Series Editor

Ashok Pandey

Centre for Innovation and Translational Research,
CSIR-Indian Institute of Toxicology Research,
Lucknow, India

Biomass, Biofuels, Biochemicals

Biodegradable Polymers and Composites - Process Engineering to Commercialization

Edited by

Parameswaran Binod
Microbial Processes and Technology Division, CSIR—National Institute for Interdisciplinary Science and Technology (CSIR—NIIST), Thiruvananthapuram, India

Sindhu Raveendran
Microbial Processes and Technology Division, CSIR—National Institute for Interdisciplinary Science and Technology (CSIR—NIIST), Thiruvananthapuram, India

Ashok Pandey
Centre for Innovation and Translational Research, CSIR—Indian Institute of Toxicology Research (CSIR—IITR), Lucknow, India

Elsevier
Radarweg 29, PO Box 211, 1000 AE Amsterdam, Netherlands
The Boulevard, Langford Lane, Kidlington, Oxford OX5 1GB, United Kingdom
50 Hampshire Street, 5th Floor, Cambridge, MA 02139, United States

Copyright © 2021 Elsevier Inc. All rights reserved.

No part of this publication may be reproduced or transmitted in any form or by any means, electronic or mechanical, including photocopying, recording, or any information storage and retrieval system, without permission in writing from the publisher. Details on how to seek permission, further information about the Publisher's permissions policies and our arrangements with organizations such as the Copyright Clearance Center and the Copyright Licensing Agency, can be found at our website: www.elsevier.com/permissions.

This book and the individual contributions contained in it are protected under copyright by the Publisher (other than as may be noted herein).

Notices
Knowledge and best practice in this field are constantly changing. As new research and experience broaden our understanding, changes in research methods, professional practices, or medical treatment may become necessary.

Practitioners and researchers must always rely on their own experience and knowledge in evaluating and using any information, methods, compounds, or experiments described herein. In using such information or methods they should be mindful of their own safety and the safety of others, including parties for whom they have a professional responsibility.

To the fullest extent of the law, neither the Publisher nor the authors, contributors, or editors, assume any liability for any injury and/or damage to persons or property as a matter of products liability, negligence or otherwise, or from any use or operation of any methods, products, instructions, or ideas contained in the material herein.

British Library Cataloguing-in-Publication Data
A catalogue record for this book is available from the British Library

Library of Congress Cataloging-in-Publication Data
A catalog record for this book is available from the Library of Congress

ISBN: 978-0-12-821888-4

For Information on all Elsevier publications
visit our website at https://www.elsevier.com/books-and-journals

Publisher: Susan Dennis
Acquisitions Editor: Kostas KI Marinakis
Editorial Project Manager: Andrea Dulberger
Production Project Manager: Swapna Srinivasan
Cover Designer: Greg Harris

Typeset by MPS Limited, Chennai, India

Contents

List of contributors xvii
Preface xxi

Part I General 1

Chapter 1: Introduction to biodegradable polymers and composites: process engineering to commercialization 3

Parameswaran Binod, Raveendran Sindhu and Ashok Pandey

   1.1 Introduction 3
   1.2 Plant-based biopolymers 4
   1.3 Microbial and insect biopolymers 6
   1.4 Biopolymer composites 7
   1.5 Process engineering and commercialization 8
   1.6 Conclusions and perspectives 9
   References 9

Chapter 2: Microplastics in aquatic and terrestrial environment 11

*Shikhangi Singh, Taru Negi, Ayon Tarafdar, Ranjna Sirohi,
Ashutosh Kumar Pandey, Mohd. Ishfaq Bhat and Raveendran Sindhu*

   2.1 Introduction 11
   2.2 Microplastic in aquatic environment 13
         2.2.1 River and lakes 14
         2.2.2 Marine 15
   2.3 Microplastics in soil 17
   2.4 Microplastic interaction with biotas 18
   2.5 Microplastic in waste 20
         2.5.1 Microplastic in solid waste 20

2.5.2 Microplastic in industrial waste 23
2.6 Conclusions and perspectives..24
Acknowledgment 25
References 25

Chapter 3: Thermoplastic starch 31

Ranjna Sirohi, Shikhangi Singh, Ayon Tarafdar, Nalla Bhanu Prakash Reddy, Taru Negi, Vivek Kumar Gaur, Ashutosh Kumar Pandey, Raveendran Sindhu, Aravind Madhavan and K.B. Arun

3.1 Introduction 31
3.2 Characterization of thermoplastic starch 34
3.3 Properties of the thermoplastic starch 36
    3.3.1 Mechanical properties 36
    3.3.2 Thermal properties 37
    3.3.3 Rheological and viscoelastic properties...............................37
    3.3.4 Crystal property 38
3.4 Biodegradability of the thermoplastic starch 39
3.5 Methods of preparing the thermoplastic starch.............................41
    3.5.1 Film blowing 42
    3.5.2 Injection blow molding 42
    3.5.3 Injection stretch blow molding..44
    3.5.4 Thermoforming 45
3.6 Applications of the thermoplastic starch 45
3.7 Conclusions and perspectives..46
Acknowledgment 47
References 47

Part II Plant - based biopolymers...51

Chapter 4: Cellulose...53

Niveditha Kulangara and Swapna Thacheril Sukumaran

4.1 Introduction 53
4.2 Biopolymers 53
    4.2.1 Biodegradable polymers and polymer composites..............54
    4.2.2 Sources of cellulose 54
    4.2.3 Plant-based cellulose...55
    4.2.4 Applications of cellulose..60
    4.2.5 Process engineering and product development 66

	4.2.6	Limitations of biopolymers and overcoming strategies 66
	4.2.7	Current status and challenges in the production of cellulose-based biopolymers 67
4.3	Conclusions and perspectives 67
Conflict of interest 68
References 68

Chapter 5: Starch 75

Susan Grace Karp, Maria Giovana Binder Pagnoncelli, Fernanda Prado, Rafaela de Oliveira Penha, Antônio Irineudo Magalhães Junior, Gabriel Sprotte Kumlehn and Carlos Ricardo Soccol

5.1	Introduction 75
5.2	Structure and properties 76
5.3	Starch-processing techniques 79
5.4	Improving mechanical and physicochemical properties 83
5.5	Starch-based materials 86
	5.5.1 Starch-based polymer blending 86
	5.5.2 Starch-based foaming 87
	5.5.3 Starch-based nanocomposites 89
5.6	Applications of starch-based materials 89
5.7	Conclusions and perspectives 93
References 94

Chapter 6: Pectin 101

Poonam Sharma, Krishna Gautam, Ashutosh Kumar Pandey, Vivek Kumar Gaur, Alvina Farooqui and Kaiser Younis

6.1	Introduction 101
6.2	Structure and classification of pectin 102
	6.2.1 Homogalacturonan 103
	6.2.2 Rhamnogalacturonan I 104
	6.2.3 Rhamnogalacturonan II 105
	6.2.4 Xylogalacturonan 105
6.3	Functional properties of pectin 105
6.4	Sources of pectin 108
6.5	Recent advances in the extraction of pectin 110
	6.5.1 Microwave-assisted extraction of pectin 112
	6.5.2 Ultrasound-assisted extraction of pectin 113

6.6	Application of pectin	115
	6.6.1	Food-processing industries	115
	6.6.2	Pharmaceutical industry	118
	6.6.3	Cosmetics	119
6.7	Current challenges and future implications	120
6.8	Conclusions and perspectives	121
	Acknowledgments	121
	References	121

Chapter 7: Xylan ... 129

Luciana Porto de Souza Vandenberghe, Kim Kley Valladares-Diestra, Gustavo Amaro Bittencourt, Ariane Fátima Murawski de Mello and Carlos Ricardo Soccol

7.1	Introduction	129
7.2	Methods for xylan extraction	130
	7.2.1	Sources and types of xylans	130
	7.2.2	Physicochemical characteristics	132
	7.2.3	Motivation for xylan extraction from different sources	132
	7.2.4	Pretreatment methods for xylan extraction	133
7.3	Bioproducts obtained from xylan	140
	7.3.1	Xylitol	142
	7.3.2	Bioethanol	146
	7.3.3	Hydrogels	147
	7.3.4	Packaging	148
	7.3.5	Xylooligosaccharides	149
	7.3.6	Enzymes	150
7.4	Advances and innovation	151
7.5	Environmental aspects	153
7.6	Conclusions and perspectives	153
	References	154

Part III Microbial-based biopolymers ... 163

Chapter 8: Production and applications of pullulan ... 165

Ashutosh Kumar Pandey, Ranjna Sirohi, Vivek Kumar Gaur and Ashok Pandey

8.1	Introduction	165
8.2	Biosynthesis of pullulan	166
	8.2.1	Mechanism of pullulan biosynthesis	167

	8.2.2 Physicochemical properties	168
	8.2.3 Factor affecting the production of pullulan	169
8.3	Microbial consortia for the production of pullulan	188
	8.3.1 Aureobasidium pullulans	188
	8.3.2 Cell morphologies of Aureobasidium pullulans	189
8.4	Production of pullulan by fermentation of agroindustrial by-products	189
8.5	Bioreactors and mode of operation for the production of pullulan	192
	8.5.1 Batch fermentation	192
	8.5.2 Fed-batch and continuous fermentation	193
	8.5.3 Immobilized cell bioreactors	194
	8.5.4 Airlift and other fermenters for the production of pullulan	195
8.6	Chemical modification of pullulan and advancement in processing	196
	8.6.1 Carboxymethylation	196
	8.6.2 Cross-linking	198
	8.6.3 Hydrophobic modification	199
	8.6.4 Grafting	199
8.7	Downstream processing	200
8.8	Applications of pullulan	203
	8.8.1 Healthcare	203
	8.8.2 Food industry	204
	8.8.3 Waste remediation	205
	8.8.4 Miscellaneous applications	205
8.9	Conclusions and perspectives	206
References		206
Further reading		220

Chapter 9: Production and application of bacterial polyhydroxyalkanoates 223

*Vivek Kumar Gaur, Poonam Sharma, Janmejai Kumar Srivastava,
Ranjna Sirohi and Natesan Manickam*

9.1	Introduction	223
9.2	Classification of polyhydroxyalkanoates	224
	9.2.1 Short chain length polyhydroxyalkanoates	227
	9.2.2 Medium chain length polyhydroxyalkanoates	227
	9.2.3 Chemical modifications of polyhydroxyalkanoates	228
9.3	Structure and properties	231
	9.3.1 Chemical structure	231
	9.3.2 Properties	231
9.4	Industrial-scale production of polyhydroxyalkanoates	235
	9.4.1 Batch fermentation	235

         9.4.2 Fed-batch fermentation 236
         9.4.3 Continuous fermentation 237
   9.5   Application of polyhydroxyalkanoates 239
         9.5.1 Polyhydroxyalkanoates in medical implants and medicines 239
         9.5.2 Polyhydroxyalkanoates in drug delivery 241
         9.5.3 Polyhydroxyalkanoates in tissue engineering 242
   9.6   Conclusions and perspectives 243
   Acknowledgment 244
   References 244

Chapter 10: Production and applications of polyglutamic acid 253

Kritika Pandey, Ashutosh Kumar Pandey, Ranjna Sirohi, Srinath Pandey, Aditya Srivastava and Ashok Pandey

   10.1 Introduction 253
   10.2 Microbial biosynthesis pathway 254
   10.3 Process parameters for production 255
         10.3.1 Substrate 255
         10.3.2 Microbial consortia 257
         10.3.3 Bioreactors mode of operation for production 262
         10.3.4 Isolation, analysis, and determination of PGA 264
         10.3.5 Structure of γ-polyglutamic acid 266
   10.4 Characterization of polyglutamic acid 266
   10.5 Commercial production 267
         10.5.1 Production cost 269
   10.6 Applications 271
         10.6.1 Healthcare 271
         10.6.2 Personal-care products 272
         10.6.3 Food industry 272
         10.6.4 Bioremediation 272
         10.6.5 Other applications 273
   10.7 Conclusions and perspectives 274
   References 274

Chapter 11: Production and applications of polyphosphate 283

Raj Morya, Bhawna Tyagi, Aditi Sharma and Indu Shekhar Thakur

   11.1 Introduction 283
   11.2 Structure and types of polyphosphate 284
         11.2.1 Pyrophosphate 284

		11.2.2	High molecular weight polyphosphate	286

- 11.2.2 High molecular weight polyphosphate 286
- 11.2.3 Cyclophosphates 286
- 11.3 Acidocalcisomes 286
- 11.4 Biogenic production of polyphosphate 287
  - 11.4.1 Prokaryotes 288
  - 11.4.2 Eukaryotes 291
- 11.5 Applications of polyphosphates 294
  - 11.5.1 Applications in environmental bioremediation 295
  - 11.5.2 Applications in industry 296
  - 11.5.3 Biotechnological applications 297
  - 11.5.4 Application in medical field 298
- 11.6 Challenges associated with polyphosphate production strategies 300
- 11.7 Strategies to improve the yield of polyphosphate 301
- 11.8 Conclusions and perspectives 302
- Conflict of interest 303
- References 303

Chapter 12: Production and applications of polylactic acid 309

Ashutosh Kumar Pandey, Ranjna Sirohi, Sudha Upadhyay, Mitali Mishra, Virendra Kumar, Lalit Kumar Singh and Ashok Pandey

- 12.1 Introduction 309
- 12.2 Substrate 310
- 12.3 Microbial production 311
  - 12.3.1 Use of bacterial strains 316
  - 12.3.2 Use of fungi and yeast 317
  - 12.3.3 Use of cyanobacteria 318
- 12.4 Strain improvement 318
- 12.5 Commercial strains 319
- 12.6 Fermentation modes and bioreactors 320
  - 12.6.1 Batch fermentation 320
  - 12.6.2 Fed-batch fermentation 322
  - 12.6.3 Continuous fermentation 323
- 12.7 Type of reactors used for production 324
  - 12.7.1 Continuous stirred tank reactor 324
  - 12.7.2 Packed-bed reactor 325
  - 12.7.3 Fluidized-bed reactor 325
  - 12.7.4 Airlift bioreactors 326
  - 12.7.5 Fibrous-bed reactors 327

12.8 Isolation, analysis, and determination technique and process 327
    12.8.1 Diffusion dialysis 328
    12.8.2 Membrane filtration 329
    12.8.3 Electrodialysis 330
    12.8.4 Reactive extraction 331
    12.8.5 Adsorption 331
12.9 Synthesis and structure of polymers 332
    12.9.1 Polylactic acid synthesis 332
    12.9.2 Structure of polymer 333
    12.9.3 Process flow diagram for production 334
    12.9.4 Properties of PLA polymers 335
12.10 Commercialization and application 336
    12.10.1 Textiles 336
    12.10.2 Biomedical and pharmaceutical applications 337
    12.10.3 Tissue engineering 338
    12.10.4 Drug-delivery system 339
    12.10.5 Packaging and service wares 339
    12.10.6 Plasticulture/agriculture 341
    12.10.7 Environmental remediation 342
    12.10.8 Other applications 343
12.11 Conclusions and perspectives 343
References 344

Chapter 13: Production and applications of bacterial cellulose 359

Fazli Wahid and Cheng Zhong

13.1 Introduction 359
13.2 A brief history of bacterial cellulose 360
13.3 Bacterial cellulose production 361
    13.3.1 Selection of bacterial strain 361
    13.3.2 Culture medium 362
    13.3.3 Cultivation methods 363
13.4 Structural and functional features of bacterial cellulose 365
    13.4.1 Mechanical properties 365
    13.4.2 Water holding/release capacity 366
    13.4.3 Structure, pore size, and morphology 366
    13.4.4 Biodegradability 367
    13.4.5 Biocompatibility 368
13.5 Applications of bacterial cellulose 369
    13.5.1 Biomedical applications 369

	13.5.2	Applications of bacterial cellulose in food 376
	13.5.3	Applications in cosmetics 378
	13.5.4	Electronics 379
	13.5.5	Water purification 379
	13.5.6	Other applications 380
13.6	Commercialization of BC-based products 381
13.7	Conclusions and perspectives 382
Acknowledgments 382
References 382

Part IV Biopolymer composites 391

Chapter 14: Biodegradable polymer composites 393

R. Reshmy, Eapen Philip, P.H. Vaisakh, Raveendran Sindhu, Parameswaran Binod, Aravind Madhavan, Ashok Pandey, Ranjna Sirohi and Ayon Tarafdar

14.1	Introduction 393
	14.1.1 Polymer composites 394
	14.1.2 Advantages of biodegradable polymer composites 394
	14.1.3 General commercialization processes 395
14.2	Types of biodegradable polymer composites 396
	14.2.1 Natural fiber composites 396
	14.2.2 Double-layer polymer composites 401
	14.2.3 Carbon nanotube-reinforced composites 404
	14.2.4 Petrochemical-based biocomposites 406
14.3	Potentials and applications 408
14.4	Conclusions and perspectives 410
Acknowledgments 411
References 411

Chapter 15: Thermal/rheological behavior and functional properties of biopolymers and biopolymer composites 413

Prachi Gaur, Vivek Kumar Gaur, Poonam Sharma and Ashok Pandey

15.1	Introduction 413
15.2	Biocomposites derived from polylactic acid 416
15.3	Biocomposites derived from polyhydroxyalkanoate 418
15.4	Thermal and rheological properties 420
	15.4.1 Polylactides and its biocomposites 420

        15.4.2   Polyhydroxyalkanoate and its biocomposites 425
   15.5 Functional properties of biopolymers and biocomposites 427
        15.5.1   Tensile strength of biopolymer 427
        15.5.2   Crystallinity of biocomposites 429
        15.5.3   Biopolymer film formation 430
   15.6 Conclusions and perspectives 432
   Acknowledgment 432
   References 432

Chapter 16: Synthesis and applications of chitosan and its composites 439

Thana Saffar, Narisetty Vivek, Sara Magdouli, Joseph Amruthraj Nagoth, Maria Sindhura John, Raveendran Sindhu, Parameswaran Binod and Ashok Pandey

   16.1 Introduction 439
   16.2 Biofunctionality of chitosan 440
        16.2.1   Extraction and chemical modification of chitosan for bio-based materials 442
   16.3 Synthesis of composite blends of chitosan 444
   16.4 Applications 447
        16.4.1   Food and packaging 448
        16.4.2   Wastewater treatment 450
        16.4.3   Bioremediation 451
        16.4.4   Drug delivery 451
        16.4.5   Medical 452
   16.5 Conclusions and perspectives 453
   Acknowledgment 453
   References 454

Chapter 17: Nanocellulose-reinforced biocomposites 461

Sam Sung Ting, Gan Pei Gie, Mohd Firdaus Omar and Muhammad Faiq Abdullah

   17.1 Introduction 461
   17.2 Cellulose 462
   17.3 Nanocellulose 463
        17.3.1   Cellulose nanofiber 463
        17.3.2   Cellulose nanocrystal 465
   17.4 Processing methods of nanocellulose-reinforced biocomposites 468
        17.4.1   Solvent casting 468
        17.4.2   Melt processing 470
        17.4.3   Electrospinning 471
        17.4.4   Layer-by-layer 473

17.5 Properties of nanocellulose-reinforced biocomposites 473
    17.5.1 Tensile properties 473
    17.5.2 Thermal properties 475
    17.5.3 Barrier property 480
    17.5.4 Biodegradation property 483
17.6 Conclusions and perspectives 485
References 486

Chapter 18: Biomedical applications of microbial polyhydroxyalkanoates 495

Aravind Madhavan, K.B. Arun, Raveendran Sindhu, Parameswaran Binod, Ashok Pandey, Ranjna Sirohi, Ayon Tarafdar and R. Reshmy

18.1 Introduction 495
18.2 Types of polyhydroxyalkaonates for biomedical application 496
18.3 Genetic-engineered strains for the production of polyhydroxyalkaonates 496
    18.3.1 Rational strategies for cost-effective, good-quality large-scale production of PHAs 497
18.4 Polyhydroxyalkanoates for drug delivery 501
18.5 Polyhydroxyalkanoates in tissue engineering 503
    18.5.1 Tissue engineering—bone 503
    18.5.2 Tissue engineering—cartilage 504
    18.5.3 Tissue engineering—nerve 504
    18.5.4 Tissue engineering—peridontal 505
    18.5.5 Tissue engineering—cardiovascular 505
18.6 Conclusions and perspectives 507
Acknowledgments 507
References 508

Part V Process engineering and commercialization 515

Chapter 19: Process engineering and commercialization of polyhydroxyalkanoates (PHAs) 517

Lalit R. Kumar, Bhoomika Yadav, Rajwinder Kaur, Sravan Kumar Yellapu, Sameer Pokhrel, Aishwarya Pandey, Bhagyashree Tiwari and R.D. Tyagi

19.1 Introduction 517
19.2 Types and properties of polyhydroxyalkanoates 518
19.3 Applications of polyhydroxyalkanoates 518
19.4 Process development at lab scale 518
    19.4.1 Upstream processing 519

xvi Contents

   19.4.2 Downstream processing 529
   19.4.3 Other methods 532
 19.5 Scale-up from lab scale to pilot scale 532
   19.5.1 Scale-up parameters for equipment 532
   19.5.2 Process validation 533
   19.5.3 Good manufacturing practices 534
   19.5.4 Problems and challenges encountered in scale-up 535
 19.6 Commercialization of polyhydroxyalkanoates 538
   19.6.1 Social factors affecting polyhydroxyalkanoate commercialization 538
   19.6.2 Technoeconomic studies 539
   19.6.3 Environmental assessment 540
   19.6.4 Commercial production of polyhydroxyalkanoates 542
   19.6.5 Challenges in the commercialization of polyhydroxyalkanoates 543
 19.7 Conclusions and perspectives 543
 Conflicts of Interest 544
 Acknowledgment 544
 References 544

Chapter 20: Lignin production in plants and pilot and commercial processes 551

Ayyoub Salaghi, Long Zhou, Preety Saini, Fangong Kong, Mohan Konduri and Pedram Fatehi

 20.1 Introduction 551
   20.1.1 Occurrence and formation of lignin 552
 20.2 Lignin extraction methods at laboratory and pilot scales 561
   20.2.1 Lignin extraction using milling methods 561
   20.2.2 Lignin production by novel extraction technologies at pilot scales 564
 20.3 Methods for commercial lignin production 572
   20.3.1 Lignosulfonate production 572
   20.3.2 Kraft lignin production 573
   20.3.3 Organosolv and soda lignin 576
   20.3.4 Thermomechanical pulp-bio lignin 577
 20.4 Opportunities and challenges in the commercialization of lignin production ... 578
 20.5 Conclusions and perspectives 580
 Acknowledgments 581
 References 581

Index 589

List of contributors

Muhammad Faiq Abdullah School of Bioprocess Engineering, Universiti Malaysia Perlis, Arau, Malaysia

Gustavo Amaro Bittencourt Federal University of Paraná, Department of Bioprocess Engineering and Biotechnology, Centro Politécnico, Curitiba, Brazil

K.B. Arun Rajiv Gandhi Center for Biotechnology, Thiruvananthapuram, India

Parameswaran Binod Microbial Processes and Technology Division, CSIR—National Institute for Interdisciplinary Science and Technology (CSIR—NIIST), Thiruvananthapuram, India

Rafaela de Oliveira Penha Department of Bioprocess Engineering and Biotechnology, Federal University of Paraná, Curitiba, Brazil

Alvina Farooqui Department of Bioengineering, Integral University, Lucknow, India

Pedram Fatehi Green Processes Research Centre and Chemical Engineering Department, Lakehead University, Thunder Bay, ON, Canada; State Key Laboratory of Biobased Material and Green Papermaking, Qilu University of Technology, Jinan, P.R. China

Prachi Gaur Institute of Information Management and Technology, Aligarh, India

Vivek Kumar Gaur Environmental Biotechnology Division, Environmental Toxicology Group, CSIR—Indian Institute of Toxicology Research, Lucknow, India; Amity Institute of Biotechnology, Amity University, Lucknow, India

Krishna Gautam Academy of Scientific and Innovative Research (AcSIR), CSIR—Indian Institute of Toxicology Research, Lucknow, India

Gan Pei Gie School of Bioprocess Engineering, Universiti Malaysia Perlis, Arau, Malaysia

Mohd. Ishfaq Bhat Department of Post Harvest Process and Food Engineering, G.B. Pant University of Agriculture and Technology, Pantnagar, India

Maria Sindhura John School of Biosciences and Veterinary Medicine, University of Camerino, Camerino, Italy

Susan Grace Karp Department of Bioprocess Engineering and Biotechnology, Federal University of Paraná, Curitiba, Brazil

Rajwinder Kaur Centre Eau Terre Environnement, Institut National de la Recherche Scientifique, Québec City, QC, Canada

Kim Kley Valladares-Diestra Federal University of Paraná, Department of Bioprocess Engineering and Biotechnology, Centro Politécnico, Curitiba, Brazil

Mohan Konduri Bio-Economy Technology Centre, FPInnovations, Thunder Bay, ON, Canada

Fangong Kong State Key Laboratory of Biobased Material and Green Papermaking, Qilu University of Technology, Jinan, P.R. China

Niveditha Kulangara Department of Biotechnology, University of Kerala, Thiruvananthapuram, India

Lalit R. Kumar Centre Eau Terre Environnement, Institut National de la Recherche Scientifique, Québec City, QC, Canada

Virendra Kumar Department of Biotechnology, Dr. Ambedkar Institute of Technology for Handicapped, Kanpur, India

Gabriel Sprotte Kumlehn Department of Bioprocess Engineering and Biotechnology, Federal University of Paraná, Curitiba, Brazil

Aravind Madhavan Rajiv Gandhi Center for Biotechnology, Thiruvananthapuram, India

Antônio Irineudo Magalhães Junior Department of Bioprocess Engineering and Biotechnology, Federal University of Paraná, Curitiba, Brazil

Sara Magdouli Centre Technologique des Résidus Industriels, University of Quebec in Abitibi Témiscamingue, Quebec, Canada

Natesan Manickam Environmental Biotechnology Division, Environmental Toxicology Group, CSIR—Indian Institute of Toxicology Research, Lucknow, India

Mitali Mishra Department of Biotechnology, Dr. Ambedkar Institute of Technology for Handicapped, Kanpur, India

Raj Morya School of Environmental Sciences, Jawaharlal Nehru University, New Delhi, India

Ariane Fátima Murawski de Mello Federal University of Paraná, Department of Bioprocess Engineering and Biotechnology, Centro Politécnico, Curitiba, Brazil

Joseph Amruthraj Nagoth School of Biosciences and Veterinary Medicine, University of Camerino, Camerino, Italy

Taru Negi Department of Food Science and Technology, College of Agriculture, G.B. Pant University of Agriculture and Technology, Pantnagar, India

Mohd Firdaus Omar Center of Excellence Geopolymer and Green Technology, Universiti Malaysia Perlis (UniMAP), Kangar, Perlis, Malaysia; School of Material Engineering, Universiti Malaysia Perlis, Arau, Malaysia

Maria Giovana Binder Pagnoncelli Department of Chemistry and Biology, Federal University of Technology, Curitiba, Brazil

Aishwarya Pandey Centre Eau Terre Environnement, Institut National de la Recherche Scientifique, Québec City, QC, Canada

Ashok Pandey Centre for Innovation and Translational Research, CSIR—Indian Institute of Toxicology Research (CSIR—IITR), Lucknow, India; Centre for Energy and Environmental Sustainability, Lucknow, India

Ashutosh Kumar Pandey Centre for Energy and Environmental Sustainability, Lucknow, India

Kritika Pandey Department of Biotechnology, Dr. Ambedkar Institute of Technology for Handicapped, Kanpur, India

Srinath Pandey Department of Biotechnology, Naraina Group of Institution, Kanpur, India

Eapen Philip Post-Graduate and Research Department of Chemistry, Bishop Moore College, Mavelikara, India

Sameer Pokhrel Centre Eau Terre Environnement, Institut National de la Recherche Scientifique, Québec City, QC, Canada

Luciana Porto de Souza Vandenberghe Federal University of Paraná, Department of Bioprocess Engineering and Biotechnology, Centro Politécnico, Curitiba, Brazil

Fernanda Prado Department of Bioprocess Engineering and Biotechnology, Federal University of Paraná, Curitiba, Brazil

Nalla Bhanu Prakash Reddy Department of Post-Harvest Process and Food Engineering, G. B. Pant University of Agriculture and Technology, Pantnagar, India

R. Reshmy Post-Graduate and Research Department of Chemistry, Bishop Moore College, Mavelikara, India

Thana Saffar Centre Technologique des Résidus Industriels, University of Quebec in Abitibi Témiscamingue, Quebec, Canada

Preety Saini Green Processes Research Centre and Chemical Engineering Department, Lakehead University, Thunder Bay, ON, Canada

Ayyoub Salaghi Green Processes Research Centre and Chemical Engineering Department, Lakehead University, Thunder Bay, ON, Canada

Aditi Sharma School of Environmental Sciences, Jawaharlal Nehru University, New Delhi, India

Poonam Sharma Department of Bioengineering, Integral University, Lucknow, India

Raveendran Sindhu Microbial Processes and Technology Division, CSIR—National Institute for Interdisciplinary Science and Technology (CSIR—NIIST), Thiruvananthapuram, India

Lalit Kumar Singh Department of Biochemical Engineering, Harcourt Butler Technical University, Kanpur, India

Shikhangi Singh Department of Post-Harvest Process and Food Engineering, G.B. Pant University of Agriculture and Technology, Pantnagar, India

Ranjna Sirohi Department of Post-Harvest Process and Food Engineering, G.B. Pant University of Agriculture and Technology, Pantnagar, India

Carlos Ricardo Soccol Department of Bioprocess Engineering and Biotechnology, Federal University of Paraná, Curitiba, Brazil; Federal University of Paraná, Department of Bioprocess Engineering and Biotechnology, Centro Politécnico, Curitiba, Brazil

Aditya Srivastava Department of Biotechnology, Dr. Ambedkar Institute of Technology for Handicapped, Kanpur, India

Janmejai Kumar Srivastava Amity Institute of Biotechnology, Amity University, Lucknow, India

Swapna Thacheril Sukumaran Department of Botany, University of Kerala, Thiruvananthapuram, India

Ayon Tarafdar Department of Food Engineering, National Institute of Food Technology, Entrepreneurship and Management, Sonipat, India; Division of Livestock Production and Management, ICAR-Indian Veterinary Research Institute, Izatnagar, Bareilly, India

Indu Shekhar Thakur School of Environmental Sciences, Jawaharlal Nehru University, New Delhi, India

Sam Sung Ting School of Bioprocess Engineering, Universiti Malaysia Perlis, Arau, Malaysia; Center of Excellence Geopolymer and Green Technology, Universiti Malaysia Perlis (UniMAP), Kangar, Perlis, Malaysia

Bhagyashree Tiwari Centre Eau Terre Environnement, Institut National de la Recherche Scientifique, Québec City, QC, Canada

Bhawna Tyagi School of Environmental Sciences, Jawaharlal Nehru University, New Delhi, India

R.D. Tyagi   BOSK Bioproducts, Québec City, QC, Canada

Sudha Upadhyay   Department of Chemical Engineering, Indian Institute of Technology, Guwahati, India

P.H. Vaisakh   Post-Graduate and Research Department of Chemistry, Bishop Moore College, Mavelikara, India

Narisetty Vivek   Department of Biochemical Engineering and Biotechnology, Indian Institute of Technology Delhi, New Delhi, India

Fazli Wahid   State Key Laboratory of Food Nutrition & Safety, Tianjin University of Science & Technology, Tianjin, P.R. China; Key Laboratory of Industrial Fermentation Microbiology (Ministry of Education), Tianjin University of Science & Technology, Tianjin, P.R. China

Bhoomika Yadav   Centre Eau Terre Environnement, Institut National de la Recherche Scientifique, Québec City, QC, Canada

Sravan Kumar Yellapu   Centre Eau Terre Environnement, Institut National de la Recherche Scientifique, Québec City, QC, Canada

Kaiser Younis   Department of Bioengineering, Integral University, Lucknow, India

Cheng Zhong   State Key Laboratory of Food Nutrition & Safety, Tianjin University of Science & Technology, Tianjin, P.R. China; Key Laboratory of Industrial Fermentation Microbiology (Ministry of Education), Tianjin University of Science & Technology, Tianjin, P.R. China

Long Zhou   Green Processes Research Centre and Chemical Engineering Department, Lakehead University, Thunder Bay, ON, Canada

Preface

The book titled *Biodegradable Polymers and Composites—Process Engineering to Commercialization* is a part of the Elsevier comprehensive book series on *Biomass, Biofuels, Biochemicals* (Editor-in-Chief: Ashok Pandey). This book intends to cover different aspects of biodegradable polymers, proving state-of-art information on their production and applications and perspectives for future developments.

The constant growth of population, technological improvements, modernization of existing infrastructures, expansion of industries, and other businesses have led to a steep surge in demand for basic commodities, including polymers. The global demand for the plastics is rising every year with scientific advancements in their manufacture and applications. However, the fossil-derived plastics or synthetic plastics have led to very serious environmental concerns globally. The increasing industrial developments have also led to increasing levels of pollution due to the plastics contamination of water, soil, and sediment environments. Therefore there has been a paradigm shift/transition from the infrastructure and industrial practices reliant on fossil-derived plastics toward a post-fossil, sustainable future. In other words, a transition for a livable, sustainable, post-fossil carbon society, in which these plastics could be replaced by the biodegradable biopolymers (bioplastics).

The constant increase in the demand for bio-based polymers is mainly due to strong public concerns about the waste generation and its disposal. All over the world, people have realized the menace of conventional petroleum-based plastics, and a commercially viable, competitive, and biodegradable polymer as an alternative to plastics is an extreme need of the hour. It is not only the biodegradability of polymer that matters but how the individual monomer of the biopolymer or biopolymer composite affects the environment also needs to be evaluated. The major bottleneck in the commercialization of biodegradable polymers is the cost factor. The biopolymer production process has to be competitive enough to the equipment currently available in the market and also, the product cost has to be on par with the conventional plastics. However, in spite of enormous global efforts to make the biodegradable polymers competitive with the well-established petro-based polymers, it has still not met the benchmark regarding the economic competitiveness and characteristics. The production cost mainly depends on the cost of raw materials as well as the production strategies. The raw material has to be very cheap, easily available, and should be sustainable. The downstream processing and polymer blending is required and has to be simple and should be devoid of any harmful solvents. The treatment of effluent generated during the process also needs to be considered. The technoeconomic and life-cycle analysis of the biopolymer is also very important to study the long-term effect of the biopolymer to the environment and on life forms.

This book covers the recent advances and development in the area of biopolymers and composites, focusing on process engineering to commercialization. There are a total of 19 chapters in the book, which have been organized in five sections: Part 1: General, Part 2: Plant-based Biopolymers, Part 3: Microbial-based Biopolymers, Part 4: Biopolymer Composites, and Part 5: Process Engineering and Commercialization.

The General section includes three chapters, which basically cover general introduction to biopolymers and importance of biopolymers in the environment. The environmental impact and pollution created due to the presence of microplastics in aquatic and terrestrial environments are described in one of the chapters, and this sheds light on the importance and need for replacing the conventional plastics with biodegradable plastics to save the ecosystem. Thermoplastic starch is a potential polymer to produce the biodegradable packaging material, and detailed updates on the development of research in this area are described in a chapter.

The section on plant-based biopolymers gives detailed information on the biopolymers from plant origin such as cellulose, starch, pectin, and xylan. Plant-based polymers have immense possibilities in many fields such as food industry, paper making, water purification, packaging, and the biomedical industry.

The third section is on the biopolymers produced by the microbes. Polymers such as pullulan, polyhydroxyalkanoates (PHAs), polyglutamic acid, polyphosphate, polylactic acid, and bacterial cellulose are described in this section.

Biodegradable polymer composites have enormous potential in the industrial sector due to their biodegradability and other mechanical properties. The fourth section in the book elaborates on various biopolymer composites and critically evaluates the strengths and weaknesses of biocomposites as well as thermal and functional properties of natural material-reinforced biopolymers, their potential, and future opportunities. The biopolymers such as chitin and chitosan are also used for the development of various functional biocomposites and blends. Nanocellulose has attracted a great level of interest in scientific and industrial fields as a reinforcing filler in polymeric composite due to its abundant availability, good renewability, good biodegradability, lightweight, excellent mechanical properties, and high surface area. The chapter on nanocellulose-reinforced biocomposites provides updated information on the processing methods and various applications in response to innovative and evolving societal requirements, including circular bioeconomy.

One chapter in this section specifically discusses the application of PHAs and their composites for biomedical applications.

The process development and commercialization are the challenging tasks in biopolymer research. The success of a commercial process depends on the minute analysis of various unit operations in terms of yield and cost. A chapter in this section deals on the process development for PHA production, process scale-up from laboratory to pilot scale, PHA commercialization, and challenges observed during different stages. A comprehensive appraisal of the pilot scale to commercialization of lignin based on various industrial process are also discussed in this section.

Overall, the book provides a wide source of knowledge in the area of biopolymers. The editors would like to express their sincere gratitude to all the authors who contributed in this book. We would like to express our special thanks to all the experts, who reviewed various chapters of this book and provided very valuable suggestions to the authors to improve their manuscripts. The editors also express their sincere thanks to Elsevier team comprising Dr. Kostas Marinakis, Senior Book Acquisition Editor, Andrea Dulberger, Editorial Project Manager, and the entire production team for their cooperation and support in the publication of this book.

We are confident that the readers will find this book useful.

Parameswaran Binod
Sindhu Raveendran
Ashok Pandey

PART I

General

CHAPTER 1

Introduction to biodegradable polymers and composites: process engineering to commercialization

Parameswaran Binod[1], Raveendran Sindhu[1] and Ashok Pandey[2]

[1]Microbial Processes and Technology Division, CSIR—National Institute for Interdisciplinary Science and Technology (CSIR—NIIST), Thiruvananthapuram, India, [2]Centre for Innovation and Translational Research, CSIR—Indian Institute of Toxicology Research (CSIR—IITR), Lucknow, India

1.1 Introduction

Biopolymers are those polymers that are derived from living organisms. It includes polymers derived or produced from plants, animals, and microorganisms. Living organisms are able to synthesize a wide range of polymers and these polymers contribute to the major portion of their cellular dry matter. In most cases, these polymers are essential for these organisms as they provide their cellular structures. These polymers are helpful to these organisms in many ways such as the storehouse of carbon and energy and during stress, helping the organism to thrive the condition. They also conserve and express the genetic information. These polymers in some microorganisms help in defending and protecting them against the attack of other organisms, hazardous environmental factors, and sensing biotic and abiotic factors. It also help in communication with the environment and other organisms. In some cases, they help in the adhesion of the cells to the surfaces of other organisms or non-living materials.

The biopolymers are synthesized by the enzymatic process in the cytoplasm or in various organelles of the cells or at the cell surfaces or synthesized extracellular. Many of the biopolymers produced by these living organisms are useful to human beings in one way or another. The environmental problem associated with the impact of persistent plastic wastes can be avoided by replacing conventional plastic with bioplastics. A number of polymers produced by the living organisms have either been adopted as commercial product or have the potential for commercialization. There are different methods and strategies that are used to produce biopolymers from living organisms to make them available for various applications. Plants and algae are among the excellent sources of biopolymers.

Microorganisms such as bacteria and fungi are also good sources of the biopolymers. The fermentative production of biopolymers from these microbes is followed in many industries. In vitro synthesis of biopolymers with isolated enzymes in cell-free system also offers another possibility to produce biopolymers.

1.2 Plant-based biopolymers

Plant-based polymers have immense possibilities in many fields such as the food industry, paper making, water purification, packaging as well as biomedical industry. Such polymers (e.g., cellulose, starch, pectin, xylan, lignin, and even protein) are renewable resources that attracted the scientific community due to the diverse uses and the eco-friendly nature. Cellulose can be modified in macro-, micro-, and nanoforms for enhancement in mechanical strength, water retention, change in degradation rate, etc. Cellulose has attracted significant attention due to its excellent properties such as biocompatibility, biodegradability, mechanical properties, and thermal and chemical stability. Its high strength and stiffness make it a suitable material as reinforcing filler in composites with structural requirements (Vlaia, Coneac, Olariu, Vlaia, & Lupuleasa, 2016). It can also be converted into two different nanoforms: cellulose nanocrystal and cellulose nanofiber. Nanocellulose offers numerous attractive characteristics such as great surface area, high tensile strength, high modulus elasticity, and high crystallinity. Besides, nanocellulose is capable to attain a similar enhancement as cellulose, which requires higher loading to improve the properties of composites. Therefore nanocellulose can be considered as a significant reinforcing material for the development of high-performance composites (Khoo, Ismail, & Chow, 2016).

Starch is a promising alternative in the field of bio-based materials. It is abundantly available in nature and is relatively cheap and highly biodegradable. However, there are some challenges to overcome for its use as a substitute for fossil-based polymers such as poor mechanical properties, high vapor permeability due to its hydrophilic nature, and the tendency of undergoing retrogradation. Improvement of starch-based materials can be accomplished by physical, chemical, and biological strategies. Chemical modifications may include crosslinking and addition of functional groups, which can be accomplished both by chemical and enzymatic reactions. Genetic modifications in the botanical source can be a tool to obtain starch with modified properties through the modification of amylose and amylopectin ratios. Another interesting strategy to improve the mechanical behavior of starch-based materials is the manufacture of starch composites. The addition of fibers, particles, or flakes in a starch matrix allows the combination of the different mechanical properties of the matrix and the inclusions, yielding various types of materials. Besides, the addition of plasticizer agents and the formation of starch blends with other polymers also increase the range of possibilities for starch-based materials (Ivanovic, Milovanovic, & Zizovic, 2016).

The major areas of application of biopolymers are in packaging and disposable materials. Several biopolymers are known to have the property to use as a packaging material. Among these, thermoplastic starch is one of the potential materials that can produce biodegradable packaging material. One of the major issues associated with the use of thermoplastic starch is the poor mechanical and water-resistant properties (Carvalho, Curvelo, & Agnelli, 2001). However, this property can be improved by blending it with other polymers and plasticizers. Various research-and-development activities are going on across the globe to achieve the commercialization of products using thermoplastic starch.

Pectin is another plant-derived polymer ubiquitously present between the cell wall of terrestrial plants. This polymer is proven to be an excellent biomaterial in various sectors such as food processing, food packaging, nutraceuticals, pharmaceutics, and cosmetics. In the food processing industry, it has found multiple utilities due to its excellent gelling, emulsifying, and texture modifying properties. It is a suitable biopolymer to replace synthetic packaging material with natural substitute (Mahalik & Nambiar, 2010). Owing to these properties, it is used in nutraceuticals and as a promising delivery agent for drugs and probiotics.

Xylans and mannans are constituents of hemicellulose, which are abundant plant cell wall polysaccharides. They are present in hardwoods and some plants such as grasses, cereals, and herbs. These biopolymers can also be recovered from the side-streams of the agriculture and forestry industry. Xylan can be extracted from hemicellulose through different techniques, including chemical, physicochemical and biological pretreatments. After extraction, xylan can be employed in the generation of commercial bio-based products such as xylitol, biofuels, hydrogels for drug delivery and biomedical products, enzymes, xylooligosaccharides, packaging materials, etc. (Deutschmann & Dekker, 2012).

Research on understanding lignin and its market potential in various fields, including novel chemical production, value-added material manufacturing, and bioenergy generation is being exploited extensively. Many research papers and review articles have been published on lignin; however, little attention has been given to collectively address the research gaps, opportunities, and challenges encountered by the researchers and the market off-takers in commercializing lignin-based products. Currently, the commercially proven technologies for the use of lignin do not match the quantity of lignin production, implying that the production and extraction of lignin overbalance the applications of value-added lignin-based products. Moreover, the quality of the lignin-based products used in well-established fields of the industry, for example, adhesives, adsorbents, and dispersants, still needs improvement to compete against the existing commercial products (Smolarski, 2012). Therefore it is essential to develop various applications utilizing technical lignin-based products that can be exploited on an industrial scale.

1.3 Microbial and insect biopolymers

Microorganisms are capable of producing vast diversity of polymers and it offers an excellent application area. Pullulan is an edible bacterial exopolysaccharide that displays many potential applications owing to its peculiar characteristics such as stability and nontoxicity. Its applications in food, pharmaceutical, medical, and environmental remediation have expanded more since the introduction of chemical modifications in the polysaccharide structure (Dionísio et al., 2016). Although many attempts have been successfully made to optimize the downstream processing of pullulan, more engineering innovations and use of improved production strains can enhance the productivity and purity of pullulan, further widening its range of utilization and application.

Polyhydroxyalkanoates (PHAs) have emerged as a promising alternative to synthetic petroleum-based plastics. Replacing synthetic plastics offers advantages in protecting the environment and reducing the use of fossil fuels. Beginning from 1959 to till date, the market of PHAs is continuously increasing. However, PHA has not attained its possible peak in production and commercialization, mainly due to lack of low yields and complexity in purification. With the current understanding and following diverse strategies such as mixed cultures and two-stage processes, large-scale production is made more economical with improved yield. PHAs are widely applicable in medical implants, tissue engineering, and drug delivery agents (Hazer & Steinbüchel, 2007).

Poly-glutamic acid (PGA) is a biopolymer that has gained immense consideration over the past with an increasing number of studies. Its properties such as biodegradability, high solubility, and nontoxicity make it of interest for applications in food, medicine, waste treatment, and many others. However, the high cost and low productivity are still major concerns regarding its commercial production and use. Also, for specific applications, molecular mass and enantiomer conformation of the polymer are important for the efficient use of γ-PGA in different industries. Its biosynthesis pathway has been decoded but efforts related to the proper selection of substrates and microorganisms for efficient production are still required. Production mechanism for γ-PGA changes with diverse microorganisms and commercial production requires the selection of high producing strains. Modification in medium composition, genetic engineering, and advancements in molecular biology has led to increased yields with enhanced durability of the microorganisms in adverse fermentation conditions. Fermentation using renewable biomass has been established as the best approach for the commercial production of γ-PGA (Zhu et al., 2014). Current studies are mainly aimed at enhancing the yields with economic production and recovery of the biopolymer.

Polyphosphates are the linear polymers of the orthophosphates produced by the living cells. The application of this polymer broadly varies from environmental bioremediation (as heavy metal and enhanced biological phosphate removal) to industrial and biotechnological applications (as food additives, detergent additives, and enzyme technology)

(Keasling, Van Dien, & Trelstad, 2000). Polyphosphates also have applications in the medicinal field for therapeutic drug design and drug delivery.

Poly-lactic acid (PLA) is a biodegradable polymer made up of lactic acid monomers. It is being increasingly produced today because of its potential applications in textile industries, pharmaceutical, packaging, bioremediation, and many more. The commercial production of lactic acid enantiomers is mostly done using renewable material such as lignocellulosic and starchy biomass along with the milk processing industry by-products such as whey. Fermentation has been commercially performed in batch, fed-batch, and continuous mode with the use of techniques such as membrane filtration, reactive extraction, and others for separation from the broth (Abdel-Rahman, Tashiro, & Sonomoto, 2011). PLA can be composed of either pure L-isomer or D-isomer or both D,L-lactic acid depending upon the requirement and use. Polymerization processes such as polycondensation and ring-opening polymerization and direct methods like azeotropic dehydration and enzymatic polymerization are used to form PLA from lactic acid monomers.

Cellulose produced by the bacteria is considered an important material for the development of bio-based products. Bacterial cellulose is a unique biopolymer, which has a variety of applications in various fields, including food, water purification, cosmetics, paper making, biomedical, and various other applications. Compared to plant cellulose, bacterial cellulose exhibits higher mechanical properties, higher water retention, higher purity, and better biocompatibility. It consists of pure cellulose fibers arranged in a 3D network structure spun by the bacteria. Because of its distinctive properties, bacterial cellulose could be promising biomass for the development of numerous industrial products.

Chitin and chitosan are natural linear polysaccharides with exceptional properties such as the biocompatible, renewable, biodegradable, chelating agent, as a drug carrier, etc. Similar to the numerous applications, even the source of chitin is also abundant in the form of crustaceans, molluscs, insects, fungi, and algae. Chitin and chitosan are insoluble in organic and aqueous solutions but can be chemically modified to be soluble desirable to the application. These flexible characteristics of the biopolymer enhance the physical and mechanical properties and lead to the development of various functional biocomposites and blends. These blends are used in as biodegradable films in food packaging such as wound dressings, drug carriers in pharmaceutical and medical, as porous membranes or coagulating agents in wastewater treatments, and as nitrogen source or encapsulating material in agriculture and bioremediation applications (Kyzas, Siafaka, Pavlidou, Chrissafis, & Bikiaris, 2015; Ngah, Endud, & Mayanar, 2002; Ribeiro et al., 2009).

1.4 Biopolymer composites

Biopolymer composites also offer various industrial applications. These composites have a wide range of properties and other characters, which can be designed based on end-applications.

They can be designed to have high mechanical resistance, high thermogravimetric properties, high oxygen barrier, high chemical resistance, and high biodegradable properties. There is no single natural biopolymer that can achieve such a wide range of properties as that of biopolymer composites. The biopolymers can be functionalized for better compatibility during the preparation of composites. The production of composites can be done by combining two or more chemically different materials or phases such as polymers in an orderly manner or merely mixing, which play a key role.

PHA- and PLA-based composites extend great interest as food packaging material with excellent gas and moisture barrier properties. Strengths and weaknesses of PLA- and PHA-based biocomposites as well as thermal and functional properties of natural material reinforced biopolymers and their potentials are important to explore and exploit further. Functional properties such as film formation, crystallinity, tensile strength, flexural properties of PLA, and PHA polymer can be modified to exponential level using different concentrations of natural filler material. The enhancement in the functional and thermal behavior of these biocomposites opens doorway toward their increased applications in medical and industrial aspects.

1.5 Process engineering and commercialization

The main limitations in the development of these biopolymers are the lack of efficient extraction and purification together with lower yields. However, new developments in the area of biotechnology and other research fields could improve the process efficiencies, which may lead to the commercial feasibility of the production of biopolymers. The research on the production and application of biopolymers requires interdisciplinary knowledge. Successful commercialization of biopolymers requires integrated efforts by biotechnologists, microbiologists, polymer scientists, geneticists, design engineers, chemists, botanists, medical scientists, venture capitalists, and government agencies. There has to be a shift in the approach of chemical companies manufacturing plastics to move toward bioplastic for sustainability. As far the market is concerned, there is no dearth of their potential as the consumers have increased liking for biopolymers. However, merely reducing costs and increasing yields will not guarantee bioplastics survival in the market. Rather, innovative business models should be developed to create a sustainable value proposition. For successful commercialization the scale-up of each equipment must be carefully designed and tested such that conditions at a larger scale should be similar to that of conditions on a smaller scale. During the scale-up of downstream processing equipment, fluid dynamics, material properties, and thermodynamics need to be taken into account. In this regard, techno-economic and life-cycle assessment have shown that the use of biowaste as a substrate during fermentation reduces the production cost and greenhouse gas emission (Kookos, Koutinas, & Vlysidis, 2019).

1.6 Conclusions and perspectives

Biopolymers are an excellent source of material having wide applications. Both plants and microbes produce a wide variety of biopolymers. The blending of polymers to make composite also offers wide applications and depending on the kind of application, the polymer property can be changed, which would suit the application. There are several challenges, which still need to be addressed to commercialize many of the biopolymers. The processing machinery used for conventional plastic processing cannot be used in many biopolymers, mainly due to the low melt temperatures. Researches are going on to address these issues. Due to the environmental issues associated with conventional plastics and microplastics, there is a huge demand for biodegradable polymers, and in near future, biopolymers can make a new revolution by replacing conventional plastics.

References

Abdel-Rahman, M. A., Tashiro, Y., & Sonomoto, K. (2011). Lactic acid production from lignocellulose-derived sugars using lactic acid bacteria: Overview and limits. *Journal of Biotechnology*.

Carvalho, A. J. F., Curvelo, A. A. S., & Agnelli, J. A. M. (2001). A first insight on composites of thermoplastic starch and kaolin. *Carbohydrate Polymers*, 45, 189–194.

Deutschmann, R., & Dekker, R. F. H. (2012). From plant biomass to bio-based chemicals: Latest developments in xylan research. *Biotechnology Advances*, 30, 1627–1640.

Dionísio, M., Braz, L., Corvo, M., Lourenço, J. P., Grenha, A., & Rosa da Costa, A. M. (2016). Charged pullulan derivatives for the development of nanocarriers by polyelectrolyte complexation. *International Journal of Biological Macromolecules*. Available from https://doi.org/10.1016/j.ijbiomac.2016.01.054.

Hazer, B., & Steinbüchel, A. (2007). Increased diversification of polyhydroxyalkanoates by modification reactions for industrial and medical applications. *Applied Microbiology and Biotechnology*, 74(1), 1–12.

Ivanovic, J., Milovanovic, S., & Zizovic, I. (2016). Utilization of supercritical CO_2 as a processing aid in setting functionality of starch-based materials. *Starch*, 68, 821–833. Available from https://doi.org/10.1002/star.201500194.

Keasling, J. D., Van Dien, S. J., Trelstad, P., et al. (2000). Application of polyphosphate metabolism to environmental and biotechnological problems. *Biochemistry C/C of Biokhimiia (Moscow, Russia)*, 65, 324–331.

Khoo, R. Z., Ismail, H., & Chow, W. S. (2016). Thermal and morphological properties of poly(lactic acid)/nanocellulose nanocomposites. *Procedia Chemistry*, 19, 788–794.

Kookos, I. K., Koutinas, A., & Vlysidis, A. (2019). Life cycle assessment of bioprocessing schemes for poly(3-hydroxybutyrate) production using soybean oil and sucrose as carbon sources. *Resources, Conservation and Recycling*, 141, 317–328.

Kyzas, G. Z., Siafaka, P. I., Pavlidou, E. G., Chrissafis, K. J., & Bikiaris, D. N. (2015). Synthesis and adsorption application of succinyl-grafted chitosan for the simultaneous removal of zinc and cationic dye from binary hazardous mixtures. *Chemical Engineering Journal*, 259, 438–448.

Mahalik, N. P., & Nambiar, A. N. (2010). Trends in food packaging and manufacturing systems and technology. *Trends in Food Science & Technology*, 21(3), 117–128.

Ngah, W. W., Endud, C., & Mayanar, R. (2002). Removal of copper (II) ions from aqueous solution onto chitosan and cross-linked chitosan beads. *Reactive and Functional Polymers*, 50(2), 181–190.

Ribeiro, M. P., Espiga, A., Silva, D., Baptista, P., Henriques, J., Ferreira, C., et al. (2009). Development of a new chitosan hydrogel for wound dressing. *Wound Repair and Regeneration*, 17(6), 817–824.

Smolarski, N. (2012). *High-value opportunities for lignin: Unlocking its potential*. Frost & Sullivan.

Vlaia, L., Coneac, G., Olariu, I., Vlaia, V., & Lupuleasa, D. (2016). Cellulose-derivatives-based hydrogels as vehicles for dermal and transdermal drug delivery. *Emerging Concepts in Analysis and Applications of Hydrogels*, 2, 64.

Zhu, F., Cai, J., Zheng, Q., Zhu, X., Cen, P., & Xu, Z. (2014). A novel approach for poly-γ-glutamic acid production using xylose and corncob fibres hydrolysate in *Bacillus subtillis* HB-1. *Journal of Chemical Technology and Biotechnology (Oxford, Oxfordshire: 1986)*, 89(4), 616−622.

CHAPTER 2

Microplastics in aquatic and terrestrial environment

Shikhangi Singh[1], Taru Negi[2], Ayon Tarafdar[3], Ranjna Sirohi[1], Ashutosh Kumar Pandey[4], Mohd. Ishfaq Bhat[1] and Raveendran Sindhu[5]

[1]Department of Post Harvest Process and Food Engineering, G.B. Pant University of Agriculture and Technology, Pantnagar, India, [2]Department of Food Science and Technology, College of Agriculture, G.B. Pant University of Agriculture and Technology, Pantnagar, India, [3]Division of Livestock Production and Management, ICAR-Indian Veterinary Research Institute, Izatnagar, Bareilly, India, [4]Centre for Energy and Environmental Sustainability, Lucknow, India, [5]Microbial Processes and Technology Division, CSIR—National Institute for Interdisciplinary Science and Technology (CSIR—NIIST), Thiruvananthapuram, India

2.1 Introduction

Plastics are the synthetic organic materials, which are in use by mankind over long time due to their durability, inexpensive nature, and buoyancy properties, with their mass production starting from the 1950s. Around 322 million metric tons of plastics were produced in 2015 globally, while in 1950 only 350,000 metric tons of plastics were produced (Habib, Thiemann, & Al Kendi, 2020). Raw plastics are produced by the industries and are transported to application-based industries for making individual manufactured goods or components. After the end-of-life, these plastics can be recycled for reuse or incinerated for energy production. For example, in European countries where 49 million metric tons of plastics were produced, out of which 5.5 million tons were recycled and 7.6 million tons were incinerated (Zmak & Hartmann, 2017). Nevertheless, after all these conversion strategies, most of the nonrecycled plastics end up as the waste. These are scattered in all over the world. Many of them are on simply disposed on the land but most of them end up into the oceans. Accumulation of these plastics also occurs on closed bays, gulfs, and seas, which are mostly surrounded by the coastal lines and watersheds. There is rising concern globally due to the harmful effects on marine and terrestrial life caused by these plastics, which has become most serious environmental concern for public at large, including national and international governments.

The demand for plastics has consistently increased over recent years, which has led to the production of a broad variety of plastics (Table 2.1). Plastics have good barrier properties,

Table 2.1: Classification of polymers found in marine ecosystem.

Polymer name	Recycled annually	Decomposing time	Uses
LDPE	6%	500–1000 years	Plastic bags, six-pack rings, bottles, netting, drinking straws
HDPE	30%–35%	100 years	Milk and juice jug
PVC	<1%	Never	Plastic film, bottles, cups
PET	36%	5–10 years	Plastic beverage bottles
PP	3%	20–30 years	Rope, bottle, caps, netting
PS	34%	50 years	Plastic utensils, food containers

they are strong, light weighted, buoyant, and transparent, which make them widely used in various application. They come in many forms and played an important tool in our life, from a toothbrush to a lunch container. All plastic items are used and eventually thrown out. However, some of the plastic products are designed only for single use, for instance, plastic water/drink bottle which end up in oceans if they are not properly managed or disposed (Anderson, Park, & Palace, 2016).

With regard to plastic pollution, since a few decades, a new associated threat has appeared, which is microplastics. Small bits of plastics are known as "microplastics," which are defined as the plastics having particle size in between 5 mm and 100 nm in size (Welden & Lusher, 2020); 5 mm can be the longest axis in measuring the two dimensions (length and width). A 5 mm particle is very different from 500 µm and 50 µm particle. The particle size between 5 mm and 500 µm can be easily digested by the human digestive tract, whereas 50 µm particles are small enough to cross the cell boundaries. The outline for new definition of plastic has been recently recommended by the global experts by considering all the biological influences and other cross-disciplines and concluded that the size is not the most important evocative factor (New Plastic Economy, 2020). In fact, there has been inconsistent use of the term "plastics" for nano, meso, and macro, which have been referred as micro (Hartmann et al., 2019). Therefore SI scale should be considered as the convenient option to distinguish these measurements such as nanoplastics (1–1000 nm), microplastics (1–1000 µm), milliplastics (1–10 mm), centiplastics (1–10 cm), and deciplastics (1–10 dm) (Corona, Martin, Marasco, & Duarte, 2020).

Microplastics are categorized into two types, namely, primary and secondary microplastics. Primary microplastics are those particles that have been intentionally produced into resin pellets or in capsules in size range of microplastics by the plastic industries whereas secondary microplastics are those particles, which are fragmented in surroundings or broken during the processing (Hartmann et al., 2019). Other sources of primary microplastics are deliberately designed small, known as "microbeads," a kind of microplastics that are added as an exfoliants in many beauty and health products such as some cleansers and toothpastes. Microbeads used in the cosmetics can easily pass through water filtration system and then enters into the waterways, including rivers and oceans. According to one of the statistical survey studies, 8 trillion microbeads worldwide passing into the water bodies through wastewater treatments plants (Wright & Kelly, 2017). Other primary sources of microplastics are the cleaning products and the domestic waste, which are disposed directly in wastewater drains.

There are two sources of secondary microplastics—terrestrial and aquatic. Two different phases of secondary production of microplastics are as follows: in-use fragmentation and postuse fragmentation. In-use fragmentation refers to the formation of microfibers during washing of clothes; tire and road wear material and wear out of fishing gear, which can be degraded by the action of microorganisms or by the chemical and physical means (Davidson & Dudas, 2016). On the other hand, postuse fragmentation is referred as the breakdown of used and disposed plastics. The small plastic debris enter the environment through domestic and industrial wastes, wastewater system and carrier being wind and water bodies. These plastic are impacted by the sun rays (get exposed to UV rays), wave action of water and wind and lose their durability, break down and fragmented into many smaller pieces. Interestingly, birds and aquatic animals mistakenly consider them as a food and eat, which very harmfully affect their health. However, some reports have suggested that meso- and macroplastics could be ingested by the larger organisms, whereas microplastics affects marine animals such as fish, crustaceans, planktons, etc. (Welden & Cowie, 2017). Whatever be the case, there are strong signs that microplastics have potential to transfer into humans too, although there is no evidence found yet. But the ubiquitous presence of microplastics in nature indicates that all forms of lives are exposed with it.

The thwarting issue is that microplastics can carry pathogen, chemical pollutants, and organic materials which are high in toxicity such as hexachlorocyclohexane, chlorinated benzenes, chlorinated diphenyl, and polycyclic aromatic hydrocarbons (Dawson et al., 2018). These organic matters can easily get adsorbed on the surface of microplastics due to hydrophobicity and large surface area of microplastic, which helps the adsorption. According to United Nations Environment Program, these microplastics first appeared as plastic microbeads in personal-care products about 50 years ago and there has been constant increase in the products showing their presence (Thompson, Moore, Vom Saal, & Swan, 2009).

2.2 Microplastic in aquatic environment

Microplastics have become an emerging environmental restraint as it poses risk to the aquatic ecosystems. An increasing number of studies have been done on the effects of microplastics in aquatic environments. Widespread concerned was increased when a plastic particle was first found in the seas back in 1972. Later, it became a hot topic for research interests. The toxic chemical transportation of the plastic pellets and long residence times in marine world may increase the chances of ingestion of these pellets by biota. One of the most thorough estimates suggested that there could be around 5—51 trillion particles in the marine ecosystem, which is equivalent to 90,000—263,000 tons of the plastics. In the following section, we look at various microplastics in marine and freshwater system (river and lakes) and their indirect routes through which they enter the aquatic environment.

Widespread occurrence of microplastic has led to possible risk to aquatic environment. Ingestion of plastic particles by marine fishes was first reported a decade ago (Ferreira, Thompson, Paris, Rohindra, & Rico, 2020). As microplastics are buoyant and flexible, they sink to remote water areas. It has been estimated that nearly 80% of the plastic pollution is from terrestrial source and other 18% from the fishery industry or aquaculture (Mai, Bao, Shi, Wong, & Zeng, 2018). Marine ecosystem is a highly complex three-dimensional system and the formation of cluster and movement of the microplastics are due to thermohaline circulation, storms and wind driven from land toward the sea.

2.2.1 River and lakes

Microplastics in the freshwater system have become an increasingly serious issue with limited studies suggesting the risk to the biotas due to high contamination levels in the rivers and lakes. Lesser attention has been given to freshwater system as compared to the effects of microplastics in marine water, despite the fact that most of the garbage litter are scattered onshore and riverside and later enter into the oceans. The sediments and water sample near the continental area are affected by the microplastic pollution. The three main ways through which microplastics can contaminate these freshwater systems are:

- discharge of effluents from the wastewater treatment systems,
- overflow of sewers during high rainfall, and
- run off from sludge applied to agricultural land.

The major sources of microplastics are the microbeads from the cosmetics industries and microfibers from the textile industries. Personal-care products such as the toothpaste, shampoos, facial cleanser, shaving gel, and other cosmetic products are the source of primary microplastics. One of recent studies has suggested that one shirt made of polyester produces 1900 microplastic fibers with each wash (Cheung & Fok, 2016; Hernandez, Nowack, & Mitrano, 2017). These are abundantly found in different shape, size, and color. Exclusion of these microfibers from the aquatic ecosystem is a difficult task as it also removes all plankton-size organisms. Another study in Canada showed that six brands of facial scrubs produced 4594—94,500 microbeads, which were around 164—327 μm in diameter and were released into the wastewater system after the use of the product (Dris et al., 2015).

There are only a few studies on the microplastics in river sediments. A river in Canada, *Saint Lawrence River*, has shown the high presence of microbeads of size 0.5–2 mm (Dris et al., 2015). One of the studies in the Netherlands proved that the presence of microplastics in Rhine River has reduced the high concentration of family of worms *Naididae*. Worms play a vital role in the environment. They increase the amount of air and water that get into the soil. Worms also compost the organic matter, making the soli fertile. According to Besseling et al., (Besseling, Redondo-Hasselerharm, Foekema, & Koelmans, 2019) worms ingest the plastic debris, which remain within them and reduces their ability to eat and grow, resulting in declination of their population. It was the first evidence showing that minuscule worms are vital for the environments found at the bottom of the lakes, rivers, ponds, and canals. Rivers are the common route of domestic and industrial waste, surface runoff and release stored water, which later merges into the seas and oceans. Sediments from the river Thames of the United Kingdom signify the presence of 12–22 synthetic microfibers per 100 g; river Rhine had near about 9,000,00 particles of microplastics per square kilometer in the urban areas, whereas Los Angeles River, San Gabriel River, and Coyote Creek showed the presence of about 13,000 fragments per cubic meter of plastic debris (Welden & Lusher, 2020).

Lakes are the temporary features in the aquatic ecosystem and act as a reservoir for microplastics. They are found abundantly where the movement of water is low. Great lakes are the series of interconnected freshwater lakes, which connects Atlantic Ocean through *Saint Lawrence River*. The concentration of microplastic in this lake is found to be 43,000 fragments per square kilometer, which is similar in concentration of Lake Geneva in Europe (48,000 fragments per square kilometer) (Welden & Lusher, 2020). Composition and quantity of microplastics in Lakes are influenced by the urbanization, industrialization, and wastewater treatment system and textile industries. In the United Kingdom, a single semiurban pond reveals that it had 5–80 fibers per kg of sediment of microfibers concentration. A total of 7% of the renewable freshwater of worlds is remained in the rivers and lakes; therefore an effective monitoring is required to limit the concentration of microplastics (Besseling et al., 2019).

2.2.2 Marine

Marine ecosystem provides important services, including tourism and transportation that are crucial to the economy of any country. Marine biodiversity and ecosystem are, therefore, important for the society as they provide feed for livestock and, provide food security (seafood for human consumption), medicines, and building material. Marines are the habitat of many species and serve in controlling the pollution and flood, protecting from storm, and shoreline stabilization (Free et al., 2014). Although the marines hold an important place, it is also the most neglected in terms of maintenance which has converted it into a sea of microplastic pollution. Beach debris and terrestrial sediments extracted by the marine are the major sources of microplastic pollution.

Plastic particles of size from few μm to 500 μm reside into seawater (Chatterjee & Sharma, 2019). Microplastics are not visible from naked eyes and even mesoplastics mixed with sand are not recognizable. In order to check the microplastics in the ocean, several new technologies have been introduced to identify microplastics. Some of these include optical microscopy, electron microscopy, Raman and FTIR spectroscopy, etc. (Osterlund, Renberg, Nordqvist, & Viklander, 2019) (Fig. 2.1).

Several studies have shown that unidirectional flow of freshwater system has driven the plastic debris into the oceans. These debris, which are made of plastic monofilament line along with the net, which are made of nylon, are buoyant in nature and are drifted deep in the ocean. The nylon nets act as a deathly weapon when biotas residing in oceans get entangled with them, known as "ghost fishing" (Andrady, 2011; Lozano & Mouat, 2009). Back in 1970, it was observed that the marine vessels (ships, containers, boats, and yachts) were the main source for marine littering and contributed around dumping 26,000 tons of plastic packaging materials globally. Thus a ban was implemented with international agreement (MARPOL 73/78 Annex V) to reduce or limit the disposing of wastes in the oceans. It was also believed that the lack of education and guidance made shipping an influential source of marine littering (Cole, Lindeque, Halsband, & Galloway, 2011).

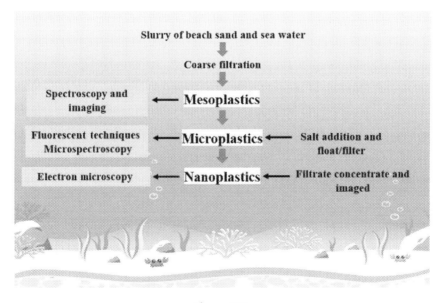

Figure 2.1
Sampling approach designed to isolate microplastics.

2.3 Microplastics in soil

While much focus has been given on the studies on microplastics prevalence and pollution in aquatic environment, there are not much known about this for soil. This could be due to the fact that microplastic filaments were easily obtained and computed from the sea water than from land (Corcoran et al., 2015), although there are reports describing the sources of plastic debris in terrestrial soil, including agricultural field, cities, industrial sites, and remote areas (Fuller & Gautam, 2016; Piehl et al., 2018; Rillig, 2012). The most affected biota in the soil due to microplastics is earthworms; the presence of plastic debris affects the soil biophysical properties such as bulk density and water-holding capacity (Scheurer & Bigalke, 2018). They also affect the growth of plants due to adversely affected soil parameters (Fig. 2.2).

Around 67% of global river input of plastics into land is contributed by the top 20 polluting rivers of the world with major rivers from Asia (Wan, Wu, Xue, & Hui, 2019; Blettler, Abrial, Khan, Sivri, & Espinola, 2018). Sewage sludge is the main source for the scattering of plastic debris on the land. Wastewater systems receive microplastics from the toothpastes, personal-care products, cosmetic particles (facial scrubs), and industrial wastes. Storm drains in the United Kingdom receive most of the synthetic fibers from the textile industries

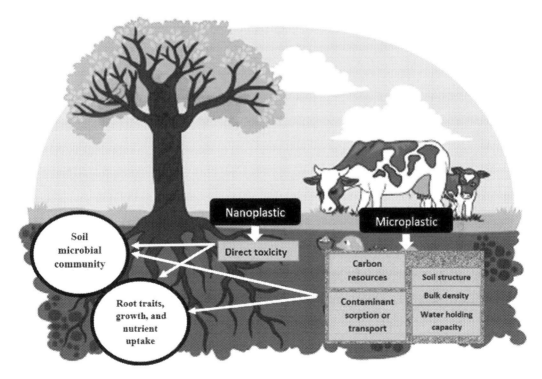

Figure 2.2
Mechanism through which microplastic affects the soil.

(Jambeck et al., 2015). These fibers originate from the domestic washing machines and reach the oceans. Wastewater system plant removes around 99% of the sludge containing microplastics. This sludge then ends up in meeting the requirement as a fertilizer in agricultural lands. It was believed that incorporating sludge in fields will improve the soil fertility and is therefore adopted and legally approved to be used as a fertilizer. This is how microplastics got their route to agricultural soil (Horton, Svendsen, Williams, Spurgeon, & Lahive, 2017; de Souza Machado, Kloas, Zarfl, Hempel, & Rillig, 2018).

Microplastics could be conceptually viewed as soil contaminant as microfibers can lower the bulk density of soil, causing reduction in the penetration of plant roots, which can result in soil drying (Scheurer & Bigalke, 2018). It also has negative impact on soil structure as it alters the composition and functioning of many soil microbes, thus hindering the growth of the plants. Changes in the soil structure could also affect the soil aggregation, which directly influences soil aeration and root growth. Microfibers negatively affect the soil aggregation but when entangled with soil particles could also act as a positive aid in soil aggregate formation (de Souza Machado et al., 2018).

Plastic particles do not get decomposed due to their high content of carbon, which is relatively inert (de Souza Machado et al., 2018).

Some contaminants can induce the hydrophobicity in the soil. Microplastics contain these phytotoxic substances at the time of manufacturing and could be transported into soil with the microplastics (Galloway, Cole, Lewis, Atkinson, & Allen, 2017). As discussed, the toxicity adversely affects the plant roots and plant growth. On the safer side, some contaminants are absorbed on the microplastic surface making them less available for the soil and terrestrial ecosystem. Studies have shown decreased plant performance parameter in the presence of biodegradable plastics, which was believed to cause immobilization of microbes (Rillig, Lehmann, de Souza Machado, & Yang, 2019). Plant performances are mainly dependent on the soil biota and their diversity, including root colonizing microorganism such as N_2 fixers, pathogens, and mycorrhizal fungi (Qi et al., 2018).

Microplastics cannot be taken up by the roots, but nanoplastics could be. Decreased particle size could affect the biota to become more toxic. Underlying mechanism of nanoplastics adsorbed strongly on to soil surfaces show the alteration of cell membrane and intercellular molecules with the generation of oxidative stress (Powell & Rillig, 2018). Nanoplastics could also be absorbed by the plant, which are intended for human and livestock consumption, thus entering the food chain.

2.4 Microplastic interaction with biotas

The abundant presence of microplastics and their routine interaction with the organisms in nature drew attention of ecotoxicologists on its safety and toxicity. Small plastic particles get adhered on the surface of biotas and get entangled. The studies of adherence on organisms are underexplored. But the cases of entanglement of microplastics are less likely to be observed as compared to ingestion. Ingestion of plastic debris has been recorded in numerous terrestrial and marine species of varied size and feeding mode. Microplastics act as carrier of hydrophobic organic contaminants, which release some chemical in the organism when adsorbed. These contaminants are more likely absorbed directly or indirectly by the prey animals, for example, by grazing herbivores such as *Littorina* and get transferred by the uptake in gill membranes in marine environment. Around 250 marine species were believed to be affected by the plastic ingestion. Microplastics are regularly taken by larger group of invertebrates due to smaller size of microplastic and its suspension in the water column. Feeding type and development drive the ingestion of microplastics in marine invertebrates such as free swimming and planktonic filterers. Microplastics reduce the mechanism of filtration and induce an inflammatory response in these organisms (Navarro et al., 2008).

Studies have shown that copepods ingest microplastic, which result in increased mortality rate, decreased fertility rate, weight loss, and poor feed intake, followed by the accumulation of polychlorinated biphenyl (Anbumani & Kakkar, 2018; Lee, Shim, Kwon, & Kang, 2013). The transfer of microplastic in marine food caused adverse health effects from monomers and plastic additives that are capable of inducing carcinogenesis and endocrine disruption. It was believed that major portion of microplastics was accumulated in the digestive tract of many consumers by the consumption of some small pelagic fish species. Moreover, nonbiodegradability of these plastics causes them to bioaccumulate in organisms through ingestion in their respective body tissues.

Information regarding the ecotoxicity of microplastic in freshwater is limited. It has been observed that the freshwater sponge *Hydra attenuate* has the potential of ingesting plastic particles of size less than 400 μm while other water flea, *Daphnia magna* can ingest plastic particles in gut epithelia of size up to 0.01 and 1 mm (Besseling, Wegner, Foekema, Van Den Heuvel-Greve, & Koelmans, 2013; Murphy & Quinn, 2018). Daphnids are also likely to be ingesting the microfibers from the textile weathering and washing of fibers size between 300 and 1400 μm. Ogonowshi et al. (Rosenkranz, Chaudhry, Stone, & Fernandes, 2009) reported the exposure of *Daphnia magna* with primary and secondary plastic accumulation in digestive tract and noted its performance on feeding rate and reproductivity. The effect of fluorescent polystyrene beads ingestion of size 2 μm and 100 nm by *Daphnia magna* has been reported (Ogonowski, Schur, Jarsen, & Gorokhova, 2016). Results on feeding rate and reproductivity performance showed that particle sizes of 2 μm were five times higher ingestion at the end of 21-day exposure. Survival rate, body length, and feeding rate of sediment dwelling organism *Caenorhabditis elegans* of freshwater pelagic and benthic ecosystem were studied and which showed reduced particle size induces oxidative stress (Rist et al., 2016). Microplastic uptake by *Mytilus edulis* and lug worm, *Arenicola marina* (deposit feeder) showed 20–90 and 15- to 500-μm-size microplastic particles present in feces and 15–100 and 35- to 1000-μm-size particles in the tissues (Lei et al., 2018). High-density polyethylene microplastics were observed in *M. edulis* at cell and tissue level leading to uptake in gills and digestive gland (Van Cauwenberghe, Claessens, Vandegehuchte, & Janssen, 2015).

2.5 Microplastic in waste

Wastes are residual part of raw material that is generally undesirable or useless for consumer after primary use (Banerjee et al., 2018; Von Moos, Burkhardt-Holm, & Kohler, 2012). It can be classified based on various categories such as physical state (solid, liquid, and gas); source (household/domestic waste, industrial waste, agricultural waste, commercial waste, demolition and construction waste, and mining waste); and environmental (hazardous waste and nonhazardous waste) (Basu, 2009). These wastes are the result of various human activities. In the sixteenth century, considerable increase in the volume of waste generation was seen. The increase in waste was associated with the initiation of migration of people from rural areas to urban areas in search of employment in industry which results in expansion of population in urban areas. This finally led to the current volume and variety available in waste (Amasuomo & Baird, 2016). Solid and industrial wastes constitute the major fraction of the waste generated from urban areas and contain huge amount of microplastic. In the following sections, microplastic from solid and industrial waste is discussed.

2.5.1 Microplastic in solid waste

Solid waste is the useless, unwanted, and discarded material, resulting from day-to-day activities in the community (Wilson, 2007). It mainly comprises of village, hospital, agriculture, and municipal solid wastes (MSWs) (Fig. 2.3). Among these, majority of solid

Figure 2.3
Different kinds of solid waste.

wastes is generated from the municipalities. Rise in the living standard in the developing countries, burgeon economy, rapid expansion in population, and urbanization have greatly speed up the rate, amount, and quality of MSW (Mishra, Mishra, & Tiwari, 2014). Solid waste disposal can be done by various methods such as landfills, controlled dumps, and open-pit dumps. Out of these, land filling is the most attractive and cheap method to manage MSW (Abdel-Shafy & Mansour, 2018). About 90% of MSW is dumped in open field unscientifically, which creates environmental deterioration and adversely affects human health (Madera & Valencia-Zuluaga, 2009). MSW includes waste produced from the commercial, domestic, industrial, institutional, demolition, construction, and municipal services (Sharholy, Ahmad, Mahmood, & Trivedi, 2008). MSW is one of the main sources of microplastics.

Both primary and secondary microplastics can occur in the environment. Primary microplastics, such as microbeads (cosmetics), capsules (medical), fibers (textiles), and pellets (in plastic manufacturing), are created intentionally using certain chemicals. Bisphenol A (BPA) is a chemical, which is used in the manufacture of polycarbonate plastics and epoxy resins. BPA is responsible for the endocrine disturbance, which is found in the leachates from MSW landfill and considered as the primary source of contamination in landfill leachate (Barnabas et al., 2017; Horton, Walton, Spurgeon, Lahive, & Svendsen, 2017). Leachates are polluted water produced due to the leaching of rainwater or inherent water present in the wastes (Fig. 2.4A,B) (Xu, Zhang, He, & Shao, 2011). Microplastics containing leachate come into the environment through leachate leakage or by the discharge of leachate collecting and treatment systems (Renou, Givaudan, Poulain, Dirassouyan, & Moulin, 2008).

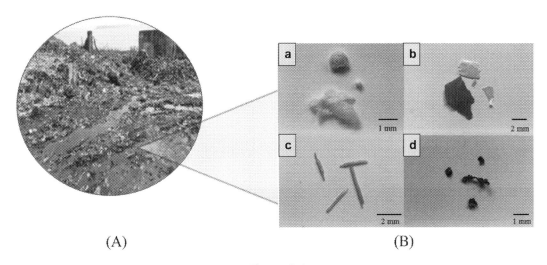

Figure 2.4
(A) Leachates from landfill polluting the environment. (B) Microplastics from the leachate samples: (a) polystyrene foams and polymethyl methacrylate pellet; (b) polyethlene flakes; (c) polypropylene lines; and (d) polyethylene fragments.

In contrast, the secondary microplastics are the result of the breakdown of large pieces of plastic into smaller fragments due to physical damage/stress and prolonged exposure to sunlight. Secondary microplastics are considered the "big contributor" of microplastic pollution in the environment. It occurs from human activity, such as laundry washing, littering, and MSW collection and disposal processes. Laundry washing machines discharge large amount of microplastic into wastewaters. Household dusts also contain fragments of plastic. All these microplastics drift to land and aquatic environment by wind whereas denser plastic fragments are buried deeper in the soil layers (Fig. 2.5) (He, Chen, Shao, Zhang, & Lü, 2019; Laskar & Kumar, 2017).

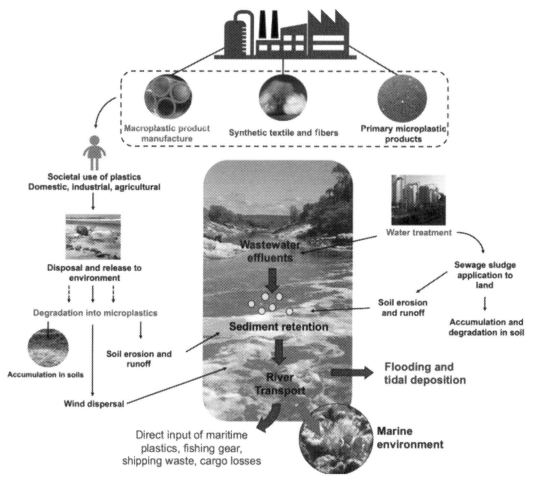

Figure 2.5
Microplastic sources and flows throughout and between anthropogenic, terrestrial, freshwater, and marine environmental compartments.

Paint particles are also described as microplastics because of the presence of resin in it. It can enter into the environment through weathering, sanding of old paint layers, and through rinsing brushes and rollers. Another source of microplastic is road vehicle tire tread as it consists of rubber polymers. Further, these microplastics become part of solid waste and create problems in environment (Akdogan & Guven, 2019).

In agriculture and horticulture field, plastic mulching and polytunnels play an important role as it has various advantages such as enhancing crop yield, early harvesting, improving fruit quality, control temperature, and increase water-use efficiency. It also helps to reduce the weed and, therefore, less amount of herbicide spray is needed. Apart from the benefits, however, these anthropogenic activities are the main cause of environmental pollution also. These mulches are not degradable in nature that leads to environmental pollution through incineration emissions, landfill leaching, or microplastic residues. Application of sewage sludge for soil amendment and polymer seed coating are other anthropogenic activities which are responsible for microplastic pollution in soil. All these plastics debris ultimately get fragmented as secondary microplastics. Earth worm also grinds the plastic debris in their gizzard that adds up to secondary microplastics. Anecic earth worms have the habit to make vertical burrows in soil and therefore, deposit the surface plastic debris into the soil. Other soil mesofauna (such as collembola or mites) as well as digging mammals (gophers) are also responsible for the fragmentation of plastic into microplastic that further accumulate in the soil detrital food web (Fuller & Gautam, 2016; Verschoor, Pooter, Droge, Kuenen, & De Valk, 2020).

2.5.2 Microplastic in industrial waste

Microplastics are tiny particles, which are easy to incorporate in products; hence it increases the volume, weight, viscosity, and adhesive quality of the product (Steinmetz et al., 2016). Because of these properties, microplastic beads are used in the manufacturing of soap, shampoo, deodorant, toothpaste, wrinkle creams, moisturizers, shaving cream, sunscreen lotion, facial masks, lipstick, eye shadow, children's bubble bath, and in textile industries. Mainly polyethylene microbeads are used in the manufacturing of personal-care products (Raj, Kumar, & Galhotra, 2018). Wastes from these industries, which are the main source of microplastics, enter the wastewater treatment plants and then river catchments area (Bhattacharya, 2016; Brate, Blazquez, Brooks, & Thomas, 2018). Between 4594 and 94,500 microbeads are released from exfoliate wash in a single use (Kay, Hiscoe, Moberley, Bajic, & McKenna, 2018). Several countries, including Canada, Ireland, the United Kingdom, and the United States proposed to ban these microbeads. Due to this, companies such as Crest, Johnson and Johnson, and L'Óreal have stopped using microbeads in their products.

Industries that manufacture the raw material for plastic industry such as pellets, spherules, granules, discs, etc. are also responsible for microplastic pollution (Napper, Bakir, Rowland,

& Thompson, 2015). These plastic pellets (also called nurdles and mermaid tears) are used further for the manufacture of various products like plastic bottles. Pellets are sold and shipped to companies that manufacture product from these pellets. Pellets can be spilled into the environment throughout this process. Glitter, contact lens cleaners, small buttons, and jewelry are other consumer product that creates microplastic pollution. Nondomestic source of microplastic includes plastic particles used in air blasting to clean paint and engines; preproduction pellets lost during transport or manufacturing; lost Styrofoam used in fillings or shipping; fibers from synthetic textile industries; and dust from drilling and cutting plastics. Microplastics from these nondomestic sources enter the sewage system and pollute the environment (Lechner & Ramler, 2015). Furthermore, abrasive blasting media as well as oil and gas exploration, drilling fluids based on plastic microbeads are also other sources of microplastic (Prata, 2018).

Textiles industry materials are not only used for making clothing but also for carpets, household textiles, vehicles, and a wide range of technical products like the filter for improving the quality of air and water. Synthetic fiber is one of the most common forms of microplastic which is used by these industries. It can enter the aquatic environment from various pathways, for example, industrial wastewater, laundry washing machine, fishing nets, and disposed clothing. Synthetic fiber does not decompose in environment and therefore contaminate sewage sludge during wastewater treatment (Sundt, Schulze, & Syversen, 2014; Dris, Gasperi, Saad, Mirande-Bret, & Tassin, 2016). Fisheries and aquaculture industries also generate waste during their operations. These industries use polyamide and polyolefin for the manufacture of ropes, lines, and fishing nets. Waste coming out from these industries is dumped in the ocean which create a problem for aquatic environment (Andrady, 2011; Toben, 2017).

2.6 Conclusions and perspectives

Microplastics are the contaminants of very serious global concern. The presence of microplastics in the environment has become a major issue since the last few decades and has gained major attention from the researchers all over the world. In recent years, tremendous understanding has been on basic aspects of their spread in the environment, more specifically in aquatic life and how the life there is impacted due to their presence. However, much is to be understood still to curb the microplastics footprint. Some of them are listed below:

1. A detailed protocol should be prepared for the isolation, identification, characterization, and validation of instrumental analysis for the analysis of microplastic in the environmental samples.
2. Till-date most attention has been paid on microplastic in aquatic environment; studies on the presence of microplastic in air and terrestrial ecosystem are lacking and should

be carried out to compliment the spread and impact of microplastic in the environment as a whole.
3. Effect of microplastic on the growth of the plants should be studied, including the aspects related with the factors affecting plant species, plant performance, root and plant growth, and likely other mechanism.
4. Effects of microplastics on the invertebrates should be studied as they are possibly the mostly affected species due to the allocation of plastics to higher tropics.
5. Research should focus in developing alternative microbeads for use in cosmetic products, which should be biodegradable in nature.
6. Studies should also be undertaken on the impact of microplastics on human health through sea food as these are also currently not available.

To sum up, the research focus should be on developing sound information on the contamination in aquatic and terrestrial biodiversity. The presence of microplastics in all kinds of environment has become a serious concern for the public at large, including scientists, academicians, and governments. As the plastic usage is continuing to rise in our daily lives, the sources of microplastics will rise accordingly. Therefore, to provide a maximum integrated view, a more scientometric approach is required to quantify and assess the knowledge related to microplastics.

Acknowledgment

One of the authors (Ranjna Sirohi) acknowledges CSIR, New Delhi for providing a fellowship under direct SRF scheme bearing the grant no. 09/171(0136)/19. Raveendran Sindhu acknowledges the Department of Science and Technology, New Delhi for sanctioning a project under DST WOS-B scheme.

References

Abdel-Shafy, H. I., & Mansour, M. S. M. (2018). Solid waste issue: Sources, composition, disposal, recycling, and valorization. *Egyptian Journal of Petroleum*, *27*, 1275–1290.

Akdogan, Z., & Guven, B. (2019). Microplastics in the environment: A critical review of current understanding and identification of future research needs. *Environmental Pollution*, *254*, 11301.

Amasuomo, E., & Baird, J. (2016). The concept of waste and waste management. *Journal of Management and Sustainability*, *6*(4), 88–96.

Anbumani, S., & Kakkar, P. (2018). Ecotoxicological effects of microplastics on biota: A review. *Environmental Science and Pollution Research*, *25*(15), 14373–14396.

Anderson, J. C., Park, B. J., & Palace, V. P. (2016). Microplastics in aquatic environments: Implications for Canadian ecosystems. *Environmental Pollution*, *218*, 269–280.

Andrady, A. L. (2011). Microplastics in the marine environment. *Marine Pollution Bulletin*, *62*, 1596–1605.

Banerjee, P., Hazra, A., Ghosh, P., Ganguly, A., Murmu, N. C., & Chatterjee, P. K. (2018). Solid waste management in India: A brief review. In S. K. Ghosh (Ed.), *Waste Management and Resource Efficiency* (pp. 1027–1049). Springer Publishing.

Barnabas, S. G., Sivakumar, G. D., Pandian, G. S., Vasantha Geethan, K. A., Kumar, S. P., & Kumar, P. D. (2017). Solid waste management across the world—A review. *Ecology, Environment and Conservation*, *23*, 339–348.

Basu, R. (2009). Solid waste management—A model study. *SIES Journal of Management, 6*, 20−24.

Besseling, E., Redondo-Hasselerharm, P., Foekema, E. M., & Koelmans, A. A. (2019). Quantifying ecological risks of aquatic micro- and nanoplastic. *Critical Reviews in Environmental Science & Technology, 49*(1), 32−80.

Besseling, E., Wegner, A., Foekema, E. M., Van Den Heuvel-Greve, M. J., & Koelmans, A. A. (2013). Effects of microplastic on fitness and PCB bioaccumulation by the lugworm *Arenicola marina* (L.). *Environmental Science & Technology, 47*(1), 593−600.

Bhattacharya, P. (2016). A review on the impacts of microplastic beads used in cosmetics. *Acta Biomedica Scientia., 3*, 47−52.

Blettler, M. C., Abrial, E., Khan, F. R., Sivri, N., & Espinola, L. A. (2018). Freshwater plastic pollution: Recognizing research biases and identifying knowledge gaps. *Water Research, 143*, 416−424.

Brate, I. L. N., Blazquez, M., Brooks, S. J., & Thomas, K. V. (2018). Weathering impacts the uptake of polyethylene microparticles from toothpaste in Mediterranean mussels (*M. galloprovincialis*). *Science of the Total Environment, 626*, 1310−1318.

Van Cauwenberghe, L., Claessens, M., Vandegehuchte, M. B., & Janssen, C. R. (2015). Microplastics are taken up by mussels (*Mytilus edulis*) and lugworms (*Arenicola marina*) living in natural habitats. *Environmental Pollution, 199*, 10−17.

Chatterjee, S., & Sharma, S. (2019). Microplastics in our oceans and marine health. *Field Actions Science Reports. The Journal of Field Actions, 19*, 54−61.

Cheung, P. K., & Fok, L. (2016). Evidence of microbeads from personal care product contaminating the sea. *Marine Pollution Bulletin, 109*(1), 582−585.

Cole, M., Lindeque, P., Halsband, C., & Galloway, T. S. (2011). Microplastics as contaminants in the marine environment: A review. *Marine Pollution Bulletin, 62*(12), 2588−2597.

Corcoran, P. L., Norris, T., Ceccanese, T., Walzak, M. J., Helm, P. A., & Marvin, C. H. (2015). Hidden plastics of Lake Ontario, Canada and their potential preservation in the sediment record. *Environmental Pollution, 204*, 17−25.

Corona, E., Martin, C., Marasco, R., & Duarte, C. M. (2020). Passive and active removal of marine microplastics by a mushroom coral (*Danafungia scruposa*). *Frontiers in Marine Science, 7*, 128.

Davidson, K., & Dudas, S. E. (2016). Microplastic ingestion by wild and cultured Manila clams (*Venerupis philippinarum*) from Baynes Sound, British Columbia. *Archives of Environmental Contamination and Toxicology, 71*(2), 147−156.

Dawson, A. L., Kawaguchi, S., King, C. K., Townsend, K. A., King, R., Huston, W. M., & Nash, S. M. B. (2018). Turning microplastics into nanoplastics through digestive fragmentation by Antarctic krill. *Nature Communications, 9*(1), 1−8.

Dris, R., Gasperi, J., Saad, M., Mirande-Bret, C., & Tassin, B. (2016). Synthetic fibers in atmospheric fallout: A source of microplastics in the environment? *Marine Pollution Bulletin, 104*(1−2), 290−293, 10.1016/j.marpolbul.2016.01.006.hal-01251430.

Dris, R., Imhof, H., Sanchez, W., Gasperi, J., Galgani, F., Tassin, B., et al. (2015). Beyond the ocean: Contamination of freshwater ecosystems with (micro-) plastic particles. *Environmental Chemistry, 12*(5), 539−550.

Ferreira, M., Thompson, J., Paris, A., Rohindra, D., & Rico, C. (2020). Presence of microplastics in water, sediments and fish species in an urban coastal environment of Fiji, a Pacific small island developing state. *Marine Pollution Bulletin, 153*, 110991.

Free, C. M., Jensen, O. P., Mason, S. A., Eriksen, M., Williamson, N. J., & Boldgiv, B. (2014). High-levels of microplastic pollution in a large, remote, mountain lake. *Marine Pollution Bulletin, 85*(1), 156−163.

Fuller, S., & Gautam, A. (2016). A procedure for measuring microplastics using pressurized fluid extraction. *Environmental Science & Technology, 50*(11), 5774−5780.

Galloway, T. S., Cole, M., Lewis, C., Atkinson, A., & Allen, J. I. (2017). Interactions of microplastic debris throughout the marine ecosystem. *Nature Ecology & Evolution, 1*, 0116.

Habib, R. Z., Thiemann, T., & Al Kendi, R. (2020). Microplastics and waste water treatment plants—A review. *Journal of Water Resource and Protection, 12*(01), 1.

Hartmann, N. B., Huffer, T., Thompson, R. C., Hassellov, M., Verschoor, A., Daugaard, A. E., ... Wagner, M. (2019). Are we speaking the same language? Recommendations for a definition and categorization framework for plastic debris. *Environmental Science & Technology*, *53*, 1039–1047.

He, P., Chen, L., Shao, L., Zhang, H., & Lü, F. (2019). Municipal solid waste (MSW) landfill: A source of microplastics?—Evidence of microplastics in landfill leachate. *Water Research*, *159*, 38–45.

Hernandez, E., Nowack, B., & Mitrano, D. M. (2017). Polyester textiles as a source of microplastics from households: A mechanistic study to understand microfiber release during washing. *Environmental Science & Technology*, *51*(12), 7036–7046.

Horton, A. A., Svendsen, C., Williams, R. J., Spurgeon, D. J., & Lahive, E. (2017). Large microplastic particles in sediments of tributaries of the River Thames, UK—Abundance, sources and methods for effective quantification. *Marine Pollution Bulletin*, *114*(1), 218–226.

Horton, A. A., Walton, A., Spurgeon, D. J., Lahive, E., & Svendsen, C. (2017). Microplastics in freshwater and terrestrial environments: Evaluating the current understanding to identify the knowledge gaps and future research priorities. *Science of the Total Environment*, *586*, 127–141.

Jambeck, J. R., Geyer, R., Wilcox, C., Siegler, T. R., Perryman, M., Andrady, A., & Law, K. L. (2015). Plastic waste inputs from land into the ocean. *Science (New York, N.Y.)*, *347*(6223), 768–771.

Kay, P., Hiscoe, R., Moberley, I., Bajic, L., & McKenna, N. (2018). Wastewater treatment plants as a source of microplastics in river catchments. *Environmental Science and Pollution Research*, *25*, 20264–20267.

Laskar, N., & Kumar, U. (2017). Plastics and microplastics: A threat to environment. *Environmental Technology and Innovation*, *14*, 100352.

Lechner, A., & Ramler, D. (2015). The discharge of certain amounts of industrial microplastic from a production plant into the River Danube is permitted by the Austrian legislation. *Environmental Pollution*, *200*, 159–160.

Lee, K. W., Shim, W. J., Kwon, O. Y., & Kang, J. H. (2013). Size-dependent effects of micro polystyrene particles in the marine copepod *Tigriopus japonicus*. *Environmental Science & Technology*, *47*(19), 11278–11283.

Lei, L., Wu, S., Lu, S., Liu, M., Song, Y., Fu, Z., & He, D. (2018). Microplastic particles cause intestinal damage and other adverse effects in zebra fish *Danio rerio* and nematode *Caenorhabditis elegans*. *Science of the Total Environment*, *619*, 1–8.

Lozano, R. L., & Mouat, J. (2009). *Marine litter in the North-East Atlantic region: Assessment and priorities for response*. London, UK: OSPAR Commission. Available from https://qsr2010.ospar.org/media/assessments/p00386, ISBN 978-1-906840-26-6.

Madera, C., Valencia-Zuluaga, V. (2009). Landfill leachate treatment: One of the bigger and underestimated problems of the urban water management in developing countries. *9th World Wide Workshop for Young Environmental Scientists WWW-YES-Brazil-2009: Urban Waters: Resource or Risks*, No. 12, Belo Horizonte, MG, Brazil. https://hal.archives-ouvertes.fr/hal-00593299.

Mai, L., Bao, L. J., Shi, L., Wong, C. S., & Zeng, E. Y. (2018). A review of methods for measuring microplastics in aquatic environments. *Environmental Science and Pollution Research*, *25*(12), 11319–11332.

Mishra, A. R., Mishra, S. A., & Tiwari, A. V. (2014). Solid waste management—Case study. *International Journal of Research in Advent Technology*, *2*(1), 396–399.

Von Moos, N., Burkhardt-Holm, P., & Kohler, A. (2012). Uptake and effects of microplastics on cells and tissue of the blue mussel *Mytilus edulis* L. after an experimental exposure. *Environmental Science & Technology*, *46*(20), 11327–11335.

Murphy, F., & Quinn, B. (2018). The effects of microplastic on freshwater *Hydra attenuata* feeding, morphology & reproduction. *Environmental Pollution*, *234*, 487–494.

Napper, I. E., Bakir, A., Rowland, S. J., & Thompson, R. C. (2015). Characterization, quantity and sorptive properties of microplastics extracted from cosmetics. *Marine Pollution Bulletin.*, *99*, 178–185.

Navarro, E., Baun, A., Behra, R., Hartmann, N. B., Filser, J., Miao, A. J., & Sigg, L. (2008). Environmental behavior and ecotoxicity of engineered nanoparticles to algae, plants, and fungi. *Ecotoxicology (London, England)*, *17*(5), 372–386.

New Plastic Economy. (2020). Retrieved from https://www.ellenmacarthurfoundation.org/assets/downloads/13319-Global-Commitment-Definitions.pdf on 10.09.2020.

Ogonowski, M., Schur, C., Jarsen, A., & Gorokhova, E. (2016). The effects of natural and anthropogenic microparticles on individual fitness in Daphnia magna. *PLoS One*, *11*(5).

Osterlund, H., Renberg, L., Nordqvist, K., Viklander, M. (2019). Micro litter in the urban environment: Sampling and analysis of undisturbed snow. In *Novatech 2019 10th international conference*. Retrieved from urn:nbn:se:ltu:diva-77479 on 30.04.2020.

Piehl, S., Leibner, A., Loder, M. G. L., Dris, R., Bogner, C., & Laforsch, C. (2018). Identification and quantification of macro- and microplastics on an agricultural farmland. *Scientific Reports*, *8*, 17950.

Powell, J. R., & Rillig, M. C. (2018). Biodiversity of arbuscular mycorrhizal fungi and ecosystem function. *New Phytologist*, *220*(4), 1059−1075.

Prata, J. C. (2018). Microplastics in wastewater: State of the knowledge on sources, fate and solutions. *Marine Pollution Bulletin*, *129*, 262−265.

Qi, Y., Yang, X., Pelaez, A. M., Lwanga, E. H., Beriot, N., Gertsen, H., & Geissen, V. (2018). Macro-and micro-plastics in soil-plant system: Effects of plastic mulch film residues on wheat (*Triticum aestivum*) growth. *Science of the Total Environment*, *645*, 1048−1056.

Raj, U., Kumar, P., & Galhotra, A. (2018). Microplastics—All we know till now and the way out. *Indian Journal of Community and Family Medicine*, *4*, 19−21.

Renou, S., Givaudan, J. G., Poulain, S., Dirassouyan, F., & Moulin, P. (2008). Landfill leachate treatment: Review and opportunity. *Journal of Hazardous Materials*, *150*, 468−493.

Rillig, M. C. (2012). Microplastic in terrestrial ecosystems and the soil? *Environmental Science & Technology*, *46*, 6453−6454.

Rillig, M. C., Lehmann, A., de Souza Machado, A. A., & Yang, G. (2019). Microplastic effects on plants. *New Phytologist*, *223*(3), 1066−1070.

Rist, S. E., Assidqi, K., Zamani, N. P., Appel, D., Perschke, M., Huhn, M., & Lenz, M. (2016). Suspended micro-sized PVC particles impair the performance and decrease survival in the Asian green mussel *Perna viridis*. *Marine Pollution Bulletin*, *111*(1−2), 213−220.

Rosenkranz, P., Chaudhry, Q., Stone, V., & Fernandes, T. F. (2009). A comparison of nanoparticle and fine particle uptake by Daphnia magna. *Environmental Toxicology and Chemistry: An International Journal*, *28*(10), 2142−2149.

Scheurer, M., & Bigalke, M. (2018). Microplastics in Swiss floodplain soils. *Environmental Science & Technology*, *52*(6), 3591−3598.

Sharholy, M., Ahmad, K., Mahmood, G., & Trivedi, R. C. (2008). Municipal solid waste management in Indian cities—A review. *Waste Management*, *28*, 459−467.

de Souza Machado, A. A., Kloas, W., Zarfl, C., Hempel, S., & Rillig, M. C. (2018). Microplastics as an emerging threat to terrestrial ecosystems. *Global Change Biology*, *24*(4), 1405−1416.

de Souza Machado, A. A., Lau, C. W., Till, J., Kloas, W., Lehmann, A., Becker, R., & Rillig, M. C. (2018). Impacts of microplastics on the soil biophysical environment. *Environmental Science & Technology*, *52*(17), 9656−9665.

Steinmetz, Z., Wollmann, C., Schaefer, M., Buchmann, C., David, J., Troger, J., ... Schaumann, G. E. (2016). Plastic mulching in agriculture. Trading short-term agronomic benefits for long-term soil degradation. *Science of the Total Environment*, *550*, 690−705.

Sundt, P., Schulze, P.E., Syversen, F. (2014). Sources of microplastic-pollution to the marine environment. Mepex Report for the Norwegian Environment Agency, pp. 1−108. https://www.miljodirektoratet.no/globalassets/publikasjoner/M321/M321.pdf Retrieved on 01.05.2020.

Thompson, R. C., Moore, C. J., Vom Saal, F. S., & Swan, S. H. (2009). Plastics, the environment and human health: Current consensus and future trends. *Philosophical Transactions of the Royal Society B: Biological Sciences*, *364*(1526), 2153−2166.

Toben, M. (2017). Microplastic pollution originating from textiles and paints: Environmental impacts and solutions (CCB technical report), Uppsala, Sweden. Retrieved from www.ccb.se on 05.05.2020.

Verschoor, A., Pooter, D., Droge, L., Kuenen, R., De Valk, E. (2016). National Institute for Public Health and the Environment. Report on the emission of microplastics and potential mitigation measures. Retrieved from https://www.rivm.nl/bibliotheek/rapporten/2016-0026.pdf on 01.07.2020.

Wan, Y., Wu, C., Xue, Q., & Hui, X. (2019). Effects of plastic contamination on water evaporation and desiccation cracking in soil. *Science of the Total Environment, 654*, 576−582.

Welden, N. A., & Cowie, P. R. (2017). Degradation of common polymer ropes in a sublittoral marine environment. *Marine Pollution Bulletin, 118*(1-2), 248−253.

Welden, N. A., & Lusher, A. (2020). *Microplastics: From origin to impacts. In Plastic Waste and Recycling* (pp. 223−249). Academic Press.

Wilson, D. C. (2007). Development drivers for waste management. *Waste Management & Research the Journal of the International Solid Wastes & Public Cleansing Association Iswa, 25*(3), 198−207.

Wright, S. L., & Kelly, F. J. (2017). Plastic and human health: A micro issue? *Environmental Science & Technology, 51*(12), 6634−6647.

Xu, S. Y., Zhang, H., He, P. J., & Shao, L. M. (2011). Leaching behaviour of bisphenol A from municipal solid waste under landfill environment. *Environmental Technology, 32*, 1269−1277.

Zmak, I., & Hartmann, C. (2017). Current state of the plastic waste recycling system in the European Union and in Germany. *Tehnicki Glasnik, 11*(3), 138−142.

CHAPTER 3

Thermoplastic starch

Ranjna Sirohi[1], Shikhangi Singh[1], Ayon Tarafdar[2], Nalla Bhanu Prakash Reddy[1], Taru Negi[3], Vivek Kumar Gaur[4], Ashutosh Kumar Pandey[5], Raveendran Sindhu[6], Aravind Madhavan[7] and K.B. Arun[7]

[1]Department of Post-Harvest Process and Food Engineering, G. B. Pant University of Agriculture and Technology, Pantnagar, India, [2]Division of Livestock Production and Management, ICAR-Indian Veterinary Research Institute, Izatnagar, Bareilly, India, [3]Department of Food Science and Technology, College of Agriculture, G. B. Pant University of Agriculture and Technology, Pantnagar, India, [4]Environmental Biotechnology Division, Environmental Toxicology Group, CSIR—Indian Institute of Toxicology Research, Lucknow, India, [5]Centre for Energy and Environmental Sustainability, Lucknow, India, [6]Microbial Processes and Technology Division, CSIR—National Institute for Interdisciplinary Science and Technology (CSIR—NIIST), Thiruvananthapuram, India, [7]Rajiv Gandhi Center for Biotechnology, Thiruvananthapuram, India

3.1 Introduction

Plastics are synthetic polymer of high molecular mass obtained mostly from the fossil fuels such as coal and oil (crude oil). Due to their characteristics such as good strength, low cost, light weight, easy, and energy-efficient processability, plastics have become an essential commodity of the modern life, whose demand has been rapidly increasing. The numerous properties stated above along with other properties such as ease of formation, heat sealing capability, durability, force of effect, shear strength, and most importantly, good water vapor, and oxygen-barrier properties have made plastic the most versatile material for the packaging, medical utilities, automobile parts, daily-use household items, etc. Some of its specific attribute make it suitable for food packaging also (Laftah, 2017). Polymers that are mostly used for food packaging are nonbiodegradable petroleum-based polymers, which have low density and occupy more volume when discarded in the landfills and dumps (Khan, Khan Niazi, Samin, & Jahan, 2016). Disposal in the environment causes large ecological impact due to their nonbiodegradable nature as well as nonrecyclability. Therefore, many countries have imposed laws the banning their disposal in the environment or even regulating their industrial manufacturing. This has led to search and develop eco-friendly renewable natural polymer, which can meet the demand of customers as well as industries and are environmental-friendly.

Major renewable polymers occurring in the nature are starch, cellulose, protein, and chitin. Out of these, starch has been gaining considerably attention since several decades due to its high productivity from renewable resources and simple processing ability with conventional plastic-processing equipment. Starch is well known for its functional properties and is mostly employed by food industries as a hydrocolloid. Wheat, rice, maize, cassava, and potatoes are some of the major sources of plant-based starch (Oliveira, Fidalgo, Valera, & Demarquette, 2014).

Starch is polysaccharide made of glucose units with basic chemical formula $C_6H_{10}O_5$. It has two components: amylose and amylopectin. Amylose is a straight-chain polymer made of alpha-1,4-glycosidic bonds, while amylopectin is a branched polymer of alpha-1,4- and alpha-1,6-glycosidic bonds joined at branching points (Fig. 3.1). Most of the starch contain 20%−30% amylose and 70%−80% amylopectin. Around 200−20,000 glucose units are connected with the alpha-1,4-glycosidic bonds in amylose molecule. The framework of amylose is linear, which closely resembles with the behavior of conventional synthetic polymers, though it changes during the thermal processing. The very reason behind is that the native starch cannot fall under the category of thermoplastic polymer as they exhibit a powerful intra- and intermolecular H-bonding in the macromolecular chains of amylose as well as amylopectin. The free hydroxyl group present in the native starch absorbs water, resulting in the swelling of the starch granules. Starch granules after heating swell and the intermittent H-bonding in between the glucose unit changed damaging the crystalline nature. This results in the breakdown of starch granules to some extent and is referred to as "de-structured starch." Variations in size and form of starch granules is because of their plant origin from which they are derived. The botanical sources of starch and their characterization, including amylose content, granule size, and gelatinization temperature are shown in Table 3.1.

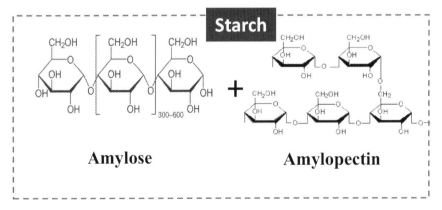

Figure 3.1
Structure of amylose and amylopectin.

Table 3.1: Characterization of different source of starch (Perez-Pacheco et al., 2016).

Starch source	Amylose (%)	Granule dimension (μm)	Gelatinization temperature (°C)
Potato	20	10–100	57–65
Rice	0–33	2–7	55–79
Maize	55–75	6–40	63–67
Wheat	25	Spherical–6 and flattened–25	58–64
Mango	9–16	7–28	77–80
Cassava	16.8–21.5	5–35	60–80
Pea	25.3	10–70	65–70
Sorghum	16.1–55.8	5.5–30	70–75

Source: *Modified from Perez-Pacheco, E., Canto-Pinto, J.C., Moo-Huchin, V.M., Estrada-Mota, I.A., Estrada-León, R.J., & Chel-Guerrero, L. (2016). Thermoplastic starch (TPS)-cellulosic fibers composites: Mechanical properties and water vapour barrier: A review.* Composites from renewable and sustainable materials *(pp. 85–105). Inteach.*

High moisture sensitivity and the low mechanical strength are the limitation of starch-based polymers. Starch has two main disadvantages over the synthetic polymeric materials: first, it is a hydrophilic polymer because of the existence of hydroxy groups. In excess water, amylopectin loses its crystallinity and starch granules swell due to hydrolytic degradation, which ultimately diffuse the linear amylose molecules into solution. When starch granules are exposed to moisture, it disintegrates and loses its properties (Carvalho, Curvelo, & Agnelli, 2001). Secondly, it cannot be considered as a real thermoplastic polymer until starch is processed under the action of heat, water/plasticizer, and shear, a process known as gelatinization (Carvalho, 2008; Garcia, Ribba, Dufresne, Aranguren, & Goyanes, 2011).

Thermoplastic starch (TPS) is a gelatinized starch instead of a granular starch. Several methods can be used to convert the granular starch to TPS. Preparation process of TPS under high temperature and shear breaks the strong intramolecular H-bond and is replaced by intermolecular H-bond between the polysaccharide chain and plasticizer. Starch with low moisture content (10%–30%) and plasticizers (glycerol, water, sorbitol, etc.) is extruded under shear at high temperature (90°C–180°C). Due to the impact of temperature, the starch (native form) instantaneously melts and starts flowing, therefore is utilized as an injection molding, or blowing material, in the same way as most of the conventional synthetic thermoplastic polymers are employed to achieve thermoplastic melt leaving no crystallinity of residual starch.

TPS has the potential to increase the ductility of granular starch. Native starch shows high fragility and low mechanical properties. However, the presence of plasticizers or blends with the plasticizer enhances the innate characteristics of TPS. Plasticizers play the role in breaking the H-bond in starch-starch granules and lowering the melting temperature below their breakdown temperature (230°C). Some TPSs are transformed into commercial polymers such as packaging materials (loose fillers and films), fertilizer bags, glazes, disposable nappies, and bark films. TPS made films and coatings are used in seafood products, meats and poultry, grain, vegetables and fruits, and candy-processing plants. However, high amylose content in starch leads to decrease in the flexibility as compared to high amylopectin during TPS development. Moreover, TPS developed from the native starch undergoes structural modifications, exhibiting increased weakness or stiffness depending on the amount of plasticizer added.

3.2 Characterization of thermoplastic starch

TPS formation requires structural disruption of starch granules to form the amorphous or semicrystalline material that contains any one or a mixture of plasticizers. The source and amount of plasticizer used determines the processing conditions and thermal and mechanical features of the finished product. Amylopectin is accountable for imparting semicrystalline (20%–45%) property in starch. Starch properties such as the mechanical strength and elasticity depends on the ratio of amylose and amylopectin, plant type, and degree of branching (Perez-Pacheco et al., 2016). For TPS, it is important to know the characteristics of amylose and amylopectin (Table 3.2) and, to eliminate the crystalline behavior by the thermal or mechanical processing (Thomas & Atwell, 1999). Glass transition temperature of sheer amylose and amylopectin is around 500K ± 10K, which is much higher than the thermal degradation temperature of most of the polymers.

The role of plasticizer is vitally important in damaging the crystalline structure of starch particles in the manufacturing process of TPS. Plasticizers lower the melting point of starch up to its temperature of decomposition. The most important and easily accessible plasticizer is water, but it makes TPS very brittle and inelastic at room temperature. Water as a plasticizer also lowers the mechanical strength due to the formation of steam bubbles and empty spaces inside TPS.

The gelatinization temperature of starch is increased by adding nonvolatile water-replacing organic solvents such as sugars and glycerol. The plasticizers of high molecular weight increase the penetration in starch granules and they also can increase the available space of amorphous regions. This reduces its activities and the volume concentration of water as plasticizer. Starch granules structure is depicted in Fig. 3.2 (Kainuma, 1984).

Table 3.2: Characterization of amylose and amylopectin (Thomas & Atwell, 1999).

Feature	Amylose	Amylopectin
Structure	Linear	Branched
Bond	Alpha-1,4	Alpha-1,4 and alpha-1,6
Molecular weight	Usually less than 10.5×10^6	$10 \times 10^6 - 500 \times 10^6$
Film intensity	Solid (Strong)	Soft (weak)
Gel formation	Strong and flexible	Neither gelating nor weak
Color by iodine	Blue	Red

Source: *Modified from Thomas, D.J., & Atwell, W.A. (1999). Starch analysis methods. Starches, 13–24.*

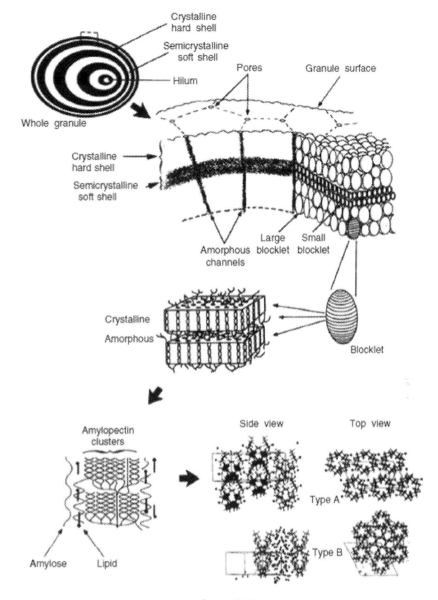

Figure 3.2
Starch granule structure (Kainuma, 1984).

TPS holds low permeability for the gases, weak barrier property of water vapor, high hygroscopicity as well as low tensile strength, which resist their use in packaging (Muller, Laurindo, & Yamashita, 2009). The use of strengthening agents and several types of biodegradable reinforcements in starch to influence the quality of the finished product by increasing the mechanical strength, flexibility, and improving color. The main and common feature of all these auxiliary substances is to provide neutral impact on the environment. Ingredients such as cellulose fibers, whiskers, and cellulosic nanofibers are added to make efficient and cheap starch biocomposites. Although the mechanical and tensile strength of TPS increases with the addition of auxiliary substances but results in decrease in elongation.

Water absorption by the TPS from the environment during manufacturing and storage is one of the major problems, as it changes the mechanical properties and it is hard to retain constant and stable environment necessary in the production and storage. Therefore, to reduce the effects of water intake, addition of some of the hygroscopic substances are advised to form a coating or a cover, which act as a moisture barrier on the surface of processed TPS.

3.3 *Properties of the thermoplastic starch*

3.3.1 *Mechanical properties*

The mechanical properties of TPS depend upon the process temperature, moisture in the starch, and nature and quantity of the plasticizer used. Plasticizers have the greatest impact on providing the mechanical strength to TPS as they have similar hydroxy radicals as starch. Commonly used plasticizers are sorbitol, glycerol, agar, glycol, and polylactic acid (PLA). Studies have shown that on increasing glycerol content up to 35%, the tensile strength of TPS decreases. This is due to weak intermolecular bond between the starch molecules, which are replaced with the glycerol that bind to starch molecules. Some of the auxiliary materials such as cellulose fibers, linen, clay, and pectin can improve the mechanical strength but decrease the elongation (De Graaf, Karman, & Janssen, 2003). However, on increasing the water content up to 35%, increase in the tensile strength and elongation of the TPS were observed (Hulleman, Janssen, & Feil, 1998). Tensile modulus and strength both show direct proportionality with PLA percentage as it enhances the rigidity of the blend due to its strengthening effect (Shirai et al., 2013). Tensile modulus and strength of TPS are also improved by the polar interaction among PLA carboxyl and starch hydroxy group (Wittaya, 2012).

There is a decline in plasticizer content during the storage due to relocation of the plasticizer in matrix to surface, thus influencing the mechanical characteristics, for instance, increase the tensile strength and a reduction in elongation (Finkenstadt & Willett, 2004). Extruded materials containing high glycerol content absorbs extra water during the storage, thus raising the rate of retrograde (Van Soest & Vliegenthart, 1997). Plasticizers permeable to various gases such as oxygen, carbon dioxide as well as water steam have been subjected to a selection of plasticizer. The permeability of plasticizers increases with an increase in their proportion. This increase in permeability has been found higher in hydrophilic compounds when compared to hydrophobic ones.

3.3.2 Thermal properties

Crystalline starch is heated for as long as the granular crystalline melting temperature is reached in front of the solvent to break the hydrogen bonds, which holds together the starch molecules under high shear conditions. When starch and solvent are heated at the high temperature, the solvent interaction with the hydroxy groups of starch reduces the H-bonding in the starch-starch molecules. This tendency is known as gelatinization and the temperature is called as "gelatinized temperature." Gelatinization allows individual starch chain to relocate easily relative to one another, thereby forming a melt-processed starch. The biggest problem with the starchy material is their brittleness which is brought on by the high temperature. TPS thermal properties are measured by the differential scanning calorimetry and dynamic mechanical analysis. A glass transition temperature (T_g) of TPS vary from $-10°C$ to $0°C$. In glass transition state, a material changes from a rubbery state into a highly glassy state at a given heating temperature. T_g verifies the change in the dimensional stability and mechanical property of TPS when processed at a high temperature. Melting temperature (T_m) determines the processability of the polymer. With increase in the concentration of glycerol or sorbitol, T_g of TPS declines. Table 3.3 shows that thermoplastics with different starch origin reduce the T_g with increase in glycerol content in the solution (Janssen & Moscicki, 2006; Mitrus, 2004; Moscicki, 2011; Van Soest, Hulleman, De Wit, & Vliegenthart, 1996; Gaudin, Lourdin, Le Botlan, Ileri, & Colonna, 1999).

3.3.3 Rheological and viscoelastic properties

Like other biopolymers, melted TPS also displays a non-Newtonian flow characteristic of pseudoplastic fluids, regardless the type and volume of plasticizer used. Studies on the evaluation of the viscosity of TPS made from the water and plasticizer have shown that

Table 3.3: Characteristics of thermoplastic with different blends (Moscicki, 2011; Janssen & Moscicki, 2006; Van Soest et al., 1996; Gaudin et al., 1999; Mitrus, 2004).

Type of TPS	Moisture content (%)	Glycerol content (%)	$T_g(°C)$	Reference
TP maize starch	10	25–35	71–83	Janssen & Moscicki (2006)
TP potato starch	11	26	40	Van Soest et al. (1996a)
TP potato starch	13	15	25	Gaudin et al. (1999)
TP potato starch	13	25	0	Mitrus (2004)

Source: Modified from Moscicki, L. (Ed.). (2011). Extrusion-cooking techniques: Applications, theory and sustainability. Wiley-VCH Verlag & Co. KGaA. Janssen, L.P.B.M., & Moscicki, L. (2006). Thermoplastic starch as packaging material. Acta Scientiarum Polonorum Technica Agraria, 5(1), 19–25. Van Soest, J.J., Hulleman, S.H.D., De Wit, D., & Vliegenthart, J.F.G. (1996). Crystallinity in starch bioplastics. Industrial Crops and Products, 5(1), 11–22. Gaudin, S., Lourdin, D., Le Botlan, D., Ileri, J.L., & Colonna, P. (1999). Plasticisation and mobility in starch-sorbitol films. Journal of Cereal Science, 29, 273. Mitrus, M. (2004). The effect of barothermal treatment on changes in the physical properties of biodegradable starch biopolymers (PhD thesis). Lubin: Lublin Agricultural University.

bubbles start to form at the exist point of the extruder at higher temperatures up to 130°C and 100°C for the materials constituting 30% of moisture content (Wittaya, 2012). The viscous behavior of TPS depends on the total volume of plasticizer content. Generally, glycerol content between 36% and 40% results within the 20% decrease in the viscosity (Rodriguez-Gonzalez, Ramsay, & Favis, 2004). Decrease in the viscosity of TPS corresponds to the separation of starch chains (amylose or amylopectin). This process increases the viscosity of TPS as the amylose leaches out from the superstructure forming transparent gelatinous mass. In contrast to the conventional polymers, viscosity of TPS shows exponential dependency on the temperature. Under glass transition temperature ($-50°C$ to $-10°C$), TPS can flow in shear as it is heated and then gradually crystallizes. Dynamic mechanical thermal analysis is achieved to test the viscoelastic property of TPS. This approach investigates the alterations in E' (storage modulus), E'' (loss modulus), and the internal friction coefficient {$\tan \delta = E''$ (loss modulus)/E' (storage modulus)} is a function of the temperature in different mechanical stress frequencies. E' measures the flexibility and elasticity of a material and E'' measures the energy dispersed in the material in the form of heat.

3.3.4 Crystal property

The crystal property of starch varies depending on as per plant origin. Under elevated temperature processing of TPS, the crystallinity of starch is damaged caused by their location in starch chain. Crystallinity of two kinds is acknowledged in TPS after they are processed. First type is *residual-type crystallinity* (native A-, B-, or C-) that are formed by the partially melt-processed starch (Fig. 3.2). The type and total plasticizer used has a direct impact on the residual crystallinity, for example, the low content of glycerol reduces the residual crystallinity. Process conditions such as temperature and shear strength play major role in identifying the amount of residual crystallinity. The second type of crystallinity is *processing-induced-type crystallinity* (V_H, V_A, or E_H) that are formed during the time of thermomechanical processing due to amylose. Crystal structure of the V_H-type shape is by amylose recrystallization into a single helical structure. The addition of a casting agent such as calcium stearate or the availability of lysophospholipids helps to shape the crystalline systems of amylose. E_H-type crystal structures are detected simply after extrusion and are produced during the processing.

The residual A- and B-type crystallinity occurs during low input energy, during the processing for complete melt of native crystals and same phenomena occurs with type C crystallinity, whereas processing at high temperatures does not result in any residual crystallinity, some crystallinity of E_H type is observed because of amylose crystallization.

On increasing the residence period, the quantity of E_H-type crystallinity also rises due to the increment of shear applied as well as the degree of disruption of the starch granule.

Applied shear stress on the blend also influenced by the speed of extruder which shows the variation of amylose-type crystallinity. Variation in formation of amylose crystallinity is also observed with different starch origin. Table 3.4 illustrates the extrusion molding that has been accomplished by starch–glycerol in proportion of 100:30 and 10%–15% (w/w) water content for extrusion and 10%–35% (w/w) for the compression molding (Van Soest, 1996).

3.4 Biodegradability of the thermoplastic starch

The biodegradation of polymers eliminates the plastics and its fragments from the surroundings in an environment-friendly way by breaking down the polymers in gases, salt, and biomass. These polymers are referred to as biopolymers as they are able to decay by the microbial and enzymatic activity. Some of the commercially available biopolymers are cellulose, polysaccharides, and polypeptides. The two mechanisms responsible for the occurrence of biodegradation are hydrolysis or photodegradation and mechanistic route of biological processes, which involve the aerobic and anaerobic breakdown of polymers. Biodegradation mechanism of the plastic under aerobic conditions is shown in Fig. 3.3.

Hydrolysis is the process where the water molecules rupture the chemical bonds, leading to either partial or complete degradation. Photodegradation is carried under sunlight or air and is usually same as oxidation or hydrolysis, whereas biological process involves either presence or absence of oxygen to form smaller fragments of polymers. Following reactions take place during the aerobic and anaerobic degradation:

$$\text{Aerobic degradation} \rightarrow \text{Polymer} + O_2 = \text{Residue} + \text{Biomass} + CO_2 + H_2O$$

$$\text{Anaerobic degradation} \rightarrow \text{Polymer} = \text{Residue} + \text{Biomass} + CO_2 + CH_4 + H_2O$$

Factors which affect the degradability of biopolymers are type of chemical bonds, solubility, and type of the polymers. One of the main factors is the environment constraints such as temperature, humidity, presence of microorganisms, and pH. Then biodegradability of polymers is assessed by change in molecular weight and morphology, mechanical

Table 3.4: Influence of extrusion molding on different source of starch for the formation of amylose crystallinity (Van Soest, 1996).

Source	Crystalline behavior
Waxy corn	No V-type
Potato	E_H and V_H type
Corn	E_H and V_H type
Rice	V_H and V_A type
Wheat	E_H and V_H type

Source: *From Van Soest, J.J.G. (1996). Starch plastics: Structure-property relationships (PhD thesis). Utrecht: Utrecht University.*

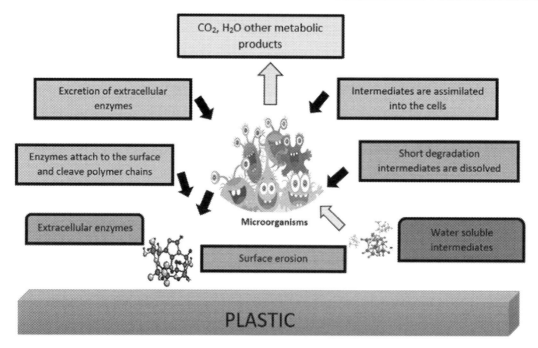

Figure 3.3
General mechanism of plastic biodegradation under aerobic conditions.

properties, and radiation of CO_2. These assessments indicate the biodegradation and fragmentation of the materials. The carbon content present in the polymers is converted to biomass carbon, minerals, and dissolved organic carbon. The mineralization of carbon is about 85%–90% for most of the materials.

The biodegradability of starch granules treated by the enzyme demonstrate that the crystalline behavior of starch has a great impact on the biodegradation rate as enzyme catalyzes the anhydrous glucose units in amorphous starch as disputed in the crystalline areas (Imam et al., 2006). During starch-restructuring process, the crystalline structure is broken down due to the plasticizer molecules move into the starch granule. The replacement of starch granule by the plasticizer–starch hydrogen bonding results in the disappearance of type-A crystallinity of starch. Also, the rate of biodegradation of TPS is typically higher than the native starch because of its amorphous structure. The biodegradation of starch blends is mostly dependent on the total starch content. Higher the starch content, higher will be microorganism's activity, increasing biodegradation rates (Thakore, Desai, Sarawade, & Devi, 2001). Higher starch content also stimulates significant pathways for moisture penetration which is essential for microbial activity and effective degradation (Wool, Raghavan, Wagner, & Billieux, 2000).

3.5 Methods of preparing the thermoplastic starch

To make starch a thermoplastic, mechanical power, friction, heat, and plasticizers (glycerin, water, or other polyols) must change its crystalline structure. Conversion of this starch down in the TPS is possible only in the presence of plasticizers (water, sugars, glycol, urea, sorbitol, amide, and quaternary amine) (Wang, Yang, & Wang, 2003). It is know that the starch processing into TPS results in the reduction of starch molecular weight. To prepare the TPS, starch is liquefied by the operations such as injection molding, extrusion, and pressure molding along with the help of comparatively small quantities of plasticizers. For certain cases, the water content during processing is <20%. In some cases, water content is replaced by the addition of glycol in low concentration.

TPS can also be prepared in a two-stage process. In the first stage, premix containing starch, water, and glycerol is retained for a particular time (1 h) to get plasticizing and swelling of starch granules. Then this mixture is transferred to a mixer where the starch gelatinization occurs at 110°C for a time period of 25 min at a roller speed of 100 rpm (Park et al., 2002). The biggest downside of TPS prepared material is water sensitivity. Generally, the plasticizers have hydrophilic nature and they can be easily washed out with water. Prior to extrusion processing, the addition of water into the starch is essential. Depending on a relative humidity in atmosphere, starch may be able to absorb considerable quantities of water. Hence it is comparatively difficult to process native starch owing to its high viscosity, hydrophilic behavior, and weak mechanical characteristics and is therefore used often as composites or mixtures rather than in its natural form. It is often coated or blended or encapsulated with polymer, polystyrene, polyethylene, polypropylene, and polyvinyl alcohol and PLA blends which are used in various applications where a greater attention has to be given to water and oxygen transmissibility (Khan et al., 2016).

On the completion of processing, the mixture of TPS is cooled and cut in granules so that it would be suitable for further processing such as the blending with polymer materials that enhances the required properties of the end-use material. The most regular process used in TPS preparation has been casting in which dilute slurry is poured into predefined molds. But this approach may be costly and inadequate for large-scale production. Some of the alternatives for this method are as follows (Platt, 2006):

1. Film blowing
2. Injection blow molding
3. Injection stretch blow molding
4. Thermoforming

In all the above processes, care must be given to moisture exclusion in order to prevent the hydrolytic degradation during the drying of polymers. Monomer formation may occur because of thermodynamic equilibrium between the polymerization reaction and reverse reaction as they have been synthesized mostly by ring-opening polymerization. Excessively higher processing temperatures possibly results in the formation of monomer during the process of molding or extrusion. Existence of an extra monomer possibly acts like a plasticizer leading alteration in mechanical characteristics and cause of hydrolysis in material, consequently changing the kinetics of degradation. So, these substances must be processed at least-possible temperatures (Bastioli, 2005).

3.5.1 Film blowing

Two processes used in general for film making using TPS include casting and blowing, the latter being one of the more common methods of film manufacturing. Film blowing is also referred as tubular film extrusion in many places. It is one of the regularly used methods for production of self-reliant plastic films in which hollow tube is extruded first and then expanded by increasing the inside pressure of tube (Fig. 3.4A). Various properties including tensile property play a major role in melt calendaring as well as film blowing. Meager melt persistence has been recognized as one of the restrictions during the processing and extruding of TPS (Thunwall, Boldizar, & Rigdahl, 2006). Material goes through a circular die during the process of extrusion and accompanied by bubble-type enlargement.

The entire film-blowing process consists of four major steps namely, mixing, compounding, conditioning, and film blowing. In the first step, the starch and plasticizers are mixed by hand or any mixer. The mixture is fed into an extruder or kneader where that mixture gets compounded and extruded into strands, after which they are cut into pellets of required size before sending them for further conditioning (Thunwall, Kuthanova, Boldizar, & Rigdahl, 2008).

In conditioning processing, salt solutions will be added (optional) and kept for conditioning for some days depending on categories of product film. The process of conditioning is followed with film blowing wherein the process is achieved by compact extruder. The mixture of pellets is fed into the extruder that in general is equipped with controllable temperature film-blowing die and film-blowing tower. Film formed by plastic melt in extrusion through an annular slit die, usually vertical in direction. Simultaneously, pressurized air is blown through film-blowing tower consisting of a hole in the center of the annular slit of the die which blows up the tube upward followed by continuous cooling of the film (Dang & Yoksan, 2015; Thunwall et al., 2008).

3.5.2 Injection blow molding

This type of molding is mostly employed in shaping synthetic polymers. It is usually employed for the manufacture of hollow materials in bulky quantities like bottles, jars, etc. TPS is liquefied in a screw device or cylinder, later pressed in a definite hollow or definite

(A) Film blowing

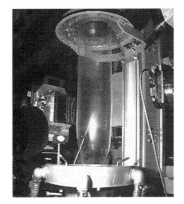

(B) Injection blow molding

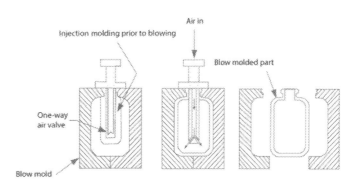

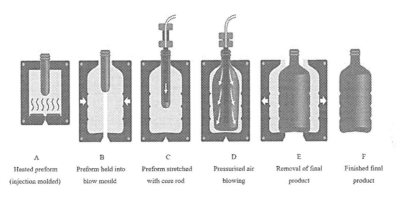

(C) Injection stretch blow molding

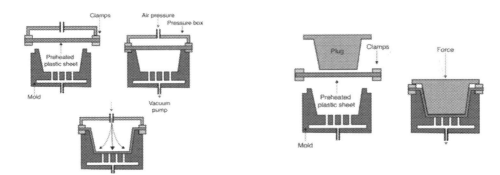

(D) Vacuum thermoforming **(E) Mechanical thermoforming**

Figure 3.4
Manufacturing/molding process. (A) Film blowing; (B) Injection blow molding; (C) Injection stretch blow molding; (D) Vacuum thermoforming; (E) mechanical thermoforming.

shaped chamber (Fig. 3.4B). After cooling and solidification, liquefaction is carried out under continuous pressure to gain shrinkage and contraction of that material, and later removed when solidified (Pielichowski, Puszyński, & Pielichowski, 1994). Greater dimensional and visual quality bottles are created with injection blow-molding technique. The procedure of this technique consists of three phases (Platt, 2006; Rosato, Rosato, & Rosato, 2004).

3.5.2.1 Injection

The liquefied polymer is placed in a manifold and inserted over the nozzle in a hollow preform-heated mold, which forms the outside shape. The interior shape of preform is made by fastening it around the core rod. It is composed of a completely formed object along with a dense polymer tube attached, which forms the body of the required object.

3.5.2.2 Blowing

The preform mold is opened and the core rod is then rotated. The core rod when opened, allows compressed air to pass into the preformed that expands into the shape of the final required object.

3.5.2.3 Ejection

Once the cooling period is over, the blow mold is opened, and the core rod is rotated that will make the final object to be in an ejectable position. To speed up the production, numerous similar molds are placed in one cycle during the same extruder unit. This whole process is completely different from the glass bottle production in which molten material is enforced inserted in a mold along with atmospheric pressure.

3.5.3 Injection stretch blow molding

Injection stretch blow molding (ISBM) is employed in order to produce superior quality as well as greater clarity containers. The ISBM process has four stages (Fig. 3.4C).

3.5.3.1 Injection

The prefrom along with the core pin is produced by feeding the molten polymer into an injection cavity. The core pin helps in producing the internal and external diameter by cavity. Once the set time is over, the core pins and injection molds get apart followed by the preform being held in a carrier that is rotated perpendicularly to loosen and removal of the object.

3.5.3.2 Conditioning

The preform is rapidly cooled in the injection station due to variable temperatures throughout the wall thickness of preform. To ensure pioneer quality of the final object, maintaining uniform temperature is a key factor, so to attain specific level of conditioning, heating is required.

3.5.3.3 Stretching

After conditioning to the required level of temperature, the preform is stretched and blew into the desired shape.

3.5.3.4 Blowing

When the preform is placed inside the blow mold and closed, a stretch rod later put inside the assembly to stretch the preform longitudinally. By maintaining the air pressure in two levels, the preform is blown circumferentially.

3.5.4 Thermoforming

Thermoforming has handy resemblances to vacuum forming; apart from the fact that larger use is of atmospheric pressure and softer surface forming plug-assisted (Mohan & Kanny, 2015; Platt, 2006). This process is comparatively faster and automated than that in the process of vacuum forming (Fig. 3.4D). The principal application for thermoformed objects is for pharmaceuticals, toiletries, food packaging, and electronics packaging. Accurate properties for particular applications are provided by coextrusion. Roll-fed machines are being used for major part of the production whereas sheet-fed machines are used for small-scale operations (Fig. 3.4E). In the heating station, the polymer sheet is softened; it is then guided toward the forming station.

The forming process of the particular sheet is achieved by combining atmospheric pressure and male core plugs. Greater precision of cut of the final product can be attained by this method as the article that is produced, and the scrap, need not be repositioned.

3.6 Applications of the thermoplastic starch

As TPS is plasticized starch that is (in general, with temperature or pressure) meant for damaging the starch crystalline structure completely to bring up an amorphous TPS (Halley et al., 2007). TPS, as previously described also, is developed from starch treated at elevated temperatures in the presence of plasticizers, namely, water, glycerol, and sorbitol (Pervaiz, Oakley, & Sain, 2014).

TPSs, containing starch and polyols, potentially can be used for preparation of disposable as well as reusable biodegradable plastic materials that could provide alternative for traditional plastic materials. For example, biomaterial-based TPS-containing cinnamon oil emulsion is found to increase the mechanical resistance (27.4%) and elongational property at break (44.0%) in active packaging of food (Díaz-Galindo, Nesic, Bautista-Baños, Dublan García, &

Cabrera-Barjas, 2020). Nowadays, one of the most promising reinforce material cellulose nanofibrils (CNFs) is used to make a biodegradable composite film because of its improving mechanical and moisture absorption property, low weight, high strength, and biodegradability. CNF features a high lignin content which is hydrophobic in nature and thus capable of increasing the properties of the films like water vapor barrier and thermal stability (Zhang et al., 2020). The CNF (15% w/w) increases the tensile strength (319%) and modulus of composites (800%) as compared to pure TPS films (Zhang et al., 2020). To restrict the production economics, it is always a better option to go for natural resources for starch and other composites of final blend. Disadvantages of utilization of pure starch for packaging are both the brittleness and the quick retrogradation that is caused by recrystallization with time (Bastioli, 2002). Nowadays, some of the degradable plastics existing in the market that are used in particular applications where in one of the considerable factors is biodegradability, some of them are composting, fast food tableware (straws, plates, cups, cutlery, etc.), packaging (film wrapping, industrial packaging, laminated paper, food containers, etc.), agriculture (plant labels, mulch film, nursery pots, etc.), hygiene (cotton swab, nappy back sheet, etc.), and pharmaceutical sectors (Bastioli, 2005).

Starch-filled polyethylene consisting of prooxidants are generally used in making mulch film that is employed in agricultural operations, in bags and in six-pack yoke packaging (Bastioli, 2002; Otey, Mark, Mehltretter, & Russell, 1974).

In the present scenario, biopolymers are largely used for food items like fresh juices or meat. Biopolymer packaging can also pack fat-rich items. In the food industry, starch is used in nonplasticized form, for modification in the functional properties of foodstuffs, such as texture adjustment or thickening. It is also used in the form of a pregelatinized starch (heat-modified); this is a component of the so-called starch. For those extended shelf life applications where high oxygen/water vapor barriers are not needed (Auras, Harte, Selke, & Hernandez, 2003). Due to immanent nature of biodegradability and potential biocompatibility, TPSs do have applications in extensive ways like biomedical, pharmaceutical, food and other like packaging, etc. (De Carvalho & Trovatti, 2016; Robert & Ing, 2012).

3.7 Conclusions and perspectives

TPS has the potential to produce biodegradable packaging material. The mechanical and vapor-resistance properties of the TPS are still a big challenge but can be improved by the blending of plasticizers, polymers, nanoparticles, nanocellulose, etc. New kinds of blends with TPS may fulfill the versatile demands of the industries. Various research-and-development activities are going on across the globe to achieve this.

Usage of starch-based materials in the form of films, bags, sachets, and pouches in horticulture and agricultural field may improve the economics of farmers. They not only reduce the risk factors such as toxins released from the synthetic polymers, but also limit the environmental pollution. Consumers rely on the synthetic polymers because they are generally cheap and fulfill various needs. However, in order to make TPS more acceptable with desired properties, research work must focus toward the drawbacks of TPS such as its sensitivity to moisture and retrogradation process. Thus future investigations on the starch, plasticizers, cellulose, and nanofibers are required that mainly should concentrate on reducing the water absorption and retrogradation process, which directly improves the mechanical properties of TPS. Replacement of the synthetic plastics is need of an hour, especially in packaging applications, and thus economical production of TPS with unique properties is highly desirable.

Acknowledgment

One of the authors, Ranjna Sirohi acknowledges CSIR for providing fellowship under direct SRF scheme bearing the grant no. 09/171(0136)/19. Raveendran Sindhu acknowledges Department of Science and Technology for sanctioning a project under DST WOS-B scheme. Aravind Madhavan acknowledges Department of Science and Technology for SERB Post-Doctoral fellowship. K. B. Arun acknowledges Department of Health Research for sanctioning fellowship under Young Scientist Scheme.

References

Auras, R. A., Harte, B., Selke, S., & Hernandez, R. (2003). Mechanical, physical, and barrier properties of poly (lactide) films. *Journal of Plastic Film & Sheeting*, *19*(2), 123–135.

Bastioli, C. (2005). *Handbook of biodegradable polymers*. Rapra Technology Ltd.

Bastioli, C. (2002). Starch-polymer composites. In G. Scott (Ed.), *Degradable polymers*. Springer.

Carvalho, A. J. (2008). Starch: Major sources, properties and applications as thermoplastic materials. In M. N. Belgacem, & A. Gandini (Eds.), Monomers, polymers and composites from renewable resources (pp. 321–342). Elsevier.

Carvalho, A. J. F., Curvelo, A. A. S., & Agnelli, J. A. M. (2001). A first insight on composites of thermoplastic starch and kaolin. *Carbohydrate Polymers*, *45*, 189–194.

De Carvalho, A. J. F., & Trovatti, E. (2016). Biomedical applications for thermoplastic starch. *Biodegradable and biobased polymers for environmental and biomedical applications* (pp. 1–23). Wiley-Scrivener Publication.

Dang, K. M., & Yoksan, R. (2015). Development of thermoplastic starch blown film by incorporating plasticized chitosan. *Carbohydrate Polymers*, *115*, 575–581.

Díaz-Galindo, E. P., Nesic, A., Bautista-Baños, S., Dublan García, O., & Cabrera-Barjas, G. (2020). Cornstarch-based materials incorporated with cinnamon oil emulsion: Physico-chemical characterization and biological activity. *Foods*, *9*(4), 475.

Finkenstadt, V. L., & Willett, J. L. (2004). A direct-current resistance technique for determining moisture content in native starches and starch-based plasticized materials. *Carbohydrate Polymers*, *55*, 149–154.

Garcia, N. L., Ribba, L., Dufresne, A., Aranguren, M., & Goyanes, S. (2011). Effect of glycerol on the morphology of nanocomposites made from thermoplastic starch and starch nanocrystals. *Carbohydrate Polymers*, *84*(1), 203–210.

Gaudin, S., Lourdin, D., Le Botlan, D., Ileri, J. L., & Colonna, P. (1999). Plasticisation and mobility in starch-sorbitol films. *Journal of Cereal Science*, *29*, 273.

De Graaf, R. A., Karman, A. P., & Janssen, L. P. B. M. (2003). Material properties and glass transition temperatures of different thermoplastic starches after extrusion processing. *Starch*, *55*, 80.

Halley, P. J., Truss, R. W., Markotsis, M. G., Chaleat, C., Russo, M., Sargent, A. L., & Sopade, P. A. (2007). A review of biodegradable thermoplastic starch polymers. *Polymer durability and radiation effects, ACS Symposium Series* (978, pp. 287−300). ACS Publications.

Hulleman, S. H. D., Janssen, F. H. P., & Feil, H. (1998). The role of water during plasticization of native starches. *Polymer*, *39*, 2043.

Imam, S. H., Gordon, S. H., Mohamed, A., Harry-O'kuru, R., Chiou, B. S., Glenn, G. M., & Orts, W. J. (2006). Enzyme catalysis of insoluble cornstarch granules: Impact on surface morphology, properties and biodegradability. *Polymer Degradation and Stability*, *91*(12), 2894−2900.

Janssen, L. P. B. M., & Moscicki, L. (2006). Thermoplastic starch as packaging material. *Acta Scientiarum Polonorum Technica Agraria*, *5*(1), 19−25.

Kainuma, K. (1984). Starch oligosaccharides: Linear, branched, and cyclic. *Starch: Chemistry and technology* (pp. 125−152). Academic Press.

Khan, B., Khan Niazi, M. B., Samin, G., & Jahan, Z. (2016). Thermoplastic starch: A possible biodegradable food packaging material—A review. *Journal of Food Process Engineering*, *40*(3), 12447.

Laftah, W. A. (2017). Starch based biodegradable blends: A review. *International Journal of Engineering, Research and Technology*, *6*, 1151−1168.

Mitrus, M. (2004). *The effect of barothermal treatment on changes in the physical properties of biodegradable starch biopolymers* (PhD thesis). Lubin: Lublin Agricultural University.

Mohan, T., & Kanny, K. (2015). Thermoforming studies of corn starch-derived biopolymer film filled with nanoclays. *Journal of Plastic Film and Sheeting*, *32*(2), 163−188.

Moscicki, L. (Ed.), (2011). *Extrusion-cooking techniques: Applications, theory and sustainability*. Wiley-VCH Verlag & Co. KGaA.

Muller, C. M., Laurindo, J. B., & Yamashita, F. (2009). Effect of cellulose fibers on the crystallinity and mechanical properties of starch-based films at different relative humidity values. *Carbohydrate Polymers*, *77*(2), 293−299.

Oliveira C., Fidalgo N.N., Valera, T.S., & Demarquette N.R. (2014). Thermoplastic starch: The preparation method. *ANTEC 2014—Proceedings of the technical conference & exhibition* (pp. 431−434). Las Vegas, Nevada, USA, April 28−30.

Otey, F. H., Mark, A. M., Mehltretter, C. L., & Russell, C. R. (1974). Starch-based film for degradable agricultural mulch. *Industrial & Engineering Chemistry Product Research and Development*, *13*(1), 90−92.

Park, H. M., Li, X., Jin, C. Z., Park, C. Y., Cho, W. J., & Ha, C. S. (2002). Preparation and properties of biodegradable thermoplastic starch/clay hybrids. *Macromolecular Materials and Engineering*, *287*(8), 553−558.

Perez-Pacheco, E., Canto-Pinto, J. C., Moo-Huchin, V. M., Estrada-Mota, I. A., Estrada-León, R. J., & Chel-Guerrero, L. (2016). Thermoplastic starch (TPS)-cellulosic fibers composites: Mechanical properties and water vapour barrier: A review. *Composites from renewable and sustainable materials* (pp. 85−105). Inteach.

Pervaiz, M., Oakley, P., & Sain, M. (2014). Development of novel wax-enabled thermoplastic starch blends and their morphological, thermal and environmental properties. *International Journal of Composite Materials*, *4*(5), 204−212.

Pielichowski, K., Puszyński, A., & Pielichowski, J. (1994). Thermal analysis of selectively-brominated polystyrene. *Polymer Journal*, *26*(7), 822−827.

Platt, D. K. (2006). *Biodegradable polymers: Market report*. Smithers Rapra Limited.

Robert, S., & Ing, K. (2012). Thermoplastic Starch. *Thermoplastic Elastomers*, 95−116.

Rodriguez-Gonzalez, F. J., Ramsay, B. A., & Favis, B. D. (2004). Rheological and thermal properties of thermoplastic starch with high glycerol content. *Carbohydrate Polymers*, *58*(2), 139−147.

Rosato, D. V., Rosato, D. V., & Rosato, M. V. (2004). *Injection molding. Plastic product material and process selection handbook* (pp. 192−226).

Shirai, M. A., Grossmann, M. V. E., Mali, S., Yamashita, F., Garcia, P. S., & Muller, C. M. O. (2013). Development of biodegradable flexible films of starch and poly (lactic acid) plasticized with adipate or citrate esters. *Carbohydrate Polymer, 92*, 19–22.

Van Soest, J. J., Hulleman, S. H. D., De Wit, D., & Vliegenthart, J. F. G. (1996). Crystallinity in starch bioplastics. *Industrial Crops and Products, 5*(1), 11–22.

Van Soest, J.J.G. (1996). *Starch plastics: Structure-property relationships* (PhD thesis). Utrecht: Utrecht University.

Van Soest, J. J. G., & Vliegenthart, J. F. G. (1997). Crystallinity in starch plastics: Consequences for material properties. *Trends Biotechnology, 15*, 208–213.

Thakore, I. M., Desai, S., Sarawade, B. D., & Devi, S. (2001). Studies on biodegradability, morphology and thermo-mechanical properties of LDPE/modified starch blends. *European Polymer Journal, 37*(1), 151–160.

Thomas, D. J., & Atwell, W. A. (1999). Starch analysis methods. *Starches*, 13–24.

Thunwall, M., Boldizar, A., & Rigdahl, M. (2006). Compression molding and tensile properties of thermoplastic potato starch materials. *Biomacromolecules, 7*(3), 981–986.

Thunwall, M., Kuthanova, V., Boldizar, A., & Rigdahl, M. (2008). Film blowing of thermoplastic starch. *Carbohydrate Polymers, 71*(4), 583–590.

Wang, X. L., Yang, K. K., & Wang, Y. Z. (2003). Properties of starch blends with biodegradable polymers. *Journal of Macromolecular Science, Part C: Polymer Reviews, 43*(3), 385–409.

Wittaya, T. (2012). Rice starch-based biodegradable films: Properties enhancement. *Structure and Function of Food Engineering, 5*, 103–134.

Wool, R. P., Raghavan, D., Wagner, G. C., & Billieux, S. (2000). Biodegradation dynamics of polymer–starch composites. *Journal of Applied Polymer Science, 77*(8), 1643–1657.

Zhang, C. W., Nair, S. S., Chen, H., Yan, N., Farnood, R., & Li, F. Y. (2020). Thermally stable, enhanced water barrier, high strength starch bio-composite reinforced with lignin containing cellulose nanofibrils. *Carbohydrate Polymers, 230*, 115626.

PART II
Plant - based biopolymers

CHAPTER 4

Cellulose

Niveditha Kulangara[1] and Swapna Thacheril Sukumaran[2]

[1]Department of Biotechnology, University of Kerala, Thiruvananthapuram, India, [2]Department of Botany, University of Kerala, Thiruvananthapuram, India

4.1 Introduction

The high use of petrochemicals in the 21st century has resulted in increased environmental pollution. Nonbiodegradable synthetic polymers significantly enhanced this pollution and the use of such synthetic polymer is very high because of its stability. Around 140 million tons of synthetic polymers per year are synthesized all around the world, and these nondegradable synthetic polymers accumulated on earth act as the primary source of environmental pollution. Hence, eco-friendly alternative methods such as degradable biopolymers should be promoted worldwide to reduce pollution (Dassanayake, Acharya, & Abidi, 2018). Substitution of a synthetic polymer with a natural polymer such as cellulose $[(C_6H_{10}O_5)_n]$, which is the most abundant sustainable resource on earth, can help to decrease the pollution and health hazards created by synthetic polymer to an extent (Anjana, Hinduja, Sujitha, & Dharani, 2020). Many researchers have focused on developing materials made from biomass, such as cellulose-based biocomposite, for diverse applications in various fields (Li, Wang, Ma, & Wang, 2018).

4.2 Biopolymers

Biopolymers are obtained from natural sources such as plants. Some microorganisms, for example, bacteria and fungi, can also produce polymeric compounds. A major limitation of biopolymer production from renewable resources is the low yield. Synthetic polymers have been produced from nonrenewable resources in massive amounts. Both polymers consist of numerous monomeric units, which are covalently linked to form larger structures (Hallinan & Balsara, 2013). Biopolymers, especially bioplastics, are an excellent substitute for conventional plastics (Vroman & Tighzert, 2009), due to their biodegradable properties. Soil microbes, while degrading these biopolymers, release CO_2 and water, which are quickly reabsorbed by the plants and reduce the atmospheric CO_2 levels (Huang, Huang, Lin, Xu, & Cen, 2010).

4.2.1 Biodegradable polymers and polymer composites

Biodegradable polymers or resins from renewable resources (Mahalakshmi et al., 2019), when combined with the natural reinforcing materials, are called biopolymer matrix composites (Lalit, Mayank, & Ankur, 2018). Cellulose polymer matrix/binder combined with reinforcement material such as natural fibers can form a composite material having properties of the matrix as well as reinforcement material added gives improved mechanical properties for the matrix. Cellulose acts as a load-bearing constituent when combines with reinforcing materials such as fibers and whisker. Silicate clays, carbon nanotubes, and cellulose nanocrystals are common types of reinforcement used (Miao & Hamad, 2013), and these composites have several applications in the industries (Rudin & Choi, 2012).

4.2.2 Sources of cellulose

Payen (1838) discovered cellulose in plants and reported its hydrophilic property. Cellulose and starch are polymers commonly derived from renewable sources such as corn, wheat, and different agricultural residues (Tanase, Rapa, & Popa, 2014), but cellulose can also be produced by microorganisms (Fig. 4.1) and tunicates. Tunicates belong to marine invertebrates and are the only known animals with cellulose biosynthesis ability (Klemm, Heublein, Fink, & Bohn, 2005). In the case of tunicates, cellulose forms their skeletal structure, and like plants, they synthesize cellulose using cellulose synthase enzyme present in the plasma membrane. It also helps in metamorphosis and normal development in tunicates (Kimura & Itoh, 2007). Crystalline cellulose in tunicates was not found in its pure

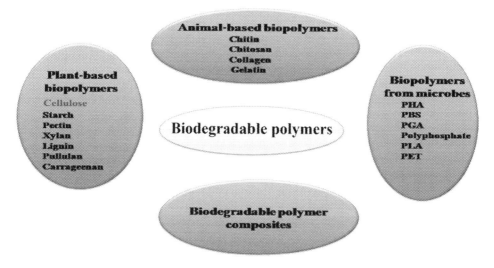

Figure 4.1
Types of biodegradable polymers.

form but present along with various membrane components such as mucopolysaccharides, glycans, proteins, and lipids in the embedded form. For isolating cellulose from tunicate tissue, other components have to be removed by a suitable method such as hydrolysis and bleaching. Like plant cellulose, industries use purified fibrous cellulose from tunicates. Cellulose produced from different strains has different properties. For example, cellulose obtained from tunicates like *Ascidia* sp. (order Phelbobranchia) is soft, but from *Styela plicata*, it is hard due to the structural variations. Pure cellulose fibers from tunicates were thermally stable in nature and in the microfibrillar form (Zhao & Li, 2014).

Cellulose is also obtained from some microbes such as bacteria, fungi, and algae. Cellulose isolated from bacteria *Acetobacter xylinum* (*Gluconacetobacter xylinum*) was initially reported by Brown et al. in 1886. Other bacterial strains used for cellulose production and isolation include *Rhizobium*, *Acetobacter*, *Agrobacterium*, and *Pseudomonas* sp. (Brown, 1886; Jonas & Farah, 1998). Cellulose from different microorganisms will be having different structural and morphological properties with diverse applications. For example, carboxymethylcellulose (CMC) and hydroxypropyl methylcellulose, both produced from different bacterial strains, have different applications in the food industries itself. CMC is added to flour dough for regulating rheology. On the other hand, hydroxypropyl methylcellulose is added to whipped cream for enhancing the texture of the cream. Cellulose synthase catalyzes the cellulose production in bacteria. The most commonly used strains for industrial-scale production of bacterial cellulose include *A. xylinum* or *G. xylinum*, *Acetobacter hansenii*, and *Acetobacter pasteurianus* (Wang, Tavakoli, & Tang, 2019).

Overexpression of ces A gene for the enzyme cellulose synthase enhances the extracellular production of cellulose in bacteria. Zhao et al. (2015) reported that in Cyanobacterium *Synechococcus*, overexpression of multiple genes such as *cmc*, *bgl*, *ces* AB, and *ces* C resulted in enhanced production of type I cellulose. *Synechococcus* has cellulose in between the two membrane layer, and silencing of *ces* A and low salt level in the medium supports the enhanced production. Cyanobacterial system is also useful for lignin-free cellulose production on an industrial scale (Zhao et al., 2015). Cellulose can be modified in macro and microforms for enhancement in mechanical strength, water retention, change in degradation rate, etc. (Vroman & Tighzert, 2009).

4.2.3 Plant-based cellulose

Cellulose is a profusely available natural resource on earth, which is present in the cell wall of plants. The yield of cellulose is about 1.5×10^{12} t annually, but only 2% is recovered industrially (Dassanayake et al., 2018). Cellulose rarely occurs in pure form, since it is usually present along with pectin, lignin, and hemicellulose (Moon, Martini, Nairn, Simonsen, & Youngblood, 2011). In agriculture residues, about 30%–35% of cellulose, 25%–30% of hemicellulose, 15%–28% lignin, and 4%–7% of ash are available, but in fibers of cotton, cellulose is estimated as 94% (Marques, Rencoret, Gutiérrez, Alfonso, & del Río, 2010).

Biocomposites based on cellulose have significant advantages due to the exceptional features of cellulose in a wide array of fields (Qiu & Hu, 2013). The mechanical properties of cellulose make it an efficient and acceptable biopolymer or composite material (Hu, Chen, Yang, Li, & Wang, 2014). Cellulose is a polymer of D-glucose. Each monomeric glucose unit was linked by β-1,4-glycosidic bonds formed between the oxygen atom and hydroxyl groups of the nearby glucose molecule. The linear structure of the cellulose was illustrated in Fig. 4.2. Cellulose linear structure is formed by such glycosidic linkages between adjacent glucose molecules, but each linear chain is further polymerized to form polysaccharides by van der Waals interactions and hydrogen bonding. Hydrogen bonds permit a high degree of crosslinking and help to build parallel fibers. These glucan chains interact with each other side by side, forming microfibrils of cellulose, and stacking of microfibrils gives cellulose a crystalline structure (Brett, 2000). The number of monomers in the polymeric chain can extend from 100 as in wood to 6000 as in cotton.

On the hydrolysis, cellulose produces individual D-glucose monomeric units. Cellulose can be cleaved with β-glucosidase. Polysaccharide monooxygenases can cleave cellulose to glucuronic acid. Generally, 500–1000 units of glucuronic acid contain one glucuronic acid A (GlcA). The degree of polymerization (DP) changes based upon the source of cellulose. In α-cellulose, DP is less than 150, in β-cellulose between 10 and 150, and in γ-cellulose less than 10. Hydrogen bonding allows a high degree of crosslinking, which regulates the molecular structure (Klemm et al., 2005).

In cotton fibers, microfiber of 300 nm, along with microfiber of 2–4 nm, are included in the polysaccharide chain. In wood fibers, lignin is interspaced with elementary fibers of cellulose. Cellulose I, with all parallel chains with a degree of crystallinity of 60%, when processed with NaOH, can be modified into a new allelomorph, cellulose II with antiparallel chains, and low free energy structure, by a process called mercerization. This allomorph is stable than cellulose I thermodynamically, and the changes are irreversible (Klemm et al., 2005). Mercerization causes increased stiffness and shiny texture with

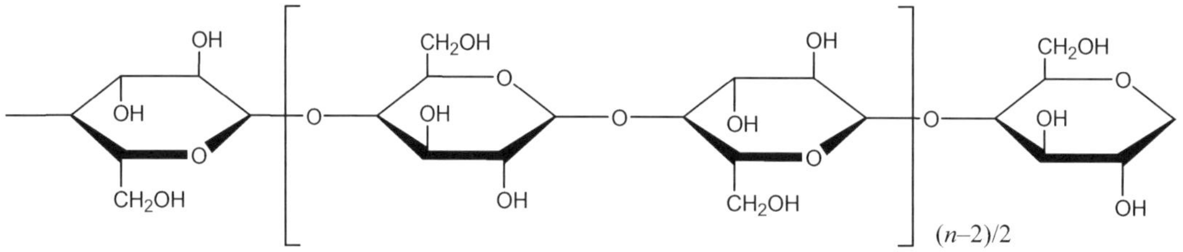

Figure 4.2
Linear structure of cellulose.

changes in dyeing property. Between cellulose I and cellulose II transition, many intermediates such as IIA, IIB, III, and IV can be formed. Generally, all plants are an excellent and abundant source of cellulose, but industrial production of cellulose is mainly from wood (40%−45% by weight). Various plant sources of cellulose and the percentage of cellulose in them were represented in Table 4.1.

For the industrial production of cellulose from wood, cellulose has to be separated from hemicellulose and lignin. The most preferred technique for the separation of cellulose from wood chips is chemical treatment with sodium hydroxide and sodium sulfide solution along with the simultaneous application of heat under high pressure (Nawrath, Poirier, & Somerville, 1995). Other extraction methods are acid hydrolysis, enzymatic hydrolysis, and mechanical process. This processed cellulose could be used for various purposes since the hard crystalline structure gives extra stability and strength (Rinaudo, 2006). Strong bonding makes the cellulose insoluble in many solvents, so acetal and ester derivatives are used for various applications. Various applications and production of cellulose-based biopolymer were illustrated in Fig. 4.3.

Cellulose synthesis is catalyzed by cellulose synthase complex (CSC) comprising three different cellulose synthases, located on the plasma membrane of plant cells. The presence of endo-1,4-β-D-glucanase also supports CSC activity. Uridine diphosphate glucose supply, lipid raft composition, cortical microtubules, and various other membrane compositions influence the speed or catalysis rate of CSC. This enzyme complex interacts with proteins such as korrigan, cellulose synthase interacting protein 1 (CSI 1), and glycosylphosphatidylinositol, as well as lipid anchored proteins such as CTL1. Korrigan, which is endoglucanase attached to the plasma membrane, has a direct or indirect influence on cellulose biosynthesis. CSI helps CSC tethering with cortical microtubules. This complex of proteins, along with cellulose synthase enzyme, is responsible for the synthesis of cellulose in the apoplast to form crystals (McFarlane, Doring, & Persson, 2014).

Cellulose fiber from plants can be obtained from woody (softwood and hardwood) or nonwoody (agricultural waste, native plants, and nonwood plant fibers) biomass (Arvanitoyannis & Biliaderis, 1999; Ma, Hu, & Wang, 2016; Park, Weller, Vergano, & Testin, 1993). Thus cellulose is one of the major constituents of plant fiber, with 20%−90%

Table 4.1: List of natural cellulose fibers and their percentages in the composition.

Sl. no.	Source	% of cellulose	References
1	Banana leaf	60−65	Thomas, Paul, Pothan, and Deepa (2011)
2	Cotton seed	82−95	Fernandes, Pires, Mano, and Reis (2013)
3	Maize straw	28−44	Rehman et al. (2014)
4	Rice husk straw	25−35	Matias, Cruz, Garcia, and Gonzalez (2019)
5	Wheat straw stalk	30−35	Fernandes et al. (2013)

Evaluation of rice straw yield, fiber composition, and collection under Mediterranean conditions.

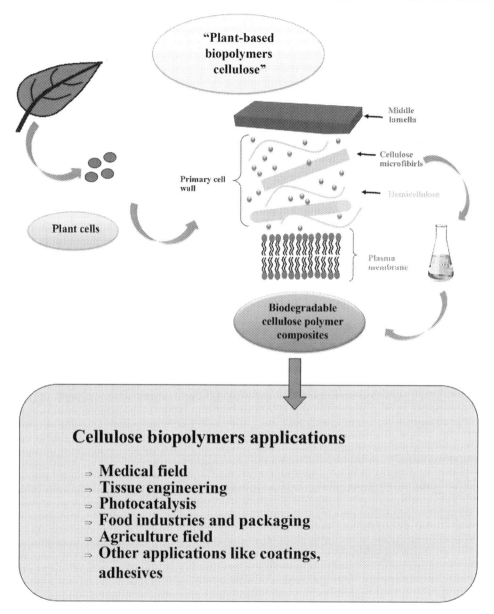

Figure 4.3
Production and application plant-based cellulose.

crystallinity. Normally cellulose is not found in pure form and is usually associated with hemicellulose and lignin (Dassanayake et al., 2018; Khalil et al., 2017). This lignin matrix gives protection from stress conditions like wind and biological attacks. On the other hand, hemicellulose has compatibility with cellulose (Kalia et al., 2011). Generally, pulp is prepared from plant fiber, which is later used to produce boards, paper, etc., and through treatments such as pulping and hydrolysis, pure cellulose, which is greater than 90% cellulose content, can be produced. Such a pure form of cellulose can be processed further through mechanical or chemical means to produce micro or nanoscale cellulose and derivatives of cellulose. Such derivatives are usually ether and ester forms which are used in composites such as hydrogels and thin polymer films. The hydrogel is made up of polymers having hydrophilic properties. These polymers form a three-dimensional (3D) network by crosslinking with nearby polymeric chains. Hydrogels had a water-holding capacity without altering their standard structural features (Anwar, Gulfraz, & Irshad, 2014; Ashori & Nourbakhsh, 2010; Fink, Weigel, Purz, & Ganster, 2001; Tye, Lee, Abdullah, & Leh, 2016).

Hydroxyl group present in the cellulose has the ability to form intermolecular and intramolecular hydrogen bonds, which gives it a semicrystalline nature with a crystalline and amorphous domain.

Cellulose nanocomposites can be synthesized from hydrolysis of cellulose I by cleaving the amorphous region (O'Sullivan, 1997; Wada, Okano, & Sugiyama, 2001). Cellulose and chitosan nanocomposites thus prepared due to their biocompatibility can be used in biomedical research purposes, especially as a scaffold material for creating biomedical devices (Domingues et al., 2015; Li et al., 2015; Lin & Dufresne, 2014; Sunasee, Hemraz, & Ckless, 2016). Chitosan is a polymer of NAG/*N*-acetyl glucosamine. Due to its structural similarity with cellulose, many researchers worldwide focused on making composites of chitosan and cellulose with altered degradation rates in the human body. Chitosan and cellulose exhibit unique properties, such as biocompatibility, degradability, and physiochemical properties. Biocomposites by these polymers can form films, and they exhibit antimicrobial activity along with improved mechanical properties (Mesquita, Donnici, Teixeira, & Pereira, 2012).

Cellulose synthase enzyme complex present in the membrane catalyzes the glucan chain polymerization reaction. (Mohan, Oluwafemi, Kalarikkal, Thomas, & Songca, 2016). DP of cellulose in the primary cell wall is about 8000, whereas the secondary cell wall is more than 15,000 (Donaldson, 2016). In wood, 60%−70% cellulose is crystalline in nature, the remaining portion having less-ordered amorphous behavior (Rowell, 2012). The amorphous region absorbs water more quickly due to the greater accessibility. So the crystalline and amorphous components critically affect the final properties such as mechanical performance, water absorption, and durability. Crystalline polymers were ordered in structure, which provides a rigid nature. But amorphous polymers have disordered orientation, which gives opposite properties. Natural cellulose has eight hydrogen bonds per glucose unit in its crystalline form, but in amorphous cellulose, it is only 5.3 H_2 bonds (Wei & McDonald, 2016).

Cellulose, having a complex fibrillar morphology, can be altered by strategies such as chemical alteration and crosslinking. Crosslinking leads to changes in the physical and chemical properties of cellulose. This is because of the pillaring effects during a net reduction of one OH group in each crosslinking. Chemical treatment methods such as sodium hydroxide treatment can modify the surface of cellulose (Mohamed, Udoetok, Wilson, & Headley, 2015; Oyama, Udoetok, Wilson, & Headley, 2018; Udoetok, Wilson, & Headley, 2018).

Various methods were identified for cellulose composite synthesis. For example, cellulose/grapheme oxide composite is prepared by mixing graphene oxide and cellulose at 100°C followed by casting step and washing in deionized water. For any biocomposite preparation the selection of reinforcing material is a crucial step (Hao et al., 2019). Based on the applications, combinations of cellulose composite materials can be used. Cellulose polymer can hold about 99% of water due to its porous structure (Klemm et al., 2006). The hydrated form of cellulose (Czaja, Romanovicz, & Malcolm Brown, 2004) can be used for hydrogel-based applications, especially in the medical field (Caccavo, Cascone, Lamberti, & Barba, 2018; Iguchi, Yamanaka, & Budhiono, 2000).

4.2.4 Applications of cellulose

Nanostructures can exhibit higher specific surface areas, density, and surface energy and can lead to the production of materials with new and enhanced properties due to synergistic effects. Thus cellulose and its bio-nanocomposites are useful in medicine, pharmaceutics, agriculture, forestry, food packaging, electronics, transport, construction, etc. (DiezPascual, 2019).

4.2.4.1 Cellulose for medical applications

Cellulose composites are used for the preparation of different materials, especially scaffold materials mimicking our skin in biomedical fields due to the biocompatible nature. Artificial skins, thus made from these composite materials, are a boon to patients suffering from chronic wounds/burns, which indicates the immense possibilities in medical as well as tissue-engineering fields (Calo & Khutoryanskiy, 2015).

For drug-delivery applications, cellulose derivatives are used as a jelling agent. Derivatives of cellulose used for this purpose include methylcellulose, hydroxyl ethyl cellulose, hydroxyl propyl cellulose, and sodium carboxyl methylcellulose which enable controlled drug delivery with very few side effects. Hydrogel based on cellulose is used in ointments for transdermal action as the base material because of its transparency, high water content, bioadhesiveness, and biocompatibility (Vlaia, Coneac, Olariu, Vlaia, & Lupuleasa, 2016).

For drug delivery, cellulose nanocomposites can also be used (Ciolacu & Suflet, 2018). Jackson et al. reported the use of nanocrystalline cellulose for the binding, followed by the controlled release of drugs. Molecules such as cellulose and chitosan can be used as binding agents for delivering drug molecules. Due to the biocompatibility of cellulose fibers, it is an excellent vehicle to deliver drugs and hence widely applied in the therapeutical field (Jackson et al., 2011) including scaffold material preparation (Mathew, Oksman, Pierron, & Harmand, 2012).

Cellulose can also be used for making superabsorbent polymers due to its water-absorbing ability. At the same time, after polymerization, cellulose polymers used as insulation shows fungal resistance. High porosity enables vapors to move out, thereby causing low humidity, which favors fungal resistance (Guan, Li, Zhang, & Yu, 2017; Peng et al., 2016). Peng et al. (2016) introduced cell-based hydrogel that has applications in disposable diapers. CMC and hydroxyethylcellulose linked with divinyl sulfone have good water uptake and swelling capacity compared to synthetic material. After crosslinking, quaternized cellulose-based hydrogels have the property of hydrogel as well as cellulose. Crosslinking provides space for hydrogels for improved absorption as hydrogel itself has water-holding capacity and swelling ratio. The reason for the enhanced swelling ratio in a composite having quaternized cellulose is due to the electrostatic repulsion of quarternary ammonium groups. Properties of native cellulose differ from quaternized cellulose due to the absence of 4° ammonium groups. In diapers the use of quaternized cellulose-based hydrogel composite improves water-holding and swelling properties when compared to native cellulose. Hence, such natural polymer is an excellent alternative to the synthetic polymers (Sannino, Mensitieri, & Nicolais, 2004). Cellulose biofilms are also used for wound dressing because of the excellent wound drainage absorption and oxygen permeability, which enhances wound healing (Portela, Leal, Almeida, & Sobraj, 2019).

4.2.4.2 Cellulose for tissue-engineering applications

Tissue engineering is a merging of life sciences and engineering for developing alternatives for replacing or repairing damaged tissues or organs by regenerating tissues that need a suitable scaffold. Tissue-engineering triads include cells, materials, and growth factors. Using the different combinations of triads, organs can be modified or even can replace the entire tissue/organ. Cellulose is used in tissue engineering as natural scaffold materials. Natural polymer scaffolds are far better than synthetic scaffolds because of their biocompatibility and hence reduce the risk of immunological reactions. Cellulose biopolymer is widely used in regenerative medicines (Salgado et al., 2013). It is proved that isolated cells seeded on the cellulose-based scaffold material promote cell attachment, growth, and proliferation (Taboas, Maddox, Krebsbach, & Hollister, 2003; Dhandayuthapani, Yoshida, Maekawa, & Kumar, 2011).

Halib, Ahmad, Grassi, and Grassi (2019) reported that the bacterial cellulose fibers and associated proteins mimic extracellular matrix (ECM). In tissue engineering, porous nature of cellulose provides a suitable environment for cell attachment, growth, and multiplication

(Halib et al., 2019). ECM in organisms consists of proteoglycans, glycosaminoglycans, and fibrous proteins such as collagen, laminin, fibronectin, and elastin. In the case of bacterial cellulose fibers, their structure shows the similarity between collagen fibers. Bacterial cellulose, such as collagen fibers, can also provide mechanical support to tissues (Hoseny et al., 2015). Not only cellulose, but hemicellulose, lignin, etc. are also proved as good scaffold material in tissue engineering, which provides an ideal environment for the growth of seeded cells with sufficient porosity for nutrient and oxygen supply (Cooper, Hausman, & Hausman, 2000).

Cellulose-based hydrogel composites are used for culturing cells since they mimic natural ECM with a swollen 3D network. These cellulose fibers and associated proteins used in artificial ECM help to culture cells for their growth and proliferation. Alginate used in cellulose composite material has viscoelastic properties, which makes it ideal for different tissue-engineering fields. There are reports on mesenchymal stem cells (MSCs) seeded on the cellulose-based artificial ECM support growth of the cell. MSCs attached and differentiated into osteogenic cells in bone tissue engineering (Chaudhuri et al., 2016). These ECM-mimicking scaffolds support the growth of new tissue from seeded MSCs. Even in the artificial skin, cellulose (mimicking collagen) was added as a component to create artificial ECM (Hoseny et al., 2015).

The use of highly elastic and viscous hydrogels promotes the growth of MSCs when attached to the scaffold forming a collagen-rich matrix. Artificial ECM created from natural materials encourages cell growth and proliferation. Artificial blood vessels were created using cellulose-based scaffolds. 3D cellulose-based scaffolds are helpful for nerve suture cuffing and direct implant of artificial blood vessels (Klemm, Schumann, Udhardt, & Marsch, 2001; Lv et al., 2018; Yang et al., 2014). Cellulose can be used in biomedical fields, including skin therapy, wound dressings, and drug delivery (Portela et al., 2019). Artificial skin replacement therapy has disadvantages such as graft rejections and immunological reactions. These hurdles can be overcome using scaffolds designed from natural sources (Klemm et al., 2001).

Studies have highlighted promising and potential biomedical applications for cellulose in tissue-engineering fields (Meyer, 2009). For tissue-engineering purposes, pure cellulose should be isolated on a large scale, and a steady supply of high-quality cellulose has to be ensured using plants since they are the most abundant natural resources in the earth (Huang et al., 2014).

4.2.4.3 Cellulose for photocatalysis and photovoltaics/electronics

Cellulose biopolymer can also be used for photocatalysis and photovoltaics/electronics. Photovoltaics are simply is a technique in which solar energy is used for power generations. Cellulose can act as photocatalyst for utilizing solar energy into power and nanocellulose is used for the preparation of solar cells (Khan, Nair, Colmenares, & Glaser, 2018).

Cellulose sheets are used as a semitransparent electrode for photovoltaic cells in electrochemical devices. Cellulose in its nanoform used in such devices shows better performance and is proved to be having excellent tensile strength and electrical and thermal conductivity. Nanostructures and bond strength between carbon atoms were responsible for such properties. Cellulose nanofibers and nanocrystals are now available for the production of ultrathin storage devices of renewable energy, which is also used in flow batteries. Carbon nanotube/cellulose composite has applications in this field. Metal, mineral, and carbon nanomaterials can be incorporated into a cellulose substrate to enhance electrical, catalytic, and optical properties (Wang et al., 2016; Wei, Rodriguez, Renneckar, & Vikesland, 2014).

4.2.4.4 Cellulose for food industries and packaging applications

Cellulose is widely used in food industries as the packaging of food material, salad dressings, desserts, cake batters, whipped toppings, stabilizers, beverages, etc. The material used as a covering membrane should remain chemically inert so that it will not react with the components of food. Food-packaging material should be able to withstand the processing temperatures too since preservation processing of many food materials uses heat. Cellulose membranes produced from *A. xylinum* are used for clarification of apple juice where the membrane is used as a filter bed. For food processing purposes, cellulose biopolymer–based membrane-like dicarboxy methylcellulose was also used. These low-cost films were having gas barrier properties and are easy to prepare from natural sources (Berk, 2018). Since cellulose can adsorb compounds such as phenols and pigments, the wine industry is hesitant to use these membranes. Nonresistance to disinfectants, as well as detergents, is also a concern.

Cellulose itself and in composite form or film blend composites are far better than synthetic packaging materials. For example, cellophane and cellulose acetate are the biodegradable films used as an excellent material for food-packaging purposes. Cellophane shows more mechanical properties than cellulose acetate. Cellophane is a transparent film showing low permeability to microbes, oxygen, and water vapor. Biodegradable laminate from cellulose–chitosan and polycaprolactone composite material is used in modified atmosphere packaging in various types of food industries. These are transparent thin films in nature, which are biodegradable also. Biofilms in the blended form are highly useful in food packaging (Ahmadi, Ghanbarzadeh, Ayaseh, & Pezeshki, 2017).

Bioplastic polymers are the best substitute for nondegradable materials used for packaging, thereby reducing serious health hazards caused by nondegradable packaging materials. The use of cellulose composite film for packaging purposes is indeed an eco-friendly approach (Vartiainen, Vähä-Nissi, & Harlin, 2014). A comparison of the properties of cellulose-based packaging materials has provided its advantages over commercial packaging materials.

Such cellulose blended film provides better insulation and barrier properties, and hence, it is an excellent material for food packaging (Johansson et al., 2012).

Good-quality cellulose is obtained from pulping technique, and byproducts from paper industries can be utilized for manufacturing cellulose-based composite materials. Cellulose polymer can form paper and cardboard, carton, which is used as the exterior packaging layer for the storage of various perishable and semiperishable food products. The shelf life of such commodities is improved while packed in such cellulose-based materials. For example, in the case of perishable food materials such as lettuce and tomatoes, shelf life can be improved up to 4–6 days under 10°C–25°C storage by the use of biopackaging by cellulose materials (Ballinas-Casarrubias et al., 2016a).

4.2.4.5 Cellulose for agriculture

In agriculture, the proper supply of nutrients gives better productivity. Nutrients fabricated with cellulose-based hydrogel are easier to assimilate by plants; hence, the cultivation of plants with nutrients provided in polymer improves growth and food productivity. The hydrogel can also be used for the controlled and smooth release of agrochemicals such as pesticides and herbicides (Kabir et al., 2018). Because of the hydrogel structure, these chemicals are retained inside for some time and prevented from fast release (Ismail, Irani, & Ahmad, 2013).

Cellulose-based hydrogels are used in agricultural fields due to its nontoxic nature, biodegradability, and its natural origin (Cannazza et al., 2014; Demitri, Scalera, Madaghiele, Sannino, & Maffezzoli, 2013; Montesano, Parente, Santamaria, Sannino, & Serio, 2015; Sanchez-Orozco, Timoteo-Cruz, Torres-Blancas, & Ureña-Nunez, 2017). CMC is used for the controlled release of an herbicide like acetochlor based on the diffusion mechanism (Li, Lai, & Luo, 2009; Li, Li, & Dong, 2008).

Cellulose obtained from palm oil–free fruit and hydrogels such as polyacrylic acid together form composites and proved that they improve soil–water-retaining capacity as well as low rate of soil urea leaching (Laftah & Hashim, 2013). The use of hydrogels including methylcellulose can enhance the controlled release of fertilizer with the help of ammonia volatilization (Bortolin, Aouada, Mattoso, & Ribeiro, 2013). Plant-based hydrogels are employed as nutrient carriers for improving productivity in agricultural fields (Guilherme et al., 2015) instead of synthetic polyvinyl pyrrolidone hydrogel composites commonly used in agricultural areas (Wang & Wang, 2010).

4.2.4.6 Other applications

Cellulose-based composites are used in the textile industries. During the processing of textile materials, wastewater treatment is highly significant. Water-absorbing hydrogels absorb excess water (Shen, Shamshina, Berton, Gurau, & Rogers, 2016) and are used as flocculants and electrolyte membranes. Many derivatives of cellulose are known with diverse properties. Some of the cellulose derivatives and their applications (Smitha, Sridhar, & Khan, 2004) are given in Table 4.2.

Instead of using synthetic nondegradable polymers, natural polymers obtained from plants such as cellulose, lignin, and chitin can reduce the environmental pollution rate. Biopolymers derived from natural sources can also be used for bioremediation purposes. Cellulose–starch-based biodegradable polymers are used for the covering of greenhouse and controlled release of fertilizer materials (Shokri & Adibkia, 2013) instead of traditional films to avoid the time-consuming, noneconomic methods of disposal of synthetic material which may lead to environmental pollution (Dilara & Briassoulis, 2000).

The utilization efficiency of fertilizers is a crucial element for the development of agricultural productions. However, due to surface run off, leaching, and vaporization, the fertilizers escape to the environment to cause an adverse effect on the economy and environmental problems (Bohlmann & Toki, 2004; Dave, Mehta, Aminabhavi, Kulkarni, & Soppimath, 1999). Cellulose can be used as the fertilizer controlled release matrices to release the fertilizers slowly or in a controlled way. As a result, the loss of fertilizers and environmental pollution can be avoided or reduced (Guo, Liu, Zhan, & Wu, 2005; Kumbar, Kulkarni, Dave, & Aminabhavi, 2001).

Biomass waste–containing cellulose produced in different types of industries or byproducts like cellulose acetate can be purified and used (Chen, Xie, Zhuang, Chen, & Jing, 2008). It can also be used for the preparation of bioethanol production (Binod et al., 2010); microporous membrane filters and cellulose-coated tables with aperture act as monolithic osmotic pumps for controlled drug release, thickening and stabilizing agents, etc. (Shokri & Adibkia, 2013). Natural rubber contains cellulose in which cellulose acts as reinforcing material (Ballinas-Casarrubias et al., 2016b).

Cellulose has applications in different fields; it can be used as biofuels, cellulose-based biodegradable plastics, high-quality paper, electroconductive films, optically transparent films, magnetically responsive films, nanomaterial composites, transparent coatings,

Table 4.2: Cellulose derivatives and its applications (Bortolin et al., 2013).

Sl. No.	Cellulose derivatives	Applications
1	Alkyl cellulose	Tablet, cosmetics
2	Carboxymethylcellulose	Ion exchange chromatography, medical field as tear substitute
3	Cellulose acetate	Acetate silk, films, plastics
4	Cellulose nitrate	Munition, celluloid film, glossy sheets
5	DEAE cellulose	Ion exchange chromatography
6	Xanthogenate	Textile industry, cellophane production

DEAE cellulose, Diethylaminoethyl cellulose.

vacuum insulation, electronic sensors, construction, filtration, barrier film, packaging (biofilm), composites, paints and coatings, paper and pulp, facial scrub, mask, chromatographic column for separation and purification of metabolites, etc. (Thakore 2012; Shaghaleh, Xu, & Wang, 2018; Neg & Vibha, 2017).

4.2.5 Process engineering and product development

Processing of cellulose-based composite and product development has invited the attention of researchers. Sustainable development is aimed at the use of cellulose-based biopolymers, since they are an excellent substitute for synthetic polymers. Process development and engineering strategies to improve new cellulose-based composites are mainly based on the source of cellulose from which is isolated. Processing techniques also vary based on the type of cellulose. Such cellulose-based green materials are indeed eco-friendly and reduce the pollution rate drastically. Purified cellulose is used in different combinations with other materials, and a lot of research indeed is vital for engineering good cellulose-based composites (Moohan et al., 2020).

4.2.6 Limitations of biopolymers and overcoming strategies

Bio-based polymers are gaining more demands these days mainly because of their biodegradable properties, which are becoming more critical due to strong public concerns about the waste generation and its disposal. All over the world, people realized the menace of conventional petroleum-based plastics, and a commercially viable, competitive, and biodegradable polymer as an alternative to plastics is a vital need of the hour. It is not only the biodegradability of polymer matters but how the individual monomer of the biopolymer or biopolymer composite affects the environment is also needed to be evaluated. Biopolymers are completely degradable or completely recyclable. The major problem in the commercialization of biodegradable polymers is the cost factor. The biopolymer production process has to be competitive enough to currently available processes and products in the market.

Moreover, the product cost has to be on par with conventional plastics. Although several efforts were undertaken all over the world to make biodegradable polymers competitive with the well-established petro-based polymers, it still has not met the benchmark regarding economic competitiveness and characteristics. The production cost mainly depends on the cost of raw materials as well as the production strategies. The raw material has to be very cheap, readily available, and it should be sustainable. Plant-based cellulose is a possible solution to the problem. The downstream processing and polymer blending are required and have to be simple and should be devoid of any harmful solvents. The treatment of effluent generated during the process also needs to be considered (Faruk, Bledzki, Fink, & Sain,

2014). The techno-economic and life-cycle analysis of the biopolymer is also fundamental to study the long-term effect of the biopolymer on the environment and on life forms.

4.2.7 Current status and challenges in the production of cellulose-based biopolymers

Nowadays, lots of researches are carried out worldwide to replace synthetic polymers with natural biopolymers such as cellulose. The usage of sustainable natural polymeric products creates environment-friendly approaches with a low cost of production than synthetic polymer because of the availability of abundant natural resources. Polymer composites are categorized into three types: partially degradable type, completely degradable type, and nondegradable type materials. In partially degradable types, only biofiller part of the composite is degradable. All parts including the matrix are made up of degradable materials in completely recyclable polymers (Buggy, 2006). These challenges involved in engineering cellulose–based composites are needed to be solved in the upcoming years.

Other challenges such as strategies to develop less time-consuming and easy procedures for the synthesis of cellulose composites are still ahead. By using ideas from material science, chemistry, and process engineering, these challenges can be overcome (Shaghaleh et al., 2018). This chapter aims to give readers interest in cellulose-based polymer composites and ideas about their wide range of applications. To produce cost-effective and eco-friendly natural cellulose–based biopolymers from plants, more researches are required for the natural methods for the development of cellulose polymer and its commercialization.

4.3 Conclusions and perspectives

This chapter explains the application of the substitution of synthetic polymers by plant-based cellulose polymers. Various issues related to the production and commercialization of plant cellulose-based biodegradable polymers and their composites are also mentioned here, and the adoption of novel techniques is the need of the hour in this area, which is addressing in this chapter. The possibilities of production and commercialization of cellulose from plants described in this chapter give a definite idea about how to use cellulose as a natural biopolymer. Cellulose can be a good substitute for the synthetic polymers used in different applications in various fields.

Biopolymers are defined as the polymeric materials derived from biological sources. Polysaccharides from plants, especially cellulose, are an excellent alternative to synthetic polymers due to their biodegradability, abundance, highly sustainable, functional, and low cost of production. We can conclude that it is worth to investigate further the creation of cost-effective plant-based cellulose polymer and its composite materials having a wide range of applications.

Conflict of interest

The authors declare that they have no conflict of interest.

References

Ahmadi, R., Ghanbarzadeh, B., Ayaseh, A., & Pezeshki, A. (2017). Nanotechnology and biopolymers in food packaging: Focused on carboxymethyl cellulose (CMC) as an edible biopolymer in biobased films and nanocomposites. In: *24th Iranian food science and technology congress* (pp. 1–9).

Anjana, K., Hinduja, M., Sujitha, K., & Dharani, G. (2020). Review on plastic wastes in marine environment—Biodegradation and biotechnological solutions. *Marine Pollution Bulletin, 150*, 110733.

Anwar, Z., Gulfraz, M., & Irshad, M. (2014). Agro-industrial lingo cellulosic biomass a key to unlock the future bio-energy: A brief review. *Journal of Radiation Research and Applied Sciences, 7*(2), 163–173.

Arvanitoyannis, I., & Biliaderis, C. G. (1999). Physical properties of polyol-plasticized edible blends made of methyl cellulose and soluble starch. *Carbohydrate Polymers, 38*(1), 47–58.

Ashori, A., & Nourbakhsh, A. (2010). Performance properties of microcrystalline cellulose as a reinforcing agent in wood plastic composites. *Composites Part B: Engineering, 41*(7), 578–581.

Ballinas-Casarrubias, L., Camacho-Davila, A., Gutierrez-Mendez, N., Ramos-Sanchez, V. H., Chavez-Flores, D., Manjarrez-Nevárez, L., Zaragoza-Galán, G., & González-Sanchez, G. (2016a). *Biopolymers from waste biomass—Extraction, modification and ulterior uses. Recent advances in biopolymers* (pp. 1–16). IntechOpen.

Ballinas-Casarrubias, L., Camacho-Davila, A., Gutierrez-Mendez, N., Ramos-Sanchez, V. H., Chavez-Flores, D., Manjarrez-Nevárez, L., Zaragoza-Galan, G., & Gonzalez-Sanchez, G. (2016b). *Biopolymers from waste biomass—Extraction, modification and ulterior uses. Recent advances* in biopolymers (pp. 3–15). IntechOpen.

Berk, Z. (2018). *Food process engineering and technology*. Academic Press.

Binod, P., Sindhu, R., Singhania, R. R., Vikram, S., Devi, L., Nagalakshmi, S., Kurien, N., Sukumaran, R. K., & Pandey, A. (2010). Bioethanol production from rice straw: An overview. *Bioresource Technology, 101*(13), 4767–4774.

Bohlmann, G., & Toki, G. (2004). *Chemical economics handbook*. SRI International, Menlo Park (47, pp. 777–780).

Bortolin, A., Aouada, F. A., Mattoso, L. H., & Ribeiro, C. (2013). Nanocomposite PAAm/methyl cellulose/montmorillonite hydrogel: Evidence of synergistic effects for the slow release of fertilizers. *Journal of Agricultural and Food Chemistry, 61*(31), 7431–7439.

Brett, C. T. (2000). Cellulose microfibrils in plants: Biosynthesis, deposition, and integration into the cell wall. *International Review of Cytology, 199*, 161–199.

Brown, A. J. (1886). XLIII. On an acetic ferment which forms cellulose. *Journal of the Chemical Society, Transactions, 49*(432–439), 1886.

Buggy, M. (2006). Review of natural fibers, biopolymers, and biocomposites. In A. K. Mohanty, M. Misra, & L. T. Drzal (Eds.), *Polymer International* (p. 1462).

Caccavo, D., Cascone, S., Lamberti, G., & Barba, A. A. (2018). Hydrogels: Experimental characterization and mathematical modelling of their mechanical and diffusive behaviour. *Chemical Society Reviews, 47*(7), 2357–2373.

Calo, E., & Khutoryanskiy, V. V. (2015). Biomedical application of hydrogels: A review of patents and commercial products. *European Polymer Journal, 65*, 252–267.

Cannazza, G., Cataldo, A., De Benedetto, E., Demitri, C., Madaghiele, M., & Sannino, A. (2014). Experimental assessment of the use of a novel superabsorbent polymer (SAP) for the optimization of water consumption in the agricultural irrigation process. *Water, 6*(7), 2056–2069.

Chaudhuri, O., Gu, L., Klumpers, D., Darnell, M., Bencherif, S. A., Weaver, J. C., Huebsch, N., Lee, H. P., Lippens, E., Duda, G. N., & Mooney, D. J. (2016). Hydrogels with tunable stress relaxation regulate stem cell fate and activity. *Nature Materials, 15*(3), 326.

Chen, L., Xie, Z., Zhuang, X., Chen, X., & Jing, X. (2008). Controlled release of urea encapsulated by starch-g-poly(L-lactide). *Carbohydrate Polymers, 72*(2), 342–348.

Ciolacu, D. E., & Suflet, D. M. (2018). *Cellulose-based hydrogels for medical/pharmaceutical applications. Biomass as renewable raw material to obtain bioproducts of high-tech value* (pp. 401–439). Elsevier.

Cooper, G. M., Hausman, R. E., & Hausman, R. E. (2000). *The cell: A molecular approach* (10). Washington, DC: ASM Press.

Czaja, W., Romanovicz, D., & Malcolm Brown, R. (2004). Structural investigations of microbial cellulose produced in stationary and agitated culture. *Cellulose, 11*(3–4), 403–411.

Dassanayake, R. S., Acharya, S., & Abidi, N. (2018). *Biopolymer-based materials from polysaccharides: properties, processing, characterization and sorption applications. Advanced sorption process applications.* IntechOpen.

Dave, A. M., Mehta, M. H., Aminabhavi, T. M., Kulkarni, A. R., & Soppimath, K. S. (1999). A review on controlled release of nitrogen fertilizers through polymeric membrane devices. *Polymer-Plastics Technology and Engineering, 38*(4), 675–711.

Demitri, C., Scalera, F., Madaghiele, M., Sannino, A., & Maffezzoli, A. (2013). Potential of cellulose-based superabsorbent hydrogels as water reservoir in agriculture. *International Journal of Polymer Science, 2013*.

Dhandayuthapani, B., Yoshida, Y., Maekawa, T., & Kumar, D. S. (2011). Polymeric scaffolds in tissue engineering application: A review. *International Journal of Polymer Science, 2011*.

DiezPascual, A. M. (2019). Synthesis and applications of biopolymer composites. *International Journal of Molecular Sciences, 20*, 2321.

Dilara, P. A., & Briassoulis, D. (2000). Degradation and stabilization of low-density polyethylene films used as a greenhouse covering materials. *Journal of Agricultural Engineering Research, 76*(4), 309–321.

Domingues, R. M., Silva, M., Gershovich, P., Betta, S., Babo, P., Caridade, S. G., Mano, J. F., Motta, A., Reis, R. L., & Gomes, M. E. (2015). Development of injectable hyaluronic acid/cellulose nanocrystals bionanocomposite hydrogels for tissue engineering applications. *Bioconjugate Chemistry, 26*(8), 1571–1581.

Donaldson, L. A. (2016). Lignification and lignin topochemistry—An ultrastructural view. *Phytochemistry, 57*(6), 859–873.

Faruk, O., Bledzki, A. K., Fink, H. P., & Sain, M. (2014). Progress report on natural fiber reinforced composites. *Macromolecular Materials and Engineering, 299*(1), 9–26.

Fernandes, E. M., Pires, R. A., Mano, J. F., & Reis, R. L. (2013). Bionanocomposites from lignocellulosic resources: Properties, application and future trends for their use in the biomedical field. *Progress in Polymer Science, 38*(10–11), 1415–1441.

Fink, H. P., Weigel, P., Purz, H. J., & Ganster, J. (2001). Structure formation of regenerated cellulose materials from NMMO solutions. *Progress in Polymer Science, 26*(9), 1473–1524.

Guan, H., Li, J., Zhang, B., & Yu, X. (2017). Synthesis, properties, and humidity resistance enhancement of biodegradable cellulose-containing superabsorbent polymer. *Journal of Polymers*.

Guilherme, M. R., Aouada, F. A., Fajardo, A. R., Martins, A. F., Paulino, A. T., Davi, M. F., Rubira, A. F., & Muniz, E. C. (2015). Superabsorbent hydrogels based on polysaccharides for applications in agriculture as a soil conditioner and nutrient carrier: A review. *European Polymer Journal, 72*, 365–385.

Guo, M., Liu, M., Zhan, F., & Wu, L. (2005). Preparation and properties of a slow-release membrane-encapsulated urea fertilizer with superabsorbent and moisture preservation. *Industrial & Engineering Chemistry Research, 44*(12), 4206–4211.

Halib, N., Ahmad, I., Grassi, M., & Grassi, G. (2019). The remarkable three dimensional network structure of bacterial cellulose for tissue engineering applications. *International Journal of Pharmaceutics, 566*, 631–640.

Hallinan, D. T., Jr, & Balsara, N. P. (2013). Polymer electrolytes. *Annual Review of Materials Research, 43*, 503–525.

Hao, Y., Cui, Y., Peng, J., Zhao, N., Li, S., & Zhai, M. (2019). Preparation of graphene oxide/cellulose composites in ionic liquid for Ce (III) removal. *Carbohydrate Polymers, 208*, 269–275.

Hoseny, S. M., Basmaji, P., de Olyveira, G. M., Costa, L. M. M., Alwahedi, A. M., da Costa Oliveira, J. D., & Francozo, G. B. (2015). Natural ECM-bacterial cellulose wound healing—Dubai study. *Journal of Biomaterials and Nanobiotechnology*, 6(04), 237.

Huang, J., Huang, L., Lin, J., Xu, Z., & Cen, P. (2010). *Organic chemicals from bioprocesses in China. Biotechnology in China II* (pp. 43–71). Berlin, Heidelberg: Springer.

Huang, Y., Zhu, C., Yang, J., Nie, Y., Chen, C., & Sun, D. (2014). Recent advances in bacterial cellulose. *Cellulose*, 21(1), 1–30.

Hu, W., Chen, S., Yang, J., Li, Z., & Wang, H. (2014). Functionalized bacterial cellulose derivatives and nanocomposites. *Carbohydrate Polymers*, 101, 1043–1060.

Iguchi, M., Yamanaka, S., & Budhiono, A. (2000). Bacterial cellulose – A masterpiece of nature's arts. *Journal of Materials Science*, 35(2), 261–270.

Ismail, H., Irani, M., & Ahmad, Z. (2013). Starch-based hydrogels: Present status and applications. *International Journal of Polymeric Materials and Polymeric Biomaterials*, 62(7), 411–420.

Jackson, J. K., Letchford, K., Wasserman, B. Z., Ye, L., Hamad, W. Y., & Burt, H. M. (2011). The use of nanocrystalline cellulose for the binding and controlled release of drugs. *International Journal of Nanomedicine*, 6, 321–330.

Johansson, C., Bras, J., Mondragon, I., Nechita, P., Plackett, D., Simon, P., Svetec, D. G., Virtanen, S., Baschetti, M. G., Breen, C., & Aucejo, S. (2012). Renewable fibers and bio-based materials for packaging applications—A review of recent developments. *BioResources*, 7(2), 2506–2552.

Jonas, R., & Farah, L. F. (1998). Production and application of microbial cellulose. *Polymer Degradation and Stability*, 59(1–3), 101–106.

Kabir, S. F., Sikdar, P. P., Haque, B., Bhuiyan, M. R., Ali, A., & Islam, M. N. (2018). Cellulose-based hydrogel materials: Chemistry, properties and their prospective applications. *Progress in Biomaterials*, 7(3), 153–174.

Kalia, S., Dufresne, A., Cherian, B. M., Kaith, B. S., Averous, L., Njuguna, J., & Nassiopoulos, E. (2011). Cellulose-based bio- and nanocomposites: A review. *International Journal of Polymer Science*.

Khalil, H. P. S., Tye, Y. Y., Saurabh, C. K., Leh, C. P., Lai, T. K., Chong, E. W. N., Fazita, M. R., Hafiidz, J. M., Banerjee, A., & Syakir, M. I. (2017). Biodegradable polymer films from seaweed polysaccharides: A review on cellulose as a reinforcement material. *Express Polymer Letters*, 11(4), 244–265.

Khan, A., Nair, V., Colmenares, J. C., & Glaser, R. (2018). Lignin-based composite materials for photocatalysis and photovoltaics. *Topics in Current Chemistry*, 376(3), 20.

Kimura, S., & Itoh, T. (2007). Biogenesis and function of cellulose in the tunicates. In R. M. Brown, Jr, & I. Saxena (Eds.), *Cellulose: molecular and structural biology* (pp. 217–236). Dordrecht: Springer.

Klemm, D., Heublein, B., Fink, H. P., & Bohn, A. (2005). Cellulose: Fascinating biopolymer and sustainable raw material. *Angewandte Chemie International Edition*, 44(22), 3358–3393.

Klemm, D., Schumann, D., Kramer, F., Heber, N., Hornung, M., Schmauder, H. P., & Marsch, S. (2006). *Nanocelluloses as innovative polymers in research and applications. Polysaccharides II* (pp. 49–96). Berlin, Heidelberg: Springer.

Klemm, D., Schumann, D., Udhardt, U., & Marsch, S. (2001). Bacterial synthesized cellulose—Artificial blood vessels for microsurgery. *Progress in Polymer Science*, 26(9), 1561–1603.

Kumbar, S. G., Kulkarni, A. R., Dave, A. M., & Aminabhavi, T. M. (2001). Encapsulation efficiency and release kinetics of solid and liquid pesticides through urea formaldehyde crosslinked starch, guar gum, and starch + guar gum matrices. *Journal of Applied Polymer Science*, 82(11), 2863–2866.

Laftah, W. A., & Hashim, S. (2013). *Preparation and possible agricultural applications of polymer hydrogel composite as soil conditioner, . Advanced materials research* (626, pp. 6–10). Trans Tech Publications.

Lalit, R., Mayank, P., & Ankur, K. (2018). Natural fibers and biopolymers characterization: A future potential composite material. *Strojnicky Casopis—Journal of Mechanical Engineering*, 68(1), 33–50.

Lin, N., & Dufresne, A. (2014). Nanocellulose in biomedicine: Current status and future prospect. *European Polymer Journal*, 59, 302–325.

Li, H., Lai, F., & Luo, R. (2009). Analysis of responsive characteristics of ionic-strength-sensitive hydrogel with consideration of the effect of equilibrium constant by a chemo-electro-mechanical model. *Langmuir: The ACS Journal of Surfaces and Colloids*, 25(22), 13142−13150.

Li, J., Li, Y., & Dong, H. (2008). Controlled release of herbicide acetochlor from clay/carboxylmethylcellulose gel formulations. *Journal of Agricultural and Food Chemistry*, 56(4), 1336−1342.

Li, W., Lan, Y., Guo, R., Zhang, Y., Xue, W., & Zhang, Y. (2015). In vitro & in vivo evaluation of a novel collagen/cellulose nanocrystals scaffold for achieving the sustained release of basic fibroblast growth factors. *Journal of Biomaterials Applications*, 29(6), 882−893.

Li, Y. Y., Wang, B., Ma, M. G., & Wang, B. (2018). Review of recent development on preparation, properties, and applications of cellulose-based functional materials. *International Journal of Polymer Science*, 18.

Lv, X., Feng, C., Liu, Y., Peng, X., Chen, S., Xiao, D., Wang, H., Li, Z., Xu, Y., & Lu, M. (2018). A smart bilayered scaffold supporting keratinocytes and muscle cells in micron/nano-scale for urethral reconstruction. *Theranostics*, 8(11), 3153−3163.

Mahalakshmi, M., Selvanayagam, S., Selvasekarapandian, S., Moniha, V., Manjuladevi, R., & Sangeetha, P. (2019). Characterization of biopolymer electrolytes based on cellulose acetates with magnesium perchlorate (Mg $(ClO_4)_2$) for energy storage devices. *Journal of Science: Advanced Materials and Devices*, 276−284.

Marques, G., Rencoret, J., Gutiérrez, A., Alfonso, J. E., & del Río, J. C. (2010). Evaluation of the chemical composition of different non-woody plant fibers used for pulp and paper manufacturing. *The Open Agriculture Journal*, 4(1).

Mathew, A. P., Oksman, K., Pierron, D., & Harmand, M. F. (2012). Fibrous cellulose nanocomposite scaffolds prepared by partial dissolution for potential use as ligament or tendon substitutes. *Carbohydrate Polymers*, 87(3), 2291−2298.

Matias, J., Cruz, V., Garcia, A., & Gonzalez, D. (2019). Evaluation of rice straw yield, fibre composition and collection under Mediterranean conditions. *Acta Technologica Agriculturae*, 22(2), 43−47.

Ma, Q., Hu, D., & Wang, L. (2016). Preparation and physical properties of gum film reinforced with cellulose nanocrystals. *International Journal of Biological Macromolecules*, 86, 606−612.

McFarlane, H. E., Doring, A., & Persson, S. (2014). The cell biology of cellulose synthesis. *Annual Review of Plant Biology*, 65, 69−94.

Mesquita, J. P., Donnici, C. L., Teixeira, I. F., & Pereira, F. V. (2012). Bio-based nanocomposites obtained through the covalent linkage between chitosan and cellulose nanocrystals. *Carbohydrate Polymers*, 90(1), 210−217.

Meyer, U. (2009). *The history of tissue -engineering and regenerative medicine in perspective. Fundamentals of tissue engineering and regenerative medicine* (pp. 5−12). Berlin, Heidelberg: Springer.

Miao, C., & Hamad, W. Y. (2013). Cellulose reinforced polymer composites and nanocomposites: A critical review. *Cellulose*, 20(5), 2221−2262.

Mohamed, M. H., Udoetok, I. A., Wilson, L. D., & Headley, J. V. (2015). Fractionation of carboxylate anions from aqueous solution using chitosan cross-linked sorbent materials. *RSC Advances*, 5(100), 82065−82077.

Mohan, S., Oluwafemi, O. S., Kalarikkal, N., Thomas, S., & Songca, S. P. (2016). *Biopolymers − Application in nanoscience and nanotechnology. Recent advances in biopolymers* (pp. 47−66). IntechOpen.

Montesano, F. F., Parente, A., Santamaria, P., Sannino, A., & Serio, F. (2015). Biodegradable superabsorbent hydrogel increases water retention properties of growing media and plant growth. *Agriculture and Agricultural Science Procedia*, 4, 451−458.

Moohan, J., Stewart, S. A., Espinosa, E., Rosal, A., Rodríguez, A., Larrañeta, E., Donnelly, R. F., & Domínguez-Robles, J. (2020). Cellulose nanofibers and other biopolymers for biomedical applications. A review. *Applied Sciences*, 10(1), 65.

Moon, R. J., Martini, A., Nairn, J., Simonsen, J., & Youngblood, J. (2011). Cellulose nanomaterials review: Structure, properties and nanocomposites. *Chemical Society Reviews*, 40(7), 3941−3994.

Nawrath, C., Poirier, Y., & Somerville, C. (1995). Plant polymers for biodegradable plastics: Cellulose, starch and polyhydroxyalkanoates. *Molecular Breeding*, 1(2), 105−122.

Negi, S., & Vibha, K. (2017). Amylolytic enzymes: Glucoamylases. In A. Pandey, S. Nedi, & C. R. Soccol (Eds.), *Current developments in biotechnology & bioengineering: Production, isolation and purification of industrial products* (pp. 25–41). Elsevier.

O'Sullivan, A. C. (1997). Cellulose: The structure slowly unravels. *Cellulose*, 4(3), 173–207.

Oyama, T., Udoetok, I. A., Wilson, L. D., & Headley, J. V. (2018). Cross-linked polymer synthesis. *Encyclopedia of Polymeric Nanomaterials*, 496–505.

Park, H. J., Weller, C. L., Vergano, P. J., & Testin, R. F. (1993). Permeability and mechanical properties of cellulose-based edible films. *Journal of Food Science*, 58(6), 1361–1364.

Payen, A. (1838). Study of the composition of the natural tissue of plants and of lignin. *Comptes Rendus*, 7, 1052–1056.

Peng, N., Wang, Y., Ye, Q., Liang, L., An, Y., Li, Q., & Chang, C. (2016). Biocompatible cellulose-based superabsorbent hydrogels with antimicrobial activity. *Carbohydrate Polymers*, 137, 59–64.

Portela, R., Leal, C. R., Almeida, P. L., & Sobraj, R. G. (2019). Bacterial cellulose: A versatile biopolymer for wound dressing applications. *Microbial Biotechnology*, 12(4), 586–610.

Qiu, X., & Hu, S. (2013). "Smart" materials based on cellulose: A review of the preparations, properties, and applications. *Materials*, 6(3), 738–781.

Rehman, N., de Miranda, M. I. G., Rosa, S. M., Pimentel, D. M., Nachtigall, S. M., & Bica, C. I. (2014). Cellulose and nano cellulose from maize straw: An insight on the crystal properties. *Journal of Polymers and the Environment*, 22(2), 252–259.

Rinaudo, M. (2006). Chitin and chitosan: Properties and applications. *Progress in Polymer Science*, 31(7), 603–632.

Rowell, R. M. (2012). *Handbook of wood chemistry and wood composites*. CRC Press.

Rudin, A., & Choi, P. (2012). *The elements of polymer science and engineering*. Academic Press.

Salgado, A. J., Oliveira, J. M., Martins, A., Teixeira, F. G., Silva, N. A., Neves, N. M., Sousa, N., & Reis, R. L. (2013). *Tissue engineering and regenerative medicine: Past, present and future, . International review of neurobiology* (108, pp. 1–33). Academic Press.

Sanchez-Orozco, R., Timoteo-Cruz, B., Torres-Blancas, T., & Ureña-Nunez, F. (2017). Valorization of superabsorbent polymers from used disposable diapers as soil moisture retainer. *International Journal of Research*, 5(4), 115–117.

Sannino, A., Mensitieri, G., & Nicolais, L. (2004). Water and synthetic urine sorption capacity of cellulose-based hydrogels under a compressive stress field. *Journal of Applied Polymer Science*, 91(6), 3791–3796.

Shaghaleh, H., Xu, X., & Wang, S. (2018). Current progress in production of biopolymeric materials based on cellulose, cellulose nanofibers, and cellulose derivatives. *RSC Advances*, 8(2), 825–842.

Shen, X., Shamshina, J. L., Berton, P., Gurau, G., & Rogers, R. D. (2016). Hydrogels based on cellulose and chitin: Fabrication, properties, and applications. *Green Chemistry*, 18(1), 53–75.

Shokri, J., & Adibkia, K. (2013). *Applications of cellulose and cellulose derivatives in pharmaceutical industries. Cellulose—Medical, pharmaceutical and electronic applications*. IntechOpen.

Smitha, B., Sridhar, S., & Khan, A. A. (2004). Polyelectrolyte complexes of chitosan and poly (acrylic acid) as proton exchange membranes for fuel cells. *Macromolecules*, 37(6), 2233–2239.

Sunasee, R., Hemraz, U. D., & Ckless, K. (2016). Cellulose nanocrystals: A versatile nanoplatform for emerging biomedical applications. *Expert Opinion on Drug Delivery*, 13(9), 1243–1256.

Taboas, J. M., Maddox, R. D., Krebsbach, P. H., & Hollister, S. J. (2003). Indirect solid free form fabrication of local and global porous, biomimetic and composite 3D polymer-ceramic scaffolds. *Biomaterials*, 24(1), 181–194.

Tanase, E. E., Rapa, M., & Popa, O. (2014). Biopolymers based on a renewable resource—A review. *Scientific Bulletin, Series F. Biotechnologies*, 18, 188–195.

Thakore, S. I. (2012). *Role of biopolymers in green nanotechnology* (pp. 119–140). Intech Open Access Publisher.

Thomas, S., Paul, S. A., Pothan, L. A., & Deepa, B. (2011). *Chapter 1: Natural fibres: Structure, properties and applications. Bio- and nano-polymer composites* (pp. 1–42). Springer-Verlag.

Tye, Y. Y., Lee, K. T., Abdullah, W. N. W., & Leh, C. P. (2016). The world availability of non-wood lignocellulosic biomass for the production of cellulosic ethanol and potential pretreatments for the enhancement of enzymatic saccharification. *Renewable and Sustainable Energy Reviews*, *60*, 155–172.

Udoetok, I. A., Wilson, L. D., & Headley, J. V. (2018). "Pillaring effects" in cross-linked cellulose biopolymers: A study of structure and properties. *International Journal of Polymer Science*.

Vartiainen, J., Vähä-Nissi, M., & Harlin, A. (2014). Biopolymer films and coatings in packaging applications – A review of recent developments. *Materials Sciences and Applications*, *5*(10), 708–718.

Vlaia, L., Coneac, G., Olariu, I., Vlaia, V., & Lupuleasa, D. (2016). Cellulose-derivatives-based hydrogels as vehicles for dermal and transdermal drug delivery. *Emerging Concepts in Analysis and Applications of Hydrogels*, *2*, 64.

Vroman, I., & Tighzert, L. (2009). Biodegradable polymers. *Materials*, *2*(2), 307–344.

Wada, M., Okano, T., & Sugiyama, J. (2001). Allomorphs of native crystalline cellulose I evaluated by two equatorial *d*-spacings. *Journal of Wood Science*, *47*(2), 124–128.

Wang, J., Tavakoli, J., & Tang, Y. (2019). Bacterial cellulose production, properties and application with different culture methods—A review. *Carbohydrate Polymers*, *219*, 63–76.

Wang, W., & Wang, A. (2010). Synthesis and swelling properties of pH-sensitive semi-IPN superabsorbent hydrogels based on sodium alginate-*g*-poly(sodium acrylate) and polyvinylpyrrolidone. *Carbohydrate Polymers*, *80*(4), 1028–1036.

Wang, W., Wang, J., Shi, X., Yu, Z., Song, Z., Dong, L., Jiang, G., & Han, S. (2016). Synthesis of mesoporous TiO_2 induced by nano-cellulose and its photocatalytic properties. *BioResources*, *11*(2), 3084–3093.

Wei, L., & McDonald, A. (2016). A review on grafting of biofibers for biocomposites. *Materials*, *9*(4), 303–328.

Wei, H., Rodriguez, K., Renneckar, S., & Vikesland, P. J. (2014). Environmental science and engineering applications of nanocellulose based nanocomposites. *Environmental Science: Nano*, *1*(4), 302–316.

Yang, J., Lv, X., Chen, S., Li, Z., Feng, C., Wang, H., & Xu, Y. (2014). In situ fabrication of a microporous bacterial cellulose/potato starch composite scaffold with enhanced cell compatibility. *Cellulose*, *21*(3), 1823–1835.

Zhao, C., Li, Z., Li, T., Zhang, Y., Bryant, D. A., & Zhao, J. (2015). High-yield production of extracellular type-I cellulose by the cyanobacterium *Synechococcus* sp. PCC 7002. *Cell Discovery*, *1*, 15004.

Zhao, Y., & Li, J. (2014). Excellent chemical and material cellulose from tunicates: Diversity in cellulose production yield and chemical and morphological structures from different tunicate species. *Cellulose*, *21*(5), 3427–3441.

… CHAPTER 5

Starch

Susan Grace Karp[1], Maria Giovana Binder Pagnoncelli[2], Fernanda Prado[1], Rafaela de Oliveira Penha[1], Antônio Irineudo Magalhães Junior[1], Gabriel Sprotte Kumlehn[1] and Carlos Ricardo Soccol[1]

[1]Department of Bioprocess Engineering and Biotechnology, Federal University of Paraná, Curitiba, Brazil, [2]Department of Chemistry and Biology, Federal University of Technology, Curitiba, Brazil

5.1 Introduction

Bio-based polymers are becoming a subject of great interest in materials science and technology. Decades of extensive utilization of fossil-based nonbiodegradable materials have caused serious environmental damages, and today the demand for environmentally friendly materials comes from many sectors of society such as governments and consumers. Consequently the legislative scenario is also changing, favoring bio-based products and affecting the perspectives for the industrial sector. This can be seen in practice though initiatives such as the Lead Market Initiative (European Union) and the BioPreferred (USA).

Among the sources of biodegradable materials, starch appears as a promising biopolymer for being a rich natural source of overwhelming abundance, being relatively cheap, highly biodegradable, and with annual renewability. Its decomposition in nature results from hydrolytic reactions that can be catalyzed by enzymes. Additionally, starch can be utilized as a carbon source by a great variety of microorganisms, which favors its biodegradation.

The abundance of hydroxyl groups in its constituent monomer, D-glucose, associated with its lower crystallinity degree as compared, for example, to cellulose, confers hydrophilic characteristics to the starch molecules depending on their physicochemical environment. The behavior of starch is in great extent dependent on the content of water and temperature. Because of these peculiar characteristics in terms of gaining and losing water, the use of starch as a structural material is a challenging task.

Starch is extracted industrially from various vegetable sources like corn, wheat, rice, potatoes, and cassava. The main objective is to release the starch from the cell structure and separate it from other components such as proteins and insoluble fiber. The processing steps depend on the raw material and on the intended product, for example, unmodified starch, modified starch, gelatinized starch, and D-glucose from starch. After extraction, the processing of starch granules to produce starch-based materials can be accomplished by many of the traditional methods, such as extrusion, injection molding, compression molding, casting, and foaming.

In order to use starch as a substitute for petroleum-based materials, many obstacles need to be overcome, such as poor mechanical properties, high vapor permeability due to its hydrophilic nature, and the phenomenon of retrogradation. The physicochemical and mechanical properties of starch-based materials can be improved by physical, chemical, or enzymatic treatments or genetic modifications. Examples of physical treatments are shear application, fluidized bed heating and deep freezing and thawing cycles. Chemical modifications may include cross-linking and addition of functional groups, which can be accomplished both by chemical and enzymatic reactions. Finally, genetic modifications in the botanical source can be a tool to obtain starch with enhanced properties, through the modification of amylose and amylopectin ratios.

Another interesting strategy to improve the mechanical behavior of starch-based materials is the manufacture of starch composites. The addition of fibers, particles, or flakes in a starch matrix allows the combination of the different mechanical properties of the matrix and the inclusions, yielding various types of materials. Besides, the addition of plasticizer agents and the formation of starch blends with other polymers also increase the range of possibilities for starch-based materials.

This chapter presents an overview of starch properties, extraction and processing methods, and strategies for starch-based materials development and applications. A patent survey on starch-based materials is also presented, especially focusing on starch-based polymer blending and compositing, foaming, and nanocomposites.

5.2 Structure and properties

Starch is a carbohydrate produced abundantly by photosynthesis and is the main carbon reserve in higher plants. It is a polysaccharide composed exclusively of α-D-glucose residues and its basic chemical formula is $(C_6H_{10}O_5)_n$. This homopolymer is formed by two components in different proportions, according to the species of the plant: amylose and amylopectin (Fig. 5.1). Amylose is a linear chain consisting of about 500–2000 glucose units linked by α-(1→4) linkages, and eventually some branches of α-(1→6) linkages (0.3%–0.5%). Amylopectin is a highly branched molecule consisting of over 1,000,000 glucose units linked by α-(1→4) and α-(1→6) bonds, the branch points make up approximately 4%–6% of the total linkages in the molecule. Amylopectin is a very large and highly branched molecule, reaching a molecular mass of up to 6×10^9, while the

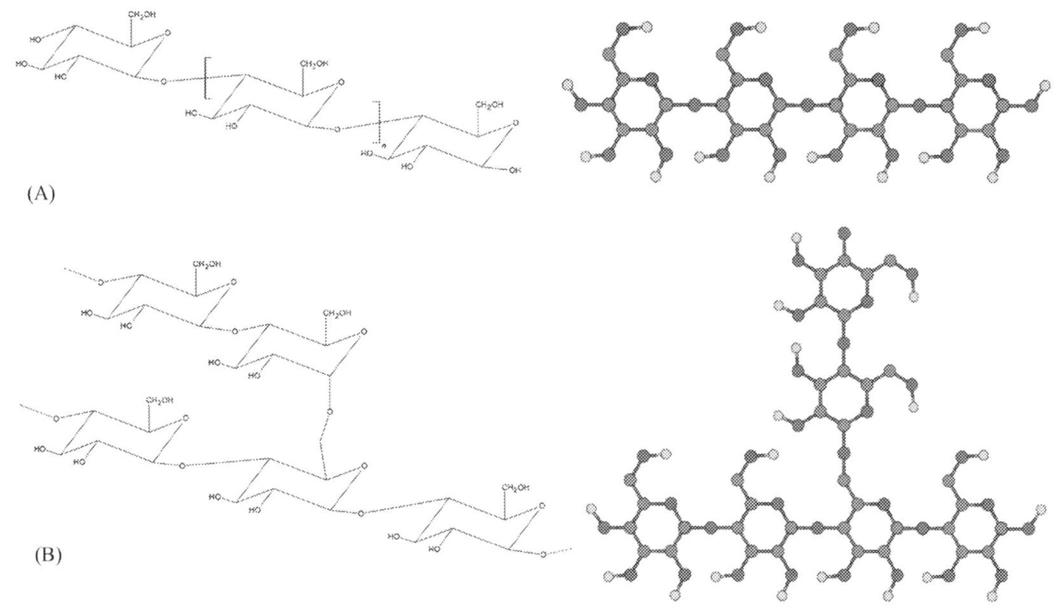

Figure 5.1
Structures of amylose (A) and amylopectin (B).

molecular mass of an amylose molecule from different botanical sources of starch may vary between 10^5 and 10^6 (Robyt, 2008; Egharevba, 2019).

D-glucose contains groups of hydroxyls capable of forming hydrogen bonds with one or more water molecules. This feature confers to glycans a strong affinity with water, enabling hydration when these groups are available. Another important feature is the tetrahedral chemistry of these molecules. Due to the spatial arrangement of glycosidic bonds between the D-glucose residues, amylose and amylopectin can reveal a particular spatial configuration, which gives them certain characteristics and properties. The amylose forms an overall spiral or helical shape, with predominance of hydrogen atoms inside the spiral (hydrophobic region) and the hydroxyl groups are positioned on the outside. Whereas the model which is widely accepted until today for the amylopectin is the "cluster" model proposed by French in 1972 and modified by Robin et al. in 1974. In this model the branch points are located in the clusters and the external branch chains are presented in a double helical crystalline structure. The presence of crystallinity makes the polysaccharide insoluble and resistant to rupture (Jane, 2009; Huber & BeMiller, 2019).

Starch is distinguished from other carbohydrates as it occurs in nature like crystalline particles. The amylose and the amylopectin are arranged in granules forming a structure of layers which overlap around a point called hilum. The layers that surround the hilum are the result of starch deposits of different degrees of hydration due to the presence of amylose and amylopectin

(Leonel & Lima, 2016). The structures of these granules are defined by the length of glucose chains, the amylose/amylopectin ratio, and the branching degree of amylopectin. The starch granules produced by each plant species have specific structures and compositions. Furthermore the amylose content can vary within the same botanical variety due to differences in geographical origin and culture conditions. In natural starches, that is, those from natural maize, rice, wheat, and potato, the percentage of amylose ranges from 11% to 37%, however there are mutants composed of 100% amylopectin (waxy mutants) and high-amylose starches that contain from 53% to 70% amylose. The granules also contain water of hydration, ranging from 10% to 15% (Robyt, 2008; Shannon, Garwood, & Boyer, 2009).

The starch granules can occur in all shapes and sizes (spheres, ellipsoids, polygons, platelets, and irregular tubules); their long dimensions range from 0.1 μm to at least 200 μm. Amylose and amylopectin are the major macromolecular components of starch granules (98%–99%). The phosphate groups, protein and fat contents may also vary significantly in granules from different botanical sources. The size and shape of the starch granules are among the important features in determining its applications. The stacks of crystalline and amorphous lamellae build up a semicrystalline region that alternates with more amorphous regions to create the "growth rings," and these concentric rings radiate out from the central hilum to the surface of the granule. The crystalline region is less susceptible to enzymatic hydrolysis, water penetration, and other chemical reactions than the amorphous region. Amylopectin has a lower tendency to gelatinization, retrogradation, and syneresis because of the branched structure (Jane, 2009; Pérez, Baldwin, & Gallant, 2009; Hamaker, Tuncil, & Shen, 2019).

Starch granules are insoluble in cold water, but when the water is heated in excess, they undergo a transition phase called gelatinization. During the gelatinization process, the heating promotes increased motion of the molecules, and as a result the hydrogen and hydrophobic bonds break and the water molecules start interacting with the amylose and amylopectin hydroxyl groups, causing an increase in the size of the granules and partial solubilization of the starch, due to melting of the crystallites and unfolding of the double helix structures. Under these conditions the crystalline structure of the granules is irreversibly broken down, usually keeping a certain amount of structural integrity. The changes in properties are water uptake, granular swelling, crystallite melting, birefringence loss, starch solubilization, and viscosity development. The gelatinization temperature (60°C–70°C) is a characteristic of the starch source, however, as the temperature is increased, the granules could burst, releasing the amylose and amylopectin molecules into the aqueous solution. Cooling of the solution results in the formation of a viscoelastic firm gel (Robyt, 2008; Leonel & Lima, 2016; Hamaker et al., 2019; Biliaderis, 2009).

As the temperature decreases, the formation of association zones between molecules occurs in the gelatinized starch, so they gradually realign and become less soluble. This phenomenon is called retrogradation, and occurs much faster with amylose than with amylopectin. The rate of retrogradation of a starch gel depends on the ratio of amylose to amylopectin, the structure of amylose and amylopectin molecules, the temperature and the starch concentration (Robyt, 2008; Leonel & Lima, 2016; Hamaker et al., 2019; Biliaderis, 2009).

The characteristic of starch in gaining or losing water can result in significant changes in physical and mechanical properties of starch-based materials. Also, the branched nature of amylopectin yields a polymer with poor properties relative to linear polymers. Nowadays several approaches have been used for incorporating starch into plastics and they will be discussed further (Jiang, Duan, Zhu, Liu, & Yu, 2020).

5.3 Starch-processing techniques

The technique used for starch extraction depends on the raw material chosen for processing. Corn, wheat, rice, potatoes, and cassava are the main feedstocks processed to produce starch, but others can also be used on a smaller scale, such as oats, sorghum, rye, and barley (Alcázar-Alay & Meireles, 2015; Blennow, 2004). Corn and rice starches are prepared from the whole seed, but most cereal starches are extracted from flour. The feedstock composition will also make it possible to choose the starch-processing technique, which may involve several steps and procedures. The main objective in starch processing is to release starch from the cell structure free of other components such as proteins, gluten, and fibers.

Starch granules, in their natural state, have limited capacity to absorb water (Haroon et al., 2016). When mixed with water, starch forms a suspension. The concentration of starch in the suspension should be as high as possible to improve the efficiency of extraction processes. The temperature and concentration of some substances, such as potassium salts and sulfurous acid, can influence the suspension of starch causing swelling of the granules (Alcázar-Alay & Meireles, 2015; Schirmer, Jekle, & Becker, 2015). When the suspension is heated, the starch granules gelatinize to obtain a thick starch paste. Heating a suspension of starch causes the breakdown of hydrogen bonds, hydration of the hydroxyl groups of the glycoses, and the granule swells (Alcázar-Alay & Meireles, 2015). The starch gelatinization temperature is of $52°C-63°C$ for wheat, $56°C-66°C$ for potatoes, $58°C-70°C$ for cassava, $62°C-72°C$ for corn, and $61°C-78°C$ for rice (Early, 1997). After gelatinization, the hydrogen bonds continue to be broken, the swelling of the granules continues until the break, the granular structure ceases to exist, and the viscosity decreases abruptly (Kumar & Khatkar, 2017). The gelatinization temperature range, viscosity, and stability of the starch paste are of great importance in the food industry and determine the use of starch (Ashogbon & Akintayo, 2014). During the cooling of gelatinized starch pastes, the crystallization of the starch molecules or "retrogradation" occurs, with the formation of hydrogen bonds (Alcázar-Alay & Meireles, 2015).

Starch processing is carried out using different methods, such as steeping, grinding, and crushing, often with the aid of chemical, biochemical, and biological reactions (Ashogbon & Akintayo, 2014). The extracted starch grains are separated due to their density and dimension in equipment such as sieves, filters, hydrocyclones, and centrifuges. The choice of processing varies according to the type of product, such as unmodified starch, modified starch, gelatinized starch, and D-glucose from starch. Thus different configurations are found for the processing of starch. One of the main techniques used in grains is steeping with immersion in sulfur dioxide (SO_2), as in the processing of corn and rye starch, or alkaline solution, as in the rice starch extraction method (Autio & Eliasson, 2009a; Eckhoff & Watson, 2009; Mitchell, 2009). In the processing of barley and oat starch, the release of the starch granules is carried out by treatment in a solution containing enzymes, such as cellulases and hemicellulases (Autio & Eliasson, 2009a; Hoover & Vasanthan, 2009). The most common industrial processes for extracting starch will be described in the following paragraphs.

The main objective in corn processing is to extract starch from the seed endosperm, ensuring that the corn germ, used for oil production, is not damaged. Thus at the end of the process, starch, fiber, germ, and gluten are produced. The first stage involves cleaning and steeping of the grains. Steeping requires immersing the corn in a solution of 0.1%–0.2% SO_2, in a countercurrent flow 48°C–55°C for 24–36 hours (Eckhoff & Watson, 2009). Corn steeping has more specific stages and occurs in three stages between 8 and 12 hours each. In the first stage, lactic acid is produced by *Lactobacillus* spp. through the fermentation of sugars present in the steeping water. This reduces the pH of the medium and promotes ideal conditions for hydration and separation of the grain components, in addition to softening the seeds, releasing the starch granules from the protein matrices by breaking the sulfur bonds and inhibiting the growth of deteriorating microorganisms. The release of starch in the endosperm occurs after the diffusion of SO_2 in the core of the corn seed. Then, the seeds absorb 0.02%–0.04% (w/w) SO_2 with water absorption of about 45% in wet basis (Eckhoff & Watson, 2009). At the end of the steeping, the germ is released intact and free of adhesion from the endosperm or pericarp, being easily removed as a white flake. Then, the water-soaked corn is wet milled to separate the germ, while a finer grind followed by washing removes the fibers. The washing broth is sent to hydrocyclones that separate starch and gluten. Concentration, purification, dehydration, and drying steps are used to finish the starch processing. These finishing steps are, in general, common among all starch-processing methods.

The steeping in SO_2 in the extraction of rye starch is carried out under intense agitation of 0.5%–0.15% SO_2 between 30°C and 48°C for 5–24 hours, and the incubation of lactobacilli, streptococci, sourdough extract, pasta of sour culture, or a set of microbial enzymes reduces the viscosity of the suspension (Autio & Eliasson, 2009b). The steeping of rice occurs in a solution of sodium hydroxide (NaOH) between 0.3% and 0.5% at 20°C–50°C for 10–24 hours for the solubilization of glutelin, which constitutes about 80% of the protein in the grain (Mitchell, 2009). Then, the paste is filtered to remove the cell wall, washed with water to remove the protein, neutralized, and dried. Rice steeping can also be carried out without adding an alkaline solution. In this process, the protein is removed by physical separation after the mechanical release of the granules and starch proteins from the endosperm. Products containing 0.25%–7% protein can be produced using this method (Mitchell, 2009). If not, protein removal steps are necessary. The main advantage of the mechanical process is the variety of starches with concentrations of proteins that can be produced.

The production of wheat starch involves the separation of the flour components, starch A, starch B, gluten, and pentosans, by wet milling in warm water in a short period between 10 and 12 minutes (Maningat, Seib, Bassi, Woo, & Lasater, 2009). Enzymes, such as hemicellulases and pentosanases, can be used to improve the separation of starch and gluten. Then, the dough is vigorously stirred with water to concentrate the proteins and agglomerate the gluten strands. The dough can be separated by the difference in density of the dough components, with starch A being the densest component, followed by starch B, gluten and pentosans using a settling centrifuge, hydrocyclones, or high-pressure equipment.

The extraction of cassava starch is carried out by crushing the cassava into 1–2 cm pieces using a cutting blade and feeding to a saw tooth scraper for an intense attraction in a pulp paste (Breuninger, Piyachomkwan, & Sriroth, 2009). The scraped cassava paste is pumped through a series of thick and thin extractors, vertical or horizontal, where the fiber is removed by screens arranged conically in continuous centrifuges. Also, the starch paste passes through hydrocyclones to ensure complete removal of the sand. The concentration of the starch paste is carried out using a bi- or three-phase separator or a series of hydrocyclones. The starch can be recovered from the final stream and dried, or the stream can be used to provide a chemical modification process. To produce native starch, the starch paste is dehydrated in a horizontal centrifuge to produce a starch cake with a moisture content of 35%–40%.

High-pressure filtration or filter presses are used in some starch plants to reduce the loss of starch in the liquid stream. The high moisture starch cake is then subjected to a pneumatic conveying dryer, known as an instant dryer, to decrease the moisture content to 13% before packaging (Breuninger et al., 2009). Starch dust lost in cyclones can be recovered by being retained with venturi washers or filter bags. The basic screening and density separation processes remain common to all machine variations used in the production of starch with high quality.

Potato starch is extracted by grinding and crushing, which forms a mixture of starch granules, broken cell walls, and cell debris. The separation of these three components can be done in different ways according to the factors chosen. Cellular residues contain soluble proteins, amino acids, sugars, and salts. Some small particles of fiber and proteins that remain in the raw starch are removed by centrifugal sieves and classification steps in a continuous disk-type centrifuge (Grommers & Krogt, 2009). After the extraction of the fibers and classification, a paste containing starch with high concentration is obtained. Figure 5.2 represents the main steps involved in starch processing from different raw materials.

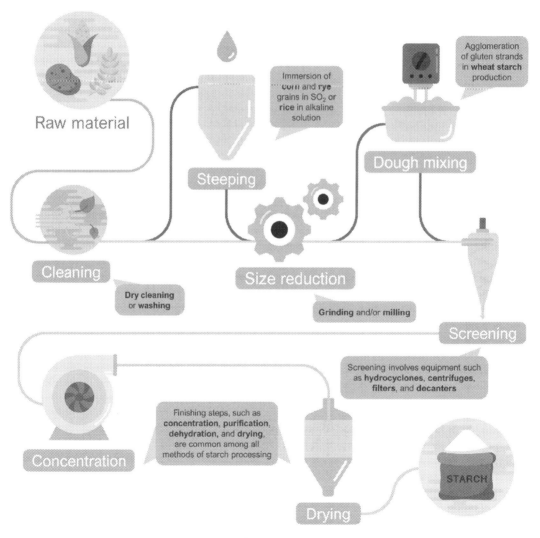

Figure 5.2
Flowchart of the main steps in starch processing from different raw materials.

After extraction, the processing of starch granules to produce starch-based materials can be accomplished by many of the traditional methods, such as extrusion, injection molding, compression molding, casting, and foaming. Extrusion involves melting and solidification processes, that for starch granules can result in phase changes like swelling, loss of birefringence, melting, and solubilization. In conditions of high shear and high pressure, gelatinization can occur at lower water contents, and starch degradation can also take place. Some alternatives that reduced the losses of starch during extrusion included the adjustment of speed, temperature, and moisture contents and the addition of glycerol. Compression molding and injection molding have also been evaluated for starch processing, resulting in preservation of amylopectin crystals with formation of brittle materials, and increase of the amorphous phase with formation of materials with improved mechanical properties, respectively. Starch films or sheets are usually produced by extrusion casting (Jiang et al., 2020).

Foaming is another interesting processing technique for starch-based materials. The desired mechanical characteristics of cellular solids or foams can be achieved with starch when process conditions such as water content are properly adjusted. The use of starch foams has been largely stimulated due to their biodegradability. Starch can also be applied in the manufacture of biomaterials in a process called starch consolidation, where starch gel is incorporated in a ceramic matrix and then heated until total volatilization, leaving void spaces in a porous material.

5.4 Improving mechanical and physicochemical properties

Even though starch has several qualities (renewable, biodegradable, cheap, and abundant), to be an ideal solution for the replacement of petroleum-based materials, many shortcomings need to be addressed. Even after the addition of plasticizer agents, transforming it into thermoplastic starch, it still shows poor mechanical properties, high vapor permeability due to its hydrophilic nature, and suffers retrogradation. To overcome these aspects, many techniques and methods have been developed to modify starch and tailor its properties, which can be broadly classified into four main areas: physical, chemical, enzymatic, and genetic modifications (Ribba, Garcia, D'Accorso, & Goyanes, 2017).

Either through biotechnology or traditional plant breeding, the ratio between amylose and amylopectin can be altered, modifying the composition of starch. Several sources of starch have been modified biosynthetically to yield both high amylose and high amylopectin starches. When compared to natural starch, both modifications resulted in higher stress at break, although due to the molecules intrinsic characteristics, the mechanical properties of the resulting materials are still not comparable to conventional plastics (Johnson, Baumel, Hardy, & White, 1999; Menzel et al., 2015; Lourdin, Della Valle, & Colonna, 1995). Physical modification methods are basically different processing techniques, and can be very environmentally friendly, since there are no chemicals involved. Friction, collision, impingement, shear, extrusion or fluidized bed heating, deep freezing, and thawing cycles are all physical methods to disrupt the crystalline structure of starch (Che, Li, Wang, Chen, & Mao, 2007; Szymońskaa, Krokb, Komorowska-Czepirskac, & Rebilas, 2003; González, Carrara, Tosi, Añón, & Pilosof, 2007).

The alteration of the molecular structure of starch has been vastly studied and can be achieved with either chemical or enzymatic reactions. These reactions can introduce new functional groups to the amylose and amylopectin chains through derivatization or decomposition. For example, hydroxypropylated starch, produced with propylene oxide in alkaline conditions exhibited lower solubilization temperatures, allowing film formation in milder conditions. Characteristics such as flexibility and transparency can also be enhanced. In addition, most enzymatic modifications significantly reduced retrogradation rates (Tarvainen, Sutinen, Peltonen, Tiihonen, & Paronen, 2002; López, García, & Zaritzky, 2008; Woggum, Sirivongpaisal, & Wittaya, 2015; Lafargue, Pontoire, Buléon, Doublier, & Lourdin, 2007; Auh et al., 2006).

Another chemical modification method is cross-linking. The promoting chemicals must be chosen carefully so no toxic residues are formed. Citric acid is a good example of a nontoxic, inexpensive cross-linking agent, which improves mechanical properties and reduces starch solubility in water. In general, cross-linked starch has higher tensile strength than native starch. Retrogradation rates are also greatly reduced, when not avoided altogether (Reddy & Yang, 2009; González Seligra, Medina Jaramillo, Famá, & Goyanes, 2016; Kim & Lee, 2002).

In order to produce even more specific materials, methods can be combined in several different ways, like cross-linking while adding functional groups, derivatization coupled with enzymatic hydrolysis, microwave-assisted catalyzed decomposition, and so on (Deetae et al., 2008; Karim, Sufha, & Zaidul, 2008; Rajan, Sudha, & Abraham, 2008).

The use of different plasticizers also plays a major role on the engineering of novel thermoplastic starch materials. These agents function by disrupting intermolecular bonds between the polymeric chains, resulting in the reduction of the melting temperature, allowing it to be processed. The ideal plasticizer is a small molecule, which is hydrophilic, polar, and compatible with starch. For the production of films, it also needs to have a high boiling point so it will not compromise the polymer's structure when dried. Water and glycerol are mainly used, and citric acid can also act as a plasticizer agent, depending on the concentration (Souza & Andrade, 2002; Pushpadass, Marx, & Hanna, 2008; Pushpadass et al., 2009; Ghanbarzadeh, Almasi, & Entezami, 2011).

Natural fiber reinforcement has proven its value along the years thanks to excellent specific properties such as high strength, low weight, and stiffness. A considerable amount of lignocellulosic materials have been studied along the years such as cotton, straw, bamboo, corn and wheat hulls, walnut shells, and even laver, and edible alga (Jiang et al., 2020).

Composites can be produced by the addition of nano- and microsized components known as fillers into the starch matrix. They can be introduced in several shapes, like fibers, particles, and flakes. Overall they improve mechanical and barrier properties, and some nanofillers have demonstrated to improve thermal and electrical capabilities also. Furthermore, nanofillers have a higher interfacial area per volume, which is a crucial factor in composite development, and do not scatter light, maintaining the material's transparency. Filler benefits increase with concentration, until a certain limit, when particles start to agglomerate, compromising the composite. Particles made from starch itself can be used as fillers, and studies have shown that the resulting material tensile strength almost doubled with only 6% starch nanocrystals (Siqueira, Bras, & Dufresne, 2009; Kudus, Akil, Mohamad, & Loon, 2011; Dai, Qiu, Xiong, & Sun, 2015).

Other natural polymers have been used as fillers as well. Cellulose nanofiber incorporation has been evaluated, and even though the composite strength was higher, it became more rigid, with reduced flexibility. Chitin fillers, both in nanofibrils and nanocrystals, also resulted in a more rigid and fragile material due to intermolecular bonding between starch hydroxyl radicals and chitin residual amine groups (Jonoobi, Mathew, Abdi, Davoodi Makinejad, & Oksman, 2012; Salaberria, Diaz, Labidi, & Fernandes, 2015).

The addition of well-dispersed fillers often results in increased water stability. Their presence causes a more tortuous diffusion path, reducing water vapor permeation. Older techniques involved the use of lipid microdroplets on the starch matrix to reduce hydrophilicity. Nowadays lipid monolayers are capable of cutting down water vapor permeation by almost half (Chen, Liu, Chang, Cao, & Anderson, 2009; Slavutsky & Bertuzzi, 2016).

Another way to overcome starch-based material's shortcomings is combining them with different polymers. Blends with biodegradable polymers result in fully compostable materials while with petroleum-based plastics allow long-term applications. Mechanical properties of the blended composites are influenced by the affinity and adhesion between starch and the other polymer(s). Compatibility can be enhanced by the addition of compatibilizer agents that increase miscibility. However, low compatibility can also be explored as an advantage. Since thermoplastic starch and biodegradable polyester are immiscible, after film formation, phase separation occurs, resulting in a multilayered structure (Pushpadass, Weber, Dumais, & Hanna, 2010; Averous & Fringant, 2001).

High molecular weight chitosan incorporation has the same effect of the chitin fillers, reducing water affinity, while low molecular weight chitosan promotes extensible materials. Chitosan can also be introduced into starch in the form of microcapsules for drug-delivery systems development. These microcapsules also act as moisture barriers, increasing the water resistance of the composite (Bof, Bordagaray, Locaso, & García, 2015; Huo et al., 2016).

Starch composites can be even more functional, by blending with antimicrobial components. Natural antimicrobial oils and extracts can be dispersed into a starch matrix, producing biodegradable, edible, and safe packaging for the food industry.

5.5 Starch-based materials

5.5.1 Starch-based polymer blending

Starch-based polymer blending has been well studied for developing composites. Recently, a technology has been described on marine biodegradable plastics with a blend of polyester and a starch-based polymeric material. Some polyester plastics may exhibit some biodegradability when on land with specific conditions. However, almost no biodegradability is shown in a marine environment. The blend between polyester plastics and starch-based polymeric material led to a complete marine biodegradability of the entire composite after 400 days. A fully biodegradable starch-based composite has also been developed (Li & Li, 2010). A blend of starch and polyester was created with a thermoplastic hydroxylation polyester as an interface reinforcing agent. In this case, glycerol was used as plasticizer. Apparently the composite presents good mechanical properties and high starch content (up to 70% of the total weight of the composite).

Composites of starch and other biodegradable polymers can also be found in protected technologies. For example, a blend of polyhydroxyalkanoate and starch was used to create laminates and films (Bond & Noda, 2003). Since natural starch generally has a granular structure, it needs to be destructured before it can be melt processed. The destructuring process can be achieved by dissolving the starch in water. Suitable naturally occurring starches can include corn, potato, sweet potato, wheat, sago palm, tapioca, rice, soybean, arrow root, bracken, lotus, cassava, waxy maize, and high amylose corn starches. Blends of starch may also be used. If a material comprising the presented composition is deposited in a solid-waste treatment facility or sanitary landfill, biodegradation occurs more rapidly than the materials with different composition.

Another example of a blended material is the use of starch and polylactic acid. A high strength plastic from reactive blending of starch and polylactic acid has been developed (Sun, Seib, & Wang, 2001). Basically, the method comprises forming and heating a blended mixture of polylactic acid, a starch, and a linkage group (diphenylmethylene diisocyanate, hexamethylene diisocyanate, and isophorone diisocyanate) for joining or copolymerizing the polylactic acid and starch. The material showed a high tensile strength, modulus of elasticity, percent elongation, and thermal stability, which made it suitable to be formed into a desired final plastic product.

Blends containing hydroxypropyl methylcellulose (HPMC) and starch are gaining growing attention as materials in medicinal capsules because of the much lower cost when compared with pure-HPMC materials. HPMC is known to have good mechanical properties, water solubility, and plant origin as advantages. However, starch is a cheaper source and, when blended with HPMC, it presents good properties for application (Zhang et al., 2014). Based on this, a hydrophilic controlled release formulation was reported that can prevent dose dumping (Vandecruys & Jans, 2000). The formulation comprises pregelatinized starch and hydrophilic polymers that can include HPMC. A more recent technology described that the combination of hydroxypropyl starch, starch, mannitol, HPMC, sorbitol, and polyvidone can create stable capsules with a greatly shortened production time. The gelatine capsules, however, are a subject of conflict for some religions and vegetarians. In this sense, plant-based capsules are an alternative to the traditional gelatine capsules.

Other examples of starch-blended polymers include polylactic acid/starch/fibrilia bio-based degradable composite material (Chen, Li, Wang, & Tang, 2015), cellulose acetate and starch-based biodegradable injection-molded plastics (Mayer & Elion, 1993), biodegradable thermoplastic composition based on protein and starch (Wang & Wang, 1999), fully degradable polymer composed by starch and vegetable fiber (Zhu, Yuan, & Zai, 2009), material with increased strength and biodegradability with a starch-based polymeric material, and a sustainable polymeric material, for example, green polyethylene, green polypropylene, and green polyethylene terephthalate (LaPray, Quan, & Allen, 2017). More information about the origin of starches and the blended materials can be found in Table 5.1.

5.5.2 Starch-based foaming

Starch-based foam like materials have been described since the nineteenth century (Graefe, 1982). However, the resulting products were usually unstable, with a very expensive production, or were not a proper foam material. In 1982, an innovation on starch-based foam technology was developed (Graefe, 1982) for preparing foamed gelatinized starch products, heating and pressing granular or pulverized starch or starch-containing materials in an extruder at temperatures of $60°C-220°C$ in the presence of $10\%-30\%$ (w/w) of water, and a gas-forming or gas-generating expanding agent. The content is then extruded to a final product similar to synthetic foams that can be used for technical purposes or as food.

Recently, several other technologies were developed. For example, a starch composite foam constituted mainly of starch, ethylene vinyl acetate, and elastomer was prepared, which beneficial to environmental protection, and with it the shortage phenomenon of plastic raw material could be alleviated (Wu & Zheng, 2011). In another technology,

Table 5.1: Origin of starches and blended materials, as retrieved from patent documents.

Patent document	Starch origin	Blended material
US20190276664A1 LaPray, Allen, and Miura (2019)	Corn, tapioca, cassava, wheat, potato, rice, sorghum	Polyesters: PBAT, PLA, PBS, PCL, PHA
CN101928411B Li and Li (2010)	Not disclosed	Polyesters: PCL, PLA, PHB, PHB/V, PBS
US20030108701A1 Bond and Noda (2003)	Corn, potato, sweet potato, wheat, sago palm, tapioca, rice, soybean, arrow root, bracken, lotus, cassava, waxy maize, high amylose corn starch, and commercial amylose powder	PHA
US6211325B1 Sun et al. (2001)	Corn, wheat, sorghum, potato, tapioca	Polylactic acid
CA2371940C Vandecruys and Jans (2000)	Drum-dried waxy maize starch is preferred	HPC, HPMC
CN105030723A Shuai, Wang, and Zhang (2015)	Tapioca, sweet potato, corn, barley	Mannitol, HPMC, sorbitol, and polyvidone
CN104693707A Chen et al. (2015)	Tapioca, sweet potato, yam, wheat, water caltrop, rice, *Rhizoma Nelumbinis*	Polylactic acid, fibrilia
US5288318A Mayer and Elion (1993)	Corn, tapioca, potato, sago, wheat, rye, pea, sorghum, rice, arrow root	Cellulose acetate
JP4302318B2 Wang and Wang (1999)	Corn, potato, sweet potato, wheat, rice, tapioca, sorghum	Soy protein, cellulose fiber
CN101885231A Zhu et al. (2009)	Corn, wheat, tapioca	Vegetable fiber: wood, peanut hull, bamboo
US20180100060A1 LaPray et al. (2017)	Potato, corn, tapioca	Polyethylene, polypropylene, or polyethylene terephthalate

Note: *HPC*, Hydroxypropyl cellulose; *HPMC*, hydroxypropyl methylcellulose; *PBAT*, polybutylene adipate terephthalate; *PBS*, polybutylene succinate; *PCL*, polycaprolactone; *PHA*, polyhydroxyalkanoate; *PHB*, polyhydroxybutyrate; *PHV*, polyhydroxyvalerate; *PLA*, polylactic acid.

starch-lignin-based foams presented a substitution of 10% of the starch for lignin causes no deleterious effects on foam density, morphology, compressive strength, or resiliency when compared to a starch-extruded foam, with retention of its integrity after immersion for 24 hours in water. When the substitution achieves 20%, there are important detriments in the foam's properties. The addition of cellulose fibers restores these properties; however, it results in increased density (Stevens, 2016).

There are several other innovations described for starch-based foams such as fully degradable foam material with starch and cellulose as base material (Wang, 2011), low-cost starch foams using specialized lignin (Changfeng, Aldi, Lansing, & Tudman, 2017), foamed starch sheet (Bastioli, Bastioli, Lombi, & Salvati, 2001), and process and device for producing a foamed product or foam made of unmodified starch (Kustner, 1994).

5.5.3 Starch-based nanocomposites

Despite of the commercial advantages of using starch-based materials (low cost, availability, and biodegradability), fundamental properties such as mechanical properties and moisture sensitivity of plasticized starch-based materials, are not always competitive with traditional petroleum-based plastics. One possible solution is the development of nanobiocomposites. These technologies are basically dispersions of nanosized fillers into a starch biopolymer matrix (Xie, Pollet, Halley, & Avérous, 2013).

An innovation has been described focused on a material based on a chemically modified plasticized starch and a chemically modified starch nanoclay. The use of nanoclay provides an increased mechanical strength, flame retardancy, gas barrier properties, and improved barrier to water vapor (Narayan et al., 2004). Other example of the use of nanoclay involves the material composing starch and nanoclay ceramic powder, combined by electrostatic attractive force. The inventors created an environmentally friendly flame-retardant multilayer thin film which is excellent in flame retardancy, is flexible and easy to process (Choi, Kim, Park, & Seo, 2016).

Many other nanofillers are used in starch-based nanocomposites, such as graphite (Weng, Lohse, Mota, Silva, & Kresge, 2008), nanotubes (Cîrciumaru, Andrei, Dima, & Murărescu, 2011), cellulose (Yang, Wang, & Liu, 2015), and chitosan (Yang, Yang, & Liu, 2015). Nevertheless, the development of new technologies using different nanofillers is essential to the development of promising starch-based nanocomposites with excellent performance, new functionalities, and market competitiveness (Xie et al., 2013).

5.6 Applications of starch-based materials

Starch-based materials find applications in many technological fields such as aerogels, biodegradable packages and films, nonbiodegradable green plastics, micro- and nanoparticles, and biomedical devices. Some of these applications will be reviewed in this section.

Aerogels consist of a porous and synthetic ultralight material derived from a gel, in which the liquid is replaced by a gas (Alemán et al., 2007). Starch-based aerogels are becoming particularly attractive and suitable for different uses. Some processes are necessary for the manufacture of aerogels, namely gelatinization, retrogradation, solvent exchange, and drying. The solvent exchange is necessary to form alcogel, in which the liquid component is an alcohol, for example, ethanol. The drying process can be supercritical or lyophilization, giving rise to the final product (Fig. 5.3) (Ivanovic, Milovanovic, & Zizovic, 2016; Ganesan et al., 2018).

One of the main applications of aerogels is the encapsulation and controlled release of bioactive compounds, being used as carriers to improve the solubility and bioavailability of

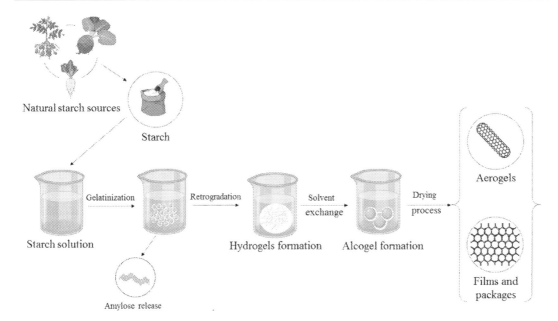

Figure 5.3
Formation of aerogels, films, and starch-based packages.

these compounds. Studies show that the use of starch airgel in in vitro and in vivo tests significantly increased drug loading, controlled release pattern, and dissolution. Some examples include: the use of ketoprofen-loaded aerogels, where the ketoprofen release is more sustained than in the pure sample (García-Gonzales & Smirnova, 2013); foods with phytosterols precipitated in the airgel system present greater solubility in the gastrointestinal fluid, higher impregnation rate, and better distribution (Ubeyitogullari & Ciftci, 2016); incorporation of aerogels increases pore connection and matrix porosity, thus promoting the release of ketoprofen and the growth of bone tissue (Goimil et al., 2017); quercetin in the aerogels matrix is released slowly and the release pattern shows that the systems can be used for active food packaging, while preserving the product's nutritional and organoleptic properties (Franco, Aliakbarian, Perego, Reverchon, & De Marco, 2018).

In the application of active food packaging, edible starch-based films have a great capacity to form films with good mechanical resistance. Some characteristics that made them attract so much attention are the low cost, abundance, permeability to O_2 and CO_2, insolubility, organoleptic properties like absence of taste and odor, and optical properties like transparency and absence of color. They are used to involve food products to avoid rapid degradation, moisture loss, and unwanted oxidation (Acosta, Chiralt, Santamarina, & Rosello, 2016; Galindez, Daniel, Alexander, Homez-Jara, & Eim, 2019). Different starch sources are used in the composition of edible films, like maize, cassava, sweet potato, potato, wheat, and corn starch. Their physical and mechanical properties depend on the botanical source, starch concentration, type of plasticizer, length of the chain, hydrophobic tail of the surfactant, and mechanical treatment (Galindez et al., 2019; Santacruz, Rivadeneira, & Castro, 2015).

For the production of these packages, the material goes through different hydrothermal processes, such as gelatinization (thermal) and retrogradation (postthermal). These become the basis of film and coating formation, as they use intermolecular associations−dissociations of starch units (Molina-Boisseau, Dole, & Dufresne, 2006; Thakur et al., 2019). Mixing starch with other mineral or natural polymers of different compatibilities is seen as a strategy to improve the properties and stability of starch-based films, giving rise to edible, biodegradable films and creating extra features. These may gain antimicrobial properties against foodborne pathogens when incorporated into natural bioactive compounds, thus protecting food products from microbial contamination (Sung, Tin, Tan, & Vikhraman, 2013). They also act as a barrier for gases and volatile compounds, protecting fresh food against dehydration, controlling loss of flavor during storage (Ferreira et al., 2015), and maintaining the quality of food products for distant marketing (Vásconez, Flores, Campos, Alvarado, & Gerschenson, 2009). Combining nanotechnology and starch-based packages results in extended shelf life, better traceability of food products, healthier foods, and safer packaging. Thus the use of these active films results in the development of new products, such as individual packaging of particulate foods, nutritional supplements, and carriers of different additives (Mohammadi, Tabatabaei, Pashania, Rajabi, & Karim, 2013; Nouri, Mohammadi, & Karim, 2014).

Currently, starch-based nanoparticles (SNPs) have also gained prominence for their controllable release capacity, better water solubility, and good distribution of active compounds in food and in the human body (Wang et al., 2018). They are promising nanoscale carriers due to their biodegradability, nontoxicity, and hydrophilicity (Ayadi, Bayer, Marras, & Athanassiou, 2016). Different applications are provided to SNPs. They can be directed to the food industry for bioactives and nutraceuticals delivery, as they promote bioavailability and maintain controlled transport and release of bioactive substances to target environments (Rostamabadi, Falsafi, & Jafari, 2019).

In drug-delivery systems, starch-based materials become attractive because they show the ability to control the rate and increase the bioavailability of bioactives, control time, and direct drug release to specific regions of the gastrointestinal tract (Gombotz & Wee, 2012; Qiu et al., 2020).

The bioavailability of many hydrophobic drugs is restricted due to their low bioaccessibility, low chemical stability, and limited absorption, thus, SNPs are designed and the slow release of the drug, provided by the hydrophobic character of some SNPs, ensures an improved concentration gradient and a potentiated effect of the medication (Santander-Ortega et al., 2010; Wang et al., 2019). In the gastrointestinal system, the release of these bioactive SNPs is mediated through the simulation of different environments, and the theory of digestion of these comprises diffusion, swelling, erosion, and disintegration—enzymatic degradation by amylase in the mouth, stomach, and small intestine (Fig. 5.4) (Walkey, Olsen, Guo, Emili, & Chan, 2012). The functional performance of SNPs can be modified through SNPs−protein interactions. After the drug's nanoparticles enter the human body, their surfaces are quickly surrounded by a layer of protein, forming a "protein corona," which determines the subsequent destination of the nanoparticles within the body (Pyrgiotakis, Blattmann, Pratsinis, & Demokritou, 2013; McClements, 2014).

Another applicability of starch nanoparticles is the use for water treatment, which has been one of the greatest challenges in environmental science due to the large quantities and types of contaminants. Starch has excellent swelling capacity, providing fast and efficient kinetics in removing pollutants (Jackson & Hillmyer, 2010). Due to its functional groups, starch can be easily modified and adapted to target specific pollutants (Jiang et al., 2020).

Besides SNPs, microscale starch-based particles also find various applications in different technological fields. Studies suggest that bioadhesives in starch microparticles work by increasing the systemic absorption of drugs and polypeptides through the nasal mucosa. Another area in which they are applied is in dry powder inhalation formulations acting on the deep pulmonary release of various agents. These, when reach the lungs, are rapidly phagocyted by alveolar macrophages. In the administration of pulmonary drugs,

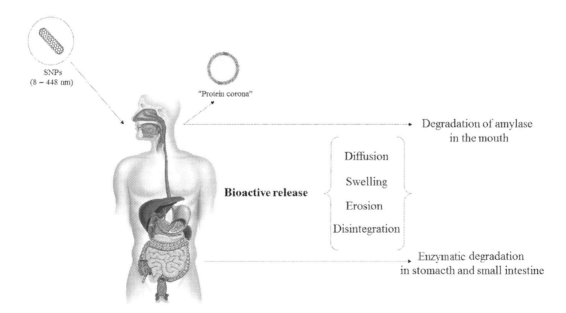

Figure 5.4
Mechanism of drug-delivery systems (DDS) using starch-based nanoparticles (SNPs).

microparticles decrease the frequency and magnitude of the dose, maintaining the concentration and adherence of the medication to the patient (Rodrigues & Emeje, 2012).

The advantages of using starch microparticles as carriers in tissue engineering include preparation at room temperature and adequateness for controlled drug release. They allow the adhesion of biologically active agents and are suitable vehicles for the incorporation and release of the growth factor, acting on delivery to living cells and enhancing bone regeneration (Balmayor, Tuzlakoglu, Azevedo, & Reis, 2009; Malafaya, Stappers, & Reis, 2006; Silva, Coutinho, Ducheyne, Shapiro, & Reis, 2007). In the production of packages, they can work as a vehicle for active compounds as antioxidants and antimicrobials (Fonseca et al., 2015; Sanyang, Sapuan, Jawaid, Ishak, & Sahari, 2016). And as emulsion stabilizers, they adsorb particles and generate repulsive steric interactions, thus stabilizing the system and avoiding the phase separation between oil and water (Agama-Acevedo & Bello-Perez, 2017).

It can be concluded that the applications of starch have been the focus of extensive research and technological advances. Features such as compatibility with the environment and ease of chemical modification allow for numerous applications for specific end uses. Thus in the global search for sustainability, starch becomes one of the most promising natural polymers for the replacement of petrochemical-derived materials.

5.7 Conclusions and perspectives

The commercial interest in biodegradable polymers, evidenced by the profile of applied science research papers and patent applications, is motivated by the negative impacts caused by the disposal of fossil-based, nonbiodegradable plastics in the environment, which led to changes in regulations and consumer habits. There is a tendency of gradual replacement of these petroleum-based materials by natural, renewable, nontoxic, and biodegradable alternatives.

Starch stands out for being abundantly available and relatively cheap. However, its native form has some characteristics such as insolubility in cold water, hygroscopicity, texture, tendency to retrograde, and loss of viscosity and thickening power, which limit its industrial applications. In this sense, processing strategies are needed, whether chemical, enzymatic, or physical, to enhance its properties and allow its technological applications in different areas.

Since the starch extraction methods are traditional and consolidated industrial processes, the greatest technological challenges are related to the optimization of processing methods (extrusion, injection molding, compression molding, casting, and foaming) and postprocessing modifications to overcome the limitations in mechanical and physicochemical properties. Besides the physical treatments and chemical or enzymatic modifications in the starch molecule, the development of starch-based composites and blends stands out as a key target for scientific research and technological development.

References

Acosta, S., Chiralt, A., Santamarina, P., & Rosello, J. (2016). Antifungal films based on starch-gelatin blend, containing essential oils. *Food Hydrocolloids*, *61*, 233–240. Available from: https://doi.org/10.1016/j.foodhyd.2016.05.008.

Agama-Acevedo, E., & Bello-Perez, L. A. (2017). Starch as an emulsions stability: The case of octenyl succinic anhydride (OSA) starch. *Current Opinion in Food Science*, *13*, 78–83. Available from: https://doi.org/10.1016/j.cofs.2017.02.014.

Alcázar-Alay, S. C., & Meireles, M. A. A. (2015). Physicochemical properties, modifications and applications of starches from different botanical sources. *Food Science and Technology*, *35*, 215–236. Available from: https://doi.org/10.1590/1678-457X.6749.

Alemán, J., Chadwick, A. V., He, J., Hess, M., Horie, K., Jones, R. G., ... Jones, R. G. (2007). Definitions of terms relating to the structure and processing of sols, gels, networks and inorganic-organic hybrid. *Pure and Applied Chemistry*, *79*, 1801–1829. Available from: https://doi.org/10.1351/pac200779101801.

Ashogbon, A. O., & Akintayo, E. T. (2014). Recent trend in the physical and chemical modification of starches from different botanical sources: A review. *Starch*, *66*, 41–57. Available from: https://doi.org/10.1002/star.201300106.

Auh, J. H., Chae, H. Y., Kim, Y. R., Shim, K. H., Yoo, S. H., & Park, K. H. (2006). Modification of rice starch by selective degradation of amylose using alkalophilic *Bacillus* cyclomaltodextrinase. *Journal of Agricultural and Food Chemistry*, *54*(6), 2314–2319.

Autio, K., & Eliasson, A.-C. (2009a). *Rye starch. Starch: Chemistry and technology* (pp. 579–587). Oxford, UK: Elsevier Inc. Available from: https://doi.org/10.1016/B978-0-12-746275-2.00014-8.

Autio, K., & Eliasson, A.-C. (2009b). *Oat starch. Starch: Chemistry and technology* (pp. 589–599). Oxford, UK: Elsevier Inc. Available from: https://doi.org/10.1016/B978-0-12-746275-2.00015-X.

Averous, L., & Fringant, C. (2001). Association between plasticized starch and polyesters: Processing and performances of injected biodegradable systems. *Polymer Engineering and Science*, *41*(5), 727–734.

Ayadi, F., Bayer, I. S., Marras, S., & Athanassiou, A. (2016). Synthesis of water dispersed nanoparticles from different polysaccharides and their application in drug release. *Carbohydrate Polymers*, *136*, 282–291. Available from: https://doi.org/10.1016/j.carbpol.2015.09.033.

Balmayor, E. R., Tuzlakoglu, K., Azevedo, H. S., & Reis, R. L. (2009). Preparation and characterization of starch-poly-ε-caprolactone microparticles incorporating bioactive agents for drug delivery and tissue engineering applications. *Acta Biomaterialia*, *5*, 1035–1045. Available from: https://doi.org/10.1016/j.actbio.2008.11.006.

Bastioli, A., Bastioli, C., Lombi, R., Salvati, P. (2001). Foamed starch sheet. *EP Patent No. 1127914B1*.

Biliaderis, C. G. (2009). Structural transitions and related physical properties of starch. In J. N. BeMiller, & R. L. Whistler (Eds.), *Starch: Chemistry and technology* (3rd ed., pp. 293–372). Academic Press.

Blennow, A. (2004). Starch bioengineering. *Starch in food* (pp. 97–127). Elsevier. Available from: https://doi.org/10.1533/9781855739093.1.97.

Bof, M. J., Bordagaray, V. C., Locaso, D. E., & García, M. A. (2015). Chitosan molecular weight effect on starch-composite film properties. *Food Hydrocolloids*, *51*, 281–294.

Bond, E., Noda, I. (2003). Polyhydroxyalkanoate copolymer/starch compositions for laminates and films. *U.S. Patent Application No. 20030108701(A1)*.

Breuninger, W. F., Piyachomkwan, K., & Sriroth, K. (2009). *Tapioca/cassava starch: Production and use. Starch: Chemistry and technology* (pp. 541–568). Elsevier Inc. Available from: https://doi.org/10.1016/B978-0-12-746275-2.00012-4.

Changfeng, G.E., Aldi, R., Lansing, B., Tudman, S. (2017). Starch foams using specialized lignin. *U.S. Patent Application No. 20180002451(A1)*.

Che, L., Li, D., Wang, L., Chen, X. D., & Mao, Z. (2007). Micronization and hydrophobic modification of cassava starch. *International Journal of Food Property, 10*(3), 527–536.

Chen, H., Li, P., Wang, X., Tang, Z. (2015). Polylactic acid/starch/fibrilia bio-based degradable composite material and preparation method thereof. *China Patent No. 104693707A*.

Chen, Y., Liu, C., Chang, P. R., Cao, X., & Anderson, D. P. (2009). Bionanocomposites based on pea starch and cellulose nanowhiskers hydrolyzed from pea hull fibre: Effect of hydrolysis time. *Carbohydrate Polymers, 76*(4), 607–615.

Choi, K.H., Kim, J., Park, Y.T., Seo, S.M. (2016). Eco-friendly flame-retardancy multilayer structure and method of manufacturing the same. *Korean Patent No. 101891572B1*.

Cîrciumaru, A., Andrei, G., Dima, D., Murărescu, M. (2011). Epoxy matrix additivated with carbon nanotubes and starch. *Romania Patent No. 128730B*.

Dai, L., Qiu, C., Xiong, L., & Sun, Q. (2015). Characterisation of corn starch-based films reinforced with taro starch nanoparticles. *Food Chemistry, 174*, 82–88.

Deetae, P., Shobsngob, S., Varanyanond, W., Chinachoti, P., Naivikul, O., & Varavinit, S. (2008). Preparation, pasting properties and freeze–thaw stability of dual modified crosslink-phosphorylated rice starch. *Carbohydrate Polymers, 73*(2), 351–358.

Early, R. (1997). *The technology of dairy products* (2nd ed.). Chapman and Hall.

Eckhoff, S. R., & Watson, S. A. (2009). Corn and sorghum starches: Production. *Starch: Chemistry and technology* (pp. 373–439). Oxford, UK: Elsevier Inc. Available from: https://doi.org/10.1016/B978-0-12-746275-2.00009-4.

Egharevba, H. O. (2019). Chemical properties of starch and its application in the food industry. In M. Emeje (Ed.), *Chemical properties of starch* (pp. 1–26). IntechOpen.

Ferreira, M. S. L., Fai, E. C., Andrade, C. T., Paulo, H., Azero, G., & Gonçalves, É. C. (2015). Edible films and coatings based on biodegradable residues applied to acerolas (*Malpighia punicifolia* L.). *Journal of the Science of Food and Agriculture, 96*(5), 1634–1642. Available from: https://doi.org/10.1002/jsfa.7265.

Fonseca, L. M., Gonçalves, J. R., El Halal, S. L. M., Pinto, V. Z., Dias, A. R. G., Jacques, A. C., . . . da, R. (2015). Oxidation of potato starch with different sodium hypochlorite concentrations and its effect on biodegradable films. *LWT—Food Science Technolology, 60*, 714–720. Available from: https://doi.org/10.1016/j.lwt.2014.10.052.

Franco, P., Aliakbarian, B., Perego, P., Reverchon, E., & De Marco, I. (2018). Supercritical adsorption of quercetin on aerogels for active packaging applications. *Industrial & Engineering Chemistry Research, 57*(44), 15105–15113. Available from: https://doi.org/10.1021/acs.iecr.8b03666.

Galindez, A., Daniel, L., Alexander, H., Homez-Jara, A., & Eim, V. S. (2019). Characterization of ulluco starch and its potential for use in edible films prepared at low drying temperature. *Carbohydrate Polymers, 215*, 143–150. Available from: https://doi.org/10.1016/j.carbpol.2019.03.074.

Ganesan, K., Budtova, T., Ratke, L., Gurikov, P., Baudron, V., Preibisch, I., . . . Milow, B. (2018). Review on the production of polysaccharide aerogel particles. *Materials, 11*, 1–37. Available from: https://doi.org/10.3390/ma11112144.

García-Gonzales, C. A., & Smirnova, I. (2013). Use of supercritical fluid technology for the production of tailor-made aerogel particles for delivery systems. *Journal of Supercritical Fluids, 79*, 152–158. Available from: https://doi.org/10.1016/j.supflu.2013.03.001.

Ghanbarzadeh, B., Almasi, H., & Entezami, A. A. (2011). Improving the barrier and mechanical properties of corn starch-based edible films: Effect of citric acid and carboxymethyl cellulose. *Industrial Crops and Products, 33*(1), 229–235.

Goimil, L., Braga, M. E. M., Dias, A. M. A., Gómez-Amoza, J. L., Concheiro, A., Alvarez-Lorenzo, C., . . . García-González, C. A. (2017). Supercritical processing of starch aerogels and aerogel-loaded poly (e-caprolactone) scaffolds for sustained release of ketoprofen for bone. *Biochemical Pharmacology, 18*, 237–249. Available from: https://doi.org/10.1016/j.jcou.2017.01.028.

Gombotz, W. R., & Wee, S. F. (2012). Protein release from alginate matrices. *Advanced Drug Delivery Reviews*, *64*, 194−205. Available from: https://doi.org/10.1016/j.addr.2012.09.007.

González, R., Carrara, C., Tosi, E., Añón, M. C., & Pilosof, A. (2007). Amaranth starch-rich fraction properties modified by extrusion and fluidized bed heating. *LWT—Food Science Technology*, *40*(1), 136−143.

González Seligra, P., Medina Jaramillo, C., Famá, L., & Goyanes, S. (2016). Biodegradable and non-retrogradable eco-films based on starch−glycerol with citric acid as crosslinking agent. *Carbohydrate Polymers*, *138*, 66−74.

Graefe, J.E. (1982). A process for preparing foamed gelatinized starch products. DE Patent No. 3206751C2.

Grommers, H. E., & Krogt, D. A. V. D. (2009). Potato starch: Production, modifications and uses. *Starch: Chemistry and technology* (pp. 511−539). Elsevier Inc. Available from: https://doi.org/10.1016/B978-0-12-746275-2.00011-2.

Hamaker, B. R., Tuncil, Y. E., & Shen, X. (2019). Carbohydrates of the kernel. In S. O. Serna-Saldivar (Ed.), *Corn: Chemistry and technology* (pp. 305−318). Elsevier.

Haroon, M., Wang, L., Yu, H., Abbasi, N. M., Zain-ul-Abdin., Saleem, M., ... Wu, J. R. (2016). Chemical modification of starch and its application as an adsorbent material. *RSC Advances*, *6*, 78264−78285. Available from: https://doi.org/10.1039/C6RA16795K.

Hoover, R., & Vasanthan, T. (2009). Barley starch: Production, properties, modification and uses. *Starch: Chemistry and technology* (pp. 601−628). Oxford, UK: Elsevier Inc. Available from: https://doi.org/10.1016/B978-0-12-746275-2.00016-1.

Huber, K. C., & BeMiller, J. N. (2019). Carboidratos. In S. Damodaran, & K. L. Parkin (Eds.), *Química de alimentos de fennema* (pp. 91−174). Artmed.

Huo, W., Xie, G., Zhang, W., Wang, W., Shan, J., Liu., & Zhou, X. (2016). Preparation of a novel chitosan-microcapsules/starch blend film and the study of its drug-release mechanism. *International Journal of Biological Macromolecules*, *87*, 114−122.

Ivanovic, J., Milovanovic, S., & Zizovic, I. (2016). Utilization of supercritical CO_2 as a processing aid in setting functionality of starch-based materials. *Starch*, *68*, 821−833. Available from: https://doi.org/10.1002/star.201500194.

Jackson, E. A., & Hillmyer, M. A. (2010). Nanoporous membranes derived from block copolymers: From drug delivery to water filtration. *ACS Nano*, *4*(7), 3548−3553. Available from: https://doi.org/10.1021/nn1014006.

Jane, J. (2009). Structural features of starch granules II. In J. N. BeMiller, & R. L. Whistler (Eds.), *Starch: Chemistry and technology* (3rd ed., pp. 193−236). Academic Press.

Jiang, T., Duan, Q., Zhu, J., Liu, H., & Yu, L. (2020). Starch-based biodegradable materials: Challenges and opportunities. *Advanced Industrial and Engineering Polymer Research*, *3*, 8−18. Available from: https://doi.org/10.1016/j.aiepr.2019.11.003.

Johnson, L. A., Baumel, C. P., Hardy, C. L., & White, P. J. (1999). *Identifying valuable corn quality traits for starch production*. Iowa State University Extension.

Jonoobi, M., Mathew, A. P., Abdi, M. M., Davoodi Makinejad, M., & Oksman, K. (2012). A comparison of modified and unmodified cellulose nanofiber reinforced polylactic acid (PLA) prepared by twin screw extrusion. *Journal of Polymers and the Environment*, *20*(4), 991−997.

Karim, A. A., Sufha, E. H., & Zaidul, I. S. M. (2008). Dual modification of starch via partial enzymatic hydrolysis in the granular state and subsequent hydroxypropylation. *Journal of Agricultural and Food Chemistry*, *56*(22), 10901−10907.

Kim, M., & Lee, S. J. (2002). Characteristics of crosslinked potato starch and starch-filled linear low-density polyethylene films. *Carbohydrate Polymers*, *50*(4), 331−337.

Kudus, M. H. A., Akil, H. M., Mohamad, H., & Loon, L. E. (2011). Effect of catalyst calcination temperature on the synthesis of MWCNT−alumina hybrid compound using methane decomposition method. *Journal of Alloys and Compounds*, *509*(6), 2784−2788.

Kumar, R., & Khatkar, B. S. (2017). Thermal, pasting and morphological properties of starch granules of wheat (*Triticum aestivum* L.) varieties. *Journal of Food Science and Technology*, *54*, 2403−2410. Available from: https://doi.org/10.1007/s13197-017-2681-x.

Kustner, F. (1994). Process and device for producing a foamed product or foam made of unmodified starch. *CA Patent No. 2119688C*.

Lafargue, D., Pontoire, B., Buléon, A., Doublier, J. L., & Lourdin, D. (2007). Structure and mechanical properties of hydroxypropylated starch films. *Biomacromolules*, 8(12), 3950–3958.

LaPray, B., Allen, D.R., Miura, S. (2019). Marine biodegradable plastics comprising a blend of polyester and a carbohydrate-based polymeric material. *U.S. Patent Application No. 20190276664(A1)*.

LaPray, B., Quan, W., Allen D.R. (2017). Articles formed with renewable and/or sustainable green plastic material and carbohydrate-based polymeric materials lending increased strength and/or biodegradability. *U.S. Patent Application No. 20180100060(A1)*.

Leonel, M., & Lima, G. P. P. (2016). Carboidratos e fibras alimentares. In C. O. Silva, E. M. M. Tassi, & G. B. Pascoal (Eds.), *Ciência dos alimentos: Princípios de bromatologia* (pp. 53–72). Rubio.

Li, Y., Li, B. (2010). Starch-based biodegradable combination as well as preparation method and application thereof. *China Patent No. 101928411B*.

López, O. V., García, M. A., & Zaritzky, N. E. (2008). Film forming capacity of chemically modified corn starches. *Carbohydrate Polymers*, 73(4), 573–581.

Lourdin, D., Della Valle, G., & Colonna, P. (1995). Influence of amylose content on starch films and foams. *Carbohydrate Polymers*, 27(4), 261–270.

Malafaya, P. B., Stappers, F., & Reis, R. L. (2006). Starch-based microspheres produced by emulsion crosslinking with a potential media dependent responsive behavior to be used as drug delivery carriers. *Journal of Materials Science: Materials in Medicine*, 17, 371–377. Available from: https://doi.org/10.1007/s10856-006-8240-z.

Maningat, C. C., Seib, P. A., Bassi, S. D., Woo, K. S., & Lasater, G. D. (2009). Wheat starch: Production, properties, modification and uses. *Starch: Chemistry and technology* (pp. 441–510). Oxford, UK: Elsevier Inc. Available from: https://doi.org/10.1016/B978-0-12-746275-2.00010-0.

Mayer, J.M., Elion, G.R. (1993). Cellulose acetate and starch based biodegradable injection molded plastics compositions and methods of manufacture. *U.S. Patent Application No. 5288318(A)*.

McClements, D. J. (2014). *Nanoparticle-and microparticle-based delivery systems: Encapsulation, protection and release of active compounds* (1st ed., p. 572) CRC Press.

Menzel, C., Andersson, M., Andersson, R., Vázquez-Gutiérrez, J. L., Daniel, G., Langton, M., & Koch, M. (2015). Improved material properties of solution-cast starch films: Effect of varying amylopectin structure and amylose content of starch from genetically modified potatoes. *Carbohydrate Polymers*, 130, 388–397.

Mitchell, C. R. (2009). Rice starches: Production and properties. *Starch: Chemistry and technology* (pp. 569–578). Oxford, UK: Elsevier Inc. Available from: https://doi.org/10.1016/B978-0-12-746275-2.00013-6.

Mohammadi, A., Tabatabaei, R. H., Pashania, B., Rajabi, H. Z., & Karim, A. A. (2013). Effects of ascorbic acid and sugars on solubility, thermal, and mechanical properties of egg white protein gels. *International Journal of Biological Macromolecules*, 62, 397–404. Available from: https://doi.org/10.1016/j.ijbiomac.2013.09.050.

Molina-Boisseau, S., Dole, P., & Dufresne, A. (2006). Thermoplastic starch-waxy maize starch nanocrystals nanocomposites. *Biomacromolecules*, 61, 531–539. Available from: https://doi.org/10.1021/bm050797s.

Narayan, R., Balakrishnan, S., Nabar, Y., Shin, B.-Y., Dubois, P., Raquez, J.-M. (2004). Chemically modified plasticized starch compositions by extrusion processing. *U.S. Patent Application No. 7153354(B2)*.

Nouri, L., Mohammadi, A., & Karim, A. A. (2014). Phytochemical, antioxidant, antibacterial, and α-amylase inhibitory properties of different extracts from betel leaves. *Industrial Crops and Products*, 62, 47–52. Available from: https://doi.org/10.1016/j.indcrop.2014.08.015.

Pérez, S., Baldwin, P. M., & Gallant, D. J. (2009). Structural features of starch granules I. In J. N. BeMiller, & R. L. Whistler (Eds.), *Starch: Chemistry and technology* (3rd ed., pp. 149–192). Academic Press.

Pushpadass, A., Weber, R. W., Dumais, J. J., & Hanna, M. A. (2010). Biodegradation characteristics of starch–polystyrene loose-fill foams in a composting medium. *Bioresource Technology*, 101(19), 7258–7264.

Pushpadass, H. A., Kumar, A., Jackson, D., Wehling, R. L., Dumais, J. J., & Hanna, M. A. (2009). Macromolecular changes in extruded starch-films plasticized with glycerol, water and stearic acid. *Starch*, *61*(5), 256–266.

Pushpadass, H. A., Marx, D. B., & Hanna, M. A. (2008). Effects of extrusion temperature and plasticizers on the physical and functional properties of starch films. *Starch*, *60*(10), 527–538.

Pyrgiotakis, G., Blattmann, C. O., Pratsinis, S., & Demokritou, P. (2013). Nanoparticle–nanoparticle interactions in biological media by atomic force microscopy. *Langmuir: The ACS Journal of Surfaces and Colloids*, *29*(36), 11385–11395. Available from: https://doi.org/10.1021/la4019585.

Qiu, C., Wang, C., Gong, C., Julian, D., Jin, Z., & Wang, J. (2020). Advances in research on preparation, characterization, interaction with proteins, digestion and delivery systems of starch-based nanoparticles. *International Journal of Biological Macromolecules*, *152*, 117–125. Available from: https://doi.org/10.1016/j.ijbiomac.2020.02.156.

Rajan, A., Sudha, J. D., & Abraham, T. E. (2008). Enzymatic modification of cassava starch by fungal lipase. *Industrial Crops and Products*, *27*(1), 50–59.

Reddy, N., & Yang, Y. (2009). Preparation and properties of starch acetate fibers for potential tissue engineering applications. *Biotechnology and Bioengineering*, *103*(5), 1016–1022.

Ribba, L., Garcia, N. L., D'Accorso, N., & Goyanes, S. (2017). Disadvantages of starch-based materials, feasible alternatives in order to overcome these limitations. *Starch-based materials in food packaging* (1st ed., pp. 37–76). Elsevier.

Robyt, J. F. (2008). Starch: Structure, properties, chemistry, and enzymology. In B. O. Fraser-Reid, K. Tatsuta, J. Thiem, et al. (Eds.), *Glycoscience: Chemistry and chemical biology* (pp. 1437–1472). Springer.

Rodrigues, A., & Emeje, M. (2012). Recent applications of starch derivatives in nanodrug delivery. *Carbohydrate Polymers*, *87*, 987–994. Available from: https://doi.org/10.1016/j.carbpol.2011.09.044.

Rostamabadi, H., Falsafi, S. R., & Jafari, S. M. (2019). Starch-based nanocarriers as cutting-edge natural cargos for nutraceutical delivery. *Trends in Food Science & Technology*, *88*, 397–415. Available from: https://doi.org/10.1016/j.tifs.2019.04.004.

Salaberria, A. M., Diaz, R. H., Labidi, J., & Fernandes, S. C. M. (2015). Role of chitin nanocrystals and nanofibers on physical, mechanical and functional properties in thermoplastic starch films. *Food Hydrocolloids*, *46*, 93–102.

Santacruz, S., Rivadeneira, C., & Castro, M. (2015). Edible films based on starch and chitosan. Effect of starch source and concentration, plasticizer, surfactant's hydrophobic tail and mechanical treatment. *Food Hydrocolloids*, *49*, 89–94. Available from: https://doi.org/10.1016/j.foodhyd.2015.03.019.

Santander-Ortega, M. J., Stauner, T., Loretz, B., Ortega-Vinuesa, J. L., Bastos-González, D., Wenz, G., . . . Lehr, C. M. (2010). Nanoparticles made from novel starch derivatives for transdermal drug delivery. *Journal of Controlled Release: Official Journal of the Controlled Release Society*, *141*, 85–92. Available from: https://doi.org/10.1016/j.jconrel.2009.08.01.

Sanyang, M. L., Sapuan, S. M., Jawaid, M., Ishak, M. R., & Sahari, J. (2016). Development and characterization of sugar palm starch and poly(lactic acid) bilayer films. *Carbohydrate Polymers*, *146*, 36–45. Available from: https://doi.org/10.1016/j.carbpol.2016.03.051.

Schirmer, M., Jekle, M., & Becker, T. (2015). Starch gelatinization and its complexity for analysis. *Starch*, *67*, 30–41. Available from: https://doi.org/10.1002/star.201400071.

Shannon, J. C., Garwood, D. L., & Boyer, C. D. (2009). Genetics and physiology of starch development. In J. N. BeMiller, & R. L. Whistler (Eds.), *Starch: Chemistry and technology* (3rd ed., pp. 23–82). Academic Press.

Shuai, F., Wang, X., Zhang, J. (2015). Starch capsule and method for preparing same. *China Patent No. 105030723A*.

Silva, G. A., Coutinho, O. P., Ducheyne, P., Shapiro, I. M., & Reis, R. L. (2007). Starch-based microparticles as vehicles for the delivery of active platelet-derived growth factor. *Tissue Engineering*, *13*, 1259–1268. Available from: https://doi.org/10.1089/ten.2006.0194.

Siqueira, G., Bras, J., & Dufresne, A. (2009). Cellulosic bionanocomposites: A review of preparation and properties of nanocomposites. *Polymers*, *2*(4), 728–765.

Slavutsky, A. M., & Bertuzzi, M. A. (2016). Improvement of water barrier properties of starch films by lipid nanolamination. *Food Packaging and Shelf Life*, 7, 41−46.

Souza, R. C. R., & Andrade, C. T. (2002). Investigation of the gelatinization and extrusion processes of corn starch. *Advances in Polymer Technology*, 21(1), 17−24.

Stevens, E.S. (2016). Extruded starch-lignin foams. *U.S. Patent Application No. 10400105(B2)*.

Sun, X.S., Seib, P., Wang, H. (2001). High strength plastic from reactive blending of starch and polylactic acids. *U.S. Patent Application No. 6211325(B1)*.

Sung, S., Tin, L., Tan, A., & Vikhraman, M. (2013). Antimicrobial agents for food packaging applications. *Trends in Food Science & Technology*, 33, 110−123. Available from: https://doi.org/10.1016/j.tifs.2013.08.001.

Szymońskaa, J., Krokb, F., Komorowska-Czepirskac, F., & Rebilas, K. (2003). Modification of granular potato starch by multiple deep-freezing and thawing. *Carbohydrate Polymers*, 52(1), 1−10.

Tarvainen, M., Sutinen, R., Peltonen, S., Tiihonen, P., & Paronen, P. (2002). Starch acetate—A novel film-forming polymer for pharmaceutical coatings. *Journal of Pharmaceutical Sciences*, 91(1), 282−289.

Thakur, R., Pristijono, P., Scarlett, C. J., Bowyer, M., Singh, S. P., & Vuong, Q. V. (2019). Starch-based films: Major factors affecting their properties. *International Journal of Biological Macromolecules*, 132, 1079−1089. Available from: https://doi.org/10.1016/j.ijbiomac.2019.03.190.

Ubeyitogullari, A., & Ciftci, O. N. (2016). Phytosterol nanoparticles with reduced crystallinity generated using nanoporous starch aerogels. *RSC Advances*, 108319−108327. Available from: https://doi.org/10.1039/c6ra20675a.

Vandecruys, R.P.G., Jans, E.M. (2000). Pregelatinized starch in a controlled release formulation. *CA Patent No. 2371940C*.

Vásconez, M. B., Flores, S. K., Campos, C. A., Alvarado, J., & Gerschenson, L. N. (2009). Antimicrobial activity and physical properties of chitosan—Tapioca starch based edible films and coatings. *Food Research International*, 42, 762−769. Available from: https://doi.org/10.1016/j.foodres.2009.02.026.

Walkey, C. D., Olsen, J. B., Guo, H., Emili, A., & Chan, W. C. (2012). Nanoparticle size and surface chemistry determine serum protein adsorption and macrophage uptake. *Journal of the American Chemical Society*, 134(4), 2139−2147. Available from: https://doi.org/10.1021/ja2084338.

Wang, H. (2011). Fully-degradable foam material using starch as base material and preparation method for fully-degradable foam material. *China Patent No. 102206361A*.

Wang, H., Feng, T., Zhuang, H., Xu, Z., Ye, R., & Sun, M. (2018). A review on patents of starch nanoparticles: Preparation, applications, and development. *Recent Patents on Food, Nutrition & Agriculture*, 9, 23−30. Available from: https://doi.org/10.2174/2212798410666180321101446.

Wang, L., Zhao, X., Yang, F., Wu, W., Wu, M., Li, Y., & Zhang, X. (2019). Loading paclitaxel into porous starch in the form of nanoparticles to improve its dissolution and bioavailability. *International Journal of Biological Macromolecules*, 138, 207−214. Available from: https://doi.org/10.1016/j.ijbiomac.2019.07.083.

Wang, S.H., Wang, H. (1999). Biodegradable thermoplastic composition based on protein and starch. *Japan Patent No. 4302318B2*.

Weng, W., Lohse, D., Mota, M.O., Silva, A.S., Kresge, E.M. (2008). Graphite nanocomposites. *U.S. Patent Application No. 7923491(B2)*.

Woggum, T., Sirivongpaisal, P., & Wittaya, T. (2015). Characteristics and properties of hydroxypropylated rice starch based biodegradable films. *Food Hydrocolloids*, 50, 54−64.

Wu, B., Zheng, Y. (2011). Starch composite foamed material and its production method. *China Patent No. 102226015A*.

Xie, F., Pollet, E., Halley, P. J., & Avérous, L. (2013). Starch-based nano-biocomposites. *Progress in Polymer Science*, 38, 1590−1628.

Yang, F., Yang, R., Liu, Q. (2015). Nano chitosan base quaternary alkyl ammonium starch ether composite antibacterial paper and its preparation and application. *China Patent No. 105040509B*.

Yang, R., Wang, W., Liu, Q. (2015). Nano-cellulose base oxidized starch compound bio latex and its preparation and application. *China Patent No. 105461971B*.

Zhang, L., Wang, Y., Yu, L., Liu, H., Simon, G., Zhang, N., & Chen, L. (2014). Rheological and gel properties of hydroxypropyl methylcellulose/hydroxypropyl starch blends. *Colloid and Polymer Science, 293*, 229−237.

Zhu, J., Yuan, Y., Zai, J. (2009). Preparation method of fully-degradable polymer wood plastic composite. *China Patent No. 101885231A*.

CHAPTER 6

Pectin

Poonam Sharma[1], Krishna Gautam[2], Ashutosh Kumar Pandey[3], Vivek Kumar Gaur[4,5], Alvina Farooqui[1] and Kaiser Younis[1]

[1]Department of Bioengineering, Integral University, Lucknow, India, [2]Academy of Scientific and Innovative Research (AcSIR), CSIR—Indian Institute of Toxicology Research, Lucknow, India, [3]Centre for Energy and Environmental Sustainability, Lucknow, India, [4]Environmental Biotechnology Division, Environmental Toxicology Group, CSIR—Indian Institute of Toxicology Research, Lucknow, India, [5]Amity Institute of Biotechnology, Amity University, Lucknow, India

6.1 Introduction

Pectin derived its name from the Greek word πηχτες (pektes), coined by Henri Braconnot in 1825 meaning "Coagulated material." It is a naturally occurring complex family of heteropolysaccharide, which constitutes the main component of primary cell wall of nonwoody plants and plays a central role in their growth and development (Ciriminna, Chavarría-Hernández, Inés Rodríguez Hernández, & Pagliaro, 2015; Smith, Moxon, & Morris, 2016). In higher plants, approximately one-third weight of the dry cell wall is contributed by pectin. The concentrations of pectin decrease from primary cell wall to plasma membrane. Its highest concentration is in the middle lamella of cell wall (Sriamornsak, 2003). In plants, pectin imparts flexibility and strength to the cell wall, along with its crucial role in various fundamental biological functions (cell signaling, differentiation, cell adhesion, and cell proliferation) (Ciriminna et al., 2015). It consists of up to 300–1000 saccharide units with an approximate molecular weight of 150 kDa. In majority, it consists of polymeric units of galacturonic acid (GalA) residues (Fig. 6.1), with different ratio of the acid groups such as methoxy esters and some amount of neutral sugars in side chains such as glucose, xylose, mannose, arabinose, fucose, or galactose (Voragen, Coenen, Verhoef, & Schols, 2009). As per the constitutional arrangement of subunits, pectin possesses three polysaccharide structures, namely, homogalacturonan (HGA), rhamnogalacturonan II (RG-II), and rhamnogalacturonan I (RG-I) (Maneerat, Tangsuphoom, & Nitithamyong, 2017).

Pectin has a myriad of nutritive and functional uses within the food and nonfood applications such as cosmetics, personal-care products (paints, toothpaste, and shampoos), and pharmaceutical (drug delivery, wound healing, cholesterol-lowering, film-coated dose form, matrix tablets, and gel beads) (Sriamornsak, 2003; Sundar Raj, Rubila, Jayabalan, & Ranganathan, 2012).

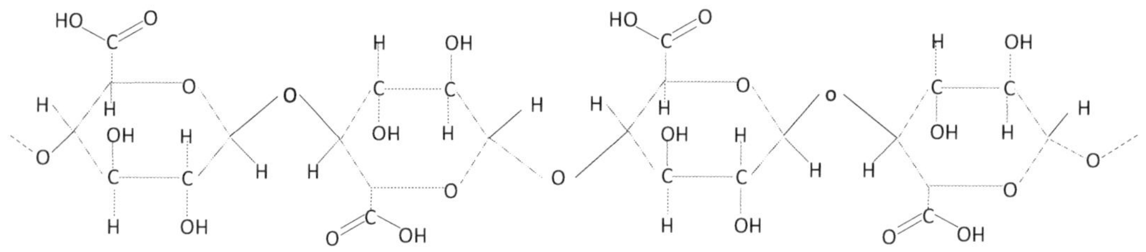

Figure 6.1
Typical structure of pectin polymer.

It has been used for its functional properties in foods for many years such as jam, jellies, yogurt drinks, fruity milk drinks, acidic milk drinks, bread and dough, salad dressings, and ice cream (Ajibade & Ijabadeniyi, 2019; Brummel & Lee, 1990; Maneerat et al., 2017; Sharma, Naresh, Dhuldhoya, Merchant, & Merchant, 2006). It is highly valued food additive as a gelling agent, thickening agent, and viscosifier in beverages and soft drinks. Due to high-ester bonds, it finds applications as a mouth-feel improver (Adetunji, Adekunle, Orsat, & Raghavan, 2017; Dranca & Oroian, 2018; Flutto, 2003). The European Commission and Joint FAO/WHO Committee have recommended pectin as an additive having no limitation for the acceptable daily intake other than as prescribed by the good manufacturing practice. It is recommended that pectin must contain a minimum of 65% GalA (Müller-Maatsch et al., 2016).

The production of pectin was started in the 1900s in Germany, utilizing apple pomace produced by apple juice–manufacturing industry as a waste. By 1930 the demand of pectin as a gelling agent increased significantly; therefore the process of pectin manufacturing was industrialized by companies such as Opetka, Obipektin, and Herbstreith & Fox. Pectin is a highly value product from the industrial point of view as the substrate required for its extraction is mostly by-products or wastes of other industries, which offers an economic as well as environment-friendly approach to produce pectin at a large scale. Pectin can also be extracted from a wide array of other sources such as citrus peel, apple pomace, and sugar beet pulp. The multivariate applications of pectin in several industries, including food processing, pharmaceutical, neutraceuticals, and cosmetology, warrant more comprehensive researches for the improved yield, novel sources, and better extraction procedures and functional properties of pectin.

6.2 Structure and classification of pectin

Pectin is a naturally occurring carbohydrate of high molecular weight found in all complex groups of plants, which contributes to the formation of the cell wall (Harris & Smith, 2006). It is the main constituents of middle lamella and the primary cell wall, providing mechanical strength, turgidity, and resistance to abiotic environmental stresses and also in maintaining the physical structure of the plant (Sirisomboon, Tanaka, Fujita, & Kojima, 2000). The texture and consistency of the fruit are also one of the significant contributory parts of the pectin, as its content may vary from one species to other (Prasanna, Prabha, & Tharanathan, 2007). It is involved in the ripening of the tissues, which makes the fruit squashy and pulpy due to its degradation and dissolution (Sakai, Sakamoto, Hallaert, & Vandamme, 1993).

Although pectin is known since two centuries, yet its chemical and structural properties are an enormous research subject because of the inhomogeneity. Several variables can be seen during the plant growth cycle, influencing the pectin structure and differentiation in structure. Pectin is made up of a polymeric chain of GalA, joined by α-1,4-glycosidic bonds and some of these galacturonic chains are partially or totally esterified with methanol to form methyl esters (Harris & Smith, 2006). This formation of methyl esters can be calculated as the degree of esterification (DE); these ester molecules interact with other molecules or ions that lead to changes in local electrostatic charge density. On this basis, pectin can be classified as pectin polysaccharides, protopectin, galactans, arabinogalactan, and arabinans (Ovodov, 2009). Pectin's complex chain comprises the 17 most abundant groups of monosaccharides, such as GalA, followed by D-galactose, L-arabinose, L-rhamnose, and some others with approximately 20 different types of linkages (Ovodov, 2009; Voragen et al., 2009). The residues of 1,4-linked α-D-GalA form the backbone of pectin with three different arrangements of GalA in the chain, that is, O-acetylated GalA, methylated GalA, and GalA (Chan, Choo, Young, & Loh, 2017; Ovodov, 2009). The structural analysis of pectin shows the presence of various polysaccharide domains, that is, RG-I, RG-II, and HGA (Fig. 6.2), linked by a covalent bond and crossed with other pectin chains to form a branched network or structure throughout the primary cell wall (O'neill, Albersheim, & Darvill, 1990; Harholt, Suttangkakul, & Scheller, 2010; Gawkowska, Cybulska, & Zdunek, 2018). Among these, HGA constitutes the major fraction in pectin molecule and is stored in the cell wall in the form of GalA remnants (70%–80%) (O'neill et al., 1990). However, the fragments of RG-I fluctuate to a notable extent in the diverse group of pectin and its backbone comprises of 1,4-linked GalA and 1,2-linked rhamnose residues, which is partly replaced by the residues of galactose accompanied by 1,4 linkages of residues of rhamnose. The RG-I subunits have extended side chains of monosaccharides of galactose and arabinose (Ovodov, 2009). Several methods have been developed to explore the extremely complex and diverse pectin assembly (Willats, Knox, & Mikkelsen, 2006) whose composition differs not only between the different plant classes but also within the same plant cell wall.

6.2.1 Homogalacturonan

HGA is an extensively found linear homopolymer of α-1,4-linked GalA, contributing approximately 60%–65% of total pectin (Voragen et al., 2009). The linear pectin polymer undergoes methylation and acetylation reactions during the polymerization process, resulting in three distinct pectin groups, that is, pectin consisting of >75% of the methylated carboxyl

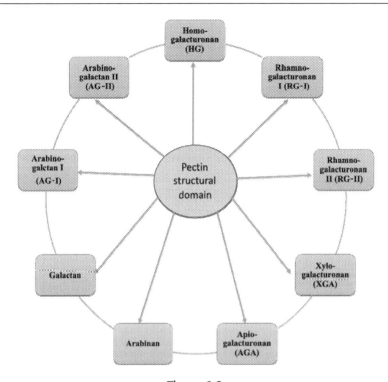

Figure 6.2
Structural domains of pectin present in various plant matrixes (Voragen et al., 2009).

group, pectin consisting of <75% of methylation, and pectin without methyl-esterified carboxyl group (Pedrolli, Monteiro, Gomes, & Carmona, 2009). It is estimated that around 100–200 units of GalA are present in HGA (Bonnin, Dolo, Le Goff, & Thibault, 2002). One or two α-L-rhamnopyranose residues by 1,2-linkage join the 1,4-linked α-D-galactopyranosyluronic acid residues, which constitute the linear region of HGAs. Besides, two additional domains of pectin, apiogalacturonan or xylogalacturonan (XGA), are formed by replacing the GalA units with apiose or xylose residues at C-2 and C-3 positions (Ovodov, 2009).

6.2.2 Rhamnogalacturonan I

RG-I is another domain of pectin, which accounts for more than 30% in total amount of pectin and comprises of repeating units of α-1,2-linked-L-rhamnose-α-1,4-D-GalA, in which 50%–78% residues are of RG-I (Renard, Crépeau, & Thibault, 1995). About 20%–80% of rhamnose-GalA residues are connected to neutral side sugar chains, primarily xylose, galactose, and arabinose, which form arabin, arabinogalactans, and galactans at C-4 (Mohnen, 2008; Valdés, Burgos, Jiménez, & Garrigós, 2015). In addition, there is also the possibility of linking the neutral sugar side chain at position C-2 or C-3 (Chan et al., 2017).

6.2.3 Rhamnogalacturonan II

RG-II consists of a branched-chain of HGA as a backbone with a variety of glycosyl residues and is known to be a hairy component of pectin structure (Valdés et al., 2015). The RG-II structure is extremely dense and consists of about nine units of α-1,4-linked GalA with four separate polymeric side chains and some unusual sugars such as aceric acid, apiose, 2-*O*-methyl-D-xylose, and 3-deoxy-D-manno-octulosonic acid (Yapo, 2011). RG-II is only a pectin domain that has no significant structural variation in its composition.

6.2.4 Xylogalacturonan

The substitution with β-D-xylopyranose (1→3) in the HGA forms XGA and this process is known as xylosidation (Voragen et al., 2009). In the hairy regions of pectin, this biopolymer occurs as RG-I side chain (Vincken et al., 2003). NMR spectroscopy of methyl-esterified XGA fractions shows these methyl esters uniformly divided between the unsubstituted and substituted galacturonosyl residues (Schols, Bakx, & Voragen, 1995). The presence of XGA in various tissues of *Arabidopsis thaliana* has been demonstrated (Zandleven et al., 2007).

6.3 Functional properties of pectin

The remarkable functional flexibility of pectin can be attributed to its high differential structure. Pectins are certainly the most biologically versatile biopolymers, exhibiting a broad variety of biological (plant cell proliferation, protection against prey, resistance, etc.), technofunctional (gelling, thickening, emulsification, film-forming, etc.), and health benefits (immunomodulating, biosorbent, prebiotic, antitumor, antioxidant, anticomplementary, antimicrobial, etc.) (Yapo & Gnakri, 2015). The three main important functional properties of pectin are (1) gelling, (2) emulsifying, and (3) anticancerous. Such functional properties are closely related to their composition, including molecular weight, protein quantity, and degree of methylation, based on the origin of the plant and the process of extraction (Geerkens et al., 2015).

As a polysaccharide with positive health results, pectin attracts expanded focus due to its lowering of cholesterol, lowering of blood glucose and anticancer properties (Leclere, Van Cutsem, & Michiels, 2013; Liu, Dong, Yang, & Pan, 2016). As compared to normal sugar beet pectin, alkali-treated sugar beet pectin shows enhanced rate of apoptosis in cancerous cell lines (Maxwell et al., 2016). pH or heat-modified forms of pectin (modified citrus pectin) exhibit strong antimetastatic properties, antitumor, and chemopreventive activities against various malignancies, aggressive and repetitive cancers by silencing the overexpression of pleiotropic galectin-3 protein (Leclere et al., 2013). Papaya pectin extracted after 3 days of harvesting showed enhanced cell detachment and promoted apoptosis/necroptosis activity by disrupting cancer cells and extracellular matrix proteins interaction (do Prado et al., 2017). As an anticancerous compound, pectin reduces the incidences of chemoresistance associated with the conventional chemotherapeutic drugs (Leclere et al., 2013).

Increased cholesterol level in the body increases the risk of developing cardiovascular disease. Dietary intervention is, therefore, needed to address the risk associated with the development of the disease. The viscosity, DE, and molecular weight of pectin decide the cholesterol-lowering capacity of pectin. Pectin derives the cholesterol-lowering effect from its capacity to form a viscous gel that facilitates its attachment to cholesterol and bile acids, thus decreases the reabsorption and facilitating excretion (Naqash, Masoodi, Rather, Wani, & Gani, 2017).

Pectin is a versatile natural food ingredient used mainly for the purposes such as an emulsifier, thickener, gelling agent, drug transport agent, and binder (Begum, Yusof, Aziz, & Uddin, 2017). Food industries use pectin as stabilizer and gelling agent for mass-produced food because of their emulsifying and gelation effects (Funami, 2011). There is a growing need for pectin with differing properties to stabilize the food products, which has necessitated the studies for new pectin sources such as sugar beet residues, papaya peel, tomato waste, and cocoa pod husks (Chen, Fu, & Luo, 2015; Grassino et al., 2016; Koubala, Christiaens, Kansci, Van Loey, & Hendrickx, 2014; Vriesmann, Teófilo, & de Oliveira Petkowicz, 2012). Pectin from some sources exhibits immunomodulatory, antihypertensive, antioxidant, cytoprotective, prebiotic, hypoglycemic, hypocholesterolemic, and other functions that offer a compelling context for their use in a broad variety of functional foods (Nara, Yamaguchi, Maeda, & Koga, 2009; Torkova et al., 2018; Wang, Hu, Nie, Yu, & Xie, 2016).

Lack of dietary fiber in regular meals adversely impacts human health by increasing the risk of diseases such as cancer (Kaczmarczyk, Miller, & Freund, 2012; Kushi et al., 2012; Pietrzyk, Torres, Maciejewski, & Torres, 2015). The antioxidant function of the polysaccharides can be impaired by their structural properties or by the existence of noncarbohydrate elements, like phenolic compounds (Košťálová, Hromádková, & Ebringerová, 2010; Wang et al., 2016; Yang, Zhao, Shi, Yang, & Jiang, 2008). In addition to its own antioxidant influence, pectin has the purpose of transporting dietary antioxidants (vitamin C, phenolic compounds, and carotenoids) into the gastrointestinal tract and preventing them from breakdown in the acidic environment of digestive system (Dikeman & Fahey, 2006; Saura-Calixto, 2011). Besides, in terms of beneficial effects on the gastrointestinal tract, pectin has a broad range of functional properties (Brownlee, 2011). In the human body the intestinal epithelial cells are also critical sites for the antioxidant activity of pectic polysaccharides and serve as an ideal model for measuring the antioxidant property of pectin (Torkova et al., 2018).

Since pectin is persistently present in fruits and vegetables, it contributes to the textural consistency of the products produced from them (Wicker et al., 2014). Neutral pectin is a functional component of food and acts in the fruits and vegetables as a soluble dietary fiber (Zhang, Xu, & Zhang, 2015). This is primarily used in jams and jellies as a gelling agent and is often used to efficiently preserve fruit juices and acidified milk beverages (Srivastava & Malviya, 2011), high-quality fruit drinks, and improved antioxidant products (Wicker et al., 2014). Pectin is also used in low-calorie foods as thickeners, stabilizers, water binders, and textured fat substitutes to imitate the mouthfeel of lipids (Maran, Sivakumar, Thirugnanasambandham, & Sridhar, 2014).

Pectin displays a variety of roles in the therapeutic arena, from reducing blood fat to fighting various types of cancers. The therapeutic function of pectin relieves discomfort, lowers the occurrence of cardiac failure, decreases blood cholesterol, prevents the production of lipase, development and metastasis of cancer cells, and causes apoptosis, which leads to its extensive use in the pharmaceutical field (Bagherian, Ashtiani, Fouladitajar, & Mohtashamy, 2011; Jackson et al., 2007; Kumar & Chauhan, 2010; Liu, Cao, Huang, Cai, & Yao, 2010). Tumor causes almost 7.6 million deaths worldwide per year (Ferlay et al., 2010). Research into pectin, its derivatives, especially modified pectin, has proven its function in inhibiting metastasis. The antitumor effects associated with the dietary pectin are largely due to its role in the defense of the immune system and its capacity to promote the development of probiotics, suppress tumor development and control oncogenes (Flint, Bayer, Rincon, Lamed, & White, 2008). Pectin-related antitumor function affects both colon cancers and cellular immunogenesis (Jeon et al., 2011). Improved antitumor properties of pectin modified with additives, radiation, heating, and enzymatic possesses have been documented relative to natural pectin. Zhang et al. (2015) have thoroughly studied the role of pectin in cancer therapy. Several researchers (Maxwell et al., 2016; Mahalik & Nambiar, 2010) found that apoptosis-induced colorectal cancer cell activity in modified sugar beet pectin. Modified pectin and its association with the neutral side chain of sugar showed an important role in inhibiting or destroying the cancer cells via apoptosis.

In terms of food preservation, pectin has several applications as carrier molecule for antioxidants, antimicrobials, and in packaging. Increasing health issues and competition for environmental-friendly and renewable products have moved research priorities into organic food-packaging materials (Espitia, Du, de Jesús Avena-Bustillos, Soares, & McHugh, 2014). Since pectin exhibits strong insoluble polymers as it interacts with multivalent metal cations such as calcium, it is used in edible packaging (Falguera, Quintero, Jiménez, Muñoz, & Ibarz, 2011; Medeiros, Pinheiro, Carneiro-da-Cunha, & Vicente, 2012). It is also used as a component of nanomultilayer coating with chitosan. This nanomultilayer device on whole mangoes called "Tommy Atkins" stopped their degradation. The gas-barrier and antimicrobial properties of chitosan and decreased oxygen permeability of pectin have aided in prolonging the shelf life of mangoes (Medeiros et al., 2012).

The above text well illustrates the multifold applications of pectin and special attention that scientists from multidisciplinary area need to continue paying by creating effective techniques with the intention of unraveling the interesting and challenging structures of certain pectic biopolymers and complex natural pectin. It would certainly be a significant move toward a greater understanding of the structure, the functions to open up a new field of functionality and application.

6.4 Sources of pectin

As per a report of FAO in 2016, it has been estimated that around 1.6 billion tons of food waste is generated holding a major proportion of fruits and vegetables (Gaur et al., 2020; Medeiros et al., 2012). Food waste generation is a detrimental loss of crop produce, economy as well as environment [BIO-Intelligence Service, Food and Agriculture Organization of the United Nations (FAO), 2013]. Valorization of fruits and vegetable waste as a substrate of pectin extraction would be of greater choice. Waste generated from the food-processing industry such as pomegranate peel, kiwifruit peel, and melon rind shows the pectin yields from the kiwifruit (KP), pomegranate (PP), and melon rinds (MP), as 8.03%, 6.54%, and 6.13%, respectively (Güzel & Akpınar, 2019).

Commercially pectin production is prominently from citrus peels, lemon, limes, orange, and apple pomace, which contain about 85%, 56%, 30%, 13%, and 14% pectin, respectively, and a minor fraction is obtained from sugar beet (Ciriminna et al., 2015; Pacheco, Villamiel, Moreno, & Moreno, 2019). The source of extraction greatly alters the yield, physicochemical properties, and chemical structure of the pectin in addition to its gelling properties, emulsion stabilities, emulsion activities, and effect on complex food matrices (Müller-Maatsch et al., 2016; Sundar Raj et al., 2012). The ability of pectin to form the gel depends on the DE and molecular size. Pectin obtained from apple peel produces harder and more viscous gel, therefore finds greater choice in bakery products, whereas citrus peel pectin is suitable for the candies and jellies due to its lighter color. The market demand of pectin cannot be fulfilled only by using these few sources, therefore finding alternative sources for pectin recovery is necessary to compete with qualitative properties needed for commercial-grade pectin and its production cost.

Several food waste and agriwaste streams have been used for pectin extraction such as jackfruit, cactus, soy hull, onion hull, carrot, tomato, potatoes, grapes, citrus, sugar beet flakes, pea pods, cauliflower, pumpkin without kernels, apple pomace, pumpkin kernel cake, cabbage, sour cucumber, sea buckthorn pulp, sea buckthorn seed, parsley, hop, sabal, rapeseed press cake, papaya, banana stem, jackfruit, watermelon rind, olive pomace, orange peels, grape pomace, berries, whole pears, whole apples, Belgian endive leaves and root, leek, papaya peel, sisal waste, carrot, peach pomace, and soy hull (Jafari, Khodaiyan, Kiani, & Hosseini, 2017; Müller-Maatsch et al., 2016; Venkatanagaraju, Bharathi, Sindhuja, Chowdhury, & Sreelekha, 2019). A wide array of fruit and vegetable by-products such as guava peel, citrus fruits, banana peel, cocoa pods, cocoa pod husk, orange peel, lemon peel, durian peel, durian rind, mango peel, horse eye bean peel, lemon pomace, grapes, pomegranate peel, sweet potato peel, and apple pomace have also been explored for the extraction of pectin (Sandarani, 2017). Extraction of pectin is a critical unit operation for the recovery from sidestreams of different plant-based food-processing units (Table 6.1). Extraction can be done using acids, solvents, including water and irradiation, but water-based extraction method is an environment-friendly and economic approach to obtain pectin from food waste, for example, from orange peel. Alcohol precipitation has been used for banana stem, leaf, peel, orange peel. The ultrasound-microwave HCl-assisted treatment has been applied on potato pulp to obtain branched RG-I (Yang, Mu, & Ma, 2019). Pectin

Table 6.1: Yield of pectin from different sources employing various extraction techniques.

Natural source	Extraction method	Extraction parameters	Pectin yield	References
Ambarella or golden apple (*Spondias cytherea*)	Acid extraction	Hydrochloric acid (HCl) (0.03 M) pH 1.5, for 1 h at 85°C	19.4 mg g^{-1}	Koubala et al. (2008)
	Acid extraction	Ammonium oxalate/oxalic acid (0.25%), pH 4.6, for 1 h at 85°C	22 mg g^{-1}	Koubala et al. (2008)
	Water	Deionized water for 1 h at 75°C	15.6 mg g^{-1}	Koubala et al. (2008)
Lime (*Citrus latifolia*)	Acid extraction	HCl (0.03 M) pH 1.5, for 1 h at 85°C	26.9 mg g^{-1}	Koubala et al. (2008)
	Acid extraction	Ammonium oxalate/oxalic acid (0.25%), pH 4.6, for 1 h at 85°C	8.8 mg g^{-1}	Koubala et al. (2008)
	Water	Deionized water for 1 h at 75°C	29.7 mg g^{-1}	Koubala et al. (2008)
Durian rind (*Durio zibethinus*)	Acid extraction	HCl, pH 2.8, 43 min, S:L 1:10 g mL^{-1}, 86°C	9.1%	Valdés et al. (2015)
Cacao pod husk (*Theobroma cacao*)	Acid extraction	Nitric acid pH 1.5 at 100°C for 30 min	3.7%–8.6%	Vriesmann et al. (2012)
Cocoa peel (*Theobroma cacao*)	Microwave-assisted extraction	Citric acid solution (pH 1.5), microwave power 300 W, 30 min	42.3%	Sarah, Hanum, Rizky, and Hisham (2018)
Banana peel	Acid extraction	Citric acid solution, pH 2.0, 87°C, 160 min	13.89%	Valdés et al. (2015)
		Citric acid and HCl, pH 1.5, 90°C, 4 h	16.54%	Valdés et al. (2015)
		HCl (pH 1.5) and water (pH 6.0) at 90°C ± 5°C, for 30–120 min	7%–11%	Maneerat et al. (2017)
	Microwave-based extraction	Microwave power: 700 W; irradiation time: 2 min; pH 1.5	12.2%	Kazemi, Khodaiyan, and Hosseini (2019)

(*Continued*)

Table 6.1: (Continued)

Natural source	Extraction method	Extraction parameters	Pectin yield	References
Brinjal waste (calyxes and peel)	Microwave-based extraction	Microwave power: 700 W; irradiation time: 2 min; pH 1.5	29.17% and 18.36%, respectively	Kazemi et al. (2019)
Sugar beet pulp	Acid extraction	Citric acid, pH 1, 99°C, 166 min	23.95%	Valdés et al. (2015)
Jackfruit waste (*Artocarpus heterophyllus* Lam) fruit peel	Optimized acid extraction	Oxalic acid at 90°C 60 min	38.45%	Venkatanagaraju et al. (2019)
	Ultrasound-based extraction	pH 1.6, temperature 60°C, and sonication time 24 min	14.5%	Moorthy et al. (2015)
Watermelon seed	Acid extraction	HCl, pH 2, 60 min, 85°C	19.75%	Valdés et al. (2015)
Tomato waste	Ultrasound-assisted extraction	Ultrasound treatment 37 kHz, temperatures 60°C and 80°C	15.1–35.7	Venkatanagaraju et al. (2019)
Tomato peel	Acid extraction	Ammonium oxalate and oxalic acid, 90°C	32.0	Valdés et al. (2015)
Pomegranate peel	Ultrasound-assisted extraction	1:17.52 g mL^{-1} of solid–liquid ratio, 28.31 min of extraction time, 1.27 pH, at 61.90°C	23.87%	Moorthy et al. (2015)
Sunflower head	Acid extraction	Sodium citrate, 85°C, 3.5 h	16.90	Valdés et al. (2015)
Orange peel	UHP	500 MPa, 55°C for 10 min	20.44%	Guo et al. (2012)
	Microwave	Microwave power 500 W, 80°C, 21 min	18.13%	Guo et al. (2012)
Apple pomace	MAE	Microwave power 499.4 W, 20.8 min, pH 1.01	0.315 g from 2 g dried apple pomace	Wang et al. (2007)

MAE, Microwave-assisted extraction; *UHP*, ultra-high pressure.

extracted from waste could be a promising food additive as gelling agent, texturizer, viscosifier, stabilizer, fat replacer, and nutraceutical, which finds its applicability in various food products to modify their technological properties as discussed before.

Although the process of extraction is well established, yet growing demand of pectin in different food and nonfood-based industries shows certain shortcomings in the traditional methods of extraction, thus prompting toward the need of amelioration of the process by adopting novel technologies.

6.5 Recent advances in the extraction of pectin

Several advances and rapid methods to extract pectin from different plant sources have been made (Fig. 6.3). The use of an effective pectin extraction method is important to improve the quality and optimize the yield of the product. The processing and separation of pectin depend on the rate of hydrolysis of protopectin (begins at a temperature of 80°C−85°C) and the solubilization of resulting pectin so that the extraction solvent reaches a saturation state (Adetunji et al., 2017). The extraction processes are, therefore, regulated by the concepts of mass transfer. This whole method of pectin isolation is outlined in three broad categories, that is, pretreatment of the raw material, extraction, and postextraction (Gentilini et al., 2014).

Traditionally, pectin is typically extracted by the continuous stirring in water acidified with 0.05−2 M sulfuric, phosphoric, acetic, hydrochloric, or nitric acid for 1 h between 80°C and 100°C (Georgiev, Ognyanov, Yanakieva, Kussovski, & Kratchanova, 2012). The traditional extraction process depends on numerous factors such as pH range, temperature, solvent properties, particle size, dry solids, solvent ratio, and diffusion speed. The conventional extraction (CE) strategy employs the use of mineral acids has, however, some drawbacks such as increased manufacturing costs, loss of volatile compounds, and environmental issues, as well as the degradation of valuable, targeted compounds found in plant materials. For this purpose, several sustainable and faster alternatives have emerged to extract natural compounds from the biological sources, which have resulted reduced extraction time, lower temperature, higher yield, lower solvent usage, and even improved quality of extracts as compared to the conventional methods (Shams et al., 2015). The fundamentals of conventional methods vary from advanced extraction techniques as the extraction takes place owing to the changes in the cell structure induced by electromagnetic or ultrasound waves. The most promising of all emerging pectin-related extraction techniques are ultrasound-assisted extraction (Vinatoru, 2001), microwave-assisted extraction (Kaufmann & Christen, 2002), supercritical fluid

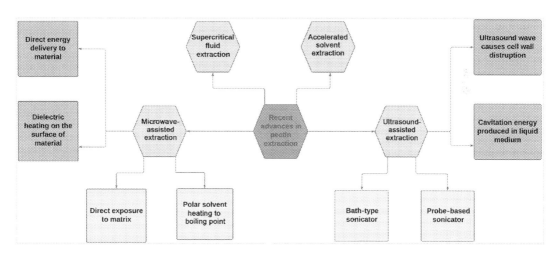

Figure 6.3
Schematic representation of different extraction processes.

extraction (Meireles, 2003), and accelerated solvent extraction (Smith, 2002). These techniques can work at high temperatures and pressure, which significantly reduce the time of extraction. As a result, pectin with different structural features can be obtained using different extraction methods. This section provides details of some of the most innovative and relevant approaches for the extraction of pectin.

6.5.1 Microwave-assisted extraction of pectin

Microwaves are electromagnetic waves nonionizing radiation having frequency between 0.3 and 300 GHz and wavelengths from 1 mm to 30 cm (Aguilar-Reynosa et al., 2017). Microwave irradiation is possibly one of the most powerful techniques for noncontact heating. Microwave energy has an advantage of direct delivery of energy to materials that absorbs microwave thereby allowing the volumetric heating. Owing to these properties, microwaves have been attractively used in industries alternatively to conventional processing methods. Dielectric heating (i.e., dipole rotation millions of times per second) induced by the microwave radiation inside to outside in the cell produces a pressure on the cell wall during the extraction process. As a result of this pressure, the plant tissue ruptures and the contents are released into the solvent (Dhobi, Mandal, & Hemalatha, 2009). Two extraction principles remain in the use of microwave energy (Letellier & Budzinski, 1999), namely, (1) direct exposure to the matrixes, leading to separation of the target compound into solvent and (2) boiling point heating of polar solvent leads to the extraction of the compound.

In comparison to the traditional heating methods, which require ample amounts of solvents with extended treatment time, microwave-assisted extraction is favored due to shortened processing time, substantial yield increases, lower energy usage, and lower solvent use. In addition to these characteristics, during microwave-based extractions, particle size and moisture content of the plant matrixes to be extracted are also considered. Microwave energy is mainly absorbed by the water present in biological and food materials; thus the faster it responds to microwave heating, the higher is the moisture content of the biomaterial (Venkatesh & Raghavan, 2004).

The microwave-based pectin extraction has been applied as a novel approach for a wide range of plant materials such as pomelo peel, lime, orange peel (Fishman, Chau, Hoagland, & Hotchkiss, 2006; Maran, Sivakumar, Thirugnanasambandham, & Sridhar, 2013; Quoc et al., 2015), and lime albedo, flavedo, and pulp. The exposure time ranged from 1 to 10 min and the final optimized heating time was 3 min, followed by a decrease in molar mass, viscosity, gyration radius, and hydrated radius. Depending on the lime fraction obtained after 3-min microwave exposure, the intrinsic weight average viscosity ranged from around 9.5 to 13 dL g^{-1} and the average molar mass ranged from 310,000 to 515,000 Da (Fishman et al., 2006). Similarly, optimizing the pectin extraction process using microwave, the highest yield was reported by apple pomace (i.e., 0.315 g/2 g dried apple pomace), under optimum conditions: 20.8 min (time); 0.069 (solid:liquid ratio); 1.01 (pH); and 499.4 W (microwave power) (Wang et al., 2007). This time for traditional heating was comparatively lower than those previously published (Cho & Hwang, 2000; Minkov, Minchev, & Paev, 1996). The reasons for the initial improvement in the efficiency of extraction can be attributed to the decrease in surface tension and viscosity, which accelerates the protopectin and pectin hydrolysis and solubility, respectively. The time of exposure to the microwave energy, also termed as irradiation time, may be the most significant factor on pectin yield and performance. As a result, increased microwave energy will increase the temperature inside the device and lead to increased pectin yields (Maran et al., 2013). Therefore the duration of microwave exposure should be considered very carefully in the optimization of the extraction process. With the aid of microwave and tartaric acid, pectin extraction from pomelo (*Citrus maxima*) peels greatly affected pectin properties and its extraction efficiency. The optimal condition for pectin extraction was pH (1.5), material/solvent rate (1/40), irradiation power (660 W), and exposure time (9 min), yielding 23.83% pure pectin. The obtained pectin contained high methoxyl pectin with more than 90% DE value and low viscosity (Quoc et al., 2015). It can be used in the food industry, especially the processing of beverages and jams.

Hylocereus polyrhizus is one of the tropical fruits in India that belongs to the cactus family. It is a very good source of pectin, better known as dragon fruit. Microwave-based pectin extraction from dragon fruit peel exhibited the highest pectin yield of 7.5% under the parameters, viz. microwave exposure of 400 W at 45°C, 120 s, and a solid–liquid ratio of 24 g mL^{-1} (Wicker et al., 2014). Due to the simplicity of method and low cost, microwave-assisted extraction is also comparable with other CE techniques, namely, supercritical fluid extraction. Thus overall the microwave-assisted extraction method offers an environmental-friendly processing solution for pectin isolation (Chemat et al., 2017).

6.5.2 Ultrasound-assisted extraction of pectin

Ultrasound-assisted extraction is a technique used to extract target compounds from different plant matrices using acoustic energy and solvents. Sound waves (SWs) consist of mechanical vibrations that can be generated with frequencies greater than 20 kHz in solid water or gas treatments (Roselló-Soto et al., 2015). These waves are different from the electromagnetic waves in the term that they require a medium, that is, solid, liquid, or gas, for their propagation, involve a sequence of expansion pulling molecules apart and compression cycles, bringing them together. There are two types of ultrasonic systems available for separating the targeted molecules: (1) bath and (2) probe-based. On the basis of reproducibility in the extraction process, probe-based systems are considered better in comparison to the bath system, since their energy can be concentrated on specific sample zone, resulting in more efficient cavitation of liquids.

Cavitation is a process in liquid by which bubbles form, grow, and collapse (Luque-Garcia & Castro, 2003). In the effective and efficient implementation of ultrasound-based product extraction, ultrasound device conditions usually involve temperature, sonication time, frequency, and pressure. Solvent-biomass ratio is also significant because ultrasound intensity attenuation occurs with increasing content of solid particles (Wang & Weller, 2006). Since pectin is a soluble fiber present in the cell walls of plant, cell disruption and cavitation caused by ultrasound waves enhances the transfer of mass from the solid phase to the solvent, thereby improve pectin extraction. The cavitation cycle lasts approximately 400 ms during which high temperatures and pressures can be measured, approximately 5000°C and 1000 atm, respectively (Azmir et al., 2013). However, the yield of pectin may be reduced by the decomposition and structural destruction that may have caused by the longer ultrasound treatment during extraction and the heating effect (Maran & Priya, 2015).

Maran and Priya (2015) used Response Surface Methodology (RSM) model to test the influence of ultrasound and temperature on the process of pectin extraction from fruit peel. Wave exposure time over the dried peel/extractant at 1:30 ratio was 10 min. Maximum yield was obtained by
644 W cm^{-2} energy strength and 85°C temperature. The content of GalA, the DE, and yield, were 66.65%, 60.36%, and 12.67%, respectively. The same criteria were followed for the traditional processes for comparison. The results showed that the use of ultrasound facilitated a better pectin extraction yield as compared to the conventional method (de Oliveira et al., 2016). In 2016 a study was conducted to explore the use of tomato waste using ultrasound technique and conventional technique for the commercial production of pectin. The waste from tomato was treated with ammonium oxalate/oxalic acid by the CE under the influence of reflux and SWs at 37 kHz and a temperature of 60°C and 80°C. The obtained pectin was analyzed and compared in terms of structure, yield, and chemical properties. The esterified structure of isolated pectin was studied using NMR and FTIR spectroscopy, which showed that the quality of pectin was superior with the highest yield among the methods tested. Conventional 24-h solvent extraction yielded equal quantity of pectin compared with 15 min of ultrasound therapy. Thus it could be concluded that SWs significantly reduced the extraction time (Moorthy, Maran, Muneeswari, Naganyashree, & Shivamathi, 2015).

Several studies have been done on the extraction of pectin from the by-products of food using different frequencies and ultrasound system power to achieve maximum yield in a short period of time. In a study to extract pectin from the *Punica granatum* (Pomegranatum) by-product, immersed sonotrode at 20 kHz frequency and ultrasound system power of 130 W was used with 1.27 pH, 28.31-min processing time, and temperature at 61.9°C. In this the SWs were produced using a flat tip 2-cm-diameter probe. The highest pectin extraction was obtained with a solid−liquid ratio of 1:17.52 g mL^{-1} (Grassino et al., 2016). Wang et al. (2007) also used grapefruit peel to extract pectin by RSM and the conditions optimized were 12.56 W cm^{-1} power frequency, 66.7°C extraction temperature, and 27.95-min sonic time. The quantitative analysis revealed that in approximately 38% less time, there was an increase of 16% yield as compared to the conventional method. FTIR analysis showed insignificant variations with better color and more loosened microstructure in the chemical structure. This makes ultrasound-based extraction method as an industrial potential procedure for future applications (Wang et al., 2015).

6.6 Application of pectin

6.6.1 Food-processing industries

6.6.1.1 Dairy industry

In the competitive era of market survival, quality parameters of dairy products are needed to be amplified each day with a simultaneous increase in the demand of newer products. Pectin as a food additive improves several quality characteristics of foods such as texture, stability, flavor retention, fat reduction, easy digestibility, bioavailability of amino acids, and shelf life. Advance processing practices employs incorporation of multiple dairy and nondairy products to alter the quality parameters of the products (Korhonen, 2011). In the same context, plant hydrocolloids like pectin alginates, carrageenans, gum arabic, guar gum, and locust bean gum, are used in dairy industry to improve the stability, taste, and texture of food products such as flavored ice cream, yogurt drinks, milk, cream, puddings, and custards.

Pectin is a polysaccharide that interacts with milk proteins and forms protein−polysaccharide complexes with strong interaction (covalent bonding) or weak interactions (hydrophobic bonding, electrostatic, and van der Waals). The formation of weak or strong interaction depends on aqueous environmental conditions such as ionic strength and pH, charge density, and molecular weight of both the reacting macromolecules. Pectin addition improves textural, nutritional properties, and stability of dairy products (Mouécoucou et al., 2003).

Flavors are volatile components associated with fat molecules for stability in a food matrix. They are key players in deciding the acceptability and likelihood of food. Along with greater concern toward low calories diet and certain health aliments skimmed milk is generally preferred for manufacturing of various delicious desserts (Lubbers, Decourcelle, Martinez, Guichard, & Tromelin, 2007). With the reduction of fat content, there will be imbalances in flavor release like perception of flavor, mouthfeel, appearance, and structure of food product. Pectin is incorporated in the food prior to fat separation, and being a fat substitute, it traps volatile flavor components within them (Lubbers et al., 2007). Viscosity is a vital parameter accounting for consistency and sensory attributes (appearance and mouthfeel) of a food product. Incorporation of increasing concentration of pectin in fat-free yoghurt increases viscosity along with flavor retention. This practice of introducing pectin in different food system in varied concentration enhances acceptability of food product (Lubbers et al., 2007).

In Greek-style yoghurt manufacturing, whey is removed mechanically to achieve desired texture in terms of solidity which generates large quantities of acid whey as a waste, causing serious economic and environmental concern (Gyawali & Ibrahim, 2018). Owing to the excellent hygroscopic, gelling, thickening, stabilizing, and syneresis controlling properties, pectin significantly improves the water-holding capacity of yoghurt and concurrently reduces the amount of acid whey that is generated during concentration step of yoghurt production. Pectin inhibits the formation of whey protein derived large oligomeric aggregates by forming protein network by interacting with milk proteins, thereby inhibiting free movement of water and reduces syneresis (Gyawali & Ibrahim, 2018).

Pectin enhances the nutritional value of food by enhancing digestibility of food component such as of β-lactoglobulin (β-lg), a globular protein of lipocalin family. β-lg is a bioactive component present in whey protein of bovine milk and colostrum. It is resistant to digestive enzymes in gastrointestinal tract (Korhonen, 2011). It contributes around 50% of whey protein in cow milk (Mouécoucou et al., 2003). β-lg is a versatile material that is used as an ingredient owing to diverse food and biochemical applications due to its multipotent functional, nutritional properties, and bioactivities, such as antimicrobial, antioxidative, antihypertensive, immunomodulatory, anticarcinogenic, opioid, and hypocholesterolemic (Mouécoucou et al., 2003). Pectin is resistant to digestive enzymes present in gastrointestinal tract, thus it reaches intestinal mucosa in almost intact form. From there, it gets absorbed in blood stream and induces allergenic reactions (Mouécoucou et al., 2003). Several treatments like sterilization, hydrostatic pressure, or boiling partially increases whey protein digestibility. As compared to the protein alone, the digestibility of β-lg was significantly enhanced by the incorporation of low methoxy pectin (50 wt.%).

6.6.1.2 Bakery industry

The incorporation of pectin in fiber-depleted diets, containing refined carbohydrates, silences/eradicates diabetes-causing factors from the diet. After ingestion, unrefined carbohydrate-containing foods leads to controlled rise in blood sugar level as compared to refined carbohydrates. Pectin fulfills the requirement of fibrous content in diet therefore plays a significant role in diabetes control. It also promotes nutritional quality of food in the form of short-chain fatty acids (Ho, Matia-Merino, & Huffman, 2015; Jenkins, Leeds, Gassull, Cochet, & Alberti, 1977).

Besides providing nutritional benefits, pectin also modifies functional and textural properties of dough used in bakery industry. Involvement of pectin in popular staple foods like bread enhances the content of dietary fiber, which is of high consumer demand due to its nutritional and quality characteristics. Pectin imparts water binding or holding capacity to dough due to its hydration effect of flour components and encounters self-breakdown process of protein strands therefore retards retrogradation of bread. Addition of higher amount of pectin in bread dough results in dilution of gluten protein in whole formulation, thereby causing reduced loaf volume in bread (Sivam, Sun-Waterhouse, Waterhouse, Quek, & Perera, 2011).

Pectin forms hydrophilic complexes with gluten proteins through electrostatic interactions which contributes in dough stability, elevates dough height, and greatly improves fermentation properties of dough. Addition of 0.2%–0.6% pectin significantly enhances gas-retention capacity and extensibility of dough (Li, Zhu, Yadav, & Li, 2019).

In bread manufacturing the gluten protein content of wheat flour is an essential component that is helpful in providing fine bread crumb structure. Pectin in combination with suitable emulsifying agents can be helpful in the manufacturing of bread with substitution of different crop flours which are deficient in gluten protein content. Pectin is responsible to improve specific loaf volume and firmness of composite bread (Eduardo, Svanberg, & Ahrné, 2016). Pectin improves the baking performance and acts as a fat replacer. Incorporation of pectin extracted from Yuja (*Citrus junos*) pomace in cake batter yield greater viscosity and lesser shear-thinning behavior. Pectin concentration highly correlates to specific gravity of cake batter and volume raise after baking. In baked products, pectin can replace up to 10% of shortening content. This substitution of pectin in baked products causes increased softness, lighter surface color, a significant reduction in calories and fat content of the products (Lim, Ko, & Lee, 2014).

6.6.1.3 Confectionery

Pectin is recommended as safe food additive by a Joint FAO/WHO Committee, requiring no limit on acceptable daily intake. Depending on the age of men and women, the daily requirement of dietary fiber was suggested to be 30–38 and 21–25 g, respectively, by the Food and Nutrition Board, Institute of Medicine of the USA National Academies recommendations (Figueroa & Genovese, 2018). Owing to its fibrous potential, pectin is widely used in confectionery products (jam, jellies, and marmalades), calorie foods, frozen foods, etc. (Porta, Mariniello, Di Pierro, Sorrentino, & Giosafatto, 2011). A wide variety of lucrative confectionery items such as jelly snakes, jelly babies, and jelly bear has been developed using pectin as a gelling and stabilizing agent. It is acid stable in nature therefore, is highly preferred as gelling agent for acidic fruit gels (Burey, Bhandari, Rutgers, Halley, & Torley, 2009). For jam and jelly preparations, both high and low methoxyl pectin (commercial and amidated types) are used. Commercial pectin contains sugar, which help in preventing clumps that are formed during wetting of pectin and standardizing viscosity (Saha & Bhattacharya, 2010). Gel-formation tendency or gelation of pectin depends upon the accuracy of pH and sugar content of mixture. Upon hydration, pectin absorbs water and the mixture rapidly thickens to produces gel confectionery in continuous manner (Burey et al., 2009).

Jams and marmalade are filler-matrix composite gels, in which matrix is made by pectin networking, while filler consists insoluble fruit particles/pieces or pulp. Pectin forms cross-linking network in gel by involving a composite action of hydrophobic interactions and hydrogen bonds between pectin molecules. However, other factor that influences the gelation in pectin is the degree of methoxylation (charge density and molecular size, pH, and other solutes) (Porta et al., 2011).

According to the Argentine Food Code, to receive "source of fiber" label, a food product should have dietary fiber at a concentration of 3 g/100 g of food product. Commercial fruit jams contain lower than 3 g/100 g of dietary fiber. Therefore other confectionery items like gels are prepared by incorporating the required amount of pectin as fiber in sugar acid, which claims the label "source of fiber" as a healthy confectionery item (Figueroa & Genovese, 2018).

Pectin has the ability to be casted as edible films with the potential of commercial usage. Biodegradable edible films with desired gas-barrier properties and mechanical characteristics are used for food wrapping and coating. It provides multiple desired characteristics such as physical protection to food material, reduction in moisture loss, restricts oxygen absorption, reduced migration of lipids, improved mechanical handling features, and ability to be applied directly for food packaging. Purified pectin/soy flour films and pectin-containing fruit puree/soy flour films are produced to wrap food products (Porta et al., 2011).

6.6.2 Pharmaceutical industry

Owing to a variety of biological properties and numerous applications of pectin in pharmacology, an increasing interest has been focused to the research of natural polysaccharides (Noreen et al., 2017). The bioactivity of pectin polysaccharides includes pharmacological applications, namely, antibacterial, antioxidant, immunoregulatory, hypoglycemic, antitumor activities, and antiinflammatory. The pectin carbohydrate chain influences the function of immune suppression. Pectin was reported to contain more than 80% of GalA residues which suppress macrophage activity and inhibit the hypersensitivity reaction of the delayed-type. Furthermore, the pectin macromolecule branched region mediates phagocytosis stimulation and increased antibodies production (Yu, Shen, Song, & Xie, 2018). Similarly, the physicochemical properties of lemon pectin, such as the extent of polymerization and the degree of methyl esterification, can affect their immunostimulatory characteristics and may, therefore, be relevant when using pectin to enhance the immune status (Popov & Ovodov, 2013). Citrus pectin which is low on methyl esterification inhibits systemic and local inflammation, whereas highly esterified pectin inhibits inflammation of the intestines as evaluated by oral administration in mice (Vogt et al., 2016). Drugs based on medicinal and dietary plants do not exhibit side effects, which promise pectin to be incorporated as a new class of hypoglycemic drugs. Previously, pectin polysaccharides suppress the elevated hyperglycemic activity of streptozotocin-induced diabetic mice by probably slowing the reaction of the peroxidation chain and could be utilized as a powerful new diabetes therapy (Zhang, Sun, & Jiang, 2018). The antibacterial role of pectin is of considerable interest as it has a Gram-positive and Gram-negative bactericidal impact on microorganisms (Minzanova et al., 2018). A green approach has been involved in synthesizing silver (Ag) nanoparticles (NPs) that use citrus pectin as the reducing and capping agent. The Ag NPs showed potent antibacterial activity against *Escherichia coli* and *Staphylococcus aureus* bacteria (Zhang, Zhao, Jiang, & Zhou, 2017). The rapidly growing area of the practical application of pectin is anticancer therapy. Various in vivo and in vitro studies of native and modified pectin related to the antitumor activity showed a decrease in adhesion and proliferation of cells, as well as apoptosis initiation and migration (Bush, 2014). Different forms of pectin can also be used for drug delivery in combination with other molecules in target-based therapies. The ability to get dissolved in the upper gastrointestinal tract, mucoadhesivity, degradation stability against proteases and amylases, and ease of gel formation in acid environments makes pectin ideal in different forms of drug delivery such as beads, pellets, microparticles, and microspheres (Marras-Marquez, Peña, & Veiga-Ochoa, 2015; Zhang, Xiang, Zheng, Yan, & Min, 2018). The wide range of applications and the growing number of studies on pectin hydrogels indicate the potential of pectin as novel and flexible biomaterials.

6.6.3 Cosmetics

Characteristics of outstanding gelling properties, excellent biocompatibility, and nontoxicity, as well as biodegradability, allow pectin to be an enticing new biopolymer product that can be used in the cosmetic and health-promotion applications. Pectin serves as an emulsion stabilizer in cosmetic products. It is used to manufacture body and hand lotions, make-up formulations, shampoos, hair conditioners, and cosmetics for personal cleanliness (Augustine, Venugopal, Snigdha, Kalarikkal, & Thomas, 2015). Pectin offers a significant advantage in the cosmetic industry in the form of gelation, viscosity, absorption of moisture, emulsification, esterification, adhesion, and chelation. Microencapsulation of lipophilic materials is a way of formulating a liquid material in a solid dosage form to promote its handling, boosting its stability, and preserving its release (Turchiuli et al., 2005). Incorporating these microcapsules/microspheres into a cosmetic gel base will improve the esthetic appeal of these products and provide a slow release of the active components while preserving product stability and efficiency throughout its shelf life. The findings from a study revealed that the 5:1.5 ratio of pectin alginate microspheres is best suited for incorporation into a gel base. Such microspheres can be filled with vitamin E or any other antioxidant or scent to improve product stability and enhance formulation efficacy (Shalaka, Naik, Amruta, & Parimal, 2009). If rubbed between the palms, these microspheres will rupture releasing vitamin E. Besides, physically disrupted pectin gels were used in lotions and cream formulation without the use of surfactants (Bouyer, Mekhloufi, Rosilio, Grossiord, & Agnely, 2012). High methoxyl citrus and apple pectin can replace other hydrocolloids or emulsifiers in whole or in part (Endreß & Rentschler, 1999). Hydrated pectin upon swelling makes smooth, wet particle system that gives skin creams a fat-like texture. The unesterified form when used in skincare products, protects the tissue from the effects of ultraviolet radiation, delays the photoaging process and exhibits moisturizing properties. The gel-forming unique ability of pectin and its effective bioadhesiveness is a landmark in the production of many skincare products (Endreß & Rentschler, 1999).

6.7 Current challenges and future implications

The global demand of pectin is increasing due to its applications as a technological adjuvant in various industries. Pectin can be extracted from a wide array of plants tissues, yet the demand cannot be matched because the pectin extracted from different source varies in its structural configurations at backbone chain level and distribution of methyl ester groups and different carbohydrate moieties (Bouyer et al., 2012; Endreß & Rentschler, 1999). It is challenging to fully exploit the functional properties of the structurally diverse pectin, as differing structures alter the gelling, emulsifying, and stabilizing properties. The potential evaluation of techno- and biofunctionality of pectin from different sources is still largely incomplete and needs to be well defined (Bouyer et al., 2012; Vanitha & Khan, 2019). There is a vast gap of understanding about the structural−functional relationship and its impact on the interactions of pectin with different food products (Bouyer et al., 2012).

For the extraction of pectin, it is necessary to maintain highly acidic conditions, which is a serious environmental concern; it also causes corrosion of the reactors (Endreß & Rentschler, 1999). Technological advancement in the extraction methods still seek improvements in terms of yield potential, extraction time, solvent toxicity, and capital investment. Modern method utilizing supercritical fluid extraction integrated with subcritical water hydrolysis for the recovery of pectin could be attractive and environmental-friendly approach but this also poses challenges like slow biomass hydrolysis rates and bonding between the cell wall composition and its monosaccharide structure (May, 1990).

There is huge projection in the market demand of pectin, proliferating exponentially every year. Such demands warrant improvement in well-established pectin extraction process in terms of operating speed, reproducibility, and predictability of physicochemical properties of extracted pectin. Pectin can successfully substitute a broad group of food additives and other technological adjuvants in nonfood industry, still its application needs are needed to be explored (Adetunji et al., 2017).

6.8 Conclusions and perspectives

Pectin is a natural biodegradable polymer present ubiquitously in the cell wall of plants and can also be obtained from the waste streams of fruits and vegetable-processing industries. Source and process of extraction greatly influence the functional properties of pectin and a wide array of plant sources have been exploited using different procedures from traditional to advance methods using enzymes, microwave, and ultrasound to achieve the maximum yield and modifying quality attributes. Pectin is a versatile food additive due to its potential to be used as emulsifying, gelling, stabilizing, and thickening agent in various food products, along with enhancing fibrous content of the diet.

Acknowledgments

Vivek Kumar Gaur acknowledges the Council of Scientific and Industrial Research (CSIR), New Delhi for Senior Research Fellowship. This chapter bears Integral University communication number: IU/R&D/2020-MCN000958.

References

Adetunji, L. R., Adekunle, A., Orsat, V., & Raghavan, V. (2017). Advances in the pectin production process using novel extraction techniques: A review. *Food Hydrocolloids*, *62*, 239–250.

Aguilar-Reynosa, A., Romani, A., Rodriguez-Jasso, R. M., Aguilar, C. N., Garrote, G., & Ruiz, H. A. (2017). Microwave heating processing as alternative of pretreatment in second-generation biorefinery: An overview. *Energy Conversion and Management*, *136*, 50–65.

Ajibade, B. O., & Ijabadeniyi, O. A. (2019). Effects of pectin and emulsifiers on the physical and nutritional qualities and consumer acceptability of wheat composite dough and bread. *Journal of Food Science and Technology*, *56*(1), 83–92.

Augustine, R., Venugopal, B., Sajeendra Babu, S., Kalarikkal, N., & Thomas, S. (2015). Polyuronates and their application in drug delivery and cosmetics. *Green polymers and environmental pollution control*. Oakville: Apple Academic Press, (pp. 239–269).

Azmir, J., Zaidul, I. S. M., Rahman, M. M., Sharif, K. M., Mohamed, A., Sahena, F., ... Omar, A. K. M. (2013). Techniques for extraction of bioactive compounds from plant materials: A review. *Journal of Food Engineering*, *117*(4), 426–436.

Bagherian, H., Ashtiani, F. Z., Fouladitajar, A., & Mohtashamy, M. (2011). Comparisons between conventional, microwave- and ultrasound-assisted methods for extraction of pectin from grapefruit. *Chemical Engineering and Processing: Process Intensification*, *50*(11–12), 1237–1243.

Begum, R., Yusof, Y. A., Aziz, M. G., & Uddin, M. B. (2017). Structural and functional properties of pectin extracted from jackfruit (*Artocarpus heterophyllus*) waste: Effects of drying. *International Journal of Food Properties*, *20*(Suppl. 1), S190–S201.

BIO-Intelligence Service, Food and Agriculture Organization of the United Nations (FAO). (2013). *Food wastage footprint. Impacts on natural resources. Summary report 9* (p. 63) Food and Agriculture Organization of the United Nations (FAO).

Bonnin, E., Dolo, E., Le Goff, A., & Thibault, J. F. (2002). Characterisation of pectin subunits released by an optimised combination of enzymes. *Carbohydrate Research, 337*(18), 1687–1696.

Bouyer, E., Mekhloufi, G., Rosilio, V., Grossiord, J. L., & Agnely, F. (2012). Proteins, polysaccharides, and their complexes used as stabilizers for emulsions: Alternatives to synthetic surfactants in the pharmaceutical field? *International Journal of Pharmaceutics, 436*(1–2), 359–378.

Brownlee, I. A. (2011). The physiological roles of dietary fibre. *Food Hydrocolloids, 25*(2), 238–250.

Brummel, S. E., & Lee, K. (1990). Soluble hydrocolloids enable fat reduction in process cheese spreads. *Journal of Food Science, 55*(5), 1290–1292.

Burey, P., Bhandari, B. R., Rutgers, R. P. G., Halley, P. J., & Torley, P. J. (2009). Confectionery gels: A review on formulation, rheological and structural aspects. *International Journal of Food Properties, 12*(1), 176–210.

Bush, P. L. (2014). *Pectin: Chemical properties, uses and health benefits.* Nova Science Publishers, Inc.

Chan, S. Y., Choo, W. S., Young, D. J., & Loh, X. J. (2017). Pectin as a rheology modifier: Origin, structure, commercial production and rheology. *Carbohydrate Polymers, 161*, 118–139.

Chemat, F., Rombaut, N., Meullemiestre, A., Turk, M., Perino, S., Fabiano-Tixier, A. S., & Abert-Vian, M. (2017). Review of green food processing techniques. Preservation, transformation, and extraction. *Innovative Food Science & Emerging Technologies, 41*, 357–377.

Chen, H. M., Fu, X., & Luo, Z. G. (2015). Properties and extraction of pectin-enriched materials from sugar beet pulp by ultrasonic-assisted treatment combined with subcritical water. *Food Chemistry, 168*, 302–310.

Cho, Y. J., & Hwang, J. K. (2000). Modeling the yield and intrinsic viscosity of pectin in acidic solubilization of apple pomace. *Journal of Food Engineering, 44*(2), 85–89.

Ciriminna, R., Chavarría-Hernández, N., Inés Rodríguez Hernández, A., & Pagliaro, M. (2015). Pectin: A new perspective from the biorefinery standpoint. *Biofuels, Bioproducts and Biorefining, 9*(4), 368–377.

de Oliveira, C. F., Giordani, D., Lutckemier, R., Gurak, P. D., Cladera-Olivera, F., & Marczak, L. D. F. (2016). Extraction of pectin from passion fruit peel assisted by ultrasound. *LWT—Food Science and Technology, 71*, 110–115.

Dhobi, M., Mandal, V., & Hemalatha, S. (2009). Optimization of microwave assisted extraction of bioactive flavonolignan-silybinin. *Journal of Chemical Metrology, 3*(1), 13.

Dikeman, C. L., & Fahey, G. C., Jr (2006). Viscosity as related to dietary fiber: A review. *Critical Reviews in Food Science and Nutrition, 46*(8), 649–663.

do Prado, S. B. R., Ferreira, G. F., Harazono, Y., Shiga, T. M., Raz, A., Carpita, N. C., & Fabi, J. P. (2017). Ripening-induced chemical modifications of papaya pectin inhibit cancer cell proliferation. *Scientific Reports, 7*(1), 1–17.

Dranca, F., & Oroian, M. (2018). Extraction, purification and characterization of pectin from alternative sources with potential technological applications. *Food Research International, 113*, 327–350.

Eduardo, M., Svanberg, U., & Ahrné, L. (2016). Effect of hydrocolloids and emulsifiers on the shelf-life of composite cassava-maize-wheat bread after storage. *Food Science & Nutrition, 4*(4), 636–644.

Endreß, H. U., & Rentschler, C. (1999). Chances and limits for the use of pectin as emulsifier: Part I. *European Food and Drink Review*, 49–53.

Espitia, P. J. P., Du, W. X., de Jesús Avena-Bustillos, R., Soares, N. D. F. F., & McHugh, T. H. (2014). Edible films from pectin: Physical-mechanical and antimicrobial properties—A review. *Food Hydrocolloids, 35*, 287–296.

Falguera, V., Quintero, J. P., Jiménez, A., Muñoz, J. A., & Ibarz, A. (2011). Edible films and coatings: Structures, active functions and trends in their use. *Trends in Food Science & Technology, 22*(6), 292–303.

Ferlay, J., Shin, H. R., Bray, F., Forman, D., Mathers, C., & Parkin, D. M. (2010). Estimates of worldwide burden of cancer in 2008: GLOBOCAN 2008. *International Journal of Cancer, 127*(12), 2893–2917.

Figueroa, L. E., & Genovese, D. B. (2018). Pectin gels enriched with dietary fibre for the development of healthy confectionery jams. *Food Technology and Biotechnology*, *56*(3), 441–453.

Fishman, M. L., Chau, H. K., Hoagland, P. D., & Hotchkiss, A. T. (2006). Microwave-assisted extraction of lime pectin. *Food Hydrocolloids*, *20*(8), 1170–1177.

Flint, H. J., Bayer, E. A., Rincon, M. T., Lamed, R., & White, B. A. (2008). Polysaccharide utilization by gut bacteria: Potential for new insights from genomic analysis. *Nature Reviews Microbiology*, *6*(2), 121–131.

Flutto, L. (2003). *Pectin| Food use. Encyclopedia of food sciences and nutrition* (pp. 4449–4456). Elsevier.

Funami, T. (2011). Next target for food hydrocolloid studies: Texture design of foods using hydrocolloid technology. *Food Hydrocolloids*, *25*(8), 1904–1914.

Gaur, V. K., Sharma, P., Sirohi, R., Awasthi, M. K., Dussap, C. G., & Pandey, A. (2020). Assessing the impact of industrial waste on environment and mitigation strategies: A comprehensive review. *Journal of Hazardous Materials*, 123019.

Gawkowska, D., Cybulska, J., & Zdunek, A. (2018). Structure-related gelling of pectins and linking with other natural compounds: A review. *Polymers*, *10*(7), 762.

Geerkens, C. H., Nagel, A., Just, K. M., Miller-Rostek, P., Kammerer, D. R., Schweiggert, R. M., & Carle, R. (2015). Mango pectin quality as influenced by cultivar, ripeness, peel particle size, blanching, drying, and irradiation. *Food Hydrocolloids*, *51*, 241–251.

Gentilini, R., Bozzini, S., Munarin, F., Petrini, P., Visai, L., & Tanzi, M. C. (2014). Pectins from aloe vera: Extraction and production of gels for regenerative medicine. *Journal of Applied Polymer Science*, *131*(2).

Georgiev, Y., Ognyanov, M., Yanakieva, I., Kussovski, V., & Kratchanova, M. (2012). Isolation, characterization and modification of citrus pectins. *Journal of BioScience & Biotechnology*, *1*(3).

Grassino, A. N., Brnčić, M., Vikić-Topić, D., Roca, S., Dent, M., & Brnčić, S. R. (2016). Ultrasound assisted extraction and characterization of pectin from tomato waste. *Food Chemistry*, *198*, 93–100.

Guo, X., Han, D., Xi, H., Rao, L., Liao, X., Hu, X., & Wu, J. (2012). Extraction of pectin from navel orange peel assisted by ultra-high pressure, microwave or traditional heating: A comparison. *Carbohydrate Polymers*, *88*(2), 441–448.

Güzel, M., & Akpınar, Ö. (2019). Valorisation of fruit by-products: Production characterization of pectins from fruit peels. *Food and Bioproducts Processing*, *115*, 126–133.

Gyawali, R., & Ibrahim, S. A. (2018). Addition of pectin and whey protein concentrate minimises the generation of acid whey in Greek-style yogurt. *Journal of Dairy Research*, *85*(2), 238–242.

Harholt, J., Suttangkakul, A., & Scheller, H. V. (2010). Biosynthesis of pectin. *Plant Physiology*, *153*(2), 384–395.

Harris, P. J., & Smith, B. G. (2006). Plant cell walls and cell-wall polysaccharides: Structures, properties and uses in food products. *International Journal of Food Science & Technology*, *41*, 129–143.

Ho, I. H., Matia-Merino, L., & Huffman, L. M. (2015). Use of viscous fibres in beverages for appetite control: A review of studies. *International Journal of Food Sciences and Nutrition*, *66*(5), 479–490.

Jackson, C. L., Dreaden, T. M., Theobald, L. K., Tran, N. M., Beal, T. L., Eid, M., ... Mohnen, D. (2007). Pectin induces apoptosis in human prostate cancer cells: Correlation of apoptotic function with pectin structure. *Glycobiology*, *17*(8), 805–819.

Jafari, F., Khodaiyan, F., Kiani, H., & Hosseini, S. S. (2017). Pectin from carrot pomace: Optimization of extraction and physicochemical properties. *Carbohydrate Polymers*, *157*, 1315–1322.

Jenkins, D. J., Leeds, A. R., Gassull, M. A., Cochet, B., & Alberti, K. G. M. (1977). Decrease in postprandial insulin and glucose concentrations by guar and pectin. *Annals of Internal Medicine*, *86*(1), 20–23.

Jeon, C., Kang, S., Park, S., Lim, K., Hwang, K. W., & Min, H. (2011). T cell stimulatory effects of Korean Red Ginseng through modulation of myeloid-derived suppressor cells. *Journal of Ginseng Research*, *35*(4), 462.

Kaczmarczyk, M. M., Miller, M. J., & Freund, G. G. (2012). The health benefits of dietary fiber: Beyond the usual suspects of type 2 diabetes mellitus, cardiovascular disease and colon cancer. *Metabolism*, *61*(8), 1058–1066.

Kaufmann, B., & Christen, P. (2002). Recent extraction techniques for natural products: Microwave-assisted extraction and pressurised solvent extraction. *Phytochemical Analysis: An International Journal of Plant Chemical and Biochemical Techniques*, *13*(2), 105−113.

Kazemi, M., Khodaiyan, F., & Hosseini, S. S. (2019). Utilization of food processing wastes of eggplant as a high potential pectin source and characterization of extracted pectin. *Food Chemistry*, *294*, 339−346.

Korhonen, H. J. (2011). Bioactive milk proteins, peptides and lipids and other functional components derived from milk and bovine colostrum. In *Functional foods* (pp. 471−511). Woodhead Publishing.

Košťálová, Z., Hromádková, Z., & Ebringerová, A. (2010). Isolation and characterization of pectic polysaccharides from the seeded fruit of oil pumpkin (*Cucurbita pepo* L. var. *Styriaca*). *Industrial Crops and Products*, *31*(2), 370−377.

Koubala, B. B., Christiaens, S., Kansci, G., Van Loey, A. M., & Hendrickx, M. E. (2014). Isolation and structural characterisation of papaya peel pectin. *Food Research International*, *55*, 215−221.

Koubala, B. B., Mbome, L. I., Kansci, G., Mbiapo, F. T., Crepeau, M. J., Thibault, J. F., & Ralet, M. C. (2008). Physicochemical properties of pectins from ambarella peels (*Spondias cytherea*) obtained using different extraction conditions. *Food Chemistry*, *106*(3), 1202−1207.

Kumar, A., & Chauhan, G. S. (2010). Extraction and characterization of pectin from apple pomace and its evaluation as lipase (steapsin) inhibitor. *Carbohydrate Polymers*, *82*(2), 454−459.

Kushi, L. H., Doyle, C., McCullough, M., Rock, C. L., Demark-Wahnefried, W., Bandera, E. V. . . . American Cancer Society 2010 Nutrition and Physical Activity Guidelines Advisory Committee. (2012). American Cancer Society Guidelines on nutrition and physical activity for cancer prevention: Reducing the risk of cancer with healthy food choices and physical activity. *CA: A Cancer Journal for Clinicians*, *62*(1), 30−67.

Leclere, L., Van Cutsem, P., & Michiels, C. (2013). Anti-cancer activities of pH-or heat-modified pectin. *Frontiers in Pharmacology*, *4*, 128.

Letellier, M., & Budzinski, H. (1999). Microwave assisted extraction of organic compounds. *Analusis*, *27*(3), 259−270.

Li, J., Zhu, Y., Yadav, M. P., & Li, J. (2019). Effect of various hydrocolloids on the physical and fermentation properties of dough. *Food Chemistry*, *271*, 165−173.

Lim, J., Ko, S., & Lee, S. (2014). Use of yuja (*Citrus junos*) pectin as a fat replacer in baked foods. *Food Science and Biotechnology*, *23*(6), 1837−1841.

Liu, L., Cao, J., Huang, J., Cai, Y., & Yao, J. (2010). Extraction of pectins with different degrees of esterification from mulberry branch bark. *Bioresource Technology*, *101*(9), 3268−3273.

Liu, Y., Dong, M., Yang, Z., & Pan, S. (2016). Anti-diabetic effect of citrus pectin in diabetic rats and potential mechanism via PI3K/Akt signaling pathway. *International Journal of Biological Macromolecules*, *89*, 484−488.

Lubbers, S., Decourcelle, N., Martinez, D., Guichard, E., & Tromelin, A. (2007). Effect of thickeners on aroma compound behavior in a model dairy gel. *Journal of Agricultural and Food Chemistry*, *55*(12), 4835−4841.

Luque-Garcia, J. L., & De Castro, M. L. (2003). Ultrasound: A powerful tool for leaching. *TrAC Trends in Analytical Chemistry*, *22*(1), 41−47.

Mahalik, N. P., & Nambiar, A. N. (2010). Trends in food packaging and manufacturing systems and technology. *Trends in Food Science & Technology*, *21*(3), 117−128.

Maneerat, N., Tangsuphoom, N., & Nitithamyong, A. (2017). Effect of extraction condition on properties of pectin from banana peels and its function as fat replacer in salad cream. *Journal of Food Science and Technology*, *54*(2), 386−397.

Maran, J. P., & Priya, B. (2015). Ultrasound-assisted extraction of pectin from sisal waste. *Carbohydrate Polymers*, *115*, 732−738.

Maran, J. P., Sivakumar, V., Thirugnanasambandham, K., & Sridhar, R. (2013). Optimization of microwave assisted extraction of pectin from orange peel. *Carbohydrate Polymers*, *97*(2), 703−709.

Maran, J. P., Sivakumar, V., Thirugnanasambandham, K., & Sridhar, R. (2014). Microwave assisted extraction of pectin from waste *Citrullus lanatus* fruit rinds. *Carbohydrate Polymers*, *101*, 786−791.

Marras-Marquez, T., Peña, J., & Veiga-Ochoa, M. D. (2015). Robust and versatile pectin-based drug delivery systems. *International Journal of Pharmaceutics*, *479*(2), 265–276.

Maxwell, E. G., Colquhoun, I. J., Chau, H. K., Hotchkiss, A. T., Waldron, K. W., Morris, V. J., & Belshaw, N. J. (2016). Modified sugar beet pectin induces apoptosis of colon cancer cells via an interaction with the neutral sugar side-chains. *Carbohydrate Polymers*, *136*, 923–929.

May, C. D. (1990). Industrial pectins: Sources, production and applications. *Carbohydrate Polymers*, *12*(1), 79–99.

Medeiros, B. G. D. S., Pinheiro, A. C., Carneiro-da-Cunha, M. G., & Vicente, A. A. (2012). Development and characterization of a nanomultilayer coating of pectin and chitosan—Evaluation of its gas barrier properties and application on 'Tommy Atkins' mangoes. *Journal of Food Engineering*, *110*(3), 457–464.

Meireles, M. A. A. (2003). Supercritical extraction from solid: Process design data (2001–2003). *Current Opinion in Solid State and Materials Science*, *7*(4–5), 321–330.

Minkov, S., Minchev, A., & Paev, K. (1996). Modelling of the hydrolysis and extraction of apple pectin. *Journal of Food Engineering*, *29*(1), 107–113.

Minzanova, S. T., Mironov, V. F., Arkhipova, D. M., Khabibullina, A. V., Mironova, L. G., Zakirova, Y. M., & Milyukov, V. A. (2018). Biological activity and pharmacological application of pectic polysaccharides: A review. *Polymers*, *10*(12), 1407.

Mohnen, D. (2008). Pectin structure and biosynthesis. *Current Opinion in Plant Biology*, *11*(3), 266–277.

Moorthy, I. G., Maran, J. P., Muneeswari, S., Naganyashree, S., & Shivamathi, C. S. (2015). Response surface optimization of ultrasound assisted extraction of pectin from pomegranate peel. *International Journal of Biological Macromolecules*, *72*, 1323–1328.

Mouécoucou, J., Sanchez, C., Villaume, C., Marrion, O., Frémont, S., Laurent, F., & Méjean, L. (2003). Effects of different levels of gum arabic, low methylated pectin and xylan on in vitro digestibility of β-lactoglobulin. *Journal of Dairy Science*, *86*(12), 3857–3865.

Müller-Maatsch, J., Bencivenni, M., Caligiani, A., Tedeschi, T., Bruggeman, G., Bosch, M., ... Sforza, S. (2016). Pectin content and composition from different food waste streams. *Food Chemistry*, *201*, 37–45.

Naqash, F., Masoodi, F. A., Rather, S. A., Wani, S. M., & Gani, A. (2017). Emerging concepts in the nutraceutical and functional properties of pectin—A review. *Carbohydrate Polymers*, *168*, 227–239.

Nara, K., Yamaguchi, A., Maeda, N., & Koga, H. (2009). Antioxidative activity of water soluble polysaccharide in pumpkin fruits (*Cucurbita maxima* Duchesne). *Bioscience, Biotechnology, and Biochemistry*, *73*(6), 1416–1418.

Noreen, A., Akram, J., Rasul, I., Mansha, A., Yaqoob, N., Iqbal, R., ... Zia, K. M. (2017). Pectins functionalized biomaterials; a new viable approach for biomedical applications: A review. *International Journal of Biological Macromolecules*, *101*, 254–272.

O'neill, M., Albersheim, P., & Darvill, A. (1990). The pectic polysaccharides of primary cell walls. In *Methods in plant biochemistry* (Vol. 2, pp. 415–441). Academic Press.

Ovodov, Y. S. (2009). Current views on pectin substances. *Russian Journal of Bioorganic Chemistry*, *35*(3), 269.

Pacheco, M. T., Villamiel, M., Moreno, R., & Moreno, F. J. (2019). Structural and rheological properties of pectins extracted from industrial sugar beet by-products. *Molecules*, *24*(3), 392.

Pedrolli, D. B., Monteiro, A. C., Gomes, E., & Carmona, E. C. (2009). Pectin and pectinases: Production, characterization and industrial application of microbial pectinolytic enzymes. *Open Biotechnology Journal*, 9–18.

Pietrzyk, L., Torres, A., Maciejewski, R., & Torres, K. (2015). Obesity and obese-related chronic low-grade inflammation in promotion of colorectal cancer development. *Asian Pacific Journal of Cancer Prevention*, *16*(10), 4161–4168.

Popov, S. V., & Ovodov, Y. S. (2013). Polypotency of the immunomodulatory effect of pectins. *Biochemistry (Moscow)*, *78*(7), 823–835.

Porta, R., Mariniello, L., Di Pierro, P., Sorrentino, A., & Giosafatto, C. V. L. (2011). Transglutaminase crosslinked pectin-and chitosan-based edible films: A review. *Critical Reviews in Food Science and Nutrition*, *51*(3), 223–238.

Prasanna, V., Prabha, T. N., & Tharanathan, R. N. (2007). Fruit ripening phenomena—An overview. *Critical Reviews in Food Science and Nutrition*, 47(1), 1–19.

Quoc, L. P. T., Huyen, V. T. N., Hue, L. T. N., Hue, N. T. H., Thuan, N. H. D., Tam, N. T. T., ... Duy, T. H. (2015). Extraction of pectin from pomelo (*Citrus maxima*) peels with the assistance of microwave and tartaric acid. *International Food Research Journal*, 22(4), 1637.

Renard, C. M., Crépeau, M. J., & Thibault, J. F. (1995). Structure of the repeating units in the rhamnogalacturonic backbone of apple, beet and citrus pectins. *Carbohydrate Research*, 275(1), 155–165.

Roselló-Soto, E., Galanakis, C. M., Brnčić, M., Orlien, V., Trujillo, F. J., Mawson, R., ... Barba, F. J. (2015). Clean recovery of antioxidant compounds from plant foods, by-products and algae assisted by ultrasounds processing. Modeling approaches to optimize processing conditions. *Trends in Food Science & Technology*, 42(2), 134–149.

Saha, D., & Bhattacharya, S. (2010). Hydrocolloids as thickening and gelling agents in food: A critical review. *Journal of Food Science and Technology*, 47(6), 587–597.

Sakai, T., Sakamoto, T., Hallaert, J., & Vandamme, E. J. (1993). Pectin, pectinase and protopectinase: Production, properties, and applications. *Advances in Applied Microbiology*, 39, 213–294.

Sandarani, M. D. J. C. (2017). A review: Different extraction techniques of pectin. *Journal of Pharmacognosy & Natural Products*, 3(3), 1–5.

Sarah, M., Hanum, F., Rizky, M., & Hisham, M. F. (2018). Microwave-assisted extraction of pectin from cocoa peel. *IOP Conference Series: Earth and Environmental Science*, 122, 012079.

Saura-Calixto, F. (2011). Dietary fiber as a carrier of dietary antioxidants: An essential physiological function. *Journal of Agricultural and Food Chemistry*, 59(1), 43–49.

Schols, H. A., Bakx, E. J., & Voragen, D. S. A. (1995). A xylogalacturonan subunit present in the modified hairy regions of apple pectin. *Carbohydrate Research*, 279, 265–279.

Shalaka, D., Naik, S. R., Amruta, A., & Parimal, K. (2009). Vitamin E loaded pectin alginate microspheres for cosmetic application. *Journal of Pharmacy Research*, 2(6).

Shams, K. A., Abdel-Azim, N. S., Saleh, I. A., Hegazy, M. E. F., El-Missiry, M. M., & Hammouda, F. M. (2015). Green technology: Economically and environmentally innovative methods for extraction of medicinal & aromatic plants (MAP) in Egypt. *Journal of Chemical and Pharmaceutical Research*, 7(5), 1050–1074.

Sharma, B. R., Naresh, L., Dhuldhoya, N. C., Merchant, S. U., & Merchant, U. C. (2006). Xanthan gum—A boon to food industry. *Food Promotion Chronicle*, 1(5), 27–30.

Sirisomboon, P., Tanaka, M., Fujita, S., & Kojima, T. (2000). Relationship between the texture and pectin constituents of Japanese pear. *Journal of Texture Studies*, 31(6), 679–690.

Sivam, A. S., Sun-Waterhouse, D., Waterhouse, G. I., Quek, S., & Perera, C. O. (2011). Physicochemical properties of bread dough and finished bread with added pectin fiber and phenolic antioxidants. *Journal of Food Science*, 76(3), H97–H107.

Smith, A. M., Moxon, S., & Morris, G. A. (2016). Biopolymers as wound healing materials. In *Wound healing biomaterials* (pp. 261–287). Woodhead Publishing.

Smith, R. M. (2002). Extractions with superheated water. *Journal of Chromatography A*, 975(1), 31–46.

Sriamornsak, P. (2003). Chemistry of pectin and its pharmaceutical uses: A review. *Silpakorn University International Journal*, 3(1–2), 206–228.

Srivastava, P., & Malviya, R. (2011). Sources of pectin, extraction and its applications in pharmaceutical industry—An overview. *Indian Journal of Natural Products and Resources*, 2, 10–18.

Sundar Raj, A. A., Rubila, S., Jayabalan, R., & Ranganathan, T. V. (2012). A review on pectin: Chemistry due to general properties of pectin and its pharmaceutical uses. *Scientific Reports*, 1, 550–551.

Torkova, A. A., Lisitskaya, K. V., Filimonov, I. S., Glazunova, O. A., Kachalova, G. S., Golubev, V. N., & Fedorova, T. V. (2018). Physicochemical and functional properties of *Cucurbita maxima* pumpkin pectin and commercial citrus and apple pectins: A comparative evaluation. *PLoS One*, 13(9).

Turchiuli, C., Fuchs, M., Bohin, M., Cuvelier, M. E., Ordonnaud, C., Peyrat-Maillard, M. N., & Dumoulin, E. (2005). Oil encapsulation by spray drying and fluidised bed agglomeration. *Innovative Food Science & Emerging Technologies*, 6(1), 29–35.

Valdés, A., Burgos, N., Jiménez, A., & Garrigós, M. C. (2015). Natural pectin polysaccharides as edible coatings. *Coatings, 5*(4), 865–886.

Vanitha, T., & Khan, M. (2019). Role of pectin in food processing and food packaging. In *Pectins—Extraction, purification, characterization and applications*. IntechOpen.

Venkatanagaraju, E., Bharathi, N., Sindhuja, R. H., Chowdhury, R. R., & Sreelekha, Y. (2019). Extraction and purification of pectin from agro-industrial wastes. In *Pectins—Extraction, purification, characterization and applications*. IntechOpen.

Venkatesh, M. S., & Raghavan, G. S. V. (2004). An overview of microwave processing and dielectric properties of agri-food materials. *Biosystems Engineering, 88*(1), 1–18.

Vinatoru, M. (2001). An overview of the ultrasonically assisted extraction of bioactive principles from herbs. *Ultrasonics Sonochemistry, 8*(3), 303–313.

Vincken, J. P., Schols, H. A., Oomen, R. J., Beldman, G., Visser, R. G., & Voragen, A. G. (2003). Pectin—The hairy thing. In *Advances in pectin and pectinase research* (pp. 47–59). Dordrecht: Springer.

Vogt, L. M., Sahasrabudhe, N. M., Ramasamy, U., Meyer, D., Pullens, G., Faas, M. M., . . . de Vos, P. (2016). The impact of lemon pectin characteristics on TLR activation and T84 intestinal epithelial cell barrier function. *Journal of Functional Foods, 22*, 398–407.

Voragen, A. G., Coenen, G. J., Verhoef, R. P., & Schols, H. A. (2009). Pectin, a versatile polysaccharide present in plant cell walls. *Structural Chemistry, 20*(2), 263.

Vriesmann, L. C., Teófilo, R. F., & de Oliveira Petkowicz, C. L. (2012). Extraction and characterization of pectin from cacao pod husks (*Theobroma cacao* L.) with citric acid. *LWT—Food Science and Technology, 49*(1), 108–116.

Wang, J., Hu, S., Nie, S., Yu, Q., & Xie, M. (2016). Reviews on mechanisms of in vitro antioxidant activity of polysaccharides. *Oxidative Medicine and Cellular Longevity, 2016*.

Wang, L., & Weller, C. L. (2006). Recent advances in extraction of nutraceuticals from plants. *Trends in Food Science & Technology, 17*(6), 300–312.

Wang, S., Chen, F., Wu, J., Wang, Z., Liao, X., & Hu, X. (2007). Optimization of pectin extraction assisted by microwave from apple pomace using response surface methodology. *Journal of Food Engineering, 78*(2), 693–700.

Wang, W., Ma, X., Xu, Y., Cao, Y., Jiang, Z., Ding, T., . . . Liu, D. (2015). Ultrasound-assisted heating extraction of pectin from grapefruit peel: Optimization and comparison with the conventional method. *Food Chemistry, 178*, 106–114.

Wicker, L., Kim, Y., Kim, M. J., Thirkield, B., Lin, Z., & Jung, J. (2014). Pectin as a bioactive polysaccharide—Extracting tailored function from less. *Food Hydrocolloids, 42*, 251–259.

Willats, W. G., Knox, J. P., & Mikkelsen, J. D. (2006). Pectin: New insights into an old polymer are starting to gel. *Trends in Food Science & Technology, 17*(3), 97–104.

Yang, B., Zhao, M., Shi, J., Yang, N., & Jiang, Y. (2008). Effect of ultrasonic treatment on the recovery and DPPH radical scavenging activity of polysaccharides from longan fruit pericarp. *Food Chemistry, 106*(2), 685–690.

Yang, J. S., Mu, T. H., & Ma, M. M. (2019). Optimization of ultrasound-microwave assisted acid extraction of pectin from potato pulp by response surface methodology and its characterization. *Food Chemistry, 289*, 351–359.

Yapo, B. M. (2011). Pectin rhamnogalacturonan II: On the "small stem with four branches" in the primary cell walls of plants. *International Journal of Carbohydrate Chemistry, 2011*.

Yapo, B. M., & Gnakri, D. (2015). Pectic polysaccharides and their functional properties. *Polysaccharides: Bioactivity and Biotechnology*, 1729–1749.

Yu, Y., Shen, M., Song, Q., & Xie, J. (2018). Biological activities and pharmaceutical applications of polysaccharide from natural resources: A review. *Carbohydrate Polymers, 183*, 91–101.

Zandleven, J., Sørensen, S. O., Harholt, J., Beldman, G., Schols, H. A., Scheller, H. V., & Voragen, A. J. (2007). Xylogalacturonan exists in cell walls from various tissues of *Arabidopsis thaliana*. *Phytochemistry, 68*(8), 1219–1226.

Zhang, T., Xiang, J., Zheng, G., Yan, R., & Min, X. (2018). Preliminary characterization and anti-hyperglycemic activity of a pectic polysaccharide from okra (*Abelmoschus esculentus* (L.) Moench). *Journal of Functional Foods, 41*, 19–24.

Zhang, W., Xu, P., & Zhang, H. (2015). Pectin in cancer therapy: A review. *Trends in Food Science & Technology, 44*(2), 258–271.

Zhang, W., Zhao, X. J., Jiang, Y., & Zhou, Z. (2017). Citrus pectin derived silver NPs and their antibacterial activity. *Inorganic and Nano-Metal Chemistry, 47*(1), 15–20.

Zhang, Y., Sun, T., & Jiang, C. (2018). Biomacromolecules as carriers in drug delivery and tissue engineering. *Acta Pharmaceutica Sinica B, 8*(1), 34–50.

CHAPTER 7

Xylan

Luciana Porto de Souza Vandenberghe, Kim Kley Valladares-Diestra, Gustavo Amaro Bittencourt, Ariane Fátima Murawski de Mello and Carlos Ricardo Soccol

Federal University of Paraná, Department of Bioprocess Engineering and Biotechnology, Centro Politécnico, Curitiba, Brazil

7.1 Introduction

Lignocellulosic biomass is composed by three basic polymers: lignin, cellulose, and hemicellulose. The hemicellulose composition varies depending on biomass type and provenience with cellulose ranging from 45% to 55%, hemicellulose from 25% to 35%, and lignin from 20% to 30%. Hardwood species, which include grasses and agroindustrial by products (cereals straws, sugarcane bagasse, corn stover, wood sawdust), are mostly composed of xylan, formed by β-(1→4)-linked D-xylopyranoside monomer units. Softwoods are mainly composed by glucomannans, including xylans that are smaller proportion, consisting of 5%—15% of plant dry weight (Deutschmann & Dekker, 2012).

Xylan is a type of hemicellulose that consists of β-D-xylopyranosyl (xylose) residues that are linked through β-1,4 glycosidic bonds (Van Dyk & Pletschke, 2012). This polysaccharide is considered the second most abundant polymer in the plant kingdom (Limayem & Ricke, 2012). The presence of xylan is not only restricted to the plant kingdom, it can also be found in other organisms such as algae. The structural diversity of xylan is related to its functionality within plant tissue. Besides, this diversity also serves to discriminate the type of plant, especially at the level of order classification within plants (Mussatto & Teixeira, 2010; Naidu, Hlangothi, & John, 2018).

The main purpose of this chapter is to present an overview of the latest research developments on xylan exploitation. As this plant polysaccharide has been extensively studied for its interesting characteristics, different extraction and purification technologies will be presented. Potential applications and developed processes of industrial products will also be shown, together with the new technologies that are being proposed by the scientific community and industrial sector.

7.2 Methods for xylan extraction

7.2.1 Sources and types of xylans

Hemicellulose is a heteropolymer composed of different monosaccharides such as pentose, hexose, acetylated sugars, and uronic acids. Different than cellulose, hemicellulose is less polymerized and does not have crystalline structures, on the contrary, it presents branched chains (Mussatto & Teixeira, 2010). Consequently the presence of these ramifications causes a wide variety of hemicellulose subclassification (Limayem & Ricke, 2012).

Xylan consists of a linear polymer of β-(1,4)-linked xylose unities switched by acetyl, glucuronic acid (GA), 4-O-methylglucuronic acid (MGA), and arabinose (A) residues (Rennie & Scheller, 2014). Xylan's chemical and structural composition and concentration vary according to different plants. In flowering plants, xylan is the predominant hemicellulose in secondary cell walls. In flowering and seeds'-producing plants, the ratio of GA and MGA substitutions is around two, and MGA substitutions are found at a ratio of 1:8 xylose residues. Other minor components that can be linked to the xylan structure are D-galactose (G) and ferulic acid (FEA) moieties linked to L-arabinose residues (Deutschmann & Dekker, 2012) (Fig. 7.1).

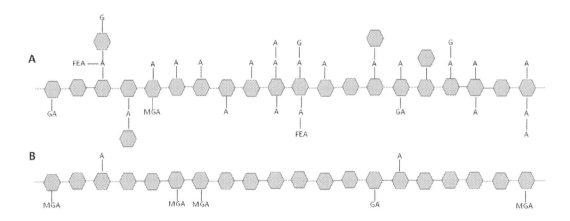

Figure 7.1
Chemical structures of xylans: (A) grasses and cereals, (B) softwoods, and (C) hardwoods (X—xylose, A—arabinose, G—galactose, GA—glucuronic acid, MGA—4-O-methyl-glucuronic acid, FEA—ferulic acid).

Within woody plant tissues such as dicotyledons and lignified tissues such as monocotyledons, xylan can represent 25%–35% of dry biomass, while in the case of seeds (agricultural crops) the percentage of xylan can reach 50% of dry biomass (Ebringerova & Heinze, 2000). In recent years, agricultural residues have shown great potential as sources of xylan, especially by-products such as wheat straw, sugarcane bagasse, corn cobs, soybean hulls, as well as waste from the wood industry.

The diversity of compounds that can bind to the main skeleton of xylan fiber form a large number of branch types, which are classified into four xylan subcategories: homoxylan, arabinoxylan, glucuronoxylan, and arabinoglucuronoxylan.

- Homoxylan: Consists of xylose residues, which are linearly arranged or branched, linked by glycosidic bonds and regularly distributed in pentameric form into the cell wall (Chakdar, Kumar, Pandiyan, & Singh, 2016; Ebringerová & Thomas, 2005). This type of xylan is found mainly in the cell wall of primitive plants such as green algae (Ebringerova & Heinze, 2000).
- Arabinoxylan: It is the largest fraction of hemicellulose, which is found in cereal crops, this polymer presents arabinose monosaccharides in addition to xylose. Arabinose is normally attached to the carbon 2 or 3 position of the xylose units that form the main chain of xylan (monosubstituted or disubstituted), reaching a degree of polymerization that varies from 50 to 185 monomer units. This type of xylan can also have substitutions on carbon 2 or 3 by acetyl groups (Ebringerová & Thomas, 2005; Zhou, Li, Mabon, & Broadbelt, 2016). One of the factors that most affect the degree of branching, and consequently, the physicochemical characteristics of arabinoxylan is the arabinose/xylose ratio, which can directly influence solubility against water or other solvents (Zhang, Smith, & Li, 2014).
- Glucuronoxylans: Like arabinoxylans, this type of xylan can have acetylated substitutions at positions 2 and 3 in the carbon of xylose units. It may also have substitutions at position 2 of the carbon of the main chain by GA residues. In most of the cases, xylose units are attached only to one GA residue (Gröndahl, Teleman, & Gatenholm, 2003). Substitution by chemical groups is responsible for different properties of this polysaccharide such as solubility, when substituted by acetyl, and resistance to alkaline substances, when substituted by GA (Van Dyk & Pletschke, 2012). The average degree of polymerization of glucuronoxylan can achieve from 150 to 200 monomer units, being considered the largest hemicellulose component in hardwoods and reaching 15%–30% of the dry mass weight (Gröndahl et al., 2003; Hilpmann et al., 2016).
- Arabinoglucuronoxylans: The substitutions found in this type of xylan are arabinofuranose and glucuronic acid, these residues are bonded on positions 2 or 3 of the backbone of xylose. The main chain may have some acetylation (Peng, Peng, Xu, & Sun, 2012). The frequency of substitution is given in approximately 1 unit of glucuronic

acid for every 5–6 units of xylose, while 1 arabinofuranose unit can be found for every 5–12 units of xylose (Yamasaki et al., 2012). Other residues, such as ferulic acid, can also be found attached to arabinose residues by means of an ester bond (Kozlova, Mikshina, & Gorshkova, 2012). This type of xylan is present in softwoods and mainly found in nonwoody plants such as grasses, where it reaches 40%–50% of the total hemicellulose (Sorek, Yeats, Szemenyei, Youngs, & Somerville, 2014).

- Xyloglucan: In this polysaccharide the backbone is formed by xylose and glucose residues, presenting branches of arabinose and galactose residues that are linked to xyloses (Gírio, Fonseca, Carvalheiro, Duarte, & Marques, 2010). This type of xylan is found mainly in nongraminaceous plants, presenting up to 75% of ramifications throughout the main chain. The mainly function of xyloglucans is its interaction with cellulose preserving the structural integrity of the cell wall. It represents 20% of the total dry mass of the cell wall composition (Gírio et al., 2010; Naidu et al., 2018).

7.2.2 Physicochemical characteristics

Xylan's solubility is mainly influenced by the intra- and intermolecular hydrogen bridge bonds. These bonds can undergo great modifications in the extraction and storage process, changing the xylan solubility. The strongest interaction between xylan fibers can occur mainly in chains with low frequency of substitutions, which present a greater amount of free hydrogens for the formation of bonds (Ebringerova & Heinze, 2000). For its part, the molecular weight and size of the xylan is also diverse, not having a determined pattern. It varies depending on the source and the recovery methodology used; however, the molecular weight distribution can be calculated by size exclusion chromatography.

Xylan's rheological behavior depends on the type and degree of substitution present in the main chain, for example, arabinoxylan containing phenolic acids can generally form gels after an oxidation process. This fact is due to the formation of cross bonds by the ferulic acid residues that are present in the side chains of different xylan molecules. On the other hand, the thermal behavior of cell wall xylan is comparable to the behavior of cellulose in higher plants, with thermal decomposition at 2008°C. Other studies also showed that arabinoxylan can inhibit the formation of ice, acting as a thermal protector in cereals. Regarding its biological activity, xylan can be used as a diet rich in fibers, improving some metabolic processes within the human or animal organism. In addition to xylan can act as hypocholesterolemic and immunological agent, controlling liver disorders, and even indirectly inhibiting the growth of tumors (Ebringerova & Heinze, 2000).

7.2.3 Motivation for xylan extraction from different sources

The agroindustrial sector generates 5 billion tons of waste per year (United Nations Environment Programme), which are often destined to nonsustainable uses, such as dumping to decompose, which requires land that could be used for different purposes, and burning for energy generation that releases carbon dioxide and other toxic substances to the environment (Naidu et al., 2018). However, these agroindustrial residues present great amounts of carbohydrates and other nutrients that can be used for the fabrication of high added value products (Bharathiraja, Suriya, Krishnan, Manivasagan, & Kim, 2017; Deutschmann & Dekker, 2012; Ravindran, Hassan, Williams, & Jaiswal, 2018). Therefore the generated waste is availed and enriched, and the productive chain is closed and sustainable.

To use and isolate the nutrients of this biomass, which is composed by lignocellulose, extraction methods can be applied. As previously mentioned, the hemicellulose fraction (that usually contains xylan) can be extracted by well-established methods. These methods can involve chemicals, like organic solvents and alkaline substances, and can be physical, such as steam explosion, ultrasonication, and screw extrusion (Naidu et al., 2018). Once the xylan is extracted, it can be used in several industries, as it will be described below.

7.2.4 Pretreatment methods for xylan extraction

Biomass has complex structural mechanisms and different chemical bonds that serve to prevent microbial and enzymatic degradation, generally known as biomass recalcitrance (Himmel et al., 2007). Many researchers have extensively studied this natural resistance, related to different biomass constituents, including cellulose, hemicellulose, and lignin, and developed different pretreatment strategies with and efficiencies, depending on employed feedstocks and process objectives. Being one of the most important stage to efficient conversion of biomass into products, the choice of pretreatment type and configuration can greatly improve overall cost (Prasad, Sotenko, Blenkinsopp, & Coles, 2016). Concerning the hemicellulose, the main factors that impede its extraction are the hydrogen bonds between individual hemicellulose units and its ether and ester linkages with lignin (Ebringerova & Heinze, 2000). Thus to explore the xylan fraction of biomass, it is necessary to preserve the hemicellulose polymers fraction while achieving maximum conversion yields, with low energy and less formation of toxic compounds.

In an integrated biorefinery facility, all the different biomass constituents of the feedstocks are processed in order to maximize the overall profitability. Hence major final product, source materials, conversion yields, and produced wastes affect upon possible other high-value bioproducts that can be manufactured (Deutschmann & Dekker, 2012). In this way, different pretreatments can be carried out with the use of chemical reagents, such as alkaline, acid, and organic solvent pretreatment, or specific equipment, which employ severe pressure and temperature, such as hydrothermal pretreatment. Each method has advantages and disadvantages related to the specific process necessity and used feedstock, therefore some strategies may present higher costs depending on the used equipment or chemicals and their recycling.

7.2.4.1 Alkaline pretreatment

In alkaline pretreatment, ester linkages between hemicellulose and lignin can be hydrolyzed, including aryl-ether and C—C bonds, liberating the hemicellulosic heteropolymers from the biomass matrix (Saratale & Oh, 2015). Sodium hydroxide, calcium hydroxide, potassium hydroxide, and ammonium hydroxide can be used, with the variation of reaction time, temperature, and reagent concentration. In comparison with other traditional pretreatment methods, alkaline pretreatment requires mild operating conditions, such as lower temperatures, room pressures, and has no necessity for reactors with corrosion resistance (Kumar, Bhardwaj, Agrawal, Chaturvedi, & Verma, 2020).

The alkaline pretreatment has different effectiveness depending on the biomass species, which is linked to the capacity to remove acetyl groups from the hemicellulose. It is most effective for hardwoods, where the largest hemicellulose component is glucuronoxylans, consisting of 15%—30% of its dry mass weight (Gröndahl et al., 2003). The disadvantages of alkaline pretreatment are longer periods of reaction time and necessity of neutralization, besides the potential formation of salts, which can be impregnated in the biomass and may hinder its recovery, causing extra costs of post-pretreatment steps (Singh, Suhag, & Dhaka, 2015; Wan, Zhou, & Li, 2011).

An et al. (2019) proposed a dilute hydrochloric acid and alkaline treatment (aqueous ammonia) wet oxidation of corn stover, in two steps, so as to surpass these disadvantages and improve sugar recovery (82.8% xylan and 71.5% glucan) and lignin removal (86.1%), with short reaction time (80 min). Despite the multiple steps and technologies in pretreatment strategies, one step of alkaline extraction can be suitable for xylan exploitation. Banerjee, Patti, Ranganathan, and Arora (2019) reached 96% of maximum hemicellulose recovery using 15% NaOH reaction, during 16 h, 45°C, and liquid—solid ratio (LSR) of 10 in a pineapple peel biorefining. The pretreated sample was utilized to produce xylose-rich liquor using nitric acid hydrolysis, with 91.3% of conversion yield. The same authors found similar results with less quantity of sodium hydroxide and shorter period of reaction time, implementing a one-step hydrothermal-assisted alkaline pretreatment. With a reaction time of 1.5 h, 121°C, 15 psi, and 10% NaOH they reached 87.6% of hemicellulose recovery. In an eco-friendly process, an enzymatic hydrolysis reached 25.7% of xylooligosaccharides (XOSs) from the hydrothermal-assisted alkaline pretreated pineapple peel.

Sporck et al. (2017) submitted sugarcane bagasse to an alkaline-sulfite pretreatment step (5% NaOH and 10% Na_2SO_3), testing different extraction methods, which the best condition (40% NaOH, 60°C, 2 h) reached 53% of xylan recovery. Svärd, Brännvall, and

Edlund (2015) extracted xylan rapeseed straw to produce rapeseed xylan films with strain-to-break values >60% without any added plasticizers. The authors compared extractions in conditions with and without sodium hydroxide, changing reaction times and temperatures, in the context of hydrothermal and alkaline pretreatment. Therefore extractions in the presence of sodium hydroxide reached higher yields of primarily xylan (33.2%–57.5%). Hydrothermal extraction, without the presence of sodium hydroxide, yielded fractions rich in galactoglucomannan and lesser concentrations of xylose (3.0%–9.3%). The authors also studied 0.1 mol NaOH L^{-1} concentration level, yielding 11.7%–16.9% of xylose. Thus it is possible to observe the importance of alkali reactants in xylan extraction processes.

7.2.4.2 Acid pretreatment

The acid pretreatment is the most extensively studied extraction technique for ethanol production processes. High concentration acid processes promote the hydrolysis of both hemicellulose and cellulose, using acid concentrations such as 72% sulfuric acid, 40% hydrochloric acid, and 100% trifluoracetic (Gírio, Carvalheiro, Duarte, & Bogel-Łukasik, 2012). Besides the low selectivity, concentrated acid pretreatments promote highly toxic compounds formation and highly equipment corrosion, engendering other schemes for acid pretreatment biorefinery (Gírio et al., 2012; Kumar et al., 2020).

Organic acids, such as acetic, formic, and oxalic acids, and inorganic acids, such as hydrochloric, maleic, nitric, nitrous, and sulfuric acid, are used in diluted reactions (Bai, Lant, Jensen, Astals, & Pratt, 2016). Higher temperatures, reaction times, and acid concentrations are factors that can be varied to improve pretreatment efficiency and prevent process inhibitory compounds formation and undesirable reactions. An opportune factor design may minimize furfural and hydroxymethyl furfural generation, as well as the formation of acetic acid and aromatic compounds (Kumari & Singh, 2018).

Thinking about the fact that acid pretreatment solubilizes mainly hemicellulose, it is a suitable strategy as first-step pretreatment in a multiple-step-designed lignocellulosic biorefinery, to fractionate hemicellulose before targeting process to major cellulosic product. Lai et al. (2019) used poplar sawdust to evaluate a sequencing pretreatment strategy. An acetic acid pretreatment was used as first step to produce XOSs, followed by an alkaline-assisted poly(ethylene glycol) diglycidyl ether pretreatment of the resulted solid material. The results showed that 59.4% of xylan was extracted (37.6% as XOS and 21.8% of xylose) with 5% acetic acid and 170°C during 30 min.

As long as lignin is more efficiently treated in alkaline media and organic solvents, acid medium pretreatments can be most efficient for the conversion of xylans to xylose bioproducts, combining effectiveness and short reaction times (Chandel et al., 2012; Vallejos & Area, 2017). Hemicellulose is composed of many different sugar chains with amorphous and branched structure that are easily degraded in acid medium with recovery of around 70%–95% of sugars, such as xylose, glucose, arabinose, galactose, and mannose, depending on reaction conditions (Delgado Arcaño, Valmaña García, Mandelli, Carvalho, & Magalhães Pontes, 2020). Mazlan et al. (2019) used nitric acid pretreatment to evaluate the influence of factors affecting xylan recovery, such as temperature, solid loading, pretreatment time, acid concentration, and agitation, in a factorial design. The results recommended that mild acid concentration conditions through increased time incubations improve xylan recovery. Their best condition (37°C, 24 h, 0.01% HNO_3) resulted in an acid pretreated oil palm frond bagasse with 27.63% of xylan presence. Cortivo, Hickert, Rosa, and Ayub (2020) applied a sulfuric acid pretreatment on a mixture of soybean and oat hulls to produce ethanol and xylitol, resulting in 28.2 g L^{-1} of xylose, 0.26 g L^{-1} of 5-hydroxymethylfurfural (HMF), and 0.09 g L^{-1} of furfural.

Liu et al. (2016) developed a two-step process of diluted acid pretreatment (0.7% HCl, 120°C, 40 min) and wet milling (15 min) of corn stover, resulting in 81% xylose, 64% glucose with the use of cellulase at 3 FPU g^{-1} of substrate. These results were attained because of the improved accessibility of cellulase after hemicelluloses degradation and increase in pores' volume.

7.2.4.3 Organosolv pretreatment

Organosolv pretreatment utilizes organic solvent such as short-chain aliphatic alcohols (ethanol, methanol), polyhydric alcohols (ethylene glycol, glycerol, and triethylene glycol), alkylene carbonates, or mixtures of those with or without catalysts addition (acids or bases) (Zhang et al., 2016). The organic solvents solubilize lignin and hemicellulose structure's bonds. When the addition of catalysts takes place, the rate and extent of hemicellulose hydrolysis and lignin α- and β-aryl ether linkages cleavages are enhanced, achieving high rates of hemicellulose recovery and pure lignin extraction (Ferreira & Taherzadeh, 2020). The main disadvantages of organosolv pretreatment is the solvent volatility and the necessity to recycle it, so as to be a cost-effective process, besides its inflammability, which has safety concerns bringing extra costs to preventions (Kumar et al., 2020).

However, valorization strategies of organosolv pretreated hemicellulosic fractions include mainly studies toward production of enzymes, ethanol, and methane, out of biorefinery concept (Lú-chau et al., 2018; Nozari, Mirmohamadsadeghi, & Karimi, 2018; Smichi, Messaoudi, & Gargouri, 2018; Teramura et al., 2018). Smichi et al. (2018) and Teramura et al. (2018) employed a three-stage pretreatment strategy on new feedstocks as *Juncus maritimus* and *Retama raetam*. In the first step, the authors used 80% (vol.:vol.) phosphoric acid medium during 24 h; in the second step, 71.4% acetone was added to the mixture; and in the third step, the resulted sample was mixed with 200 mL acetone, before targeted the resulted solid residue to ethanol fermentation process. The hemicellulosic fraction reached 100% and was used to produce xylanases reaching 0.44 U mL^{-1} of enzyme activity. Due to many steps to produce a hemicellulosic-rich fraction and the safety risks related to the solvents, the entirely process may be expensive for industrial applications. Cebreiros, Clavijo, Boix, Ferrari, and Lareo (2020) evaluated ethanol organosolv pretreatment in eucalyptus sawdust, at 180°C, with a reaction time of 45 min, LSR of 8% and 50% ethanol concentration. Xylose extraction reached 42.8% in organosolv liquor; after lignin precipitation, 6.6 g of XOS production was achieved from 100 g of eucalyptus sawdust. The same authors compared this result to autohydrolysis pretreatment, reaching higher results of xylan extraction of 49.4% (32.9% of xylose and 16.5% of XOS).

7.2.4.4 Hydrothermal pretreatment

Hydrothermal pretreatment uses water as solvent for hydrolysis reactions with no need for chemical addition. The process uses high temperatures (150°C–220°C) and, frequently, pressures (145–290 psi) generated from water steam in proper reactors. The increased temperature of the reaction leads to pKw water decrease leading to the formation of hydronium ions that catalyze the hydrolysis of hemicellulose acetyl groups. The consequent decrease of pH reaction promotes hemicellulose's ether bonds cleavage, releasing sugars from it (Meighan et al., 2017). Differently from steam explosion, which releases the reaction pressure abruptly, autohydrolysis employs high pressures to avoid water evaporation at high temperatures (Marques, Silva, Lomonaco, de F. Rosa, & Leitão, 2020).

Cheng, Li, Feng, Zhan, and Xie (2014) employed autohydrolysis (160°C, 210 min, LSR of 10) resulting in 70.2% of xylan extracted, which was mainly composed of XOS. The authors observed that higher temperatures promoted similar results of xylan recovery in shorter periods of time (in the range of 150°C–170°C), even though higher severity factors resulted information of furfural and HMF, diminishing the xylose production. Bittencourt et al. (2019) pretreated sugarcane bagasse with autohydrolysis (183°C, 41 min, LSR of 3.9) reaching 73% of hemicellulose extraction with 20.9 g L^{-1} of XOS and 15.3 g L^{-1} of xylose in the extracted liquor, which was used to produce biomethane.

7.2.4.5 Steam explosion pretreatment

Comparatively to hydrothermal pretreatment, steam explosion works with a sudden discharge of pressure reaction causing an explosion that disrupts biomass cell wall structure and separate its individual fibers (Pielhop, Amgarten, Von Rohr, & Studer, 2016). This enable shorter periods of time reactions and increased surface area and porosity level of solid residues. Typically, steam exploded biomasses produce xylan extracts in oligomers and monomers forms and have great results of hemicellulose valorization (Silveira, Chandel, Vanelli, Sacilotto, & Cardoso, 2018).

Bhatia et al. (2020) evaluated *Miscanthus* (*Miscanthus sinensis* and *M. sacchariflorus*) to XOS production through steam explosion pretreatment (200°C, 218 psi, 10 min) leading to 52% conversion yield related to initial xylan. Mihiretu, Chimphango, & Görgens (2019) employed a two-stage pretreatment in sugarcane trash and aspenwood with an alkaline impregnation of 5% NaOH before steam explosion (204°C, 10 min). Maximum xylan recovery of 51% and 24% was achieved for sugarcane trash and aspenwood, respectively. The authors concluded that this pretreatment strategy was suitable to produce xylan economically and with reduced formation of monomeric sugars and degradation products thereof.

7.2.4.6 Subcritical or supercritical fluids

The pretreatment of lignocellulosic biomass by means of subcritical or supercritical fluids can be carried out with or without the use of catalysts (Fu et al., 2019). Usually, this method destabilizes the xylan and lignin fractions, making them soluble in liquid media, for this reason most of the products obtained by these methodologies are derived from the main polysaccharides of plant biomass. In the case of xylan, the main derivatives obtained by this methodology are small fibers, XOS, and fermentable sugars (mainly xylose) (Coelho, Rocha, Saraiva, & Coimbra, 2014; Liu, Wei, & Wu, 2019).

Coelho et al. (2014) used brewers' spent grain to obtain arabinoxylan and arabinoxylo-oligosaccharides (AXOS), by means of a sequential three-step thermal degradation process with the following conditions: (1) 140°C 2 min at 4 bar pressure; (2) 180°C, 2 min at 8 bar pressure, and (3) 0.1 M KOH 180°C at 27 bar pressure; obtaining 62% of arabinoxylan and AXOS. They also observed that the efficiency of obtaining AXOS was directly related to the increase in temperature, however higher temperatures promoted depolymerization, deesterification, and generated the formation of brown products. Liu et al. (2019) obtained oligosaccharides and fermentable sugars from corn straw through the application of autohydrolysis assisted by subcritical CO_2 (170°C 5 MPa CO_2 for 40 min) achieving an XOS yield of 90.2% in the prehydrolysate, and the functional XOS was approximately 40% in total XOS, with generation of few fermentation inhibitors.

7.2.4.7 Ionic liquids and deep eutectic solvent

Ionic liquids (ILs) are solvents that are integrally composed of ions, not volatile and easily dissolve in lignocellulosic biomass under mild conditions. Due to their both ionic and organic properties, these solvents can effectively reduce the intractability of biomass by altering the lignin/carbohydrate complex and decreasing the recalcitrance of plant biomass. IL are included within green solvents (Fu et al., 2019). On the other hand, deep eutectic solvents (DESs) have great potential as a method to be applied in the extraction of lignocellulosic fractions, since they have ideal physical and chemical properties for dissolution of polysaccharides, such as cellulose and hemicellulose. They are considered an alternative to IL (Vigier, Chatel, & Jérôme, 2015). DES generally have a component that acts as hydrogen-bond acceptor and another component that acts as hydrogen-bond donor. The main characteristics of DES are: easily synthesized, renewable in nature, biocompatible, biodegradable, and profitable (Loow et al., 2018).

Wang, Gräsvik, Jönsson, and Winestrand (2017) found that the imidazole-based ionic liquid with an anionic component [HSO_4] generated greater hydrolysis of xylan in the pretreatment of xylano-rich hardwood, generating the hydrolysis of more than a half of the xylan in products as xylose or XOS. Morais et al. (2018) used DES (choline chloride and urea) to extract xylan from hardwood, demonstrating the successful extraction of xylan from *Eucalyptus globulus* wood. They also found that uronic acid fractions were excised from the xylan skeleton during pretreatment process.

7.2.4.8 Microwave-assisted extraction

Microwave-assisted extraction (MAE) treatment is an alternative method for the extraction and separation of lignocellulosic fractions, which can generate selective isolation of lignocellulosic fractions, presenting high extraction performance, low cost, without significant changes in molecular structures (Fu et al., 2019). Due its selective heating capacity, MAE leads to high solubility of polymers, creation of hot spots, that when in alkaline solutions, generate alkali and fibers interaction. This generates an explosive effect that separates the fibers from lignocellulosic biomass (Panthapulakkal, Kirk, & Sain, 2015; Singh, Bhuyan, Banerjee, Muir, & Arora, 2017).

Panthapulakkal et al. (2015) demonstrated that, under alkaline extraction (NaOH), a low microwave power (110 W) applied to wood birch in different treatment times, promotes an extraction efficiency of 60% xylan, with different degrees polymerization. This process demonstrated the same xylan extraction efficiency compared to alkaline methods, however xylan recovery can take less time (10-fold) when applying MAE. Singh et al. (2017) showed that the efficiency recovery of hemicellulose in areca nut hulk increased proportionally with the application of microwave power, highlighting the importance of the pretreatment of biomass in alkaline medium, obtaining a total recovery of 52% of hemicellulose. Despite the low efficiency of hemicellulose recovery, the application of MAE method shows a significant delignification of the material ($\sim 82\%$) and, above all, a significant saving of time in the pretreatment process.

7.2.4.9 Biopulping treatment

Biopulping methods are generally applied to lignocellulosic biomass to release cellulose fractions and, in a few cases, hemicellulose. The process consists mainly in lignin and hemicellulose fractions degradation by the action of wood-degrading fungi, and, in some cases, bacteria or filamentous fungi are also employed (Pathak, Kaur, & Bhardwaj, 2016). These organisms can degrade the cell wall selectively, promoting the obtainment of pulp rich in cellulose and/or hemicellulose (Bajpai, 2018). Once the lignocellulosic substrate is colonized, the production and release of different enzymes starts for substrate hydrolysis. The composition and type of enzymes will depend on the employed organism (Singh, 2018). However, this process needs the maintaining of optimal conditions for organisms' growth and takes long time for an efficient biomass treatment. On the other hand, the direct enzyme treatment of lignocellulosic materials can be more efficient and take less time than biopulping. This methodology is based on the use of pure or mixed enzymes, which will act according to the objective to be achieved, having the advantage of being more specific to each treated biomass (Lin, Wu, Zhang, Liu, & Nie, 2018).

The main enzymes that can be used in the xylan extraction process are laccase, lignin peroxidase, and manganese peroxidase. These enzymes are capable of degrading lignocellulosic materials through the oxidation of the different components of lignin (Bharathiraja et al., 2017). On the other hand, the use of cellulases could hydrolyze the cellulose remaining fraction allowing the xylan to be released. However, this enzymatic hydrolysis method is not widely studied, since xylan represents a low percentage of the total biomass and its complex structure turns difficult its efficient degradation and release of cellulose fraction (Queiroz, Awan, & Tasic, 2016). Even so, enzymatic treatments reduce the use of toxic and aggressive chemicals and present high extraction efficiencies under optimal conditions. For this reason, the combination of physicochemical and enzymatic treatments is outlined with great potential in the recovery process of lignocellulosic fractions, allowing the progressive reduction of the use of chemical products to find a balance between performance, productivity, and reduction of environmental impacts. Some methods for xylan extraction from different sources are presented in Table 7.1.

7.3 Bioproducts obtained from xylan

The recovery of agroindustrial by-products is certainly a sustainable way to produce a large diversity of industrial products (Table 7.2). Hemicellulose, which is composed of xylan, can be transformed into different products with commercial applications. With the increasing demand for sustainable sources of energy, xylan can be applied in the production of biofuels. For this application, the polysaccharide needs to be hydrolyzed with acids or enzymes into monomers that will be fermented by microorganisms, resulting in bioethanol (Beltrán-Ramírez et al., 2019). Lactic acid and furfural can also be produced from xylan. Lactic acid follows the same route as bioethanol, with microorganisms fermenting the monomers obtained from hydrolysis of xylan. For furfural production, the xylose molecule needs to be dehydrated (Naidu et al., 2018).

In the food industry, xylan can be biotransformed to XOSs and xylitol. XOSs are oligomers of xylose that presents great organoleptic characteristics for incorporation in food. Besides, XOSs are classified as prebiotics due to its biological properties—such as favoring the growth of the gastrointestinal microbiota. An innovative approach is to combine XOS with probiotics, generating synbiotics (Sedlmeyer, 2011; Singh et al., 2015). Xylitol presents sweetness and can substitute common sugar in food and it's obtained by the conversion of xylose with chemical or fermentative routes. For the manufacture of these products, xylan

Table 7.1: Methods for xylan extraction and their final products.

Source	Methods	Conditions	Yield	Final product	Reference
Corn stover	Acid and alkaline wet oxidation	120°C, 40 min, 1% HCl and 130°C, 12.6% NH_4OH, 3 MPa O_2, 40 min	82.8%	Hemicellulose	An et al. (2019)
Pineapple peel	Alkaline pretreatment	45°C, 16 h, 15% NaOH	91.3%	Xylose	Banerjee et al. (2019)
Pineapple peel	Hydrothermal-assisted alkaline pretreatment	121°C, 1.5 h, 15 psi, 10% NaOH	25.7%	XOS	Banerjee et al. (2019)
Sugarcane bagasse	Two-stage alkaline pretreatment	5% NaOH, 10% Na_2SO_3 and 60°C, 2 h 40% NaOH	53%	Xylan recovery	Sporck et al. (2017)
Rapeseed straw	Alkaline pretreatment	105°C, 20 min, 0.5 M NaOH	57.3%	Plastic films	Svärd et al. (2015)
Oil palm frond bagasse	Acid	37°C, 24 h, 0.01% HNO_3; LSR of 20	27.63%[a]	Xylan recovery	Mazlan et al. (2019)
Poplar sawdust	Acid	170°C, 30 min, 5% CH_3COOH	37.6%	XOS	Lai et al. (2019)
Soybean and oat hulls mixture	Acid	121°C, 20 min, 1% H_2SO_4, LSR of 10	28.2[b]	Xylitol	Cortivo et al. (2020)
Juncus maritimus and *Retama raetam*	Three-stage acid/organosolv strategy	Complex conditions on each step	100%	Hemicellulose extraction	Smichi et al. (2018)
Eucalyptus sawdust	Organosolv	180°C, 45 min, 50% EtOH, LSR of 8	42.8%	Xylose	Cebreiros et al. (2020)
Eucalyptus sawdust	Hydrothermal	180°C, 45 min, LSR of 8	49.3%	Xylan recovery	Cebreiros et al. (2020)
Corn stover	Hydrothermal	160°C, 210 min, LSR of 10	70%	Xylose	Cheng et al. (2014)
Sugarcane bagasse	Hydrothermal	183°C, 41 min, LSR of 3.9	73%	Hemicellulose extraction	Bittencourt et al. (2019)
Sugarcane stalks	Steam explosion	204°C, 10 min, LSR of 10	51%	XOS	Mihiretu et al. (2019)
Aspenwood	Steam explosion	204°C, 10 min, LSR of 10	24%	XOS	Mihiretu et al. (2019)
Miscanthus	Steam explosion	200°C, 218 psi, 10 min, LSR of 10	52%	XOS	Bhatia et al. (2020)
Brewers' spent grain	Supercritical water and MAVE	140°C at 4 bar; 180°C at 8 bar and 0.1 M KOH 180°C at 27 bar	62%	Arabinoxylan and AXOS	Coelho et al. (2014)
Corn straw	Subcritical CO_2	170°C, 5 MPa CO_2, 40 min	40%	XOS and fermentable sugar	Liu et al. (2019)
Hardwood (aspen)	Ionic liquid $[C_4C_1im][HSO_4]$	1/20 sample/IL for 20 h at 100°C	~50%	XOS and fermentable sugar	Wang et al. (2017)

(Continued)

Table 7.1: (Continued)

Source	Methods	Conditions	Yield	Final product	Reference
Eucalyptus globulus	Deep eutectic solvent	50% w/v DES in water for 24 h at 90°C solid/liquid 1/25	~90%	Xylan	Morais et al. (2018)
Birchwood	MAVE	3 g wood fiber 1/10 solid–liquid ratio 4% NaOH 110 W	60%	Xylan	Panthapulakkal et al. (2015)
Arecanuthusk	MAVE	900 W 3 min 1/10 solid liquid ratio 15% NaOH	52%	Hemicellulose	Singh et al. (2017)

[a]Global yield, related to total composition of raw material.
[b]Results in g/L, mass balance and yields not showed.

needs to be hydrolyzed to smaller molecules (Venkateswar Rao, Goli, Gentela, & Koti, 2016). The C5 and C6 sugar fractions can be used in fermentation processes to produce different commercially important biomolecules with adapted strains such as biofuels and organic acids.

In the pharmaceutical industry, xylan can be applied in hydrogels that are composed by cross-linked polymer chains and can be applied for drug-delivery systems since xylan is not toxic and presents sensibility to pH alterations, which is advantageous for a targeted release. Xylan can also form biofilms and be used in packaging of oils and fats due to its hydrophilicity (da Silva et al., 2012; Deutschmann & Dekker, 2012; Naidu et al., 2018) (Fig. 7.2).

7.3.1 Xylitol

Xylitol ($C_5H_{12}O_5$) is a polyol is the most widely produced xylan-derived product with great applicability (Table 7.3). It possesses a high sweetening power with 40% less calories than sucrose (De Albuquerque, Da Silva, De MacEdo, & Rocha, 2014) that is why it is largely employed against dental caries and a usual component of chewing gums, toothpastes, and diabetic products (Deutschmann & Dekker, 2012). In the Second World War the lack of sugars supply raised the interest for xylitol production and commercialization as a substitute of sugars (Ur-Rehman, Mushtaq, Zahoor, Jamil, & Murtaza, 2015). Xylitol exists in fruits and vegetables at concentrations lower than 1%. That is why the direct extraction is economically inviable. The richest sources are strawberries, raspberries, berries, corn husks, oats, mushrooms, hardwoods, wood shavings, peel cottonseed, nuts, straw, stems, corncobs, or sugarcane bagasse that contain around 20%–35% xylan (Delgado Arcaño et al., 2020).

A fourfold increase of xylitol production was observed in the last four decades. In 2016, the estimated global market for xylitol was 190.9 thousand metric tons, with a market value of US$725.9 million. For 2022, the estimated production may reach 266.5 thousand metric

Table 7.2: Examples of production/preparation processes of xylan-based products—sources and techniques.

Product	Substrate	Microorganism	Production/Preparation process	Reference
Bioethanol	Oil palm empty fruit bunch	*Saccharomyces cerevisiae* INVSc1	Batch fermentation mode	Liu, Peng, Huang, & Geng (2020)
	Sugarcane bagasse	*Pichia stipitis*	Batch fermentation mode	Phaiboonsilpa, Chysirichote, & Champreda (2020)
	Corn stover	*S. cerevisiae*SyBE00	Batch fermentation mode	Li et al. (2018)
	Khejurer Rosh (an overnight natural fermented date palm)	*Saccharomyces cerevisiae*	Batch fermentation mode	Talukdera et al. (2019)
Xylanases	Sugarcane bagasse	Microorganism consortia	Batch fermentation mode	Evangelista, Kadowaki, Mello, & Polikarpov (2018)
	Corncob	*Penicillium chrysogenum* QML-2	Batch fermentation mode	Zhang and Sang (2015)
	Wheat bran	*Aspergillus niger* CCUG33991	Solid-state fermentation (tray bioreactor)	Khanahmadi, Arezi, Amiri, & Miranzadeh (2018)
	Wheat bran	*T. lanuginosus* VAPS24	Batch fermentation mode	Kumar and Shukla (2018)
	Oil palm empty fruit bunch	*Aspergillus niger*	Solid-state fermentation	Ajijolakewu, Peng, Nadiah, & Abdullah (2017)
	Raw oil palm frond leaves	*Rhizopus oryzae* UC2	Solid-state fermentation	Ezeilo, Abdul, & Arafat (2020)
Hydrogels	Wheat straw		Extraction with toluene, alkali treatment, and redox initiation system	Sun et al. (2013)
	Beech wood xylan		Cross-linking copolymerization of xylan with N-isopropylacrylamide (NIPAm) and acrylic acid (AA) using N,Nı́-methylenebis-acrylamide (MBA)	Gao et al. (2016)
	Wheat straw		Fe_3O_4 nanoparticles prepared using chemical coprecipitation method	Sun, Liu, Jing, & Wang (2015)
	Xylan		Polymerized hydrogels were prepared using EGDE cross-linker with a varying molar feed composition of xylan and βCD	Gami, Kundu, Dileep, Seera, & Banerjee (2020)

(Continued)

Table 7.2: (Continued)

Product	Substrate	Microorganism	Production/Preparation process	Reference
Packing material—xylan films	Xylan obtained from oat spelt, with glycerol or sorbitol as plasticizer		Solvent casing	Mikkonen et al. (2009)
	Xylan		Long-chain anhydrides using solvent casting method	Zhong, Peng, Yang, Cao, & Sun (2013)
XOS	Kenaf stem (Hibiscus cannabinus)		Enzymatic hydrolysis xylanase: arabinofuranosidase (Xyn2: AnabfA)	Izyan et al. (2016)
	Hazelnut shell		Autohydrolysis 190°C and 30 min of holding time	Surek and Buyukkileci (2017)
	Meranti wood sawdust		Xylanase was immobilized by a combination of entrapment and covalent binding techniques with 60 h of hydrolysis	Sabrina, Sukri, & Sakinah (2018)
	Poplar		Poplar was pretreated by hydrogen peroxide–acetic acid (HPAC) with H_2SO_4 as catalyst to remove lignin, and the solid residues were used to produce xylooligosaccharides (XOS) by two-step xylanase and cellulase hydrolysis	Hao et al. (2020)
Xylitol	Vegetable waste	Candida athensensis SB18	Batch mode (bioreactor)	Zhang, Geng, Yao, Lu, & Li (2012)
	Corncob	Candida tropicalis CCTCC M2012462	Fed-batch mode (bioreactor)	Ping, Ling, Song, & Ge (2013)
	Sugarcane bagasse	Debaryomyces hansenii	Batch mode (immobilized cells)	Prakash, Varma, Prabhune, Shouche, & Rao (2011)
	Sunflower stalks	Hansenula polymorpha ATCC 34438	Batch mode (bioreactor)	Lourdes, Sánchez, & Bravo (2019)
	Corn stover	Pichia stipitis NRRL Y-30785	Batch mode (Erlenmeyer flasks)	Valderez et al. (2014)

tons, with a value of US$1 billion (Chandel et al., 2012). Xylitol production is concentrated in Asia Pacific, Europe, the United States, and Australia. The expansion of its production in other countries is limited probably due to the costly manufacturing. Xylitol can be produced from lignocellulosic biomasses. However, at industrial level, it is produced through

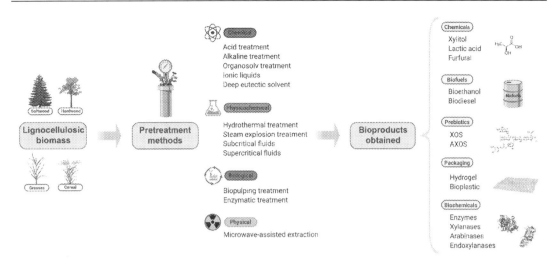

Figure 7.2
Biobased products obtained from xylan.

Table 7.3: Physical and chemical properties of xylitol (De Albuquerque et al., 2014).

Formula	$C_5H_{12}O_5$
Molecular weight	152.15 g mol^{-1}
Odor	none
Appearance	White, crystalline
Solubility at 20°C	169 g/100 g H_2O
pH in water	5–7
Melting point	93°C–94.5°C
Boiling point (at 760 mmHg)	216°C
Density (15°C)	1.50 g L^{-1}
Viscosity	10%—1.23 cP; 40%—4.18 cP; 50%—8.04 cP; 60%—20.63 cP
Caloric value (4.06 cal g^{-1})	16.88 J g^{-1}
Heat of solution	Endothermic 36.61 cal g^{-1} (−157.1 kJ kg^{-1})
Hygroscopicity	More hygroscopic than sucrose
Optical activity	Optically active
Moisture adsorption (%)	At 60% relative humidity 0.55
4 days, 20°C–22°C	At 92% relative humidity 90
Relative sweetness	Equal to sucrose, greater than sorbitol and mannitol

Source: *Modified from De Albuquerque T.L., I.J. Da Silva, G.R. De MacEdo, M.V.P. Rocha. (2014). Biotechnological production of xylitol from lignocellulosic wastes: A review. Process Biochemistry, 49, 1779–1789.*

chemical processes including the acid hydrolysis to xylose; purification of the obtained hydrolysate; catalytic hydrogenation of xylose and, finally the purification of xylitol. All these steps certainly lead to high production costs and, consequently, a high price of final product (Delgado Arcaño et al., 2020).

Chemical production of xylitol is carried out through the catalytic hydrogenation of xylose, which is obtained from the hydrolysis of birch wood (Delgado Arcaño et al., 2020). Concerns about the environment have forced researchers to find alternatives to produce xylitol and use xylan resources. In fact, biotechnological synthesis of xylitol is a very attractive way to replace chemical processes because it occurs under mild conditions. Besides, it can be based on sugars from lignocellulosic hydrolysates, xylan, to reduce energy and costs of substrate purification (Dasgupta, Bandhu, Adhikari, & Ghosh, 2017).

Xylitol is synthetized via glycolysis pathway in living organisms including plants and animals. Biotechnological conversion is mostly accomplished by bacteria, fungi, and yeast that are capable of fermenting xylose. An alternative for xylitol production is cell free enzymatic reduction or immobilized enzyme reaction (Dasgupta et al., 2017). Xylitol production from yeasts presents some bottlenecks that are related to low product yields (Sun, Byung, Dae, & Doo, 2005). However, selective gene manipulation to increase productivity has been studied with the use of increased knowledge about metabolic mapping including the optimization of selective xylose transporter system to increase xylose uptake rate, overexpression of xylose reductase to maximize reduction of xylose into xylitol, and others (Dasgupta et al., 2017).

7.3.2 Bioethanol

Bioethanol and biofuels are still the main products of biorefineries. First-generation bioethanol production is carried through direct fermentation of sucrose from sugarcane and sugarbeet juices. The enzymatic conversion of starchy sources, such as corn starch, to glucose with further bioethanol production is also largely employed. However, second-generation (2G) biofuels market has grown 50% between 2014 and 2020 and, in 2020, its value has been estimated in US$ 23.9 billion (UNCTAD, 2016). The global installed capacity of 2G ethanol is 1390.48 (million liters). The main players are the USA (35%), China (24%), Canada (22%), European Union (9%), and Brazil (9%).

Research has been conducted on 2G ethanol production from cellulose and hemicellulose, which are obtained from agricultural and forestry waste residues. Some technical difficulties of saccharification and fermentation of these polysaccharides including have been subject of intense research (Deutschmann & Dekker, 2012). In general, hemicelluloses are first partially hydrolyzed with major part of lignin. After pretreatment step, cellulose fibrils are accessible to subsequent enzymatic saccharification. The most employed pretreatment methods include dilute or concentrated acids treatment, alkalis treatment; treatments with SO_2, H_2O_2 or steam-explosion/autohydrolysis, ammonia fiber explosion (AFEX), and CO_2 explosion processes. Each pretreatment must be studied and optimized and tested for each biomass source and present some advantages and disadvantages. According to the type of lignocellulosic residue and treatment, some compounds of degradation may be generated, such as furfural, hydroxymethyl furfural, organic acids, and phenolics, which may inhibit ethanol fermentation process (Almeida, Runquist, Sànchez Nogué, Lidén, & Gorwa-Grauslund, 2011; Chandra et al., 2007; Deutschmann & Dekker, 2012).

Saccharomyces cerevisiae and *Zymomonas mobilis* strains are usually employed in alcoholic fermentation in the direct conversion of glucose into ethanol. However, these species are not able to consume xylose and other pentoses (Canio, Bari, & Patrizi, 2011). Besides, xylose consumer microorganisms are moderate producers of ethanol and they are also sensitive to fermentation subproducts and may have a slow fermentation rate (Rao, Bhadra, & Shivaji, 2008). Research must then be focused on: new genetic engineering technologies to develop new strains with these capacities and higher productivities, use of C5 sugars, either for fermentation or upgrading to valuable coproducts, use of lignin as value-adding energy carrier or material feedstock, feedstock handling, and processing in cellulosic plants.

7.3.3 Hydrogels

Hydrogels are polymer chains that are cross-linked through chemical or physical reactions. These bioderived and biocompatible materials present good perspectives of application by different industrial sectors such as cosmetics, tissue engineering, drug delivery, insulation, and gas storage (Aaltonen & Jauhiainen, 2009). The global market for hydrogels reached $15.6 billion in 2016 with an annual growth rate of 6.3% from 2017 to 2022, it should reach a total of $22.3 billion by 2022 (Report Buyer, 2021).

The formation of porous three-dimensional structures (gels and foams) is achieved with the help of polymeric material that are cross-linked or are capable of establishing a strong network. Xylan can be used to ameliorate some properties of the gel such as increased porosity, higher thermal stability, and improved mechanical properties. In most cases, xylan is mixed with other hydrophilic polymers (cellulose or other polysaccharides), or modified and cross-linked to another polymeric compound (Deutschmann & Dekker, 2012; Fonseca Silva, Habibi, Colodette, & Lucia, 2011).

Hydrogels have the ability to retain large amounts of water such as living tissue making them suitable for different biomedical applications such as drug delivery for drugs' encapsulation because they present some characteristics and may act on immunological defense, inhibition of cell mutation, and present anticancer and antioxidant properties (Gao et al., 2016). So they can be employed for targeted release of drugs (Chimphango, Van Zyl, & Görgens, 2012; Mihiretu et al., 2019) as they are resistant to digestion in the human stomach and are only hydrolyzed by enzymes of the human colon (Naidu et al., 2018; Sun, Wang, Jing, & Mohanathas, 2013). The use of hydrogels loaded with drugs was carried out for in vitro tests and revealed that the pH sensitivity and biodegradability proved their suitability as oral drug carriers with varying amounts of N-isopropylacrylamide, acrylic acid, and N,N′-methylene bis-acrylamide (Gao et al., 2016). Hydrogels that are composed of polysaccharides, which are obtained from renewable resources, have several advantages over synthetic ones due to the variety in their chemical structure that cannot be reproduced in laboratory. Xylan on chitosan-based hydrogels were applied in bone tissue regeneration where the hydrogel was liquid at room temperature and then injected into mice and a rat. The addition of xylan to the chitosan hydrogel improved, compared to chitosan alone, the rate of recovery of rodents that suffered from various bone injuries (Bush, Liang, Dickinson, & Botchwey, 2016; Naidu et al., 2018). *Propionibacterium acidipropionici* cells were immobilized in a xylan-based disulfide-cross-linked hydrogel matrix reinforced with cellulose nanocrystals. The study was performed with continuous cultivation using different dilution rates with better productivities of propionic acid (Wallenius et al., 2015).

7.3.4 Packaging

It is a fact that the use of bioplastics is increasing due to environmental legislations and conscious concerns, waste management, and attractive properties. The packaging industry represents almost 70% (1.2 million tons) of the total bioplastics market, remaining the single largest field of application for bioplastics (Naidu et al., 2018). In the last decades, the development of renewable and biodegradable materials for packaging films highly increased due to their interesting physical properties such as low oxygen permeability, water resistance, mechanical strength, and flexibility (Deutschmann & Dekker, 2012).

Hemicelluloses obtained from agricultural subproducts are interesting alternatives for food-packaging sector when compared to oil-based polymers (Deutschmann & Dekker, 2012). However, xylan's poor film-forming abilities may be a limiting factor for some applications. Hence it needs to be combined with other components to achieve certain abilities (Naidu et al., 2018; Ren et al., 2015), such as chitosan, cellulose nanofibers and other fibers, glucomannans, or polyvinyl alcohol, which promote film formation and raise permeability barrier properties (Stevanic, Bergström, Gatenholm, Berglund, & Salmén, 2012). The film is formed even if alkali-soluble lignin is not completely removed during purification (Goksu, Karamanlioglu, Bakir, Yilmaz, & Yilmazer, 2007), but its brownish color and almost complete solubility in water may limit the spectrum of applications. The film-forming capacity of xylan depends on the degree of polymerization, heterogeneity, and chemical structure of the blending compounds, which have a strong influence on the crystallinity of samples that is dependent on the regularity of xylan chain (Deutschmann & Dekker, 2012).

Xylan extracted from barley husk presents film-forming capacity without the addition of a plasticizer (Höije, Gröndahl, Tømmeraas, & Gatenholm, 2005), which suggests that xylan's structure strongly influences its film-forming abilities. Xylans are hydrophilic, but do not act as good barriers for water (Stevanic et al., 2011; Stevanic et al., 2012). Other properties of xylan are low oxygen permeability, aroma permeability, and high light transmittance that favor its application for packaging (Mikkonen & Tenkanen, 2012).

The combination of PVA and xylan in a 3:1 ratio was tested with the addition of ammonium zirconium carbonate. Higher ammonium zirconium concentrations promoted increased hydrophobicity and decreased solubility of films with stable water vapor permeability and increased elongation at break. These characteristics are necessary for packaging applications (Chen, Ren, & Meng, 2015). Stevanic et al. (2011) studied the combination of xylan with nanofibrillated cellulose at different concentrations to form films with improvement of moisture sorption, tensile properties, and oxygen barrier properties with higher concentrations of this polymer.

Interesting characteristics of xylan- and mannan-based films such as better oxygen, grease, and aroma barrier were observed by Mikkonen and Tenkanen (Mikkonen & Tenkanen, 2012). This material could be applied for packaging of low-moisture foods, coatings on fruit, cheese, or paper. In fact, the major drawback of xylans and mannans' films is their sensitivity to moisture. Alekhina, Mikkonen, Alén, Tenkanen, and Sixta (2014) studied the synthesis of carbomethyl xylan from bleached birch Kraft pulp with the use of sodium monochloroacetate in different ratios. Some characteristics of the material were observed depending on the concentration of sodium monochloroacetate, such as oxygen permeability. Xylan and chitosan films were composed at different ratios using the solvent casting technique. Higher xylan contents led to increased hydrophilicity, decreased oxygen permeability, tensile strength, and water vapor transmission rate of the films (Luo, Pan, Ling, Wang, & Sun, 2014).

Wang, Ren, Li, Sun, and Liu (2014) tested the addition of citric acid to PVA and xylan films. The effect of different citric acid concentration on some physical properties of films was observed. The higher citric acid content, the lower was the degree of swelling and a slight increase in the solubility of films and elongation at break. Lower water vapor permeability and tensile strength were reported with increasing loading of citric acid. The improved characteristics of xylan films with citric acid make them suitable for packaging applications (Naidu et al., 2018).

7.3.5 Xylooligosaccharides

XOSs are short saccharides chain, derived from xylan and formed by xylose residues joined through beta-1,4 bonds. XOSs' structures vary according to the degree of polymerization (number of xylose residues in the main chain), degree of substitution (arabinose/xylose ratio), monomer units, and type of bond. The number of residues involved in the formation of XOS can vary from 2 to 10 xyloses [such as xylobiose ($C_{10}H_{18}O_9$), xylotriose ($C_{15}H_{26}O_{13}$), xylotetraose ($C_{20}H_{34}O_{17}$), and others] (Broekaert et al., 2011; De Maesschalck et al., 2015; Morgan, Wallace, Bedford, & Choct, 2017).

Xylan's enzymatic hydrolysis allows the natural generation of XOS, which are intermediate products, which can be produced at industrial scale using arabinoxylan extracted from vegetable biomass (Rantanen et al., 2007; Rivière et al., 2014). However, fibers from plants' cell wall have a significant amount of lignin that can interfere with extraction processes and production of oligosaccharides leading to additional extraction processes, for example, the combination of physicochemical methods with enzymatic hydrolysis (Moore & Jung, 2001). For this reason, lignocellulosic biomass delignification process is crucial for higher degree of enzyme digestibility (Vermaas et al., 2015).

The worldwide market for XOS is expected to grow 4.1% reaching 120 million USD in 2024, from 94 million USD in 2019 (MarketWatch, 2020). Among different oligosaccharides, XOSs show great pharmaceutical potential and may be applied as additives in animal feed formulations (Carvalho, de O Neto, da Silva, & Pastore, 2013). In fact, XOS consumption promotes healthy and functional maintenance of the digestive system of monogastric organisms. The selective fermentation of XOS brings positive effects to the composition and activity of the gastrointestinal microbiota, so they can be defined as high value prebiotics and food additive ingredients. These prebiotic properties are strongly influenced by the XOS production method, degree of polymerization and substitution. Numerous studies have demonstrated the protective effect of XOS as prebiotics in animals against *Salmonella*, *E. coli*, and *Listeria monocytogenes* (de M de LeBlanc, Castillo, & Perdigon Gabriela, 2010; Gill, Shu, Lin, Rutherfurd, & Cross, 2001; Silva et al., 2004). The antimicrobial activity promoted an increase in resistance against infectious bacteria, which contributes to the prevention of diseases and a greater survival of animals (Licht, Ebersbach, & Frøkiær, 2012). Also, XOSs have the ability to contribute with a selective growth of probiotic bacteria. Some studies have shown that XOSs are the most effective prebiotics required for the growth and development of *Bifidobacterium lactis* Bb12, *Lactobacillus plantarum*, and *L. paracasei* (Adamberg et al., 2014; Shin, Lee, Pestka, & Ustunol, 2000). The fermentation of XOS by the genus *Bifidobacterium* and *Lactobacillus* may stimulate the synthesis of secondary metabolites that are benefic for the host (Belghith, Dahech, Belghith, & Mejdoub, 2012).

XOSs are can be used in the production of low-calorie sweeteners without changing other organoleptic characteristics. In addition, due to the stability over a wide temperature and pH range and prebiotic properties, XOS can be exploited by the food industry (Lasrado & Gudipati, 2014).

7.3.6 Enzymes

The global enzymes market reached a value of USD 9.9 billion in 2019. It is expected to grow 7.1% from 2020 to 2027 due to increasing demand from the food and beverage, biofuel, and animal feed industries (Grand View Research, 2020). Due to the great complexity of xylan, different hydrolytic enzymes can be produced such as α-arabinofuranosidase (EC3.2.1.55), acetylxylan esterase (EC3.1.1.72), α-glucuronidase (EC3.2.1.139), β-xylosidase (EC3.2.1.37), and endoxylanase (EC3.2.1.8) (Rahman, Sugitani, Hatsu, & Takamizawa, 2003; Selvarajan & Veena, 2017). The cited enzymes act in synergy for xylan conversion into simpler and easily metabolizable sugar, with xylanases being the most active enzymes in xylan degradation (Hu, Arantes, & Saddler, 2011; Su et al., 2013). Among xylanases, endoxylanases are very important because they act on the breakage of glycosidic bonds and the release of small chains of XOS (Dey & Roy, 2018).

Xylanases' production is carried out by a high variety of organisms from bacteria, fungi, and protozoa to higher organisms, such as marine algae, insects, gastropods, arthropods, and plants (Andrade, Santana, & Motta, 2013). Fungi and bacteria are the most employed for their production where filamentous fungi stands out with xylan as an inducer. The process's cost of xylanases production from pure xylan is a critical factor at industrial scale. For this reason, lignocellulosic biomass, with high concentrations of xylan and without previously chemical treatment, can be used in xylanases production. Sources rich in natural xylan such as sugarcane bagasse, corncob, and wheat bran are used in xylanases production (Di Marco, Soraire, Romero, Villegas, & Martínez, 2017; Elegbede & Lateef, 2018; Lal, Dutt, Kumar, & Gautam, 2015). This allows the development of simple, efficient, and economically viable strategies for xylanases production.

Xylanases have a wide range of applications in different industrial fields such as processing of pulp and fibers, saccharification of vegetable biomass, flour improvement for bakery products, food additives, and improvement of the nutritive properties in animal feed, extraction of vegetable oils, biobleaching of wood pulp, as an alternative for the treatment of textile-cellulosic residues, and industrial and municipal residues (Polizeli et al., 2005).

7.4 Advances and innovation

In order to reveal the innovation tendencies and the state of art of xylan extraction and application, patents search, and analysis were conducted. Lens.org database was used for patents search and terms associated with each topic (extraction and application) were combined using Boolean language in order to obtain the best result. The analyzed period was from 2000 to April 2020. The graphics were generated using Python programming language with Anaconda3 software. Therefore the following analysis was done: how many patents documents were applied per year and per country, which documents had more citations, which are the most recent, what is the profile of the depositor, and which are the enterprises that play a significant role in this area are shown hereafter. The refined databases of xylan application and extraction contained 228 and 112 entries, respectively.

As it is observed in Fig. 7.3A, there is a peak of applied documents in 2011 justified (31 documents) by some actions of Tate & Lyle in previous years such as of the opening

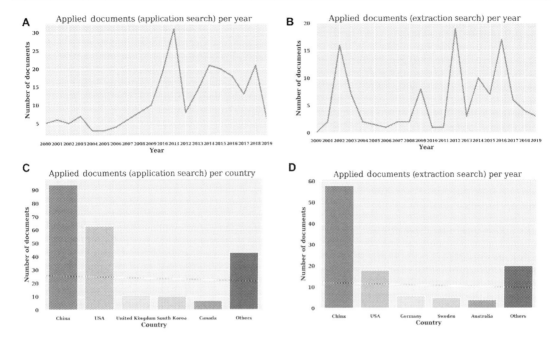

Figure 7.3
Number of patents and depositions: (A) xylan application by year; (B) xylan extraction by year; (C) xylan application by country; and (D) xylan extraction by country.

industrial plants with DuPont in Tennessee, R&D centers in Shanghai (2007), and an Innovation Center in Lille, France (2008). Besides, in 2011, Evonik invested on polymers for medical and pharmaceutical applications. In 2013 inaugurated an Innovation Center for research in cosmetics justifying the growth between 2012 and 2014. In Fig. 7.3B, peaks can be observed. In 2002 (16 documents) the peak is justified by the production of bioethanol by Lantamännen that started in 2001. CH Bioforce, in 2011, developed a new process for hemicellulose extraction (using vacuum), maintaining its polymeric form and, in 2016 joined forces with Oy Chemec Ab in order to replace fossil and food-based materials by biobased products. These facts explain the peaks in 2012 (19 documents) and 2016 (17 documents), respectively. In both graphics, there is a decay from 2019 to 2020 related to the time that the document stays in secrecy (18 months).

In Figs. 7.3C,D, the five countries that had more documents are shown (China, USA, United Kingdom, South Korea, Canada, and others). This result matches the expected since these countries are great technology developers (World Intellectual Property Organization) and express interest in reducing environmental impact and in evolving sustainable processes (Morgan et al., 2017). As it can be observed, the number of obtained documents of application of xylan is higher when compared to the number of obtained documents of extraction methods. This happens because the methods of extraction, as previously mentioned, are well established and present high yields. Therefore there are more research efforts on the application of xylan and more cooperation between inventors and sectors of society, since each of them can contribute, particularly, for technology development. Besides, inventors tend to deposit in regional or world offices and use systems that make the deposit process easier, rather than depositing in individual offices of each desired country.

The most cited patent in xylan application is US20010020091 (Buchanan, Buchanan, Debenham, Shelton, & Matthew, 2010) that presents methods for extracting xylan from corn cobs and its transformation in esters and ethers, while the most recent patent [GB 2576122 (Wei & Shannon, 2020)] applies hemicellulose for a tissue sheet. For xylan extraction, the most cited patent is US20040108085 (Kettenbach & Stein, 2007) that proposes a biomass treatment with an aqueous solution using complex substances, making the hemicellulose soluble. The most recent document is US20200040110 (Vilaplana & Ruthes, 2016) that comprises a hydrothermal enzymatic method for hemicellulose extraction.

7.5 Environmental aspects

Global waste generation reaches 12.73 kg capita^{-1} day^{-1} of industrial wastes, 3.35 kg capita^{-1} day^{-1} of agriculture wastes, and 0.74 kg capita^{-1} day^{-1} of municipal wastes (Usmani et al., 2020). Part of this large quantities of solid residues is burnt or transported to dump sites for disposal. While these residues decompose, they would still take up space that could be used for other purposes. Burning of agricultural waste on the other hand generates unnecessary carbon dioxide and often leads to the release of chemicals such as polycyclic aromatic hydrocarbons and dioxins (Naidu et al., 2018). Biobased biorefineries with the use of agricultural or food-processing wastes streams have attracted attention. In fact, new possibilities erase from these subproducts with the development of processes for the production of value-added products. This means that lignocellulosic biomass can be used as clean and renewable source of feedstock for the production of chemicals and biocomposites. Xylan appears as a potential component for the production of different biobased products with significant reduction of environmental impact.

7.6 Conclusions and perspectives

Recent developments have proven the interest of the global scientific and economic sectors to find solutions for the reuse of lignocellulosic biomass, which is a serious environmental and economic problem all over the world. Xylan, one of its constituents, appears as a great source for potential production of different value-added biobased products, chemicals, packaging materials, drug delivery, and biomedical products. Different chemical, physicochemical, and biological pretreatments were developed and are well stablished for xylan extraction. That is the reason why new technologies are now more focused on xylan fractions' application. This fact can be observed by the high number of deposed patents by the academy and industrial sectors, mainly by big technology developers, which express the interest in reducing the environmental impact and in establishing new sustainable processes.

References

Aaltonen, O., & Jauhiainen, O. (2009). The preparation of lignocellulosic aerogels from ionic liquid solutions. *Carbohydrate Polymers*, 75, 125−129.

Adamberg, S., Sumeri, I., Uusna, R., Ambalam, P., Kondepudi, K. K., Adamberg, K., ... Ljungh, Å. (2014). Survival and synergistic growth of mixed cultures of bifidobacteria and lactobacilli combined with prebiotic oligosaccharides in a gastrointestinal tract simulator. *Microbial Ecology in Health and Disease*, 25.

Ajijolakewu, A. K., Peng, C., Nadiah, W., & Abdullah, W. (2017). Biocatalysis and agricultural biotechnology optimization of production conditions for xylanase production by newly isolated strain *Aspergillus niger* through solid state fermentation of oil palm empty fruit bunches. *Biocatalysis and Agricultural Biotechnology*, 11, 239−247.

Alekhina, M., Mikkonen, K. S., Alén, R., Tenkanen, M., & Sixta, H. (2014). Carboxymethylation of alkali extracted xylan for preparation of bio-based packaging films. *Carbohydrate Polymers*, 100, 89−96.

Almeida, J. R. M., Runquist, D., Sànchez Nogué, V., Lidén, G., & Gorwa-Grauslund, M. F. (2011). Stress-related challenges in pentose fermentation to ethanol by the yeast Saccharomyces cerevisiae. *Biotechnology Journal*, 6, 286−299.

An, S., Li, W., Liu, Q., Xia, Y., Zhang, T., Huang, F., ... Chen, L. (2019). Combined dilute hydrochloric acid and alkaline wet oxidation pretreatment to improve sugar recovery of corn stover. *Bioresource Technology*, 271, 283−288.

Andrade, C. C. P., Santana, M. H. A., & Motta, F. L. (2013). A review of xylanase production by the fermentation of xylan: Classification, characterization and applications. In A. K. Chandel, & S. S. da Silva (Eds.), *Sustainable degradation of lignocellulosic biomass—Techniques, applications and commercialization* (pp. 251−275). Croatia: IntechOpen.

Bai, X., Lant, P. A., Jensen, P. D., Astals, S., & Pratt, S. (2016). Enhanced methane production from algal digestion using free nitrous acid pre-treatment. *Renewable Energy*, 88, 383−390.

Bajpai, P. (2018). *Biotechnology for pulp and paper processing*. Singapore: Springer.

Banerjee, S., Patti, A. F., Ranganathan, V., & Arora, A. (2019). Hemicellulose based biorefinery from pineapple peel waste: Xylan extraction and its conversion into xylooligosaccharides. *Food and Bioproducts Processing*, 117, 38−50.

Belghith, K. S., Dahech, I., Belghith, H., & Mejdoub, H. (2012). Microbial production of levansucrase for synthesis of fructooligosaccharides and levan. *International Journal of Biological Macromolecules*, 50, 451−458.

Beltrán-Ramírez, F., Orona-Tamayo, D., Cornejo-Corona, I., Luz Nicacio González-Cervantes, J, de Jesús Esparza-Claudio, J., & Quintana-Rodríguez, E. (2019). Agro-industrial waste revalorization: The growing biorefinery. *Biomass bioenergy—Recent trends and future challenges*. IntechOpen.

Bharathiraja, S., Suriya, J., Krishnan, M., Manivasagan, P., & Kim, S.-K. (2017). Production of enzymes from agricultural wastes and their potential industrial applications. In S.-K. Kim, & F. Toldrá (Eds.), *Marine enzymes biotechnology: Production and industrial applications, Part III—Application of marine enzymes* (pp. 125−148). Academic Press, Inc.

Bhatia, R., Winters, A., Bryant, D. N., Bosch, M., Clifton-brown, J., Leak, D., & Gallagher, J. (2020). Pilot-scale production of xylo-oligosaccharides and fermentable sugars from Miscanthus using steam explosion pretreatment. *Bioresource Technology*, 296, 122285.

Bittencourt, G. A., da, E., Barreto, S., Brandão, R. L., Baêta, B. E. L., & Gurgel, L. V. A. (2019). Fractionation of sugarcane bagasse using hydrothermal and advanced oxidative pretreatments for bioethanol and biogas production in lignocellulose biorefineries. *Bioresource Technology, 292*, 121963.

Broekaert, W. F., Courtin, C. M., Verbeke, K., van de Wiele, T., Verstraete, W., & Delcour, J. A. (2011). Prebiotic and other health-related effects of cereal-derived arabinoxylans, arabinoxylan-oligosaccharides, and xylooligosaccharides. *Critical Reviews in Food Science and Nutrition, 51*, 178−194.

Buchanan C. M., Buchanan N. L., Debenham J. S., Shelton M. C., Wood M. D. (2010). Corn fiber for the production of advanced chemicals and materials:arabinoxylan and arabinoxylan derivatives made therefrom. *U.S. Patent & Application No. 6388069(B1)*.

Bush, J. R., Liang, H., Dickinson, M., & Botchwey, E. A. (2016). Xylan hemicellulose improves chitosan hydrogel for bone tissue regeneration. *Polymers for Advanced Technologies, 27*, 1050−1055.

Canio, D., Bari, D., & Patrizi, R. (2011). Latest frontiers in the biotechnologies for ethanol production from lignocellulosic biomass. *Biofuel production—Recent developments and prospects*. IntechOpen.

Carvalho, A. F. A., de O Neto, P., da Silva, D. F., & Pastore, G. M. (2013). Xylo-oligosaccharides from lignocellulosic materials: Chemical structure, health benefits and production by chemical and enzymatic hydrolysis. *Food Research International, 51*, 75−85.

Cebreiros, F., Clavijo, L., Boix, E., Ferrari, M., & Lareo, C. (2020). Integrated valorization of eucalyptus sawdust within a biorefinery approach by autohydrolysis and organosolv pretreatments. *Renewable Energy, 149*, 115−127.

Chakdar, H., Kumar, M., Pandiyan, K., & Singh, A. (2016). Bacterial xylanases: Biology to biotechnology. *3 Biotech, 6*(2), 150.

Chandel, A. K., Antunes, F. A. F., de Arruda, P. V., Milessi, T. S. S., da Silva, S. S., & das G de A Felipe, M. (2012). Dilute acid hydrolysis of agro-residues for the depolymerization of hemicellulose: State-of-the-art. *D-Xylitol: Fermentive production, application and commercialization* (pp. 1−345). Springer.

Chandra, R. P., Bura, R., Mabee, W. E., Berlin, A., Pan, X., & Saddler, J. N. (2007). Substrate pretreatment: The key to effective enzymatic hydrolysis of lignocellulosics? *Advances in Biochemical Engineering/Biotechnology, 108*, 67−93.

Chen, X. F., Ren, J. L., & Meng, L. (2015). Influence of ammonium zirconium carbonate on properties of poly (vinyl alcohol)/xylan composite films. *Journal of Nanomaterials, 2015*.

Cheng, H., Li, J., Feng, Q., Zhan, H., & Xie, Y. (2014). Hot water extraction of corn stover: Hemicellulose fractionation and its effect on subsequent soda-AQ pulping. *BioResources, 9*, 2671−2680.

Chimphango, A. F. A., Van Zyl, W. H., & Görgens, J. F. (2012). In situ enzymatic aided formation of xylan hydrogels and encapsulation of horse radish peroxidase for slow release. *Carbohydrate Polymers, 88*, 1109−1117.

Coelho, E., Rocha, M. A. M., Saraiva, J. A., & Coimbra, M. A. (2014). Microwave superheated water and dilute alkali extraction of brewers' spent grain arabinoxylans and arabinoxylo-oligosaccharides. *Carbohydrate Polymers, 99*, 415−422.

Cortivo, P. R. D., Hickert, L. R., Rosa, C. A., & Ayub, M. A. Z. (2020). Conversion of fermentable sugars from hydrolysates of soybean and oat hulls into ethanol and xylitol by *Spathaspora hagerdaliae* UFMG-CM-Y303. *Industrial Crops and Products, 146*.

Dasgupta, D., Bandhu, S., Adhikari, D. K., & Ghosh, D. (2017). Challenges and prospects of xylitol production with whole cell bio-catalysis: A review. *Microbiological Research, 197*, 9−21.

da Silva, A. E., Marcelino, H. R., Gomes, M. C. S., Oliveira, E. E., Nagashima-Jr, T., & Egito, E. S. T. (2012). *Xylan, a promising hemicellulose for pharmaceutical use. Products and applications of biopolymers* (p. 220) IntechOpen.

De Albuquerque, T. L., Da Silva, I. J., De MacEdo, G. R., & Rocha, M. V. P. (2014). Biotechnological production of xylitol from lignocellulosic wastes: A review. *Process Biochemistry, 49*, 1779−1789.

de M de LeBlanc, A., Castillo, N. A., & Perdigon Gabriela, G. (2010). Anti-infective mechanisms induced by a probiotic *Lactobacillus* strain against *Salmonella enterica* serovar *Typhimurium* infection. *International Journal of Food Microbiology, 138*, 223−231.

De Maesschalck, C., Eeckhaut, V., Maertens, L., De Lange, L., Marchal, L., Nezer, C., ... Van Immerseel, F. (2015). Effects of xylo-oligosaccharides on broiler chicken performance and microbiota. *Applied and Environmental Microbiology, 81*, 5880–5888.

Delgado Arcaño, Y., Valmaña García, O. D., Mandelli, D., Carvalho, W. A., & Magalhães Pontes, L. A. (2020). Xylitol: A review on the progress and challenges of its production by chemical route. *Catalysis Today, 344*, 2–14.

Deutschmann, R., & Dekker, R. F. H. (2012). *From plant biomass to bio-based chemicals: Latest developments in xylan research, Biotechnology Advances* (30, pp. 1627–1640).

Dey, P., & Roy, A. (2018). Molecular structure and catalytic mechanism of fungal family G acidophilic xylanases. *3 Biotech, 8*, 1–13.

Di Marco, E., Soraire, P. M., Romero, C. M., Villegas, L. B., & Martínez, M. A. (2017). Raw sugarcane bagasse as carbon source for xylanase production by *Paenibacillus* species: A potential degrader of agricultural wastes. *Environmental Science and Pollution Research, 24*, 19057–19067.

Ebringerova, A., & Heinze, T. (2000). Xylan and xylan derivatives—Biopolymers with valuable properties, 1. Naturally occurring xylans structures, isolation procedures and properties. *Macromolecular Rapid Communications, 21*(9), 542–556.

Ebringerová, A., & Thomas, H. (2005). *Hemicellulose* (pp. 1–67). Springer.

Elegbede, J. A., & Lateef, A. (2018). Valorization of corn-cob by fungal isolates for production of xylanase in submerged and solid state fermentation media and potential biotechnological applications. *Waste and Biomass Valorization, 9*, 1273–1287.

Evangelista, D. E., Kadowaki, M. A. S., Mello, B. L., & Polikarpov, I. (2018). Biochemical and biophysical characterization of novel GH10 xylanase prospected from a sugar cane bagasse compost-derived microbial consortia. *International Journal of Biological Macromolecules, 109*, 560–568.

Ezeilo, U. R., Abdul, R., & Arafat, N. (2020). Optimization studies on cellulase and xylanase production by *Rhizopus oryzae* UC2 using raw oil palm frond leaves as substrate under solid state fermentation. *Renewable Energy, 156*, 1301–1312.

Ferreira, J. A., & Taherzadeh, M. J. (2020). Improving the economy of lignocellulose-based biorefineries with organosolv pretreatment. *Bioresource Technology, 299*.

Fonseca Silva, T. C., Habibi, Y., Colodette, J. L., & Lucia, L. A. (2011). The influence of the chemical and structural features of xylan on the physical properties of its derived hydrogels. *Soft Matter, 7*, 1090–1099.

Fu, G.-Q., Hu, Y.-J., Bian, J., Li, M.-F., Peng, F., & Sun, R.-C. (2019). Isolation, purification, and potential applications of xylan. *Production of materials from sustainable biomass resources*. Springer.

Gami, P., Kundu, D., Dileep, S., Seera, K., & Banerjee, T. (2020). Chemically crosslinked xylan—β-Cyclodextrin hydrogel for the in vitro delivery of curcumin and 5-fluorouracil. *International Journal of Biological Macromolecules, 158*, 18–31.

Gao, C., Ren, J., Zhao, C., Kong, W., Dai, Q., Chen, Q., ... Sun, R. (2016). Xylan-based temperature/pH sensitive hydrogels for drug controlled release. *Carbohydrate Polymers, 151*, 189–197.

Gill, H. S., Shu, Q., Lin, H., Rutherfurd, K. J., & Cross, M. L. (2001). Protection against translocating *Salmonella typhimurium* infection in mice by feeding the immuno-enhancing probiotic *Lactobacillus rhamnosus* strain HN001. *Medical Microbiology and Immunology, 190*, 97–104.

Gírio, F. M., Carvalheiro, F., Duarte, L. C., & Bogel-Łukasik, R. (2012). Deconstruction of the hemicellulose fraction from lignocellulosic materials into simple sugars. *D-Xylitol: Fermentive production, application and commercialization* (pp. 3–37). Springer.

Gírio, F. M., Fonseca, C., Carvalheiro, F., Duarte, L. C., & Marques, S. (2010). Bioresource technology hemicelluloses for fuel ethanol: A review. *Bioresource Technology, 101*, 4775–4800.

Goksu, E. I., Karamanlioglu, M., Bakir, U., Yilmaz, L., & Yilmazer, U. (2007). Production and characterization of films from cotton stalk xylan. *Journal of Agricultural and Food Chemistry, 55*, 10685–10691.

Grand View Research (2020). Enzymes Market Size, Share. Global Industry Report, 2020–2027. <https://www.grandviewresearch.com/industry-analysis/enzymes-industry> (Accessed 11.07.20).

Gröndahl, M., Teleman, A., & Gatenholm, P. (2003). Effect of acetylation on the material properties of glucuronoxylan from aspen wood. *Carbohydrate Polymers, 52*, 359–366.

Hao, X., Wen, P., Wang, J., Wang, J., You, J., & Zhang, J. (2020). Production of xylooligosaccharides and monosaccharides from hydrogen peroxide-acetic acid-pretreated poplar by two-step enzymatic hydrolysis. *Bioresource Technology, 297*, 122349.

Hilpmann, G., Becher, N., Pahner, F., Kusema, B., Mäki-arvela, P., & Lange, R. (2016). Acid hydrolysis of xylan, *Catalysis Today* (259, pp. 376−380).

Himmel, M. E., Ding, S. Y., Johnson, D. K., Adney, W. S., Nimlos, M. R., Brady, J. W., & Foust, T. D. (2007). Biomass recalcitrance: Engineering plants and enzymes for biofuels production. *Science, 315*, 804−807.

Höije, A., Gröndahl, M., Tømmeraas, K., & Gatenholm, P. (2005). Isolation and characterization of physicochemical and material properties of arabinoxylans from barley husks. *Carbohydrate Polymers, 61*, 266−275.

Hu, J., Arantes, V., & Saddler, J. N. (2011). The enhancement of enzymatic hydrolysis of lignocellulosic substrates by the addition of accessory enzymes such as xylanase: Is it an additive or synergistic effect? *Biotechnology for Biofuels, 4*, 36.

Izyan, N., Azelee, W., Jahim, J., Fauzi, A., Fatimah, S., Mohamad, Z., ... Illias, R. (2016). High xylooligosaccharides (XOS) production from pretreated kenaf stem by enzyme mixture hydrolysis. *Industrial Crops and Products, 81*, 11−19.

Kettenbach, G., & Stein, A. (2007). Method for separating hemicelluloses from abomass containing hemicelluloses and biomass and hemicelluloses obtained by said method. *U.S. Patent Application No. 7198695(B2)*.

Khanahmadi, M., Arezi, I., Amiri, M., & Miranzadeh, M. (2018). Bioprocessing of agro-industrial residues for optimization of xylanase production by solid- state fermentation in flask and tray bioreactor. *Biocatalysis and Agricultural Biotechnology, 13*, 272−282.

Kozlova, L. V., Mikshina, P. V., & Gorshkova, T. A. (2012). Glucuronoarabinoxylan extracted by treatment with endoxylanase from different zones of growing maize root. *Biochemistry (Moscow), 77*, 395−403.

Kumar, B., Bhardwaj, N., Agrawal, K., Chaturvedi, V., & Verma, P. (2020). Current perspective on pretreatment technologies using lignocellulosic biomass: An emerging biorefinery concept. *Fuel Processing Technology, 199*.

Kumar, V., & Shukla, P. (2018). Extracellular xylanase production from *T. lanuginosus* VAPS24 at pilot scale and thermostability enhancement by immobilization. *Process Biochemistry, 71*, 53−60.

Kumari, D., & Singh, R. (2018). Pretreatment of lignocellulosic wastes for biofuel production: A critical review. *Renewable and Sustainable Energy Reviews, 90*, 877−891.

Lai, C., Jia, Y., Wang, J., Wang, R., Zhang, Q., & Chen, L. (2019). Co-production of xylooligosaccharides and fermentable sugars from poplar through acetic acid pretreatment followed by poly (ethylene glycol) ether assisted alkali treatment. *Bioresource Technology, 288*.

Lal, M., Dutt, D., Kumar, A., & Gautam, A. (2015). Optimization of submerged fermentation conditions for two and their biochemical characterization. *Cellulose Chemistry and Technology, 49*, 5−6.

Lasrado, L. D., & Gudipati, M. (2014). Antioxidant property of synbiotic combination of *Lactobacillus* sp. and wheat bran xylo-oligosaccharides. *Journal of Food Science and Technology, 52*, 4551−4557.

Li, W., Li, X., Zhu, J., Qin, L., Li, B., & Yuan, Y. (2018). Improving xylose utilization and ethanol production from dry dilute acid pretreated corn stover by two-step and fed-batch fermentation, *Energy, 157*, 877−885.

Licht, T. R., Ebersbach, T., & Frøkiær, H. (2012). Prebiotics for prevention of gut infections. *Trends in Food Science & Technology, 23*, 70−82.

Limayem, A., & Ricke, S. C. (2012). Lignocellulosic biomass for bioethanol production: Current perspectives, potential issues and future prospects. *Progress in Energy and Combustion Science, 38*, 449−467.

Lin, X., Wu, Z., Zhang, C., Liu, S., & Nie, S. (2018). Enzymatic pulping of lignocellulosic biomass. *Industrial Crops and Products, 120*, 16−24.

Liu, Q., Li, W., Ma, Q., An, S., Li, M., Jameel, H., & Chang, H. M. (2016). Pretreatment of corn stover for sugar production using a two-stage dilute acid followed by wet-milling pretreatment process. *Bioresource Technology, 211*, 435−442.

Liu, T., Peng, B., Huang, S., & Geng, A. (2020). Recombinant xylose-fermenting yeast construction for the co-production of ethanol and cis, cis-muconic acid from lignocellulosic biomass. *Bioresource Technology Reports*, *9*, 100395.

Liu, X., Wei, W., & Wu, S. (2019). Subcritical CO_2-assisted autohydrolysis for the co-production of oligosaccharides and fermentable sugar from corn straw, *Cellulose*, *26*, 7889−7903.

Loow, Y. L., Wu, T. Y., Yang, G. H., Ang, L. Y., New, E. K., Siow, L. F., . . . Teoh, W. H. (2018). Deep eutectic solvent and inorganic salt pretreatment of lignocellulosic biomass for improving xylose recovery. *Bioresource Technology*, *249*, 818−825.

Lourdes, M., Sánchez, S., & Bravo, V. (2019). Production of xylitol and ethanol by *Hansenula polymorpha* from hydrolysates of sunflower stalks with phosphoric acid. *Industrial Crops and Products*, *40*, 160−166.

Lú-chau, T. A., Martínez-Patiño, J. C., Gullón, B., García-Torreiro, M., Moreira, M. T., Lema, J. M., & Eibes, G. (2018). Scale-up and economic analysis of the production of ligninolytic enzymes from a side-stream of the organosolv process. *Journal of Chemical Technology and Biotechnology (Oxford, Oxfordshire: 1986)*, *93*, 3125−3134.

Luo, Y., Pan, X., Ling, Y., Wang, X., & Sun, R. (2014). Facile fabrication of chitosan active film with xylan via direct immersion. *Cellulose.*, *21*, 1873−1883.

Market Watch. (2020). Xylooligosaccharides (XOS) market share and size 2020: Financial matrix, growth figures, advanced strategies, analysis and forecast 2024—MarketWatch. <https://www.marketwatch.com/press-release/xylooligosaccharides-xos-market-share-and-size-2020-financial-matrix-growth-figures-advanced-strategies-analysis-and-forecast-2024-2020-07-08> (Accessed 12.07.20).

Marques, F. P., Silva, L. M., Lomonaco, D., de F. Rosa, M., & Leitão, R. C. (2020). Steam explosion pretreatment to obtain eco-friendly building blocks from oil palm mesocarp fiber. *Industrial Crops and Products*, *143*.

Mazlan, N. A., Samad, K. A., Yahya, N. D., Samah, R. A., Jahim, J., & Yussof., H. W. (2019). Factorial analysis on nitric acid pretreatment of oil palm frond bagasse for xylan recovery. *Materials Today: Proceedings*, *19*, 1189−1198.

Meighan, B. N., Lima, D. R. S., Cardoso, W. J., Baêta, B. E. L., Adarme, O. F. H., Santucci, B. S., . . . Gurgel, L. V. A. (2017). Two-stage fractionation of sugarcane bagasse by autohydrolysis and glycerol organosolv delignification in a lignocellulosic biorefinery concept. *Industrial Crops and Products*, *108*, 431−441.

Mihiretu, G. T., Chimphango, A. F., & Görgens, J. F. (2019). Steam explosion pre-treatment of alkali-impregnated lignocelluloses for hemicelluloses extraction and improved digestibility. *Bioresource Technology*, *294*.

Mikkonen, K. S., Heikkinen, S., Soovre, A., Peura, M., Serimaa, R., Hyvo, L., . . . Hele, H. (2009). Films from oat spelt arabinoxylan plasticized with glycerol and sorbitol. *Journal of Applied Polymer Science*, *114*(1), 457−466.

Mikkonen, K. S., & Tenkanen, M. (2012). Sustainable food-packaging materials based on future biorefinery products: Xylans and mannans. *Trends in Food Science & Technology*, *28*, 90−102.

Moore, K. J., & Jung, H.-J. G. (2001). Lignin and fiber digestion. *Journal of Range Management*, *54*, 420.

Morais, E. S., Mendonça, P. V., Coelho, J. F. J., Freire, M. G., Freire, C. S. R., Coutinho, J. A. P., & Silvestre, A. J. D. (2018). Deep eutectic solvent aqueous solutions as efficient media for the solubilization of hardwood xylans. *ChemSusChem.*, *11*, 753−762.

Morgan, N. K., Wallace, A., Bedford, M. R., & Choct, M. (2017). Efficiency of xylanases from families 10 and 11 in production of xylo-oligosaccharides from wheat arabinoxylans. *Carbohydrate Polymers*, *167*, 290−296.

Mussatto, S., & Teixeira, J. (2010). Lignocellulose as raw material in fermentation processes. *Applied Microbiology and Biotechnology*, *2*, 897−907.

Naidu, D. S., Hlangothi, S. P., & John, M. J. (2018). Bio-based products from xylan: A review. *Carbohydrate Polymers*, *179*, 28−41.

Nozari, B., Mirmohamadsadeghi, S., & Karimi, K. (2018). Bioenergy production from sweet sorghum stalks via a biorefinery perspective. *Bioenergy and Biofuels*, *102*, 3425−3438.

Panthapulakkal, S., Kirk, D., & Sain, M. (2015). Alkaline extraction of xylan from wood using microwave and conventional heating. *Journal of Applied Polymer Science, 132*, 1−10.

Pathak, P., Kaur, P., & Bhardwaj, N. K. (2016). *Microbial biotechnology*. CRC Press.

Peng, F., Peng, P., Xu, F., & Sun, R. (2012). Fractional purification and bioconversion of hemicelluloses. *Biotechnology Advances, 30*, 879−903.

Phaiboonsilpa, N., Chysirichote, T., & Champreda, V. (2020). Fermentation of xylose, arabinose, glucose, their mixtures and sugarcane bagasse hydrolyzate by yeast *Pichia stipitis* for ethanol production. *Energy Reports, 6*, 710−713.

Pielhop, T., Amgarten, J., Von Rohr, P. R., & Studer, M. H. (2016). Steam explosion pretreatment of softwood: The effect of the explosive decompression on enzymatic digestibility. *Biotechnology for Biofuels, 9*, 1−13.

Ping, Y., Ling, H., Song, G., & Ge, J. (2013). Xylitol production from non-detoxified corncob hemicellulose acid hydrolysate by *Candida tropicalis*. *Biochemical Engineering Journal, 75*, 86−91.

Polizeli, M. L. T. M., Rizzatti, A. C. S., Monti, R., Terenzi, H. F., Jorge, J. A., & Amorim, D. S. (2005). Xylanases from fungi: Properties and industrial applications. *Applied Microbiology and Biotechnology, 67*, 577−591.

Prakash, G., Varma, A. J., Prabhune, A., Shouche, Y., & Rao, M. (2011). Microbial production of xylitol from D-xylose and sugarcane bagasse hemicellulose using newly isolated thermotolerant yeast *Debaryomyces hansenii*. *Bioresource Technology, 102*, 3304−3308.

Prasad, A., Sotenko, M., Blenkinsopp, T., & Coles, S. R. (2016). Life cycle assessment of lignocellulosic biomass pretreatment methods in biofuel production. *International Journal of Life Cycle Assessment, 21*, 44−50.

Queiroz, V. L., Awan, A. T., & Tasic, L. (2016). *Low-cost enzymes and their applications in bioenergy sector*. Elsevier Inc.

Rahman, A. K. S., Sugitani, N., Hatsu, M., & Takamizawa, K. (2003). A role of xylanase, α-L-arabinofuranosidase, and xylosidase in xylan degradation. *Canadian Journal of Microbiology, 49*, 58−64.

Rantanen, H., Virkki, L., Tuomainen, P., Kabel, M., Schols, H., & Tenkanen, M. (2007). Preparation of arabinoxylobiose from rye xylan using family 10 *Aspergillus aculeatus* endo-1,4-β-d-xylanase. *Carbohydrate Polymers, 68*, 350−359.

Rao, R. S., Bhadra, B., & Shivaji, S. (2008). Isolation and characterization of ethanol-producing yeasts from fruits and tree barks. *Letters in Applied Microbiology, 47*, 19−24.

Ravindran, R., Hassan, S. S., Williams, G. A., & Jaiswal, A. K. (2018). A review on bioconversion of agro-industrial wastes to industrially important enzymes. *Bioengineering., 5*, 1−20.

Ren, J., Wang, S., Gao, C., Chen, X., Li, W., & Peng, F. (2015). TiO_2-containing PVA/xylan composite films with enhanced mechanical properties, high hydrophobicity and UV shielding performance. *Cellulose., 22*, 593−602.

Rennie, E. A., & Scheller, H. V. (2014). Xylan biosynthesis. *Current Opinion in Biotechnology, 26*, 100−107.

Report Buyer. (2021). Hydrogels: Applications and global markets to 2022. <https://www.reportbuyer.com/product/5200527/hydrogels-applications-and-global-markets-to-2022.html> (Accessed 12.07.20).

Rivière, A., Moens, F., Selak, M., Maes, D., Weckx, S., & De Vuyst, L. (2014). The ability of bifidobacteria to degrade arabinoxylan oligosaccharide constituents and derived oligosaccharides is strain dependent. *Applied and Environmental Microbiology, 80*, 204−217.

Sabrina, S., Sukri, M., & Sakinah, A. M. M. (2018). Production of high commercial value xylooligosaccharides from Meranti wood sawdust using ommobilised xylanase. *Applied Biochemistry and Biotechnology, 184*, 278−290.

Saratale, G. D., & Oh, M. K. (2015). Improving alkaline pretreatment method for preparation of whole rice waste biomass feedstock and bioethanol production. *RSC Advances, 5*, 97171−97179.

Sedlmeyer, F. B. (2011). Xylan as by-product of biorefineries: Characteristics and potential use for food applications. *Food Hydrocolloids, 25*, 1891−1898.

Selvarajan, E., & Veena, R. (2017). Recent advances and future perspectives of thermostable xylanase. *Biomedical and Pharmacology Journal, 10*, 261−279.

Shin, H., Lee, J., Pestka, J., & Ustunol, Z. (2000). Growth and viability of commercial *Bifidobacterium* spp. in honey-sweetened skim milk. *Journal of Food Science, 65*, 884−886.

Silva, A. M., Barbosa, F. H. F., Duarte, R., Vieira, L. Q., Arantes, R. M. E., & Nicoli, J. R. (2004). Effect of *Bifidobacterium longum* ingestion on experimental salmonellosis in mice. *Journal of Applied Microbiology, 97*, 29−37.

Silveira, M., Chandel, A., Vanelli, B., Sacilotto, K., & Cardoso, E. (2018). Production of hemicellulosic sugars from sugarcane bagasse via steam explosion employing industrially feasible conditions: Pilot scale study. *Bioresource Technology Reports, 3*, 138−146.

Singh, S. (2018). White-rot fungal xylanases for applications in pulp and paper industry. In S. Kumar, P. Dheeran, M. Taherzadeh, & S. Khanal (Eds.), *Fungal biorefineries* (pp. 47−63). Springer.

Singh, J., Suhag, M., & Dhaka, A. (2015). Augmented digestion of lignocellulose by steam explosion, acid and alkaline pretreatment methods: A review. *Carbohydrate Polymers, 117*, 624−631.

Singh, R. D., Bhuyan, K., Banerjee, J., Muir, J., & Arora, A. (2017). Hydrothermal and microwave assisted alkali pretreatment for fractionation of arecanut husk. *Industrial Crops and Products, 102*, 65−74.

Smichi, N., Messaoudi, Y., & Gargouri, M. (2018). Lignocellulosic biomass fractionation: Production of ethanol, lignin and carbon source for fungal culture. *Waste and Biomass Valorization, 9*, 947−956.

Sorek, N., Yeats, T. H., Szemenyei, H., Youngs, H., & Somerville, C. R. (2014). The implications of lignocellulosic biomass chemical composition for the production of advanced biofuels. *Bioscience, 64*, 192−201.

Sporck, D., Reinoso, F. A. M., Rencoret, J., Gutiérrez, A., Rio, J. C., Ferraz, A., & Milagres, A. M. F. (2017). Xylan extraction from pretreated sugarcane bagasse using alkaline and enzymatic approaches. *Biotechnology for Biofuels, 10*.

Stevanic, J. S., Bergström, E. M., Gatenholm, P., Berglund, L., & Salmén, L. (2012). Arabinoxylan/nanofibrillated cellulose composite films. *Journal of Materials Science, 47*, 6724−6732.

Stevanic, J. S., Joly, C., Mikkonen, K. S., Pirkkalainen, K., Serimaa, R., Rémond, C., . . . Salmén, L. (2011). Bacterial nanocellulose-reinforced arabinoxylan films. *Journal of Applied Polymer Science, 122*, 1030−1039.

Su, X., Han, Y., Dodd, D., Moon, Y. H., Yoshida, S., Mackie, R. I., & Cann, I. K. O. (2013). Reconstitution of a thermostable xylan-degrading enzyme mixture from the bacterium *Caldicellulosiruptor bescii*. *Applied and Environmental Microbiology, 79*, 1481−1490.

Sun, M. P., Byung, I. S., Dae, W. P., & Doo, H. P. (2005). Electrochemical reduction of xylose to xylitol by whole cells or crude enzyme of *Candida peltata*. *Journal of Microbiology (Seoul, Korea), 43*, 451−455.

Sun, X., Liu, B., Jing, Z., & Wang, H. (2015). Preparation and adsorption property of xylan/poly (acrylic acid) magnetic nanocomposite hydrogel adsorbent. *Carbohydrate Polymers, 118*, 16−23.

Sun, X. F., Wang, H. H., Jing, Z. X., & Mohanathas, R. (2013). Hemicellulose-based pH-sensitive and biodegradable hydrogel for controlled drug delivery. *Carbohydrate Polymers, 92*, 1357−1366.

Surek, E., & Buyukkileci, A. O. (2017). Production of xylooligosaccharides by autohydrolysis of hazelnut (*Corylus avellana* L.) shell. *Carbohydrate Polymers, 174*, 565−571.

Svärd, A., Brännvall, E., & Edlund, U. (2015). Rapeseed straw as a renewable source of hemicelluloses: Extraction, characterization and film formation. *Carbohydrate Polymers, 133*, 179−186.

Talukdera, A. A., Adnana, N., Siddiqaa, A., Miaha, R., Tulia, J. F., Khanb, S. T., . . . Yamadad, M. (2019). Fuel ethanol production using xylose assimilating and high ethanol producing thermosensitive Saccharomyces cerevisiae isolated from date palm juice in Bangladesh. *Biocatalysis and Agricultural Biotechnology, 18*, 101029.

Teramura, H., Sasaki, K., Oshima, T., Kawaguchi, H., Ogino, C., Sazuka, T., & Kondo, A. (2018). Effective usage of sorghum bagasse: Optimization of organosolv pretreatment using 25% 1-butanol and subsequent nanofiltration membrane separation. *Bioresource Technology, 252*, 157−164.

UNCTAD. (2016). *Second generation biofuel markets: State of play, trade and developing country perspectives.* United Nations Publication.

Ur-Rehman, S., Mushtaq, Z., Zahoor, T., Jamil, A., & Murtaza, M. A. (2015). Xylitol: A review on bioproduction, application, health benefits, and related safety issues. *Critical Reviews in Food Science and Nutrition, 55*, 1514–1528.

Usmani, Z., Sharma, M., Karpichev, Y., Pandey, A., Chandra Kuhad, R., Bhat, R., ... Gupta., V. K. (2020). Advancement in valorization technologies to improve utilization of bio-based waste in bioeconomy context. *Renewable and Sustainable Energy Reviews, 131*, 109965.

Vallejos, M. E., & Area, M. C. (2017). *Area and forest biorefinery*. Elsevier Inc.

Valderez, M., Rocha, P., Helena, T., Rodrigues, S., Lima, T., Albuquerque, D., ... Macedo, D. (2014). Evaluation of dilute acid pretreatment on cashew apple bagasse for ethanol and xylitol production. *Chemical Engineering Journal, 243*, 234–243.

Van Dyk, J. S., & Pletschke, B. I. (2012). A review of lignocellulose bioconversion using enzymatic hydrolysis and synergistic cooperation between enzymes—Factors affecting enzymes, conversion and synergy. *Biotechnology Advances, 30*, 1458–1480.

Venkateswar Rao, L., Goli, J. K., Gentela, J., & Koti, S. (2016). Bioconversion of lignocellulosic biomass to xylitol: An overview. *Bioresource Technology, 213*, 299–310.

Vermaas, J. V., Petridis, L., Qi, X., Schulz, R., Lindner, B., & Smith, J. C. (2015). Mechanism of lignin inhibition of enzymatic biomass deconstruction. *Biotechnology for Biofuels, 8*.

Vigier, K. D. O., Chatel, G., & Jérôme, F. (2015). Contribution of deep eutectic solvents for biomass processing: Opportunities, challenges, and limitations. *ChemCatChem, 7*, 1250–1260.

Vilaplana, F., & Ruthes, A. C. (2016). International application published under the patent cooperation treaty (PCT).

Wallenius, J., Pahimanolis, N., Zoppe, J., Kilpeläinen, P., Master, E., Ilvesniemi, H., ... Ojamo, H. (2015). Continuous propionic acid production with *Propionibacterium acidipropionici* immobilized in a novel xylan hydrogel matrix. *Bioresource Technology, 197*, 1–6.

Wan, C., Zhou, Y., & Li, Y. (2011). Liquid hot water and alkaline pretreatment of soybean straw for improving cellulose digestibility. *Bioresource Technology, 102*, 6254–6259.

Wang, S., Ren, J., Li, W., Sun, R., & Liu, S. (2014). Properties of polyvinyl alcohol/xylan composite films with citric acid. *Carbohydrate Polymers, 103*, 94–99.

Wang, Z., Gräsvik, J., Jönsson, L. J., & Winestrand, S. (2017). Comparison of $[HSO_4]^-$, $[Cl]^-$ and $[MeCO_2]^-$ as anions in pretreatment of aspen and spruce with imidazolium-based ionic liquids. *BMC Biotechnology, 17*, 1–10.

Wei, N., & Shannon, T. G. (2020). Tailored hemicellulose in non-wood fibers for tissue products. *Patent No. GB2576122*.

Yamasaki, T., Enomoto, A., Kato, A., Ishii, T., Kameyama, M., Anzai, H., & Shimizu, K. (2012). Enzymatically derived aldouronic acids from *Cryptomeria japonica*. *Carbohydrate Polymers, 87*, 1425–1432.

Zhang, H., & Sang., Q. (2015). Production and extraction optimization of xylanase and β-mannanase by *Penicillium chrysogenum* QML-2 and primary application in saccharification of corn cob. *Biochemical Engineering Journal, 97*, 101–110.

Zhang, J., Geng, A., Yao, C., Lu, Y., & Li, Q. (2012). Xylitol production from D-xylose and horticultural waste hemicellulosic hydrolysate by a new isolate of *Candida athensensis* SB18. *Bioresource Technology, 105*, 134–141.

Zhang, Z., Harrison, M. D., Rackemann, D. W., Doherty, W. O. S., O'Hara, I. M. O., & Harrison, M. D. (2016). Organosolv pretreatment of plant biomass for enhanced enzymatic saccharification. *Green Chemistry, 18*, 360–381.

Zhang, Z., Smith, C., & Li, W. (2014). Extraction and modification technology of arabinoxylans from cereal by-products: A critical review. *Food Research International, 65*, 423–436.

Zhong, L., Peng, X., Yang, D., Cao, X., & Sun, R. (2013). Long-chain anhydride modification: A new strategy for preparing xylan films. *Journal of Agricultural and Food Chemistry, 61*(3), 655–661.

Zhou, X., Li, W., Mabon, R., & Broadbelt, L. J. (2016). A critical review on hemicellulose pyrolysis. *Energy Technology, 5*, 216.

PART III
Microbial-based biopolymers

CHAPTER 8

Production and applications of pullulan

Ashutosh Kumar Pandey[1], Ranjna Sirohi[2], Vivek Kumar Gaur[3] and Ashok Pandey[4]

[1]Centre for Energy and Environmental Sustainability, Lucknow, India, [2]Department of Post-Harvest Process & Food Engineering, G.B. Pant University of Agriculture and Technology, Pantnagar, India, [3]Environmental Biotechnology Division, Environmental Toxicology Group, CSIR—Indian Institute of Toxicology Research, Lucknow, India, [4]Centre for Innovation and Translational Research, CSIR—Indian Institute of Toxicology Research, Lucknow, India

8.1 Introduction

Pullulan is one of the commonly known neutral polymers that are water-soluble and produced microbially as a polysaccharide in large quantities by fermentation (Pollock, Thorne, & Armentrout, 1992; Taguchi, Kikuchi, Sakano, & Kobayashi, 1973). Sugar nucleotide-lipid carrier intermediates associated with the cell membrane act as mediators for the biosynthesis of pullulan (Catley & McDowell, 1982; Taguchi et al., 1973). The carbon and hydrogen in pullulan are present as $(C_6H_{10}O_5)_n$; its reaction with iodine yields no color and forms complexes with Cu^{2+} (Ueda, Fujita, Komatsu, & Nakashima, 1963). *Aureobasidium pullulans* produces pullulan as an amorphous slime matter with maltotriose repeating units joined by α-1,6-linkages and α-1,4-glycosidic bonds among glucose units of maltotriose (Catley, Ramsay, & Servis, 1986) (Fig. 8.1). The molecular weight of pullulan ranges from 45 to 600 kDa, varying with changing cultivation parameters (Lee & Yoo, 1993). Temperature (McNeil & Kristiansen, 1990), initial pH of the medium (Imshenetskii, Kondrat'eva, & Smut'ko, 1981a; Lacroix, LeDuy, Noel, & Choplin, 1985; Ono, Yasuda, & Ueda, 1977), oxygen supply (Rho, Mulchandani, Luong, & LeDuy, 1988; Wecker & Onken, 1991), the concentration of nitrogen (Auer & Seviour, 1990), and the carbon source (Badr-Eldin, El-Tayeb, El-Masry, Mohamad, & El-Rahman, 1994) are the major factors that influence the production of pullulan. Culture conditions and type of strain affect the molecular weight of pullulan (Pollock et al., 1992; Slodki & Cadmus, 1978). Pullulan synthesis in *A. pullulans* has been described by Leathers (2003). The main focus in this chapter is to present the concepts, methods, and strategies regarding pullulan biosynthesis, utilization of industrial by-products as substrates, industrial production process, applications, and to discuss the scope of chemical modifications in the expansion of the applications of pullulan derivatives.

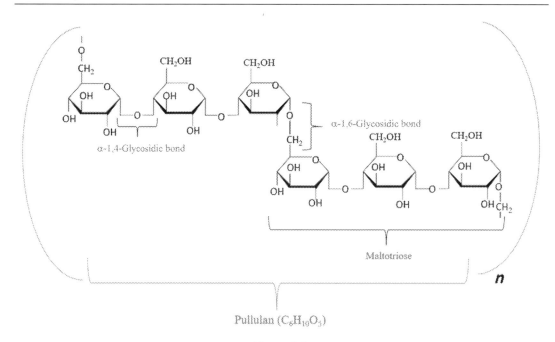

Figure 8.1
Structure of pullulan.

8.2 Biosynthesis of pullulan

One of the functions of microbial exopolysaccharide (EPS) is to protect the host microbe (Kumar, Mody, & Jha, 2007). The cells of microbes are surrounded by layers of these EPS, which also protects them from being desiccated and predated. The diffusion potency of the cells is also determined by these EPSs (Dudman & Sutherland, 1977). Pullulan is a type of EPS produced inside the cell wall of microbes and found in the form of slime layers over the cell surface (Simon, Caye-Vaugien, & Bouchonneau, 1993). Several studies have been carried out on the cytology and physiology of *A. pullulans*, still, the biosynthesis pathway of the production of pullulan is poorly understood. According to Duan, Chi, Wang, and Wang (2008), glucose and three keys enzymes (phosphoglucose mutase, uridine diphosphoglucose pyrophosphorylase, and glucosyltransferase) play important role in the biosynthesis of pullulan in *A. pullulans*. Besides, different sugars like mannose, sucrose, galactose, fructose, maltose, and wastes enriched in carbohydrates can also be used by *A. pullulans* (Catley, 1971; Leathers, 2003; Madi, McNeil, & Harvey, 1996).

To synthesize uridine diphosphate glucose (UDPG), a precursor of pullulan, the activities of isomerase and hexokinase play an important role. These enzymes facilitate the formation of UDPG from the carbon sources. UDPG mediates the addition of D-glucose to lipid molecule, that is, lipid hydroperoxide by forming phosphodiester bond (Catley & McDowell, 1982)

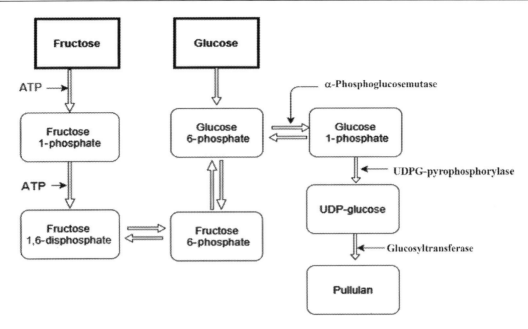

Figure 8.2
Biosynthesis pathway for pullulan.

(the transfer of D-glucose from UDPG forms lipid-linked with isomaltose). This isomaltose after reacting with glucose linked with lipid yields isopropyl residues. These residues act as monomers and polymerize to form pullulan. Figure 8.2 summarizes the biosynthesis pathway of pullulan.

Pullulans are also synthesized when sucrose is metabolized by *A. pullulans* enzymes in the presence of ATP and UDPG (Ono et al., 1977). As ADPG cannot replace the UDPG in the reaction, this implies that the precursors of pullulan are derived from the UDPG. In addition to medium having maltose, the precursors needed for pullulan formation are panose and/isomaltose. Both precursors are synthesized in *A. pullulans* by a reaction mediating glucose transfer (Hayashi, Hayashi, Takasaki, & Imada, 1994). A study proposed that accumulation of sugars occurs inside the cells and in late stages of the growth cycle these reserved sugars are used for pullulan synthesis (Simon, Bouchet, Bremond, Gallant, & Bouchonneau, 1998). This was further proved by the study of Simon et al. (1998), according to which there is an inverse correlation present between stored glycogen and pullulan concentration.

8.2.1 Mechanism of pullulan biosynthesis

Many studies have been carried out to understand the biosynthesis mechanism of EPSs (Degeest & De Vuyst, 2000), still, the mechanism of biosynthesis of pullulan is poorly understood. To increase the productivity of pullulan by *A. pullulans*, it is very critical to explore the synthesis and regulatory pathways in *A. pullulans*.

Substrates for pullulan are ATP and UDPG (Shingel, 2004). High pullulan concentration can be linked to the concentration of UDPG, and bioactivity of glucosyltransferase enzyme present in strain Y68 of *A. pullulans* (Chi et al., 2009). To investigate the factors affecting the yield of pullulan, the effects of UDPG, catalytic activities of enzymes like UDPG-pyrophosphorylase, glucosyltransferase, and phosphoglucose mutase were deciphered (Duan et al., 2008). High production of this polysaccharide by the yeast was obtained in the medium supplemented with glucose rather than the medium containing other sugars. Also, UDPG concentration decreased in the cells of *A. pullulans* Y68 with a high percentage of pullulan showing that in *A. pullulans* Y68 cells grown in media containing different sugars, the yield of pullulan was positively correlated with high catalytic activities of UDP-pyrophosphate, glucosyltransferase, and phosphoglucose mutase (Chi et al., 2009). The mechanism of biosynthesis of pullulan in *A. pullulans* Y68 was suggested based on other studies (Chi et al., 2009). The reason behind the low yield of pullulans in the medium inoculated with *A. pullulans* Y68 and supplemented by fructose and xylose is because of the long pathway involved in the synthesis of UDPG from xylose and fructose. Due to the high activity of glucosyltransferase, the maximum of the UDPG get utilized in the synthesis of pullulan, therefore the cells become deprived of UDPG. Thus the high catalytic activity of glucosyltransferase in the yeast cells makes them potent in producing a higher concentration of pullulans. An orchestral activity of phosphoglucomutase and UDPG-pyrophosphorylase along with higher activities of glucosyltransferase ensures the continual supplementation of precursors via utilizing the UDPG so that high yield of pullulans can be obtained by *A. pullulans* Y68 grown on media having high glucose concentration. However, higher accumulation of UDPG in yeast cells was obtained when they were cultured in medium supplemented with fructose, xylose, respectively. This may be due to the low activities of glucosyltransferase in the cells (Prajapati, Jani, & Khanda, 2013).

8.2.2 Physicochemical properties

Pullulan is a heat stable, white to off-white polysaccharide, which forms a stable, viscous solution when dissolved in water. Its viscosity varies with heat treatment depending on the molecular weight. When heated at 90°C for an hour, the viscosity decreases by about 10% for a pullulan with a molecular weight of 3×10^6 Da, whereas trivial decrease is observed for 10^5 Da pullulan (Tsujisaka & Mitsuhashi, 1993). The equilibrium moisture content of pullulan, when present in an atmosphere with a relative humidity of less than 70%, is about 10%–15% without any hygroscopicity (Tsujisaka & Mitsuhashi, 1993). In terms of viscosity, pullulan resembles gum Arabic while the surface tension of its solution is 0.074 N m^{-1}, which is slightly higher than that of water at 25°C. pH does not impart large variation in the viscosity of pullulan solution when the range of pH values is <2 to >11

(Tsujisaka & Mitsuhashi, 1993). This polysaccharide is nontoxic, nonmutagenic, edible, tasteless, and odorless (Kimoto, Shibuya, & Shiobara, 1997) with high mechanical strength, adhesiveness, film formability, and degrades in the presence of enzymes (Shingel, 2004). Its solution when applied to paper or wood imparts strength and shows adhesion to glass, metal, and concrete. It is highly stable in the presence of most metal ions and salts like sodium chloride. Enzymes that actively hydrolyze pullulan include pullulanase and isopullulanase (4-alpha pullulan hydrolase) (Kikuchi, Taguchi, Sakano, & Kobayashi, 1973). Also, it is slightly susceptible to degradation in presence of many alpha-amylases such as salivary alpha-amylase which causes very slow hydrolysis affecting the maltotriose regions of pullulan suggesting toward very little hydrolysis of pullulan by gut digestive enzymes, which also makes it suitable for oral drug delivery (Tsujisaka & Mitsuhashi, 1993). Decomposition of pullulan due to temperature begins at 250°C and char is formed at 280°C. Pullulan is insoluble in alcohol and other organic solvents, except for dimethyl sulfoxide and formamide and high solubility has been observed in water as well as dilute alkali (Prajapati et al., 2013). The molecular structure of pullulan has the property of random coiling due to its highly flexible nature. It forms a straight unbranched chain with a molecular weight ranging from 5000 to 9,000,000 g mol^{-1} (Prajapati et al., 2013). Chemical modifications such as esterification, etherification, or cross-linking can be done to alter the solubility and degradability of pullulan.

8.2.3 Factor affecting the production of pullulan

8.2.3.1 Effect of carbon source

The potential of *A. pullulans* to grow on different kinds of substrates, including agricultural wastes can be credited to its multienzyme system, which enables the conversion of complex plant fibers into glucose units to be used as a carbon source for the cell growth (Duan et al., 2008; Leathers, 2003; Singh & Saini, 2008). Many parameters require optimization for appropriate microbial growth and pullulan production (Fig. 8.3). The proper selection of carbon, nitrogen source and supplements is important for the microbial pullulan production.

Various carbon sources, including glucose, sucrose, mannose, galactose, fructose, etc. have been utilized for pullulan synthesis (Duan et al., 2008; Singh & Saini, 2008). Utilization of glucose and sucrose (22 and 20 g L^{-1}, respectively) by a wild strain of *A. pullulans* with the yield of 1.3 g of pullulan per gram of dry cells during 5 days of production was first reported (Bender, Lehmann, & Wallenfels, 1959). In comparison to glucose, sucrose has been reported as a superior carbon source both in terms of yield and titer and a small amount of sucrose stimulates the enzymatic action, which is responsible for pullulan synthesis (Bender et al., 1959; LeDuy & Mian Boa, 1983; Roukas, 1999a). For *A. pullulans*, the starch hydrolysate is a preferred carbon source. No differences are perceived with starch

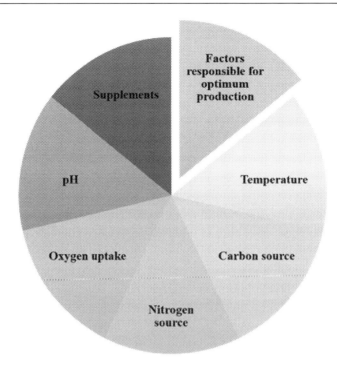

Figure 8.3
Responsible factors for optimum production.

syrups regardless of the conversion system used, acidic or enzymatic. Pullulan yield is based on the dextrose equivalent (DE) of the syrups. Use of a starch syrup with a DE in the range of 40–50 gives the highest pullulan yield against saccharides. Further investigations show that carbon concentration in the range of 10%–15% gives the excessively high pullulan yield against saccharide (Tsujisaka & Mitsuhashi, 1993). If the concentration of the carbon source is very low, more saccharide is used for cell growth, which leads to a lower pullulan yield. When the concentration exceeds 20%, agitation or stirring efficiency becomes low (due to the increased viscosity of culture broth by pullulan production), cell growth is reduced, cultivation period extends, and lowers the pullulan yield against saccharide (Tsujisaka & Mitsuhashi, 1993). Corn syrup as a carbon source has also been used in pullulan fermentation medium (Singh & Saini, 2008; Wu, Jin, Tong, & Chen, 2009). An *Aureobasidium* sp. NRRLY 12974 was reported by Leathers and Gupta to grow well on basal medium with wet milled corn fiber or corn condensed distiller's soluble but not on thin stillage (by-products from sugar and starch processes) (Leathers & Gupta, 1994).

The use of xylose or lactose as a carbon source in the medium results in poor cell growth as well as low pullulan yield (Duan et al., 2008; Leathers, 2003; LeDuy & Mian Boa, 1983). Other wastes from the food and agricultural industries such as whey, molasses, sugarcane juice, sweet potato, and peat hydrolysate are also regarded as economical and efficient for the production of pullulan (Kim, Kim, Lee, Lee, & Kim, 2000; McNeil, Kristiansen, & Seviour, 1989; Shin, Kim, Lee, Cho, & Byun, 1989).

The amount of carbon source above 5% exert an inhibitory effect on the production of pullulan (Cheng, Demirci, & Catchmark, 2010c; Reeslev, StrØm, Jensen, & Olsen, 1997). This inhibition could be attributed to the suppression effect caused by the sugars on enzymes associated with pullulan production such as α-phosphoglucose mutase, UDPG-pyrophosphorylase, and glycosyltransferase (Duan et al., 2008). The use of fed-batch and continuous fermentation with proper strain selection and enhancement of cultivation methods are some strategies that can be employed to control this suppression effect (Bulmer, Catley, & Kelly, 1987; Cheng et al., 2010c; Gibbs & Seviour, 1992). Optimization of carbon and nitrogen sources' concentration was done by Cheng et al. (2010c), which resulted in 60.7 g L^{-1} pullulan when 100 g L^{-1} of sucrose was applied.

8.2.3.2 Effect of nitrogen source

For pullulan production, nitrogen depletion acts as a trigger for fermentative EPS formation by *A. pullulans* (Mehta, Prasad, & Choudhury, 2014; Orr, Zheng, Campbell, McDougall, & Seviour, 2009; Reed-Hamer & West, 1994; Schuster, Wenzig, & Mersmann, 1993; Thirumavalavan, Manikkadan, & Dhanasekar, 2009). Ammonium ion (NH_4^+) as a nitrogen source has an important role in pullulan production. Auer and Seviour (1990) found ammonium salts like ammonium nitrate to be conducive nitrogen sources for pullulan synthesis. The concentration of ammonium ions significantly influences the diversion of glucose from assimilation into cellular material to that in polysaccharide formation (Catley et al., 1986). NH_4^+ may affect the activity of pullulan degrading enzymes, managing the flow of carbon within the cell (Catley et al., 1986). Furthermore, a surplus supply of nitrogen may result increased biomass but cannot increase the production of polysaccharide (Orr et al., 2009). Studies have shown the role of certain complex nitrogen sources, including soybean pomace, corn steep liquor, Jatropha seed cake, and soybean hydrolysate for the production of pullulan (Campbell, McDougall, & Seviour, 2003; Duan et al., 2008; Wiederschain, 2007; Zheng, Campbell, McDougall, & Seviour, 2008). One of the reasons responsible for low pullulan yield is the detrimental activity of a pullulan degrading enzyme in the late stages of fermentation (Catley, 1971; Cheng, Demirci, & Catchmark, 2011a; Pollock et al., 1992). This occurrence is only the result of the consumption of the carbon source and a high starting nitrogen concentration (Campbell et al., 2003). It has also been found stated that for EPS production, a 10:1 carbon/nitrogen ratio is the most conducive condition (Kumar et al., 2007; Wang, Chen, Wei, Jiang, & Dong, 2015). This has been supported by a study on the medium composition of sucrose, ammonium sulfate, and yeast extract (Cheng et al., 2011a), where 75 g L^{-1} sucrose, 3.0 g L^{-1} of yeast extract, and 5.0 g L^{-1} of ammonium sulfate were used for the production of pullulan achieving pullulan concentration of 25.8 g L^{-1} with 94.5% purity after a week of cultivation. The influence of nitrogen source on the activity of UDPG-pyrophosphorylase and production of pullulan was studied by Wu, Chen, and Pan (2012). The effect of nitrogen limitation on the production of pullulan as compared to nonlimiting conditions directed toward the enhanced overproduction of pullulan in nitrogen limitation due to an increase in the pursuit of α-phosphoglucose mutase and glucosyltransferase (Wang et al., 2015).

8.2.3.3 Effect of supplements

The effect of uracil on pullulan production, biomass yield, and activity of uridine phosphorylase (UPase) showed the positive impact of uracil on pullulan yield as Sheng, Zhu, and Tong (2014) observed an increase in pullulan yield from 37.72 to 49.07 g L^{-1} due to UPase activity in the presence of 5 mM uracil at 48 h. Varying concentrations of iodoacetic acid were also tested to decode its influence on the activity of enzymes involved in pullulan synthesis, and hence on pullulan production and biomass concentration (Sheng et al., 2014). Roukas (Shabtai & Mukmenev, 1995) achieved a pullulan concentration of 32 g L^{-1} at 0.635 g pullulan g^{-1} substrate from beet molasses medium incorporated with supplements such as Tween 80 and olive oil. Supplementing beet molasses medium with olive oil increased the concentration of pullulan to 49 g L^{-1} by *A. pullulans* P 56 (Lazaridou, Biliaderis, Roukas, & Izydorczyk, 2002). The use of production medium with Tween 80 as a supplement enhances the yield and productivity of pullulan (Sheng, Zhu, & Tong, 2013). A production medium with brewery waste, L-glutamic acid, and K$_2$HPO$_4$ was developed by Roukas and Serris (Madi, Harvey, Mehlert, & McNeil, 1997) with olive oil and Tween 80 as supplements resulting in pullulan concentration of 11 g L^{-1} at an initial solution pH of 7.5. Pullulan production in submerged fermentation with medium supplemented with olive oil and Tween 80 yielded 11.0 ± 0.5 g L^{-1} pullulan (Roukas, 1999b). The use of soybean oil as a nitrogen supplement has also been studied for the production of pullulan (Catley, 1973; Ronen, Guterman, & Shabtai, 2002). The literature regarding the type of microorganism, raw materials utilized, condition of fermentation, yield, and productivity of pullulan are compiled in Table 8.1.

8.2.3.4 Effect of pH and temperature

The optimum pH for the production of pullulan is generally in the range of 5.5–7.5 (Lazaridou, Biliaderis, et al., 2002; Lee & Yoo, 1993; Shingel, 2004). At relatively low pH values, pullulan synthesis is suppressed while the production of insoluble glucan is promoted (Madi et al., 1997). The first study on the effect of pH by Catley (1971) showed that the optimal pH required for the synthesis of pullulan differed from that required for cell mass growth. A two-stage pH profile was proposed by Lacroix et al. (1985) in which an initial low pH of 2.0 promoted biomass growth and the pH was later increased to 5.0 for enhancing pullulan synthesis. This two-stage regulation resulted in an increase in pullulan from 17 to 26 g L^{-1} after 13 days of cultivation. Very few studies state acidic pH for the

Table 8.1: Microbial production of pullulan from different strain and carbon sources.

S. no.	Substrate (carbon and nitrogen source)	Strains	Mode of fermentation	Fermentation conditions	Pullulan production yield (g L^{-1})	Reference
1	Beet molasses and L-glutamic acid	*Aureobasidium pullulans* P56	Submerge fermentation (100 mL batch culture)	Temp.—28°C, impeller speed—200 rpm, fermentation time—120 h	32 g L^{-1}; 6.4 g L^{-1} h^{-1}	Roukas (1998)
2	Beet molasses with ammonium sulfate and yeast extract	*A. pullulans* P 56	Submerge fermentation (6 L working volume)	Temp.—28°C, pH 5, impeller speed—700 rpm, sparging rate—0.23 vvm	49 g L^{-1}	Lazaridou, Roukas, et al. (2002)
3	Beet molasses with ammonium sulfate and yeast extract	*A. pullulans* P56	Airlift reactor (1.4 L working volume)	Temp.—28°C, pH 6.5, fermentation time—120 h, sparging rate—2 vvm	18.5 g L^{-1}; 3.7 g L^{-1} h^{-1}	Roukas (1998)
4	Brewery wastes and L-glutamic acid	*A. pullulans* P56	Submerge fermentation (100 mL batch culture)	pH 7.5, fermentation time—72 h, supplements—K$_2$HPO$_4$ + olive oil + Tween 80	11 g L^{-1}; 3.67 g L^{-1} h^{-1}	Roukas (1999)
5	Carob pod and ammonium sulfate	*A. pullulans* M18	Submerge fermentation (100 mL batch culture)	Temp.—25°C–30°C, pH 6.5, impeller speed—200 rpm	6.5 g L^{-1}	Roukas and Biliaderis (1995)
6	Cassava bagasse and sodium glutamate	*A. pullulans* MTCC 1991	Solid-state fermentation (40 mL batch culture)	Temp.—28°C, pH 6.5, fermentation time—7 days, moisture content—1–2	18.6 g L^{-1}; 2.66 g L^{-1} h^{-1}	Ray and Moorthy (2007)
7	Coconut milk and yeast extract	*A. pullulans* MTCC 2195	Submerge fermentation (50 mL batch culture)	Temp.—28°C, fermentation time—144 h, pH 7, impeller speed—200 rpm	58.0 g L^{-1}; 10.83 g L^{-1} h^{-1}	Thirumavalavan et al. (2009)
8	Coconut water and yeast extract	*A. pullulans* MTCC 2195	Submerge fermentation (50 mL batch culture)	Temp.—28°C, pH 7, fermentation time—144 h, impeller speed—200 rpm	38.3 g L^{-1}; 6.38 g L^{-1} h^{-1}	Thirumavalavan et al. (2009)
9	Corn syrup	*A. pullulans* ATCC 201253 RP-I entrapped with agar	Immobilized cell bioreactor (50 mL batch medium)	Time—7 days, temp.—30°C	6.8 ± 0.6 g L^{-1}; 0.97 g L^{-1} h^{-1}	West and Strohfus (1998)
10	Corn syrup (2.5%) and ammonium sulfate	*A. pullulans* ATCC 201253 immobilized with chitosan	Submerge fermentation (50 mL batch culture)	Temp.—30°C, fermentation time—168 h, pH 6.0, impeller speed—125 rpm, chitosan beads—1%	5.32 g L^{-1}; 0.75 g L^{-1} h^{-1}	West (2011)

(*Continued*)

Table 8.1: (Continued)

S. no.	Substrate (carbon and nitrogen source)	Strains	Mode of fermentation	Fermentation conditions	Pullulan production yield (g L^{-1})	Reference
11	Corn syrup	A. pullulans ATCC 201253 entrapped with agarose	Immobilized cell bioreactor (50 mL batch medium)	Fermentation time—168 h, temp.—30°C	—	West (2000)
12	Corn syrup	A. pullulans ATCC 201253 entrapped with alginate	Column bioreactor	Temp.—25°C, pH 6.5, sparging rate—500 mL min^{-1}, fermentation time—168 h	5.93 ± 0.55 g L^{-1}, 0.85 g L^{-1} h^{-1}	West and Strohfus (2001b)
13	Corn syrup and ammonium sulfate	A. pullulans ATCC 201253 immobilized with TEAE cellulose	Immobilized cell bioreactor	Temp.—30°C, pH 6.5, impeller speed—125 rpm, fermentation time—120 h	6 g L^{-1}, 1.2 g L^{-1} h^{-1}	West (2012)
14	Corn syrup and ammonium sulfate	A. pullulans ATCC 42023 immobilized on epichlorohydrin triethanolamine cellulose at pH 2	Submerge fermentation (50 mL batch culture)	Initial pH 6.0, fermentation time—120 h	6 g L^{-1}, 1.2 g L^{-1} h^{-1}	West (2010)
15	Corn syrup with ammonium sulfate and yeast extract	A. pullulans ATCC 42023	Submerge fermentation (50 mL)	Temp.—30°C, impeller speed—200 rpm, fermentation time—7 days, supplement—MnCl$_2$ (10 μM)	5.63 g L^{-1}, 0.8 g L^{-1} h^{-1}	West and Strohfus (1997)
16	Corn syrup with ammonium sulfate and yeast extract	A. pullulans NYSRP-1	Submerge fermentation (50 mL batch culture)	Temp.—30°C, impeller speed—200 rpm, fermentation time—7 days	7 g L^{-1}, 1 g L^{-1} h^{-1}	West and Strohfus (2001a)
17	Deproteinized whey with L-glutamic acid	A. pullulans P56	Submerge fermentation (100 mL batch culture)	Temp.—30°C, pH 6.5, impeller speed—200 rpm, fermentation time—10 days	11.0 ± 0.5 g L^{-1}, 1.1 g L^{-1} h^{-1}	Roukas (1999a)
18	Dextrose and corn steep liquor	A. pullulans RBF 4A3	Submerge fermentation (5 L batch culture)	Temp.—20°C, fermentation time—120 h, impeller speed—250 rpm, sparging rate—1 vvm	88.59 g L^{-1}, 17.72 g L^{-1} h^{-1}	Sharma et al. (2013)
19	Fuel ethanol fermentation stillage with yeast extract and sodium nitrate	Aureobasidium sp. NRRL 12,974	Submerge fermentation (50 mL batch culture)	Temp.—28°C, impeller speed—200 rpm, fermentation time—9 days	4.5 ± 0.2 g L^{-1}, 0.5 g L^{-1} h^{-1}	Leathers and Gupta (1994)

#	Substrate	Organism	Culture type	Conditions	Yield	Reference
20	Glucose and ammonium sulfate	A. pullulans	Proton type reactor, 800 mL (continuous mode) working volume	Temp.—27°C, sparging rate—3/4 vvm, impeller speed—1000 rpm, pH 5	3.6 g L^{-1}, 1.2 g L^{-1} h^{-1}	Reeslev, Nielsen, Olsen, Jensen, and Jacobsen (1991)
21	Glucose and ammonium sulfate	A. pullulans ATCC 9348	Airlift/batch	pH 4.5, temp.—26°C, impeller speed—500 rpm, sparging rate—0.3 vvm	18.0 g L^{-1}	Orr et al. (2009)
22	Glucose and ammonium sulfate	A. pullulans P56	Submerge fermentation (5 L chemostat)	Dilution rate—0.05 h^{-1}, initial pH 4.5, impeller speed—400 rpm, sparging rate—1 vvm	0.35 g L^{-1} h^{-1}	Schuster et al. (1993)
23	Glucose and ammonium sulfate	A. pullulans CCTCC M2012223	Submerge fermentation (3 L batch culture)	Fermentation time—60 h, supplement—Tween 80 (1%)	28.8 g L^{-1}, 11.52 g L^{-1} h^{-1}	Tu, Wang, Ji, and Zou (2015)
24	Glucose and NaNO$_2$	A. pullulans AP329	Submerge fermentation (3 L batch culture)	Temp.—28°C, impeller speed—800 rpm, sparging rate—4 lpm, pH 6.5, fermentation time—4	33.83 g L^{-1}, 8.46 g L^{-1} h^{-1}	Jiang et al. (2011)
25	Glucose and NaNO$_3$	A. pullulans ATCC 9348	Submerge fermentation (5 L CSTR)	Temp.—26°C, impeller speed—500 rpm, sparging rate—0.3 vvm, constant pH 4.5	7.02 g L^{-1}	Zheng et al. (2008)
26	Glucose and soybean cake hydrolysate	Yeast strain Y68	Submerge fermentation (50 mL batch culture)	Initial pH 6, temp.—30°C, impeller speed—180 rpm, fermentation time—60 h	59 g L^{-1}, 23.6 g L^{-1} h^{-1}	Chi and Zhao (2003)
27	Glucose and soybean cake hydrolysate	A. pullulans Y68	Submerge fermentation (3 L batch fermenter)	pH 7, impeller speed—300 rpm	52.47 ± 0.09 g L^{-1}	Duan et al. (2008)
28	Glucose and soybean pomace	A. pullulans HP-2001	Submerge fermentation (100 mL batch culture)	Impeller speed—200 rpm, temp.—30°C, fermentation time—120 h	7.6 g L^{-1}, 1.52 g L^{-1} h^{-1}	Seo et al. (2004)
29	Glucose and yeast extract	A. pullulans HP-2001	Submerge fermentation	Initial pH 5.5, fermentation time—72 h, temp.—30°C	11.49 g L^{-1}, 3.83 g L^{-1} h^{-1}	Gao, Chung, Li, and Lee (2011)
30	Glucose and yeast extract	A. pullulans SZU 1001 (mutant)	Submerge fermentation (50 mL batch culture)	Temp.—30°C, impeller speed—200 rpm, fermentation time—72 h	25.65 g L^{-1}, 8.55 g L^{-1} h^{-1}	Yu, Wang, Wei, and Dong (2012)
31	Glucose and yeast extract	A. pullulans HP-2001	Submerge fermentation (100 mL of batch medium)	Fermentation time—72 h, temp.—30°C, impeller speed—200 rpm	13.2 g L^{-1}, 4.4 g L^{-1} h^{-1}	Gao et al. (2011)

(Continued)

Table 8.1: (Continued)

S. no.	Substrate (carbon and nitrogen source)	Strains	Mode of fermentation	Fermentation conditions	Pullulan production yield (g L^{-1})	Reference
32	Glucose and deoiled Jatropha seed cake	A. pullulans RBF 4A3	Submerge fermentation (3.5 L batch culture)	Temp.–28°C, impeller speed–200 rpm, pH 6.0, time–96 h	83.98 g L^{-1}, 20.99 g L^{-1} h^{-1}	Choudhury et al. (2012)
33	Glucose and glucosamine with yeast extract	A. pullulans ATCC 42023	Submerge fermentation (50 mL batch culture)	Initial pH 6.5, impeller speed–200 rpm	8 g L^{-1}	Kim et al. (2000)
34	Glucose with ammonium sulfate and yeast extract	A. pullulans ATCC 42023	Submerge fermentation (3 L batch culture)	Initial pH 6.5, impeller speed–500 rpm, DO–50%	25.5 g L^{-1}	Lee et al. (2002)
35	Glucose with ammonium sulfate and yeast extract	A. pullulans NRM2	Submerge fermentation (100 mL batch culture)	Temp.–30°C ± 2°C, impeller speed–150 rpm, fermentation time–7 days, pH 6.5	25.1 g L^{-1}	Prasongsuk et al. (2007)
36	Glucose with ammonium sulfate and yeast extract	A. pullulans ATCC 42023	Submerge fermentation (1.5 L batch culture)	pH 6.5, temp.–28°C, impeller speed–500 rpm, sparging rate–0.5 vvm, fermentation speed–5 days	15 g L^{-1}, 3 g L^{-1} h^{-1}	Lee et al. (2001)
37	Glucose with ammonium sulfate and yeast extract	A. pullulans	Submerge fermentation (4.5 L fed-batch culture)	Sparging rate–0.5 vvm, temp.–46°C, initial pH 6.5, fermentation time–95 h, fermentation time–72 h, impeller speed–40 rpm	45 g L^{-1}, 30 g L^{-1} h^{-1}	Gaur and Singh (2010)
38	Glucose with ammonium sulfate and sodium nitrate	A. pullulans ICCF-68	Fed-batch reactor (60 L working volume)	Impeller speed–340 rpm, sparging rate–1 vvm	101 g L^{-1}	Moscovici et al. (1996)
39	Glucose with ammonium sulfate	A. pullulans ATCC 9348	Submerge fermentation (10 L batch culture)	Constant pH 4.5, temp.–28°C, sparging rate–0.4 plum	11.27 g L^{-1}	Gibbs and Seviour (1996)

40	Glucose with yeast extract and ammonium sulfate	A. pullulans CCTCC M2012259	Submerge fermentation (3 L batch culture)	Temp.—30°C, pH at 3.8, impeller speed—200 rpm, fermentation time—72 h	26.8 g L^{-1}, 8.93 g L^{-1} h^{-1}	Wang, Yu, and Gongyuan (2013)
41	Glucose with yeast extract and peptone	A. pullulans RBF 4A3	Submerge fermentation (50 mL batch culture)	Fermentation time—96 h, temp.—30°C	66.79 g L^{-1}, 16.7 g L^{-1} h^{-1}	Choudhury et al. (2011)
42	Glucose with yeast extract and ammonium sulfate	A. pullulans CCTCC M2012259	Submerge fermentation (3 L batch culture)	Temp.—30°C, impeller speed—350 rpm, sparging rate—3 lpm	25.3 ± 0.5 g L^{-1}	Wang et al. (2016)
43	Glucose with yeast extract and ammonium sulfate	A. pullulans CCTCC M2012259	Submerge fermentation (3 L batch culture)	Temp.—30°C, sparging rate—1.0 vvm, initial pH 6.8, impeller speed (500 rpm at 18 HR, 400 rpm till 72 h)	27.4 g L^{-1}	Wang et al. (2015)
44	Glucose with yeast extract ammonium sulfate	A. pullulans ATCC 9348	Submerge fermentation (50 L working volume)	DO—50%, sparging rate—1 vvm, initial pH 6.5	23.5 g L^{-1}	Wecker and Onken (1991)
45	Glucose with yeast extract and ammonium sulfate	A. pullulans ATCC 201253	Submerge fermentation (1.5 L batch culture)	Fermentation time—7 days, two-stage pH control (pH 2 for 72 h, pH 5 up to 168 h)	60.7 g L^{-1}	Cheng et al. (2010c)
46	Glucose with yeast extract and ammonium sulfate	A. pullulans CTCC M 2012259	Submerge fermentation (50 mL batch culture)	Temp.—30°C, impeller speed—200 rpm	32.28 mg g cells^{-1} h^{-1}	Ju, Wang, Zhang, Cao, and Wei (2014)
47	Grape skin pulp extract	A. pullulans NRRLY-6220	Submerge fermentation (120 mL batch culture)	Temp.—28°C, impeller speed—200 rpm, fermentation time—7 days	—	Israilides et al. (1994)
48	Hemicellulose hydrolysate with ammonium sulfate and yeast extract	A. pullulans AY82	Submerge fermentation (3 L batch culture)	Two-stage pH control (pH 3 in 1st day, pH 5 up to 7th day), sparging rate—1.5 vvm, impeller speed—400 rpm, fermentation time—7 days	12.65 g L^{-1}, 1.81 g L^{-1} h^{-1}	Chen et al. (2014)

(*Continued*)

Table 8.1: (Continued)

S. no.	Substrate (carbon and nitrogen source)	Strains	Mode of fermentation	Fermentation conditions	Pullulan production yield (g L^{-1})	Reference
49	Hydrolyzed beet molasses and yeast extract	A. pullulans P56	Submerge fermentation (50 mL batch culture)	Temp.—28°C, impeller speed—200 rpm, pH 7.5, time—96 h (beet molasses)	16.9 g L^{-1}, 4.23 g L^{-1} h^{-1}	Goksungur, Ucan, and Guvenc (2004)
50	Hydrolyzed potato starch waste and yeast extract	A. pullulans P56	Submerge fermentation (50 mL batch culture)	Fermentation time—111.8 h, impeller speed—200 rpm, pH 7.26	19.2 g L^{-1}, 4.12 g L^{-1} h^{-1}	Göksungur, Uzunoullari, and Dağbağli (2011)
51	Hydrolyzed potato starch and ammonium sulfate	A. pullulans NRRLY-6220	Submerge fermentation (50 mL batch culture)	Temp.—29°C, impeller speed—200 rpm	58 g L^{-1}	Barnett, Smith, Scanlon, and Israilides (1999)
52	Jackfruit seed with yeast extract and ammonium sulfate	A. pullulans NCIM 1049	Solid-state fermentation	Moisture content—47.9 wt.%	—	Sugumaran et al. (2013)
53	Jackfruit seed powder and yeast extract	A. pullulans MTCC 2195	Submerge fermentation (50 mL batch culture)	Temp.—30°C, impeller speed—200 rpm, fermentation time—168 h	17.95 g L^{-1}, 2.56 g L^{-1} h^{-1}	Sharmila, Muthukumaran, Nayan, and Nidhi (2013)
54	Jaggery and yeast extract	A. pollutants CFR-77	Submerge fermentation (200 mL batch)	Final pH 3.89, fermentation time—72–96 h, temp.—30°C, impeller speed—150 rpm	51.9 g L^{-1}, 17.3 g L^{-1} h^{-1}	Vijayendra et al. (2001)
55	Jaggery with corn steep liquor and deoiled Jatropha seed cake	A. pullulans RBF 4A3	Submerge fermentation (20 mL batch culture)	Temp.—28°C, impeller speed—250 rpm, fermentation time—72 h	61.17 g L^{-1}, 20.39 g L^{-1} h^{-1}	Mehta et al. (2014)
56	Molasses with peptone and yeast extract	A. pullulans MTCC 2195	Submerge fermentation (100 mL batch culture)	Impeller speed—150 rpm, temp.—35°C, fermentation time—5 days, pH 6.4	45 g L^{-1}, 9 g L^{-1} h^{-1}	Srikanth et al. (2014)

57	Peat hydrolysate and ammonium sulfate	A. pullulans 140B	Submerge fermentation (50 mL batch culture)	Temperature—20°C–25°C, impeller speed—145 rpm, fermentation time—5 days	16 g L^{-1}; 3.2 g L^{-1} h^{-1}	Boa and LeDuy (1984)
58	Peat hydrolysate with yeast extract and ammonium sulfate	A. pullulans 2552	Submerge fermentation	pH 6.5, impeller speed—150 rpm, temp.—25°C, fermentation time—7 days	14 g L^{-1}; 2 g L^{-1} h^{-1}	LeDuy and Mian Boa (1983)
59	Potato starch hydrolysate and sucrose with yeast extract and ammonium sulfate	A. pullulans 201253	Submerge fermentation (working volume—7 L)	Temp.—28°C, impeller speed—500 rpm, inocum size—5.0% (v/v)	54.57 g L^{-1}	An, Ma, Chang, and Xue (2017)
60	Rice hull hydrolysate with ammonium sulfate and yeast extract	A. pullulans CCTCC M2012259	Submerge fermentation (3 L batch culture)	Temp.—28°C, impeller speed—400 rpm, sparging rate—3 lpm, fermentation time—148 h, constant pH 3.8	22.2 g L^{-1}; 3.6 g L^{-1} h^{-1}	Wang, Ju, et al. (2014)
61	Sucrose (100 g L^{-1})	A. pullulans ATCC 201253	Biofilm reactor (1.5 L)	pH 5.0, temp.—30°C, impeller speed—200 rpm, sparging rate—1.5 vvm, time—168 h	60.7 g L^{-1}; 8.67 g L^{-1} h^{-1}	Cheng, Demirci, and Catchmark (2010a)
62	Sucrose	A. pullulans NRRLY-2311-1	Submerge fermentation (5 L reactor)	Sparging rate—1 vvm, DO—100%, impeller speed—300–500 rpm, fermentation time—4 days	26.2 g L^{-1}; 6.6 g L^{-1} h^{-1}	Gibson and Coughlin (2002)
63	Sucrose	A. pullulans ATCC 42023	—	Initial pH 6, fermentation time—168 h, temp.—30°C	10.28 g L^{-1}, 1.47 g L^{-1} h^{-1}	West (2011)
64	Sucrose and ammonium sulfate	A. pullulans P 56	Submerge fermentation (batch—6 L, fed-batch—3 L)	pH 7, impeller speed—650 rpm, airflow rate flow—3 lpm	31.3 g L^{-1} (batch), 24.5 g L^{-1} (fed-batch)	Youssef et al. (1999)
65	Sucrose and ammonium sulfate	A. pullulans	Immobilized cell bioreactor (polyurethane foam) (continuous mode)	Temp.—42°C, pH 5.5, sparging rate—0.5 vvm	37 g L^{-1}	Singh, Saini, and Kennedy (2011)
66	Sucrose and ammonium sulfate	A. pullulans ATCC 42023	Submerge fermentation (100 mL batch culture)	Temp.—28°C–30°C, impeller speed—250 rpm, fermentation time—7 days, fermentation time—5th day	22.6 g L^{-1}; 4.52 g L^{-1} h^{-1}	Lin, Zhang, and Thibault (2007)

(Continued)

Table 8.1: (Continued)

S. no.	Substrate (carbon and nitrogen source)	Strains	Mode of fermentation	Fermentation conditions	Pullulan production yield (g L^{-1})	Reference
67	Sucrose and glutamic acid	A. pullulans ATCC 42023	Submerge fermentation (12 l working volume batch)	Fermentation time—60 h, temp.—28°C, pH 4.0, impeller speed—300 rpm, supplement—soybean oil	35 g L^{-1}, 14 g L^{-1} h^{-1}	Shabtai and Mukmenev (1995)
68	Sucrose and sodium nitrate	A. pullulans YS67	Submerge fermentation (100 mL batch culture)	Temp.—28°C, impeller speed 200 rpm	40.1 g L^{-1}	Ravella et al. (2010)
69	Sucrose and sodium nitrate	A. pullulans KCTC 6081	Fed-batch (3 L working volume)	pH 4, two-stage impeller speed—(200 for 40 h, 600 rpm for 120 h)	47 g L^{-1}, 9.4 g L^{-1} h^{-1}	Seo et al. (2014)
70	Sucrose and yeast extract	A. pullulans AP329	Submerge fermentation (3 L batch culture)	Two-stage (pH 2.6 for 18 HR, 5.5 for 30 h), impeller speed—800 rpm, airflow rate—4 lpm, fermentation time—60 h	31 g L^{-1}, 12.4 g L^{-1} h^{-1}	Xia, Wu, and Pan (2011)
71	Sucrose and yeast extract	A. pullulans var. melanogenium P16	Submerge fermentation (6.5 L batch culture)	Impeller speed—300 rpm, sparging rate—6.5 lpm, temp.—28°C, fermentation time—132 h	65.3 g L^{-1}, 11.87 g L^{-1} h^{-1}	Ma, Chi, Geng, Zhang, and Chi (2012)
72	Sucrose and yeast extract	A. pullulans IFO 4464	Submerge fermentation (2 L Jar fermenter)	Two-stage pH control (pH 6.0→3.0), temp.—23°C, sparging rate—1 vvm	22 g L^{-1}	Lee and Yoo (1993)
73	Sucrose and yeast extract	A. pullulans CJ001	Submerge fermentation (3 L batch culture)	Constant pH 5.5, optimum fermentation time—4 days	32.5 g L^{-1}, 8.13 g L^{-1} h^{-1}	Pan et al. (2013)
74	Sucrose and yeast extract	A. pullulans ATCC 42023	Submerge fermentation (12 L batch culture)	Impeller speed—150 rpm, pH 5.3, fermentation time—102 h	23.25 g L^{-1}, 0.3 g L^{-1} h^{-1}	Lin and Thibault (2013)
75	Sucrose and yeast extract	A. pullulans N3.387	Submerge fermentation, 400 mL (batch culture)	Impeller speed—180 rpm, temp.—28°C, fermentation time—72 h	20.78 ± 0.57 g L^{-1}	Kang et al. (2011)
76	Sucrose and yeast extract	A. pullulans CGMCC1234	Submerge fermentation (3 L batch culture)	pH 6.5, temp.—28°C, sparging rate—3 lpm, impeller speed—300 rpm, supplement—1% Tween 80	55 g L^{-1}	Sheng et al. (2013)
77	Sucrose and yeast extract	A. pullulans CGMCC1234	Submerge fermentation (50 mL batch culture)	Impeller speed—200 rpm, pH 6.5, two-stage temp. (32°C for first 2 days, 26°C for last 2 days)	27.39 g L^{-1}, 6.82 g L^{-1} h^{-1}	Wu, Jin, Tong, et al. (2009)

#	Substrate	Organism	Fermentation	Conditions	Yield	Reference
78	Sucrose and yeast extract	A. pullulans SK1002	Submerge fermentation (50 mL batch culture)	Temp.—28°C, pH 5.5, fermentation time—5 days	30.28 g L^{-1}, 6.06 g L^{-1} h^{-1}	Jiang (2010)
79	Sucrose and yeast extract	A. pullulans FB-1	Submerge fermentation (100 mL batch culture)	Initial pH 6.5, temp.—30°C, impeller speed—150 rpm, fermentation time—7 days	23.1 g L^{-1}, 3.3 g L^{-1} h^{-1}	Singh and Saini (2008)
80	Sucrose and yeast extract	A. pullulans CJ001	Submerge fermentation (50 mL batch culture)	Fermentation time—4 days, pH 5.0, temp.—22°C	31.25 g L^{-1}, 7.81 g L^{-1} h^{-1}	Wu et al. (2012)
81	Sucrose and yeast extract	A. pullulans DSM2404	Airlift reactor (1.3 L working volume)	pH 7.5, temp.—28°C, fermentation time—5.36 days, sparging rate—1.93 vvm	39.2 g L^{-1}, 7.36 g L^{-1} h^{-1}	Özcan et al. (2014)
82	Sucrose and yeast extract	A. pullulans FB-1	Submerge fermentation (100 mL batch culture)	Initial pH 6.5, fermentation time—7 days, temp.—25°C	—	Singh and Saini (2008)
83	Sucrose and ammonium sulfate	A. pullulans RG5	Submerge fermentation (6 L batch reactor)	Temp.—42°C, pH 5.5, airflow rate—0.5 vvm	37.1 g L^{-1}	Sheoran, Dubey, Tiwari, and Singh (2012)
84	Sucrose and hydrolyzed soya extract	A. pullulans MTCC 1991 (mutant)	Submerge fermentation (3.5 L batch culture)	Impeller speed—180 rpm, pH 5.9, air supply—1.5 vvm	125.7 g L^{-1}, 17.95 g L^{-1} h^{-1}	Sheoran et al. (2012)
85	Sucrose (15 g L^{-1}) and ammonium sulfate	A. pullulans ATCC 201253	Biofilm reactor with continuous model (1.5 L)	Dilution rate—0.16 h^{-1}	8.3 g L^{-1} and 1.33 g L^{-1} h^{-1}	Cheng et al. (2010a)
86	Sucrose with ammonium sulfate and yeast extract	A. pullulans 2552, 140B	Submerge fermentation (9 L batch culture)	Two-stage pH control (pH 2.0 for 4 days→4.0 for 5.07 days), sparging rate—0.8 vvm	27 g L^{-1}, 5.33 g L^{-1} h^{-1}	Lacroix et al. (1985)
87	Sucrose with ammonium sulfate and yeast extract	Eurotium chevalieri MTCC 9614	Submerge fermentation (250 mL batch culture)	Initial pH 5.5, temp.—35°C, impeller speed—30 rpm, fermentation time—65 h	38 g L^{-1}, 14.07 g L^{-1} h^{-1}	Gaur and Singh (2010)
88	Sucrose with ammonium sulfate and yeast extract	A. pullulans FB-1	Submerge fermentation (50 mL batch culture)	Impeller speed—150 rpm, fermentation time—7 days, pH 6.5	44.4 g L^{-1}, 6.34 g L^{-1} h^{-1}	Singh et al. (2009)
89	Sucrose with ammonium sulfate and yeast extract	A. pullulans P56	Submerge fermentation (5 L batch culture)	Initial pH 7.5, temp.—28°C, impeller speed—345 rpm, sparging rate—2.36 vvm	17.2 g L^{-1}	Göksungur, Dağbağli, Uçan, and Güvenç (2005)

(Continued)

Table 8.1: (Continued)

S. no.	Substrate (carbon and nitrogen source)	Strains	Mode of fermentation	Fermentation conditions	Pullulan production yield (g L^{-1})	Reference
90	Sucrose with ammonium sulfate and yeast extract	A. pullulans P56	Immobilization/batch and repeated batch fermentation	Initial pH 7.3, impeller speed—191.5 rpm, fermentation time—101.2 h	21.1 g L^{-1}	Ürküt, Dağbağli, and Göksungur (2007)
91	Sucrose with ammonium sulfate and yeast extract	A. pullulans P 56	Submerge fermentation (100 mL batch culture)	Temp.—28°C, time—168 h, impeller speed—200 rpm, supplement—olive oil (25 mL L^{-1})	51.5 g L^{-1}, 7.36 g L^{-1} h^{-1}	Youssef et al. (1998)
92	Sucrose with ammonium sulfate and yeast extract	A. pullulans CJ001	Submerge fermentation (50 mL batch culture)	Impeller speed—200 rpm, temp.—28°C, initial pH 5.5, fermentation time—96 h	26.13 g L^{-1}, 6.53 g L^{-1} h^{-1}	Chen, Wu, and Pan (2012)
93	Sucrose with ammonium sulfate and yeast extract	A. pullulans NRRL Y-2311-1	Submerge fermentation (100 mL batch culture)	Fermentation time—96th hour, temp.—26°C, impeller speed—250 rpm, fermentation time—120 h	26.24 g L^{-1}, 5.25 g L^{-1} h^{-1}	Sena et al. (2006)
94	Sucrose with ammonium sulfate and yeast extract	A. pullulans W518	Submerge fermentation (5 L batch reactor)	Temp.—28°C ± 1.5°C, pH 6.5, optimum time—132 h, optimum aerate rate—1.0 vvm, impeller speed—500 rpm	35 g L^{-1}, 6.36 g L^{-1} h^{-1}	Ma et al. (2012)
95	Sucrose with ammonium sulfate and yeast extract	A. pullulans IMI 145194	Oscillatory baffled fermenter (working volume of 2.5 dm^3)	Sparging rate—1 vvm, initial pH 4.5, temp.—28°C, fermentation time—52 h	17 g L^{-1}, 7.84 g L^{-1} h^{-1}	Gaidhani, McNeil, and Ni (2005)
96	Sucrose with yeast extract and ammonium sulfate	A. pullulans ATCC 42023	Submerge fermentation (12 L batch culture)	Impeller speed—150 rpm, sparging rate—0.92 vvm	23.25 g L^{-1}	Lin and Thibault (2013)

97	Sucrose with yeast extract and ammonium sulfate	A. pullulans P56	Submerge fermentation (2 L airlift reactor)	pH 7.5, fermentation time—144 h, sparging rate—2 vvm, impeller speed—200 rpm	30 g L^{-1}, 5 g L^{-1} h^{-1}	Roukas and Mantzouridou (2001)
98	Sucrose with yeast extract and ammonium sulfate	A. pullulans 2552	Submerge fermentation (1.5 L batch culture)	Pressure—0.6 MPa, sparging rate—2 vvm, initial pH 5.5	13.8 g L^{-1}	Dufresne et al. (1990)
99	Sucrose with yeast extract and ammonium sulfate	A. pullulans	Submerge fermentation (12 L batch fermenter)	Sparging rate—1 vvm, temp.—28°C, initial pH 4.5, impeller—750 rpm, sparging rate—1 vvm	19.5 g L^{-1}	McNeil et al. (1989)
100	Sucrose with yeast extract and ammonium sulfate	A. pullulans IFO 4464	Submerge fermentation (3 L fed-batch culture)	Fed-batch mode started after 3 days, temp.—27°C, impeller speed—400 rpm, sparging rate—1 vvm	58.0 g L^{-1}	Shin et al. (1989)
101	Sucrose with yeast extract and ammonium sulfate	A. pullulans CGMCC1234	Submerge fermentation (3 L batch culture)	Temp.—28°C, sparging rate—3 lpm, impeller speed—300 rpm, initial pH 6.5, time—96 h, supplement—uracil (5 mM)	49.07 g L^{-1}, 12.27 g L^{-1} h^{-1}	Sheng et al. (2013)
102	Sweet potato and yeast extract	A. pullulans AP329	Submerge fermentation (50 mL batch culture)	pH 6.5, temp.—28°C, fermentation time—96 h, impeller speed—200 rpm	29.4 g L^{-1}, 7.35 g L^{-1} h^{-1}	Wu et al. (2009)
103	Glucose	Aureobasidium pullulans CCTCC M2012259	Submerge fermentation	Temp.—30°C, impeller speed—200 rpm, supplement—copper sulfate (0.1 mg L^{-1})	20 g L^{-1}	Wang et al. (2016)

(Continued)

Table 8.1: (Continued)

S. no.	Substrate (carbon and nitrogen source)	Strains	Mode of fermentation	Fermentation conditions	Pullulan production yield (g L^{-1})	Reference
104	Beet molasses	A. pullulans P 56	Submerged	Temp.—28°C, fermentation time—8 days, impeller speed—200 rpm	24 g L^{-1}	Lazaridou, Biliaderis, et al. (2002)
			Shake-flask fermentationSubmerged stirred tank reactor fermentation	Temp.—28°C, fermentation time—8 days, impeller speed—700 rpm, sparging rate—1.4 dm^3 min^{-1}	49 g L^{-1}	Youssef et al. (1998)
			Submerged shake-flask fermentation	Temp.—28°C, fermentation time—5 days, impeller speed—200 rpm	28.2 g L^{-1}	Roukas and Serris (1999)
			Submerged airlift reactor fermentation	Temp.—28°C, 5 days, shear rate—42 s^{-1}, sparging rate—2 vvm	27 g L^{-1}	Roukas and Liakopoulou-Kyriakides (1999)
			Submerged stirred tank reactor fermentation	Temp.—28°C, fermentation time—5 days, impeller speed—650 rpm, sparging rate—1 vvm	23 g L^{-1}	Goksungur et al. (2004)
			Submerged shake-flask fermentation	Temp.—28°C, fermentation time—5 days, impeller speed—200 rpm	16.7 g L^{-1}	Goksungur et al. (2004)
			Submerged stirred tank reactor fermentation	Temp.—28°C, fermentation time—5 days, impeller speed—400 rpm, sparging rate—2 vvm	6.6 g L^{-1}	Srikanth et al. (2014)
		A. pullulans MTCC 2195	Submerged shake-flask fermentation	Temp.—35°C, impeller speed—150 rpm, fermentation time—5 days	45 g L^{-1}	
105	Cassava bagasse	A. pullulans MTCC 1991	Submerged solid-state fermentation in flasks	Temp.—28°C, fermentation time—7 days	32 g L^{-1}	Ray and Moorthy (2007)
		A. pullulans MTCC 2670	Submerged	Temp.—30°C, fermentation time—5 days	19 g L^{-1}	Sugumaran and Ponnusami (2017a)

106	Corn steep liquor	*A. pullulans* RBF 4A3	Submerged shake-flask fermentation	Temp.—28°C, fermentation time—5 days, impeller speed—200 rpm	88.59 g L^{-1}	Sharma et al. (2013)
107	Coconut milk	*A. pullulans* ATCC 42023	Submerged shake-flask fermentation	Temp.—28°C, fermentation time—5 days, impeller speed—210 rpm	65.3 g L^{-1}	Abdel Hafez et al. (2007)
108	Coconut water	*A. pullulans* MTCC 2195	Submerged shake-flask fermentation	Temp.—28°C, fermentation time—7 days, impeller speed—200 rpm	58 g L^{-1}	Thirumavalavan et al. (2009)
109	Deoiled rice bran	*A. pullulans* MTCC 2195	Submerged shake-flask fermentation	Temp.—28°C, fermentation time—7 days, impeller speed—200 rpm	38.8 g L^{-1}	Thirumavalavan et al. (2009)
110	Grape skin pulp extract	*A. pullulans* MTCC 6994	Submerged shake-flask fermentation	Temp.—30°C, fermentation time—7 days, impeller speed—150 rpm	54.8 g L^{-1}	Singh and Kaur (2019)
111	Jackfruit seeds	*A. pullulans* NRRLY-6220	Submerged shake-flask fermentation	Temp.—28°C, fermentation time—5 days, impeller speed—200 rpm	22.30 g L^{-1}	Israilides et al. (1998)
		A. pullulans NCIM 1049	Submerged solid-state fermentation in flasks	Temp.—28°C, fermentation time—7 days	34.22 g L^{-1}	Sugumaran et al. (2013)
112	Jatropha seed cake	*A. pullulans* MTCC 2195	Submerged shake-flask fermentation	Temp.—30°C, fermentation time—7 days, impeller speed—200 rpm	18.76 g L^{-1}	Sharmila et al. (2013)
		A. pullulans RBF 4A3	Submerged shake-flask fermentation	Temp.—28°C, fermentation time—5 days, impeller speed—200 rpm	83.98 g L^{-1}	Choudhury et al. (2012)
113	Palm kernel	*A. pullulans* MTCC 2670	Submerged solid-state fermentation in flasks	Temp.—30°C, fermentation time—7 days	18.43 g L^{-1}	Sugumaran et al. (2013)
114	Potato starch water	*A. pullulans* NRRLY-6220	Submerged shake-flask fermentation	Temp.—29°C, fermentation time—7 days, impeller speed—200 rpm	69 g L^{-1}	Barnett et al. (1999)
		A. pullulans 201253	Submerged stirred tank reactor fermentation	Temp.—28°C, fermentation time—5 days, impeller speed—500 rpm	54.57 g L^{-1}	An et al. (2017)
		A. pullulans P 56	Submerged shake-flask fermentation	Temp.—28°C, fermentation time—5 days, impeller speed—200 rpm	19.2 g L^{-1}	Göksungur et al. (2011)

(*Continued*)

Table 8.1: (Continued)

S. no.	Substrate (carbon and nitrogen source)	Strains	Mode of fermentation	Fermentation conditions	Pullulan production yield (g L^{-1})	Reference
115	Rice hull	A. pullulans CCTCCM 2012259	Submerged stirred tank reactor fermentation	Temp.—28°C, fermentation time—3 days, impeller speed—400 rpm, sparging rate—3 L min^{-1}	22.2 g L^{-1}	Wang et al. (2014)
116	Soybean pomace	A. pullulans MTCC 1991	Submerged stirred tank reactor fermentation	Temp.—27°C, fermentation time—7 days, impeller speed—210 rpm, sparging rate—1.25 vvm	125.7 g L^{-1}	Sheoran et al. (2012)
		A. pullulans HP-2001	Submerged shake-flask fermentation	Temp.—30°C, Fermentation time—4 days, impeller speed—200 rpm	7.5 g L^{-1}	Seo et al. (2004)
117	Spent grain liquor	A. pullulans P 56	Submerged shake-flask fermentation	Temp.—28°C, fermentation time—5 days, impeller speed—200 rpm	11 g L^{-1}	Roukas (1999b)
118	Sugarcane bagasse	A. pullulans LB83	Submerged shake-flask fermentation	Temp.—28°C, fermentation time—4 days, impeller speed—200 rpm	15.7 g L^{-1}	Terán Hilares et al. (2017)
		A. pullulans LB83	Submerged shake-flask fermentation	Temp.—25.3°C, fermentation time—4 days, impeller speed—232 rpm	25.19 g L^{-1}	Terán Hilares et al. (2019)
119	Sugarcane molasses	A. pullulans MTCC 2195	Submerged shake-flask fermentation	Temp.—35°C, fermentation time—5 days, impeller speed—150 rpm	45 g L^{-1}	Srikanth et al. (2014)
120	Sweet potato hydrolysate	A. pullulans AP329	Submerged shake-flask fermentation	Temp.—28°C, fermentation time—4 days, impeller speed—200 rpm	29.43 g L^{-1}	Wu et al. (2009)
121	Whey	A. pullulans ATCC 42023	Submerged shake-flask fermentation	Temp.—28°C, fermentation time—5 days, impeller speed—210 rpm	12 g L^{-1}	Abdel Hafez et al. (2007)

production of pullulan (Cheng et al., 2011a; LeDuy & Mian Boa, 1983; Lee et al., 2001). Many studies have reported the elongation of polysaccharide in the production medium to be associated with a particular morphology of the microbial cells (Imshenetskiĭ, Kondrat'eva, & Smut'ko, 1981b; Pan, Yao, Chen, & Wu, 2013; Roukas & Biliaderis, 1995; Simon et al., 1993). The yeast-like cells at neutral pH make pullulan at a very high molecular weight greater than 2,000,000 (Lee et al., 2001), while the integrated cultivation of mycelial and yeast-like cellular forms give an elevated concentration of pullulan (Roukas & Biliaderis, 1995). The molecular weight of pullulan is also altered by the pH of the medium (Singh, Saini, & Kennedy, 2008; Wu, Chen, Jin, & Tong, 2010). Pan et al. (2013) explained the influence of pH of production medium on the production of pullulan by utilizing the activity of UDPG-pyrophosphorylase. The maximum production of pullulan and enzyme activity was accomplished with a persistent medium pH of 5.5 (Pan et al., 2013).

Temperature is an important factor to be considered for the higher synthesis of pullulan with optimum value in the range of 24°C–30°C (LeDuy & Mian Boa, 1983; McNeil & Kristiansen, 1990; Roukas & Mantzouridou, 2001; Tsujisaka & Mitsuhashi, 1993; Zheng et al., 2008). To enhance both the cell growth and pullulan production, a two-stage temperature control strategy had been reported (Wu et al., 2010). The optimum temperature was 32°C and 26°C for the cell growth and pullulan production, respectively (Wu et al., 2010). Cell growth is promoted by high temperature but does not provide the same support to pullulan synthesis. Just like pH, the temperature of the production medium also influences the form and structure of cells, which in turn affects pullulan production. *A. pullulans* have been cultivated batch-wise on media comprising starch hydrolysates, peptone, phosphate, and basal salts, and the initial culture pH was 6.5 for commercial production (Tsujisaka & Mitsuhashi, 1993). Optimal pullulan yield was obtained in 100 h at 30°C (Singh et al., 2008).

8.2.3.5 Oxygen uptake

The production of pullulan is an aerobic process; hence, cells require adequate aeration for their growth and pullulan production (Leathers, Nofsinger, Kurtzman, & Bothast, 1988). Enhanced pullulan concentration due to increased cell accumulation as a result of profound aeration of about 2.0 vvm (volume of airflow per volume of medium per minute) has been reported (Roukas & Mantzouridou, 2001). According to Wecker and Onken (1991), a decrease in dissolved oxygen concentration till a constant amount increased pullulan production due to decreased rate of shear in the cells. α-Amylase is the main enzyme responsible in the determination of the molecular weight of pullulan (Leathers, 1987) and its activity could alter with intense aeration of fermentation broth. Hence proper caution should be taken while providing aeration to the fermentation broth as the molecular weight of pullulan produced will decrease under intense aeration conditions (Audet, Lounes, & Thibault, 1996).

During the production of pullulan using *A. pullulans*, Cheng et al. (2011a) reported the highest pullulan concentration of 25.8 g L^{-1} when aeration was 1.5 vvm. Also, at this aeration rate, both productivity and yield of pullulan were increased. The oxygen transfer rate and pullulan production by *A. pullulans* increased when partial air pressure was increased up to a critical value of 0.5−0.75 MPa (Dufresne, Thibault, Leduy, & Lencki, 1990). The spontaneous clumping of cells was observed as the pressure exceeded the critical value and subsequently pullulan production stopped.

8.3 Microbial consortia for the production of pullulan

Many microorganisms can be used for the production of pullulan among which *A. pullulans* is the most widely used due to its high pullulan production and yield (Table 8.1) (Singh et al., 2008). Besides *A. pullulans*, other reported microorganisms include a mycoparasitic saprophyte fungus *Tremella mesenterica* (Fraser & Jennings, 1971), obligate parasitic fungi like *Cytariaharioti* and *C. darwinii* (Chi & Zhao, 2003; Reis, Tischer, Gorin, & Iacomini, 2002), a fungal agent of chestnut blight, *Cryphonectria parasitica* (Delben, Forabosco, Guerrini, Liut, & Torri, 2006; Forabosco et al., 2006), and *Teloschistes flavicans* (Reis et al., 2002). Yeast strains like *Rhodotorulabacarum* (Chi & Zhao, 2003) and *Rhodosporidium paludigenum* as pullulan producers (Singh, Kaur, Sharma, & Rana, 2018).

8.3.1 Aureobasidium pullulans

A. pullulans which is also called *Pullularia pullulans* is widely found in the environment and isolated from forest soil, fresh and seawater, and as saprophytes from decaying leaf litter, wood plant, animal tissues, etc. (Jiao, Fu, & Jiang, 2004; Wang et al., 2019; Wu, Jin, Kim, Tong, & Chen, 2009). Bauer in 1938 was a pioneer to discover the microbial production of pullulan by *A. pullulans* (Bauer, 1938). Strains of *A. pullulans* can be color differentiated as typical strains appear off-white to black and are named "black yeast" producing black melanin pigment while other strains produce lesser amount of melanin and appear brightly red, yellow, pink, or purple pigmented (Constantin, Bucătariu, Stoica, & Fundueanu, 2017; Dufresne et al., 1990). The melanin produced by the microorganism provides resistance to the cells toward phagocytosis in the host (Chabasse, 2002).
A. pullulans are considered to be among the deuteromycetes (Jiao et al., 2004; Kachhawa, Bhattacharjee, & Singhal, 2003). It has been observed that the fungus is omnivorous with appreciable biotechnological importance and produces various degradative enzymes, including amylases, proteases, pectinases, esterases, etc. (Chi et al., 2009). Different varieties of *A. pullulans* are distinguished by their molecular characteristics, nutritional physiology, composition, and structure of EPS produced (Jiang, Wu, & Kim, 2011; Wu, Jin, Kim, Tong, & Chen, 2009). The major EPS produced by *A. pullulans* is pullulan (Bauer,

1938) while another EPS aubasidan is produced by *A. pullulans* var. aubasidsani (Elinov, Neshatayeva, Dranishnikov, & Matveyeva, 1974).

There is structural similarity of pullulan produced by different *A. pullulans* strains and the structure has been accepted as a linear polysaccharide with maltotriose repeating units joined by α-1,6-linkages (Kikuchi et al., 1973). Simultaneous production of melanin with pullulan at large scale creates a major problem due to the discoloration of polysaccharide because of the pigment (Pollock et al., 1992). The produced melanin is removed from the fermentation broth by adsorption or solvent combinations. However, several strains with reduced pigmentation have also been identified for pullulan production eliminating the need for excess pigment removal (Bender et al., 1959; Elinov et al., 1974).

8.3.2 Cell morphologies of Aureobasidium pullulans

The yeast-like polymorphic fungus *A. pullulans* shows many cell morphologies during its life-cycle, including the formation of elongated branched septate, large chlamydospores, young blastospores, swollen blastospores, and mycelia (Ronen et al., 2002). The morphological studies have shown that the production of pullulan is not based on any specific morphology of the fungus (Seviour, Kristiansen, & Harvey, 1984). There are different morphological states for the major pullulan producers, which include blastospore cells (Gibbs & Seviour, 1992; Pan et al., 2013), yeast-like cells (Imshenetskiĭ et al., 1981b; Roukas & Biliaderis, 1995), and hyphal cells (Yurlova & Hoog, 1997). However, studies by Cheng et al. (2011a) and LeDuy and Mian Boa (1983) showed that the increment in hyphal cells after the attachment on solid support did not affect the total productivity by *A. pullulans*.

8.4 Production of pullulan by fermentation of agroindustrial by-products

In industries, pullulan is commonly produced by the nonpathogenic and nontoxic strain of *A. pullulans*. The efficacy of production generally depends upon the fermentation process parameters, morphological state of microorganism, and the type of substrate used. The fermentation medium used for commercial pullulan production consists of peptone, phosphate, and basal salts with pH adjusted to 6.5 and starch hydrolysates at 10%–15% concentration as the carbon source (Tsujisaka & Mitsuhashi, 1993). As the fermentation proceeds, the maximum growth of culture media is observed within 75 h while the pH of the medium decreases to ~3.5 during the initial 24 h and the optimal pullulan yield is obtained within 100 h (Tsujisaka & Mitsuhashi, 1993). For promoting the proper growth of cultures and metabolite formation, aeration and agitation are done and the temperature is maintained at 30°C throughout the process. Many strains used for pullulan production release melanin, imparting color to the fermentation broth, which is removed during the downstream process to reduce the hindrance in pullulan purification. Various strains of *A. pullulans* have been used for the production of pullulan using sucrose and glucose in shake flasks and stirred tank fermenter achieving productivity of 4.5 g L^{-1} day^{-1} with maximum pullulan concentration at 31.3 g L^{-1} (Youssef, Biliaderis, & Roukas, 1998).

Many pullulan productions studies have reported the use of various organic waste extracts and by-products such as molasses, starch waste, deoiled seed cakes as the substrates of fermentation as mentioned in Table 8.1. A two-stage fermentation process for the production of pullulan without melanin colorization by fungal strain *A. pullulans* ATCC 42023 was developed wherein the first-stage fermentation used soybean oil and glutamate as carbon and nitrogen source, respectively, at pH 4.5 producing 15 g L^{-1} dry cell weight without any pigment formation in the culture and low concentration of pullulan (<2 g L^{-1}) (Shabtai & Mukmenev, 1995). When the former nutrient sources were exhausted, the cells were transferred to the production medium in the second stage using sucrose as a carbon source with nitrogen limitation, resulting in the production of 35 g L^{-1} pullulan in 50 h with highly viscous broth than the first stage (Shabtai & Mukmenev, 1995).

Molasses is a by-product of sugar-processing industry released during the conversion of sugarcane or sugar beet juice to sugar granules. It is a dark brown viscous liquid consisting of high amounts (48%−60%) of glucose and fructose as fermentable sugars, total solids (70%−85%), organic content (9%−12%), and inorganic ash (10%−15%) (Singh, Kaur, & Kennedy, 2019). Although molasses contains a high content of sugars, which could be used as a carbon source for pullulan production (Israilides, Bocking, Smith, & Scanlon, 1994), the presence of heavy metals such as copper and zinc inhibits the microbial growth and product yield (Roukas, 1998). This hindrance caused by the heavy metals could be overcome by pretreatment of molasses by the addition of 1 N sulfuric acid, followed by centrifugation (Roukas, 1998). The use of pretreated molasses in the fermentation medium significantly reduces the pullulan production cost (Srikanth et al., 2014) and the efficiency of production is also enhanced as compared to the use of raw molasses. The use of pretreated molasses improved pullulan production in stirred tank reactors by 49.0 g L^{-1} (Lazaridou, Roukas, Biliaderis, & Vaikousi, 2002) and 23.0 g L^{-1} (Roukas & Liakopoulou-Kyriakides, 1999) and by 18.5 g L^{-1} in an airlift reactor (Roukas & Serris, 1999). Besides heavy metals, the removal of other coloring substances and amino acids can also be achieved by the treatment with activated carbon along with sulfuric acid, further increasing the pullulan production capacity (Orr et al., 2009; Xi et al., 1996).

Whey is also a by-product released from the dairy industries during the processing of cheese. The coagulation of milk fats and proteins releases whey as a liquid by-product, which contains several nutrients, making it an efficacious and cost-effective substrate for pullulan production. It contains about 4.5% lactose, 1.0% salts, 0.8% proteins, and 0.1%−0.8% lactic acid (Yang, Zhu, Li, & Hong, 1994). Whey has been used after deproteinization as a substrate for fermentative pullulan production by *A. pullulans* in shake flask (Roukas, 1999b). To increase its utility in fermentation processes, whey is deproteinized by heating at 90°C for about 20 minutes after which precipitated proteins are filtered to obtain whey with approximately 80% of lactose (Roukas, 1999b). Increased pullulan yield in shake-flask fermentation has also been observed using co-cultures of *A. pullulans* and *Ceratocystis ulmi* and a lactose-based medium (LeDuy & Mian Boa, 1983).

In another study, coconut by-products were used for pullulan production using *A. pullulans* where the seed medium consisted of sucrose, potassium dihydrogen phosphate, yeast extract, ammonium sulfate, and sodium chloride and the highest concentration of pullulan obtained was 54 g L^{-1} in coconut milk after 36 h (Mehta et al., 2014; Yuen, 1974). Wastes from the brewery industries such as spent grain liquor have been used for pullulan production due to their high content of organic nutrients, which is the reason for its high biochemical oxygen demand, causing pollution if released untreated. Spent grain liquor is a by-product formed when the wort is separated from the spent grains during the production of beer and other brewery products. It comprises (w/v%) hemicelluloses (40%), cellulose (12%), starch (2.7%), proteins (14.2%), lignin (11.5%), lipids (13%), and ash (3.3%) (Xiros, Topakas, Katapodis, & Christakopoulos, 2008). Roukas used spent grain liquor as a substrate for pullulan production by *A. pullulans* where the production improved when the medium was supplemented with K_2HPO_4 (0.5%, w/v), L-glutamic acid (1%, v/v), olive oil (2.5%, v/v), and Tween 80 (0.5%, v/v) (Roukas & Serris, 1999).

Carob is a leguminous tree whose seeds are present in a pod-like structure, which is edible, sweet, and rich in vitamins and minerals. The extracts of carob pods have been used for the production of pullulan. Batch fermentation of carob pod extract by *A. pullulans* for the production of extracellular polysaccharides was done at optimum conditions of initial sugar at 25 g L^{-1}, initial pH 6.5, and temperature in a range of 25°C–30°C. At these initial conditions with lower limit of temperature (25°C) and 10% inoculum (v/v), a maximum polysaccharide concentration, polysaccharide productivity, total biomass concentration, and polysaccharide yield were obtained, respectively, as 6.5 g L^{-1}, 2.16 g L^{-1} day^{-1}, 6.3 g L^{-1}, and 30% (Roukas & Biliaderis, 1995). Also, at similar conditions of initial sugar, pH, and 30°C, the highest fermentation efficiency (89%) and maximum proportion of pullulan (70% of total polysaccharides) were achieved (Prajapati et al., 2013).

Grape pomace is a waste product of wine and juice industries obtained after the processing of grape skin and juice-free pulp. Total sugar content in grape pomace is about 85.2% in which reducing sugars contribute to 3.4% with 1.28% glucose (Singh, Saini, & Kennedy, 2009; Yuen, 1974). It also contains proteins (7.8%), some pigments, acids, and salts. Grape pomace extract has been used to produce homogenous, high molecular weight pullulan with improved yield (Bambalov & Jordanov, 1993; Israilides et al., 1994). Israilides et al. (1998) used grape pomace extract (prepared by dissolving grape pomace in hot water and filtration) for fermentative pullulan production by *A. pullulans* and achieved a concentration of 22.3 g L^{-1} in shake flask.

Jatropha seed cake has also been used for the production of pullulan. Deoiled Jatropha seed cake consists of crude protein (43.48% ± 0.83%), carbohydrates (32.54% ± 0.12%), ash (8.50% ± 0.07%), crude fiber (5.73% ± 0.07%), and ether extract (1.54% ± 0.09%) (Sánchez-Arreola et al., 2015), besides some toxic constituents such as phorbol esters and phytate. The high toxicity makes it unsuitable for use as food and feeds while its nutritional aspects could be exploited for the production of microbial metabolites like pullulan. It has been used in submerged fermentation as a substrate for pullulan production by *A. pullulans* under optimized conditions in shake flask with 8% deoiled seed cake and 15% dextrose at 28°C and pH 6.0, which resulted in 83.98 g L^{-1} pullulan with further process validation in a laboratory-scale fermenter for development of a cost-effective technology (Choudhury, Sharma, & Prasad, 2012).

8.5 Bioreactors and mode of operation for the production of pullulan

The production of pullulan is a fermentative process carried out by the action of living organisms. Pullulan was first introduced in the commercial market by Yuen (1974). The industrial production of pullulan by microorganisms depends upon the mode of fermentation, reactor type and design, operating conditions, inoculum age, and composition of the medium (Lazaridou, Biliaderis, & Kontogiorgos, 2003; Leathers, 2003; Thirumavalavan et al., 2009).

8.5.1 Batch fermentation

In the batch mode of fermentation, a batch of reactants is introduced into the reactor and the operation continues until the product is formed in the reactor. The operating conditions such as aeration, agitation, temperature, and pH are maintained as per the requirement of the ongoing process in the reactor. Mixing or stirring is performed by different types of impellers, gas bubbles, etc., while the temperature of the reactor broth is generally maintained by cooling jackets and reflux condensers. Two-stage temperature, pH, and agitation control strategies in submerged fermentation operating in batch mode have been used for improving the growth of microbes and simultaneous production (Leathers, 1987; Lee et al., 2001; Özcan, Sargin, & Göksungur, 2014). The control of agitation speed, broth behavior, and shear rate are important aspects to be considered during the submerged fermentation since high agitation speed can cause microbial damage and a high shear rate (Lazaridou, Biliaderis, et al., 2002). Yeast-like cells show an enhanced growth rate under conditions of adequate aeration, moderate agitation, and low shear rate leading to increased pullulan productivity (Wecker & Onken, 1991). Elongation of polysaccharide structure in submerged conditions causes medium viscosity to increase which in turn increases the oxygen demand of the culture medium.

Product formation in solid-state fermentation is primarily influenced by the type of substrate, the composition of the medium, moisture content of solid bed, and particle size (Prajapati et al., 2013). As compared to submerged fermentation, solid-state processes have the advantage of less water consumption, the lower requirement of energy, and the production cost is also comparatively less (Prajapati et al., 2013). The production of pullulan by solid-state fermentation has been reported only in few studies, which might be attributed to the tedious operating conditions involved in the process (Chabasse, 2002; Sugumaran, Jothi, & Ponnusami, 2014; Xiao, Lim, & Tong, 2012). Small-sized solid substrate particles provide high surface area for microbial attachment and growth but reduce the aeration throughout due to low void volume and porosity of solid substrates. Low porosity and void volume cause retention of high moisture levels leading to reduced oxygen transfer rate. As a result, the overall metabolite production is reduced. However, if the moisture ratio is low, it adversely affects the mass transfer rate of the substrate due to the insufficiency of water (Sugumaran et al., 2014). When the size of particles is large, higher void volume and greater aeration rate are achieved but the reduced specific surface area leads to the poor attachment of microorganisms (Prajapati et al., 2013).

A strain mutated by treatment with ethidium bromide was used by Pollock for pullulan production (Pollock et al., 1992). The strain showed cell growth like that of yeast and reduced pigmentation of the fermentation broth. Optimization of eight process parameters for increasing the production of pullulan led to 3.5 times higher yield as compared to production without parameter optimization (Tarabasz-Szymanska, Galas, & Pankiewicz, 1999). Another study was performed by Singh, Singh, and Saini (2009) in which the optimization of five cultivation parameters, namely, sucrose, ammonium sulfate, yeast extract, dipotassium hydrogen phosphate, and sodium chloride resulted in the production of 44.2 g L^{-1} pullulan with optimal sucrose 5.31%, ammonium sulfate 0.11%, yeast extract 0.07%, dipotassium hydrogen phosphate 0.05%, and sodium chloride 0.15% (w/v). Cheng et al. (2011a) evaluated the medium and cultivation parameters for a strain *A. pullulans* ATCC 201253, which was pigment-reduced strain; the results showed that after 7 days of cultivation, the production of 25.8 g L^{-1} (94.5% purity) pullulan was obtained with an initial glucose concentration of 75 g L^{-1}.

8.5.2 Fed-batch and continuous fermentation

In this operation, one or more reactants or substrates are fed continuously or intermittently and the product is withdrawn at the same mass flow rate as the reactant. These reactors provide control over the reaction and product quality and are preferred for highly exothermic reactions. Shin et al. (1989) performed fed-batch fermentation to remove the suppression effect of sucrose on pullulan production and obtained 58 g L^{-1} pullulan. In another study, the maximum concentration of pullulan was observed to decrease noticeably to 24.5 g L^{-1} from earlier 31.3 g L^{-1} after 7 days of cultivation when fermentation was performed in fed-batch mode (Youssef, Roukas, & Biliaderis, 1999). The high substrate concentration imparts suppressive effects on glycosyltransferase, α-phosphoglucose mutase, and UDPG-pyrophosphorylase, resulting in reduced yield of pullulan (Duan et al., 2008; Reeslev et al., 1997). Pullulan production by fed-batch fermentation showed an increase in production until the 10th day with a maximum production rate constant at 0.65 g L^{-1} h^{-1}. After day 10, further addition of sucrose showed no considerable increment in production rate (Cheng et al., 2011a). Intermittent supply of limiting substrate at low concentration to the production medium can be done to mitigate the problem caused by the suppression (Bulmer et al., 1987; Characklis & Wilderer, 1984; Lazaridou et al., 2003; Sakata & Otsuka, 2009).

A plug flow reactor is an ideal tubular flow reactor in which each molecule of reactant enters the reactor at the same velocity and has the same residence time in the reactor (plug flow). No axial or radial mixing of molecules takes place inside the reactor. Many studies have reported the production of pullulan by using continuous fermentation where productivity is mainly influenced by the dilution rate. The first attempt to produce pullulan through the chemostat system was by Schuster et al. (1993), maintaining the culture for over 1000 h and achieved a production rate increment to 0.35 g L^{-1} h^{-1} from initial 0.16 g L^{-1} h^{-1} with a dilution rate at 0.05 h^{-1}. Glucose as a substrate and optimum pH 4.0 was used by Reeslev et al. (1997) to obtain 3.6 g L^{-1} pullulan in a chemostat with limiting concentration of zinc. Relatively higher pullulan productivity of 1.17 g L^{-1} h^{-1} and a concentration of 78.4 g L^{-1} were attained by Seo et al. (2006) with a very low dilution rate of 0.015 h^{-1} in a chemostat.

A combination of increased cell density and continuous fermentation can provide a feasible way for long-term production. Sucrose, ammonium sulfate, and yeast extract concentrations were optimized in the medium to 15, 0.9, and 0.4 g L^{-1}, respectively, by Cheng, Demirci, and Catchmark (2011b) to achieve maximum pullulan concentration of 8.3 g L^{-1} at a maximum production rate of 1.33 g L^{-1} h^{-1} using a biofilm reactor with continuous production process at 0.16 h^{-1}. The nitrogen limitation in the process as at optimal conditions showed no residual inorganic nitrogen (NH^{4+}) in the medium, although there was the release of some organic nitrogen compounds from the solid support, possibly maintaining the biofilm growth.

8.5.3 Immobilized cell bioreactors

Cell immobilization bioreactors for pullulan production have been studied to increase the total biomass in the reactors by using the carriers such as agar, calcium alginate, carrageenan, and agarose to immobilize microbial cells (Catley, Robyt, & Whelan, 1966; Yurlova & Hoog, 1997). Besides the use of carriers for immobilizing the cells, a more natural form of immobilization is the biofilm growth on the solid support as in the case of *A. pullulans* where the cell attachment to the support may be due to uronic acid polymer secreted by the microbial strain (Liu et al., 2017; Vashist, Luppa, Yeo, Ozcan, & Luong, 2015). Bioreactors with cell immobilization were used to improve biomass yield by immobilizing *A. pullulans* cells on agarose and carrageenan (West, 2000). Agar and alginate were used for the immobilization of *A. pullulans* among which alginate showed higher production of pullulan during the second cycle. However, the overall pullulan yield was lower as compared to that from conventional bioreactors, attaining only 0.43 mg polysaccharide g^{-1} cells h^{-1} with 36% purity (West & Strohfus, 2001a). The lower yield of pullulan by the immobilized cell reactors can be attributed to the small pore size and void volumes of the packing material, resulting in diffusional limitation when compared to conventional bioreactors. In an immobilized bioreactor system, initial pH, agitation speed, and incubation time resulted improved yield to 19.5 g L^{-1}; however, after four consecutive batch fermentations, a drop in the activity of immobilized gel beads and subsequent pullulan productivity was reported (Mihai, Mocanu, & Carpov, 2001).

In the case of biofilm growth, polyurethane foam, polypropylene, and other plastic composite support (PCS) have been used as a matrix for biopolymer production. Besides behaving as a matrix, polypropylene also integrates the agricultural mixtures (like ground soybean hulls, soybean flour) and microbial nutrients. PCS is formed as a result of a mixture of polypropylene and nutritious materials and acts as an ideal surface for biofilm formation and growth, gradually releasing the incorporated nutrients during the fermentation process (Garg et al., 2014; Krull, Ma, Li, Davé, & Bilgili, 2016). Polyurethane foam as a support for the growth of *A. pullulans* cells was employed for pullulan production, resulting in a yield of about 18 g L^{-1} after cultivation for 96 h (Mulchandani, Luong, & LeDuy, 1989). Biofilm reactor operating in a repeated batch system with PCS support was used for pullulan production with optimized cultivation parameters and the yield of pullulan under optimized conditions reached around 60 g L^{-1} (Lazaridou, Biliaderis, et al., 2002; Lazaridou et al., 2003).

8.5.4 Airlift and other fermenters for the production of pullulan

These reactors consist of no moving parts like impellers, which reduce the energy requirement. They are fluid–fluid contacting type reactor where the mixing is achieved by circulating air bubbles created by the sparger installed inside the reactor. The draft tube inside the reactor serves as an internal heat exchanger eliminating the need for a cooling jacket for temperature control. Airlift fermentation system has been employed for pullulan production (Shabtai & Mukmenev, 1995; Sharma, Prasad, & Choudhury, 2013). Pullulan production in airlift and bubble column bioreactors showed a maximum yield of 0.402 g g^{-1} in the airlift fermentation system (Özcan et al., 2014). A reciprocating plate bioreactor for high-viscosity pullulan production was developed by Audet et al. (1996). The viscosity of medium and pullulan concentration was investigated by installment of various agitation devices in fermenter among which helical ribbon impeller and reciprocating plates were efficient for high viscous medium and high concentration of pullulan but the overall productivity reduced (Audet, Gagnon, Lounes, & Thibault, 1998).

8.6 Chemical modification of pullulan and advancement in processing

Derivatization of pullulan can be done by changing its chemical structure through the addition or removal of functional groups to or from its backbone. Each repeating unit in the structure of pullulan has nine hydroxyl groups, which are available for the substitution reaction with other chemical groups under mild reaction conditions (Dionísio et al., 2016). These modifications in the chemical structure are known to enhance the applications and potential use of pullulan in various fields like drug delivery and food preservation. The reactivity of each hydroxyl group may differ from others in the structure depending on the polarity of the solvent and reagents. A large number of derivatives have been reportedly formed by different chemical reactions with the hydroxyl groups (Table 8.2).

8.6.1 Carboxymethylation

Carboxymethylation is a reaction involving the addition of carboxymethyl group to a chemical compound, often performed with neutral polysaccharides for the chemical modification and to aid their solubility in aqueous solutions. Pullulan derivatized by this mode of reaction is often termed as carboxymethyl pullulan (CMP), which is used as a carrier for the drugs. The retention of the polymer in an organism is increased due to the addition of negative charge into CMP, which results in effective and sustained drug release (Yamaoka, Tabata, & Ikada, 1993). In an alcohol-water mixture, the carboxymethylation of pullulan prevails at the hydroxyl group at the C-2 than that at C-6 (Glinel, Paul Sauvage, Oulyadi, & Huguet, 2000). CMP in conjugation with immune depressants can be designed and used for immune-compromised patients since CMP gets selectively absorbed by the spleen and lymph nodes (Masuda et al., 2001). Conjugates of CMP and doxorubicin were studied for their effect on tumor cells, which showed a higher degree of drug release and affinity toward the tumor cells when micelles with polysaccharide-rich cores were used for drug delivery (Nogusa et al., 2000; Nogusa, Yano, Okuno, Hamana, & Inoue, 1995). Chemical conjugation of pullulan with interferon by cyanuric chloride method has actively targeted the interferon to the liver with enhanced antiviral activity and higher retention of interferon within the liver (Xi et al., 1996).

Table 8.2: Type of reaction in chemical modification of pullulan (Prajapati et al., 2013).

Type of reaction	Schematic chemical structure of substituted pullulan (P-OH)
Etherification	• P-OCH$_2$COOH (carboxymethylation) • P-O(CH$_2$)$_2$-3CH$_3$ (alkylation) • P-OCH$_3$ (permethylation) • P-O(CH2)$_2$-3CH$_2$ NH^{3+} (cationization) • P-OCH$_2$CH$_2$CN (cyanoethylation) • P-O(CH$_2$)1-4-Cl (chloroalkylation) • P-OCH$_2$CH$_2$(SO)CH$_3$ (sulfinyl) • P-OCH$_2$CH$_2$CH$_2$SO$_3$Na • P-OCH$_2$CH$_2$N(CH$_2$CH$_3$)
Oxidation	• P-COOH (C6 oxidation) • Glycosidic ring opening (periodate oxidation)
Urethane derivatives	• P-OCONHCH$_2$CH(OH)CH$_3$ • P-OCONHCH$_2$CH$_2$NH^{3+} • P-OCONHR(R = phenyl or hexyl) • P-OCONH-phenyl
Esterification	• P-OCOCH$_2$CH$_2$COOH (succinoylation) • PA-OCOCH$_2$CH$_2$CO-sulfodimethoxine • P-OCOCH$_2$CH$_2$CO-cholesterol • P-abietate • P-stearate • PA-folate • P-cinnamate • P-biotin
Chlorination	• P-CH$_2$Cl (C6 substitution)
Sulfation	• P-O SO$_3$Na
CMP/hydrazone derivative	• P-OCH$_2$CONH-doxorubicin • P-OCH$_2$CONH-antibody
Azido-pullulan	• P-CH$_2$N$_2$

8.6.1.1 Sulfation

Sulfated derivatives of pullulan have been studied to develop an anticoagulant, similar to heparin and dextran sulfate (Mihai et al., 2001). These derivatives have a comparative edge over sulfated dextran derivative since the presence of more primary hydroxyl groups in pullulan helps in attaining a higher and uniform degree of sulfation than dextran. The sulfation of hydroxyl groups in pullulan occurs independently of molecular weight and the procedure employed with substitution taking place in a defined order of carbon atoms which is C-6 > C-2 > C-3 > C-4 (Alban, Schauerte, & Franz, 2002). Although the sulfated pullulan displays similar anticoagulant activity to heparin, certain properties such as the molecular weight of the polymer, rate of substitution, and distribution of the sulfated groups on the polymer backbone alter the action of these polymers from that of heparin (Alban et al., 2002). Modification of pullulan to obtain sulfated polysaccharide is done to develop nanocarriers from these charged derivatives to demonstrate their potential in drug delivery. Dionísio et al. used sulfated pullulan to form stable nanoparticles assembly via polyelectrolyte complex formation with either carrageenan or chitosan. These sulfated nanoassemblies were further successfully tested for their capacity to associate with bovine serum albumin and also displayed nontoxic behavior on a respiratory cell line (Dionísio et al., 2016).

8.6.2 Cross-linking

Cross-linking of pullulan can be used to alter certain properties such as water absorption, swelling, tensile strength, and permeability in a manner to widen the applications of pullulan in different areas. Higher mechanical strength and thermal stability have been demonstrated by the cross-linked pullulan/gelatin gels due to stable cross-linking sites (−OCO−) and molecular linkages (−OCONH and −OCOO−) (Han & Lv, 2019). The cross-linkers selected for cross-linking should be stable, nontoxic, biocompatible, and fulfill the requirements of a particular tissue. The strength of gel can be improved by optimizing the concentration of the cross-linking agent as in the cross-linking of pullulan by sodium trimetaphosphate where gel strength initially increased with cross-linker concentration followed by the formation of a plateau highlighting alterations in intermolecular bonds (Dulong, Forbice, Condamine, Le Cerf, & Picton, 2011). Kamoun, El-Betany, Menzel, and Chen (2018) showed that during the cross-linking of pullulan by hydroxyethylmethacrylate, the cross-linking behavior varied with the concentration of photoinitiator. As the photoinitiator concentration increased from 0.1 to 0.75 mol%, an increase in cross-linking density was observed, which subsequently reduced on further increasing the photoinitiator concentration (>0.75 mol%) (Kamoun et al., 2018).

8.6.3 Hydrophobic modification

Hydrophobic modification of pullulan can provide plasticity to the construct, which could be used to develop nanoaggregates such as vesicles or micelles of desired size and composition to be used as assemblies in aqueous systems (Jung, Jeong, & Kim, 2003). These spherical and worm-like micelles and vesicles can be used to carry different concentrations of drugs to target and also minimize the free diffusion to the places other than the target site. Cholesterol-bearing pullulans are known to form nanogels with many hydrophobic domains of cholesteryl groups in aqueous systems and these stable nanogels have been studied to carry drugs, enzymes, vaccines, and other therapeutic agents (Abbas et al., 2015; Radulović et al., 2008). Also, surfactants such as sodium dodecyl sulfate can be used for the association or dissociation of these gels. Wang et al. developed amphiphilic α-tocopherol pullulan polymers to be used as carriers for anticancer drugs using 1-ethyl-3-(3-dimethylaminopropyl) carbodiimide hydrochloride (EDC) and 4-dimethylaminopyridine. Successful entrapment in the micelles and translocation of drug 10-hydroxycamptothecin was observed through the cell. However, an increase in hydrophobic α-tocopheryl succinate moiety caused a reduction in critical micelle concentration (Wang, Cui, Bao, Xing, & Hao, 2014).

8.6.4 Grafting

Grafting is another modification strategy used for enhancing the application of polysaccharides. The main idea is to produce a macromolecular structure comprising a linear backbone with several side chains. This can either be done by the attachment of polymers synthesized beforehand having complementary functional groups ("grafting onto") or by attaching a monomer over the polymer backbone covalently and allowing its subsequent polymerization as a side chain ("grafting from") (Nakashio, Fujita, Domoto, Toyota, & Sekine, 1978; Onda, Muto, & Suzuki, 1982). Dense polymerization is not obtained by the former method since the association of chains is hindered by the steric obstruction of already present side chains. In "grafting from" approach the monomer initiates redox polymerization and determines the graft efficiency which generally comes out to be higher than "grafting onto." Although the exact size and number of grafts are tedious to characterize, molecular weight and its distribution can be directed by altering the conditions of the reaction (Wang et al., 2019). The properties of graft polysaccharide vary with the type of grafts, for example, when pullulan is grafted with polyethylene glycol, its solubility increases for a range of solvents whereas its hydrophilic nature is reduced with grafting of poly(methyl acrylate) (Motozato, Ihara, Tomoda, & Hirayama, 1986; Nakatani, Shibukawa, & Nakagawa, 1996). Temperature sensitivity has been introduced in pullulan by grafting of poly(N-isopropylacrylamide) (pNIPAM) as this compound shows phase transition at physiological temperatures as was reported (Constantin et al., 2017) that at about 32°C pNIPAM grafted pullulan showed a reversible phase transition forming nanoparticles exhibiting high indomethacin loading and release efficiency (Hijiya & Shiosaka, 1975; Olmo et al., 2008). Similarly, Fundueanu, Constantin, and Ascenzi (2008) assessed the effects of temperature-sensitive units by grafting pNIPAM-*co*-acrylamide on pullulan microspheres and pH-responsive groups were attached to rest of the hydroxyl groups due to which the pullulan microspheres were able to maintain a sharp volume phase transition both below and above the pKa value of carboxylic acid at low values of exchange capacity.

8.7 Downstream processing

Downstream processing (DSP) is an important step to get pure pullulan from the fermentation broth. The DSP for any process should be simple, inexpensive but efficient to obtain the products with high purity and quality. The steps involved in DSP depend upon the target industry for which the biopolymer is being produced and the degree of purity required. However, certain steps in DSP are necessary to obtain pure pullulan from the fermentation broth. These include cell harvesting, removal of by-products such as melanin and cellular proteins, concentration, precipitation of polysaccharide, and drying these are mentioned in process flow diagram for pullulan production in Fig. 8.4. During the fermentation, pullulan is produced in late log phase or early stationary phase of microbial growth whereas melanin is reportedly produced at the end of fermentation which causes the release of the later pigment with pullulan in the broth (Israilides et al., 1994; West & Strohfus, 1998). Melanin imparts dark green to black pigmentation to the broth which demands peculiar DSP of broth when pullulan is produced using melanin-producing strains. In their work, Kachhawa et al. (2003) performed a comparative evaluation of the efficacy of activated charcoal, various solvents, blends of solvents, and combination of salts and solvent in decolorizing fermentation broth to obtain melanin-free pullulan. The results showed that complete precipitation of pullulan (yield achieved: 86%) with minimum melanin contamination was obtained by a combination of 40:60 of ethyl methyl ketone/ethanol (Kachhawa et al., 2003). Wu et al. produced a pullulan by fermentation using *A. pullulans* strain. During the DSP of the extracellular polysaccharide, the fermentation broth was centrifuged to remove biomass and the supernatant with melanin was heated at 80°C for 20 min to remove proteins followed by melanin removal using hydrogen peroxide. The supernatant was further concentrated under vacuum before precipitation with ethanol and drying at 60°C to obtain high-purity pullulan (Wu, Jin, Kim, et al., 2009). Sugumaran and Ponnusami worked on pullulan extraction in solid-state fermentation by leaching and determined the effect of solvent to solid weight ratio, system temperature, pH of solvent, agitation speed, and leaching time on the recovery of pullulan. The authors used ethanol in equal volume to the supernatant, storing the solution at 4°C for 24 h to attain precipitation after which the precipitate was

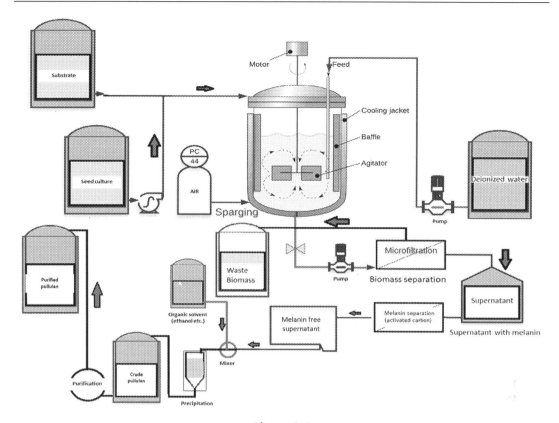

Figure 8.4
Process flow diagram along with DSP for pullulan production. *DSP*, Downstream processing.

resuspended and recentrifuged at 4400 g for 15 min and the concentration of pullulan was estimated using calibration curve (Sugumaran & Ponnusami, 2017b).

The recovery and purification of pullulan produced by strains that do not show melanin production is comparatively simpler with precipitation as a major DSP step. Use of organic solvents with higher molecular weight and lower hydrophilicity such as propyl alcohol and tetrahydrofuran for precipitation is found to be more efficacious than solvents with lower molecular weight and higher hydrophilic nature like ethanol, acetone, etc. The former solvents cause complete precipitation with adequate impurity removal at a lesser volume (Kato & Nomura, 1977). Shin et al. (1989) precipitated pullulan by treatment of broth obtained after centrifugation with a mixture of acetone and ethanol 1:1 (v/v). Centrifugation of broth followed by precipitation with methanol and further filtration has also been reported for purification of the EPS (Boa & LeDuy, 1987). Purification of pullulan from the centrifuged broth by treatment of supernatant with tetrahydrofuran and ethanol has also been reported (Dufresne et al., 1990; Zhang, Alfano, Race, & Davé, 2018). The volume of solvent used for precipitation depends upon the concentration of pullulan present in the supernatant. Various studies have reported the use of different volumes of various solvents for precipitation of pullulan. Singh et al. separated and purified pullulan from the fermentation broth by centrifugation to separate biomass and treatment of supernatant with two volumes of ice-chilled isopropyl alcohol at 4°C. The precipitates obtained were centrifuged again followed with dissolution in deionized water and reprecipitated with isopropyl alcohol after which the obtained precipitates were finally washed with acetone and deionized water and dried at 80°C until a constant weight is achieved (Singh, Saini, et al., 2009). In another study, two volumes of isopropyl alcohol were used for precipitation of supernatant obtained after centrifugation of fermentation broth, and the precipitates were washed with acetone and dried at 90°C to a constant weight to acquire purified pullulan (Vijayendra, Bansal, Prasad, & Nand, 2001). Sena, Costelli, Gibson, and Coughlin (2006) employed 2-propanol for pullulan precipitation from broth with a ratio of 1:1 (v/v) of 2-propanol to supernatant. Isopropyl alcohol for precipitation has also been reportedly used by Pollock et al. (1992) who precipitated centrifuged broth with one volume of isopropyl alcohol and the obtained precipitates were dried at 80°C till a constant weight was observed. In similar precipitation studies, Goksungur et al. and Lazaridou et al. used two volumes of ethanol at 4°C for the precipitation of pullulan to proceed for 1 h (Wickerham & Kurtzman, 1975; Xi et al., 1996).

Chi and Zhao produced extracellular pullulan by fermentation using Rhodotorula bacterium and obtained the highest pullulan yield among yeasts after 12-h precipitation with cold ethanol (absolute or 95%) at 4°C. After ethanol removal, the precipitate was dissolved and dialyzed using deionized water for 48 h, further precipitated in cold ethanol, and finally dried to a constant weight at 80°C (Chi & Zhao, 2003). Cold absolute ethanol (three volumes) was also used by Forabosco et al. (2006) for pullulan precipitation from culture filtrate keeping the solution at −20°C for 24 h to attain adequate phase separation. Haghighatpanah et al. performed centrifugation for the separation of supernatant containing pullulan from biomass and precipitation was done using cold ethanol at 2:1 (v/v) at 4°C for 16 h. The solution was again centrifuged and the supernatant was dried at 50°C for 16 h to estimate the yield of pullulan by dry-weight analysis (Haghighatpanah et al., 2020). In this study, the purification was further supplemented by deproteinization of the EPS by several method, dissolution in deionized water, and fractionation by eluting with a gradient NaCl aqueous solution followed by purification on the Sephadex G-100 gel filtration column obtaining white pure pullulan. Precipitation of polysaccharide using ethanol and fermentation supernatant in a volumetric ratio of 2:1 at 4°C for 6 h was done by Youssef et al. (1999) after which the precipitate was filtered by Whatman filter, washed with absolute ethanol thrice and dried at 105°C overnight to recover pullulan as g of polysaccharide/100 g sugar utilized. A procedure for continuous purification of pullulan was proposed by Kato and Nomura (1977) in which tanks containing organic solvent in increasing concentration were arranged serially to precipitate pullulan and the precipitated product was further dried at 80°C in an oven. Cross-flow filtration strategy was used by Yamasaki et al. with flat cellulose triacetate membrane and a nylon filter to recover pullulan from fermentation broth. The ratio of pullulan in the permeate to that in the retentate and the leakage of cells were measured to deduce that cross-flow filtration is feasible for pullulan recovery from broth (Yamasaki, Lee, Tanaka, & Nakanishi, 1993).

8.8 Applications of pullulan

8.8.1 Healthcare

Many physical and chemical properties of pullulan have confirmed its utility in pharmaceutical and biomedical areas. The high stability, biocompatibility, and biodegradability of pullulan microspheres have made them a competent candidate in the area of drug delivery (Mocanu, Mihai, Le Cerf, Picton, & Muller, 2004). pH and temperature-sensitive pullulan microspheres for controlled drug release were developed by Fundueanu et al. (2008). Initially, poly(N-isopropylacrylamide-co-acrylamide) was grafted into microspheres of pullulan to induce temperature sensitivity after which the rest of the hydroxyl groups of the polysaccharide were reacted with succinic anhydride to introduce —COOH creating more hydrophilic microspheres than the nongrafted ones. Pullulan is a naturally occurring polymer possessing the property of dissolving in aqueous media without forming a gel. Industrially it is a suitable alternative to natural gums. The solution to this polysaccharide demonstrates excellent film-forming properties (Horinaka, Hashimoto, & Takigawa, 2018). Many methods have been studied for the efficient production of pullulan films including solvent casting, melt extrusion, freeze—drying, inkjet printing, and electrospinning (Gaur & Singh, 2010; Lee et al., 2002; Prasongsuk et al., 2007). Conventionally, drugs are coated with shells derived from animal materials such as gelatin. Hard gelatin shells often result in the incomplete release of drugs and there are also some other risks associated with the use of ingredients obtained from animal sources (Sakata & Otsuka, 2009). Pullulan with molecular weight more than 2000 kDa, mechanical strength, stability, and high viscosity has the potential to replace animal-based shells, reducing the associated risks and leading to extensive recognition of such pullulan shells by vegetarians and patients with diet restrictions (Choudhury, Saluja, & Prasad, 2011; Gibbs & Seviour, 1996; Wang, Bian, Wei, Jiang, & Dong, 2016). Pullulan films allow for uniform distribution and adequate retention of drugs which has led to their utilization in the delivery of antimicrobial substances during conditions like oriental infections and wound healing (Garg et al., 2014; Xu, Weng, Gilkerson, Materon, & Lozano, 2015). Besides the drug loading and release through a single pullulan film, stacks of multiple layers have also been made with the help of hydrophobic or hydrophilic matrices in which more than one drug can be stacked at different film layers and released (Chen et al., 2014; Goksungur et al., 2004). Moreover, injectable pullulan hydrogels as microparticles or nanoparticles have also been developed for enhancing the rate of delivery and stability of drugs (Li et al., 2016). Pullulan-dextran hydrogel combined with interfacial polyelectrolyte complexation fibers to create a composite scaffold has been developed which evoked cell-specific proliferation and directed toward the potential of pullulan in tissue engineering (Cutiongco, Tan, Ng, Le Visage, & Yim, 2014). Pullulan nanogels have also been investigated to possess the properties of molecular chaperones. The catch and release mechanism of pullulan nanogels is similar to one displayed by molecular chaperones for protein refolding (Akiyoshi et al., 1998). These pullulan nanogels, like chaperones, have been reported to trap denatured proteins and release refolded complexed proteins in the presence of cyclodextrins (Nomura, Ikeda, Yamaguchi, Aoyama, & Akiyoshi, 2003). Cyclodextrins govern the binding of chaperone molecules to proteins.

8.8.2 Food industry

The utility of pullulan in the preservation of food has also been understood. Pullulan is an edible biopolymer, not easily assimilated by microorganisms. Hence it can be utilized to increase the shelf life of edible and nonedible products. Food and Drug Administration (FDA) has declared pullulan safe and conferred it as GRAS (Generally Recognized As Safe) for food applications (U.S. Food and Drug Administration Center for Food Safety and Applied Nutrition Office of Plant & Dairy Foods & Beverages, 2002). Pullulan has low-calorie content, a property that can be used to make low-calorie food and supplements (Yatmaz & Turhan, 2012). As compared to starch, pullulan has a higher moisture retention capacity which inhibits the drying of food items and can be used as a starch substitute in preparation of bakery products like cookies and cakes (Imeson, 2011). Human digestive enzymes are known to digest pullulan at a slow pace gradually converting it into glucose which makes it suitable for consumption by diabetic patients as a combination in food products made for people suffering from diabetes. Pullulan, being resistant to mammalian hydrolytic amylases, is an emerging prebiotic which promotes the growth of beneficial bacteria in the human intestine (Sugawa-Katayama, Kondou, Mandai, & Yoneyama, 1994). Viscosity and gloss can be added to food products with high salt concentration by the addition of pullulan as it is well established that the viscosity of this polysaccharide remains unaffected in presence of high sodium chloride, heat or alteration in pH to a great extent. It also has adhesive properties which help to maintain the sugar coatings on tablets and can be used as a food-binding agent (Gibson & Coughlin, 2002; Wang, Ju, Zhou, & Wei, 2014).

Pullulan is used as a food stabilizer, as gelling, glossing, and glazing agent and its coating also prevent damage in fishes due to freezing during storage. The clear film formed by the coating of pullulan solution on food materials is edible, water-soluble, and impermeable to oxygen making it both beneficial in food preservation and safe for human consumption (Lacroix & Le Tien, 2005). Hence pullulan can be adapted as a versatile packaging material both for edible and nonedible products owing to its ability to prevent oxidation. These remarkable properties of pullulan films make it an excellent alternative to petrochemical-based food packages (Tomasula et al., 2016). Prevention of moisture loss and discoloration in freshly harvested chestnut fruits was reported by Gounga, Xu, Wang, and Yang, (2008) when the fruits were coated by an edible film of whey protein isolate pullulan.

8.8.3 Waste remediation

Biosorption of heavy metals has been observed by EPSs from microbial strains revealing their role in waste remediation. The reaction between acidic functional groups of polysaccharide and metal cations leads to the removal of heavy metals (Kim, Kim, Kim, & Oh, 1996). Removal of heavy metal ions from aqueous solution with the help of *A. pullulans* has been observed by Breierová et al. (2002). The use of modified polysaccharides has shown remediation capacity with greater specificity and flocculation potential. Efficacious flocculation of pesticides has been reported with the use of cationized pullulan which further enhanced with lowering the emulsion pH and increasing the substitution rate (Ghimici & Constantin, 2015). Reduction in the concentration of several metals such as copper, iron, zinc, manganese, lead, cadmium, nickel, and chromium was reported by Radulović et al. (2008) which was attributed to the production of pullulan when *A. pullulans* CH-1 was cultured on acid hydrolysate of peat. Abbas et al. (2015) studied the highly selective sorption of more than 90% cadmium from spiked water with the use of sodium salt of succinate pullulan. The sorption process reportedly involved specific ion exchange between the acidic pullulan and divalent cadmium ion even when other divalent cations were present in water.

8.8.4 Miscellaneous applications

Hydrophilic pullulan gels are used for the immobilization of enzymes with high activity and retention capability (Hirohara, Nabeshima, Fujimoto, & Nagase, 1981). The reaction between the hydroxy group of pullulans and a bifunctional compound that could form an ether linkage is used to prepare these three-dimensional pullulan gel structures. Similarly, cyanoethylated pullulan prepared by reaction between pullulan and acrylonitrile in the presence of an alkali catalyst possess properties such as heat resistance, adhesion to metals, and formation of film marking its potential utility in electronic devices (Onda et al., 1982). Adhesion property of pullulan is used in the coating of paper which imparts better strength, ink retention, and fold resistance to it as compared to paper obtained from wood pulp (Nakashio et al., 1978). Standards for chromatography have been developed using pure pullulan samples of appropriate molecular weight and size (Kawahara, Ohta, Miyamoto, & Nakamura, 1984). Gel beads made by cross-linked pullulan of accurate size have been used commercially in gel permeation chromatography due to their moisture-resistance property, transparency, adhesiveness, and mechanical strength (Motozato et al., 1986). Capillary electrophoresis of sodium dodecyl sulfate proteins with the use of pullulan solution filled capillary and purification of human histone protein (H4) using pullulan-coated capillary has also been performed (Abdel Hafez, Abdelhady, Sharaf, & El-Tayeb, 2007; Singh & Kaur, 2019). Pullulan can be shaped by compression or extrusion into the material which resembles polystyrene or polyvinyl alcohol and possesses properties such as transparency, mechanical strength, impermeability to gas, and biodegradability (Terán Hilares et al., 2017; Terán Hilares et al., 2019).

8.9 Conclusions and perspectives

Pullulan is an edible bacterial EPS that has displayed many potential applications owing to its peculiar characteristics such as stability and nontoxicity. Its applications in food, pharmaceutical, medical, and environmental remediation have expanded more since the introduction of chemical modifications in the polysaccharide structure. Modified derivatives of pullulan possess distinctive material properties with greater stability and strength. Pullulan membranes/films are used in the coating and packaging of food items whereas pullulan composites with other biodegradable materials are used in drug delivery, tissue engineering, and waste remediation. Both immobilized and free cell bioreactor systems have been widely studied for efficient production of pullulan. The major constraint in the utilization of pullulan is its high production cost when compared to dextran and xanthan, which are other similar polysaccharides. Fermentative production of pullulan using agroindustrial wastes as a substrate could be a feasible way to reduce the upstream cost to some extent. One of the major problems encountered during the fermentative production of pullulan is the formation of melanin along with the polysaccharide, which hinders the extraction and purification process. Although many attempts have been successfully made to optimize the DSP of pullulan, more engineering innovations and use of improved production strains can enhance the productivity and purity of pullulan, further widening its range of utilization and application.

References

Abbas, A., Hussain, M. A., Amin, M., Sher, M., Tahir, M. N., & Tremel, W. (2015). Succinate-bonded pullulan: An efficient and reusable super-sorbent for cadmium-uptake from spiked high-hardness groundwater. *Journal of Environmental Sciences, 37*, 51–58. Available from https://doi.org/10.1016/j.jes.2015.04.013.

Abdel Hafez, A. M., Abdelhady, H. M., Sharaf, M. S., & El-Tayeb, T. S. (2007). Bioconversion of various industrial by-products and agricultural wastes into pullulan. *Journal of Applied Sciences Research, 32*(11), 1416–1425.

Akiyoshi, K., Kobayashi, S., Shichibe, S., Mix, D., Baudys, M., Kim, S. W., & Sunamoto, J. (1998). Self-assembled hydrogel nanoparticle of cholesterol-bearing pullulan as a carrier of protein drugs: Complexation and stabilization of insulin. *Journal of Controlled Release: Official Journal of the Controlled Release Society, 54*(3), 313–320. Available from https://doi.org/10.1016/S0168-3659(98)00017-0.

Alban, S., Schauerte, A., & Franz, G. (2002). Anticoagulant sulfated polysaccharides: Part I. Synthesis and structure−activity relationships of new pullulan sulfates. *Carbohydrate Polymers*, *47*(3), 267−276.

An, C., Ma, S.-J., Chang, F., & Xue, W.-J. (2017). Efficient production of pullulan by *Aureobasidium pullulans* grown on mixtures of potato starch hydrolysate and sucrose. *Brazilian Journal of Microbiology*, *48*(1), 180−185. Available from https://doi.org/10.1016/j.bjm.2016.11.001.

Audet, J., Gagnon, H., Lounes, M., & Thibault, J. (1998). Polysaccharide production: Experimental comparison of the performance of four mixing devices. *Bioprocess Engineering*, *19*, 45−52. Available from https://doi.org/10.1007/s004490050481.

Audet, J., Lounes, M., & Thibault, J. (1996). Pullulan fermentation in a reciprocating plate bioreactor. *Bioprocess Engineering*, *15*, 209−214. Available from https://doi.org/10.1007/BF00369484.

Auer, D. P. F., & Seviour, R. J. (1990). Influence of varying nitrogen sources on polysaccharide production by *Aureobasidium pullulans* in batch culture. *Applied Microbiology and Biotechnology*, *32*, 637−644. Available from https://doi.org/10.1007/BF00164732.

Badr-Eldin, S. M., El-Tayeb, O. M., El-Masry, H. G., Mohamad, F. H. A., & El-Rahman, O. A. A. (1994). Polysaccharide production by *Aureobasidium pullulans*: Factors affecting polysaccharide formation. *World Journal of Microbiology Biotechnology*, *10*(4), 423−426.

Bambalov, G., & Jordanov, P. (1993). Production of pullulan polysaccharide from wine-producing wastes. *Scientific Works of the HIFFI-BG*, *40*, 229−240.

Barnett, C., Smith, A., Scanlon, B., & Israilides, C. J. (1999). Pullulan production by *Aureobasidium pullulans* growing on hydrolysed potato starch waste. *Carbohydrate Polymers*, *38*(3), 203−209.

Bauer, R. (1938). Physiology of Dematium pullulans de Bary. *Zentralbl Bacteriol Parasitenkd Infekt. Hyg Abt2*, *98*, 133−167.

Bender, H., Lehmann, J., & Wallenfels, K. (1959). Pullulan, an extracellular glucan from *Pullularia pullulans* English summ TT—Pullulan, ein Extracellulares Glucan von *Pullularia pullulans* [English summ.]. *Biochimica et Biophysica Acta*, *36*, 309−316. Available from https://doi.org/10.1016/0006-3002(59)90172-6.

Boa, J. M., & LeDuy, A. (1984). Peat hydrolysate medium optimization for pullulan production. *Applied and Environmental Microbiology*, *48*(1), 26−30. Available from https://doi.org/10.1128/aem.48.1.26-30.1984.

Boa, J. M., & LeDuy, A. (1987). Pullulan from peat hydrolyzate fermentation kinetics. *Biotechnology and Bioengineering*, *30*(4), 463−470. Available from https://doi.org/10.1002/bit.260300402.

Breierová, E., Vajcziková, I., Sasinková, V., Stratilová, E., Fisera, I., Gregor, T., & Sajbidor, J. (2002). Biosorption of cadmium ions by different yeast species. *Zeitschrift für Naturforschung C*, *57*(7−8), 634−639.

Bulmer, M. A., Catley, B. J., & Kelly, P. J. (1987). The effect of ammonium ions and pH on the elaboration of the fungal extracellular polysaccharide, pullulan, by *Aureobasidium pullulans*. *Applied Microbiology and Biotechnology*, *25*, 362−365. Available from https://doi.org/10.1007/BF00252548.

Campbell, B. S., McDougall, B. M., & Seviour, R. J. (2003). Why do exopolysaccharide yields from the fungus *Aureobasidium pullulans* fall during batch culture fermentation? *Enzyme and Microbial Technology*, *33*(1), 104−112. Available from https://doi.org/10.1016/S0141-0229(03)00089-9.

Catley, B. J. (1971). Utilization of carbon sources by *Pullularia pullulans* for the elaboration of extracellular polysaccharides. *Applied Microbiology*, *22*(4), 641−649. Available from https://doi.org/10.1128/aem.22.4.641-649.1971.

Catley, B. J. (1973). The rate of elaboration of the extracellular polysaccharide, pullulan, during growth of *Pullularia pullulans*. *Journal of General Microbiology*, *78*(1), 33−38. Available from https://doi.org/10.1099/00221287-78-1-33.

Catley, B. J., & McDowell, W. (1982). Lipid-linked saccharides formed during pullulan biosynthesis in *Aureobasidium pullulans*. *Carbohydrate Research*, *103*(1), 65−75. Available from https://doi.org/10.1016/S0008-6215(82)80008-6.

Catley, B. J., Ramsay, A., & Servis, C. (1986). Observations on the structure of the fungal extracellular polysaccharide, pullulan. *Carbohydrate Research*, *153*(1), 79−86. Available from https://doi.org/10.1016/S0008-6215(00)90197-6.

Catley, B. J., Robyt, J. F., & Whelan, W. J. (1966). A minor structural feature of pullulan. *Biochemical Journal*, *100*(1), P5.

Chabasse, D. (2002). General review—Phacohyphomycetes agents of phacohyphomycosis: Emerging fungi. *Journal of Medical Mycology*, *12*(2), 65−85.

Characklis, W. G., & Wilderer, A. P. (1984). *Microbial adhesion and aggregation*. Berlin: Springer-Verlag.

Chen, J., Wu, S., & Pan, S. (2012). Optimization of medium for pullulan production using a novel strain of *Auerobasidium pullulans* isolated from sea mud through response surface methodology. *Carbohydrate Polymers*, *87*(1), 771−774. Available from https://doi.org/10.1016/j.carbpol.2011.08.062.

Chen, Y., et al. (2014). Production of pullulan from xylose and hemicellulose hydrolysate by *Aureobasidium pullulans* AY82 with pH control and DL-dithiothreitol addition. *Biotechnology and Bioprocess Engineering*, *19*, 282−288. Available from https://doi.org/10.1007/s12257-013-0715-4.

Cheng, K. C., Demirci, A., & Catchmark, J. M. (2010a). Enhanced pullulan production in a biofilm reactor by using response surface methodology. *Journal of Industrial Microbiology & Biotechnology*, *37*(6), 587−594. Available from https://doi.org/10.1007/s10295-010-0705-x.

Cheng, K. C., Demirci, A., & Catchmark, J. M. (2010c). Effects of plastic composite support and pH profiles on pullulan production in a biofilm reactor. *Applied Microbiology and Biotechnology*, *86*(3), 853−861. Available from https://doi.org/10.1007/s00253-009-2332-x.

Cheng, K. C., Demirci, A., & Catchmark, J. M. (2011a). Evaluation of medium composition and fermentation parameters on pullulan production by *Aureobasidium pullulans*. *Food Science and Technology International*, *17*(2), 99−109. Available from https://doi.org/10.1177/1082013210368719.

Cheng, K. C., Demirci, A., & Catchmark, J. M. (2011b). Continuous pullulan fermentation in a biofilm reactor. *Applied Microbiology and Biotechnology*, *90*, 921−927. Available from https://doi.org/10.1007/s00253-011-3151-4.

Chi, Z., Wang, F., Chi, Z., Yue, L., Liu, G., & Zhang, T. (2009). Bioproducts from *Aureobasidium pullulans*, a biotechnologically important yeast. *Applied Microbiology and Biotechnology*, *82*, 793−804. Available from https://doi.org/10.1007/s00253-009-1882-2.

Chi, Z., & Zhao, S. (2003). Optimization of medium and cultivation conditions for pullulan production by a new pullulan-producing yeast strain. *Enzyme and Microbial Technology*, *33*(2−3), 206−211. Available from https://doi.org/10.1016/S0141-0229(03)00119-4.

Choudhury, A. R., Saluja, P., & Prasad, G. S. (2011). Pullulan production by an osmotolerant *Aureobasidium pullulans* RBF-4A3 isolated from flowers of *Caesulia axillaris*. *Carbohydrate Polymers*, *83*(4), 1547−1552. Available from https://doi.org/10.1016/j.carbpol.2010.10.003.

Choudhury, A. R., Sharma, N., & Prasad, G. S. (2012). Deoiledjatropha seed cake is a useful nutrient for pullulan production. *Microbial Cell Factories*, *11*, 39. Available from https://doi.org/10.1186/1475-2859-11-39.

Constantin, M., Bucătariu, S., Stoica, I., & Fundueanu, G. (2017). Smart nanoparticles based on pullulan-g-poly (N-isopropylacrylamide) for controlled delivery of indomethacin. *International Journal of Biological Macromolecules*, *94*(Pt A), 698−702. Available from https://doi.org/10.1016/j.ijbiomac.2016.10.064.

Cutiongco, M. F. A., Tan, M. H., Ng, M. Y. K., Le Visage, C., & Yim, E. K. F. (2014). Composite pullulan-dextran polysaccharide scaffold with interfacial polyelectrolyte complexation fibers: A platform with enhanced cell interaction and spatial distribution. *Acta Biomaterialia*, *10*(10), 4410−4418. Available from https://doi.org/10.1016/j.actbio.2014.06.029.

Degeest, B., & De Vuyst, L. (2000). Correlation of activities of the enzymes α-phosphoglucomutase, UDP-galactose 4-epimerase, and UDP-glucose pyrophosphorylase with exopolysaccharide biosynthesis by *Streptococcus thermophilus* LY03. *Applied and Environmental Microbiology*, *66*(8), 3519−3527. Available from https://doi.org/10.1128/AEM.66.8.3519-3527.2000.

Delben, F., Forabosco, A., Guerrini, M., Liut, G., & Torri, G. (2006). Pullulans produced by strains of *Cryphonectria parasitica*—II. Nuclear magnetic resonance evidence. *Carbohydrate Polymers*, *63*(4), 545−554. Available from https://doi.org/10.1016/j.carbpol.2005.11.012.

Dionísio, M., Braz, L., Corvo, M., Lourenço, J. P., Grenha, A., & Rosa da Costa, A. M. (2016). Charged pullulan derivatives for the development of nanocarriers by polyelectrolyte complexation. *International*

Journal of Biological Macromolecules, 86, 129–138. Available from https://doi.org/10.1016/j.ijbiomac.2016.01.054.

Duan, X., Chi, Z., Wang, L., & Wang, X. (2008). Influence of different sugars on pullulan production and activities of α-phosphoglucose mutase, UDPG-pyrophosphorylase and glucosyltransferase involved in pullulan synthesis in *Aureobasidium pullulans* Y68. *Carbohydrate Polymers*, 73(4), 587–593. Available from https://doi.org/10.1016/j.carbpol.2007.12.028.

Dudman, W. F., & Sutherland, I. W. (1977). *Surface carbohydrates of the prokaryotic cell*. London and New York: Academic.

Dufresne, R., Thibault, J., Leduy, A., & Lencki, R. (1990). The effects of pressure on the growth of *Aureobasidium pullulans* and the synthesis of pullulan. *Applied Microbiology and Biotechnology*, 32, 526–532. Available from https://doi.org/10.1007/BF00173722.

Dulong, V., Forbice, R., Condamine, E., Le Cerf, D., & Picton, L. (2011). Pullulan-STMP hydrogels: A way to correlate crosslinking mechanism, structure and physicochemical properties. *Polymer Bulletin*, 67, 455–466. Available from https://doi.org/10.1007/s00289-010-0435-2.

Elinov, N. P., Neshatayeva, E. V., Dranishnikov, A. N., & Matveyeva, A. K. (1974). Effect of inorganic salts in the synthetic nutrient medium on the structure and properties of the glucan formed by *Aureobasidium pullulans* (Russian). *Prikladnaia Biokhimiia i Mikrobiologiia*, 10(4), 557–562.

Forabosco, A., Bruno, G., Sparapano, L., Liut, G., Marino, D., & Delben, F. (2006). Pullulans produced by strains of *Cryphonectria parasitica*—I. Production and characterisation of the exopolysaccharides. *Carbohydrate Polymers*, 63(4), 535–544. Available from https://doi.org/10.1016/j.carbpol.2005.10.005.

Fraser, C. G., & Jennings, H. J. (1971). A glucan from *Tremella mesenterica* NRRL-Y6158. *Canadian Journal of Chemistry*, 4(11). Available from https://doi.org/10.1139/v71-297.

Fundueanu, G., Constantin, M., & Ascenzi, P. (2008). Preparation and characterization of pH- and temperature-sensitive pullulan microspheres for controlled release of drugs. *Biomaterials*, 29(18), 2767–2775. Available from https://doi.org/10.1016/j.biomaterials.2008.03.025.

Gaidhani, H. K., McNeil, B., & Ni, X. (2005). Fermentation of pullulan using an oscillatory baffled fermenter. *Chemical Engineering Research and Design*, 83(6), 640–645. Available from https://doi.org/10.1025/cherd.04355.

Gao, W., Chung, C. H., Li, J., & Lee, J. W. (2011). Application of statistical experimental design for optimization of physiological factors and their influences on production of pullulan by *Aureobasidium pullulans* HP-2001 using an orthogonal array method. *Korean Journal of Chemical Engineering*, 28, 2184–2189. Available from https://doi.org/10.1007/s11814-011-0107-4.

Gar, R. K., Rennert, R. C., Duscher, D., Sorkin, M., Kosaraju, R., Auerbach, L. J., Lennon, J., Chung, M. T., Paik, K., Nimpf, J., Rajadas, J., Longaker, M. T., & Gurtner, G. C. (2014). Capillary force seeding of hydrogels for adipose-derived stem cell delivery in wounds. *Stem Cells Translational Medicine*, 3(9), 1079–1089. Available from https://doi.org/10.5966/sctm.2014-0007.

Gaur, R., & Singh, R. (2010). Optimization of physico-chemical and nutritional parameters for pullulan production by a mutant of thermotolerant *Aureobasidium pullulans* in fed batch fermentation process. *African Journal of Biotechnology*, 9(43). Available from https://doi.org/10.5897/AJB10.358.

Ghimici, L., & Constantin, M. (2015). The separation of the pyrethroid insecticide Fastac 10 EC by cationic pullulan derivatives. *Reactive & Functional Polymers*, 95, 12–18.

Gibbs, P. A., & Seviour, R. J. (1992). Influence of bioreactor design on exopolysaccharide production by *Aureobasidium pullulans*. *Biotechnology Letters*, 14, 491–494. Available from https://doi.org/10.1007/BF01023173.

Gibbs, P. A., & Seviour, R. J. (1996). Does the agitation rate and/or oxygen saturation influence exopolysaccharide production by *Aureobasidium pullulans* in batch culture? *Applied Microbiology and Biotechnology*, 46(5–6), 503–510.

Gibson, L. H., & Coughlin, R. W. (2002). Optimization of high molecular weight pullulan production by *Aureobasidium pullulans* in batch fermentations. *Biotechnology Progress*, 18(3), 675–678. Available from https://doi.org/10.1021/bp0200043.

Glinel, K., Paul Sauvage, J., Oulyadi, H., & Huguet, J. (2000). Determination of substituents distribution in carboxymethylpullulans by NMR spectroscopy. *Carbohydrate Research*, *328*(3), 343−354. Available from https://doi.org/10.1016/S0008-6215(00)00120-8.

Göksungur, Y., Dăgbağli, S., Uçan, A., & Güvenç, U. (2005). Optimization of pullulan production from synthetic medium by *Aureobasidium pullulans* in a stirred tank reactor by response surface methodology. *Journal of Chemical Technology and Biotechnology (Oxford, Oxfordshire: 1986)*, *80*(7), 819−827. Available from https://doi.org/10.1002/jctb.1254.

Goksungur, Y., Ucan, A., & Guvenc, U. (2004). Production of pullulan from beet molasses and synthetic medium by *Aureobasidium pullulans*. *Turkish Journal of Biology*, *28*, 23−30.

Göksungur, Y., Uzunoullari, P., & Dağbağli, S. (2011). Optimization of pullulan production from hydrolysed potato starch waste by response surface methodology. *Carbohydrate Polymers*, *83*(3), 1330−1337. Available from https://doi.org/10.1016/j.carbpol.2010.09.047.

Gounga, M. E., Xu, S., Wang, Z., & Yang, W. G. (2008). Effect of whey protein isolate−pullulan edible coatings on the quality and shelf life of freshly roasted and freeze-dried Chinese chestnut. *Journal of Food Science*, *73*(4), E155−E161.

Haghighatpanah, N., Mirzaee, H., Khodaiyan, F., Kennedy, J. F., Aghakhani, A., Hosseini, S. S., & Jahanbin, K. (2020). Optimization and characterization of pullulan produced by a newly identified strain of *Aureobasidium pullulans*. *International Journal of Biological Macromolecules*, *152*, 305−317. Available from https://doi.org/10.1016/j.ijbiomac.2020.02.226.

Han, Y., & Lv, S. (2019). Synthesis of chemically crosslinked pullulan/gelatin-based extracellular matrix-mimetic gels. *International Journal of Biological Macromolecules*, *122*, 1262−1270.

Hayashi, S., Hayashi, T., Takasaki, Y., & Imada, K. (1994). Purification and properties of glucosyltransferase from *Aureobasidium*. *Journal of Industrial Microbiology*, *13*(1), 5−9. Available from https://doi.org/10.1007/BF01569655.

Hijiya, H., & Shiosaka, M. (Mar. 18, 1975). *Shaped bodies of pullulan esters and their use*. Google patents.

Hirohara, H., Nabeshima, S., Fujimoto, M., & Nagase, T. (Jan. 27, 1981). *Enzyme immobilization with pullulan gel*. Google patents.

Horinaka, J.-i, Hashimoto, Y., & Takigawa, T. (2018). Optical and mechanical properties of pullulan films studied by uniaxial stretching. *International Journal of Biological Macromolecules*, *118*(Part A), 584−587. Available from https://doi.org/10.1016/j.ijbiomac.2018.06.127.

Imeson, A. (2011). *Food stabilisers, thickeners and gelling agents*. John Wiley & Sons.

Imshenetskii, A. A., Kondrat'eva, T. F., & Smut'ko, A. N. (1981a). Influence of the acidity of the medium, conditions of aeration, and temperature on pullulan biosynthesis by polyploid strains of *Pullularia* (*Aureobasidium*) *pullulans* [Fungi]. *Microbiologiya*, *50*, 471−475.

Imshenetskiĭ, A. A., Kondrat'eva, T. F., & Smut'ko, A. N. (1981b). Effect of carbon and nitrogen sources on pullulan biosynthesis by polyploid strains of *Pullularia pullulans*. *Mikrobiologiya*, *50*(1), 102−105.

Israilides, C., Bocking, M., Smith, A., & Scanlon, B. (1994). A novel rapid coupled enzyme assay for the estimation of pullulan. *Biotechnology and Applied Biochemistry*, *19*, 285−291.

Israilides, C. J., Smith, A., Harthill, J. E., Barnett, C., Bambalov, G., & Scanlon, B. (1998). Pullulan content of the ethanol precipitate from fermented agro-industrial wastes. *Applied Microbiology and Biotechnology*, *49*, 613−617. Available from https://doi.org/10.1007/s002530051222.

Jiang, L. (2010). Optimization of fermentation conditions for pullulan production by *Aureobasidium pullulan* using response surface methodology. *Carbohydrate Polymers*, *79*(2), 414−417. Available from https://doi.org/10.1016/j.carbpol.2009.08.027.

Jiang, L., Wu, S., & Kim, J. M. (2011). Effect of different nitrogen sources on activities of UDPG-pyrophosphorylase involved in pullulan synthesis and pullulan production by *Aureobasidium pullulans*. *Carbohydrate Polymers*, *86*(2), 1085−1088. Available from https://doi.org/10.1016/j.carbpol.2011.05.016.

Jiao, Y., Fu, Y., & Jiang, Z. (2004). The synthesis and characterization of poly(ethylene glycol) grafted on pullulan. *Journal of Applied Polymer Science*, *91*(2), 1217−1221. Available from https://doi.org/10.1002/app.13238.

Ju, X. M., Wang, D. H., Zhang, G. C., Cao, D., & Wei, G. Y. (2014). Efficient pullulan production by bioconversion using *Aureobasidium pullulans* as the whole-cell catalyst. *Applied Microbiology and Biotechnology*, 99(2), 211−220. Available from https://doi.org/10.1007/s00253-014-6100-1.

Jung, S. W., Jeong, Y.-I., & Kim, S. H. (2003). Characterization of hydrophobized pullulan with various hydrophobicities. *International Journal of Pharmaceutics*, 254(2), 109−121. Available from https://doi.org/10.1016/S0378-5173(03)00006-1.

Kachhawa, D. K., Bhattacharjee, P., & Singhal, R. S. (2003). Studies on downstream processing of pullulan. *Carbohydrate Polymers*, 52(1), 25−28. Available from https://doi.org/10.1016/S0144-8617(02)00261-8.

Kamoun, E. A., El-Betany, A., Menzel, H., & Chen, X. (2018). Influence of photoinitiator concentration and irradiation time on the crosslinking performance of visible-light activated pullulan-HEMA hydrogels. *International Journal of Biological Macromolecules*, 120(Pt B), 1884−1892. Available from https://doi.org/10.1016/j.ijbiomac.2018.10.011.

Kang, J. X., Chen, X. J., Chen, W. R., Li, M. S., Fang, Y., Li, D. S., Ren, Y. Z., & Liu, D. Q. (2011). Enhanced production of pullulan in *Aureobasidium pullulans* by a new process of genome shuffling. *Process Biochemistry*, 46(3), 792−795. Available from https://doi.org/10.1016/j.procbio.2010.11.004.

Kato, K., & Nomura, T. (Jan. 25, 1977). *Method for purifying pullulan*. Google patents.

Kawahara, K., Ohta, K., Miyamoto, H., & Nakamura, S. (1984). Preparation and solution properties of pullulan fractions as standard samples for water-soluble polymers. *Carbohydrate Polymers*, 4(5), 335−356. Available from https://doi.org/10.1016/0144-8617(84)90049-3.

Kikuchi, Y., Taguchi, R., Sakano, Y., & Kobayashi, T. (1973). Comparison of extracellular polysaccharide produced by *Pullularia pullulans* with polysaccharides in the cells and cell wall. *Agricultural and Biological Chemistry*, 37(7), 1751−1753.

Kim, J. H., Kim, M. R., Lee, J. H., Lee, J. W., & Kim, S. K. (2000). Production of high molecular weight pullulan by *Aureobasidium pullulans* using glucosamine. *Biotechnology Letters*, 22, 987−990. Available from https://doi.org/10.1023/A:1005681019573.

Kim, S.-Y., Kim, J.-H., Kim, C.-J., & Oh, D.-K. (1996). Metal adsorption of the polysaccharide produced from *Methylobacterium organophilum*. *Biotechnology Letters*, 18(10), 1161−1164.

Kimoto, T., Shibuya, T., & Shiobara, S. (1997). Safety studies of a novel starch, pullulan: Chronic toxicity in rats and bacterial mutagenicity. *Food and Chemical Toxicology*, 35(3−4), 323−329. Available from https://doi.org/10.1016/S0278-6915(97)00001-X.

Krull, S. M., Ma, Z., Li, M., Davé, R. N., & Bilgili, E. (2016). Preparation and characterization of fast dissolving pullulan films containing BCS class II drug nanoparticles for bioavailability enhancement. *Drug Development and Industrial Pharmacy*, 42(7), 1073−1085. Available from https://doi.org/10.3109/03639045.2015.1107094.

Kumar, A. S., Mody, K., & Jha, B. (2007). Bacterial exopolysaccharides—A perception. *Journal of Basic Microbiology*, 47(2), 103−117. Available from https://doi.org/10.1002/jobm.200610203.

Lacroix, C., LeDuy, A., Noel, G., & Choplin, L. (1985). Effect of pH on the batch fermentation of pullulan from sucrose medium. *Biotechnology and Bioengineering*, 27(2), 202−207. Available from https://doi.org/10.1002/bit.260270216.

Lacroix, M., & Le Tien, C. (2005). Edible films and coatings from nonstarch polysaccharides. In *Innovations in food packaging* (pp. 338−361). Elsevier.

Lazaridou, A., Biliaderis, C. G., & Kontogiorgos, V. (2003). Molecular weight effects on solution rheology of pullulan and mechanical properties of its films. *Carbohydrate Polymers*, 52(2), 151−166. Available from https://doi.org/10.1016/S0144-8617(02)00302-8.

Lazaridou, A., Biliaderis, C. G., Roukas, T., & Izydorczyk, M. (2002). Production and characterization of pullulan from beet molasses using a nonpigmented strain of *Aureobasidium pullulans* in batch culture. *Applied Biochemistry Biotechnology*, 97, 1−22. Available from https://doi.org/10.1385/ABAB:97:1:01.

Lazaridou, A., Roukas, T., Biliaderis, C. G., & Vaikousi, H. (2002). Characterization of pullulan produced from beet molasses by *Aureobasidium pullulans* in a stirred tank reactor under varying agitation.

Enzyme and Microbial Technology, 31(1−2), 122−132. Available from https://doi.org/10.1016/S0141-0229(02)00082-0.

Leathers, T. D. (1987). Host amylases and pullulan production. In D. L. Kaplan (Ed.), *First materials biotechnology symposium. Technical report NATICK/TR-88/033* (pp. 175−185). Natick, MA: US Army.

Leathers, T. D. (2003). Biotechnological production and applications of pullulan. *Applied Microbiology and Biotechnology, 62*, 468−473. Available from https://doi.org/10.1007/s00253-003-1386-4.

Leathers, T. D., & Gupta, S. C. (1994). Production of pullulan from fuel ethanol byproducts by *Aureobasidium* sp. strain NRRl Y-12,974. *Biotechnology Letters, 16*, 1163−1166. Available from https://doi.org/10.1007/BF01020844.

Leathers, T. D., Nofsinger, G. W., Kurtzman, C. P., & Bothast, R. J. (1988). Pullulan production by color variant strains of *Aureobasidium pullulans*. *Journal of Industrial Microbiology, 3*(4), 231−239. Available from https://doi.org/10.1007/BF01569581.

LeDuy, A., & Mian Boa, J. (1983). Pullulan production from peat hydrolyzate. *Canadian Journal of Microbiology, 29*(1), 143−146. Available from https://doi.org/10.1139/m83-023.

Lee, A., Ji-Hyun, A., Kim, J.-H., Kim, R.-M., Lim, S.-M., Nam, S.-W., Lee, J.-W., & Kim, S. K. (2002). Effect of dissolved oxygen concentration and pH on the mass production of high molecular weight pullulan by *Aureobasidium pullulans*. *Journal of Microbiology and Biotechnology, 12*(1), 1−7.

Lee, J. H., Lim, J.-H., Zhu, I.-H., Zhan, X.-B., Lee, J.-W., Shin, D.-H., & Kim, S.-K. (2001). Optimization of conditions for the production of pullulan and high molecular weight pullulan by *Aureobasidium pullulans*. *Biotechnology Letters, 23*, 817−820. Available from https://doi.org/10.1023/A:1010365706691.

Lee, K. Y., & Yoo, Y. J. (1993). Optimization of pH for high molecular weight pullulan. *Biotechnology Letters, 15*, 1021−1024. Available from https://doi.org/10.1007/BF00129930.

Li, X., Xue, W., Liu, Y., Li, W., Fan, D., Zhu, C., & Wang, Y. (2016). HLC/pullulan and pullulan hydrogels: Their microstructure, engineering process and biocompatibility. *Materials Science and Engineering: C, 58*, 1046−1057.

Lin, Y., & Thibault, J. (2013). Pullulan fermentation using a prototype rotational reciprocating plate impeller. *Bioprocess and Biosystems Engineering, 36*(5), 603−611. Available from https://doi.org/10.1007/s00449-012-0816-z.

Lin, Y., Zhang, Z., & Thibault, J. (2007). *Aureobasidium pullulans* batch cultivations based on a factorial design for improving the production and molecular weight of exopolysaccharides. *Process Biochemistry, 42*(5), 820−827.

Liu, N. N., Chi, Z., Wang, Q.-Q., Hong, J., Liu, G.-L., Hu, Z., & Chi, Z.-M. (2017). Simultaneous production of both high molecular weight pullulan and oligosaccharides by *Aureobasidium melanogenum* P16 isolated from a mangrove ecosystem. *International Journal of Biological Macromolecules, 102*, 1016−1024. Available from https://doi.org/10.1016/j.ijbiomac.2017.04.057.

Ma, Z. C., Chi, Z., Geng, Q., Zhang, F., & Chi, Z. M. (2012). Disruption of the pullulan synthetase gene in siderophore-producing *Aureobasidium pullulans* enhances siderophore production and simplifies siderophore extraction. *Process Biochemistry, 47*(12), 1807−1812. Available from https://doi.org/10.1016/j.procbio.2012.06.024.

Madi, N. S., Harvey, L. M., Mehlert, A., & McNeil, B. (1997). Synthesis of two distinct exopolysaccharide fractions by cultures of the polymorphic fungus *Aureobasidium pullulans*. *Carbohydrate Polymers, 32*(3−4), 307−314. Available from https://doi.org/10.1016/S0144-8617(97)00003-9.

Madi, N. S., McNeil, B., & Harvey, L. M. (1996). Influence of Culture pH and Aeration on Ethanol Production and Pullulan Molecular Weight by *Aureobasidium pullulans*. *Journal of Chemical Technology and Biotechnology (Oxford, Oxfordshire: 1986), 65*(4), 343−350. Available from https://doi.org/10.1002/(sici)1097-4660(199604)65:4<343::aid-jctb461>3.3.co;2-9.

Masuda, K., Sakagami, M., Horie, K., Nogusa, H., Hamana, H., & Hirano, K. (2001). Evaluation of carboxymethylpullulan as a novel carrier for targeting immune tissues. *Pharmaceutical Research, 18*(2), 217−223. Available from https://doi.org/10.1023/A:1011040703915.

McNeil, B., & Kristiansen, B. (1990). Temperature effects on polysaccharide formation by *Aureobasidium pullulans* in stirred tanks. *Enzyme and Microbial Technology*, *12*(7), 521−526. Available from https://doi.org/10.1016/0141-0229(90)90069-3.

McNeil, B., Kristiansen, B., & Seviour, R. J. (1989). Polysaccharide production and morphology of *Aureobasidium pullulans* in continuous culture. *Biotechnology and Bioengineering*, *33*(9), 1210−1212. Available from https://doi.org/10.1002/bit.260330918.

Mehta, A., Prasad, G. S., & Choudhury, A. R. (2014). Cost effective production of pullulan from agri-industrial residues using response surface methodology. *International Journal of Biological Macromolecules*, *64*, 252−256. Available from https://doi.org/10.1016/j.ijbiomac.2013.12.011.

Mihai, D., Mocanu, G., & Carpov, A. (2001). Chemical reactions on polysaccharides: I. Pullulan sulfation. *European Polymer Journal*, *37*(3), 541−546.

Mocanu, G., Mihai, D., Le Cerf, D., Picton, L., & Muller, G. (2004). Synthesis of new associative gel microspheres from carboxymethyl pullulan and their interactions with lysozyme. *European Polymer Journal*, *40*(2), 283−289. Available from https://doi.org/10.1016/j.eurpolymj.2003.09.019.

Moscovici, M., Ionescu, C., Oniscu, C., Fotea, O., Protopopescu, P., & Hanganu, L. D. (1996). Improved exopolysaccharide production in fed-batch fermentation of *Aureobasidium pullulans*, with increased impeller speed. *Biotechnology Letters*, *18*, 787−790. Available from https://doi.org/10.1007/BF00127889.

Motozato, Y., Ihara, H., Tomoda, T., & Hirayama, C. (1986). Preparation and gel permeation chromatographic properties of pullulan spheres. *Journal of Chromatography. A*, *355*, 434−437. Available from https://doi.org/10.1016/S0021-9673(01)97349-2.

Mulchandani, A., Luong, J. H. T., & LeDuy, A. (1989). Biosynthesis of pullulan using immobilized *Aureobasidium pullulans* cells. *Biotechnology and Bioengineering*, *33*(3), 306−312. Available from https://doi.org/10.1002/bit.260330309.

Nakashio, S., Fujita, F., Domoto, M., Toyota, N., & Sekine, N. (1978). *Paper coating material containing pullulan*. Google patents.

Nakatani, M., Shibukawa, A., & Nakagawa, T. (1996). Separatists mechanism of pullulan solution-filled capillary electrophoresis of sodium dodecyl sulfate-proteins. *Electrophoresis*, *17*(10), 1584−1586. Available from https://doi.org/10.1002/elps.1150171015.

Nogusa, H., Yamamoto, K., Yano, T., Kajiki, M., Hamana, H., & Okuno, S. (2000). Distribution characteristics of carboxymethylpullulan-peptide-doxorubicin conjugates in tumor-bearing rats: Different sequence of peptide spacers and doxorubicin contents. *Biological & Pharmaceutical Bulletin*, *23*(5), 621−626.

Nogusa, H., Yano, T., Okuno, S., Hamana, H., & Inoue, K. (1995). Synthesis of carboxymethylpullulan-peptide-doxorubicin conjugates and their properties. *Chemical and Pharmaceutical Bulletin*, *43*(11), 1931−1936.

Nomura, Y., Ikeda, M., Yamaguchi, N., Aoyama, Y., & Akiyoshi, K. (2003). Protein refolding assisted by self-assembled nanogels as novel artificial molecular chaperone. *FEBS Letters*, *553*(3), 271−276. Available from https://doi.org/10.1016/S0014-5793(03)01028-7.

Olmo, S., et al. (2008). Analysis of human histone H4 by capillary electrophoresis in a pullulan-coated capillary, LC-ESI-MS and MALDI-TOF-MS. *Analytical and Bioanalytical Chemistry*, *390*(7), 1881−1888. Available from https://doi.org/10.1007/s00216-008-1903-5.

Onda, Y., Muto, H., & Suzuki, H. (Mar. 1982). *Cyanoethylpullulan*. Google patents.

Ono, K., Yasuda, N., & Ueda, S. (1977). Effect of pH on pullulan elaboration by *Aureobasidium pullulans* S-1. *Agricultural and Biological Chemistry*, *41*(11), 2113−2118. Available from https://doi.org/10.1080/00021369.1977.10862824.

Orr, D., Zheng, W., Campbell, B. S., McDougall, B. M., & Seviour, R. J. (2009). Culture conditions affect the chemical composition of the exopolysaccharide synthesized by the fungus *Aureobasidium pullulans*. *Journal of Applied Microbiology*, *107*(2), 691−698. Available from https://doi.org/10.1111/j.1365-2672.2009.04247.x.

Özcan, E., Sargin, S., & Göksungur, Y. (2014). Comparison of pullulan production performances of air-lift and bubble column bioreactors and optimization of process parameters in air-lift bioreactor. *Biochemical Engineering Journal*, *92*, 9−15. Available from https://doi.org/10.1016/j.bej.2014.05.017.

Pan, S., Yao, D., Chen, J., & Wu, S. (2013). Influence of controlled pH on the activity of UDPG-pyrophosphorylase in *Aureobasidium pullulans*. *Carbohydrate Polymers*, *92*(1), 629−632. Available from https://doi.org/10.1016/j.carbpol.2012.08.099.

Pollock, T. J., Thorne, L., & Armentrout, R. W. (1992). Isolation of new *Aureobasidium* strains that produce high-molecular-weight pullulan with reduced pigmentation. *Applied and Environmental Microbiology*, *58*(3), 877−883. Available from https://doi.org/10.1128/aem.58.3.877-883.1992.

Prajapati, V. D., Jani, G. K., & Khanda, S. M. (2013). Pullulan: An exopolysaccharide and its various applications. *Carbohydrate Polymers*, *95*(1), 540−549. Available from https://doi.org/10.1016/j.carbpol.2013.02.082.

Prasongsuk, S., Berhow, M. A., Dunlap, C. A., Weisleder, D., Leathers, T. D., Eveleigh, D. E., & Punnapayak, H. (2007). Pullulan production by tropical isolates of *Aureobasidium pullulans*. *Journal of Industrial Microbiology & Biotechnology*, *34*, 55−61. Available from https://doi.org/10.1007/s10295-006-0163-7.

Radulović, M. Đ., Cvetković, O. G., Nikolić, S. D., Đorđević, D. S., Jakovljević, D. M., & Vrvić, M. M. (2008). Simultaneous production of pullulan and biosorption of metals by *Aureobasidium pullulans* strain CH-1 on peat hydrolysate. *Bioresource Technology*, *99*(14), 6673−6677.

Ravella, S. R., Quiñones, T. S., Retter, A., Heiermann, M., Amon, T., & Hobbs, P. J. (2010). Extracellular polysaccharide (EPS) production by a novel strain of yeast-like fungus *Aureobasidium pullulans*. *Carbohydrate Polymers*, *82*(3), 728−732. Available from https://doi.org/10.1016/j.carbpol.2010.05.039.

Ray, R. C., & Moorthy, S. N. (2007). Exopolysaccharide (pullulan) production from cassava starch residue by *Aureobasidium pullulans* strain MTTC 1991. *Journal of Scientific and Industrial Research*, *66*(3), 252−255.

Reed-Hamer, B., & West, T. P. (1994). Effect of complex nitrogen sources on pullulan production relative to carbon source. *Microbios*, *80*(323), 83−90.

Reeslev, M., Nielsen, J. C., Olsen, J., Jensen, B., & Jacobsen, T. (1991). Effect of pH and the initial concentration of yeast extract on regulation of dimorphism and exopolysaccharide formation of *Aureobasidium pullulans* in batch culture. *Mycological Research*, *95*(2), 220−226.

Reeslev, M., StrØm, T., Jensen, B., & Olsen, J. (1997). The ability of the yeast form of *Aureobasidium pullullans* to elaborate exopolysaccharide in chemostat culture at various pH values. *Mycological Research*, *101*(6), 650−652. Available from https://doi.org/10.1017/S0953756296003255.

Reis, R. A., Tischer, C. A., Gorin, P. A. J., & Iacomini, M. (2002). A new pullulan and a branched (1→3)-, (1→6)-linked β-glucan from the lichenised ascomycete *Teloschistes flavicans*. *FEMS Microbiology Letters*, *210*(1), 1−5.

Rho, D., Mulchandani, A., Luong, J. H. T., & LeDuy, A. (1988). Oxygen requirement in pullulan fermentation. *Applied Microbiology and Biotechnology*, *28*, 361−366. Available from https://doi.org/10.1007/BF00268196.

Ronen, M., Guterman, H., & Shabtai, Y. (2002). Monitoring and control of pullulan production using vision sensor. *Journal of Biochemical and Biophysical Methods*, *51*(3), 243−249. Available from https://doi.org/10.1016/S0165-022X(01)00182-8.

Roukas, T. (1998). Pretreatment of beet molasses to increase pullulan production. *Process Biochemistry*, *33*(8), 805−810. Available from https://doi.org/10.1016/S0032-9592(98)00048-X.

Roukas, T. (1999a). Pullulan production from deproteinized whey by *Aureobasidium pullulans*. *Journal of Industrial Microbiology & Biotechnology*, *22*(6), 617−621. Available from https://doi.org/10.1038/sj.jim.2900675.

Roukas, T. (1999b). Pullulan production from brewery wastes by *Aureobasidium pullulans*. *World Journal of Microbiology and Biotechnology*, *15*, 447−450. Available from https://doi.org/10.1023/A:1008996522115.

Roukas, T., & Biliaderis, C. G. (1995). Evaluation of carob pod as a substrate for pullulan production by *Aureobasidium pullulans*. *Applied Biochemistry and Biotechnology*, *55*(1), 27−44.

Roukas, T., & Liakopoulou-Kyriakides, M. (1999). Production of pullulan from beet molasses by *Aureobasidium pullulans* in a stirred tank fermentor. *Journal of Food Engineering*, *40*(1−2), 89−94. Available from https://doi.org/10.1016/S0260-8774(99)00043-6.

Roukas, T., & Mantzouridou, F. (2001). Effect of the aeration rate on pullulan production and fermentation broth rheological properties in an airlift reactor. *Journal of Chemical Technology and Biotechnology (Oxford, Oxfordshire: 1986)*, *76*(4), 371–376. Available from https://doi.org/10.1002/jctb.391.

Roukas, T., & Serris, G. (1999). Effect of the shear rate on pullulan production from beet molasses by *Aureobasidium pullulans* in an airlift reactor. *Applied Biochemistry and Biotechnology*, *80*, 77–89. Available from https://doi.org/10.1385/ABAB:80:1:77.

Sakata, Y., & Otsuka, M. (2009). Evaluation of relationship between molecular behaviour and mechanical strength of pullulan films. *International Journal of Pharmaceutics*, *374*(1–2), 33–38. Available from https://doi.org/10.1016/j.ijpharm.2009.02.019.

Sánchez-Arreola, E., Martin-Torres, G., Lozada-Ramírez, J. D., Hernández, L. R., Bandala-González, E. R., & Bach, H. (2015). Biodiesel production and de-oiled seed cake nutritional values of a Mexican edible *Jatropha curcas*. *Renewable Energy*, *76*, 143–147. Available from https://doi.org/10.1016/j.renene.2014.11.017.

Schuster, R., Wenzig, E., & Mersmann, A. (1993). Production of the fungal exopolysaccharide pullulan by batch-wise and continuous fermentation. *Applied Microbiology and Biotechnology*, *39*, 155–158. Available from https://doi.org/10.1007/BF00228599.

Sena, R. F., Costelli, M. C., Gibson, L. H., & Coughlin, R. W. (2006). Enhanced production of pullulan by two strains of *A. pullulans* with different concentrations of soybean oil in sucrose solution in batch fermentations. *Brazilian Journal of Chemical Engineering*, *23*(4), 33–38. Available from https://doi.org/10.1590/S0104-66322006000400008.

Seo, C., Lee, H. W., Suresh, A., Yang, J. W., Jung, J. K., & Kim, Y. C. (2014). Improvement of fermentative production of exopolysaccharides from *Aureobasidium pullulans* under various conditions. *Korean Journal of Chemical Engineering*, *31*, 1433–1437. Available from https://doi.org/10.1007/s11814-014-0064-9.

Seo, H. P., Jo, K.-I., Son, C.-W., Yang, J.-K., Chung, C.-H., Nam, S.-W., Kim, S.-K., & Lee, J.-W. (2006). Continuous production of pullulan by *Aureobasidium pullulans* HP-2001 with feeding of high concentrations of sucrose. *Journal of Microbiology and Biotechnology*, *16*(3), 374–380.

Seo, H. P., Son, C.-W., Chung, C.-H., Jung, D.-I., Kim, S.-K., Gross, R. A., Kaplan, D. L., & Lee, J.-W. (2004). Production of high molecular weight pullulan by *Aureobasidium pullulans* HP-2001 with soybean pomace as a nitrogen source. *Bioresource Technology*, *95*(3), 293–299. Available from https://doi.org/10.1016/j.biortech.2003.02.001.

Seviour, R. J., Kristiansen, B., & Harvey, L. (1984). Morphology of *Aureobasidium pullulans* during polysaccharide elaboration. *Transactions of the British Mycological Society*, *83*(2), 350–356. Available from https://doi.org/10.1016/s0007-1536(84)80162-x.

Shabtai, Y., & Mukmenev, I. (1995). Enhanced production of pigment-free pullulan by a morphogenetically arrested *Aureobasidium pullulans* (ATCC 42023) in a two-stage fermentation with shift from soy bean oil to sucrose. *Applied Microbiology and Biotechnology*, *43*(4), 595–603.

Sharma, N., Prasad, G. S., & Choudhury, A. R. (2013). Utilization of corn steep liquor for biosynthesis of pullulan, an important exopolysaccharide. *Carbohydrate Polymers*, *93*(1), 95–101. Available from https://doi.org/10.1016/j.carbpol.2012.06.059.

Sharmila, G., Muthukumaran, C., Nayan, G., & Nidhi, B. (2013). Extracellular Biopolymer Production by *Aureobasidium pullulans* MTCC 2195 using Jackfruit seed powder. *Journal of Polymers and the Environment*, *21*, 487–494. Available from https://doi.org/10.1007/s10924-012-0459-9.

Sheng, L., Zhu, G., & Tong, Q. (2013). Mechanism study of Tween 80 enhancing the pullulan production by *Aureobasidium pullulans*. *Carbohydrate Polymers*, *97*(1), 121–123. Available from https://doi.org/10.1016/j.carbpol.2013.04.058.

Sheng, L., Zhu, G., & Tong, Q. (2014). Effect of uracil on pullulan production by *Aureobasidium pullulans* CGMCC1234. *Carbohydrate Polymers*, *101*, 435–437. Available from https://doi.org/10.1016/j.carbpol.2013.09.063.

Sheoran, S. K., Dubey, K. K., Tiwari, D. P., & Singh, B. P. (2012). Directive production of pullulan by altering cheap source of carbons and nitrogen at 5 L bioreactor level. *ISRN Chemical Engineering*, *2012*(3–4). Available from https://doi.org/10.5402/2012/867198.

Shin, Y. C., Kim, Y. H., Lee, H. S., Cho, S. J., & Byun, S. M. (1989). Production of exopolysaccharide pullulan from inulin by a mixed culture of *Aureobasidium pullulans* and *Kluyveromyces fragilis*. *Biotechnology and Bioengineering*, *33*(1), 129–133. Available from https://doi.org/10.1002/bit.260330117.

Shingel, K. I. (2004). Current knowledge on biosynthesis, biological activity, and chemical modification of the exopolysaccharide, pullulan. *Carbohydrate Research*, *339*(3), 447–460. Available from https://doi.org/10.1016/j.carres.2003.10.034.

Simon, L., Bouchet, B., Bremond, K., Gallant, D. J., & Bouchonneau, M. (1998). Studies on pullulan extracellular production and glycogen intracellular content in *Aureobasidium pullulans*. *Canadian Journal of Microbiology*, *44*(12), 1193–1199. Available from https://doi.org/10.1139/w98-115.

Simon, L., Caye-Vaugien, C., & Bouchonneau, M. (1993). Relation between pullulan production, morphological state and growth conditions in *Aureobasidium pullulans*: New observations. *Journal of General Microbiology*, *139*(5), 979–985. Available from https://doi.org/10.1099/00221287-139-5-979.

Singh, R. S., & Kaur, N. (2019). Understanding response surface optimization of medium composition for pullulan production from de-oiled rice bran by *Aureobasidium pullulans*. *Food Science and Biotechnology*, *28*, 1507–1520. Available from https://doi.org/10.1007/s10068-019-00585-w.

Singh, R. S., Kaur, N., & Kennedy, J. F. (2019). Pullulan production from agro-industrial waste and its applications in food industry: A review. *Carbohydrate Polymers*, *217*, 46–57. Available from https://doi.org/10.1016/j.carbpol.2019.04.050.

Singh, R. S., Kaur, N., Sharma, R., & Rana, V. (2018). Carbamoylethyl pullulan: QbD based synthesis, characterization and corneal wound healing potential. *International Journal of Biological Macromolecules*, *118*(Pt B), 2245–2255. Available from https://doi.org/10.1016/j.ijbiomac.2018.07.107.

Singh, R. S., & Saini, G. K. (2008). Pullulan-hyperproducing color variant strain of *Aureobasidium pullulans* FB-1 newly isolated from phylloplane of *Ficus* sp. *Bioresource Technology*, *99*(9), 3896–3899. Available from https://doi.org/10.1016/j.biortech.2007.08.003.

Singh, R. S., Saini, G. K., & Kennedy, J. F. (2008). Pullulan: Microbial sources, production and applications. *Carbohydrate Polymers*, *73*(4), 515–531. Available from https://doi.org/10.1016/j.carbpol.2008.01.003.

Singh, R. S., Saini, G. K., & Kennedy, J. F. (2009). Downstream processing and characterization of pullulan from a novel colour variant strain of *Aureobasidium pullulans* FB-1. *Carbohydrate Polymers*, *78*(1), 89–94. Available from https://doi.org/10.1016/j.carbpol.2009.03.040.

Singh, R. S., Saini, G. K., & Kennedy, J. F. (2011). Continuous hydrolysis of pullulan using covalently immobilized pullulanase in a packed bed reactor. *Carbohydrate Polymers*, *83*(2), 672–675. Available from https://doi.org/10.1016/j.carbpol.2010.08.037.

Singh, R. S., Singh, H., & Saini, G. K. (2009). Response surface optimization of the critical medium components for pullulan production by *Aureobasidium pullulans* FB-1. *Applied Biochemistry and Biotechnology*, *152*(1), 42–53. Available from https://doi.org/10.1007/s12010-008-8180-9.

Slodki, M. E., & Cadmus, M. C. (1978). Production of microbial polysaccharides. *Advances in Applied Microbiology*, *23*, 19–54. Available from https://doi.org/10.1016/S0065-2164(08)70064-9.

Srikanth, S., Swathi, M., Tejaswini, M., Sharmila, G., Muthukumaran, C., Jaganathan, M. K., & Tamilarasan, K. (2014). Statistical optimization of molasses based exopolysaccharide and biomass production by *Aureobasidium pullulans* MTCC 2195. *Biocatalysis and Agricultural Biotechnology*, *3*(3), 7–12. Available from https://doi.org/10.1016/j.bcab.2013.11.011.

Sugawa-Katayama, Y., Kondou, F., Mandai, T., & Yoneyama, M. (1994). Effects of pullulan, polydextrose and pectin on cecal microflora. *Journal of Applied Glycoscience*, *41*(4), 413–418.

Sugumaran, K. R., Gowthami, E., Swathi, B., Elakkiya, S., Srivastava, S. N., Ravikumar, R., Gowdhaman, D., & Ponnusami, V. (2013). Production of pullulan by *Aureobasidium pullulans* from Asian palm kernel: A novel substrate. *Carbohydrate Polymers*, *92*(1), 697–703. Available from https://doi.org/10.1016/j.carbpol.2012.09.062.

Sugumaran, K. R., Jothi, P., & Ponnusami, V. (2014). Bioconversion of industrial solid waste—Cassava bagasse for pullulan production in solid state fermentation. *Carbohydrate Polymers*, *99*, 22–30. Available from https://doi.org/10.1016/j.carbpol.2013.08.039.

Sugumaran, K. R., & Ponnusami, V. (2017a). Review on production, downstream processing and characterization of microbial pullulan. *Carbohydrate Polymers, 173*, 573−591. Available from https://doi.org/10.1016/j.carbpol.2017.06.022.

Sugumaran, K. R., & Ponnusami, V. (2017b). Conventional optimization of aqueous extraction of pullulan in solid-state fermentation of cassava bagasse and Asian palm kernel. *Biocatalysis and Agricultural Biotechnology, 10*, 204−208. Available from https://doi.org/10.1016/j.bcab.2017.03.010.

Taguchi, R., Kikuchi, Y., Sakano, Y., & Kobayashi, T. (1973). Structural uniformity of pullulan produced by several strains of *Pullularia pullulans*. *Agricultural and Biological Chemistry, 37*(7), 1583−1588. Available from https://doi.org/10.1271/bbb1961.37.1583.

Tarabasz-Szymanska, L., Galas, E., & Pankiewicz, T. (1999). Optimization of productivity of pullulan by means of multivariable linear regression analysis. *Enzyme and Microbial Technology, 24*(5−6), 276−282. Available from https://doi.org/10.1016/S0141-0229(98)00117-3.

Terán Hilares, R., Orsi, C. A., Ahmed, M. A., Marcelino, P. F., Menegatti, C. R., da Silva, S. S., & dos Santos, J. C. (2017). Low-melanin containing pullulan production from sugarcane bagasse hydrolysate by *Aureobasidium pullulans* in fermentations assisted by light-emitting diode. *Bioresource Technology, 230*, 76−81. Available from https://doi.org/10.1016/j.biortech.2017.01.052.

Terán Hilares, R., Resende, J., Orsi, C. A., Ahmed, M. A., Lacerda, T. M., da Silva, S. S., & Santos, J. C. (2019). Exopolysaccharide (pullulan) production from sugarcane bagasse hydrolysate aiming to favor the development of biorefineries. *International Journal of Biological Macromolecules, 127*, 169−177. Available from https://doi.org/10.1016/j.ijbiomac.2019.01.038.

Thirumavalavan, K., Manikkadan, T. R., & Dhanasekar, R. (2009). Pullulan production from coconut by-products by *Aureobasidium pullulans*. *African Journal of Biotechnology, 8*(2), 254−258. Available from https://doi.org/10.5897/AJB2009.000-9045.

Tomasula, P. M., Sousa, A. M. M., Liou, S.-C., Li, R., Bonnaillie, L. M., & Liu, L. S. (2016). Short communication: Electrospinning of casein/pullulan blends for food-grade applications. *Journal of Dairy Science, 99*(3), 1837−1845.

Tsujisaka, Y., & Mitsuhashi, M. (1993). Pullulan. In *Industrial Gums,* Elsevier, pp. 447−460.

Tu, G., Wang, Y., Ji, Y., & Zou, X. (2015). The effect of Tween 80 on the polymalic acid and pullulan production by *Aureobasidium pullulans* CCTCC M2012223. *World Journal of Microbiology and Biotechnology, 31*(1), 219−226. Available from https://doi.org/10.1007/s11274-014-1779-9.

Ueda, S., Fujita, K., Komatsu, K., & Nakashima, Z. I. (1963). Polysaccharide produced by the genus *Pullularia*. I. Production of polysaccharide by growing cells. *Applied Microbiology, 11*(3), 211−215.

Ürküt, Z., Dağbağli, S., & Göksungur, Y. (2007). Optimization of pullulan production using Ca-alginate-immobilized *Aureobasidium pullulans* by response surface methodology. *Journal of Chemical Technology and Biotechnology (Oxford, Oxfordshire: 1986), 82*(9), 837−846. Available from https://doi.org/10.1002/jctb.1750.

U.S. Food and Drug Administration Center for Food Safety and Applied Nutrition Office of Plant & Dairy Foods & Beverages. (2002). *Detection and quantitation of acrylamide in foods.*

Vashist, S. K., Luppa, P. B., Yeo, L. Y., Ozcan, A., & Luong, J. H. T. (2015). Emerging technologies for next-generation point-of-care testing. *Trends in Biotechnology, 33*(11), 692−705. Available from https://doi.org/10.1016/j.tibtech.2015.09.001.

Vijayendra, S. V. N., Bansal, D., Prasad, M. S., & Nand, K. (2001). Jaggery: A novel substrate for pullulan production by *Aureobasidium pullulans* CFR-77. *Process Biochemistry, 37*(4), 359−364. Available from https://doi.org/10.1016/S0032-9592(01)00214-X.

Wang, D., Bian, J., Wei, G., Jiang, M., & Dong, M. (2016). Simultaneously enhanced production and molecular weight of pullulan using a two-stage agitation speed control strategy. *Journal of Chemical Technology and Biotechnology (Oxford, Oxfordshire: 1986), 91*(2), 467−475. Available from https://doi.org/10.1002/jctb.4600.

Wang, D., Chen, F., Wei, G., Jiang, M., & Dong, M. (2015). The mechanism of improved pullulan production by nitrogen limitation in batch culture of *Aureobasidium pullulans*. *Carbohydrate Polymers, 127*, 325−331. Available from https://doi.org/10.1016/j.carbpol.2015.03.079.

Wang, D., Ju, X., Zhou, D., & Wei, G. (2014). Efficient production of pullulan using rice hull hydrolysate by adaptive laboratory evolution of *Aureobasidium pullulans*. *Bioresource Technology, 164*, 12−19. Available from https://doi.org/10.1016/j.biortech.2014.04.036.

Wang, D., Yu, X., & Gongyuan, W. (2013). Pullulan production and physiological characteristics of *Aureobasidium pullulans* under acid stress. *Applied Microbiology and Biotechnology, 97*(18), 8069−8077. Available from https://doi.org/10.1007/s00253-013-5094-4.

Wang, J., Cui, S., Bao, Y., Xing, J., & Hao, W. (2014). Tocopheryl pullulan-based self assembling nanomicelles for anti-cancer drug delivery. *Material Science Engineering C, 43*, 614−621.

Wang, W., Yu, Y., Wang, P., Wang, Q., Li, Y., Yuan, J., & Fan, X. (2019). Controlled graft polymerization on the surface of filter paper via enzyme-initiated RAFT polymerization. *Carbohydrate Polymers, 207*, 239−245. Available from https://doi.org/10.1016/j.carbpol.2018.11.095.

Wecker, A., & Onken, U. (1991). Influence of dissolved oxygen concentration and shear rate on the production of pullulan by *Aureobasidium pullulans*. *Biotechnology Letters, 13*, 155−160. Available from https://doi.org/10.1007/BF01025810.

West, T. P. (2000). Exopolysaccharide production by entrapped cells of the fungus *Aureobasidium pullulans* ATCC 201253. *Journal of Basic Microbiology, 40*(5−6), 397−401. Available from https://doi.org/10.1002/1521-4028(200012)40:5/6 < 397::AID-JOBM397 > 3.0.CO;2-N.

West, T. P. (2010). Pullulan production by *Aureobasidium pullulans* cells immobilized on ECTEOLA-cellulose. *Annals of Microbiology, 60*(4), 763−766. Available from https://doi.org/10.1007/s13213-010-0115-3.

West, T. P. (2011). Pullulan production by *Aureobasidium pullulans* cells immobilized in chitosan beads. *Folia Microbiologica, 56*, 335−338. Available from https://doi.org/10.1007/s12223-011-0048-7.

West, T. P. (2012). Pullulan production by *Aureobasidium pullulans* ATCC 201253 cells adsorbed onto cellulose anion and cation exchangers. *ISRN Microbiology, 2012*, 140951. Available from https://doi.org/10.5402/2012/140951.

West, T. P., & Strohfus, B. (1997). Effect of manganese on polysaccharide production and cellular pigmentation in the fungus *Aureobasidium pullulans*. *World Journal of Microbiology and Biotechnology, 13*, 233−235. Available from https://doi.org/10.1023/A:1018554201153.

West, T. P., & Strohfus, B. (1998). Polysaccharide production by *Aureobasidium pullulans* cells immobilized by entrapment. *Microbiological Research, 153*(3), 253−256. Available from https://doi.org/10.1016/S0944-5013(98)80008-6.

West, T. P., & Strohfus, B. (2001a). Polysaccharide production by a reduced pigmentation mutant of *Aureobasidium pullulans* NYS-1. *Letters in Applied Microbiology, 33*(2), 169−172. Available from https://doi.org/10.1046/j.1472-765X.2001.00975.x.

West, T. P., & Strohfus, B. (2001b). Polysaccharide production by immobilized *Aureobasidium pullulans* cells in batch bioreactors. *Microbiological Research, 156*(3), 285−288. Available from https://doi.org/10.1078/0944-5013-00106.

Wickerham, L. J., & Kurtzman, C. P. (1975). Synergistic color variants of *Aureobasidium pullulans*. *Mycologia, 67*(2), 342−361. Available from https://doi.org/10.2307/3758426.

Wiederschain, G. Y. (2007). Polysaccharides. Structural diversity and functional versatility. *Biochemistry (Moscow), 72*, 675. Available from https://doi.org/10.1134/s0006297907060120.

Wu, S., Chen, H., Jin, Z., & Tong, Q. (2010). Effect of two-stage temperature on pullulan production by *Aureobasidium pullulans*. *World Journal of Microbiology Biotechnology, 26*, 737−741. Available from https://doi.org/10.1007/s11274-009-0231-z.

Wu, S., Chen, J., & Pan, S. (2012). Optimization of fermentation conditions for the production of pullulan by a new strain of *Aureobasidium pullulans* isolated from sea mud and its characterization. *Carbohydrate Polymers, 87*(2), 1696−1700. Available from https://doi.org/10.1016/j.carbpol.2011.09.078.

Wu, S., Jin, Z., Kim, J. M., Tong, Q., & Chen, H. (2009). Graft copolymerization of methyl acrylate onto pullulan using ceric ammonium nitrate as initiator. *Carbohydrate Polymers, 76*(1), 129−132. Available from https://doi.org/10.1016/j.carbpol.2008.10.002.

Wu, S., Jin, Z., Kim, J. M., Tong, Q., & Chen, H. (2009). Downstream processing of pullulan from fermentation broth. *Carbohydrate Polymers, 77*(4), 750−753. Available from https://doi.org/10.1016/j.carbpol.2009.02.023.

Wu, S., Jin, Z., Tong, Q., & Chen, H. (2009). Sweet potato: A novel substrate for pullulan production by *Aureobasidium pullulans*. *Carbohydrate Polymers*, 76(4), 645–649. Available from https://doi.org/10.1016/j.carbpol.2008.11.034.

Xi, K., Tabata, Y., Uno, K., Yoshimoto, M., Kishida, T., Sokawa, Y., & Ikada, Y. (1996). Liver targeting of interferon through pullulan conjugation. *Pharmaceutical Research*, 13(12), 1846–1850.

Xia, Z., Wu, S., & Pan, S. (2011). Effect of two-stage controlled pH and temperature on pullulan production by *Auerobasidium pullulans*. *Carbohydrate Polymers*, 86(4), 1814–1816. Available from https://doi.org/10.1016/j.carbpol.2011.06.087.

Xiao, Q., Lim, L. T., & Tong, Q. (2012). Properties of pullulan-based blend films as affected by alginate content and relative humidity. *Carbohydrate Polymers*, 87(1), 227–234. Available from https://doi.org/10.1016/j.carbpol.2011.07.040.

Xiros, C., Topakas, E., Katapodis, P., & Christakopoulos, P. (2008). Hydrolysis and fermentation of brewer's spent grain by *Neurospora crassa*. *Bioresource Technology*, 99(13), 5427–5435. Available from https://doi.org/10.1016/j.biortech.2007.11.010.

Xu, F., Weng, B., Gilkerson, R., Materon, L. A., & Lozano, K. (2015). Development of tannic acid/chitosan/pullulan composite nanofibers from aqueous solution for potential applications as wound dressing. *Carbohydrate Polymers*, 115, 16–24. Available from https://doi.org/10.1016/j.carbpol.2014.08.081.

Yamaoka, T., Tabata, Y., & Ikada, Y. (1993). Body distribution profile of polysaccharides after intravenous administration. *Drug Delivery*, 1(1), 75–82. Available from https://doi.org/10.3109/10717549309031345.

Yamasaki, H., Lee, M. S., Tanaka, T., & Nakanishi, K. (1993). Improvement of performance for cross-flow membrane filtration of pullulan broth. *Applied Microbiology and Biotechnology*, 39, 21–25. Available from https://doi.org/10.1007/BF00166842.

Yang, S.-T., Zhu, H., Li, Y., & Hong, G. (1994). Continuous propionate production from whey permeate using a novel fibrous bed bioreactor. *Biotechnology and Bioengineering*, 43(11), 1124–1130. Available from https://doi.org/10.1002/bit.260431117.

Yatmaz, E., & Turhan, I. (2012). Pullulan production by fermentation and usage in food industry. *GIDA-Journal Food*, 37(2), 95–102.

Youssef, F., Biliaderis, C. G., & Roukas, T. (1998). Enhancement of pullulan production by *Aureobasidium pullulans* in batch culture using olive oil and sucrose as carbon sources. *Applied Biochemistry and Biotechnology*, 74, 13–30. Available from https://doi.org/10.1007/BF02786883.

Youssef, F., Roukas, T., & Biliaderis, C. G. (1999). Pullulan production by a non-pigmented strain of *Aureobasidium pullulans* using batch and fed-batch culture. *Process Biochemistry*, 34(4), 355–366. Available from https://doi.org/10.1016/S0032-9592(98)00106-X.

Yu, X., Wang, Y., Wei, G., & Dong, Y. (2012). Media optimization for elevated molecular weight and mass production of pigment-free pullulan. *Carbohydrate Polymers*, 89(3), 928–934.

Yuen, S. (1974). Pullulan and its applications. *Process Biochemistry*, 22, 7–9.

Yurlova, N. A., & De Hoog, G. S. (1997). A new variety of *Aureobasidium pullulans* characterized by exopolysaccharide structure, nutritional physiology and molecular features. *Antonie van Leeuwenhoek*, 72(2), 141–147. Available from https://doi.org/10.1023/A:1000212003810.

Zhang, L., Alfano, J., Race, D., & Davé, R. N. (2018). Zero-order release of poorly water-soluble drug from polymeric films made via aqueous slurry casting. *European Journal of Pharmaceutical Sciences: Official Journal of the European Federation for Pharmaceutical Sciences*, 117, 245–254. Available from https://doi.org/10.1016/j.ejps.2018.02.029.

Zheng, W., Campbell, B. S., McDougall, B. M., & Seviour, R. J. (2008). Effects of melanin on the accumulation of exopolysaccharides by *Aureobasidium pullulans* grown on nitrate. *Bioresource Technology*, 99(16), 7480–7486. Available from https://doi.org/10.1016/j.biortech.2008.02.016.

Further reading

Ankareddi, I., & Brazel, C. S. (2007). Synthesis and characterization of grafted thermosensitive hydrogels for heating activated controlled release. *International Journal of Pharmaceutics*, *336*(2), 241−247. Available from https://doi.org/10.1016/j.ijpharm.2006.11.065.

Cheng, K. C., Demirci, A., & Catchmark, J. M. (2010b). Advances in biofilm reactors for production of value-added products. *Applied Microbiology and Biotechnology*, *87*(2), 445−456. Available from https://doi.org/10.1007/s00253-010-2622-3.

Cooke, W. B. (1962). A taxonomic study in the 'black yeasts'. *Mycopathologia et Mycologia Applicata*, *17* (1), 1−43.

Gorin, P. A. J., Mazurek, M., & Spencer, J. F. T. (1968). Proton magnetic resonance spectra of *Trichosporon aculeatum mannan* and its borate complex and their relationship to chemical structure. *Canadian Journal of Chemistry*, *46*(13), 2305−2310. Available from https://doi.org/10.1139/v68-374.

Gunde-Cimerman, N., Zalar, P., De Hoog, S., & Plemenitaš, A. (2000). Hypersaline waters in salterns—Natural ecological niches for halophilic black yeasts. *FEMS Microbiology Ecology*, *32*(3), 235−240. Available from https://doi.org/10.1016/S0168-6496(00)00032-5.

Heald, P. J., & Kristiansen, B. (1985). Synthesis of polysaccharide by yeast-like forms of *Aureobasidium pullulans*. *Biotechnology and Bioengineering*, *27*(10), 1516−1519. Available from https://doi.org/10.1002/bit.260271019.

Lepoittevin, B., Elzein, T., Dragoe, D., Bejjani, A., Lemée, F., Levillain, J., & Dez, I. (2019). Hydrophobization of chitosan films by surface grafting with fluorinated polymer brushes. *Carbohydrate Polymers*, *205*, 437−446. Available from https://doi.org/10.1016/j.carbpol.2018.10.044.

Li, D., Zheng, Q., Wang, Y., & Chen, H. (2014). Combining surface topography with polymer chemistry: Exploring new interfacial biological phenomena. *Polymer Chemistry*, *5*, 14−24. Available from https://doi.org/10.1039/c3py00739a.

Li, H., Chi, Z., Wang, X., & Ma, C. (2007). Amylase production by the marine yeast *Aureobasidium pullulans* N13d. *Journal of Ocean University of China*, *6*, 60−65. Available from https://doi.org/10.1007/s11802-007-0060-3.

Liu, N.-N., Chi, Z., Liu, G.-L., Chen, T.-J., Jiang, H., Hu, Z., & Chi, Z.-M. (2018). α-Amylase, glucoamylase and isopullulanase determine molecular weight of pullulan produced by *Aureobasidium melanogenum* P16. *International Journal of Biological Macromolecules*, *117*, 727−734. Available from https://doi.org/10.1016/j.ijbiomac.2018.05.235.

Miyaka, T. (1979). *Shaped matters of tobaccos and process for preparing the same. Patent no. 1 049 245*. Canadian *patent office*.

Oliva, E. M., Fernandez Cirelli, A., & de Lederkremer, R. M. (1986). Characterization of a pullulan in *Cyttaria darwinii*. *Carbohydrate Research*, *158*, 262−267. Available from https://doi.org/10.1016/0008-6215(86)84025-3.

Pometto III, A. L., Demirci, A., & Johnson, K. E. (Jan. 21, 1997). *Immobilization of microorganisms on a support made of synthetic polymer and plant material. Google patents*.

Pouliot, J. M., Walton, I., Nolen-Parkhouse, M., Abu-Lail, L. I., & Camesano, T. A. (2005). Adhesion of *Aureobasidium pullulans* is controlled by uronic acid based polymers and pullulan. *Biomacromolecules*, *6*(2), 1122−1131. Available from https://doi.org/10.1021/bm0492935.

Reeslev, M., Nielsen, J. C., Olsen, J., Jensen, B., & Jacobsen, T. (1991). Effect of pH and the initial concentration of yeast extract on regulation of dimorphism and exopolysaccharide formation of *Aureobasidium pullulans* in batch culture. *Mycological Research*, *95*(2), 220−226. Available from https://doi.org/10.1016/S0953-7562(09)81016-2.

Rosa, C. A., & Péter, G. (2006). *The yeast handbook: Biodiversity and ecophysiology of yeast*. Berlin, Heidelberg: Springer.

Shen, S., Li, H., & Yang, W. (2014). The preliminary evaluation on cholesterol-modified pullulan as a drug nanocarrier. *Drug Delivery*, *21*(7), 501−508. Available from https://doi.org/10.3109/10717544.2014.895068.

Tsuji, K., Toyota, N., & Fujita, F. (Nov. 23, 1976). *Molded pullulan type resins coated with thermosetting resin films. Google patents*.

Waksman, N., de Lederkremer, R. M., & Cerezo, A. S. (1977). The structure of an α-D-glucan from *Cyttaria harioti* Fischer. *Carbohydrate Research*, *59*(2), 505−515. Available from https://doi.org/10.1016/S0008-6215(00)83187-0.

CHAPTER 9

Production and application of bacterial polyhydroxyalkanoates

Vivek Kumar Gaur[1,2], Poonam Sharma[3], Janmejai Kumar Srivastava[2], Ranjna Sirohi[4] and Natesan Manickam[1]

[1]Environmental Biotechnology Division, Environmental Toxicology Group, CSIR—Indian Institute of Toxicology Research, Lucknow, India, [2]Amity Institute of Biotechnology, Amity University, Lucknow, India, [3]Department of Bioengineering, Integral University, Lucknow, India, [4]Department of Post-Harvest Process and Food Engineering, G.B. Pant University of Agriculture and Technology, Pantnagar, India

9.1 Introduction

Plastics have improved the quality of human life by becoming a replacement for paper or glass used in packaging, thus being an important material. Although being nonbiodegradable, they exhibit properties such as lightness, durability, and strength. Plastics first appeared in the market in the 1950s and since then it has become an indispensable material in everyday life (Grigore et al., 2019). Owing to these excellent properties and the continuous use for more than 50 years, in 2013 the production of plastic reached 299 million tons globally, which was 3.9% higher as compared to the production in 2012. This increase in production and use had led to serious environmental hazards (Grigore et al., 2019; Europe, 2017; Możejko-Ciesielska & Kiewisz, 2016). The low biodegradability of petroleum-derived plastics has contributed to their accumulation as waste (after end-of-life). Also the postconsumer plastics contribute in increasing the solid-waste management burden (Możejko-Ciesielska & Kiewisz, 2016). Furthermore, the increasing concern for rapid fossil fuel depletion urges the need to produce and use bio-based polymers for fulfilling human needs.

Polyhydroxyalkanoates (PHAs) are polyesters of hydroxyalkanoates (HAs), which comprise biodegradable biopolyesters or bioplastics, and are well known among the group of biopolymers (Możejko-Ciesielska & Kiewisz, 2016). PHAs were first discovered by Beijerinck in 1888 (Khanna & Srivastava, 2005) and reported by Maurice Lemoigne in 1926 in *Bacillus megaterium* (Lemoigne, 1926); later on, in 1958 the role of bacteria in the synthesis and storage of PHA was established by Macrae and Wilkinson (1958). Starting from 1959, many companies have started the commercialization of eco-friendly bioplastics. The first start-up, that is, W.R. Grace & Company was shut down owing to the low synthesis and purification problems in poly-3-hydroxybutyric acid P(3HB) polymer. Since 1980 many PHAs have been commercialized and sold under the trade name of Biopol, BioGreen, Nodax, Biomer, and Biocycle (Możejko-Ciesielska & Kiewisz, 2016). A joint venture by Metabolix and ADM set up in 2006 for large quantity of PHAs production and sale, however, collapsed in 2012 (Możejko-Ciesielska & Kiewisz, 2016). There is a general perception that the market of PHAs needs more time to develop and is expected to rise 10-fold by 2020 (Raza, Tariq, Majeed, & Banat, 2019; Aeschelmann & Carus, 2015). Globally, the bio-based polymers production in 2013 was 5.1 million tons, which was anticipated to rise 17 million tons by 2020 (Możejko-Ciesielska & Kiewisz, 2016; Aeschelmann & Carus, 2015).

PHAs are biodegradable linear thermoplastic polyesters, which are synthesized by various Gram-positive and Gram-negative bacterial strains (Fig. 9.1) utilizing a variety of carbon sources such as alkanes, alkenes, alkanoic acids, renewable carbon sources, and wastes such as cooking oil, volatile fatty acids, and even wastewater (Table 9.1) (Raza et al., 2019; Fernández-Dacosta, Posada, Kleerebezem, Cuellar, & Ramirez, 2015; Ruiz, Kenny, Narancic, Babu, & O'Connor, 2019; Kumar et al., 2019). About 150 different PHA congeners with different properties and structure (varying functional group and side chain) have been reported from different bacterial strains (Raza et al., 2019). Their classification is done based on their chemical unit structure; Fig. 9.2 depicts the general structure of PHAs. PHAs are used for diverse applications such as food additive, in medicines, as drug-delivery system, and in making household day-to-day useful materials (packaging materials, coating paper, electronic accessories, plastic accessories, biodegradable bottles, and garments) (Clarinval & Halleux, 2005; Chen, 2009; Greene, 2013). Interestingly, PHAs also constitute a part in renewable carbon cycle and are used as substitutes to synthetic polyesters (Raza et al., 2019; Tan et al., 2014).

9.2 Classification of polyhydroxyalkanoates

Depending on the chain length, PHAs are divided into three different classes, namely, short chain length (scl) PHAs, medium chain length (mcl) PHAs, and long chain length PHAs (lcl-PHAs) (Raza, Riaz, & Banat, 2018). lcl-PHAs refer to the group containing >14 carbon

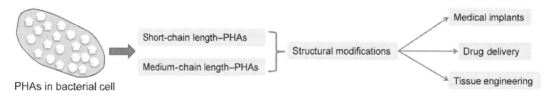

Figure 9.1
Graphical overview of bacterial PHA production and application. *PHA*, Polyhydroxyalkanoate.

Table 9.1: Polyhydroxyalkanoates (PHAs) production of by different bacterial source by utilizing diverse carbon sources.

S. no.	Name of the polymer	Source organism	Substrate	Processing condition	References
1.	Poly-3-hydroxybutyrate P(3HB)	*Bacillus mycoida* DFC1	Sucrose	pH 7.3 rpm 140 Temperature 37°C Time 48 h	Narayanan and Ramana (2012)
2.	P(3HB)	*Alcaligenes latus*	Sucrose	pH 7.0 Temperature 33°C rpm 200 Time 36 h	Gahlawat and Srivastava (2012)
3.	P(3HB)	*Caldimonas taiwanensis*	Gluconate	—	Chanprateep (2010)
4.	P(3HB)	*Bacillus megaterium uyuni* S29	Glucose	pH 7.0 Temperature 35°C rpm 500 Air inflow 5 L min^{-1} Time 18 h	López-Abelairas et al. (2015)
5.	P(HV)	*Pseudomonas oleovorans*	—	—	Ahmed et al. (2010)
6.	mcl-PHA	*Pseudomonas chlororaphis* strain PA23	Octanoic acid and nonanoic acid, vegetable oils	pH 7.0 Temperature 30°C rpm 150 Dissolved oxygen 40% Time 48 h	Sharma et al. (2017)
7.	mcl-PHA	*Pseudomonas chlororaphis* 555	Waste cooking oil	pH 6.9 Temperature 30°C rpm 500 Dissolved oxygen 20% Time 30 h	Ruiz et al. (2019)
8.	Polyhydroxyhexanoate P(HHx)	*Pseudomonas putida*	—	pH 8.0 Temperature 23°C–25°C rpm 500 Time 30 d	Albuquerque, Torres, and Reis (2010)
9.	Polyhydroxyheptanoate P(HHp)	*P. putida*	—	—	Raza et al. (2019)
10.	Polyhydroxyocantoate P(HO)	*Streptmoyces lividans*	—	—	Furutate et al. (2017)

(*Continued*)

Table 9.1: (Continued)

S. no.	Name of the polymer	Source organism	Substrate	Processing condition	References
11.	P(3HB-co-3H2MB)	Recombinant *Escherichia coli* LS5218	Tiglic acid	Temperature 37°C Time 76 h	Furutate et al. (2017)
12.	Polyhydroxynanoate P(HN)	*Alcaligenes* sp.	Plant oils	Temperature 30°C Time 72 h	Fukui and Doi (1998)
13.	Scl-mcl copolymer containing hydroxyheptanoate, 3-hydroxyvalerate, 3-hydroxyundecanoate, 3-hydroxynanoate,	*Thermus thermophilus* HB8	Whey	Temperature 70°C rpm 600 Time 48 h	Pantazaki, Papaneophytou, Pritsa, Liakopoulou-Kyriakides, and Kyriakidis (2009)
14.	P(3HB-co-4HB)	*Cupriavidus* sp. USMAA2—4	1,4-Butanediol, γ-butyrolactone	Temperature 30°C Aeration 1 w^{-1} m^{-1} Time 48 h	Vigneswari, Nik, Majid, and Amirul (2010)
15.	P(3HB-co-3HV-co-4HB)	*Cupriavidus* sp. USMAA2—4	Oleic acid, γ-butyrol actone, 1-pentanol	pH 7.0 Temperature 30°C rpm 200 Time 72 h	Aziz, Sipaut, and Abdullah (2012)
16.	P(3HB-co-3HV)	*Alcaligenes latus*	Sucrose	—	Kaur and Roy (2015)
17.	P(3H4MV)	Recombinant *Burkholderia* sp. JCM15050	Fructose, isocaproic acid	pH 7.0 Temperature 37°C rpm 200 Time 48 h	Lau, Chee, Tsuge, and Sudesh (2010)
18.	P(3HB-co-3HV-co-3HO-3HDD)	*Weutersia eutropha*	Canola oil	pH 7.0 Temperature 30°C Air inflow 5 L min^{-1} Time 36 h	López-Cuellar et al. (2011)
19.	mcl-PHA	*Pseudomonas* sp.	Grass biomass hydrolysate	Temperature 30°C rpm 200 Time 48 h	Davis et al. (2013)

atoms and is very uncommon (Grigore et al., 2019). It is prominently produced by the bacterial species such as *Aureispira marina* and *Shewanella oneidensis* (Raza et al., 2018). In addition to these, there are PHA blends such as P(3HB-co-3HV) in scl-PHA and P(3HHx-co-3HO) in mcl-PHA (Suriyamongkol, Weselake, Narine, Moloney, & Shah, 2007).

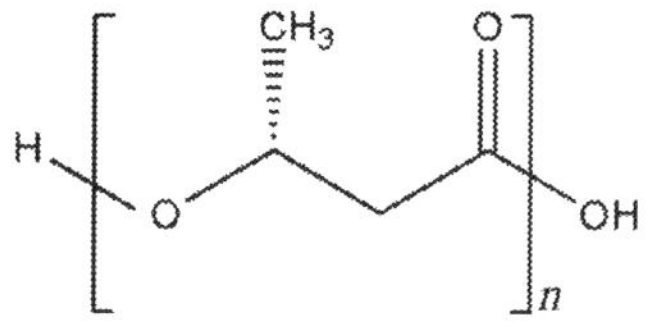

Figure 9.2
General structure of polyhydroxyalkanoates.

9.2.1 Short chain length polyhydroxyalkanoates

Polymers with five or less carbon atoms are classified as scl-PHAs. These scl-PHAs are produced by the bacterial species such as *Alcaligenes latus* and *Cupriavidus necator*. Polyhydroxybutyrate (PHB) is a brittle and crystalline polymer and categorized as scl-PHAs (and also mcl-PHA), which is produced by several bacterial strains, including *Aeromonas caviae* and *Ralstonia eutropha* and its blends have been manufactured in the plants such as in tobacco, rapeseed, cotton, and thale cress (Raza et al., 2018). The scl-PHAs and mcl-PHAs emit light at a wavelength of 590 and 575 nm, respectively (Raza et al., 2018; Wu, Sheu, & Lee, 2003).

A gene-modification study was performed by Arai et al. (Arai et al., 2002) who synthesized PHB-*co*-HV-*co*-HH copolymer by adding signal encoder (targeting peroxisome) at the carboxyl terminus (−COO) of spinach glycolate oxidase. Using *Agrobacterium*-mediated transfer, this gene was then transferred to *Arabidopsis thaliana*. This led to the expression of transgenic gene in *Arabidopsis* plant, thus producing and accumulating scl-PHAs in the tissues.

9.2.2 Medium chain length polyhydroxyalkanoates

The mcl-PHAs are polyesters of hydroxyl acids, which are primarily formed by *Pseudomonads* at times of uneven growth. In comparison to scl-PHAs the mcl-PHAs show enhanced mechanical properties such as decreased crystallinity, decreased brittleness, high melting temperatures, poor tensile strength, and low glass transition temperature. Owing to these properties, mcl-PHAs are more flexible and elastomeric, thus suitable for biomedical application such as in cardiovascular applications (Kim, Chung, & Rhee, 2007).

There are three steps involved in the synthesis of mcl-PHAs. The first step is the condensation reaction of acyl-CoA with acetyl-CoA. During the second step, the fatty acids are used as a substrate that occurs by fatty acid degradation through beta oxidation. Fatty acid biosynthesis takes place during the third step (Raza et al., 2018; Doi & Steinbüchel, 2002). 3-ketoacyl-ACP reductase and (*S*)-3-hydroxyacyl-CoA epimerase convert 3-ketoacyl-CoA and 3-hydroxyacyl-CoA molecules, respectively, to (*R*)-3-hydroxyacyl-CoA

(Steinbüchel & Hein, 2001). The (R)-specific hydration of 2-trans-enoyl-CoA (βoxidation intermediate) to (R)-3-hydroxyacyl-CoA is catalyzed by (R)-specific enoyl-CoA hydratase (PhaJ). This is an important step for PHA synthesis (Fiedler, Steinbüchel, & Rehm, 2002; Tsuge, Taguchi, & Doi, 2003). (R)-3-hydroxyacyl-ACP-CoA transferase (PhaG) plays a central role in metabolic connection of *de novo* fatty acid and mcl-PHA biosynthesis. PhaG is present in *Pseudomonas aeruginosa* and *Pseudomonas putida* and catalyzes the formation of mcl-PHAs from (R)-3-hydroxyacyl-CoA molecules (Hoffmann, Steinbüchel, & Rehm, 2000; Hoffmann & Rehm, 2004). In *Pseudomonas* sp., phaC1 and phaC2 genes have been identified as mcl-PHA synthase genes. These two genes exhibit different specificity for the substrate (Hein, Paletta, & Steinbüchel, 2002; Chen, Song, & Chen, 2006). Alkanoic acids serve as the base material for the synthesis of mcl-PHAs. The intermediates of alkanoic acid beta oxidation produce mcl-PHAs.

9.2.3 Chemical modifications of polyhydroxyalkanoates

PHAs are high impact biodegradable polymers owing to their emerging multiple applications. PHA is hydrophobic in nature with less-functional group (Raza et al., 2018), which limits their applications (Hazer & Steinbüchel, 2007; Rehm, 2010; Hazer, Kılıçay, & Hazer, 2012). Surface and chemical modification in the form of desired functional groups can be introduced in the side chain of natural PHAs to further influence the properties that may enhance their applications (Fig. 9.3) (Raza et al., 2018; Hartmann et al., 2006). These modifications can affect their mechanical properties, rate of degradation, surface structure, and amphiphilic character thereby modifying specific applications. Since the introduction of functional group is not easily accessible through the biological route, chemical modifications are required. Chemical modifications can be performed through various routes, namely, carboxylation, epoxidation, esterification-based copolymerization, blending, grafting, thiolation, metallization, and functional group incorporation through thermal degradation (Hazer & Steinbüchel, 2007; Kai & Loh, 2014).

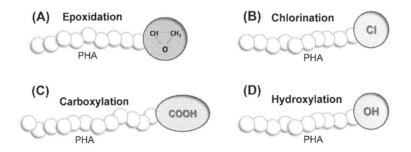

Figure 9.3
Strategies for chemical modification in PHAs: (A) epoxidation; (B) chlorination; (C) carboxylation; and (D) hydroxylation. *PHA*, Polyhydroxyalkanoate.

9.2.3.1 Epoxidation

Epoxides are the derivatives of ethylene oxide, also called as oxiranes (Parker & Isaacs, 1959). This is a highly reactive functional group and a key industrial organic intermediate used in numerous polymer reactions (Raza et al., 2018). The low molecular weights mcl-PHA produced by utilizing linseed oil contains increased side-chain unsaturation and exhibits amorphous nature and significant viscosity at room temperature (Raza et al., 2018; Ashby et al., 2000). The *meta*-chloroperoxybenzoic acid (*m*-CPBA) can be used as a cross-linker for the epoxidation of unsaturated side chains. This converts approximately 37% olefinic groups to epoxy groups, which further increase PHA cross-linking in the presence of air. The epoxidized PHA shows elevated tensile strength and Young's modulus as compared to the native one (Ashby et al., 2000). The *m*-CPBA has also been used for the epoxidation of poly(3-hydroxyoctanoate-*co*-3-hydroxy-10-undecenoate) (PHOUs). These polymers were cross-linked with succinic acid anhydride in which 2-ethyl-4-methyl imidazole was used for initiating the cross-linking reaction (Arkin, Hazer, & Borcakli, 2000). The cross-linking was indicated by an increase in glass transition temperature and sol−gel content.

A composite polymer of PHA and malic anhydride when cross-linked with the fiber of tea plant showed that the cross-linked polymer had better mechanical properties as compared to the composite polymer. This elevation in the property was due to high compatibility of polymer with tea plant fiber. The PHA/malic anhydride/tea plant fiber cross-linked polymer exhibited enhanced properties, namely, high biodegradability, low viscosity, and increase in water resistance (Wu, 2013).

9.2.3.2 Chlorination

Chlorination is the simplest method for PHAs modification, which involves the addition of chlorine to molecule or substrate. The saturated or unsaturated PHAs can undergo chlorination through the substitution reaction. For chlorination, the solution of unsaturated PHAs in chloroform is passed by chlorine gas in the presence of sunlight at room temperature (Arkin et al., 2000). The increase in the degree of chlorination changes the PHAs from sticky and soft to brittle, hard, and crystalline (Arkin et al., 2000). The chlorinated PHAs can be used for polymers blending and other modification such as carboxylation (Arkin & Hazer, 2002). Similarly, carboxylation of PHB and polyhydroxyoctanoate (PHO) was achieved using chlorine gas, which caused a reduction in the polymer weight due to hydrolysis. The melting temperature of chlorinated PHB reduced from $170°C$ to $148°C$ and the glass transition temperature increased to $10°C$ from $-20°C$, for the chlorinated PHO (Hazer & Steinbüchel, 2007). These modified PHAs can produce cross-linked polymers by reacting with benzene through Friedel−Crafts reactions (Hazer & Steinbüchel, 2007).

9.2.3.3 Carboxylation

Carboxylation is a process of adding carboxylate group to a substrate. Oxidation of unsaturated PHAs, epoxidized side chains, or chlorine groups leads to increased hydrophobicity and PHAs carboxylation (Raza et al., 2018). Addition of the carboxylic group increases the hydrophilicity of the polymer. About 70 PHA polymers have been reported with OH or -COO end group (Kai & Loh, 2014; Zook & Kee, 2016). Oxone and osmium tetroxide (OsO_4) is used for carboxylating unsaturated PHO-*co*-polyhydroxyundecenoate (PHU) and the reaction took place in warm dimethylformamide, which led to backbone degradation. The degradation was confirmed by nuclear magnetic resonance (NMR) spectroscopy and gel permeation chromatography and the carboxylation was verified by the changes in polymer solubility (Stigers & Tew, 2003).

Unsaturated PHOUs can be carboxylated using potassium permanganate at 55°C. It acts as an oxidizing agent and causes reduction in polymer molecular weight. The carboxylation is confirmed by infrared and NMR techniques. About 50% carboxylation took place in a stipulated time of 2 h and causes a significant increase in the hydrophobicity of the polymer, which was verified by testing its water solubility (Lee & Park, 2000). Carboxylation can be used to change the hydrophilicity and hydrophobicity of the PHAs, thus can be employed for hydrogel preparation.

9.2.3.4 Hydroxylation

Hydroxylation is used to modify the property of PHAs, especially its hydrophilicity and hydrophobicity (Hu et al., 2011; Zhou et al., 2012; Kwiecień, Adamus, & Kowalczuk, 2013; Andrade, Witholt, Hany, Egli, & Li, 2002; Eroğlu, Hazer, Ozturk, & Caykara, 2005). In general, the modification in PHAs through hydroxylation involves the use of acid or base in the presence of mono- or diol compounds. PHAs undergo methanolysis forming PHA methyl esters containing monohydroxy-terminated groups.

Monohydroxylated oligomers of poly(3-hydroxyoctanoate-*co*-3-hydroxy-10-undecenoate) (PHOU) and PHO can be produced through the hydrolysis catalyzed by the acid or base. Alcoholic sodium hydroxide (NaOH) was used at a pH of 10–14 to carry out basic hydrolysis and concentrated hydrochloric acid (HCl) was used to stop the reaction. For acidic hydrolysis, two different approaches were used, namely, *para*-toluene sulfonic acid monohydrate catalyzed reaction at 120°C and monohydroxylation by acidic methanolysis using sulfuric acid at 100°C, yielding respective 3-hydroxymethyl esters containing a methyl-protected carboxylic group (Timbart, Renard, Tessier, & Langlois, 2007). For producing PHO oligomers, the methanolysis and acid-catalyzed reaction is more efficient than basic hydrolysis. Also, the reduction in molecular weight of the polymer depends on the type of solvent used. Polymeric chain cleavage occurs more often in toluene than in dichloroethane. The monohydroxylation of poly-3-hydroxybutyrate-*co*-4-hydroxybutyrate is catalyzed by *p*-toluenesulfonic acid and the reaction takes place in the presence of methanol. The PHA is then acrylated, followed by grafting on poly(ethyleneimine) through addition and the produced material is used for siRNA delivery (Zhou et al., 2012).

Potassium permanganate ($KMnO_4$) is used for the hydroxylation of PHOU. It hydroxylates approximately 60% of PHOUs unsaturated side chains without considerable loss of molecular weight (Lee, Park, & Lenz, 2000). The principle of hydroboration oxidation was applied to PHOUs to completely convert the double bonds to hydroxyl group at the cost of molecular weight (Renard, Poux, Timbart, Langlois, & Guérin, 2005). The increased hydrophilicity increases the solubility of hydroxylated copolymers in polar solvents such as methanol and ethanol. The PHUs hydroxylation through hydroboration oxidation generates 100% hydroxylated polymer (Eroğlu et al., 2005), whose molecular weight is reduced considerably with increased hydrophilicity.

9.3 Structure and properties

9.3.1 Chemical structure

PHAs are mostly linear polyesters made up of monomeric units of 3-hydroxy fatty acid. The monomers are linked together by an ester bond formed by the reaction of carboxyl group of first monomer with the hydroxyl group of second, generating PHAs in rectus (R) configuration (Leja & Lewandowicz, 2010). Sinister (S) configuration of PHAs is synthesized by using stereopolymers (Kabe et al., 2012; Liong, 2015). PHB is the most common PHA, which is mostly present in R configuration (Raza et al., 2018), whereas stereocopolymer of PHBs is synthesized with butyrolactone (Kemnitzer, McCarthy, & Gross, 1992). The carbon atom of PHAs that is hydroxyl substituted, is in the R configuration apart from the ones in which chirality is absent (Liong, 2015). The general structure of PHAs is shown in Fig. 9.2; in this, the alkyl group R may vary from 1 carbon methyl (C1) to 13 carbon tridecyl (C13). The alkyl side chains may also be epoxidized aromatic, halogenated, or branched monomers (Liong, 2015). Owing to these variations in the composition and size of substituents in the side chain, PHAs are susceptible to chemical modifications, thus exhibiting diversified applications (Raza et al., 2018).

9.3.2 Properties

Based on their compositional differences, that is, homopolyester or copolyester (Bugnicourt, Cinelli, Lazzeri, & Alvarez, 2014), PHAs show different properties. The thermal and mechanical properties of PHAs can be modulated by changing the concentration of 3-hydroxyvalerate (3HV) units in the copolymer (Grigore et al., 2019). The higher content of 3HV in poly(3-hydroxybutyrate-*co*-3-hydroxyvalerate) (P(3HB-*co*-3HV)) leads to decreased crystallinity providing higher strength, elongation, and flexibility to the molecule

(Grigore et al., 2019; Leong, Show, Ooi, Ling, & Lan, 2014; López-Cuellar, Alba-Flores, Rodríguez, & Pérez-Guevara, 2011).

9.3.2.1 Thermal properties

Because of the crystalline nature of PHAs, their thermal properties are expressed in terms of glass transition temperature (T_g) and melting temperature (T_m) for the crystalline phase and amorphous phase, respectively. The increase in side-chain length from one to seven carbons decreases the T_g values, whereas increase in the content of 2-hydroxyalkanoates (2HAs) from C4 to C7 increases the T_m from 45°C to 69°C (Gopi, Kontopoulou, Ramsay, & Ramsay, 2018). mcl-PHAs have a low melting temperature ranging from 42°C and 65°C (Grigore et al., 2019). The degree of crystallinity and glass transition temperature together contribute to the elastomeric behavior of these polymers. Also, in a narrow temperature range, the mcl-PHAs behave like true elastomers owing to low melting temperature (Grigore et al., 2019). The polymer is sticky and amorphous at a temperature near or above T_m (Cerrone et al., 2014; Wecker, Moppert, Simon-Colin, Costa, & Berteaux-Lecellier, 2015; Poblete-Castro, Binger, Oehlert, & Rohde, 2014; Muhr et al., 2013). The increase in the mobility of polymer chains increases the length of group pendant, thus decreasing the T_g of mcl-PHAs (Rai, Keshavarz, Roether, Boccaccini, & Roy, 2011). The substrates used for PHAs productions have a major role in thermal properties of the produced polymer, as PHAs with different properties were synthesized by *P. putida* by utilizing different substrates (Grigore et al., 2019; Raza et al., 2018; Sharma et al., 2017).

9.3.2.2 Crystallinity

The crystallization ability of polymers is limited by an increase in the diversity of side chain, which results in differences in the degrees of crystallization in PHAs. scl-PHAs [e.g., P(3HB)] have high crystallinity as compared to mcl-PHAs (Rai et al., 2011). Bacterial P(3HB) exhibits high crystallinity of 55%–70% (Kim, Bang, Kim, & Rhee, 2000), with a relatively low rate of crystallization as compared to synthetic ones (Grigore et al., 2019). The mcl-PHAs exhibit a crystallinity of 24% (Rai et al., 2011). Many copolymers of this group do not crystallize because of the disorders in polymer structure regularity that is induced by the functional groups. The degree of crystallinity of copolymer, namely, poly(3-hydroxybutyrate-*co*-3-hydroxyhexanoate) (P(3HB-*co*-3HHx)) is determined by its 3HB component (Grigore et al., 2019). The 3HHx units in the P(3HB-*co*-3HHx) copolymer reduces the crystallization rate of P(3HB) homopolymer and the degree of crystallinity (Padermshoke et al., 2004). The homopolymer P(3HB) is rigid and fragile for many applications and the addition of 3HV monomers further reduces the fragility making it more useful for various applications (Leong et al., 2014).

The use of better degradable and less crystalline *co*- and ter-polyesters further reduces the crystallinity of these polymers. The increase in gelatin content blended with P(3HB-*co*-3HHx) enhances the viability on mouse osteoblast cells as compared to the pure P(3HB-co-3HHx). The increase in gelatin content decrease the crystallinity of the PHAs produced (Koller, 2018). The use of inorganic materials such as ceramics, nanotubes, bioactive glass, or hydroxyapatite act as nucleating agent, thus reduce the crystallite size (Xiang, Chen, Chen, Sun, & Zhu, 2017; Xiang et al., 2019; Chen, Xiang, Hu, Ni, & Zhu, 2017).

9.3.2.3 Mechanical properties

scl-PHAs are brittle and stiff with high crystallinity of 60%–80%, whereas mcl-PHAs are more elastic and flexible with T_g below room temperature, low melting temperature, crystallinity (25%), and tensile strength (Anjum et al., 2016; Meng & Chen, 2017). Blending with a copolymer or other polymers in different ratios enhances the flexibility of PHAs. This enhances their applicability in industrial and medical applications (Raza et al., 2018; Rai et al., 2011; Sanhueza et al., 2019; Zhang, Shishatskaya, Volova, da Silva, & Chen, 2018). The properties such as rigidity, hardness, and flexibility of P(3HB-co-3HHx) depend on the concentration of 3HHx units. For example, the 5.9 mol% fraction of 3HHx in P(3HB-co-3HHx) showed the highest elongation at 163% break value. Furthermore, highest tensile strength of 25.7 MPa was obtained with 2.5% 3HHx mol fraction whereas it significantly decreased to 8.8 MPa for 9.5% 3HHx (Liao, 2010). During the formation of films by mixing P(3HB-co-3HHx) and P(3HB), increasing content of P(3HB-co-3HHx) from 40% to 60% increases the break elongation from 15% to 106% with a weaken tensile strength from 23.5 to 20.9 MPa (Zhao, Deng, Chen, & Chen, 2003).

9.3.2.4 Biocompatibility

Biocompatibility is an important parameter for distinguishing the biomaterial from other materials on the basis of their ability to interact with the cells and tissues of human body without generating negative response. This is essential for the material to be used in diverse medical applications such as drug delivery, tissue engineering, and nanotechnologies (Williams, 2008; Degeratu et al., 2019; Kovalcik, Obruca, Fritz, & Marova, 2019). Because of the biocompatibility of PHAs, they have received special attention and in 2007, the use of P(4HB) in surgical sutures was approved by the Food and Drug Administration (FDA) (Hazer & Steinbüchel, 2007; Akaraonye, Keshavarz, & Roy, 2010). The biocompatibility of PHAs depends on many factors such as shape, material chemistry, surface hydrophilicity, surface energy, and surface porosity (Brandl, Bachofen, Mayer, & Wintermantel, 1995). Another factor for them to be biocompatible is that some of its monomers occur naturally in human body, namely, 3-hydroxybutyric acid in concentrations ranging from 3 to 10 mg in 100 mL blood. The hemocompatibility of P(3HB) and P(3HB-co-3HHx) based on the reduced blood platelet adhesion and thrombogenicity has also been reported in several in vitro and in vivo studies (Grigore et al., 2019; Rai, 2010; Zinn, Witholt, & Egli, 2001).

The cellular attachment, proliferation, phenotypic changes, and cell cycle in rabbit aorta smooth muscle cells (RaSMCs) were studied by Qu, Wu, Liang, Zou, and Chen (2006). They cultivated cells on P(3HB-co-3HHx) containing 0%−20% HHx (mol%) in comparison with when grown on tissue-culture plates and polylactic acid (PLA) films. After four hours of incubation, these cells exhibited strong adhesion on PHA films. Also, no difference in the attachment of these cells was observed on tissue culture plates in comparison to P(3HB-co-3HHx) films, whereas stronger adhesion was seen in the cells grown on P(3HB-co-3HHx) containing 12% HHx (mol%). After 3 days, P(3HB-co-3HHx) films with 20% HHx (mol%) exhibited highest adhesion equal to as on tissue-culture plates, which was 30%−40% higher than on PLA (Zinn et al., 2001).

9.3.2.5 Biodegradability

PHAs are biodegradable in diverse natural environments such as seawater, lake water, and soil. The rate of degradation depends on many factors such as temperature, pH, nutrient, level of hydration, and microbial load (Ojumu, Yu, & Solomon, 2004; Jendrossek & Handrick, 2002; Dai, Zou, & Chen, 2009). P(3HB) is prominently degraded by the microbes belonging to family *Micromonosporaceae, Pseudonocardiaceae, Streptomycetaceae, Streptosporangiaceae*, and *Thermomonosporaceae*. The PHA polymer is solubilized by the action of extracellular enzymes secreted by these microbes. The solubilized polymer is then absorbed/internalized through the cell wall. Some bacteria that produce PHAs degrade the intracellular polymer (Rai, 2010; Verlinden, Hill, Kenward, Williams, & Radecka, 2007). Bacteria degrade PHA by the action of enzymes such as PHA depolymerases or by PHA hydrolysis (Grigore et al., 2019). The degradation of PHAs is affected by the monomer composition and crystallinity. With the increasing crystallinity, the melting temperature also increases and this further reduces the rate of degradation. mcl-PHAs with low melting temperature and crystallinity show higher degradability as compared to scl-PHAs with high melting temperature and crystallinity (Rai et al., 2011). Thus the polymers containing P(3HB-co-3HV) or P(3HB) copolymers degrade slowly than with poly(3-hydroxybutyrate-co-4-hydroxybutyrate), P(3HB-co-4HB), and 4-hydroxybutyrate unit (4HB). When studied in buffer system, 4HB hydrolyzes faster as compared to P(3HB) (Grigore et al., 2019; Brigham & Sinskey, 2012). Carbon dioxide and water are obtained as the end products by aerobic degradation of PHAs in the environment, whereas anaerobically it leads to the production of methane (Grigore et al., 2019). Biodegradation of 10 mol% fraction of P(3HB) and 20 mol% fraction of P(3HB-co-3HV) in different environments, namely, household compost, soils, seawater, freshwater, and canal showed that both of these were completely degraded in all the conditions; also, the molecular weight was not altered by the degradation but their property of elongation at break was reduced (Qu et al., 2006).

9.4 Industrial-scale production of polyhydroxyalkanoates

The increasing applications of PHAs in various fields warrant the economic production of these biopolymers, that is, increasing yield by utilizing cheaper sources in a shorter time. The large-scale production of PHAs depends on three main factors: cost of substrate, process development, and cost of downstream processing (Kaur & Roy, 2015). In any bioprocess, the substrate cost accounts for approximately 50% of the total cost of the production. Also, the low yield of product adds up to the overall cost as more substrate is required to yield same amount of product (Kaur & Roy, 2015). If the cost of production of PHAs can be reduced to that of conventional plastics, then the PHAs application could be as vast as of plastics. PHA production process mostly yields low volumetric productivities, leading to the elevation in operational and capital costs for large bioreactors. Process economics can be improved by designing efficient bioprocess strategies focusing on overall process kinetics leading to higher PHA productivity (Kaur & Roy, 2015).

Industrial production of PHAs has been extensively reported employing Gram-negative bacteria such as *A. latus* and *C. necator*, owing to their capacity to result in high yield of PHAs (Giin-Yu, Chia-Lung, Ling, Liya, & Lin, 2014; Berezina, 2013). However, lipopolysaccharide absence in the Gram-positive bacteria makes their PHAs ideal for application in medical fields. This led to the exploration of Gram-positive strains such as *Corynebacterium glutamicum* and *Bacillus* sp. (Singh, Patel, & Kalia, 2009).

Although for more than three decades there have been continued efforts for large-scale production of PHAs, yet the two factors, namely, final quality and quantity of the PHA biopolymer define their applications in the market (Penglou, Kretza, Chatzidoukas, Parouti, & Kiparissides, 2012). In addition to this, the microorganism, its metabolic pathway, bioprocess strategy, and medium constituents are considered important factors for the quality and yield of PHAs polymer (Grothe, Moo-Young, & Chisti, 1999). Several bioreactor operating strategies such as batch fermentation, fed-batch fermentation, two-stage fermentation, and continuous fermentation have also been applied for the large-scale production of PHAs (Kaur & Roy, 2015; Blunt, Levin, & Cicek, 2018).

9.4.1 Batch fermentation

Batch fermentation is a closed system and is considered as a simple and the foremost investigation method for a bioprocess. In this system, the substrate is added in the bioreactor at the starting of the cultivation and is removed at the end, that is, no addition or removal of substrate is done in between the processing time. During the production of PHAs through batch fermentation, the maximum allowed concentration of nitrogen and carbon sources are added at the beginning of the process, thus restricting nutrient addition during the production period. This process yields low productivity (Raza et al., 2019). The batch fermentation is useful to know the influence of several process operating conditions, carbon sources bioconversion, and the use of many different types of bacterial species for the production of PHAs. Several types of carbon sources, which are very cheap such as rice straw hydrolyzate, sugar beet juice, coconut oil, etc. have been used for the production of PHAs (Kaur & Roy, 2015).

P(3HB) production is prominently done in batch fermentation, using refined or pure sugars for P(3HB) production (Narayanan & Ramana, 2012; Gahlawat & Srivastava, 2012; García et al., 2014), agroindustrial wastes such as cane molasses, rice straw hydrolysate, grass biomass hydrolysate, coconut oil, and sugar beet juice (Akaraonye, Moreno, Knowles, Keshavarz, & Roy, 2012; Sindhu, Silviya, Binod, & Pandey, 2013; Chen, Hung, Shiau, & Wei, 2013; Tripathi, Srivastava, & Singh, 2013; Davis et al., 2013; Sathiyanarayanan, Kiran, Selvin, & Saibaba, 2013; Zhang, Sun, Wang, & Geng, 2013). *Bacillus* and *Alcaligenes* sp. are predominantly used for P(3HB) production (Kaur & Roy, 2015). *Bacillus shackletoni* K5, a thermophilic bacterium, produced 72.6% biopolymer of dry cell weight in a batch process (Liu, Huang, Zhang, & Xu, 2014) and *Chelatococcus* sp. yielded 73% P(3HB) and a 4.8 g L^{-1} biomass concentration (Ibrahim & Steinbüchel, 2010). Typical concentrations of nitrogen and carbon added at the beginning of this process were 0.2–5 and 1–30 g L^{-1}, respectively (Raza et al., 2019; Koller et al., 2008). A yield below 0.4 g g^{-1} is obtained on account of hypothetical biotransformation of biogenic carbon to PHAs, thus making the batch process economically nonfeasible (Raza et al., 2019). The major drawback of this process is the nutrient limitation that inhibits bacterial growth and, thus the PHA production is restricted. Also, an increase in the fermentation period above the optimum leads to PHA degradation (Raza et al., 2019). Despite of all the above-mentioned constrains, the batch fermentation process has been the most widely used method.

9.4.2 Fed-batch fermentation

Fed-batch fermentation has been highly employed to get high cell densities, which may result in high metabolite production in bioprocess (Liu et al., 2014). In this, the substrate is added into the bioreactor during the processing but the product is recovered at the end of fermentation. Fed-batch cultures are usually fed with nutritional components for the synthesis of PHAs, which prolongs the exponential phase at an increasing rate, resulting in increased cell density (Blunt et al., 2018). Fed-batch fermentation has been used for the production of PHA copolymer by using biodiesel wastes products (unconventional substrates). Fed-batch process has been extensively used for the production of P(3HB) in which substrate pulse feeding strategy is commonly used (García et al., 2014; Sathiyanarayanan et al., 2013; Pandian et al., 2010; Pan, Perrotta, Stipanovic, Nomura, & Nakas, 2012; Rodríguez-Contreras et al., 2013). This method yielded approximately 8–11 g L^{-1} of P(3HB), depending on carbon source and microorganism used. Statistical optimization of process produced high concentration of P(3HB), that is, 19 g L^{-1} and a productivity of 0.48 g L^{-1} h^{-1} (Sathiyanarayanan et al., 2013). A significant increase in P(3HB) concentration, that is, 65.6 g L^{-1} with 1.3 g L^{-1} h^{-1} productivity was obtained by employing a novel feeding strategy on the basis of alkali addition for maintaining the substrate at a constant level (Mozumder, De Wever, Volcke, & Garcia-Gonzalez, 2014). Lower stirring speed is used to control the feeding, which reduces catabolite repression. This allows the consumption of sugars other than glucose and yields 105 g L^{-1} concentration of polymer with 1.6 g L^{-1} h^{-1} productivity (Cesário et al., 2014).

Two-stage fermentation is also one of the strategies to increase the production of PHAs. In this, fed-batch/batch or continuous mode of nutrients feeding can be involved where physical separation of the two phases takes place in which microbial growth occurs in one bioreactor and the formation of the product takes place in the second bioreactor (Kaur & Roy, 2015). *Aeromonas hydrophila* strain was used in this process for improved P(3HB) production (Mozumder et al., 2014). During stage 1, a high cell biomass density was achieved after which it was grown in a nitrogen-limiting environment during stage 2 thereby allowing its maximum accumulation. This process yielded 10.4 g L^{-1} polymer concentration and 16.8 g L^{-1} dry cell weight (Mozumder et al., 2014). *P. putida* KT2440 yielded 0.41 g g^{-1} of PHAs through fed-batch process by utilizing octanoic acid (Follonier, Riesen, & Zinn, 2015). The effect of carbon to nitrogen ratio has also been studied on product yield by fed-batch fermentation. This employs nutrient-rich medium and the pH is maintained between 7.5 and 7.8 using 1 M HCl and 1 M NaOH solution. With increase in carbon to nitrogen ratio, the cell dry mass, PHA yield, and substrate consumption decreased. Nitrogen at a concentration of 7.64, 1.95, and 0.47 g L^{-1} yielded 0.890, 0.955, and 0.970 g L^{-1} of cell dry mass, respectively (Cui, Shi, & Gong, 2017).

9.4.3 Continuous fermentation

Continuous fermentation results in high productivity with high specific growth rates which is possible by operating the bioreactor at high dilution rates and the conditions inside the bioreactor remain at steady state (Blunt et al., 2018). The dilution rate is an important factor in continuous fermentation system that can be defined as a quotient of the bioreactor working volume to the incoming flow rate. Continuous fermentation is not highly useful for the large-scale production of PHAs as it is growth associated but can be useful for the physiological parameters. Continuous fermentation in association with fed-batch cultivation of *A. latus* yielded increased amount of P(3HB-*co*-3HV), resulting 24.6 g L^{-1} of P(3HB-*co*-3HV) and a productivity of 2.18 g L^{-1} h^{-1} (Kaur & Roy, 2015). One major drawback for the industrial-level implementation of this technique is that it is considered prone to microbial contamination, which can lead to economic loss (Koller & Muhr, 2014). A three-stage continuous fermentation process was employed for PHA production by *P. putida* KT2440, which involved increased pressure in bioreactor. The stages consisted of batch cultivation (C8 carbon used), fed-batch cultivation (C8/C11:1), and lastly a continuous cultivation (C8/C11:1). Increase in pressure led to the increase in the yield and productivity of PHA content (Follonier, Henes, Panke, & Zinn, 2012). Another strategy to reduce the cost of PHAs production is the use of nonsterile cultivation process. It was employed in a two-stage continuous process. In this process, after a week of initial cultivation, the constituents of the first reactor was transferred to second stage for P(3HB) production; the second stage accounted for 65%−70% of cell dry weight with a total yield of 13 g L^{-1} (Tan, Xue, Aibaidula, & Chen, 2011).

After fermentation through any strategy, the downstream recovery process is employed to obtain the product. The process must be environmental-friendly, economic, and should avoid the use of toxic solvents (Raza et al., 2018). Downstream processing also contributes to the overall production cost of the biopolymer and is important for increasing the product purity and yield. To recover the PHA, the cell membrane and cell wall can be degraded by use of organic solvents, namely, chloroform, dichloroethane, propylene, and methylene (Ivanov, Stabnikov, Ahmed, Dobrenko, & Saliuk, 2015). However, sodium hypochlorite can also be used as a method for solvent-free recovery in which non-PHAs cellular mass get dissolved to recover the PHA. A drawback of this method is the reduced molecular weight of polymer (Raza et al., 2019). The use of enzymes, namely, lysozyme, trypsin in heat pretreatment method yielded 99% purity of PHA with higher yield (López-Abelairas, García-Torreiro, Lú-Chau, Lema, & Steinbüchel, 2015). Digestion method, utilizing alkali, acid, and strong oxidizing agent is employed to conserve PHA granules while dissolving the cell mass. This method increased the recovery of intracellular polymer and yielded 98% purity (Kapritchkoff et al., 2006). Aqueous two-phase extraction (ATPE) technique employs thermoseparating polymers such as polyethylene oxide or ethylene oxide, which upon heating makes up two layers and brings out easy separation of end products (Leong et al., 2017). The ATPE technique for downstream processing amounted to 5.77 US$/kg of isolated PHAs (Leong et al., 2017), thus this has become an industrially viable and economic technique for the purification (Raza et al., 2018).

As mentioned before, the major drawback of microbial-based plastics, that is, PHAs is their high cost owing to fermentation process (Chaudhry, Jamil, Ali, Ayaz, & Hasnain, 2011). On an average, the cost of PHA production using *Methylobacterium organophilum*, *Escherichia coli*, and *A. latus* has been 6.69, 5.37, and 2.6 US$ kg^{-1}, respectively (Raza et al., 2018; Choi & Lee, 1999). The cost of PHB was 10 Euro per kg in 2006 (Kosior, Braganca, & Fowler, 2006), whereas in 2011 the price of Mirel was approximately 1.50 Euro per kg (Chung et al., 2011). In 2010 several branded PHAs were produced, namely, Biogreen, Biomer, Enmat, Nodax, Meredian, and Green Bio with the capacity of 10,000, 50, 10,000, 2000, 272,000, and 10,000 tons, respectively. The per-kg-cost of the

PHAs from Biogreen, Biomer, Enmat, and Nodax was 2.5–3.0, 3.0–5.0, 3.26, and 3.7 Euro, respectively (Chung et al., 2011). It was estimated that by 2020 the market of bioplastic in European Union will elevate to 2–5 million tons (Chung et al., 2011).

9.5 Application of polyhydroxyalkanoates

PHA and its derived polymers are used for multifarious application as shown in Table 9.2 (Możejko-Ciesielska and Kiewisz, 2016; Ivanov et al., 2015; Kapritchkoff et al., 2006; Leong et al., 2017; Leong et al., 2017). Important medical applications of PHAs are discussed further.

9.5.1 Polyhydroxyalkanoates in medical implants and medicines

PHA has emerged as a boon in the field of medical implants. Owing to the increasing number of health issues related to cardiovascular health, bones, muscles, dental, spine, and other related injuries, the urge of natural or nature identical substitute is warranted (Singh et al., 2019; Rodriguez-Contreras, 2019). Rigorous research has intensified the abilities of PHAs as a potential material for development of various medical equipment in the few last years. There are about 150 known monomers of PHAs out of which about 8 PHAs, namely, poly-(R)-3-hydroxybutyrate (PHB), poly-(R)-3-hydroxyoctanoate (PHO), poly-4-hydroxybutyrate (P4HB), poly(R-3-hydroxybutyrate-*co*-R-3-hydroxyvalerate) (PHBV), poly(R-3-hydroxybutyrate-*co*-4-hydroxybutyrate) (P3HB4HB), poly(R-3-hydroxybutyrate-*co*-R-3-hydroxyhexanoate) (PHBHHx), poly(R-3-hydroxybutyrate-*co*-4-hydroxybutyrate-*co*-R-3-hydroxyhexanoate), and poly(R-3-hydroxy-butyrate-*co*-R-3-hydroxyvalerate-*co*-R-3-hydroxyhexanoate) are actually available in sufficient quantities for use as medical implant materials (Raza et al., 2019; Wang et al., 2010; Wei, Hu, Xie, Lin, & Chen, 2009; Chen & Zhang, 2018). These *co*- and ter-polymers exhibit tremendous potential in medical field. They exhibit desirable properties such as low toxicity, high immunotolerance, biodegradability, high biocompatibility, lesser degradation time in physiological conditions, and favorable mechanical properties (Chen & Zhang, 2018). PHA copolymers, including PHB, P4HB, PHBV, PHO, and PHBHHx composites are mostly employed for medical applications. These copolymers are used in sutures, swabs, nerve cuffs straples, repair tissue and cardiovascular patches or vascular grafts, repair devices, slings, orthopedic pins, heart valves, stents, adhesion barriers, wound dressings, guided tissue regeneration/repair devices, cartilage and tendon repair device, nerve guides, bone marrow scaffolds, bone scaffolds, bone plates, spine cages, and other products. USA-based companies involved in the production of PHA copolymers are TEPHA, Procter & Gamble Company, and Metalbolix (Misra, Valappil, Roy, & Boccaccini, 2006; Puppi, Pecorini, & Chiellini, 2019; Elmowafy et al., 2019). PHAs comprising 6–14 carbon chain are elastic in nature and are applicative in some soft tissue implants like cartilages (Chen & Zhang, 2018; Chen & Wu, 2005;

Table 9.2: Field of application and specific use of polyhydroxyalkanoates (PHA) biopolymer.

S. no.	Name of polymer	Field implication	Specific use	References
1.	PHB-TiO$_2$	Bioremediation	Removal of dyes, viz., malachite green through degradation, decolorization, and detoxification	Luef, Stelzer, and Wiesbrock (2015)
2.	P(3HB) homopolymer	Tissue engineering	Cardiovascular product fabrication, wound management, orthopaedy, and drug-delivery system	Możejko-Ciesielska and Kiewisz (2016)
3.	Copolyester of P(3HB-co-3HV)	Medical application	Fabrication of cardiovascular products, drug-delivery system, wound management, orthopaedy	Możejko-Ciesielska and Kiewisz (2016)
4.	PHBVHHx	Medical application	For bone and skin tissue engineering, proliferating umbilical cord	Chen and Zhang (2018)
5.	PHB and copolymer P(HB-HV)	Packaging	Waterproof films to be used on the back side of diaper sheets	Muhammadi, Afzal, and Hameed (2015)
6.	PHB	Wastewater treatment	Removal of organic pollutants from water	Muhammadi et al. (2015)
7.	Poly(3HB-co-3HV) + sulperazone	Drug delivery	Osteomyelitis treatment	Williams and Martin (2002)
8.	P(3HB)	Drug delivery	As a carrier for increasing drug stability	Kalia(2019)
9.	3-Hydroxyoctanoic acid (3HO)	Pharmacy industry	Antimicrobial property	Możejko-Ciesielska and Kiewisz (2016)
10.	3-Hydroxyhexanoic acid (3HHx)	Pharmacy industry	Intermediate molecule for synthesizing analogs of anti-cancer chemical namely laulimalide	Możejko-Ciesielska and Kiewisz (2016)
11.	(PHBV-CEF)	Drug delivery	Treatment of infectious diseases	Możejko-Ciesielska and Kiewisz (2016)
12.	PHBVHHx	Nerve tissue engineering	Treatment of nerve injury	Leong et al. (2017)
13.	P(3HB)	Packaging application	Used as packaging films shopping bags, containers, and paper coatings, disposable items such as razors, utensils, diapers, hygiene products for women, cosmetic containers, as well as medical surgical garments, etc.	Mathuriya and Yakhmi (2017)
14.	Hydroxybutyrate methyl ester (3HBME)	Biofuels	As fuel additive	Mathuriya and Yakhmi (2017)
15.	(P(3HB-co-4HB))	Paper finishing	Coating of brown Kraft paper	Mathuriya and Yakhmi (2017)

Dwivedi, Pandey, Kumar, & Mehrotra, 2019). Due to most elastic nature, P4HB has been approved by FDA for sutures development (Chen & Zhang, 2018). Degradable PHAs could avoid the side effects generated by permanent implantations and painful second surgery

(Chen & Zhang, 2018). Human bones are prepared by using a blend of PHB copolymer containing 8% HV and 30% (w/w) hydroxyapatite which has mechanical compressive strength similar to human bones (62 MPa) (Arai et al., 2002; Chan, Marçal, Russell, Holden, & Foster, 2011).

Coating of PHAs on medical implants opens a new horizon in combating incidences of infection caused due to postoperative bacterial proliferation in the wound or on the surface of metallic implants, improperly sterilized surgical instruments, and implant-associated infections. Implants made up of chemically stable, biologically inert biomaterial like tantalum (Ta) coated with PHA is the best substitute to human cortical bone due to its elastic modulus between 2 and 20 GPa (Rodriguez-Contreras, 2019).

9.5.2 Polyhydroxyalkanoates in drug delivery

Being natural, nontoxic, biocompatibility, and biodegradable polymer, PHA has gained much attention as a potent drug carrier. PHAs are thermoplastic or elastomeric polymers causing acute inflammatory responses (Rodriguez-Contreras, 2019; Chen & Wu, 2005; Mokhtarzadeh et al., 2016). PHBVHHx3 poly(3-hydroxybutyrate-*co*-3-hydroxyvalerate-*co*-3-hydroxyhexanoate) is the most commonly used PHA in biomedicine (Ray & Kalia, 2017). PHBHHx is used for preparation of nanoparticle and the derived nanomaterial is best suited as compared to other nanocarrier materials like liposomes micelles, silica nanoparticle, anodic alumina nanotubes, carbon nanotubes, and dendrimes (Sun et al., 2019). Drug-loaded nanoparticle of PHBHHx coupled with autophagy inhibitors like 3MA and CQ precisely target tumor site and significantly reduces volume and weight of tumors (Tavakol et al., 2019; Sun et al., 2019).

PHA act as a drug-delivery agent in several forms such as functionalized PHAs, multifunctional PHA nanoparticles, micelles, biosynthetic PHA particles, PHA-drug, and PHA-protein conjugate (Rodriguez-Contreras, 2019). Polymeric matrixes of PHAs, especially P(HB-*co*-HV) and PHB with antibiotics and drugs like Sulbactam-cefoperazone, Rifampicin, Tetracycline, Gentamicin, Sulperazone, Rubomycin, and Rhodamine B are formulated for enhancing the efficiency of drugs along with antiinflammatory, antibiofilm, antimicrobial, antifungal, and virucidal properties (Ray & Kalia, 2017). Microspheres of PHB and an antineoplastic drug Rubomycin suppress the rate of proliferation of Ehrlich's carcinoma cells (Ray & Kalia, 2017). Microsphere carrying PHB and rifampicin were used as hemoembolizing agent (Dwivedi et al., 2019). Encapsulation of Docetaxel in PHBV nanoparticles intensifies its potential as an anticancerous drug (Elmowafy et al., 2019). These drug-delivery systems release drug substance by surface erosion mechanism (Rodriguez-Contreras, 2019; Elmowafy et al., 2019). PHAs (PHB, PHBV, and PHB4HB) were formulated into water-in-oil type emulsion to be used as a dip-coating agent onto the surface of biometals as a bioactive agent (Rodriguez-Contreras, 2019).

PHAs are employed in photodynamic therapy, as a carrier of hydrophobic photosensitizers in the treatment of cancer (Elmowafy et al., 2019). PHB and PHB/poly(ethylene glycol) (PEG)-based microparticles are produced by spray-drying technique to carry antitumor drugs (Rodriguez-Contreras, 2019). PHBHHx copolymer fabricated sandwiched films with thymopentin phospholipid complexes were prepared with sustained drug release potential of over 42 days and improved immunomodulation (Rodriguez-Contreras, 2019; Peng et al., 2018). The mcl-PHAs are more effective drug-delivery agents due to lower crystallinity and melting point. They have been used for transdermal drug delivery (Ray & Kalia, 2017; Türesin, Gürsel, & Hasirci, 2001).

PHA degrade to form R-3 hydrocarboxylic acids which transforms into different HAs such as two alkylated 3HB and beta-lactones. Peptidase resistant b- and c-peptides were prepared by exploiting monomers of 3HB and 4HB which ensures longer stay of peptides along with drug in mammalian serum and, therefore, enhances its suitability as a drug carrier for cargo-drug delivery (Ray & Kalia, 2017; Philip, Keshavarz, & Roy, 2007).

9.5.3 Polyhydroxyalkanoates in tissue engineering

Treatment based on tissue engineering is heartening approach to regenerate and repair damaged tissues either from injury, disease, or a trauma (Ang, Shaharuddin, Chuah, & Sudesh, 2020). In bone tissue engineering, damaged or impaired bones are replaced by new bones, regenerated by inducing cell growth on the PHA scaffolds (Lim, You, Li, & Li, 2017). PHA can replace and heal soft and hard tissues, repair cartilages, skin patches, cardiovascular tissues heart valves, and bone and nerve conduits (Rodriguez-Contreras, 2019; Chen & Wu, 2005; Butt, Muhammad, Hamid, Moniruzzaman, & Sharif, 2018). PHA-based biomaterials are fabricated by solvent casting, phase-separation, and electrospinning method into different shapes like fibrous foam, film, and hydrogels. Electrospun films of P(3HB-co-3HHx) are ideal scaffold structures prepared by blending copolymer of 3HB and 3HHx that exhibit distinctive properties like high porosity with interconnected structure and high surface area to volume ratio (Ang et al., 2020).

The properties of PHA polymers can be modified to match biological requirements of human tissues by blending it with hydroxyapatite, ceramics, and different natural polymers like silk, collagen, and gelatin. A novel bioactive and biodegradable composite is formed by incorporating particulate hydroxyapatite into PHB as a suitable substitute for its application in hard tissue regeneration and replacement (Chen & Wu, 2005). Genetically modified *Pseudomonas entomophila* L48 produced 3-hydroxyalkanoic acids (3HA), a precursor for synthesis of antibiotics, food additive, pharmaceuticals, fragrances, and vitamins. It also produces mcl-PHA which contains 3-hydroxyoctanoate (3HO), 3-hydroxyhexanoate (3HHx), 3-hydroxydecanoate (3HD), and 3-hydroxydodecanoate (3HDD) (Chung et al., 2011).

To treat congenital heart defects, vascular grafts, and heart valves, cardiac tissue engineering is performed using P4HB as a promising biomaterial derived from PHA. Heart valve tissue engineering is an important strategy to treat cardiac disorders. Multifunctional cardiac patches based on P_3HO showed mechanical properties that are similar to those of cardiac muscle (Rodriguez-Contreras, 2019).

The performance of PHAs as biopolymer is comparable to human tissues and PHA-based scaffolds are more promising substitutes (Rodriguez-Contreras, 2019; Shijun, Junsheng, Jianqun, & Ping, 2016). It induces cell cycle progression, to sustain biocompatibility and also reduces the risk of graft reject. PHB and PHBV exhibit excellent abilities to induce the progression of cell cycle as compared to olfactory ensheathing cells and mesenchymal stem cells (MSCs) when used in nerve and bone tissues (Shijun, Junsheng, Jianqun, & Ping, 2016). PHBV has greater potential of cell adhesion and growth. PHBHHx enhances cell growth with much better cell attachment and proliferation on 3D scaffolds than PHB (Shijun et al., 2016). Researchers have developed a novel polymer PHBVHHx poly(3-hydroxybutyrate-*co*-3-hydroxyvalerate-*co*-3-hydroxyhexanoate) secreted by recombinant *A. hydrophila* 4AK4 (Wang et al., 2010). It has rough surface and is hydrophobic in nature therefore adherence of cells is better on PHBVHHx films as compared to PHBHHx films (Lim et al., 2017). PHBVHHx is a promising polymer for development of 3D supportive scaffolds for organ tissues and for the growth of stem cells (Rodriguez-Contreras, 2019). Coating of hyaluronic acid on PHA membrane improves metabolic activity of cells simultaneously declining the death rate of human MSCs (Lim et al., 2017). PHA scaffold has considerable porosity which reproduces microenvironment, enhances extracellular matrix growth in counterpart of polymer degradation to sustain scaffold mechanical properties and stability (Elmowafy et al., 2019; Butt et al., 2018).

9.6 Conclusions and perspectives

PHA biopolymers have emerged as a promising alternative to synthetic petroleum-based plastics. Replacing synthetic plastics offers advantages in protecting the environment and reducing the use of fossil fuels. Beginning from 1959 to till-date the market of PHAs is continuously increasing. However, there are major issues in the production, yield, and cost of PHAs, which has hindered their growth and market capture. Extraction and purification are also matters of concern, which affect overall economics of the production cost. With the current understanding and following diverse strategies such as mixed cultures and two-stage processes, the large-scale production could be made more economical with improved yields. Efforts should also be directed toward developing newer fields of applications of PHAs for which PHA blends with improved characteristics should be explored, specifically for applications in medical implants, tissue engineering, and drug-delivery agents.

Acknowledgment

Vivek Kumar Gaur acknowledges the Council of Scientific and Industrial Research (CSIR), New Delhi for Senior Research Fellowship.

References

Aeschelmann, F., & Carus, M. (2015). Biobased building blocks and polymers in the world: Capacities, production, and applications—Status quo and trends towards 2020. *Industrial Biotechnology, 11*(3), 154−159.

Ahmed, T., Marçal, H., Lawless, M., Wanandy, N. S., Chiu, A., & Foster, L. J. R. (2010). Polyhydroxybutyrate and its copolymer with polyhydroxyvalerate as biomaterials: Influence on progression of stem cell cycle. *Biomacromolecules, 11*(10), 2707−2715.

Akaraonye, E., Keshavarz, T., & Roy, I. (2010). Production of polyhydroxyalkanoates: The future green materials of choice. *Journal of Chemical Technology & Biotechnology, 85*(6), 732−743.

Akaraonye, E., Moreno, C., Knowles, J. C., Keshavarz, T., & Roy, I. (2012). Poly(3-hydroxybutyrate) production by *Bacillus cereus* SPV using sugarcane molasses as the main carbon source. *Biotechnology Journal, 7*(2), 293−303.

Albuquerque, M. G. E., Torres, C. A. V., & Reis, M. A. M. (2010). Polyhydroxyalkanoate (PHA) production by a mixed microbial culture using sugar molasses: Effect of the influent substrate concentration on culture selection. *Water Research, 44*(11), 3419−3433.

Andrade, A. P., Witholt, B., Hany, R., Egli, T., & Li, Z. (2002). Preparation and characterization of enantiomerically pure telechelic diols from mcl-poly[(R)-3-hydroxyalkanoates]. *Macromolecules, 35*(3), 684−689.

Ang, S. L., Shaharuddin, B., Chuah, J. A., & Sudesh, K. (2020). Electrospun poly(3-hydroxybutyrate-*co*-3-hydroxyhexanoate)/silk fibroin film is a promising scaffold for bone tissue engineering. *International Journal of Biological Macromolecules, 145*, 173−188.

Anjum, A., Zuber, M., Zia, K. M., Noreen, A., Anjum, M. N., & Tabasum, S. (2016). Microbial production of polyhydroxyalkanoates (PHAs) and its copolymers: A review of recent advancements. *International Journal of Biological Macromolecules, 89*, 161−174.

Arai, Y., Nakashita, H., Suzuki, Y., Kobayashi, Y., Shimizu, T., Yasuda, M., . . . Yamaguchi, I. (2002). Synthesis of a novel class of polyhydroxyalkanoates in Arabidopsis peroxisomes, and their use in monitoring short-chain-length intermediates of β-oxidation. *Plant and Cell Physiology, 43*(5), 555−562.

Arkin, A. H., & Hazer, B. (2002). Chemical modification of chlorinated microbial polyesters. *Biomacromolecules, 3*(6), 1327−1335.

Arkin, A. H., Hazer, B., & Borcakli, M. (2000). Chlorination of poly(3-hydroxy alkanoates) containing unsaturated side chains. *Macromolecules, 33*(9), 3219−3223.

Ashby, R. D., Foglia, T. A., Solaiman, D. K., Liu, C. K., Nuñez, A., & Eggink, G. (2000). Viscoelastic properties of linseed oil-based medium chain length poly(hydroxyalkanoate) films: Effects of epoxidation and curing. *International Journal of Biological Macromolecules, 27*(5), 355−361.

Aziz, N. A., Sipaut, C. S., & Abdullah, A. A. A. (2012). Improvement of the production of poly(3-hydroxybutyrate-*co*-3-hydroxyvalerate-*co*-4-hydroxybutyrate) terpolyester by manipulating the culture condition. *Journal of Chemical Technology & Biotechnology, 87*(11), 1607−1614.

Berezina, N. (2013). Novel approach for productivity enhancement of polyhydroxyalkanoates (PHA) production by *Cupriavidus necator* DSM 545. *New Biotechnology, 30*(2), 192−195.

Blunt, W., Levin, D. B., & Cicek, N. (2018). Bioreactor operating strategies for improved polyhydroxyalkanoate (PHA) productivity. *Polymers, 10*(11), 1197.

Brandl, H., Bachofen, R., Mayer, J., & Wintermantel, E. (1995). Degradation and applications of polyhydroxyalkanoates. *Canadian Journal of Microbiology, 41*(13), 143−153.

Brigham, C. J., & Sinskey, A. J. (2012). Applications of polyhydroxyalkanoates in the medical industry. *International Journal of Biotechnology for Wellness Industries*, *1*(1), 52−60.

Bugnicourt, E., Cinelli, P., Lazzeri, A., & Alvarez, V. A. (2014). Polyhydroxyalkanoate (PHA): Review of synthesis, characteristics, processing and potential applications in packaging. *eXPRESS Polymer Letters*, *8*(11), 791−808.

Butt, F. I., Muhammad, N., Hamid, A., Moniruzzaman, M., & Sharif, F. (2018). Recent progress in the utilization of biosynthesized polyhydroxyalkanoates for biomedical applications—Review. *International Journal of Biological Macromolecules*, *120*, 1294−1305.

Cerrone, F., Choudhari, S. K., Davis, R., Cysneiros, D., O'Flaherty, V., Duane, G., . . . O'Connor, K. (2014). Medium chain length polyhydroxyalkanoate (mcl-PHA) production from volatile fatty acids derived from the anaerobic digestion of grass. *Applied Microbiology and Biotechnology*, *98*(2), 611−620.

Cesário, M. T., Raposo, R. S., de Almeida, M. C. M., van Keulen, F., Ferreira, B. S., & da Fonseca, M. M. R. (2014). Enhanced bioproduction of poly-3-hydroxybutyrate from wheat straw lignocellulosic hydrolysates. *New Biotechnology*, *31*(1), 104−113.

Chan, R. T., Marçal, H., Russell, R. A., Holden, P. J., & Foster, L. J. R. (2011). Application of polyethylene glycol to promote cellular biocompatibility of polyhydroxybutyrate films. *International Journal of Polymer Science*, *2011*.

Chanprateep, S. (2010). Current trends in biodegradable polyhydroxyalkanoates. *Journal of Bioscience and Bioengineering*, *110*(6), 621−632.

Chaudhry, W. N., Jamil, N., Ali, I., Ayaz, M. H., & Hasnain, S. (2011). Screening for polyhydroxyalkanoate (PHA)-producing bacterial strains and comparison of PHA production from various inexpensive carbon sources. *Annals of Microbiology*, *61*(3), 623−629.

Chen, B. Y., Hung, J. Y., Shiau, T. J., & Wei, Y. H. (2013). Exploring two-stage fermentation strategy of polyhydroxyalkanoate production using *Aeromonas hydrophila*. *Biochemical Engineering Journal*, *78*, 80−84.

Chen, G. Q. (2009). A microbial polyhydroxyalkanoates (PHA) based bio- and materials industry. *Chemical Society Reviews*, *38*(8), 2434−2446.

Chen, G. Q., & Wu, Q. (2005). The application of polyhydroxyalkanoates as tissue engineering materials. *Biomaterials*, *26*(33), 6565−6578.

Chen, G. Q., & Zhang, J. (2018). Microbial polyhydroxyalkanoates as medical implant biomaterials. *Artificial Cells, Nanomedicine, and Biotechnology*, *46*(1), 1−18.

Chen, J. Y., Song, G., & Chen, G. Q. (2006). A lower specificity PhaC2 synthase from *Pseudomonas stutzeri* catalyses the production of copolyesters consisting of short-chain-length and medium-chain-length 3-hydroxyalkanoates. *Antonie Van Leeuwenhoek*, *89*(1), 157−167.

Chen, Z. Y., Xiang, H. X., Hu, Z. X., Ni, Z. G., & Zhu, M. F. (2017). Enhanced mechanical properties of melt-spun bio-based PHBV fibers: Effect of heterogeneous nucleation and drawing process. *Acta Polymerica Sinica*, *7*, 1121−1129.

Choi, J., & Lee, S. Y. (1999). Factors affecting the economics of polyhydroxyalkanoate production by bacterial fermentation. *Applied Microbiology and Biotechnology*, *51*(1), 13−21.

Chung, A. L., Jin, H. L., Huang, L. J., Ye, H. M., Chen, J. C., Wu, Q., & Chen, G. Q. (2011). Biosynthesis and characterization of poly(3-hydroxydodecanoate) by β-oxidation inhibited mutant of *Pseudomonas entomophila* L48. *Biomacromolecules*, *12*(10), 3559−3566.

Clarinval, A. M., & Halleux, J. (2005). *Classification of biodegradable polymers. Biodegradable polymers for industrial applications* (pp. 3−31). Woodhead Publishing.

Cui, Y. W., Shi, Y. P., & Gong, X. Y. (2017). Effects of C/N in the substrate on the simultaneous production of polyhydroxyalkanoates and extracellular polymeric substances by *Haloferax mediterranei* via kinetic model analysis. *RSC Advances*, *7*(31), 18953−18961.

Dai, Z. W., Zou, X. H., & Chen, G. Q. (2009). Poly(3-hydroxybutyrate-*co*-3-hydroxyhexanoate) as an injectable implant system for prevention of post-surgical tissue adhesion. *Biomaterials*, *30*(17), 3075−3083.

Davis, R., Kataria, R., Cerrone, F., Woods, T., Kenny, S., O'Donovan, A., ... Tuohy, M. G. (2013). Conversion of grass biomass into fermentable sugars and its utilization for medium chain length polyhydroxyalkanoate (mcl-PHA) production by *Pseudomonas* strains. *Bioresource Technology*, *150*, 202–209.

Degeratu, C. N., Mabilleau, G., Aguado, E., Mallet, R., Chappard, D., Cincu, C., & Stancu, I. C. (2019). Polyhydroxyalkanoate (PHBV) fibers obtained by a wet spinning method: Good in vitro cytocompatibility but absence of in vivo biocompatibility when used as a bone graft. *Morphologie: Bulletin de l'Association des Anatomistes*, *103*(341), 94–102.

Doi, Y., & Steinbüchel, A. (Eds.), (2002). *Polyesters I: Biological systems and biotechnological production*. Wiley-VCH.

Dwivedi, R., Pandey, R., Kumar, S., & Mehrotra, D. (2019). Polyhydroxyalkanoates (PHA): Role in bone scaffolds. *Journal of Oral Biology and Craniofacial Research*, *10*(1), 389–392.

Elmowafy, E., Abdal-Hay, A., Skouras, A., Tiboni, M., Casettari, L., & Guarino, V. (2019). Polyhydroxyalkanoate (PHA): Applications in drug delivery and tissue engineering. *Expert Review of Medical Devices*, *16*(6), 467–482.

Eroğlu, M. S., Hazer, B., Ozturk, T., & Caykara, T. (2005). Hydroxylation of pendant vinyl groups of poly(3-hydroxy undec-10-enoate) in high yield. *Journal of Applied Polymer Science*, *97*(5), 2132–2139.

PlasticsEurope. (2017). *Plastics—The facts 2014//2015: An analysis of European plastics production, demand and waste data*.

Fernández-Dacosta, C., Posada, J. A., Kleerebezem, R., Cuellar, M. C., & Ramirez, A. (2015). Microbial community-based polyhydroxyalkanoates (PHAs) production from wastewater: Techno-economic analysis and ex-ante environmental assessment. *Bioresource Technology*, *185*, 368–377.

Fiedler, S., Steinbüchel, A., & Rehm, B. H. (2002). The role of the fatty acid β-oxidation multienzyme complex from *Pseudomonas oleovorans* in polyhydroxyalkanoate biosynthesis: Molecular characterization of the fadBA operon from *P. oleovorans* and of the enoyl-CoA hydratase genes phaJ from *P. oleovorans* and *Pseudomonas putida*. *Archives of Microbiology*, *178*(2), 149–160.

Follonier, S., Henes, B., Panke, S., & Zinn, M. (2012). Putting cells under pressure: A simple and efficient way to enhance the productivity of medium-chain-length polyhydroxyalkanoate in processes with *Pseudomonas putida* KT2440. *Biotechnology and Bioengineering*, *109*(2), 451–461.

Follonier, S., Riesen, R., & Zinn, M. (2015). Pilot-scale production of functionalized mcl-PHA from grape pomace supplemented with fatty acids. *Chemical and Biochemical Engineering Quarterly*, *29*(2), 113–121.

Fukui, T., & Doi, Y. (1998). Efficient production of polyhydroxyalkanoates from plant oils by *Alcaligenes eutrophus* and its recombinant strain. *Applied Microbiology and Biotechnology*, *49*(3), 333–336.

Furutate, S., Nakazaki, H., Maejima, K., Hiroe, A., Abe, H., & Tsuge, T. (2017). Biosynthesis and characterization of novel polyhydroxyalkanoate copolymers consisting of 3-hydroxy-2-methylbutyrate and 3-hydroxyhexanoate. *Journal of Polymer Research*, *24*(12), 221.

Gahlawat, G., & Srivastava, A. K. (2012). Estimation of fundamental kinetic parameters of polyhydroxybutyrate fermentation process of *Azohydromonas australica* using statistical approach of media optimization. *Applied Biochemistry and Biotechnology*, *168*(5), 1051–1064.

García, A., Segura, D., Espín, G., Galindo, E., Castillo, T., & Peña, C. (2014). High production of poly-β-hydroxybutyrate (PHB) by an *Azotobacter vinelandii* mutant altered in PHB regulation using a fed-batch fermentation process. *Biochemical Engineering Journal*, *82*, 117–123.

Giin-Yu, A. T., Chia-Lung, C., Ling, L., Liya, G., & Lin, W. (2014). Start a research on biopolymer polyhydroxybutyrate (PHB). *Polymers*, *6*, 706–754.

Gopi, S., Kontopoulou, M., Ramsay, B. A., & Ramsay, J. A. (2018). Manipulating the structure of medium-chain-length polyhydroxyalkanoate (MCL-PHA) to enhance thermal properties and crystallization kinetics. *International Journal of Biological Macromolecules*, *119*, 1248–1255.

Greene, J. (2013). PHA biodegradable blow-molded bottles: Compounding and performance. *Plastics Engineering*, *69*(1), 16–21.

Grigore, M. E., Grigorescu, R. M., Iancu, L., Ion, R. M., Zaharia, C., & Andrei, E. R. (2019). Methods of synthesis, properties and biomedical applications of polyhydroxyalkanoates: A review. *Journal of Biomaterials Science, Polymer Edition*, *30*(9), 695–712.

Grothe, E., Moo-Young, M., & Chisti, Y. (1999). Fermentation optimization for the production of poly (β-hydroxybutyric acid) microbial thermoplastic. *Enzyme and Microbial Technology*, 25(1−2), 132−141.

Hartmann, R., Hany, R., Pletscher, E., Ritter, A., Witholt, B., & Zinn, M. (2006). Tailor-made olefinic medium-chain-length poly [(R)-3-hydroxyalkanoates] by *Pseudomonas putida* GPo1: Batch versus chemostat production. *Biotechnology and Bioengineering*, 93(4), 737−746.

Hazer, B., & Steinbüchel, A. (2007). Increased diversification of polyhydroxyalkanoates by modification reactions for industrial and medical applications. *Applied Microbiology and Biotechnology*, 74(1), 1−12.

Hazer, D. B., Kılıçay, E., & Hazer, B. (2012). Poly(3-hydroxyalkanoate)s: Diversification and biomedical applications: A state of the art review. *Materials Science and Engineering: C*, 32(4), 637−647.

Hein, S., Paletta, J., & Steinbüchel, A. (2002). Cloning, characterization and comparison of the *Pseudomonas mendocina* polyhydroxyalkanoate synthases PhaC1 and PhaC2. *Applied Microbiology and Biotechnology*, 58(2), 229−236.

Hoffmann, N., & Rehm, B. H. (2004). Regulation of polyhydroxyalkanoate biosynthesis in *Pseudomonas putida* and *Pseudomonas aeruginosa*. *FEMS Microbiology Letters*, 237(1), 1−7.

Hoffmann, N., Steinbüchel, A., & Rehm, B. H. (2000). The *Pseudomonas aeruginosa* phaG gene product is involved in the synthesis of polyhydroxyalkanoic acid consisting of medium-chain-length constituents from non-related carbon sources. *FEMS Microbiology Letters*, 184(2), 253−259.

Hu, D., Chung, A. L., Wu, L. P., Zhang, X., Wu, Q., Chen, J. C., & Chen, G. Q. (2011). Biosynthesis and characterization of polyhydroxyalkanoate block copolymer P3HB-b-P4HB. *Biomacromolecules*, 12(9), 3166−3173.

Ibrahim, M. H., & Steinbüchel, A. (2010). High-cell-density cyclic fed-batch fermentation of a poly(3-hydroxybutyrate)-accumulating thermophile, *Chelatococcus* sp. strain MW10. *Applied and Environmental Microbiology*, 76(23), 7890−7895.

Ivanov, V., Stabnikov, V., Ahmed, Z., Dobrenko, S., & Saliuk, A. (2015). Production and applications of crude polyhydroxyalkanoate-containing bioplastic from the organic fraction of municipal solid waste. *International Journal of Environmental Science and Technology*, 12(2), 725−738.

Jendrossek, D., & Handrick, R. (2002). Microbial degradation of polyhydroxyalkanoates. *Annual Review of Microbiology*, 56(1), 403−432.

Kabe, T., Tsuge, T., Kasuya, K. I., Takemura, A., Hikima, T., Takata, M., & Iwata, T. (2012). Physical and structural effects of adding ultrahigh-molecular-weight poly [(R)-3-hydroxybutyrate] to wild-type poly [(R)-3-hydroxybutyrate]. *Macromolecules*, 45(4), 1858−1865.

Kai, D., & Loh, X. J. (2014). Polyhydroxyalkanoates: Chemical modifications toward biomedical applications. *ACS Sustainable Chemistry & Engineering*, 2(2), 106−119.

Kalia, V. C. (Ed.), (2019). *Biotechnological applications of polyhydroxyalkanoates*. Berlin: Springer.

Kapritchkoff, F. M., Viotti, A. P., Alli, R. C., Zuccolo, M., Pradella, J. G., Maiorano, A. E., ... Bonomi, A. (2006). Enzymatic recovery and purification of polyhydroxybutyrate produced by *Ralstonia eutropha*. *Journal of Biotechnology*, 122(4), 453−462.

Kaur, G., & Roy, I. (2015). Strategies for large-scale production of polyhydroxyalkanoates. *Chemical and Biochemical Engineering Quarterly*, 29(2), 157−172.

Kemnitzer, J. E., McCarthy, S. P., & Gross, R. A. (1992). Poly(β-hydroxybutyrate) stereoisomers: A model study of the effects of stereochemical and morphological variables on polymer biological degradability. *Macromolecules*, 25(22), 5927−5934.

Khanna, S., & Srivastava, A. K. (2005). Recent advances in microbial polyhydroxyalkanoates. *Process Biochemistry*, 40(2), 607−619.

Kim, G. J., Bang, K. H., Kim, Y. B., & Rhee, Y. H. (2000). Preparation and characterization of native poly(3-hydroxybutyrate) microspheres from *Ralstonia eutropha*. *Biotechnology Letters*, 22(18), 1487−1492.

Kim, H. W., Chung, M. G., & Rhee, Y. H. (2007). Biosynthesis, modification, and biodegradation of bacterial medium-chain-length polyhydroxyalkanoates. *The Journal of Microbiology*, 45(2), 87−97.

Koller, M. (2018). Biodegradable and biocompatible polyhydroxy-alkanoates (PHA): Auspicious microbial macromolecules for pharmaceutical and therapeutic applications. *Molecules (Basel, Switzerland)*, 23(2), 362.

Koller, M., Bona, R., Chiellini, E., Fernandes, E. G., Horvat, P., Kutschera, C., ... Braunegg, G. (2008). Polyhydroxyalkanoate production from whey by *Pseudomonas hydrogenovora*. *Bioresource Technology*, *99*(11), 4854–4863.

Koller, M., & Muhr, A. (2014). Continuous production mode as a viable process-engineering tool for efficient poly (hydroxyalkanoate)(PHA) bio-production. *Chemical and Biochemical Engineering Quarterly*, *28*(1), 65–77.

Kosior, E., Braganca, R. M., & Fowler, P. (2006). Lightweight compostable packaging: Literature review. *The Waste & Resources Action Programme*, *26*, 1–48.

Kovalcik, A., Obruca, S., Fritz, I., & Marova, I. (2019). Polyhydroxyalkanoates: Their importance and future. *BioResources*, *14*(2), 2468–2471.

Kumar, G., Ponnusamy, V. K., Bhosale, R. R., Shobana, S., Yoon, J. J., Bhatia, S. K., ... Kim, S. H. A. (2019). Review on the conversion of volatile fatty acids to polyhydroxyalkonates using dark fermentative effluents from hydrogen production. *Bioresource Technology*, 121427.

Kwiecień, M., Adamus, G., & Kowalczuk, M. (2013). Selective reduction of PHA biopolyesters and their synthetic analogues to corresponding PHA oligodiols proved by structural studies. *Biomacromolecules*, *14*(4), 1181–1188.

Lau, N. S., Chee, J. Y., Tsuge, T., & Sudesh, K. (2010). Biosynthesis and mobilization of a novel polyhydroxyalkanoate containing 3-hydroxy-4-methylvalerate monomer produced by *Burkholderia* sp. USM (JCM15050). *Bioresource Technology*, *101*(20), 7916–7923.

Lee, M. Y., & Park, W. H. (2000). Preparation of bacterial copolyesters with improved hydrophilicity by carboxylation. *Macromolecular Chemistry and Physics*, *201*(18), 2771–2774.

Lee, M. Y., Park, W. H., & Lenz, R. W. (2000). Hydrophilic bacterial polyesters modified with pendant hydroxyl groups. *Polymer*, *41*(5), 1703–1709.

Leja, K., & Lewandowicz, G. (2010). Polymer biodegradation and biodegradable polymers—A review. *Polish Journal of Environmental Studies*, *19*(2).

Lemoigne, M. (1926). Products of dehydration and of polymerization of β-hydroxybutyric acid. *Bulletin de la Société de Chimie Biologique*, *8*, 770–782.

Leong, Y. K., Show, P. L., Lan, J. C. W., Loh, H. S., Lam, H. L., & Ling, T. C. (2017). Economic and environmental analysis of PHAs production process. *Clean Technologies and Environmental Policy*, *19*(7), 1941–1953.

Leong, Y. K., Show, P. L., Lan, J. C. W., Loh, H. S., Yap, Y. J., & Ling, T. C. (2017). Extraction and purification of polyhydroxyalkanoates (PHAs): Application of thermoseparating aqueous two-phase extraction. *Journal of Polymer Research*, *24*(10), 158.

Leong, Y. K., Show, P. L., Ooi, C. W., Ling, T. C., & Lan, J. C. W. (2014). Current trends in polyhydroxyalkanoates (PHAs) biosynthesis: Insights from the recombinant *Escherichia coli*. *Journal of Biotechnology*, *180*, 52–65.

Liao, Q. (2010). *Biodegradable poly(hydroxyalkanoates): Melt, solid, and foam*. Stanford University.

Lim, J., You, M., Li, J., & Li, Z. (2017). Emerging bone tissue engineering via polyhydroxyalkanoate (PHA)-based scaffolds. *Materials Science and Engineering: C*, *79*, 917–929.

Liong, M. T. (Ed.), (2015). *Beneficial microorganisms in agriculture, aquaculture and other areas* (Vol. 29). Springer.

Liu, Y., Huang, S., Zhang, Y., & Xu, F. (2014). Isolation and characterization of a thermophilic *Bacillus shackletonii* K5 from a biotrickling filter for the production of polyhydroxybutyrate. *Journal of Environmental Sciences*, *26*(7), 1453–1462.

López-Abelairas, M., García-Torreiro, M., Lú-Chau, T., Lema, J. M., & Steinbüchel, A. (2015). Comparison of several methods for the separation of poly(3-hydroxybutyrate) from *Cupriavidus necator* H16 cultures. *Biochemical Engineering Journal*, *93*, 250–259.

López-Cuellar, M. R., Alba-Flores, J., Rodríguez, J. G., & Pérez-Guevara, F. (2011). Production of polyhydroxyalkanoates (PHAs) with canola oil as carbon source. *International Journal of Biological Macromolecules*, *48*(1), 74–80.

Luef, K. P., Stelzer, F., & Wiesbrock, F. (2015). Poly(hydroxyalkanoate)s in medical applications. *Chemical and Biochemical Engineering Quarterly*, 29(2), 287−297.

Macrae, R. M., & Wilkinson, J. F. (1958). Poly-β-hyroxybutyrate metabolism in washed suspensions of *Bacillus cereus* and *Bacillus megaterium*. *Microbiology (Reading, England)*, 19(1), 210−222.

Mathuriya, A. S., & Yakhmi, J. V. (2017). Polyhydroxyalkanoates: Biodegradable plastics and their applications. *Handbook of ecomaterials* (pp. 1−29). Springer.

Meng, D. C., & Chen, G. Q. (2017). Synthetic biology of polyhydroxyalkanoates (PHA). *Synthetic biology— Metabolic engineering* (pp. 147−174). Cham: Springer.

Misra, S. K., Valappil, S. P., Roy, I., & Boccaccini, A. R. (2006). Polyhydroxyalkanoate (PHA)/inorganic phase composites for tissue engineering applications. *Biomacromolecules*, 7(8), 2249−2258.

Mokhtarzadeh, A., Alibakhshi, A., Yaghoobi, H., Hashemi, M., Hejazi, M., & Ramezani, M. (2016). Recent advances on biocompatible and biodegradable nanoparticles as gene carriers. *Expert Opinion on Biological Therapy*, 16(6), 771−785.

Mozumder, M. S. I., De Wever, H., Volcke, E. I., & Garcia-Gonzalez, L. (2014). A robust fed-batch feeding strategy independent of the carbon source for optimal polyhydroxybutyrate production. *Process Biochemistry*, 49(3), 365−373.

Możejko-Ciesielska, J., & Kiewisz, R. (2016). Bacterial polyhydroxyalkanoates: Still fabulous? *Microbiological Research*, 192, 271−282.

Muhammadi, S., Afzal, M., & Hameed, S. (2015). Bacterial polyhydroxyalkanoates-eco-friendly next generation plastic: Production, biocompatibility, biodegradation, physical properties and applications. *Green Chemistry Letters and Reviews*, 8(3−4), 56−77.

Muhr, A., Rechberger, E. M., Salerno, A., Reiterer, A., Malli, K., Strohmeier, K., ... Koller, M. (2013). Novel description of mcl-PHA biosynthesis by *Pseudomonas chlororaphis* from animal-derived waste. *Journal of Biotechnology*, 165(1), 45−51.

Narayanan, A., & Ramana, K. V. (2012). Polyhydroxybutyrate production in *Bacillus mycoides* DFC1 using response surface optimization for physico-chemical process parameters. *3 Biotech*, 2(4), 287−296.

Ojumu, T. V., Yu, J., & Solomon, B. O. (2004). Production of polyhydroxyalkanoates, a bacterial biodegradable polymers. *African Journal of Biotechnology*, 3(1), 18−24.

Padermshoke, A., Katsumoto, Y., Sato, H., Ekgasit, S., Noda, I., & Ozaki, Y. (2004). Surface melting and crystallization behavior of polyhydroxyalkanoates studied by attenuated total reflection infrared spectroscopy. *Polymer*, 45(19), 6547−6554.

Pan, W., Perrotta, J. A., Stipanovic, A. J., Nomura, C. T., & Nakas, J. P. (2012). Production of polyhydroxyalkanoates by *Burkholderia cepacia* ATCC 17759 using a detoxified sugar maple hemicellulosic hydrolysate. *Journal of Industrial Microbiology & Biotechnology*, 39(3), 459−469.

Pandian, S. R., Deepak, V., Kalishwaralal, K., Rameshkumar, N., Jeyaraj, M., & Gurunathan, S. (2010). Optimization and fed-batch production of PHB utilizing dairy waste and sea water as nutrient sources by *Bacillus megaterium* SRKP-3. *Bioresource Technology*, 101(2), 705−711.

Pantazaki, A. A., Papaneophytou, C. P., Pritsa, A. G., Liakopoulou-Kyriakides, M., & Kyriakidis, D. A. (2009). Production of polyhydroxyalkanoates from whey by *Thermus thermophilus* HB8. *Process Biochemistry*, 44(8), 847−HB853.

Parker, R. E., & Isaacs, N. S. (1959). Mechanisms of epoxide reactions. *Chemical Reviews*, 59(4), 737−799.

Peng, K., Wu, C., Wei, G., Jiang, J., Zhang, Z., & Sun, X. (2018). Implantable sandwich PHBHHx film for burst-free controlled delivery of thymopentin peptide. *Acta Pharmaceutica Sinica B*, 8(3), 432−439.

Penloglou, G., Kretza, E., Chatzidoukas, C., Parouti, S., & Kiparissides, C. (2012). On the control of molecular weight distribution of polyhydroxybutyrate in *Azohydromonaslata* cultures. *Biochemical Engineering Journal*, 62, 39−47.

Philip, S., Keshavarz, T., & Roy, I. (2007). Polyhydroxyalkanoates: Biodegradable polymers with a range of applications. *Journal of Chemical Technology & Biotechnology: International Research in Process, Environmental & Clean Technology*, 82(3), 233−247.

Poblete-Castro, I., Binger, D., Oehlert, R., & Rohde, M. (2014). Comparison of mcl-poly(3-hydroxyalkanoates) synthesis by different *Pseudomonas putida* strains from crude glycerol: Citrate accumulates at high titer under PHA-producing conditions. *BMC Biotechnology*, *14*(1), 962.

Puppi, D., Pecorini, G., & Chiellini, F. (2019). Biomedical processing of polyhydroxyalkanoates. *Bioengineering*, *6*(4), 108.

Qu, X. H., Wu, Q., Liang, J., Zou, B., & Chen, G. Q. (2006). Effect of 3-hydroxyhexanoate content in poly(3-hydroxybutyrate-*co*-3-hydroxyhexanoate) on in vitro growth and differentiation of smooth muscle cells. *Biomaterials*, *27*(15), 2944−2950.

Rai, R. (2010). *Biosynthesis of polyhydroxyalkanoates and its medical applications* (Doctoral dissertation). University of Westminster.

Rai, R., Keshavarz, T., Roether, J. A., Boccaccini, A. R., & Roy, I. (2011). Medium chain length polyhydroxyalkanoates, promising new biomedical materials for the future. *Materials Science and Engineering: R: Reports*, *72*(3), 29−47.

Ray, S., & Kalia, V. C. (2017). Biomedical applications of polyhydroxyalkanoates. *Indian Journal of Microbiology*, *57*(3), 261−269.

Raza, Z. A., Riaz, S., & Banat, I. M. (2018). Polyhydroxyalkanoates: Properties and chemical modification approaches for their functionalization. *Biotechnology Progress*, *34*(1), 29−41.

Raza, Z. A., Tariq, M. R., Majeed, M. I., & Banat, I. M. (2019). Recent developments in bioreactor scale production of bacterial polyhydroxyalkanoates. *Bioprocess and Biosystems Engineering*, *42*(6), 901−919.

Rehm, B. H. (2010). Bacterial polymers: Biosynthesis, modifications and applications. *Nature Reviews. Microbiology*, *8*(8), 578−592.

Renard, E., Poux, A., Timbart, L., Langlois, V., & Guérin, P. (2005). Preparation of a novel artificial bacterial polyester modified with pendant hydroxyl groups. *Biomacromolecules*, *6*(2), 891−896.

Rodriguez-Contreras, A. (2019). Recent advances in the use of polyhydroyalkanoates in biomedicine. *Bioengineering*, *6*(3), 82.

Rodríguez-Contreras, A., Koller, M., Miranda-de Sousa Dias, M., Calafell-Monfort, M., Braunegg, G., & Marqués-Calvo, M. S. (2013). High production of poly(3-hydroxybutyrate) from a wild *Bacillus megaterium* Bolivian strain. *Journal of Applied Microbiology*, *114*(5), 1378−1387.

Ruiz, C., Kenny, S. T., Narancic, T., Babu, R., & O'Connor, K. (2019). Conversion of waste cooking oil into medium chain polyhydroxyalkanoates in a high cell density fermentation. *Journal of Biotechnology*, *306*, 9−15.

Sanhueza, C., Acevedo, F., Rocha, S., Villegas, P., Seeger, M., & Navia, R. (2019). Polyhydroxyalkanoates as biomaterial for electrospun scaffolds. *International Journal of Biological Macromolecules*, *124*, 102−110.

Sathiyanarayanan, G., Kiran, G. S., Selvin, J., & Saibaba, G. (2013). Optimization of polyhydroxybutyrate production by marine *Bacillus megaterium* MSBN04 under solid state culture. *International Journal of Biological Macromolecules*, *60*, 253−261.

Sharma, P. K., Munir, R. I., Blunt, W., Dartiailh, C., Cheng, J., Charles, T. C., & Levin, D. B. (2017). Synthesis and physical properties of polyhydroxyalkanoate polymers with different monomer compositions by recombinant *Pseudomonas putida* LS46 expressing a novel PHA synthase (PhaC116) enzyme. *Applied Sciences*, *7*(3), 242.

Shijun, X., Junsheng, M., Jianqun, Z., & Ping, B. (2016). In vitro three-dimensional coculturing poly3-hydroxybutyrate-*co*-3-hydroxyhexanoate with mouse-induced pluripotent stem cells for myocardial patch application. *Journal of Biomaterials Applications*, *30*(8), 1273−1282.

Sindhu, R., Silviya, N., Binod, P., & Pandey, A. (2013). Pentose-rich hydrolysate from acid pretreated rice straw as a carbon source for the production of poly-3-hydroxybutyrate. *Biochemical Engineering Journal*, *78*, 67−72.

Singh, A. K., Srivastava, J. K., Chandel, A. K., Sharma, L., Mallick, N., & Singh, S. P. (2019). Biomedical applications of microbially engineered polyhydroxyalkanoates: An insight into recent advances, bottlenecks, and solutions. *Applied Microbiology and Biotechnology*, *103*(5), 2007−2032.

Singh, M., Patel, S. K., & Kalia, V. C. (2009). *Bacillus subtilis* as potential producer for polyhydroxyalkanoates. *Microbial Cell Factories*, 8(1), 38.

Steinbüchel, A., & Hein, S. (2001). *Biochemical and molecular basis of microbial synthesis of polyhydroxyalkanoates in microorganisms. Biopolyesters* (pp. 81−123). Berlin, Heidelberg: Springer.

Stigers, D. J., & Tew, G. N. (2003). Poly(3-hydroxyalkanoate) s functionalized with carboxylic acid groups in the side chain. *Biomacromolecules*, 4(2), 193−195.

Sun, X., Cheng, C., Zhang, J., Jin, X., Sun, S., Mei, L., & Huang, L. (2019). Intracellular trafficking network and autophagy of PHBHHx nanoparticles and their implications for drug delivery. *Scientific Reports*, 9(1), 1−10.

Suriyamongkol, P., Weselake, R., Narine, S., Moloney, M., & Shah, S. (2007). Biotechnological approaches for the production of polyhydroxyalkanoates in microorganisms and plants—A review. *Biotechnology Advances*, 25(2), 148−175.

Tan, D., Xue, Y. S., Aibaidula, G., & Chen, G. Q. (2011). Unsterile and continuous production of polyhydroxybutyrate by Halomonas TD01. *Bioresource Technology*, 102(17), 8130−8136.

Tan, G. Y. A., Chen, C. L., Li, L., Ge, L., Wang, L., Razaad, I. M. N., . . . Wang, J. Y. (2014). Start a research on biopolymer polyhydroxyalkanoate (PHA): A review. *Polymers*, 6(3), 706−754.

Tavakol, S., Ashrafizadeh, M., Deng, S., Azarian, M., Abdoli, A., Motavaf, M., . . . Pardakhty, A. (2019). Autophagy modulators: Mechanistic aspects and drug delivery systems. *Biomolecules*, 9(10), 530.

Timbart, L., Renard, E., Tessier, M., & Langlois, V. (2007). Monohydroxylated poly(3-hydroxyoctanoate) oligomers and its functionalized derivatives used as macroinitiators in the synthesis of degradable diblockcopolyesters. *Biomacromolecules*, 8(4), 1255−1265.

Tripathi, A. D., Srivastava, S. K., & Singh, R. P. (2013). Statistical optimization of physical process variables for bio-plastic (PHB) production by *Alcaligenes* sp. *Biomass and Bioenergy*, 55, 243−250.

Tsuge, T., Taguchi, K., & Doi, Y. (2003). Molecular characterization and properties of (R)-specific enoyl-CoA hydratases from *Pseudomonas aeruginosa*: Metabolic tools for synthesis of polyhydroxyalkanoates via fatty acid ß-oxidation. *International Journal of Biological Macromolecules*, 31(4−5), 195−205.

Türesin, F., Gürsel, I., & Hasirci, V. A. S. I. F. (2001). Biodegradable polyhydroxyalkanoate implants for osteomyelitis therapy: In vitro antibiotic release. *Journal of Biomaterials Science, Polymer Edition*, 12(2), 195−207.

Verlinden, R. A., Hill, D. J., Kenward, M. A., Williams, C. D., & Radecka, I. (2007). Bacterial synthesis of biodegradable polyhydroxyalkanoates. *Journal of Applied Microbiology*, 102(6), 1437−1449.

Vigneswari, S., Nik, L. A., Majid, M. I. A., & Amirul, A. A. (2010). Improved production of poly(3-hydroxybutyrate-co-4-hydroxbutyrate) copolymer using a combination of 1,4-butanediol and γ-butyrolactone. *World Journal of Microbiology and Biotechnology*, 26(4), 743−746.

Wang, L., Wang, Z. H., Shen, C. Y., You, M. L., Xiao, J. F., & Chen, G. Q. (2010). Differentiation of human bone marrow mesenchymal stem cells grown in terpolyesters of 3-hydroxyalkanoates scaffolds into nerve cells. *Biomaterials*, 31(7), 1691−1698.

Wang, Z., Wang, Y., Zhang, D., Li, J., Hua, Z., Du, G., & Chen, J. (2010). Enhancement of cell viability and alkaline polygalacturonate lyase production by sorbitol co-feeding with methanol in *Pichia pastoris* fermentation. *Bioresource Technology*, 101(4), 1318−1323.

Wecker, P., Moppert, X., Simon-Colin, C., Costa, B., & Berteaux-Lecellier, V. (2015). Discovery of a mcl-PHA with unexpected biotechnical properties: The marine environment of French Polynesia as a source for PHA-producing bacteria. *AMB Express*, 5(1), 1−9.

Wei, X., Hu, Y.-J., Xie, W.-P., Lin, R.-L., & Chen, G.-Q. (2009). Influence of poly(3-hydroxybutyrate-co-4-hydroxybutyrate-co-3-hydroxyhexanoate) on growth and osteogenic differentiation of human bone marrow-derived mesenchymal stem cells. *Journal of Biomedical Materials Research Part A: An Official Journal of The Society for Biomaterials, The Japanese Society for Biomaterials, and The Australian Society for Biomaterials and the Korean Society for Biomaterials*, 90(3), 894−905.

Williams, D. F. (2008). On the mechanisms of biocompatibility. *Biomaterials*, 29(20), 2941−2953.

Williams, S. F., & Martin, D. P. (2002). Applications of PHAs in medicine and pharmacy. *Biopolymers*, *4*, 91–127.

Wu, C. S. (2013). Preparation, characterization and biodegradability of crosslinked tea plant-fibre-reinforced polyhydroxyalkanoate composites. *Polymer Degradation and Stability*, *98*(8), 1473–1480.

Wu, H. A., Sheu, D. S., & Lee, C. Y. (2003). Rapid differentiation between short-chain-length and medium-chain-length polyhydroxyalkanoate-accumulating bacteria with spectrofluorometry. *Journal of Microbiological Methods*, *53*(1), 131–135.

Xiang, H., Chen, W., Chen, Z., Sun, B., & Zhu, M. (2017). Significant accelerated crystallization of long chain branched poly(3-hydroxybutyrate-*co*-3-hydroxyvalerate) with high nucleation temperature under fast cooling rate. *Composites Science and Technology*, *142*, 207–213.

Xiang, H., Chen, Z., Zheng, N., Zhang, X., Zhu, L., Zhou, Z., & Zhu, M. (2019). Melt-spun microbial poly(3-hydroxybutyrate-*co*-3-hydroxyvalerate) fibers with enhanced toughness: Synergistic effect of heterogeneous nucleation, long-chain branching and drawing process. *International Journal of Biological Macromolecules*, *122*, 1136–1143.

Zhang, J., Shishatskaya, E. I., Volova, T. G., da Silva, L. F., & Chen, G. Q. (2018). Polyhydroxyalkanoates (PHA) for therapeutic applications. *Materials Science and Engineering: C*, *86*, 144–150.

Zhang, Y., Sun, W., Wang, H., & Geng, A. (2013). Polyhydroxybutyrate production from oil palm empty fruit bunch using *Bacillus megaterium* R11. *Bioresource Technology*, *147*, 307–314.

Zhao, K., Deng, Y., Chen, J. C., & Chen, G. Q. (2003). Polyhydroxyalkanoate (PHA) scaffolds with good mechanical properties and biocompatibility. *Biomaterials*, *24*(6), 1041–1045.

Zhou, L., Chen, Z., Chi, W., Yang, X., Wang, W., & Zhang, B. (2012). Mono-methoxy-poly(3-hydroxybutyrate-*co*-4-hydroxybutyrate)-graft-hyper-branched polyethylenimine copolymers for siRNA delivery. *Biomaterials*, *33*(7), 2334–2344.

Zinn, M., Witholt, B., & Egli, T. (2001). Occurrence, synthesis and medical application of bacterial polyhydroxyalkanoate. *Advanced Drug Delivery Reviews*, *53*(1), 5–21.

Zook, E. C., & Kee, B. L. (2016). Development of innate lymphoid cells. *Nature Immunology*, *17*(7), 775.

CHAPTER 10

Production and applications of polyglutamic acid

Kritika Pandey[1], Ashutosh Kumar Pandey[2], Ranjna Sirohi[3], Srinath Pandey[4], Aditya Srivastava[1] and Ashok Pandey[5]

[1]Department of Biotechnology, Dr. Ambedkar Institute of Technology for Handicapped, Kanpur, India, [2]Centre for Energy and Environmental Sustainability, Lucknow, India, [3]Department of Post-Harvest Process and Food Engineering, G.B. Pant University of Agriculture and Technology, Pantnagar, India, [4]Department of Biotechnology, Naraina Group of Institution, Kanpur, India, [5]Centre for Innovation and Translational Research, CSIR—Indian Institute of Toxicology Research (CSIR—IITR), Lucknow, India

10.1 Introduction

Polyglutamic acid (PGA) is an anionic homopolymer of biological origin, composed of D- and L-glutamic acid units. It is found to be quite common in nature and shows properties like solubility in water, high absorbability, and high binding affinity toward metals, are even edible and does not trigger any immune response. PGA can be distinguished into two isoforms, namely, α-PGA and γ-PGA. The α isoform is not very easily synthesized via microbes but can be synthesized chemically from γ-protected N-carboxy anhydride of L-glutamic acid using nucleophile-initiated polymerization (Ogunleye et al., 2015). The γ isoform is naturally synthesized by certain microbes, including various *Bacilli*, wherein it constitutes the viscous extracellular material of various nonpathogenic species like *Bacillus subtilis*, *Bacillus licheniformis*, *Bacillus megaterium*, and *Bacillus halodurans*. It is also produced by certain archaebacteria, Gram (-) bacteria, and also some eukaryotes (Buescher & Margaritis, 2007; Schallmey, Singh, & Ward, 2004).

γ-PGA may either be formed of L-glutamate residues, or D-glutamate residues, or a mixture of both, being called γ-L-PGA, γ-D-PGA, and γ-DL-PGA, respectively. Ivanovic and Bruckner discovered γ-PGA when their nutrient medium was accidentally contaminated by *Bacillus anthracis* while autoclaving it (Hanby & Rydon, 1946). Certain food preparations, like the "Natto" of Japan and the "Chung kook Jang" of Korea, are some other common "natural" sources of γ-PGA (Shih & Van, 2001). Besides cationic cofactors, the required substrates for the biosynthesis of PGA include ATP and glutamic acid, which can either be provided in the medium itself or maybe internally produced by the microbe in the course of its own tricarboxylic acid (TCA) cycle. The overall polymerization and cross-membrane transfer of the polymer have been attributed to the genes p*gs*BCA. γ-PGA is resistant to proteases and the synthesis is independent of the ribosomal machinery since the polymer is formed by amide-bond interactions between α-amino and γ-carboxylic moieties (Bodnár et al., 2008). γ-PGA has a varying molecular weight typically from 10 to 1000 kDa which in some cases can exceed up to 2000 kDa.

γ-PGA finds a diverse array of potential applications in food, cosmetics, medicine, and bioremediation. It is being used as a carrier for drug delivery, as bioflocculant, cryoprotectant, animal feed additive, for heavy metal removal, and in various other industries as a nontoxic biodegradable polymer. The biodegradable, water-soluble, edible, and nontoxic nature of γ-PGA has substantiated it as a feasible alternative to the traditionally used chemical-based polymers. This type of replacement becomes especially important in domains where direct contact with living organisms occurs, such as medical therapies. Microbial fermentation is the most well-established mode for γ-PGA production regardless of the high cost associated including substrates and process cost. The main emphasis of this review is on the microbial production of γ-PGA and the industrial production strategies used with a focus on the parameters influencing the process costs. The role of microbes and raw materials in controlling the production costs and the utilization of γ-PGA as an efficient biopolymer has also been discussed.

10.2 Microbial biosynthesis pathway

Biosynthesis of γ-PGA occurs in the late exponential or early stationary phase during nutrient limitation for cells (Kimura, Tran, Uchida, & Itoh, 2004). Glutamate residues are primarily required for the synthesis of the polymer (Fig. 10.1). Glutamate can either be sourced internally from the TCA cycle of the bacteria or externally using extracellular L-glutamic acid. The synthesis of the latter is enzymatically catalyzed by glutamine synthetase and glutamine-2-oxoglutarate aminotransferase (when glutamine is available) or by glutamate dehydrogenase (when glutamine is unavailable). Within the microbial intracellular environment, the biosynthesis of PGA is attributed to the exogenously or endogenously derived D- and L-glutamic acid units which further polymerize into γ-PGA. Endogenous synthesis of glutamic acid occurs via TCA intermediates wherein the immediate precursor molecule for glutamic acid biosynthesis is α-ketoglutarate (Ogunleye et al., 2015; Rehm, 2009). In the case of exogenous L-glutamic acid, enzyme glutamine synthase converts the L-glutamic acid to L-glutamine, a precursor of γ-PGA (Ogunleye et al., 2015). The biosynthesis of γ-PGA can be categorized into four different stages: racemization, polymerization, regulation, and degradation. The stereochemistry of γ-PGA changes with different strains and varying culture conditions. γ-PGA is generally produced from either or both D- and L- enantiomeric forms. In γ-PGA producers, the

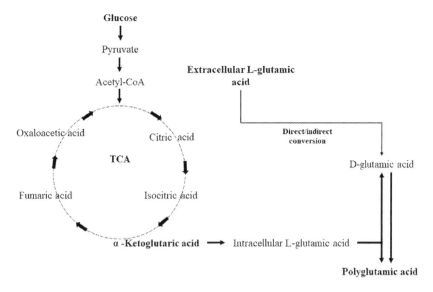

Figure 10.1
Biosynthesis pathway for polyglutamic acid.

conversion of L-glutamate into D-glutamate can either be indirect, with the help of an enzyme D-amino acid aminotransferase, or direct, with the involvement of glutamate racemase (Wu, Xu, Xu, & Ouyang, 2006). Ashiuchi et al. (2001b) brought light to the role of adenosine triphosphate on polymerization. Chain elongation occurs when amide linkage is formed between the amino group of glutamic acid and the terminal carboxylic group of γ-PGA only after the said carboxylic group has been phosphorylated by the phosphoryl group from the γ-phosphate of ATP (Ashiuchi et al., 2001b). pgs BCAE and cap BCAE are the γ-PGA synthesis genes, first recognized in *B. subtilis* and *B. anthracis* (Ashiuchi, Soda, & Misono, 1999; Candela, Mock, & Fouet, 2005). These genes are highly similar and have locations in the same γ-PGA operon. γ-PGA polymerization is regulated by the ATPase activity of pgs B and pgs C (cap B and cap C), whereas its transport is mediated by pgs A and pgs E (cap A and cap E) (Ashiuchi et al., 1999; Candela et al., 2005). In *B. subtilis*, a D, L-glutamyl hydrolase called pgdS breaks the glutamyl bond between D- and L-glutamic acids, degrading γ-PGA (Suzuki & Tahara, 2003). Hence the presence of pgs BCAE genes determine the mechanism of synthesis and transport of γ-PGA in most of the *Bacillus* species.

10.3 Process parameters for production

10.3.1 Substrate

The most widely used carbon sources for fermentative production of PGA include glucose, starch, glycerol, citric acid, and glutamic acid (Ito, Tanaka, Ohmachi, & Asada, 1996;

Ko & Gross, 1998). Other sources tested for the production of γ-PGA include cane molasses, agroindustrial wastes, rapeseed meal, xylose, soybean residue, fructose, and corncob fibers. Glycerol acts as one of the primary carbon sources for the fermentative synthesis of PGA as it not only stimulates PGA production but also decreases PGA chain length reducing the viscosity of fermentation broth (Wu, Xu, Liang, & Yao, 2010). Glycerol stimulates the enzyme polyglutamyl synthetase, which is responsible for the catalysis of glutamic acid polymerization to form PGA (Troy, 1973). Molasses, corn starch, and whey are cheaper carbon sources that can be used for PGA production. Untreated cane molasses and glutamic acid sourced from monosodium glutamate present in wastewater have also been utilized as a carbon source (Zhang, Feng, Zhou, Zhang, & Xu, 2012). The major drawback with substrates like molasses and whey is the additional cost associated with pretreatment which is necessary to avoid membrane fouling and process inhibition during production (Nayak & Pal, 2013).

The bacteria producing PGA are classified as a glutamic-acid-dependent and glutamic-acid-independent bacteria depending on whether the glutamic acid monomers have been obtained from medium or synthesized by microorganisms via TCA intermediates (Buescher & Margaritis, 2007; Shih & Van, 2001). Glutamic acid interacts with other medium components and is costlier than the rest of the medium constituents. Hence it is important to optimize its concentration to optimize PGA production in terms of both concentration and cost (Bajaj & Singhal, 2011). Many reports have suggested glycerol, citric acid, and glutamic acid as necessary medium components for γ-PGA production. Glycerol acts as cosubstrate (Du, Yang, Qu, Chen, & Lun, 2005) while citric acid and glutamic acid are precursors for polymerization (Shih & Yu, 2005; Shih, Van, & Chang, 2002). According to Du et al., glycerol as a cosubstrate greatly influences the synthesis of cell membrane phospholipids like phosphatidylethanolamine, phosphatidylserine, cardiolipin, and phosphatidylglycerol along with the reduction in ester-linked C16:1 and C18:1 fatty acids and the increment in C10:0 and C12:0 fatty acids resulting in an adequate release of γ-PGA (Du et al., 2005). When a glutamate-independent strain is employed for the fermentative production of γ-PGA using a medium with sodium glutamate, glucose, glycerol, soluble starch, alpha-lactose, maltose, sucrose, citric acid, and fructose, a concentration of about $33.84-35.34$ g L^{-1} of γ-PGA, with a molecular mass of over 10,000 kDa is recorded, significantly higher as compared to the previously used strains (Peng et al., 2015). Yeast extract has proven to be a high-quality source of nitrogen for the microbial fermentation of PGA, but its high cost has created a need to select alternative sources such as $(NH_4)_2SO_4$ or NH_4Cl (Ju, Song, Jung, & Park, 2014). Inorganic salts like Mn^{+2} also alter the stereochemical composition of γ-PGA along with enhancement in cell growth, viability, and overall γ-PGA production (Wu, Xu, & Xu, 2006).

Considering the increasing demands of γ-PGA, it is of prime importance to experts to search for cost-efficient as well as high-quality renewable sources of carbon as feedstocks.

Attempts have been made to use agroindustry wastes to produce γ-PGA through solid-state fermentation. Tang et al. utilized optimized ratio of residues of dry mushroom and monosodium glutamate with added waste glycerol as a carbon source supplement for γ-PGA production. Solid-state fermentation in batch mode resulted in a yield of 115.6 g kg^{-1} γ-PGA and further 107.7 g kg^{-1} γ-PGA was achieved by compost experiment indicating toward a novel method for efficient γ-PGA production (Tang et al., 2015a, 2015b). Lignocellulosic residues from forests and agriculture are used for the production of valuable chemicals projecting both economic and environmental significance. Certain species have been known to utilize D-xylose for γ-PGA production but in most cases use of alternative carbon source results in reduced production of target compound due to carbon catabolite repression (Kim, Block, & Mills, 2010; Yamashiro, Yoshioka, & Ashiuchi, 2011). Zhu et al. utilized xylose in batch and fed-batch fermentation using *Bacillus subtilis* HB-1 for γ-PGA production. Further, corncob fibers hydrolysate, a multiple sugar-containing alternative carbon substrate, was used for the production resulting in a high concentration of 24.92 g L^{-1} suggesting a cost-efficient, renewable lignocellulosic biomass substrate for γ-PGA production (Zhu et al., 2014).

10.3.2 Microbial consortia

Natrialba aegyptiaca, *Staphylococcus epidermidis*, *Lysinibacillus sphaericus*, *Bacillus mojavensis*, *Bacillus atrophaeus*, *B. licheniformis*, *B. subtilis*, *Fusobacterium nucleatum*, *B. anthracis*, *Bacillus thuringiensis*, *Bacillus cereus*, *Bacillus pumilus*, *Bacillus amyloliquefaciens*, and *B. megaterium* have been investigated to be chief microbial producers of PGA. γ-PGA has been known to be produced mostly by gram-positive bacteria. *Bacillus* species have been established as one of the highest γ-PGA producers giving concentration as high as 101.1 g L^{-1} (Huang et al., 2011) (Table 10.1). *B. licheniformis*, *B. subtilis*, *Bacillus pupils*, and *B. amyloliquefaciens* have been reported as the source of PGA. However, the *F. nucleatum* which is a gram-negative bacterium, some archaebacteria and eukaryotes also synthesize γ-PGA (Candela, Moya, Haustant, & Fouet, 2009; Hezayen, Rehm, Tindall, & Steinbüchel, 2001). The produced PGA is either retained by the cell or released into the environment. Halophiles such as *Planococcus halophilus*, *Sporosarcina*, and *N. aegyptiaca* have also been reported to produce PGA. They use γ-PGA to reduce the high salinity around them and survive in adverse conditions (Hezayen et al., 2001; Kandler, König, Wiegel, & Claus, 1983).

The glutamate-dependent bacteria show higher γ-PGA productivity with increased concentration of glutamate but this also increases the overall process cost, whereas glutamate-independent bacteria despite showing comparatively lower productivity are industrially more suitable. *B. subtilis* CGMCC 0833 (Xu et al., 2014) and *B. licheniformis* P-104 (Zhao et al., 2013) are glutamate-dependent while *B. subtilis* C1 (Shih & Yu, 2005)

Table 10.1: Yield of γ-PGA by *Bacillus* strains using different media components.

Strain	Key nutrients	γ-PGA yield (g L^{-1})	Reference
B. licheniformis ATCC 9945	Glutamic acid, glycerol, citric acid, NH_4Cl	23	Cromwick, Birrer, and Gross (1996); Cromwick and Gross (1995); Yoon et al. (2000); Birrer et al. (1994)
		17	
	Glutamic acid, glycerol, citric acid, NH_4Cl (Fed batch)	35	
B. subtilis ZJU-7	Glucose, l-glutamate, yeast extract, NaCl, $CaCl_2$, $MgSO_4$, $MnSO_4$	101.1	Huang et al. (2011)
B. licheniformis P-104	Glucose, sodium glutamate, sodium citrate, $(NH_4)_2SO_4$, $MnSO_4$, $MgSO_4$, K_2HPO_4	41.6	Zhao et al. (2013)
B. licheniformis NCIM 2324	Glycerol, l-glutamic acid, citric acid, $(NH_4)_2SO_4$, K_2HPO_4, $MgSO_4$, $MnSO_4$	35.75	Bajaj and Singhal (2009)
B. subtilis IFO 3335	Glutamic acid, citric acid	20	Goto and Kunioka (1992, 1994)
B. subtilis (chungkookjang)	Sucrose, $(NH_4)_2SO_4$, glutamic acid	15.60	Ashiuchi et al. (2001)
B. subtilis MJ80	Glutamic acid, starch, urea, citric acid, glycerol, NaCl, K_2HPO_4, $MgSO_4$, $MnSO_4$	68.7	Ju et al. (2014)
B. methylotrophicus SK19.001	Glucose, yeast extracts, $MgSO_4$, K_2HPO_4, $MnSO_4$	35.34	Peng et al. (2015)
B. lichenifomis CCRC 12826	Glutamic acid, glycerol, citric acid, NH_4Cl	19.80	Shih et al. (2002)
B. subtilis NX-2	Glutamic acid, glucose yeast extract, Glutamic acid, glycerol	30.20	Wu et al. (2006), Xu, Jiang, Li, Lu, and Ouyang (2005); Wu et al. (2008)
B. subtilis NX-2		31.7	
B. subtilis NX-2	Glutamate, $(NH_4)_2SO_4$, K_2HPO_4, $MgSO_4$, $MnSO_4$, and hydrolysis of rice straw	73.0	Tang et al. (2015)
B. subtilis NX-2	Glucose, glutamate, $(NH_4)_2SO_4$, K_2HPO_4, $MgSO_4$, $MnSO_4$	71.21	Xu et al. (2014)
B. subtilis NX-2	Cane molasses and monosodium glutamate waste liquor	52.1	Zhang et al. (2012)
B. subtilis (natto) MR-141	Maltose, soy sauce, sodium glutamate	35	Ogawa, Yamaguchi, Yuasa, and Tahara (1997)
B. licheniformis WBL-3	Glutamic acid, glycerol, citric acid, NH_4Cl	19.30	Du et al. (2005)
B. subtilis CGMCC 0833	Glutamic acid, $(NH_4)_2SO_4$, DMSO, Tween-80, glycerol	34.40	Wu et al. (2008)
B. subtillis HB-1	Glutamate, yeast extract, NaCl, $MgSO_4$, xylose, or corncob fibers hydrolysate	28.15	Zhu et al. (2014)
B. subtilis R 23	Glucose, citric acid, $(NH_4)Cl$, NaCl glutamic acid, a ketoglutaric acid	25.38	Bajaj, Lele, and Singhal (2009)

(*Continued*)

Table 10.1: (Continued)

Strain	Key nutrients	γ-PGA yield (g L^{-1})	Reference
B. licheniformis TISTR 1010	Glucose, citric acid, NH$_4$Cl, K$_2$HPO$_4$, MgSO$_4$, CaCl$_2$, MnSO$_4$, NaCl, Tween-80	27.5	Feng et al. (2014)
B. subtilis TAM-4	Ammonium chloride, fructose	22.1	Ito et al. (1996)
B. licheniformis A35	Glucose, MnSO$_4$, ammonium chloride	8.1	Cheng, Asada, and Aida (1989)
B. licheniformis SAB-26	Casein hydrolysate, KH$_2$PO$_4$ (NH$_4$)$_2$SO$_4$	59.90	Abdel-Fattah, Soliman, and Berekaa (2007)
B. subtilis RKY3	Glycerol, glutamic acid yeast extract, K$_2$HPO$_4$	48.5	Jeong, Kim, Wee, and Ryu (2010)
B. licheniformis NCIM 2324	Soybean meal, citric acid, glutamic acid (NH$_4$)$_2$SO$_4$, glycerol, L-glutamine, c-ketoglutaric acid	98.64	Bajaj and Singhal (2009)
B. subtilis ME714	Sodium glutamate, urea trisodium citrate, starch	75.30	Yong et al. (2011)
B. subtilis CCTCC202048	Soybean cake powder, wheat bran, glutamic acid, citric acid, NH$_4$NO$_3$	83.61	Xu et al. (2005)
B. subtilis CCTCC202048	Swine manure, soybean cake, wheat bran, glutamic acid, citric acid	60.00	Chen et al. (2005)
B. subtilis CCTCC202048	Dairy manure, wheat bran, soybean cake, glutamic acid	47.00	Yong et al. (2011)

and *B. amyloliquefaciens* LL3 (Ito et al., 1996) are some glutamate-independent bacteria. Da Silva et al. used 16s rDNA analysis to characterize *Bacillus subtilis* as the best γ-PGA producer strain among the five strains isolated (Da Silva, Cantarelli, & Ayub, 2014). γ-PGA found in the mucilage produced by *B. amyloliquefaciens* C06 enhances the bacterial colony structure, biofilm formation, and motility of cells (Liu et al., 2010). pgsB, pgsC, and pgsA genes are similar in glutamate-independent strains but similarity decreases when compared between glutamate-independent and glutamate-dependent strains.

Pathogenic bacteria like *B. anthracis*, *B. thuringiensis*, and *Staphylococcus epidermidis* are known to have surface peptidoglycan attached PGA which protects them from phagocytosis and inhibits entry of antibodies into the cells (Kocianova et al., 2005; Mesnage, Tosi-Couture, Gounon, Mock, & Fouet, 1998). However, γ-PGA attached to the peptidoglycan layer makes the recovery and purification tedious (Cachat, Barker, Read, & Priest, 2008). *B. anthracis* capsule is made up of D-enantiomer of PGA, which makes the bacteria nonimmunogenic (Candela & Fouet, 2006). To save themselves from starvation during the late stationary phase, some bacteria also use PGA bound with peptidoglycan as a source of glutamate (Kimura et al., 2004).

10.3.2.1 Strain improvement

Attempts have been made to increase γ-PGA production and to gain more information about the production mechanism using genetic engineering (Chang, Zhong, Xu, Yao, & Chen, 2013; Su et al., 2010), by inserting the genes concerned with production into glutamate-independent bacteria like *Escherichia coli* and plants such as tobacco (Cao et al., 2013; Tarui et al., 2005). Heterologous hosts have been used for the expression of γ-PGA-producing gene, *E. coli* being the most commonly employed host. Jiang et al. developed a novel host–vector system by cloning γ-PGA synthesizing genes namely pgsB, pgsC, and pgsA sourced from *B. subtilis* into *E. coli* cells using the inducible tac promoter and a derivative of the D-amino acid aminotransferase (D-AAT) gene of *Geobacillus toebii* that is a constitutive promoter P_{HCE}. The HCE promoter more efficiently increases γ-PGA production compared to trc, that too at a lower cost, providing a final γ-PGA concentration of 3.7 g L^{-1} in the fed-batch culture of metabolically engineered *E. coli* (Jiang, Shang, Yoon, Lee, & Yu, 2006). In a similar study, γ-PGA production was estimated after cloning and expression of the pgsBCA gene from *B. licheniformis* NK-03 and the racE gene from *B. amyloliquefaciens* LL3 in *E. coli* JM109 (Cao et al., 2013).

Genetic alterations to *B. subtilis* ISW 1214 were performed by creating *B. subtilis* MA41, which is a pgs BCA gene disrupting from *B. subtilis* ISW 1214. The disruptant did not produce PGA and instead, a recombinant with plasmid-mediated PGA synthesizing system was formed. The genetically engineered strain was able to produce PGA (reaching a maximum concentration of 9 g L^{-1}) in both L- and D-glutamic acid media and the production was stringently controlled by xylose (Ashiuchi, Shimanouchi, Horiuchi, Kamei, & Misono, 2006). Mutant strains of *B. subtilis* were developed by inserting sequence to control synthetic gene expression into the upstream region of ywsC-ywtAB genes of *B. subtilis* DB430 obtaining a high γ-PGA-producing mutant *B. subtilis* PGA6-2 (Yeh, Wang, Lo, Chan, & Lin, 2010). In another study, *B. subtilis* was transformed by incorporating the *Vitreoscilla* hemoglobin gene (vgb) within its chromosome to increase γ-PGA production. Expression of the *Vitreoscilla* hemoglobin gene enhanced the biomass yield by 1.26-fold under highly viscous conditions during fermentation, and γ-PGA yield by about 2.07-fold (Su et al., 2010). Modifications in the biosynthesis pathway for γ-PGA have also been done to enhance polymer production. Feng et al. (2015) engineered the metabolic pathway in *B. amyloliquefaciens* NK-1 strain by introducing several alterations such as blocking the by-product synthetic pathways, deleting the γ-PGA degrading enzyme genes, blocking the cell autoinducer synthetic pathway, or inhibiting the usage of the γ-PGA synthetic precursor. The altered NK-anti-rocG showed a 5.34-fold higher γ-PGA production than the original NK-1 strain indicating the estimated success of the modular pathway engineering approach for increasing PGA production (Feng et al., 2015). In a novel study by Jiang et al. with *B. licheniformis WX-02*, the glr gene which encodes for glutamate racemase was cloned and expressed in the bacteria to enhance γ-PGA production along with L- to D-glutamate conversion, increasing the D-glutamate proportion in the polymer from 77% to 85% (Jiang et al., 2011).

Zhang et al. increased γ-PGA production in *B. amyloliquefaciens* LL3 by analyzing the effects of the deletion of numerous genes responsible for glutamate metabolism. *gudB/rocG* double mutant resulted in a 40% rise in γ-PGA production than the wild-type strain while the *rocR* mutant showed an increase from 4.55 g L^{-1} in wild type to 5.83 g L^{-1} in the mutant (Zhang et al., 2014). Table 10.2 summarizes the genetic engineering approaches used for homologous and heterologous hosts to enhance γ-PGA production. Knockout of γ-PGA degrading enzyme coding genes in *B. subtilis*-derived strains could enhance the

Table 10.2: Genetic engineering approaches for bacterial modification and γ-PGA production.

Strains	Engineering methods	Fermentation medium	Yield (g L^{-1})	Reference
B. licheniformis WX-02	Expression of glr gene encoding glutamate racemase	Glucose, l-glutamic acid, sodium citrate, NH$_4$Cl, MgSO$_4$, K$_2$HPO$_4$, CaCl$_2$, ZnSO$_4$, MnSO$_4$	14.38	Jiang et al. (2011)
B. amyloliquefaciens LL3	Double-deletion of genes pgdS and cwlO	Sucrose, (NH$_4$)$_2$SO$_4$, MgSO$_4$, KH$_2$PO$_4$, K$_2$HPO$_4$	7.12	Feng et al. (2014)
B. amyloliquefaciens LL3	Deletion of genes roc (R, rocG, gudB, odhA)	Sucrose, (NH$_4$)$_2$SO$_4$, MgSO$_4$, KH$_2$PO$_4$, K$_2$HPO$_4$	5.68	Zhang et al. (2014)
B. subtilis ISW1214	Bearing the plasmid-borne PGA	Sucrose, NaCl, MgSO$_4$, KH$_2$PO$_4$, NaHPO$_4$, xylose	9.0	Ashiuchi et al. (2006)
E. coli BL21	Cloning and overexpressing γ-PGA biosynthesis genes	Glucose, yeast extract, l-glutamic acid, (NH$_4$)$_2$SO$_4$	3.7	Jiang et al. (2006)
E. coli JM 109	Co-expressing γ-PGA synthetase and glutamate racemase	LB medium supplemented with l-glutamate or glucose	0.65	Cao et al. (2013)
B. amyloliquefaciens	Deletions of genes (epsA-O, sac, lps, pta, pgdS, cwlO, luxS, androcG gene), expression of synthetic small regulatory RNAs which repressed the rocG and glnA gene	Sucrose, (NH$_4$)$_2$SO$_4$, MgSO$_4$, KH$_2$PO$_4$, K$_2$HPO$_4$	20.3	Feng et al. (2015)
B. subtilis PB5249	Knockout of genes (pgdS and ggt)	l-glutamic acid, citric acid, glucose, NH$_4$Cl, K$_2$HPO$_4$, MgSO$_4$·7H$_2$O, FeCl$_3$·6H$_2$O, CaCl$_2$·2H$_2$O, MnSO$_4$·H$_2$O	40	Scoffone et al. (2013)
Bacillus licheniformis WX-02	Enhanced expression of pgdS gene	Glucose, sodium glutamate, sodium citrate, NH$_4$Cl, MgSO$_4$, K$_2$HPO$_4$, CaCl$_2$, ZnSO$_4$, MnSO$_4$	20.16	Tian et al. (2014)

productivity of γ-PGA achieving concentrations comparable to that produced by wild strains (Scoffone et al., 2013).

10.3.3 Bioreactors mode of operation for production

PGA is a high molecular weight polymer released extracellularly, increasing the viscosity of the fermentation medium which as a result exhibit non-Newtonian rheology. This in turn affects the oxygen transfer rate creating the problem of oxygen limitation. It can be overcome by increasing the rate of sparging and agitation, which would not only increase the oxygen transfer rate but also enhance the uniform mixing of culture nutrients. An increase in aeration and agitation rate also reduces the molecular weight, which could be attributed to the higher activity of depolymerizes when aeration rate is higher (Birrer, Cromwick, & Gross, 1994) and also to the stronger shear forces due to the increased agitation rate (Bajaj & Singhal, 2011). The large-scale synthesis of γ-PGA has been conducted using various bioreactor operation modes including batch, fed-batch with solid-state, and submerged state fermentation strategies. Batch mode processes achieve a higher γ-PGA yield and productivity and are industrially preferred for PGA fermentations.

Richard and Margaritis analyzed the batch fermentation kinetics, major mass transfer parameters like oxygen-uptake rate, specific oxygen-uptake rate, volumetric oxygen mass transfer coefficient (kLa), and broth rheology parameters such as flow index, consistency index, and apparent viscosity during production of PGA by *B. subtilis* IFO 3335. The microorganisms showed maximum oxygen demand during the early exponential phase before an increase in broth viscosity and PGA concentration while kLa achieved a maximum of 154 h^{-1} during the exponential growth phase which reduced to 134 h^{-1} when the broth viscosity was highest. Also, the fermentation broth displayed non-Newtonian and pseudoplastic rheology during the late exponential and early stationary phase supporting the former observation (Richard & Margaritis, 2003). da Silva et al. used central composite design and response surface methodology to optimize various fermentation parameters including culture and medium composition. *B. subtilis* BL53 was found to be the most productive strain exhibiting a threefold increase in production at optimum conditions of pH 6.9, 37°C, and 1.22 mM Zn^{2+} in shaker cultivations (Da Silva et al., 2014). The concentration of PGA was further increased by 70% in a bioreactor at predetermined agitation rates and kLa. The production of PGA reflected high dependency on oxygen transfer as the cultivation time was reduced to half at kLa of 210 h^{-1} than the initial non-optimized operations at kLa 55 h^{-1} (Da Silva et al., 2014).

To overcome the problem of oxygen limitation, oxygen vectors which are hydrophobic liquids are used as a new approach to enhancing the oxygen supply and solubility. These oxygen vectors show a 15–20 times higher solubility for oxygen as compared to water (da Silva, Reis, Roseiro, & Hewitt, 2008). These vectors have been used in processes where the production of a compound increases the viscosity of fermentation broth limiting the oxygen transfer (Da Silva et al., 2006; Liu et al., 2009). Zhang et al. analyzed the effects of n-heptane and n-dodecane additives and found explicit improvement in kLa and dissolved oxygen concentrations during the PGA fermentation process of *B. subtilis* NX-2. However, in the presence of n-dodecane, intracellular NADH/NAD$^+$ and ATP levels of *B. subtilis* NX-2 were low resulting in both decreased yield and molecular weight of PGA (Zhang, Feng, Li, Chen, & Xu, 2012).

In the case of fed-batch operation, the nutrient-feeding methods are important to the success of the culture. In the case of fed-batch culture, pH and dissolved oxygen concentration increase with the depletion of carbon sources (Lee, Lee, Park, & Middelberg, 1999). However, in the fed-batch culture of certain microorganisms like *B. licheniformis*, the depletion of carbon sources display no effect on the pH and concentration of dissolved oxygen (Yoon, Do, Lee, & Chang, 2000). Fermenters with the provision of temperature control, pH control, and oxygen gas purging systems are installed for commercial production of γ-PGA (Fig. 10.2). Agitation and oxygen-level control are primarily important for efficient γ-PGA production. In the case of membrane-integrated systems,

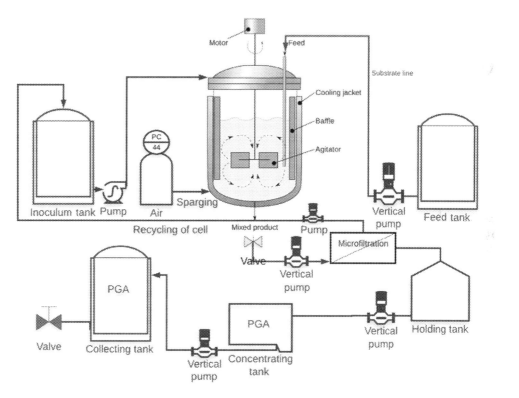

Figure 10.2
Process flow diagram for the production of polyglutamic acid.

pressure gauges are attached at the inlet and outlet of modules to maintain transmembrane pressure. Peristaltic pumps and valves are installed for the supply of inlet streams and proper removal of outlet streams. Microfiltration unit ensures clarification of permeate from cells to ease further downstream processing of the product.

Immobilization is a technique that could be employed for the enhancement of γ-PGA production in addition to enhanced biomass growth and inhibition of cell loss during fermentation processes (Zhang & Yang, 2009). Xu et al. constructed a novel aerobic plant fibrous-bed (APFB) bioreactor for immobilization of B. subtilis NX-2 to enhance γ-PGA fermentative production. The cells were immobilized in bagasse and repeated fed-batch fermentation evaluated the stability of APFB achieving a high PGA concentration of 71.21 g L^{-1} (Xu et al., 2014). The type of matrix used for cell immobilization affects the production capacity of cells. Berekaa, El Aassar, El-Sayed, and Borai (2009) studied the performance of different gel matrices for immobilization of B. licheniformis strain R cells and found maximum PGA production (36.75 g L^{-1} after 96 h) when cells were entrapped in agar-alginate gel beads mixture during batch fermentation.

PGA was produced in semicontinuous mode using trickle flow bioreactor by adsorption of B. licheniformis cells on sponge cube as a matrix which resulted in the highest cell adsorption with 43.2 g L^{-1} PGA (Berekaa et al., 2009). Jiang et al. (2016) were the first to analyze the dynamics of γ-PGA fermentation by B. subtilis NX-2 using a moving bed biofilm reactor (MBBR). Use of polypropylene TL-2 as a carrier and DO stat feeding strategy, γ-PGA concentration of 42.7 ± 0.86 g L^{-1} was achieved in a batch system (Jiang et al., 2016). Moreover, execution of repeated fed-batch cultures for γ-PGA production displayed higher stability of MBBR achieving concentration up to 74.2 g L^{-1} suggesting the potential of MBBR to be used for γ-PGA production at industrial scale (Jiang et al., 2016).

10.3.4 Isolation, analysis, and determination of PGA

The recovery and purification of PGA from fermentation broth are generally carried out in three steps, the foremost being the extraction of biomass using various techniques like filtration or centrifugation, secondly the extraction of the final product by precipitating it out from the cell-free supernatant using solvents like ethanol and methanol and lastly, the removal of low molecular weight impurities by dialysis (Birrer et al., 1994; Goto & Kunioka, 1992). Properties such as solubility, molecular weight, charge, and affinity to adsorbent are used to recover PGA from the culture broth (Hong & Bruening, 2006; Timmer, Speelmans, & Van Der Horst, 1998). The recovery of γ-PGA can be carried out either by precipitating it through complex formation or by decreasing the solvent (water) solubility of γ-PGA and then subsequently filtering it out (Buescher & Margaritis, 2007). Complex formation is done using metal ions like Cu^{2+}, Al^{3+}, Cr^{3+}, and Fe^{3+} among which

Cu^{2+} most efficiently precipitates γ-PGA (McLean, Beauchemin, Clapham, & Beveridge, 1990). The higher efficiency of Cu^{2+} is supported by Manocha and Margaritis who recovered and purified γ-PGA by Cu^{2+} precipitation followed by dialysis to obtain purified biopolymer. When compared to ethanol precipitation (82% recovery), Cu^{2+}-induced precipitation recovered 85% of γ-PGA from the broth and ethanol precipitation also leads to a higher degree of coprecipitation of other proteins along with PGA from the fermentation broth (Manocha & Margaritis, 2010).

The amount of alcohol required for precipitation decreases as the concentration of PGA increases marking the need for concentrating PGA culture broth during recovery. Concentrating 20 g L^{-1} of PGA solution to 60 g L^{-1} at pH 5 reduced the volume of ethanol required for precipitation by one-fourth (Do, Chang, & Lee, 2001). Filtration and buffer exchange is used as additional steps to efficiently separate the high molecular weight γ-PGA from the rest of the constituents in culture broth (Yoon et al., 2000).

The pH of the surrounding medium affects the separation efficiency of cells. The negative charges present on the cell surfaces significantly decrease at low pH, resulting in an aggregation of cells together. Hence the energy required for separation postacidification is not more than 17% of what is required sans acidification (Do et al., 2001). Immobilization of cells can reduce the cost associated with frequent centrifugation needed to separate the cells from the fermentation broth (Xu et al., 2014).

Membrane-based processes have not been used much for the downstream processing of γ-PGA although they have been extensively used for separation and purification of many other fermentation products (Cui & Muralidhara, 2010; Singh, Ingole, Bajaj, & Gupta, 2012). Membrane-based processes offer great flexibility of process scale and design at low capital cost. Kumar et al. developed a novel, eco-friendly, compact, and less energy-consuming hybrid reactor system with a conventional fermentation process for continuous γ-PGA production (at 0.91 g L^{-1} h^{-1}), and used membranes for downstream separation and purification (Kumar & Pal, 2015). Almost complete cell separation and recycle were attributed to the cross-flow microfiltration membrane modules (with very less fouling) and ultrafiltration membrane modules, obtaining highly concentrated γ-PGA, simultaneously ensuring the recovery and re-utilization of about 96% of the unused carbon source. This enhanced the purity of the resultant polymer, and yields as high as 0.6 g g^{-1} was observed (Kumar & Pal, 2015).

Recovery and purification of PGA after solid-state fermentation requires the removal of an insoluble substrate along with the cells. In the initial attempts for PGA recovery after solid-state fermentation, the fermented matter was mixed with distilled water using a rotary shaker (150 rpm for 1 hour) and filtered by muslin cloth. The filtrate was centrifuged at 12,000 rpm for 20 minutes separating the supernatant with PGA. The PGA in the supernatant was further precipitated using four volumes of cold ethanol and the other insoluble were removed by dialysis (Chen, Chen, Sun, & Yu, 2005). Similar extraction strategies with the mixing of fermented matter, filtration by muslin cloth, centrifugation, and ethanol precipitation have been used for the separation of γ-PGA (Jian, Shouwen, & Ziniu, 2005; Yong et al., 2011).

10.3.5 Structure of γ-polyglutamic acid

Polymerization by the γ-amide linkage between D- and/or L-glutamate residues results in the formation of γ-PGA (Ashiuchi & Misono, 2002), which is anionic and shows high optical activity due to the presence of chiral centers in each monomeric unit. γ-PGA exists in five different spatial conformations, namely, α-helix, β-sheet, helix-to-random coil transition, random coil, and enveloped aggregate (Ho et al., 2006). Factors such as pH, the concentration of polymer, and the ionic strength can influence and alter these conformations (Ho et al., 2006). Naturally produced γ-PGA often has both D- and L-glutamic acids; *B. anthracis* being an exception is known to produce only the optical pure D-form of the polymer (Bruckner, Kovács, & Dénes, 1953). The polymer generally appears as α-helix at pH 7, and at higher pH, the conformation is predominantly β-sheet (Bhat et al., 2013). The poly-[γ-D-glutamic acid] takes helical conformation in a unionized state and randomly coil in an ionized state.

The left-handed helix of the polymer is the most stable conformation due to intramolecular hydrogen bonds between side-chain carboxylic oxygen and -NH of the amide groups in the backbone. The conformation of γ-PGA can also be changed during the extraction process following fermentation. When only either L- or D-enantiomer [homopolymer] is present after fermentation, γ-PGA is soluble in ethanol whereas in cases when equal moles of L- and D-enantiomers are present, ethanol causes precipitation of γ-PGA (Candela & Fouet, 2006). The molecular mass of γ-PGA is also an important parameter that affects the property, potential, and efficacy of the polymer. γ-PGA obtained from microbial sources often has a high molecular weight, which creates hindrance during industrial applications due to high viscosity and altered fluid rheology (Shih & Van, 2001). γ-PGA produced by *B. subtilis* (natto) has a molecular weight ranging from 10 to 1000 kDa whereas γ-PGA with a molecular weight greater than 2000 kDa can be obtained from *B. subtilis* (chungkookjang) (Ashiuchi et al., 2001a). Modifications in the molecular weights of γ-PGA have been achieved by several methods such as alkaline hydrolysis, ultrasonic degradation, microbial degradation, and changing medium constitution (Shih & Van, 2001).

10.4 Characterization of polyglutamic acid

Once the γ-PGA is purified, its characterization is usually done by amino acid analysis, gel permeation chromatography for molecular mass determination, thin layer chromatography,

NMR spectroscopy, and Fourier transform infrared spectroscopy (Shih, Van, Yeh, Lin, & Chang, 2001). Irradiation with ultrasonic waves cause molecular weight reduction of naturally produced γ-PGA and decrease the polydispersity of the polymer without affecting the chemical constitution (Pérez-Camero, Congregado, Bou, & Muñoz-Guerra, 1999). Yokoi et al. performed an amino acid analysis of γ-PGA by hydrolyzing the biopolymer with 6 N HCl at 100°C in the presence of argon. The residual HCl was evaporated and the biopolymer was dissolved in distilled water followed by paper chromatography and thin-layer chromatography for analyzing amino acid composition, which was found to be only glutamic acid (Yokoi, Natsuda, Hirose, Hayashi, & Takasaki, 1995). γ-PGA produced by *B. licheniformis* was analyzed for amino acid composition by hydrolysis with 6 N HCl at 110°C in a sealed tube for 24 h and composition was determined by the amino acid analyzer (Shih et al., 2001). The component amino acid was found to be glutamic acid and the result was further validated by thin layer chromatography in which a single spot with similar R_f value to that of glutamic acid was obtained (Shih et al., 2001).

The homogeneity and degree of esterification of γ-PGA can be determined by ^{1}H- and ^{13}C-NMR spectroscopy (Birrer et al., 1994; Borbély et al., 1994), and the chemical shifts that are observed in the spectra obtained as a result are measured and compared with standard values. Fourier transform infrared spectroscopy (FTIR) produces infrared spectra that consist of peaks by specific bonds in γ-PGA, thus identifying the compound. IR spectra of the free acid form of γ-PGA and its derivative salts in pellets of potassium bromide show characteristically heightened amide absorption at $\sim 1620-1655$ cm^{-1}, a lower carbonyl C=O absorption at $\sim 1394-1454$ cm^{-1}, a higher hydroxyl absorption at $\sim 3400-3450$ cm^{-1}, and a characteristic strong C–N groups absorption in the range from 1085 to 1165 cm^{-1} (Ho et al., 2006). The absorption peaks between 2900 and 2800 cm^{-1} are characteristic of aliphatic N–H stretching, while those around 1600–1660 and 1390–1450 cm^{-1} exhibit characteristics of amide groups and C=O groups, respectively (Ho et al., 2006).

10.5 Commercial production

γ-PGA production that has been researched and published to date is inclusive of microbial, chemical, peptide, biotransformational, and fermentation methods of synthesis. Commercial production of γ-PGA in consideration of cost-effective measures has been made feasible by biomass microbial fermentation, primarily due to inexpensive raw materials and minimal hazards as far as the environmental concerns are considered. Optimization of the fermentation process may lead to better productivity, especially through the screening of efficient microbial strains. Large-scale applications of γ-PGA are only feasible when the industrial production of this soluble microbial product has new cost-effective avenues available (Chettri & Tamang, 2014).

Selection of cheaper substrates and identifying producers with high productivity and conversion efficiency could only possibly subserve the rate of industrial production of this biopolymer (Ogunleye et al., 2015). Structurally PGA exists in two isomeric forms, poly-α-glutamic acid and poly-γ-glutamic acid (Bajaj & Singhal, 2011). Poly-γ-glutamic acid is a soluble microbial product known to have random conformation at pH equal to or more than 7. The structure is primarily α-helical at low pH while β-sheets are predominant at alkaline pH levels as revealed by circular dichroism spectra and optical rotatory dispersion (Wang et al., 2017). Figure 10.3 implicates the role of microbial enzymes in the biotransformation of glutamic acid isomeric forms, the course of racemization takes over the interconversion of L-alanine to D-alanine.

The direct conversion of α-ketoglutaric acid into L-glutamic acid occurs via PGA synthetase or alanine aminotransferase. Hence the specific route to PGA synthesis is entirely based on the cost-effective process, emphasizing on cutting down the cost of substrates. *B. subtilis* ZJU-7 has proven to be most productive amongst other members of the microbial fraternity in the context of PGA production under optimized fermentation conditions (Table 10.1). Most of the cases include bacterial isolation from the soil with common media formulation ingredients, except for the use of yeast extracts and surfactants. However, the process controls have set various paradigms of PGA production; enhanced yield. *B. licheniformis* WX-02 fermentations under alkaline pH treatment regime yielded 36.26 g L^{-1} γ-PGA (Hezayen et al., 2001). A comprehensive understanding of the biological and chemical parameters can help in the optimization of the fermentation

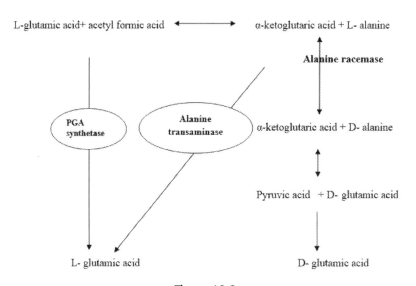

Figure 10.3
Flowchart of implicated biosynthetic routes for commercial γ-PGA synthesis.

process, especially ambient pH and dissolved oxygen concentration which is a prerequisite in aerobic fermentation and influences cellular proliferation. NAD(P)H recycling, product biosynthesis, and carbon source utilization also play a vital role in the overall γ-PGA yield (Hezayen et al., 2001).

For the adequate rates of oxygen supply, the deployment of efficacious strategies such as the provision of oxygen-deficient and oxygen-enriched air where oxygen serves as an electron acceptor, use of oxygen vectors and altering the impeller design can be useful. 0.3% n-heptane in B. subtilis batch fermentation process for the synthesis of γ-PGA, replacing gaseous oxygen as an electron acceptor, increased yield to 39.4 g L^{-1} γ-PGA with a molecular weight 19.0×10^5 Da (Zhang et al., 2012). It reduced the production costs, hence significant as far as the economics of the process is concerned. B. subtilis IFO 3335 at pH 7 have the highest biomass and γ-PGA yield (Shih & Van, 2001). B. licheniformis ATCC 9945A fermentations displayed rapid cellular proliferation at pH 6.5, indicating toward the vitality of an optimum pH level (Richard & Margaritis, 2003).

10.5.1 Production cost

Batch and fed-batch are the most conventional strategies adopted for commercial production γ-PGA (Luo et al., 2016). In cases where glutamate-dependent strains are used for γ-PGA production, the high production cost associated with the addition of exogenous L-glutamate still poses an impediment in the mass synthesis of γ-PGA. Cellular membranes of microbes possess polymerase which helps in the synthesis of γ-PGA that is excreted into the culture medium. Primarily, higher molecular weight ranging from 100 to 2000 kDa (Zhao et al., 2013) renders much of viscosity to fermentation broth, considering the nature of the polymer. This vicious nature of the polymer hinders the sedimentation serving as a grievance in the cell separation from γ-PGA. The unmanageable rheology of microbially derived γ-PGA limits industrial implications due to aggravated viscosity and difficult modification, with relatively high molecular weight (approx. Mw $\sim 10^5 - 8 \times 10^6$ Da) (Shih & Van, 2001). One of the probable options is to research renewable sources serving to replenish the necessity of raw material for enhancing productivity (müslüm altun, 2019). Agroindustrial wastes, food industry wastes, and horticultural scrap as cheaper substrates are few such substrates. For ameliorating the yield via fermentation, low-cost feathers could be utilized on a large scale, which is a waste byproduct of the processing in the poultry industry (müslüm altun, 2019). Since 90% constituent of feathers is keratinous, its recalcitrant structure can undergo degradation through their keratinolytic proteases or keratinases of bacterial origin (müslüm altun, 2019). Feather keratinous nature could be exploited to produce peptones or protein hydrolysates to supplement the need of cheaper nitrogen source vis-à-vis on the higher side as far as the conventional nitrogen sources expenses. The deployment of enzymatically synthesized feather hydrolysates has substantiated the commercial scale production of γ-PGA (müslüm altun, 2019).

Optimal process design could only be achieved via an appropriate mode of reactor operation. Continuous steady-state reactors could help achieve higher productivity due to the continuous feeding (Luo et al., 2016); however, it has its limitations associated as the probability of contamination lurks over the process and can hamper the quality of the product to a great extent. It has generally been observed that batch and fed-batch modes of fermentation are the most commonly used strategies, the batch method being more so because of the comparatively high productivity and desirable yields obtained. Thus it can be said that batch fermentation is the most feasible and promising method for the mass production of the polymer (Luo et al., 2016). Symbiotic fermentations of *B. subtilis* and *Corynebacterium glutamicum*, thus integrating bioprocesses utilizing sucrose and glucose as cocultured media, provisioning the adequate source of carbon nutrition for both the species, prevented the exogenous addition of L-glutamic acid. It concludes the fact that these coculturing fermentation strategies reduced fermentation overall duration as well as slashed down the burden of expensive substrates, reducing the production cost. The average molecular mass of γ-PGA thus produced was found to be 1.24×10^6 Da (Xu, Shi, & Cen, 2005). Downstream processing of PGA under the implication of the submerged fermentation process leads to the exogenous generation of PGA. Polymerization of N-carboxy anhydride of L-glutamic acid initiated by a nucleophile, yields to the chemical synthesis α-PGA, vis-à-vis microbial generation of γ-PGA (Ogunleye et al., 2015).

Researchers have rigorously investigated about the nutritional supplementation to ameliorate cellular proliferation, in turn, increasing the γ-PGA productivity and revealed that yeast extract could prove to be a vital nitrogen source for bacterial cell proliferation of exogenous glutamate-independent producer, thus minimizing the potential barrier of ever so burgeoning production cost of γ-PGA (Peng et al., 2015). Alteration of carbon and nitrogen sources could impact γ-PGA productivity as the incorporation of $(NH_4)_2SO_4$ or NH_4Cl can minimize production costs. Cellular proliferation with prolonged viability can be made feasible by switching to Mn^{2+}, aids in metabolizing various available carbon sources by modifying the enantiomeric and stereochemical composition of γ-PGA (Shih & Van, 2001; Wu et al., 2006). Amino acids are essential growth regulators assisting in the biosynthesis of metabolic outputs. The addition of aspartic acid and phenylalanine ameliorated the yield of γ-PGA up to 10% whereas glutamic acid addition rendered an approximate increment of 5%–10% in the yield of γ-PGA, speaking volumes of the significance of media optimization and maneuvering the growth condition according to the efficacious deployment of strategies (Zhang, Wu, & Qiu, 2018).

10.6 Applications

10.6.1 Healthcare

In recent years, γ-PGA and its derivatives have been employed for a diverse array of applications, such as metal chelators, bioadhesives, drug carriers as well as in tissue engineering. The biocompatibility and biodegradability of PGA render it important for use in drug delivery. Drugs such as chemotherapeutic agents might attach to the side-chain carboxyl groups of PGA thus becoming more soluble and easily release near tumor sites leaving the biodegradable polymer. PGA degrades to give glutamic acid which may enter the normal cellular metabolism without exhibiting any adverse effect on the system (Singer, 2005). An alternative to conventional Paclitaxel formulation was created by covalently binding the drug to PGA (PG-TXL) (molecular weight $\sim 3-6 \times 10^4$ Da) thus enhancing the pharmacokinetic profile and stability of the drug (Li et al., 1998). PG-TXL shows higher antitumor activity against mice, human tumor xenografts, and ovarian and breast cancer in animal models. Nanoparticles of PGA coupled with L-phenylalanine have been developed to test their efficiency in the treatment of retinal diseases. A detailed analysis revealed the regulatory effects of the said nanoparticles on the inflammatory phagocytic cells in the retina marking their use for delivery of drugs to the retinal cells on a long-term basis without posing any complications as caused by steroids (Ryu et al., 2011).

Potential applications of PGA nanoparticles is their use as delivering agents for tumor vaccines, as well as for antigenic proteins that are to be delivered to the antigen-presenting cells, to subsequently evoke immune responses by cytotoxic T-lymphocytes. A method to create uniform γ-PGA nanoparticles with the introduction of L-phenylalanine ethyl ester as a hydrophobic residue into the carboxyl group at alpha position γ-PGA was developed by Nakagawa. When tumor-infected mice were subcutaneously administered with ovalbumin entrapped γ-PGA nanoparticles, the growth inhibition of ovalbumin-transfected tumors was more efficient as compared to similar tests using ovalbumin emulsified with Freund's complete adjuvant, without any alterations in the overall histopathology (Nakagawa, 2008). The potential of γ-PGA nanoparticles as an antigen-delivery system is fairly established by the outcomes of this study. A similar study involving immunization with ovalbumin carrying γ-PGA nanoparticles showed expansion of antigen-specific CD^{8+} T-cells and the nanoparticles could induce either cellular or humoral immune response depending on the route of administration (Uto et al., 2009). In mice bearing major histocompatibility complex class-I deficient tumors, the oral administration of γ-PGA with molecular mass 2×10^6 Da cause induction of natural killer cell-mediated antitumor immunity (Kim et al., 2007). This suggests that γ-PGA can be efficiently used in cancer immunotherapy as it is a better stimulator of antitumor effects in mice than β-glucan (curdlan).

Biological adhesives are used for tissue adhesion, hemostasis, and to seal air and body fluid leaks during surgery. Fibrin is one such adhesive presently used which is generally unable to properly adhere to the tissues. Cross-linked gelatin and PGA is being attempted to be used as an alternative surgical adhesive and hemostatic agent known to degrade in animal models without causing any severe inflammations.

In the case of tissue engineering, chitosan is the most widely employed polymer but requires higher mechanical strength. PGA added to chitosan improved the mechanical strength, surface hydrophilicity, and water absorption rate of the biomaterial creating better tissue engineering matrices (Hsieh, Tsai, Wang, Chang, & Hsieh, 2005).

10.6.2 Personal-care products

The properties of personal-care products like creams and exfoliants can be improved with the help of PGA. The composition of PGA makes it uniformly miscible and stable when combined with different ingredients of facial creams (Ben-Zur & Goldman, 2007). It is a moisturizing hydrophilic humectant that increases the production of pyrrolidone carboxylic acid, urocanic acid, and lactic acid, which are all natural moisturizing agents. The multibag like the structure of PGA hydrogel allows it to retain moisture 5000 times its weight. PGA also aids in maintaining skin elasticity by inhibiting hyaluronidase enzyme. The water-retention capacity of PGA hydrogel allows it to protect skin and hair against external chemical effects (Ben-Zur & Goldman, 2007).

10.6.3 Food industry

γ-PGA is used in the food industry as a food supplement, oil-reducing agent, texture enhancer, agent for osteoporosis prevention, and cryoprotectant. As a food supplement, PGA has been reported to increase solubility and intestinal absorption of calcium by inhibiting the formation of insoluble calcium phosphate even in postmenopausal women (Tanimoto et al., 2007). PGA has also been reported to enhance mineral uptake in the small intestine in most humans and animals when provided in the diet. In animals, it especially enhances phosphorus uptake, results in harder eggshells, and the effects of body-fat loss are retained (Tanimoto et al., 2007). γ-PGA reduces oil in deep-fried edible products as reported by Lim et al. who analyzed a fivefold reduction in oil uptake by doughnuts with γ-PGA as compared to regular doughnut when fried for 4 minutes (Lim et al., 2012). Also, viscosity, foam stability, and emulsion stability of cake batter increased due to the addition of 0.5 g γ-PGA kg^{-1} batter (Shyu & Sung, 2010).

10.6.4 Bioremediation

PGA has the property of binding to several metal ions like Ni^{2+}, Cu^{2+}, Mn^{2+}, and Al^{3+} and can be used to remove and recover heavy metals from wastewater. Microfiltration membranes with bound PGA have been developed for the removal of heavy metals

(Bhattacharyya et al., 1998). γ-PGA obtained from *B. licheniformis* ATCC 9945 was analyzed for the removal of copper from aqueous solution (Mark, Crusberg, DaCunha, & Di Iorio, 2006). The adsorption capacity of γ-PGA for Cu^{2+} at pH 4 and 25°C was about 77.9 mg g^{-1} with a binding constant of 32 mg L^{-1} (Mark et al., 2006). The adsorption of mercury (II) on γ-PGA has also been estimated to find a way to remove mercury ions from wastewater in which pseudo-second-order kinetics described rapid adsorption of about 90% within initial 5 minutes (Inbaraj, Wang, Lu, Siao, & Chen, 2009). The complex of γ-PGA with Pb^{2+} as a novel biodegradable nanoparticle could be used as a promising sorbent for heavy metal removal from wastewater (Bodnár et al., 2008). The potential of γ-PGA/Fe_3O_4 nanoparticles in the remediation of various metallic ion pollutants like Cr^{3+}, Cu^{2+}, Pb^{2+}, and Ni^{2+} shows remarkable removal efficiency than the use of Fe_3O_4 nanoparticles or γ-PGA individually, showing the strong potential of these nanoparticles in wastewater treatment (Chang et al., 2013).

Besides bioabsorption, PGA has also been known to behave as a biodegradable bioflocculant for various organic and inorganic compounds. Shih et al. used *B. licheniformis* CCRC 12826 for the production of biopolymer flocculant later characterized as PGA (Shih et al., 2001). Similarly, a high degree of flocculation was shown by PGA from *B. subtilis* R-23 (Bajaj & Singhal, 2011). Taniguchi et al. used γ-ray to create cross-linked γ-PGA. It showed high viscosity with increasing irradiation dose and maximum water absorption capacity of 1005.6 mL g^{-1}. The flocculating and water absorption activity of cross-linked γ-PGA further increased with the addition of polyaluminum chloride up to 80°C which could help to treat polluted water (Taniguchi et al., 2005). Overall, the flocculating activity of PGA is influenced by factors like the concentration of PGA, the addition of cations and anions, pH, and temperature of reaction mixtures. PGA can be used in various water-cleansing operations including wastewater treatment, processing of potable water, and in various downstream processes that are involved in the food and fermentation industry.

10.6.5 Other applications

Preparation of different enantiomeric isomers, metal salts, and various molecular sizes of γ-PGA (produced by *B. licheniformis* CCRC 12826) was performed to analyze their antifreeze activity (Shih & Yu, 2005). The antifreeze activity of γ-PGA increases with its decreasing molecular weight and increasing ionic charge, being highest for Mg^{2+} followed by Ca^{2+}, Na^+, K^+, and inorganic chlorides, respectively (Shih, Van, & Sau, 2003). As a cryoprotectant, γ-PGA from *B. subtilis* natto ATCC 15245 shows a significant increment in the shelf life of *Lactobacillus paracasei*, as compared to the traditional treatment with 10% sucrose, without any potential toxic effects by the source bacteria *B. subtilis* var. natto (Bhat et al., 2013).

PGA has been anticipated as a suitable vector for gene delivery due to the fairly high stability of PGA and DNA complex, which could establish its potential application in gene therapy. The PGA−DNA complex has stability toward serum albumin and can simultaneously transfect 293 cells (Dekie et al., 2000). Nanoparticles of PGA can be a finer gene carrier as the PGA derivatives form polyelectrolyte complexes with DNA, reducing the surface charge and size of DNA and also protecting the DNA from DNAse I digestion (Buescher & Margaritis, 2007).

\PGA can also be used as a microencapsulation carrier. Efficient encapsulation efficiency for lycopene (76.5%) has been reported using an emulsion consisting of 10% PGA, 4.5% gelatin, and 4.8% lycopene extract with rapid lycopene release at pH 5.5 and 7 (Chiu et al., 2007). Hsieh et al. used heavy γ-PGA molecules for the microencapsulation of nattokinase and found the overall activity of encapsulated nattokinase to be higher than its free form. The microencapsulated nattokinase was found to be more storage stable and while the free form lost its activity at a temperature above 60°C and pH 5, the microencapsulated nattokinase retained its activity even after exposure to similar temperature and pH (Hsieh et al., 2009).

10.7 Conclusions and perspectives

PGA is a biopolymer that has gained immense consideration over the past years with an increasing number of researches. Its properties like biodegradability, high solubility, and nontoxicity make it of interest for applications in food, medicine, waste treatment, and many others. However, the high cost and low productivity are still a major concern regarding its commercial production and use. Molecular mass and enantiomer conformation of the polymer are important to be considered for efficient use of γ-PGA in different industries. Its biosynthesis pathway has been decoded but efforts related to the proper selection of substrates and microorganisms for efficient production are still required. The production mechanism for γ-PGA changes with diverse microorganisms and commercial production requires the selection of high producing strains. Modification in medium composition, genetic engineering, and advancements in molecular biology has led to increased yields with enhanced durability of the microorganisms in adverse fermentation conditions. Fermentation using renewable biomass has been established as the best approach for commercial production of γ-PGA. Researches are mainly aimed at enhancing the microbial yields with economic production and recovery of the biopolymer. The industrial production strategies of γ-PGA stand tested and entrenched with the recognition of some high producers. This is beneficial considering the economically and environmentally increasing value of the biopolymer.

References

Abdel-Fattah, Y., Soliman, N., & Berekaa, M. (2007). Application of Box-Behnken design for optimization of poly-γ-glutamic acid production by *Bacillus licheniformis* SAB-26. *Research Journal of Microbiology*, 2(9), 664−670.

Ashiuchi, M., Kamei, T., Baek, D.-H., Shin, S.-Y., Sung, M.-H., Soda, K., Yagi, T., & Misono, H. (2001a). Isolation of *Bacillus subtilis* (chungkookjang), a poly-γ-glutamate producer with high genetic competence. *Applied Microbiology and Biotechnology*, *57*, 764−769. Available from https://doi.org/10.1007/s00253-001-0848-9.

Ashiuchi, M., Nawa, C., Kamei, T., Song, J. J., Hong, S. P., Sung, M. H., Soda, K., Yagi, T., & Misono, H. (2001b). Physiological and biochemical characteristics of poly γ-glutamate synthetase complex of Bacillus subtilis. *European Journal of Biochemistry/FEBS*, *268*(20), 5321−5328.

Ashiuchi, M., & Misono, H. (2002). Biochemistry and molecular genetics of poly-γ-glutamate synthesis. *Applied Microbiology and Biotechnology*, *59*(1), 9−14. Available from https://doi.org/10.1007/s00253-002-0984-x.

Ashiuchi, M., Shimanouchi, K., Horiuchi, T., Kamei, T., & Misono, H. (2006). Genetically engineered poly-γ-glutamate producer from *Bacillus subtilis* ISW1214. *Bioscience, Biotechnology, and Biochemistry*, *70*(7), 1794−1797. Available from https://doi.org/10.1271/bbb.60082.

Ashiuchi, M., Soda, K., & Misono, H. (1999). A poly-γ-glutamate synthetic system of Bacillus subtilis IFO 3336: Gene cloning and biochemical analysis of poly-γ-glutamate produced by *Escherichia coli* clone cells. *Biochemical and Biophysical Research Communications*, *263*(1), 6−12.

Bajaj, I., & Singhal, R. (2011). Poly (glutamic acid)—An emerging biopolymer of commercial interest. *Bioresource Technology*, *102*(10), 5551−5561. Available from https://doi.org/10.1016/j.biortech.2011.02.047.

Bajaj, I. B., Lele, S. S., & Singhal, R. S. (2009). A statistical approach to optimization of fermentative production of poly (γ-glutamic acid) from Bacillus licheniformis NCIM 2324. *Bioresource Technology*, *100*(2), 826−832.

Bajaj, I. B., & Singhal, R. S. (2009). Enhanced production of poly (γ-glutamic acid) from Bacillus licheniformis NCIM 2324 by using metabolic precursors. *Applied Biochemistry and Biotechnology*, *159*(1), 133−141.

Bajaj, I. B., & Singhal, R. S. (2011). Flocculation properties of poly(γ-glutamic acid) produced from *Bacillus subtilis* isolate. *Food and Bioprocess Technology*, *4*, 745−752. Available from https://doi.org/10.1007/s11947-009-0186-y.

Ben-Zur, N., & Goldman, D. M. (2007). g-Polyglutamic acid: A novel peptide for skin care, *Cosmetices and Toiletries*.

Berekaa, M. M., El Aassar, S. A., El-Sayed, S. M., & Borai, A. M. E. L. (2009). Production of poly-γ-glutamate (pga) biopolymer by batch and semicontinuous cultures of immobilized bacillus licheniformis strain-r. *Brazilian Journal of Microbiology*, *40*(4), 715−724. Available from https://doi.org/10.1590/S1517-83822009000400001.

Bhat, A. R., Irorere, V. U., Bartlett, T., Hill, D., Kedia, G., Morris, M. R., Charalampopoulos, D., & Radecka, I. (2013). Bacillus subtilis natto: A non-toxic source of poly-γ-glutamic acid that could be used as a cryoprotectant for probiotic bacteria. *AMB Express*, *3*, 36. Available from https://doi.org/10.1186/2191-0855-3-36.

Bhattacharyya, D., Hestekin, J. A., Brushaber, P., Cullen, L., Bachas, L. G., & Sikdar, S. K. (1998). Novel poly-glutamic acid functionalized microfiltration membranes for sorption of heavy metals at high capacity. *Journal of Membrane Science*, *141*, 121−135. Available from https://doi.org/10.1016/S0376-7388(97)00301-3.

Birrer, G. A., Cromwick, A.-M., & Gross, R. A. (1994). γ-Poly (glutamic acid) formation by Bacillus licheniformis 9945a: Physiological and biochemical studies. *International Journal of Biological Macromolecules*, *16*(5), 265−275.

Bodnár, M., Kjøniksen, A. L., Molnár, R. M., Hartmann, J. F., Daróczi, L., Nyström, B., & Borbély, J. (2008). Nanoparticles formed by complexation of poly-gamma-glutamic acid with lead ions. *Journal of Hazardous Materials*, *153*(3), 1185−1192. Available from https://doi.org/10.1016/j.jhazmat.2007.09.080.

Borbély, M., Nagasaki, Y., Borbély, J., Fan, K., Bhogle, A., & Sevoian, M. (1994). Biosynthesis and chemical modification of poly(γ-glutamic acid). *Polymer Bulletin*, *32*, 127−132. Available from https://doi.org/10.1007/BF00306378.

Bruckner, V., Kovács, J., & Dénes, G. (1953). Structure of poly-D-glutamic acid isolated from capsulated strains of *B. anthracis*. *Nature*, *172*(4376), 508. Available from https://doi.org/10.1038/172508a0.

Buescher, J. M., & Margaritis, A. (2007). Microbial biosynthesis of polyglutamic acid biopolymer and applications in the biopharmaceutical, biomedical and food industries. *Critical Reviews in Biotechnology*, *27*(1), 1−19. Available from https://doi.org/10.1080/07388550601166458.

Cachat, E., Barker, M., Read, T. D., & Priest, F. G. (2008). A *Bacillus thuringiensis* strain producing a polyglutamate capsule resembling that of *Bacillus anthracis*. *FEMS Microbiology Letters*, 285(2), 220−226.

Candela, T., & Fouet, A. (2006). Poly−gamma−glutamate in bacteria. *Molecular Microbiology*, 60(5), 1091−1098.

Candela, T., Mock, M., & Fouet, A. (2005). CapE, a 47-amino-acid peptide, is necessary for *Bacillus anthracis* polyglutamate capsule synthesis. *Journal of Bacteriology*, 187(22), 7765−7772.

Candela, T., Moya, M., Haustant, M., & Fouet, A. (2009). *Fusobacterium nucleatum*, the first Gram-negative bacterium demonstrated to produce polyglutamate. *Canadian Journal of Microbiology*, 55(5), 627−632.

Cao, M., Geng, W., Zhang, W., Sun, J., Wang, S., Feng, J., Zheng, P., Jiang, A., & Song, C. (2013). Engineering of recombinant *Escherichia coli* cells co-expressing poly-γ-glutamic acid (γ-PGA) synthetase and glutamate racemase for differential yielding of γ-PGA. *Microbial Biotechnology*, 6(6), 675−684. Available from https://doi.org/10.1111/1751-7915.12075.

Chang, J., Zhong, Z., Xu, H., Yao, Z., & Chen, R. (2013). Fabrication of poly(γ-glutamic acid)-coated Fe_3O_4 magnetic nanoparticles and their application in heavy metal removal. *Chinese Journal of Chemical Engineering*, 21(11), 1244−1250. Available from https://doi.org/10.1016/S1004-9541(13)60629-1.

Chen, X., Chen, S., Sun, M., & Yu, Z. (2005). High yield of poly-γ-glutamic acid from *Bacillus subtilis* by solid-state fermentation using swine manure as the basis of a solid substrate. *Bioresource Technology*, 96(17), 1872−1879. Available from https://doi.org/10.1016/j.biortech.2005.01.033.

Cheng, C., Asada, Y., & Aida, T. (1989). Production of γ-polyglutamic acid by *Bacillus licheniformis* A35 under denitrifying conditions. *Agricultural and Biological Chemistry*, 53(9), 2369−2375.

Chettri, R., & Tamang, J. P. (2014). Functional properties of Tungrymbai and Bekang, naturally fermented soybean foods of North East India. *International Journal of Fermented Foods*, 3(1), 87−103.

Chiu, Y. T., Chiu, C. P., Chien, J. T., Ho, G. H., Yang, J., & Chen, B. H. (2007). Encapsulation of lycopene extract from tomato pulp waste with gelatin and poly(γ-glutamic acid) as carrier. *Journal of Agricultural and Food Chemistry*, 55(13), 5123−5130. Available from https://doi.org/10.1021/jf0700069.

Cromwick, A., Birrer, G. A., & Gross, R. A. (1996). Effects of pH and aeration on γ-poly (glutamic acid) formation by *Bacillus licheniformis* in controlled batch fermentor cultures. *Biotechnology and Bioengineering*, 50(2), 222−227.

Cromwick, A.-M., & Gross, R. A. (1995). Investigation by NMR of metabolic routes to bacterial γ-poly (glutamic acid) using 13C-labeled citrate and glutamate as media carbon sources. *Canadian Journal of Microbiology*, 41(10), 902−909.

Cui, Z. F., & Muralidhara, H. S. (2010). *Membrane technology: A practical guide to membrane technology and applications in food and bioprocessing*. Butterworth-Heinemann.

Da Silva, S. B., Cantarelli, V. V., & Ayub, M. A. Z. (2014). Production and optimization of poly-γ-glutamic acid by Bacillus subtilis BL53 isolated from the Amazonian environment. *Bioprocess and Biosystems Engineering*, 37, 469−479. Available from https://doi.org/10.1007/s00449-013-1016-1.

Da Silva, T. L., Mendes, A., Mendes, R. L., Calado, V., Alves, S. S., Vasconcelos, J. M. T., & Reis, A. (2006). Effect of n-dodecane on *Crypthecodinium cohnii* fermentations and DHA production. *Journal of Industrial Microbiology & Biotechnology*, 33(6), 408−416. Available from https://doi.org/10.1007/s10295-006-0081-8.

da Silva, T. L., Reis, A., Roseiro, J. C., & Hewitt, C. J. (2008). Physiological effects of the addition of n-dodecane as an oxygen vector during steady-state *Bacillus licheniformis* thermophillic fermentations perturbed by a starvation period or a glucose pulse. *Biochemical Engineering Journal*, 42(3), 208−216. Available from https://doi.org/10.1016/j.bej.2008.06.023.

Dekie, L., Toncheva, V., Dubruel, P., Schacht, E. H., Barrett, L., & Seymour, L. W. (2000). Poly-L-glutamic acid derivatives as vectors for gene therapy. *Journal of Controlled Release: Official Journal of the Controlled Release Society*, 65(1−2), 187−202. Available from https://doi.org/10.1016/S0168-3659(99)00235-7.

Do, J. H., Chang, H. N., & Lee, S. Y. (2001). Efficient recovery of γ-poly (glutamic acid) from highly viscous culture broth. *Biotechnology and Bioengineering*, 76(3), 219−223. Available from https://doi.org/10.1002/bit.1186.

Du, G., Yang, G., Qu, Y., Chen, J., & Lun, S. (2005). Effects of glycerol on the production of poly (γ-glutamic acid) by *Bacillus licheniformis*. *Process Biochemistry*, 40(6), 2143−2147.

Feng, J., Gao, W., Gu, Y., Zhang, W., Cao, M., Song, C., Zhang, P., Sun, M., Yang, C., & Wang, S. (2014). Functions of poly-gamma-glutamic acid (γ-PGA) degradation genes in γ-PGA synthesis and cell morphology maintenance. *Applied Microbiology and Biotechnology*, 98(14), 6397−6407. Available from https://doi.org/10.1007/s00253-014-5729-0.

Feng, J., Gu, Y., Quan, Y., Cao, M., Gao, W., Zhang, W., Wang, S., Yang, C., & Song, C. (2015). Improved poly-γ-glutamic acid production in *Bacillus amyloliquefaciens* by modular pathway engineering. *Metabolic Engineering*, 32, 106−115. Available from https://doi.org/10.1016/j.ymben.2015.09.011.

Goto, A., & Kunioka, M. (1992). Biosynthesis and hydrolysis of poly(γ-glutamic acid) from *Bacillus subtilis* IF03335. *Bioscience, Biotechnology, and Biochemistry*, 56(7), 1031−1035. Available from https://doi.org/10.1271/bbb.56.1031.

Hanby, W. E., & Rydon, H. N. (1946). The capsular substance of *Bacillus anthracis*: With an appendix by P. Bruce White. *The Biochemical Journal*, 40(2), 297.

Hezayen, F. F., Rehm, B. H., Tindall, B. J., & Steinbüchel, A. (2001). Transfer of *Natrialba asiatica* B1T to *Natrialba taiwanensis* sp. nov. and description of *Natrialba aegyptiaca* sp. nov, a novel extremely halophilic, aerobic, non-pigmented member of the Archaea from Egypt that produces extracellular poly (glutamic acid. *International Journal of Systematic and Evolutionary Microbiology*, 51(3), 1133−1142.

Ho, G. H., Ho, T. I., Hseih, K. H., Su, Y. C., Lin, P. Y., Yang, J., Yang, K. H., & Yang, S. C. (2006). γ-Polyglutamic acid produced by *Bacillus subtilis* (natto): Structural characteristics, chemical properties and biological functionalities. *Journal of the Chinese Chemical Society*, 53(6), 1363−1384. Available from https://doi.org/10.1002/jccs.200600182.

Hong, S. U., & Bruening, M. L. (2006). Separation of amino acid mixtures using multilayer polyelectrolyte nanofiltration membranes. *Journal of Membrane Science*, 280(1−2), 1−5. Available from https://doi.org/10.1016/j.memsci.2006.04.028.

Hsieh, C. W., Lu, W. C., Hsieh, W. C., Huang, Y. P., Lai, C. H., & Ko, W. C. (2009). Improvement of the stability of nattokinase using γ-polyglutamic acid as a coating material for microencapsulation. *LWT—Food Science and Technology*, 42(1), 144−149. Available from https://doi.org/10.1016/j.lwt.2008.05.025.

Hsieh, C. Y., Tsai, S. P., Wang, D. M., Chang, Y. N., & Hsieh, H. J. (2005). Preparation of γ-PGA/chitosan composite tissue engineering matrices. *Biomaterials*, 26(28), 5617−5623. Available from https://doi.org/10.1016/j.biomaterials.2005.02.012.

Huang, J., Du, Y., Xu, G., Zhang, H., Zhu, F., Huang, L., & Xu, Z. (2011). High yield and cost-effective production of poly(γ-glutamic acid) with *Bacillus subtilis*. *Engineering in Life Sciences*, 11(3), 291−297. Available from https://doi.org/10.1002/elsc.201000133.

Inbaraj, B. S., Wang, J. S., Lu, J. F., Siao, F. Y., & Chen, B. H. (2009). Adsorption of toxic mercury(II) by an extracellular biopolymer poly(γ-glutamic acid). *Bioresource Technology*, 100(1), 200−207. Available from https://doi.org/10.1016/j.biortech.2008.05.014.

Ito, Y., Tanaka, T., Ohmachi, T., & Asada, Y. (1996). Glutamic acid independent production of poly (γ-glutamic acid) by *Bacillus subtilis* TAM-4. *Bioscience, Biotechnology, and Biochemistry*, 60(8), 1239−1242.

Jeong, J.-H., Kim, J.-N., Wee, Y.-J., & Ryu, H.-W. (2010). The statistically optimized production of poly (γ-glutamic acid) by batch fermentation of a newly isolated *Bacillus subtilis* RKY3. *Bioresource Technology*, 101(12), 4533−4539. Available from https://doi.org/10.1016/J.BIORTECH.2010.01.080.

Jian, X., Shouwen, C., & Ziniu, Y. (2005). Optimization of process parameters for poly γ-glutamate production under solid state fermentation from *Bacillus subtilis* CCTCC202048. *Process Biochemistry*, 40(9), 3075−3081. Available from https://doi.org/10.1016/j.procbio.2005.03.011.

Jiang, F., Qi, G., Ji, Z., Zhang, S., Liu, J., Ma, X., & Chen, S. (2011). Expression of glr gene encoding glutamate racemase in *Bacillus licheniformis* WX-02 and its regulatory effects on synthesis of poly-γ-glutamic acid. *Biotechnology Letters*, 33(9), 1837−1840. Available from https://doi.org/10.1007/s10529-011-0631-7.

Jiang, H., Shang, L., Yoon, S. H., Lee, S. Y., & Yu, Z. (2006). Optimal production of poly-γ-glutamic acid by metabolically engineered *Escherichia coli*. *Biotechnology Letters*, 28, 1241−1246. Available from https://doi.org/10.1007/s10529-006-9080-0.

Jiang, Y., Tang, B., Xu, Z., Liu, K., Xu, Z., Feng, X., & Xu, H. (2016). Improvement of poly-γ-glutamic acid biosynthesis in a moving bed biofilm reactor by *Bacillus subtilis* NX-2. *Bioresource Technology*, *218*, 360−366. Available from https://doi.org/10.1016/j.biortech.2016.06.103.

Ju, W. T., Song, Y. S., Jung, W. J., & Park, R. D. (2014). Enhanced production of poly-γ-glutamic acid by a newly-isolated *Bacillus subtilis*. *Biotechnology Letters*, *36*(11), 2319−2324. Available from https://doi.org/10.1007/s10529-014-1613-3.

Kandler, O., König, H., Wiegel, J., & Claus, D. (1983). Occurrence of poly-γ-D-glutamic acid and poly-α-L-glutamine in the genera *Xanthobacter*, *Flexithrix*, *Sporosarcina* and *Planococcus*. *Systematic and Applied Microbiology*, *4*(1), 34−41.

Kim, J.-H., Block, D. E., & Mills, D. A. (2010). Simultaneous consumption of pentose and hexose sugars: an optimal microbial phenotype for efficient fermentation of lignocellulosic biomass. *Applied Microbiology and Biotechnology*, *88*(5), 1077−1085.

Kim, T. W., Lee, T. Y., Bae, H. C., Hahm, J. H., Kim, Y. H., Park, C., Kang, T. H., Kim, C. J., Sung, M. H., & Poo, H. (2007). Oral administration of high molecular mass poly-γ-glutamate induces NK cell-mediated antitumor immunity. *Journal of Immunology*, *179*(2), 775−780. Available from https://doi.org/10.4049/jimmunol.179.2.775.

Kimura, K., Tran, L.-S. P., Uchida, I., & Itoh, Y. (2004). Characterization of *Bacillus subtilis* gamma-glutamyltransferase and its involvement in the degradation of capsule poly-gamma-glutamate. *Microbiology (Reading, England)*, *150*(12), 4115−4123.

Ko, Y. H., & Gross, R. A. (1998). Effects of glucose and glycerol on γ-poly (glutamic acid) formation by *Bacillus licheniformis* ATCC 9945a. *Biotechnology and Bioengineering*, *57*(4), 430−437.

Kocianova, S., Vuong, C., Yao, Y., Voyich, J. M., Fischer, E. R., DeLeo, F. R., & Otto, M. (2005). Key role of poly-γ-DL-glutamic acid in immune evasion and virulence of *Staphylococcus epidermidis*. *The Journal of Clinical Investigation*, *115*(3), 688−694.

Kumar, R., & Pal, P. (2015). Fermentative production of poly (γ-glutamic acid) from renewable carbon source and downstream purification through a continuous membrane-integrated hybrid process. *Bioresource Technology*, *177*, 141−148. Available from https://doi.org/10.1016/j.biortech.2014.11.078.

Kunioka, M., & Goto, A. (1994). Biosynthesis of poly (γ-glutamic acid) from L-glutamic acid, citric acid, and ammonium sulfate in *Bacillus subtilis* IFO3335. *Applied Microbiology and Biotechnology*, *40*(6), 867−872.

Lee, J., Lee, S. Y., Park, S., & Middelberg, A. P. J. (1999). Control of fed-batch fermentations. *Biotechnology Advances*, *17*(1), 29−48. Available from https://doi.org/10.1016/S0734-9750(98)00015-9.

Li, C., Yu, D. F., Newman, R. A., Cabral, F., Stephens, L. C., Hunter, N., Milas, L., & Wallace, S. (1998). Complete regression of well-established tumors using a novel water-soluble poly(L-glutamic acid)-paclitaxel conjugate. *Cancer Research*, *58*(11), 2404−2409.

Lim, S. M., Kim, J., Shim, J. Y., Imm, B. Y., Sung, M. H., & Imm, J. Y. (2012). Effect of poly-γ-glutamic acids (PGA) on oil uptake and sensory quality in doughnuts. *Food Science and Biotechnology*, *21*, 247−252. Available from https://doi.org/10.1007/s10068-012-0032-2.

Liu, J., He, D., Li, X. Z., Gao, S., Wu, H., Gao, X., & Zhou, T. (2010). γ-Polyglutamic acid (γ-PGA) produced by *Bacillus amyloliquefaciens* C06 promoting its colonization on fruit surface. *International Journal of Food Microbiology*, *142*(1−2), 190−197. Available from https://doi.org/10.1016/j.ijfoodmicro.2010.06.023.

Liu, L., Yang, H., Zhang, D., Du, G., Chen, J., Wang, M., & Sun, J. (2009). Enhancement of hyaluronic acid production by batch culture of *Streptococcus zooepidemicus* via the addition of n-Dodecane as an oxygen vector. *Journal of Microbiology and Biotechnology*, *19*(6), 596−603. Available from https://doi.org/10.4014/jmb.0807.440.

Luo, Z., Guo, Y., Liu, J., Qiu, H., Zhao, M., Zou, W., Li, S., et al. (2016). Microbial synthesis of poly-γ-glutamic acid: Current progress, challenges, and future perspectives. *Biotechnology for Biofuels*, *9*(1), 134. Available from https://doi.org/10.1186/s13068-016-0537-7.

Manocha, B., & Margaritis, A. (2010). A novel method for the selective recovery and purification of γ-polyglutamic acid from *Bacillus licheniformis* fermentation broth. *Biotechnology Progress*, *26*(3), 734−742. Available from https://doi.org/10.1002/btpr.370.

Mark, S. S., Crusberg, T. C., DaCunha, C. M., & Di Iorio, A. A. (2006). A heavy metal biotrap for wastewater remediation using poly-γ-glutamic acid. *Biotechnology Progress*, *22*(2), 523−531. Available from https://doi.org/10.1021/bp060040s.

McLean, R. J. C., Beauchemin, D., Clapham, L., & Beveridge, T. J. (1990). Metal-binding characteristics of the gamma-glutamyl capsular polymer of *Bacillus licheniformis* ATCC 9945. *Applied and Environmental Microbiology*, *56*(12), 3671−3677.

Mesnage, S., Tosi-Couture, E., Gounon, P., Mock, M., & Fouet, A. (1998). The capsule and S-layer: Two independent and yet compatible macromolecular structures in *Bacillus anthracis*. *Journal of Bacteriology*, *180*(1), 52−58.

Müslüm, A. (2019). Bioproduction of γ-poly(glutamic acid) using feather hydrolysate as a fermentation substrate. *Trakya University Journal of Natural Science*, *20*(1), 27−34. Available from https://doi.org/10.23902/trkjnat.448851.

Nakagawa, S. (2008). Efficacy and safety of poly (γ-glutamic acid) based nanoparticles (γ-PGA NPs) as vaccine carrier. *Yakugaku Zasshi: Journal of the Pharmaceutical Society of Japan*, *128*(11), 1559−1565. Available from https://doi.org/10.1248/yakushi.128.1559.

Nayak, J., & Pal, P. (2013). Transforming waste cheese-whey into acetic acid through a continuous membrane-integrated hybrid process. *Industrial & Engineering Chemistry Research*, *52*(8), 2977−2984.

Ogawa, Y., Yamaguchi, F., Yuasa, K., & Tahara, Y. (1997). Efficient production of γ-polyglutamic acid by *Bacillus subtilis* (natto) in jar fermenters. *Bioscience, Biotechnology, and Biochemistry*, *61*(10), 1684−1687.

Ogunleye, A., Bhat, A., Irorere, V. U., Hill, D., Williams, C., & Radecka, I. (2015). Poly-γ-glutamic acid: Production, properties and applications. *Microbiology (Reading, England)*, *161*(1), 1−17.

Peng, Y., Jiang, B., Zhang, T., Mu, W., Miao, M., & Hua, Y. (2015). High-level production of poly (γ-glutamic acid) by a newly isolated glutamate-independent strain, *Bacillus methylotrophicus*. *Process Biochemistry*, *50*(3), 329−335. Available from https://doi.org/10.1016/j.procbio.2014.12.024.

Pérez-Camero, G., Congregado, F., Bou, J. J., & Muñoz-Guerra, S. (1999). Biosynthesis and ultrasonic degradation of bacterial poly(γ-glutamic acid). *Biotechnology and Bioengineering*, *63*(1), 110−115, 10.1002/(SICI)1097-0290(19990405)63:1 < 110::AID-BIT11 > 3.3.CO;2-K.

Rehm, B. H. A. (2009). *Production of biopolymers and polymer precursors: Applications and perspectives*. Horizon Scientific Press.

Richard, A., & Margaritis, A. (2003). Optimization of cell growth and poly(glutamic acid) production in batch fermentation by *Bacillus subtilis*. *Biotechnology Letters*, *25*, 465−468. Available from https://doi.org/10.1023/A:1022644417429.

Richard, A., & Margaritis, A. (2003). Rheology, oxygen transfer, and molecular weight characteristics of poly (glutamic acid) fermentation by *Bacillus subtilis*. *Biotechnology and Bioengineering*, *82*(3), 299−305. Available from https://doi.org/10.1002/bit.10568.

Ryu, M., Nakazawa, T., Akagi, T., Tanaka, T., Watanabe, R., Yasuda, M., Himori, N., Maruyama, K., Yamashita, T., Abe, T., Akashi, M., & Nishida, K. (2011). Suppression of phagocytic cells in retinal disorders using amphiphilic poly(γ-glutamic acid) nanoparticles containing dexamethasone. *Journal of Controlled Release: Official Journal of the Controlled Release Society*, *151*(1), 65−73. Available from https://doi.org/10.1016/j.jconrel.2010.11.029.

Schallmey, M., Singh, A., & Ward, O. P. (2004). Developments in the use of *Bacillus* species for industrial production. *Canadian Journal of Microbiology*, *50*(1), 1−17.

Scoffone, V., Dondi, D., Biino, G., Borghese, G., Pasini, D., Galizzi, A., & Calvio, C. (2013). Knockout of pgdS and ggt genes improves γ-PGA yield in *B. subtilis*. *Biotechnology and Bioengineering*, *110*(7), 2006−2012. Available from https://doi.org/10.1002/bit.24846.

Shih, I. L., & Van, Y. T. (2001). The production of poly-(γ-glutamic acid) from microorganisms and its various applications. *Bioresource Technology*, *79*(3), 207−225. Available from https://doi.org/10.1016/S0960-8524 (01)00074-8.

Shih, I. L., Van, Y. T., & Chang, Y. N. (2002). Application of statistical experimental methods to optimize production of poly (γ-glutamic acid) by *Bacillus licheniformis* CCRC 12826. *Enzyme and Microbial Technology*, *31*(3), 213−220.

Shih, I. L., Van, Y. T., & Sau, Y. Y. (2003). Antifreeze activities of poly(γ-glutamic acid) produced by *Bacillus licheniformis*. *Biotechnology Letters*, 25, 1709–1712. Available from https://doi.org/10.1023/A:1026042302102.

Shih, I. L., Van, Y. T., Yeh, L. C., Lin, H. G., & Chang, Y. N. (2001). Production of a biopolymer flocculant from *Bacillus licheniformis* and its flocculation properties. *Bioresource Technology*, 78(3), 267–272. Available from https://doi.org/10.1016/S0960-8524(01)00027-X.

Shih, L., & Van, Y.-T. (2001). The production of poly-(γ-glutamic acid) from microorganisms and its various applications. *Bioresource Technology*, 79(3), 207–225.

Shih, L., & Yu, Y.-T. (2005). Simultaneous and selective production of levan and poly (γ-glutamic acid) by *Bacillus subtilis*. *Biotechnology Letters*, 27(2), 103–106.

Shyu, Y. S., & Sung, W. C. (2010). Improving the emulsion stability of sponge cake by the addition of γ-polyglutamic acid. *Journal of Marine Science and Technology*, 18(6), 895–900.

Singer, J. W. (2005). Paclitaxel poliglumex (XYOTAX™, CT-2103): A macromolecular taxane. *Journal of Controlled Release: Official Journal of the Controlled Release Society*, 109(1–3), 120–126. Available from https://doi.org/10.1016/j.jconrel.2005.09.033.

Singh, K., Ingole, P. G., Bajaj, H. C., & Gupta, H. (2012). Preparation, characterization and application of β-cyclodextrin-glutaraldehyde crosslinked membrane for the enantiomeric separation of amino acids. *Desalination*, 298, 13–21. Available from https://doi.org/10.1016/j.desal.2012.04.023.

Su, Y., Li, X., Liu, Q., Hou, Z., Zhu, X., Guo, X., & Ling, P. (2010). Improved poly-γ-glutamic acid production by chromosomal integration of the Vitreoscilla hemoglobin gene (vgb) in *Bacillus subtilis*. *Bioresource Technology*, 101(12), 4733–4736. Available from https://doi.org/10.1016/j.biortech.2010.01.128.

Suzuki, T., & Tahara, Y. (2003). Characterization of the *Bacillus subtilis* ywtD gene, whose product is involved in γ-polyglutamic acid degradation. *Journal of Bacteriology*, 185(7), 2379–2382.

Tang, B., Xu, H., Xu, Z., Xu, C., Xu, Z., Lei, P., Qiu, Y., Liang, J., & Feng, X. (2015a). Conversion of agroindustrial residues for high poly (γ-glutamic acid) production by Bacillus subtilis NX-2 via solid-state fermentation. *Bioresource Technology*, 181, 351–354.

Tang, B., Lei, P., Xu, Z., Jiang, Y., Xu, Z., Liang, J., Feng, X., & Xu, H. (2015b). Highly efficient rice straw utilization for poly-(γ-glutamic acid) production by Bacillus subtilis NX-2. *Bioresource Technology*, 193, 370–376. Available from https://doi.org/10.1016/j.biortech.2015.05.110.

Taniguchi, M., Kato, K., Shimauchi, A., Xu, P., Fujita, K.-I., Tanaka, T., Tarui, Y., & Hirasawa, E. (2005). Physicochemical properties of cross-linked poly-γ-glutamic acid and its flocculating activity against kaolin suspension. *Journal of Bioscience and Bioengineering*, 99(2), 130–135. Available from https://doi.org/10.1263/jbb.99.130.

Tanimoto, H., Fox, T., Eagles, J., Satoh, H., Nozawa, H., Okiyama, A., Morinaga, Y., & Fairweather-Tait, S. J. (2007). Acute effect of poly-γ-glutamic acid on calcium absorption in post-menopausal women. *Journal of the American College of Nutrition*, 26(6), 645–649. Available from https://doi.org/10.1080/07315724.2007.10719642.

Tarui, Y., Iida, H., Ono, E., Miki, W., Hirasawa, E., Fujita, K.-i, Tanaka, T., & Taniguchi, M. (2005). Biosynthesis of poly-γ-glutamic acid in plants: Transient expression of poly-γ-glutamate synthetase complex in tobacco leaves. *Journal of Bioscience and Bioengineering*, 100(4), 443–448. Available from https://doi.org/10.1263/jbb.100.443.

Tian, G., Fu, J., Wei, X., Ji, Z., Ma, X., Qi, G., & Chen, S. (2014). Enhanced expression of pgdS gene for high production of poly-γ-glutamic aicd with lower molecular weight in *Bacillus licheniformis* WX-02. *Journal of Chemical Technology and Biotechnology (Oxford, Oxfordshire: 1986)*, 89(12), 1825–1832. Available from https://doi.org/10.1002/jctb.4261.

Timmer, J. M. K., Speelmans, M. P. J., & Van Der Horst, H. C. (1998). Separation of amino acids by nanofiltration and ultrafiltration membranes. *Seperation and Purification Technology*, 14(1–3), 133–144. Available from https://doi.org/10.1016/S1383-5866(98)00068-9.

Troy, F. A. (1973). Chemistry and biosynthesis of the poly (γ-D-glutamyl) capsule in Bacillus licheniformis I. Properties of the membrane-mediated biosynthetic reaction. *The Journal of Biological Chemistry*, 248(1), 305–315.

Uto, T., Wang, X., Akagi, T., Zenkyu, R., Akashi, M., & Baba, M. (2009). Improvement of adaptive immunity by antigen-carrying biodegradable nanoparticles. *Biochemical and Biophysical Research Communications*, 379(2), 600−604. Available from https://doi.org/10.1016/j.bbrc.2008.12.122.

Wang, L. L., Chen, J.-T., Wang, L.-F., Wu, S., Zhang, G.-z, Yu, H.-Q., Ye, X.-d, & Shi, Q.-S. (2017). Conformations and molecular interactions of poly-γ-glutamic acid as a soluble microbial product in aqueous solutions. *Scientific Reports*, 7(1). Available from https://doi.org/10.1038/s41598-017-13152-2.

Wu, Q., Xu, H., Liang, J., & Yao, J. (2010). Contribution of glycerol on production of poly (γ-glutamic acid) in *Bacillus subtilis* NX-2. *Applied Biochemistry and Biotechnology*, 160(2), 386−392.

Wu, Q., Xu, H., Shi, N., Yao, J., Li, S., & Ouyang, P. (2008). Improvement of poly (γ-glutamic acid) biosynthesis and redistribution of metabolic flux with the presence of different additives in *Bacillus subtilis* CGMCC 0833. *Applied Microbiology and Biotechnology*, 79(4), 527.

Wu, Q., Xu, H., & Xu, L. (2006). Regulation of stereochemical composition of poly-gamma-glutamic acid in *Bacillus subtilis* NX-2. *Chinese Journal of Process Engineering*, 6(3), 458.

Wu, Q., Xu, H., Xu, L., & Ouyang, P. (2006). Biosynthesis of poly (γ-glutamic acid) in Bacillus subtilis NX-2: Regulation of stereochemical composition of poly (γ-glutamic acid). *Process Biochem*, 41(7), 1650−1655.

Xu, H., Jiang, M., Li, H., Lu, D., & Ouyang, P. (2005). Efficient production of poly (γ-glutamic acid) by newly isolated *Bacillus subtilis* NX-2. *Process Biochemistry*, 40(2), 519−523.

Xu, Z., Feng, X., Zhang, D., Tang, B., Lei, P., Liang, J., & Xu, H. (2014). Enhanced poly (γ-glutamic acid) fermentation by *Bacillus subtilis* NX-2 immobilized in an aerobic plant fibrous-bed bioreactor. *Bioresource Technology*, 155, 8−14. Available from https://doi.org/10.1016/j.biortech.2013.12.080.

Xu Z., Shi F., & Cen P. (2005). Production of polyglutamic acid from mixed glucose and sucrose by co-cultivation of *Bacillus subtilis* and *Corynebacterium glutamicum*. In *The 2005 AIChE annual meeting*. Cincinnati.

Yamashiro, D., Yoshioka, M., & Ashiuchi, M. (2011). *Bacillus subtilis* pgsE (Formerly ywtC) stimulates poly-γ-glutamate production in the presence of zinc. *Biotechnology and Bioengineering*, 108(1), 226−230.

Yeh, C. M., Wang, J. P., Lo, S. C., Chan, W. C., & Lin, M. Y. (2010). Chromosomal integration of a synthetic expression control sequence achieves poly-γ-glutamate production in a *Bacillus subtilis* strain. *Biotechnology Progress*, 26(4), 1001−1007. Available from https://doi.org/10.1002/btpr.417.

Yokoi, H., Natsuda, O., Hirose, J., Hayashi, S., & Takasaki, Y. (1995). Characteristics of a biopolymer flocculant produced by *Bacillus* sp. PY-90. *Journal of Fermentation and Bioengineering*, 79(4), 378−380. Available from https://doi.org/10.1016/0922-338X(95)94000-H.

Yong, X., Raza, W., Yu, G., Ran, W., Shen, Q., & Yang, X. (2011). Optimization of the production of poly-γ-glutamic acid by *Bacillus amyloliquefaciens* C1 in solid-state fermentation using dairy manure compost and monosodium glutamate production residues as basic substrates. *Bioresource Technology*, 102(16), 7548−7554. Available from https://doi.org/10.1016/j.biortech.2011.05.057.

Yoon, S. H., Do, J. H., Lee, S. Y., & Chang, H. N. (2000). Production of poly-γ-glutamic acid by fed-batch culture of *Bacillus licheniformis*. *Biotechnology Letters*, 22(7), 585−588.

Zhang, A., & Yang, S. T. (2009). Engineering *Propionibacterium acidipropionici* for enhanced propionic acid tolerance and fermentation. *Biotechnology and Bioengineering*, 104(4), 766−773. Available from https://doi.org/10.1002/bit.22437.

Zhang, C., Wu, D., & Qiu, X. (2018). Stimulatory effects of amino acids on γ-polyglutamic acid production by *Bacillus subtilis*. *Scientific Reports*, 8(1). Available from https://doi.org/10.1038/s41598-018-36439-4.

Zhang, D., Feng, X., Li, S., Chen, F., & Xu, H. (2012). Effects of oxygen vectors on the synthesis and molecular weight of poly(γ-glutamic acid) and the metabolic characterization of *Bacillus subtilis* NX-2. *Process Biochemistry*, 47(12), 2103−2109. Available from https://doi.org/10.1016/j.procbio.2012.07.029.

Zhang, D., Feng, X., Zhou, Z., Zhang, Y., & Xu, H. (2012). Economical production of poly (γ-glutamic acid) using untreated cane molasses and monosodium glutamate waste liquor by *Bacillus subtilis* NX-2. *Bioresource Technology, 114*, 583−588.

Zhang, W., He, Y., Gao, W., Feng, J., Cao, M., Yang, C., Song, C., & Wang, S. (2014). Deletion of genes involved in glutamate metabolism to improve poly-gamma-glutamic acid production in *B. amyloliquefaciens* LL3. *Journal of Industrial Microbiology & Biotechnology, 42*(2), 297−305. Available from https://doi.org/10.1007/s10295-014-1563-8.

Zhao, C., Zhang, Y., Wei, X., Hu, Z., Zhu, F., Xu, L., . . . Liu, H. (2013). Production of ultra-high molecular weight poly-γ-glutamic acid with *Bacillus licheniformis* P-104 and characterization of its flocculation properties. *Applied Biochemistry and Biotechnology, 170*(3), 562−572. Available from https://doi.org/10.1007/s12010-013-0214-2.

Zhu, F., Cai, J., Zheng, Q., Zhu, X., Cen, P., & Xu, Z. (2014). A novel approach for poly-γ-glutamic acid production using xylose and corncob fibres hydrolysate in *Bacillus subtillis* HB-1. *Journal of Chemical Technology and Biotechnology (Oxford, Oxfordshire: 1986), 89*(4), 616−622.

CHAPTER 11

Production and applications of polyphosphate

Raj Morya, Bhawna Tyagi, Aditi Sharma and Indu Shekhar Thakur

School of Environmental Sciences, Jawaharlal Nehru University, New Delhi, India

11.1 Introduction

In the year 1890 L. Lieberman discovered a polymeric inorganic compound aggregation in yeast and named it as polyphosphate (PolyP). Arthur Meyer in 1904 reported an inorganic polymeric compound in the microorganisms and named it as "volutin." Later in 1947 J. M. Wiame confirmed the presence of PolyP in microorganisms (Kornberg, 1995). PolyP is a polymer in which the orthophosphate monomeric units are linearly bound by phosphoanhydride bonds, similar to those found in adenosine triphosphate (ATP). In nature, PolyP is formed by the dehydration of phosphate rocks at extreme temperatures in volcanoes and oceanic steam vents, which may have its genesis possibly even before the life originated on earth. PolyP accumulates in bacteria in the presence of excess phosphate and scarce sulfur. Bacterial PolyP plays a major role in oligotrophic microbial ecosystem of coral reefs and might even be related to the early earth phosphorus cycle (Zhang et al., 2015). In the present era, eutrophication is frequently occurring because of elevated accumulation of nutrients such as nitrogen and phosphorus. These findings suggest that phosphorus has always been a significant mineral, ever since prehistoric times.

Inorganic PolyPs are classified into two categories, namely, pyrophosphates (PPis) and large molecular PolyPs containing three to several hundreds of phosphate residues in a single molecule. PPi (also called diphosphate) is a simpler compound forming a P-O-P bond between two phosphorus molecules. These compounds form as a result of heating the phosphoric compounds at an elevated temperature until the condensation reaction occurs. Because of their property to constrain precipitation of calcium carbonate, they are also used as additives in detergents.

Compared to the widely studied PPis, high-molecular-weight PolyPs are not thoroughly studied. PolyP helps organism to adapt themselves in the severe conditions such as extreme levels of salinity, UV radiation, pH, osmolarity, pressure, and temperature

(Rothschild & Mancinelli, 2001; Seufferheld, Alvarez, & Farias, 2008). For primitive organisms that lived in the harsh conditions of ancient earth, these adaptations may have contributed to their survival. Polyphosphate kinases (PPKs) are the enzymes responsible for the formation of PolyPs and is present in most of the microorganisms (Kornberg, 1995; Kulaev & Vagabov, 1983). A study showed that, *ppk*1 lacking strains are more vulnerable to death because of hydrogen peroxide, salinity, and elevated temperatures than the wild-type strains (Rao & Kornberg, 2008). The views about the PolyP function in living organisms have shifted dramatically in recent years. Earlier, PolyPs were considered a "molecular fossil" or a source of phosphorous and energy for the survival of the microorganism. PolyP has various other functions in the microbes such as increasing motility, regulating gene expression, promoting translation fidelity, and virulence (Rao & Kornberg, 2008; Rao, Gomez-Garcia, & Kornberg, 2009; Itoh, Kawazoe, & Shiba, 2006). It can regulate channel activity, act as chaperone, promote cell proliferation, and ensure that genes involved in differential gene expression are explicitly stable (Dhivya et al., 2018; Gray et al., 2014; Pavlov et al., 2010; Stotz et al., 2014; Varas et al., 2018; Zakharian, Thyagarajan, French, Pavlov, & Rohacs, 2009). PolyP has been used for numerous medical purposes such as bone regeneration; it stimulates osteoblast differentiation, calcification, and inhibit bone resorption activity of osteoclasts and assist in bone tissue regeneration (Wang, Schroder, & Muller, 2014; Wang, Schroder, Schlossmacher, & Muller, 2013; Hassanian, Avan, & Ardeshirylajimi, 2017; Lui, Ao, Li, Khong, & Tanner, 2016).

11.2 Structure and types of polyphosphate

PolyPs are linear orthophosphate polymers having the general molecular formula $M_{(n+2)}P_nO_{(3n+1)}$, where n signifies the degree of polymerization between 2 and 106. The fundamental structure of the PolyP is composed of a single phosphate unit attached to an adjacent unit using an oxygen atom (Fig. 11.1). The degree of polymerization plays an important role in the existence and purity of PolyPs, those with $n = 2-5$ have a crystalline purity. Depending on the number of the orthophosphate units attached, it is classified into PPis and high molecular PolyPs.

11.2.1 Pyrophosphate

PPi/diphosphate are compounds in which two orthophosphate units are linearly linked. It is prepared by heating sodium or potassium orthophosphate salts (Fig. 11.2). In 1948 Kornberg described the first biological reaction regarding the formation of pyrophosphate (PPi) from a reversible reaction of yeast cell extract. Yeast extract mediated the formation of nicotinamide mononucleotide (NMN) and ATP using PPi and nicotinamide adenine dinucleotide (NAD)+. As a reversible reaction, one can get NAD+ and PPi by using NMN and ATP as reactants. Later in 1957 Kornberg hypothesized that pyrophosphorylases

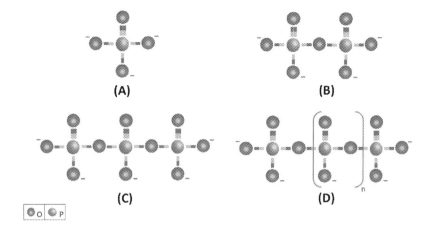

Figure 11.1
Molecular formula of (A) phosphate, (B) diphosphate, (C) tripolyphosphate, and (D) polyphosphate (Kulaev, 1979).

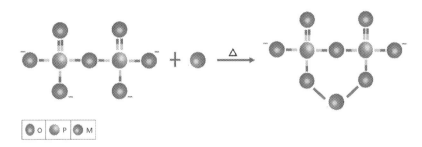

Figure 11.2
Diagrammatic representation of pyrophosphate interaction with a metal ion and heat applied for the formation of M-PPi complex (Baltscheffsky, 1967).

were the enzymes responsible for the formation of PPi resulting in the formation of other biochemical compounds. PPi is widely distributed in all the living organisms such as bacteria, lower eukaryotes, plants, animals, and humans. In humans, it is present in the blood, urine, synovial fluid, saliva, bone, teeth, cartilage, and cultured human cells. PPi is a known source of biochemical energy in bacteria, protists, plants, animals, and other eukaryotes.

To live and reproduce, all organisms need energy to meet their daily requirements and derive this biochemical energy by oxidizing a substrate (Keefe & Miller, 1996). Energy in living organisms is stored in the form of ATP and a small amount in the form of PPi. It is apparent that primitive life on the earth had a crucial reliance on low molecular weight PolyPs (PPi) (Baltscheffsky, 1967). Studies undertaken to find out the correlation between the PPi and life on primitive earth (Baltscheffsky, 1967). In addition to the biochemical source, PPi also serves as a regulator for biochemical reactions and processes. It regulates the action of few enzymes, nucleic acid synthesis, cell proliferation and iron transport, engaged in few pathological conditions, urinary stone formation, and also responsible for pseudogout (deposition of calcium pyrophosphate dihydrate crystals in knee joints) (Kornberg, Rao, & Ault-Riche, 1999).

Due to strong metal chelation property of the PPi tetra anions, it readily binds to metal ions such as Mg^{2+}, Ca^{2+}, Fe^{3+}, Mn^{2+}, and Zn^{2+} (Kornberg et al., 1999; Mekmene & Gaucheron, 2011). SunActiveFe uses a soluble ferric pyrophosphate as an iron-deficiency supplement (Rao, Nanbu, & Juneja, 2004). PPi also used as a bone-mimicking agent marketed with a tradename Technescan PYP, contains sodium metal with PPi. Triferic is a similar product used for the treatment during chronic hemodialysis patients (Fishbane et al., 2015). Apart from its nutraceutical usage, it is also used in food industry. Disodium PPi is also used as an ingredient in food processing. Food industries use ferric PPi in chocolate powders and pastas as stabilizer and emulsifier (Fidler et al., 2004). Because of its molecular structure, PPi metal complexes are commonly used in the magnetic and catalytic materials. FDK and Fujitsu developed a lithium-cobalt PPi battery which supersedes the performance of the lithium ion batteries. Replacement of the cobalt with a cheap metal ion to reach high voltage can bring down the cost of these batteries (Nishimura, Nakamura, Natsui, & Yamada, 2010).

11.2.2 High molecular weight polyphosphate

Compounds in which phosphate units (n) are linked to form a chain with $n < 10$ and $n \geq 50$ form crystallized salts (Fig. 11.1). When n increases, the degree of polymerization also increases, resulting in the formation of long chain of phosphate.

11.2.3 Cyclophosphates

Cyclophosphates are compounds with the general molecular formula of $P_nO_{3n}^{n-}$, which form a ring structure. Previously these compounds were known as metaphosphates, but IUPAC recommended the use of cyclophosphates to indicate the presence of cyclic structure (Fig. 11.3). They can be prepared by heating phosphoric acid and used as ligands for different metal ions (Montag, Clough, Müller, & Cummins, 2011).

11.3 Acidocalcisomes

Acidocalcisomes are the compounds in which PolyP is found in vicinity with calcium and other cations in an acidic environment of vacuoles. These are present in both prokaryotes and eukaryotes. In prokaryotes, their acidic nature is regulated by the vacuolar proton-pyrophosphatase/ATPase, whereas it is vacuolar-proton-ATPase in the

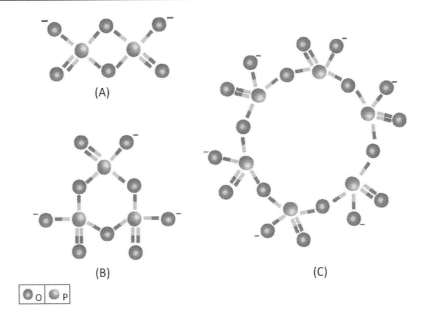

Figure 11.3
Molecular structure of cyclophosphates: (A) cyclodiphosphate, (B) cyclotriphosphate, and (C) cyclohexaphosphate (Montag et al., 2011).

case of eukaryotes (Docampo & Moreno, 2001). The role of various cations present in the acidic vacuole is to balance the charge of PolyP with the help of transporters (Huang et al., 2014). Organisms in which acidocalcisomes have been studied include *Trypanosoma cruzi*, *Phytomonas francai*, *Toxoplasma gondii*, *Chlamydomonas reinhardtii*, *Dictyostelium discoideum*, *Agrobacterium tumefaciens*, insectbasophils, and mast cells (Espiau et al., 2006; Hong-Hermesdorf et al., 2014; Marchesini, Ruiz, Vieira, & Docampo, 2002; Moreno-Sanchez, Hernandez-Ruiz, Ruiz, & Docampo, 2012; Niyogi et al., 2015; Ramos et al., 2011; Rohloff et al., 2011; Seufferheld et al., 2003).

11.4 Biogenic production of polyphosphate

PolyP is present in almost all living cells and is involved in divergent functions along the evolutionary line. Polyp found in the cytoplasm, cell surface, and plasma membrane of prokaryotes used to impart resistance to the cell against adverse conditions. The synthesis and accumulation of PolyP occurs at a higher rate in bacteria than under normal conditions; this process is called the "luxury uptake." While in yeast, the accumulation increases via cultivation on phosphate-rich media after phosphate starvation; this process is called "phosphate overplus" (Kulaev, Vagabov, & Shabalin, 1987). In yeast, PolyP localizes in cytosol as volutin granules, while in other eukaryotes it is distributed throughout different compartments of the cell. Vacuoles, the reservoir of various compounds, are also rich in

PolyP. Certain cell compartments including mitochondria, nucleic acid, and cell envelopes also accumulate a substantial amount of PolyP (Kulaev & Kulakovskaya, 2000).

PolyP was found to be present as volutin granules in the cytoplasm of the bacterium *Spirillum volutans* (Kulaev & Vagabov, 1983). PolyP biogenesis is an enzyme-mediated synthesis in all living organisms. Despite the ubiquitous presence of PolyP in all living cells, the enzymes synthesizing it are not common in all the species. There are certain enzymes that synthesize and metabolize PolyP only in prokaryotes, some only in eukaryotes and a few in both of them (Rao et al., 2009). The concentration of PolyP in the cell depends primarily on the concentration of phosphate outside the cell boundary. The presence of PolyP depends on various enzymes responsible for its formation and metabolism (Kulaev & Kulakovskaya, 2000; Wood & Clark, 1988). Enzymes provide a practical approach for research and understanding physiology. Proteomics study helps identify specific gene, its isolation, and overexpression. Overexpression results in the enhancement of targeted product and plays a vital role in the metabolism of microbes. Enzymes are the backbone of industries that help in generation of cost-effective products. PolyP, as described earlier, depends on various enzymes for metabolism (Akiyama, Crooke, & Kornberg, 1993; Wurst & Kornberg, 1994). Enzymes responsible for energy metabolism include PPK, exopolyphosphatase (PPX), polyphosphate adenosine monophosphate phosphotransferase, polyphosphate glucose-6-phosphotransferase, 1,3-diphospohglycerate polyphosphate phosphotransferase, polyphosphate glucokinase, tripolyphosphate, and endopolyphosphatase (Akiyama et al., 1993; Hsieh, Shenoy, Jentoft, & Phillips, 1993; Kornberg, Kornberg, & Simms, 1956; Kulaev, 1979; Lichko, Kulakovskaya, & Kulaev, 2010; Szymona & Ostrowski, 1964). The uptake and transport of inorganic phosphate (Pi) is assisted by phosphate transport system in bacteria. These transport systems can be classified into two categories depending on their affinity toward phosphate; low-affinity phosphate inorganic transport (Pit) and high-affinity phosphate-specific transport (Pst). Phosphate in Pit is transported as symporter of proton and metal-phosphate complex, while with Pst as $H_2PO_4^-$ and HPO_4^{2-}.

Prokaryotes have been studied considerably more than the eukaryotes for PolyP biogenesis since it accumulates in significant amount in the microbial systems. The enzyme responsible for PolyP biogenesis has been extensively researched in prokaryotes, and its homolog has been discovered in eukaryotes.

11.4.1 Prokaryotes

The location of PolyP inside the microorganism plays an important role. In bacterial cells, they are mostly concentrated in cell envelope and cytoplasm. Bacterial PolyP is known to accumulate under special stress/surplus substrate conditions. In *Burkholderia cepacia* AM19 strain, phosphate uptake increases by nearly 220% and PolyP accumulation increases by

Table 11.1: Diversity of PPK enzyme homologs found in diverse species of prokaryotes.

PPK1	PPK2	PPK1 and PPK2 both	Neither PPK1 nor PPK2
Salmonella typhi	Staphylococcus aureus	Brucella melitensis	Clostridium perfringens
Helicobacter pylori	Corynebacterium glutamicum	Bacillus anthracis	Methanococcus jannaschi
Neisseria meningitidis	Francisella tularensis	Vibrio cholerae	Mycoplasma genitalium
Burkholderia cepacia	Bordetella bronchiseptica	Agrobacterium tumefaciens	Mycoplasma pneumoniae
Mycobacterium leprae	Bordetella parapertussis	Mycobacterium tuberculosis	Pyrococcus furiosus
Bordetella pertussis	Neisseria gonorrhoeae	Pseudomonas aeruginosa	Bacillus subtilis
Yersinia pestis	Magnetococcus MC-1	Caulobacter crescentus	Halobacterium sp. NRC-1
Xylella fastidiosa	Plectonema boryanum	Ralstonia metallidurans	Saccharomyces cerevisiae
Streptomyces coelicolor	–	Synechoccus sp. WH 8102	Chlamydia muriadarum

Source: Zhang, H., Ishige, K., Kornberg, A. (2010). A polyphosphate kinase (PPK2) widely conserved in bacteria. Proceedings of the National Academy of Sciences 99, 16678–16683.

330% when the pH decreases from 7.5 to 5.5 (Mullan, Quinn, & McGrath, 2002). Several other strains including *Acinetobacter calcoaceticus, Aeromonas hydrophila*, and *Pseudomonas* sp., isolated from sewage sludge, reported for PolyP accumulation (Sidat, Bux, & Kasan, 1999). Phosphate storage is the most important function of PolyP in bacteria. In addition, it performs several other functions in the cells by affecting the expression of other enzymes. Enzymes which plays an important role in the accumulation and metabolism of phosphate inside the cell have been discussed in detail in the following sections. Most commonly known enzymes are PPK1, PPK2, and PPK3 distributed widely in the different species of bacteria (Table 11.1).

11.4.1.1 PPK1

The principal enzyme responsible for PolyP biogenesis in prokaryotes is PolyP-ADP phosphotransferase; also known as polyphosphate kinase (PPK or PPK1), discovered by Kornberg in *Escherichia coli* bacteria (Kornberg et al., 1956). ppk gene encodes PPK1. This enzyme forms tetramer with a subunit size of 80 kDa (Ahn & Kornberg, 1990). It is highly conserved in archaea and bacteria and is the most thoroughly studied enzyme for PolyP genesis, especially in the human pathogenic bacteria *Pseudomonas aeruginosa*. This enzyme generates PolyP from nucleoside diphosphates (NDPs) as well as nucleotide triphosphates (NTPs) reversibly from PolyP (Fig. 11.4). The substrate specificity for various NDPs is ADP > GDP > UDP > CDP (Kuroda & Kornberg, 1997). This shows that PPK1 has higher affinity for purine NDPs.

PPK1 catalytically relocates the terminal (γ) phosphate from ATP to an active site of histidine residue (H435) which acts as a nucleophile and attacks the phosphodiester bond of phosphate group. This step marks the initiation of the PolyP biogenesis in prokaryotes, known as autophosphorylation (Rao et al., 2009). The PolyP chain elongates by adding more phosphate groups transferred from ATP and eventually terminates when the chain length reaches about 750 units. Performance of this enzyme for PolyP genesis improved by

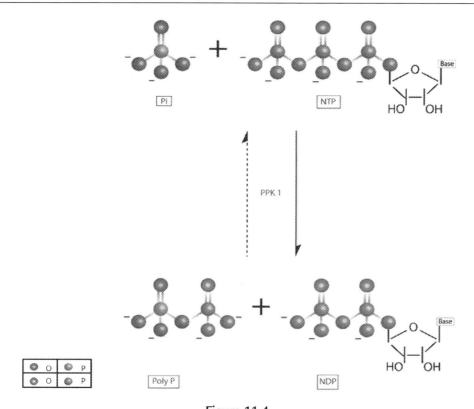

Figure 11.4
Biosynthesis of polyP from NTP with the help of the enzyme PPK1 (Kuroda & Kornberg, 1997).

divalent cation (Mg^{2+}), inducible phosphate transport systems, and exopolyphosphate gene inactivation. PolyP synthesis is downregulated by mutation in the ppk gene and in presence of fluoride ions.

11.4.1.2 PPK2

Few prokaryotes lack the enzyme PPK1 but they can still synthesize PolyP. This is possible with the help of another enzyme called PolyP-dependent nucleoside-diphosphate kinase (PNDK or PPK2). The PAO141 gene encodes this enzyme and it metabolizes PolyP instead of favoring its genesis (Kuroda & Kornberg, 1997). PPK2 can synthesize PolyP from either ATP or GTP, unlike PPK1, which uses ATP exclusively (Rao & Kornberg, 2008). Divalent manganese cation (Mn^{2+}) is preferred over Mg^{2+} for PolyP biogenesis in this case. PPK2 is supposedly an octamer with 44 kDa subunit mass, which is stable only in the presence of PolyP; otherwise it forms tetramer (Rao et al., 2009).

Some prokaryotes possess either PPK1 or PPK2, but a few possesses both of the enzymes. The bacterium *Corynebacterium glutamicum* possesses only PPK2, which favors the PolyP synthesis over the nicotinamide adenine dinucleotide phosphate synthesis from ATP or GTP as opposite to *P. aeruginosa* that possesses both PPK enzymes. The level of PPK2 is usually low during the bacterial growth cycle, except during the stationary phase, in which it increases more than 100-folds from barely detectable amount (Rao et al., 2009).

There are certain organisms in which PolyP is present intracellularly without the presence of an enzyme homologous to PPK been discovered. *Saccharomyces cerevisiae* is one such example, which possesses ample amount of PolyP in the cells but there has been no *ppk* gene or homolog enzyme discovered. Studies showed the presence of four genes (PHM1, PHM2, PHM3, and PHM4) possibly responsible for PolyP genesis (Ogawa, DeRisi, & Brown, 2000). There are some alternative PolyP-synthesizing enzymes other than PPKs, namely, polyβ-hydroxybutyrate-calcium-PolyP membrane complex and 3-phosphoglyceroylphosphate-PolyP-phosphotransferase.

11.4.1.3 PPK3

On running a blast search for PPK2 in strain *P. aeruginosa* PA0141, almost six PPK2 homologs were observed in the strain (Zhang, Ishige, & Kornberg, 2010). For further studies, PPK2 homologs were cloned from *Silicibacter pomeroyi* into *E. coli*. It was observed that one gene resembles in action to PPK1, second resembles PPK2 but the third has somehow had a different function than the conventional PPKs (Zhang et al., 2010). It has affinity for pyrimidine NDPs and was named as PPK3. It utilized PolyP as a donor to transform CDP to CTP. PPK3 can phosphorylate NDPs in the order: CDP > UDP > GDP > ADP.

Based on the above-mentioned features, PPKs can be classified as PPK1 for PolyP synthesis, PPK2 for PolyP degradation (using purine for phosphorylation), and PPK3 for PolyP degradation (using pyrimidine for phosphorylation) (Zhang et al., 2010).

11.4.2 Eukaryotes

PolyPs function as a phosphate reserve for phosphate-deficient conditions in eukaryotes. It usually accumulates in the same manner as in prokaryotes under stress conditions or surplus phosphate availability. PolyPs are wonder compound that assists eukaryotes to adapt and carry out their metabolism under harsh conditions. However, the mechanism and role of PolyP in stress response is not well known in the eukaryotic microbes. PolyP concentration depends on the stage of life and growth conditions. In fungi, PolyP is primarily present in cytoplasm, nuclei, vacuole, cell wall, and mitochondria. Compartment-specific enzymes makes the PolyP metabolism easy and fast. Different types of PolyP reserves were isolated from the yeast *S. cerevisiae* as polyP1 (acid-soluble fraction), polyP2 (salt soluble), polyP3 (weak alkali soluble), polyP4 (alkali soluble), and polyP5 (hydrolysis fraction) (Vagabov, Trilisenko, Kulakovskaya, & Kulaev, 2008). Cell compartments play a key role in the accumulation and metabolism of these heterogeneous types of PolyPs. Culture conditions affects the outcome of these PolyP chain lengths (Vagabov et al., 2008). With the shortage of phosphorous in culture media, PPN1 and PPX1 tend to degrade PolyP. The activity of the recognized enzyme PPX1 increased twice under phosphorous-limiting conditions, but no increase has been observed in the activity of PPN1 (Kulakovskaya, Andreeva, Trilisenko, Vagabov, & Kulaev, 2004). These different chain length PolyPs perform various functions like phosphorous reserve source and execute regulatory functions. A study was performed in order to check the effect of substrate availability on the production of PolyPs, hypercompensation effect was observed in which the phosphorous-starved cells were placed into fresh culture medium containing phosphorous. Different PolyP concentration were observed at different stages of growth. During first 2 h, only polyp1, polyp2, and polyp3 increased. In the later stages, only polyp1 reduced and polyp2, polyp5, and polyp3 increased. Additional phosphorous had no impact on the polyp4, this advocates the role of polyp4 for the cell wall synthesis instead of phosphorous homeostasis maintenance (Kulaev, Vagabov, & Kulakovskaya, 2004). Excess phosphorous accumulated more PolyP in mitochondria and cytoplasm than in vacuoles (Andreeva, Kulakovskaya, Kulakovskaya, & Kulaev, 2008). In *S. cerevisiae*, transformation of surplus Pi into inert PolyP controls the homeostasis of Pi and conserve the energy in phosphoester bonds. *Cryptococcus humicola* and *S. cerevisiae* both accumulated PolyP in the nitrogen-starved condition. PolyP reserve increased with the Mg^{2+} ions due to presence of Pho84p transporter.

Effect of carbon source on the PolyP production efficiency was experimented using glucose and hexadecane (Dmitriev et al., 2011). Candida maltose utilized hexadecane as sole carbon source by forming a canal-like structure to bind and oxidize hexadecane. These canals are exact cell wall molecular complexes that houses the enzymes for hydrocarbon consumption (Dmitriev et al., 2011). As compared to glucose, cells generated long-chain PolyPs during growth on hexadecane. PolyP accumulation using glucose as carbon source was 72 $\mu mol\ g^{-1}$ wet biomass, nearly three times higher than yeast grown on hexadecane (200 $\mu mol\ g^{-1}$ wet biomass).

PolyP plays a vital role in the enzymatic activities in the eukaryotes. Cell extracts of animals and yeast contains the PPK enzymatic activity. PolyP competes for binding sites with polyanionic substrates. It is one of most effective chaperones (containing ≥ 130 phosphate molecules) and supports unfolding and refolding of high-affinity proteins (Wolska-Mitaszko, 1997). PolyP inhibits the trehalase activity of vegetative yeast cells (Wolska-Mitaszko, 1997). Bg12p protein from cell wall of *S. cerevisiae* showed glucan transferase activity in the presence of PolyP. This Bg12p-PolyP complex showed transferase activity to put mannoproteins in the cell wall. The PolyP is a dual-purpose compound, which in addition to as a phosphorous-storage factory also helps in the activation of other enzymes in eukaryotic enzymatic machinery (De Venditis, Zahn, & Fasano, 1986).

Table 11.2: Difference between two types polyphosphate enzymes observed on eukaryotes (Andreeva et al., 2008).

S. no.	PPN1	PPX1
1	Depend on M^{2+}	Depend on M^{2+}
2	Exopolyphosphatase and endopolyphosphatase	Exopolyphosphatase (cleaves phosphorous from polyp end of the chain)
3	Favors long-chain polyp	Favors short-chain polyp
4	In nuclei, vacuoles, and mitochondrial membrane	Localized in mitochondria and cytoplasm
5	Decrease in polyp on overexpression of PPN1	No effect of PPX1 overexpression

11.4.2.1 Enzymes of eukaryotes

11.4.2.1.1 PPN1 and PPX1

Two genes reported to code polyphosphatases, namely, PPN1 and PPX1 hydrolyzed PolyPs in *S. cerevisiae* (Table 11.2). Their activity depends on each other, as well as on the phosphorous concentration, growth stage, and carbon source. However, they differ in their cellular localization, metabolism of PolyP, amino acid sequence, and substrate specificity. The metabolism of phosphorous by these enzymes is not well explored in higher eukaryotes.

11.4.2.1.2 DdPPK1

PolyP is present in all living cells, but very limited study has been done on the genes or enzymes responsible for PolyP accumulation in eukaryotes. Researchers have found that *D. discoideum*, a slime mold encodes the enzyme DdPPK1 by the gene homologous to the bacterial PPK1 gene (Rao et al., 2009). This enzyme has conserved sequence for autophosphorylation and ATP binding activity. However, it contained 370 amino acid N-terminal extension, which does not have any homology with the prokaryotic proteins. This N-terminal domain is important for enzyme activity, determining location within the cell, and physiological functions. DdPPK1 synthesizes PolyP chains of different lengths.

11.4.2.1.3 DdPPK2

Acidocalcisome vacuoles are present in DdPPK1 mutants, which are synchronous with the actin-like filament formation associated with PolyP synthesis. These vacuoles are responsible for calcium ion (Ca^{2+}) flux. DdPPK2 forms a tetramer of three actin-related protein (Arps), namely, Arp1, Arp2, and Arp28. Arp tetramer polymerizes into actin-like filament along with the production of PolyP in presence of ATP (Rao et al., 2009).

294   Chapter 11

11.5 Applications of polyphosphates

Inorganic biopolymer PolyP plays a key role in microbial ecology, phosphorous exchanges across sediment-water boundary, microbial stress response, biofilm production, biogeochemistry, cellular biochemistry, apatite formation, and phosphogenesis process; therefore it is a fundamental phosphorous sink present in the environment (Kulakova et al., 2011). PolyP being one of the versatile biomaterials possesses several properties such as excellent biocompatibility, structural similarity with naturally present nucleic acids, and biodegradability. Upon degradation by enzymatic hydrolysis and digestion, it gives low molecular weight products (Liu et al., 2010). PolyP has significant research potential in several environmental and biological science fields, which demarcated its environmental, medical, agricultural, and industrial applications (mineral, food, leather, and detergent) (Fig. 11.5). Applications includes the utilization of PolyP accumulating microbes in the bioremediation of toxic heavy metals and phosphate recovery from wastewater. Usage of PolyP and its materials in copper biomining. Cheap and nontoxic food additive, as an

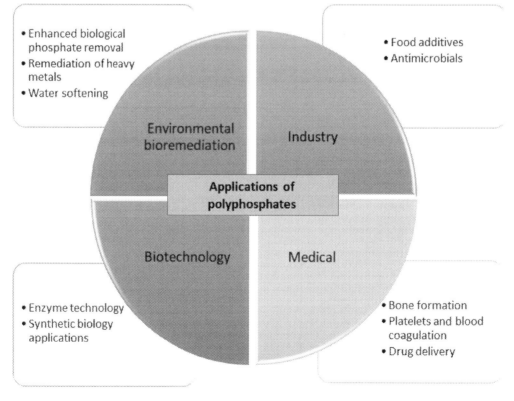

Figure 11.5
Diagrammatic representation of the application of polyphosphates in various fields.

economical source of phosphate reagents in energetic reactions, and a substitute to overpriced ATP used in biotechnology processes. It is also used in the design of therapeutic drugs for the treatment of infectious diseases, bone formation and osteoporosis, immune response, neurotransmission, regulation of Ca^{2+} level in mitochondria, cardiac ischemia, blood coagulation, apoptosis, stress-induced cell death, and also as insulating fiber used in construction process and nonflammable clothing (Rashchi & Finch, 2000). Furthermore, PolyP can be used in agriculture as a constituent of chemical fertilizers that release phosphorous under limiting condition (Alori, Glick, & Babalola, 2017).

11.5.1 Applications in environmental bioremediation

11.5.1.1 Enhanced biological phosphate removal

The increase in population growth and high agricultural demand has resulted in an imbalance of natural phosphorous cycle. Due to inappropriate regulation of pollution control and management of water bodies, recycling of phosphate from rocks in water bodies gets disturbed. Phosphate removal from water bodies is an urgent demand; otherwise, high algal bloom formation can restrict the proper use of water bodies such as lakes and bays. The PolyP granules accumulation inside the microbial and algal cells can be a potential indicator of pollution of water bodies due to high nutrient load (Eixler, Selig, & Karsten, 2005). Several studies have been conducted related to accumulation of PolyP in phosphate-utilizing microbes during wastewater treatment. A report showed the presence of PolyPs in phytoplankton of Lake Michigan (Schelske, Aldridge, & Kenney, 1999). Lake was heavily polluted due to excessive nutrient intake and this study described the impact of these compounds on pelagic community structure and dynamics. Another report described the accumulation of PolyP in sediments and phosphorus accumulation in algal cells due to high phosphorus loads in the lake (Kenney, Schelske, & Chapman, 2001). The enhanced biological phosphate removal (EBPR) method is an efficient, cheap, and eco-friendly alternative to the classical approach that requires coagulation with some metal salts such as alum, lime, or iron. In addition, the side product of the EBPR process is sludge waste, PolyP, and methane that can be managed efficiently for industrial purposes (Kulaev & Kulakovskaya, 2000). PolyP is used as a chemical dispersing agent during the ultrafiltration procedure to remove microbial contamination from drinking water as it minimizes the binding of microbes to filter surfaces due to its highly negative charge (Kulaev & Kulakovskaya, 2000).

11.5.1.2 Remediation of heavy metal

PolyP can remove heavy metals from the industrial effluents prior to discharge to a water body or wastewater treatment plant. PolyP can be used in wastewater treatment for the sequestration of alkaline earth metals as well as heavy metals such as calcium, magnesium, iron, and manganese (Li et al., 2018). PolyP is also recommended for the acid mine drainage treatment. PolyP accumulating bacteria and its heavy metal tolerance indicate the enormous possibilities toward bioremediation of heavy metals in wastewater treatment. These heavy metals precipitated on the microbial cell surface in the presence of PolyPs reserves due to high insolubility of the metal phosphates in water (Aravind, Saranya, & Kanmani, 2015). Keasling and his team found an enzyme which accumulated high amounts of PolyP in the uranyl solution under phosphate-limiting conditions (Anand & Aoyagi, 2019). The plasmid-transfected *E. coli* strain lead to a high accumulation of PolyP as well as high expression of mer genes. This recombinant strain was able to sequester phenyl mercury and Hg^{2+} from the solution, which suggested efficient mercury remediation from wastewater. Copper biomining tends to be possible as thermoacidophilic archaeon, *Sulfolobus metallicus*, has shown tolerance of up to 200 mM copper ion and sequestrated copper, mediated by PolyP accumulation (Rao et al., 2009).

11.5.1.3 Water softening

PolyP is used in wastewater treatment due to its unusual property, that is, high complexing power that controls the precipitation of metal ions with other complexing agents. This phosphate chain behaves like as "soluble ion exchanger" due to the flexibility and dimensions of the PolyP chain. PolyP also used as a potential water-softening agent in hard water treatment. PolyP chains form the corrosion inhibiting film on metal surfaces when used in low concentration in the presence of hard water to protect metal corrosion by anodic polarization process. Moreover, the application of soluble sodium chain PolyPs at a high concentration of hard water inhibits the blockage in pipes due to calcite precipitation by absorption of PolyPs by the calcite nuclei. Some of the glassy sodium PolyPs (Calgon and Gram's salt) are widely used in the detergent industry, water softening, and descaling of pipes and boilers (Rashchi & Finch, 2000).

11.5.2 Applications in industry

PolyPs such as sodium polyphosphate in low concentration are used in detergent and food industries. Due to the dispersing and de-flocculating properties, small levels (0.1%) of PolyPs are employed in making clay and titania slurries for several applications such as paper coating, preparation of emulsion paints, portland cement preparation, and treatment of oil well drilling muds (Keasling, Van Dien, Trelstad, Renninger, & McMahon, 2000).

PolyP has a key role in mineral-processing applications because of its unique complexing ability, which enables it to act as depressants, a stabilizing agent in mineral suspensions, water softening agents, complexing agents for the metal ions, and dispersing agents for slimes. Another important application of linear PolyPs is in the mineral flotation process as it prevents accidental mineral activation due to metal ions by forming a complex with the metal ions (Rauchfuss, 1990). Likewise, PolyP is used to control microbial adhesion to the soil for keeping minerals in suspension form during mineral processing.

11.5.2.1 Food additive and antimicrobials

PolyP can be used as food additive in the dairy and meat industry as it improves emulsification, water binding, flavor, antioxidant, color retention properties of the meat, and retards the rancidity of food (Bru, Jimenez, Canadell, Arino, & Clotet, 2017). PolyP mostly inhibits the growth of Gram-positive bacteria such as *B. cereus* (toxin-producing microbe) and *Staphylococcus aureus* due to its cation-chelating ability. PolyPs being a membrane permeabilizer (due to its chelating metal properties) increases the susceptibility of Gram-negative bacteria *Stenotrophomonas maltophilia* and *Acinetobacter* sp. to a range of antibiotics. Higher dose of PolyP shows cell lysis and bactericidal effect, whereas sublethal dose may lead to the formation of septum and multinucleated or filamentous cells. PolyP is being used in the meat industry as a food additive and antibacterial agent in processed fish, poultry, and meat products. PolyP is also used as a complexing agent after replacing calcium complexing agent in the meat industry to inhibit blood coagulation, as it possesses more effectiveness and ability to prevent fibrin precipitation (Kulakovskaya, Vagabov, & Kulaev, 2012). In the food industry, PolyPs such as disodium dihydrogen diphosphate ($Na_2H_2P_2O_7$) combined with sodium bicarbonate ($NaHCO_3$) are employed to leaven the bread; it is a much easy, effective, and commercial method of making bread instead of using yeast. PolyPs like calcium diphosphate ($Ca_2P_2O_7$) employed in manufacturing of fluoride toothpaste as the abrasive or polishing agent, on the other side $Na_4P_2O_7$ added in starch and flavoring agents for making instant pudding mixes (Albi & Serrano, 2016). In the leather industry, PolyPs such as sodium polyphosphate act as a pre(chrome) tanning agent on interaction with protein as albumen, and forms precipitate in low acidic solution which is beneficial for leather-making process (Seufferheld et al., 2008).

11.5.3 Biotechnological applications

11.5.3.1 Enzyme technology

As the production cost of phosphate reagents used in enzymatic ATP regenerating systems is very high, the use of ATP as a phosphorylating agent at industrial level becomes very expensive and prohibitive. Therefore an alternative ATP-regeneration system was proposed for enzyme technology applications after intensive research of two decades. This system is based on PolyP and Ec PPK1 enzyme in which a reaction mixture containing ADP and PolyP is allowed to percolate through a column. The column comprises immobilized PPK that can provide ATP equivalents at a reasonable cost. This ATP-enriched mixture is reused in the next round of reactions. The promising ATP-regeneration systems such as pyruvate kinase, acetyl phosphate, creatine phosphate, acetate kinase, phosphoenolpyruvate, and creatine kinase are used for enzyme technology applications (Achbergerova & Nahalka, 2011). A company naming GeneChem, Inc. has utilized the two ATP-regeneration systems (acetyl phosphate and acetate kinase) during the production of cytidine monophosphate-NeuAc and sialyl lactose, respectively (Kim, Yoo, Lee, Kim, & Cho, 2008). Microbial strain *C. glutamicum* utilizing nearly 37% of its cell volume for PolyP accumulation is being used for commercial-scale production of D-glutamate (1.2 million tons) and L-lysine (0.55 million tons) annually (Rao et al., 2009). The thermophilic enzymes contains a high-temperature tolerant ATP-regeneration system, which is being used for the manufacturing of economically valuable enzymes. In such applications, PPK from *Thermus thermophiles* is used, which can synthesize fructose 1,6-diphosphate enzyme for at least 1 week at 70°C (Iwamoto et al., 2007).

11.5.3.2 Synthetic biology applications

In synthetic biology approach, two areas are actively involved in which one area creates artificial life by using synthetic molecules to replicate promising and developing behaviors in natural biology, and the other area looks for identical and interchangeable units from natural biology to bring together systems that works artificially. Both areas aim to understand life in a better way and employing knowledge for commercial purposes. The primary purpose of synthetic biology is to design minimal cells. Research started to design and program, such synthetic life forms based on new DNA software, which incorporates the use of new cell components, resources, and metabolic systems. To make minimal and simple genome of a cell, PolyP represents a model energy source to power all critical functions of the synthetic cell and it can completely replace all alternative energy sources including photosynthesis, respiration, glycolysis, and others. There are only two main genes needed for PolyP utilization, PPK and PPX. A model proposed by Achbergerova, suggested the ancestral endosymbiotic origin of PolyP granules or acidocalcisomes since PolyP might be abiotically stored at high temperatures and anhydrous conditions prevailed during primitive earth formation (Achbergerova & Nahalka, 2011).

11.5.4 Application in medical field

11.5.4.1 Bone formation

PolyPs function in cells in the form of soluble salt solutions such as sodium polyphosphate (Na-PolyP) or in form of granular nanoparticles such as Ca-PolyP-microparticles (MPs). These PolyPs are used for the chelation of Ca^{2+} ions due to their negative charges. It further minimizes the free concentration of Ca^{2+} ions in the culture and results in low hydroxyapatite (HA) deposition (Wang et al., 2013). The noncrystalline calcium phosphate granules contain PolyPs in the interior of bones and cartilages. PolyP may act as an intermediate in the formation and degradation process of bones by osteoblasts and osteoclasts, respectively. An in vitro analysis on osteoblast-like MC3T3-E1 cells found that PolyP might help to promote gene expression in the bone differentiation as well as cell calcification processes (Rao et al., 2009). PolyP/Ca^{2+} complex has been suggested to be more successful than Na-PolyP in mineralization of SaOS-2 cells and expression of bone morphogenetic protein (BMP)2; therefore Ca-PolyP is being used widely in bone-regeneration-related studies. Ca-PolyP owns excellent biocompatibility, mechanical properties, bioactivity, and degradability and its composition is parallel with the inorganic matter of normal bone tissue. Besides, it remains insoluble at physiological pH and hardens as concentrations and chain lengths increases that makes it more favorable material for bone tissue engineering procedures. Sr-PolyP shows high osteogenic effect than Ca-PolyP, as it can stimulate high osteoblast activity, bone development, and can prevent bone resorption (Wang et al., 2019).

PolyP is used in bone formation, remodeling trials, and the dental implants. The natural inorganic biopolymers, whether synthesized by recombinant enzymes or chemical methods, are highly capable in the treatment of osteoporosis, illnesses, and deformities in bone as bio-PolyP. It triggers the BMP2 expression (key mediator), which further helps in differentiation of bone-forming progenitor cells to mature osteoblasts as well as retard the osteoclasts function. On the other hand, soluble PolyP (Ca-PolyP) affects the alkaline phosphatase (an enzyme responsible for phosphate metabolism in the bone) function. This enzyme generates high amounts of inorganic phosphate (Pi) in the bone matrix needed for HA crystallization (Wang et al., 2013).

Wang and his team found that alginate supplemented with PolyP (morphogenetically active additive) can be a suitable biomaterial for 3D tissue printing as it can enhance proliferation as well as differentiation of human mesenchymal stem cells (Wang et al., 2019). Further, the introduction of PolyP/Ca^{2+} complex and agarose in the alginate/gelatin hydrogel improves cells hardness and calcification and therefore can be utilized as a biomaterial for 3D tissue printing in the field of bone tissue engineering. Hence as evidenced from these studies, the composites made up of PolyP and other biomaterials showed improved mechanical properties and biological functions as compared to individual PolyP or biomaterials.

11.5.4.2 Platelets and blood coagulation

Activated platelets release PolyP, which accelerates blood clotting as it promotes clot formation and stability by stimulating the contact pathway that further activates the natural anticoagulant protein. Besides that, it also retards clot lysis by activating the natural antifibrinolytic agent (thrombin-activatable fibrinolysis inhibitor) (Anand & Aoyagi, 2019). Calcium PolyP fiber utilized as a supportive material for tendon tissue engineering in vitro (Sun & Zhao, 2002).

11.5.4.3 Drug delivery

Nowadays, a number of linear, random, and block copolymers having phosphate units being synthesized and applied to drug- and gene-delivery applications. Besides, ranges of biocompatible and biodegradable hydrogels formed by PolyPs are produced for cell encapsulations (Liu, Huang, Pang, & Yan, 2015). However, water-soluble hyperbranched polyphosphates (HBPPs) are used in drug delivery as an intracellular drug transporter due to their unique properties such as high number of surface functional hydroxyl groups, functional end groups for conjugation, lesser hydrodynamic volume, improved cellular uptake efficiency, and self-assembled nanostructures (Yan, 2016). Four types of HBPPs-based drug-delivery systems generally used, that is, nonresponsive vehicles, stimuli-responsive vehicles, drug conjugates, and self-delivery macromolecular drugs. Muller and coworkers have reported that tumor bone metastasis can be inhibited and treated by the use of Ca-PolyP-MPs embedded with anionic drug Zol (that promotes apoptosis in cancer stem cells) to make diamond-shaped Ca-PolyP-Zol-MPs after Ca^{2+} ion bonding (Muller et al., 2015). PolyP helps in the treatment of bone metastasis by participating in drug-delivery systems as it reversibly binds to drugs (Liu et al., 2015).

Phospholipid analogous polymers HPHEP alkyls are highly encouraged for drug-delivery systems. They have high biocompatibility and biodegradability along with their high in vitro anticancer effects. PolyP is used as nonresponsive carriers that facilitate the release of drug molecule from the drug-loaded nanocarriers. Because of their exceptional self-assembled structure, a number of disulfide linkages and sensitive oxidation-responsiveness, HPHSP micelles serve as drug-delivery vehicles in reactive oxygen species-mediated intracellular drug delivery in cancer therapy applications. These HBPPs-based drug-delivery systems have shown good biodegradability, outstanding biocompatibility, and less cytotoxicity than their linear analogs. These polymer-drug conjugates increase their water-solubility properties, reduce their toxicity, and inhibit their enzymatic breakdown and hydrolysis. The amphiphilic HPHEP-chlorambucil conjugates have different drug filling synthesized by linkage of terminal hydroxyls HPHEP with the carboxyl in anticancer chlorambucil for use in cancer therapy (Liu et al., 2010).

11.6 Challenges associated with polyphosphate production strategies

Although remarkable, research done for enhancing the production of PolyP, some problems still needs to be addressed. The first challenge is to optimize metabolism-related strategies such as favorable growth conditions to increase the PolyP production. There is a very limited report on PolyP accumulating organisms, metabolic reactions involved in its accumulation, and its production cost due to limitation of suitable and efficient quantification approaches involved in PolyP studies. The classical method used in PolyP accumulation analysis and quantification uses radioactivity, which makes it inefficient and inaccurate due to PolyP chain-length heterogeneity and less product recovery (Kulakova et al., 2011). Research for the quantification of PolyP production is of higher priority in current time.

Similarly, advanced research in the field of gene regulation at transcriptional and translational levels related to PolyP production needs to be done. Further, the information about the source, fate, and functions of short-chain PolyP, found in large amounts in eukaryotic cells, needs to be investigated at a large scale. The enzymes and genes involved in bioaccumulation and production of PolyP in eukaryotes are still unknown. The phosphate and toxic metals removal by the use of bioremediation technologies involving PolyP accumulating bioengineered microbes and plants need to be investigated and applied at large scale (such as wastewater treatment). The research associated with designing of successful therapeutic strategies having enzymes PPK1 and PPK2 harnessed from drug-resistant microbial pathogenic strains as *Mycobacterium tuberculosis* and *P. aeruginosa* requires utmost attention (Kulaev et al., 2004).

11.7 Strategies to improve the yield of polyphosphate

PolyP accumulated by plants, bacteria, fungi, protozoa, and mammals, is involved in several functions such as an energy source, a substitute for ATP in energy reactions, reservoir of energy and phosphate having osmotic benefits, and as DNA transfer component during the genetic transformation (Breiland et al., 2018). There are several parameters affecting microbial P-accumulation such as pH, temperature, time of anaerobic and aerobic/anoxic phases, presence of stress (organic pollutants), volatile fatty acid composition, and the presence or absence of ions (PO_4^{3-}, SO_4^{2-}, and NH_4^+) and metal ions (K^+, Na^+, Mg^{2+}, Ca^{2+}, and others) (Tarayre et al., 2016). Strategies that induce PolyP accumulation and biosynthesis in microbes are required for microbial survival and have potential biotechnological advantages. Such an approach exploits for alternative and feasible phosphate-removal method.

Generally, PolyP accumulation occurs in the bacterial cells under limited nutritional conditions. Several bacterial strains show fast and high PolyP accumulation, when Pi-starved cells were given surplus Pi in the culture media (Ohtake, Kato, Kuroda, Wu, & Ikeda, 1998). Kim and coworkers reported the extensive PolyP accumulation in *E. coli* cells when exposed to osmotic stress and limited nutrition condition of nitrogen, amino acid, or phosphate in medium (Kim et al., 1998). In a similar study by Ault-Riche, PolyP accumulation in *P. aeruginosa* mucoid strain 8830 and *Microlunatus phosphovorus* NM-1 increased during phosphate and amino acid starvation and stationary phase (Ault-Riche, Fraley, Tzeng, & Kornberg, 1998). Freshwater sponge *Ephydatia muelleri* accumulates a high amount of PolyP in presence of organic pollutants (Imsiecke et al., 1996). The microbial PolyP accumulation is also influenced and induced by exposure to acidic pH conditions. McGrath and Quinn observed a 10-fold increase in intracellular PolyP production by *Candida humicola* G-1 in acidic pH, that is, 5.5 as compared to pH 7.5 (McGrath & Quinn, 2003). Similarly, strain *B. cepacia* maximal PolyP accumulation and phosphate removal increased with 220% and 330%, respectively, at pH 5.5 in comparison to pH 7.5, furthermore the activity of PPK enzyme was also observed in microbial cells cultured at acidic pH (Mullan et al., 2002). In a report, 34% of total microbial isolates of municipal-activated sludge plants showed increased phosphate uptake and phosphate removal by 55%−124% in acidic conditions of pH 5.5 (McGrath, Cleary, Mullan, & Quinn, 2001).

The energy state of microbial cell also influences the production of PolyP from ATP or restoration of ATP from PolyP. It has been investigated that Polyp accumulates during energy-rich growth environments, whereas it degrades during energy-poor growth environments (Anand & Aoyagi, 2019). In a study, high concentration of ATP induced the in vitro synthesis of PolyP by enzyme PPK as compared to the retarded PolyP synthesis in presence of low ATP (Kornberg et al., 1999). The PolyP accumulation in *E. coli* MV1184 was improved by modification of gene regulation and supplying genes encoding PPK, acetate kinase, and the Pst system in high doses. The *E. coli* recombinant strain that was constructed by improving the dose of ppk and pstSCAB genes was accumulated a higher percent of PolyP as cell dry weight (48% as P) (Ohtake et al., 1998). The *ppk* gene has been replicated from a variety of microorganisms, including *Neisellia menigitidis*, *A. calcoaceticus*, *Synechocystis* sp. PCC9803, *E. coli*, and *Klebsiella aerogenes*.

Further, conserved sequences derived from the reported *ppk* genes could be used in synthesizing polymerase chain reaction primers for isolating the *ppk* genes from bacterial strains of different environments. Engineered recombinant microbes and genetically improved sludge bacteria may play an effective role in phosphate removal from wastewaters (Ohtake et al., 1998). However, there is limited report on PolyP accumulation by microalgae and mycetes.

11.8 Conclusions and perspectives

Phosphorus (phosphate) is considered to be present on the earth since the prehuman times. It is needed for various functions in all kinds of living organisms from prokaryotes to eukaryotes. PolyP plays a pivotal role in the cell, serving as a source of phosphorous-storage reserve. It also affects the expression of various enzymes and maintains the energy balance in the cell. It is present in phosphorylated proteins, nucleic acids, phospholipids, ATP, and carbohydrates.

PolyP is a polyanionic and neutral compound. The phosphoanhydride bond energy in the PolyP is comparable to ATP terminal phosphodiester bond. In lower eukaryotes and prokaryotes, these properties of the PolyP help in maintaining the stress response, enzyme regulation, and secondary messenger metabolism. Higher eukaryotes are generally independent of the outer environment for the production of PolyP. On the application side of the PolyP, it is clearly visible due to its polyanionic neutral nature and tendency to form the complexes with cations, it is used for several purposes. It can form complex with the heavy metals, proteins, RNAs, and polyhydroxybutyrate. This property helps PolyP to contribute in the gene regulation, enzyme expression, and membrane transport. However, despite the extensive research work in this field, not much is known on the identification of the specific genes involved in metabolism and the transcription and translational regulation, which requires further exploration. PolyP-synthesizing enzymes in eukaryotes are also still unknown (only the degradation enzymes have been reported). In the eukaryotes, only long-chain PolyP has been reported, requiring further work on the identification and characterization of short-chain PolyPs. The role of PolyPs in the biotechnological and medical industry also needs further exploration, especially for the treatment of wounds, blood clotting, and other therapeutic uses.

11.9 Conflict of interest

Authors do not have any conflict related to this publication.

References

Achbergerova, L., & Nahalka, J. (2011). Polyphosphate—An ancient energy source and active metabolic regulator. *Microbial Cell Factories*, *10*, 63.

Ahn, K., & Kornberg, A. (1990). Polyphosphate kinase from *Escherichia coli*. Purification and demonstration of a phosphoenzyme intermediate. *Journal of Biological Chemistry*, *265*, 11734−11739.

Akiyama, M., Crooke, E., & Kornberg, A. (1993). An exopolyphosphatase of *Escherichia coli*. The enzyme and its ppx gene in a polyphosphate operon. *Journal of Biological Chemistry*, *268*, 633−639.

Albi, T., & Serrano, A. (2016). Inorganic polyphosphate in the microbial world. Emerging roles for a multifaceted biopolymer. *World Journal of Microbiology and Biotechnology*, *32*, 27.

Alori, E. T., Glick, B. R., & Babalola, O. O. (2017). Microbial phosphorus solubilization and its potential for use in sustainable agriculture. *Frontiers in Microbiology*, *8*, 971.

Anand, A., & Aoyagi, H. (2019). Estimation of microbial phosphate-accumulation abilities. *Scientific Reports*, *9*, 1−9.

Andreeva, N. A., Kulakovskaya, T. V., Kulakovskaya, E. V., & Kulaev, I. S. (2008). Polyphosphates and exopolyphosphatases in cytosol and mitochondria of *Saccharomyces cerevisiae* during growth on glucose or ethanol under phosphate surplus. *Biochemistry*, *73*, 65−69.

Aravind, J., Saranya, T., & Kanmani, P. (2015). Optimizing the production of polyphosphate from *Acinetobacter towneri*. *Global Journal of Environmental Science and Management*, *1*, 63−70.

Ault-Riche, D., Fraley, C. D., Tzeng, C. M., & Kornberg, A. (1998). Novel assay reveals multiple pathways regulating stress-induced accumulations of inorganic polyphosphate in *Escherichia coli*. *Journal of Bacteriology*, *180*, 1841−1847.

Baltscheffsky, M. (1967). Inorganic pyrophosphate as an energy donor in photosynthetic and respiratory electron transport phosphorylation systems. *Biochemical and Biophysical Research Communications*, *28*, 270−276.

Breiland, A. A., Flood, B. E., Nikrad, J., Bakarich, J., Husman, M., Rhee, T. H., ... Bailey, J. V. (2018). Polyphosphate-accumulating bacteria: Potential contributors to mineral dissolution in the oral cavity. *Applied Environmental Microbiology*, 84, e02440−17.

Bru, S., Jimenez, J., Canadell, D., Arino, J., & Clotet, J. (2017). Improvement of biochemical methods of polyP quantification. *Microbial Cell*, 4, 6.

De Venditis, E., Zahn, R., & Fasano, O. (1986). Regeneration of GTP-bound from GDP-bound form of human and yeast ras proteins by nucleotide exchange, stimulatory effect of organic and inorganic polyphosphates. *European Journal of Biochemistry*, 161, 473−478.

Dhivya, S., Narayan, A. K., Kumar, R. L., Chandran, S. V., Vairamani, M., & Selvamurugan, N. (2018). Proliferation and differentiation of mesenchymal stem cells on scaffolds containing chitosan, calcium polyphosphate and pigeonite for bone tissue engineering. *Cell Proliferation*, 51, e12408.

Dmitriev, V. V., Crowley, D., Rogachevsky, V. V., Negri, C. M., Rusakova, T. G., Kolesnikova, S. A., & Akhmetov, L. I. (2011). Microorganisms form exocellular structures, trophosomes, to facilitate biodegradation of oil in aqueous media. *FEMS Microbiology Letters*, 315, 134−140.

Docampo, R., & Moreno, S. N. (2001). The acidocalcisomes. *Molecular Biochemistry and Parasitology*, 114, 151−159.

Eixler, S., Selig, U., & Karsten, U. (2005). Extraction and detection methods for polyphosphate storage in autotrophic planktonic organisms. *Hydrobiologia*, 533, 135−143.

Espiau, B., Lemercier, G., Ambit, A., Bringaud, F., Merlin, G., Baltz, T., & Bakalara, N. (2006). A soluble pyrophosphatase, a key enzyme for polyphosphate metabolism in Leishmania. *Journal of Biological Chemistry*, 281, 1516−1523.

Fidler, M. C., Walczyk, T., Davidsson, L., Zeder, C., Sakaguchi, N., Juneja, L. R., & Hurrell, R. F. (2004). A micronised, dispersible ferric pyrophosphate with high relative bioavailability in man. *Brazilian Journal of Nutrition*, 91, 107−112.

Fishbane, S. N., Singh, A. K., Cournoyer, S. H., Jindal, K. K., Fanti, P., Guss, C. D., ... Gupta, A. (2015). Ferric pyrophosphate citrate (Triferic™) administration via the dialysate maintains hemoglobin and iron balance in chronic hemodialysis patients. *Nephrology Dialysis Transplantation*, 30, 2019−2026.

Gray, M. J., Wholey, W. Y., Wagner, N. O., Cremers, C. M., Mueller-Schickert, A., Hock, N. T., ... Jakob, U. (2014). Polyphosphate is a primordial chaperone. *Molecular Cell*, 53, 689−699.

Hassanian, S. M., Avan, A., & Ardeshirylajimi, A. (2017). Inorganic polyphosphate: A key modulator of inflammation. *Journal of Thrombosis and Haemostasis*, 15, 213−218.

Hong-Hermesdorf, A., Miethke, M., Gallaher, S. D., Kropat, J., Dodani, S. C., Chan, J., ... Merchant, S. S. (2014). Subcellular metal imaging identifies dynamic sites of Cu accumulation in *Chlamydomonas*. *Nature Chemical Biology*, 10, 1034−1042.

Hsieh, P. C., Shenoy, B. C., Jentoft, J. E., & Phillips, N. F. (1993). Purification of polyphosphate and ATP glucose phosphotransferase from *Mycobacterium tuberculosis* H37Ra: Evidence that poly(P) and ATP glucokinase activities are catalyzed by the same enzyme. *Protein Expression and Purification*, 4, 76−84.

Huang, G., Ulrich, P. N., Storey, M., Johnson, D., Tischer, J., Tovar, J. A., ... Docampo, R. (2014). Proteomic analysis of the acidocalcisome, an organelle conserved from bacteria to human cells. *PLoS Pathogen*, 10, e1004555.

Imsiecke, G., Munkner, J., Lorenz, B., Müller, W. E., Schroder, H. C., & Bachinski, N. (1996). Inorganic polyphosphates in the developing freshwater sponge *Ephydatia muelleri*: Effect of stress by polluted waters. *Environmental Toxicology and Chemistry: An International Journal*, 15, 1329−1334.

Itoh, H., Kawazoe, Y., & Shiba, T. (2006). Enhancement of protein synthesis by an inorganic polyphosphate in an *E. coli* cell-free system. *Journal of Microbiological Methods*, 64, 241−249.

Iwamoto, S., Motomura, K., Shinoda, Y., Urata, M., Kato, J., Takiguchi, N., ... Kuroda, A. (2007). Use of an *Escherichia coli* recombinant producing thermostable polyphosphate kinase as an ATP regenerator to produce fructose 1, 6-diphosphate. *Applied Environmental Microbiology*, 73, 5676−5678.

Keasling, J. D., Van Dien, S. J., Trelstad, P., Renninger, N., & McMahon, K. (2000). Application of polyphosphate metabolism to environmental and biotechnological problems. *Biochemistry. Biokhimiia*, 65, 324−331.

Keefe, A. D., & Miller, S. L. (1996). Potentially prebiotic synthesis of condensed phosphates. *Origin Life Evolution and Biosphere, 26*, 15−25.

Kenney, W. F., Schelske, C. L., & Chapman, A. D. (2001). Changes in polyphosphate sedimentation: A response to excessive phosphorus enrichment in a hypereutrophic lake. *Canadian Journal of Fisheries and Aquatic Sciences, 58*, 879−887.

Kim, D. U., Yoo, J. H., Lee, Y. J., Kim, K. S., & Cho, H. S. (2008). Structural analysis of sialyltransferase PM0188 from *Pasteurella multocida* complexed with donor analogue and acceptor sugar. *BMB Reports, 41*, 48−54.

Kim, H. Y., Schlictman, D., Shankar, S., Xie, Z., Chakrabarty, A. M., & Kornberg, A. (1998). Alginate, inorganic polyphosphate, GTP and ppGpp synthesis co-regulated in *Pseudomonas aeruginosa*: Implications for stationary phase survival and synthesis of RNA/DNA precursors. *Molecular Microbiology, 27*, 717−725.

Kornberg, A. (1995). Inorganic polyphosphate: Toward making a forgotten polymer unforgettable. *Journal of Bacteriology, 177*, 491−496.

Kornberg, A., Kornberg, S. R., & Simms, E. S. (1956). Metaphosphate synthesis by an enzyme from *Escherichia coli*. *Biochimica et Biophysica Acta, 20*, 215−227.

Kornberg, A., Rao, N. N., & Ault-Riche, D. (1999). Inorganic polyphosphate: A molecule of many functions. *Annual Review of Biochemistry, 68*, 89−125.

Kulaev, I., & Kulakovskaya, T. (2000). Polyphosphate and phosphate pump. *Annual Reviews in Microbiology, 54*, 709−734.

Kulaev, I. S. (1979). *Biochemistry of inorganic polyphosphates*. Chichester: Wiley and Sons.

Kulaev, I. S., & Vagabov, V. M. (1983). Polyphosphate metabolism in microorganisms. *Advancement in Microbial Physiology, 24*, 83−171.

Kulaev, I. S., Vagabov, V. M., & Kulakovskaya, T. V. (2004). *The biochemistry of inorganic polyphosphates*. John Wiley & Sons.

Kulaev, I. S., Vagabov, V. M., & Shabalin, Y. A. (1987). New data in biosynthesis of polyphosphates in yeast. In A. Torriani-Gorini, F. G. Rothman, S. Silver, A. Wright, & E. Yagil (Eds.), *Phosphate metabolism and cellular regulation in microorganisms*. Washington DC: American Society for Microbiology, (pp. 233−238).

Kulakova, A. N., Hobbs, D., Smithen, M., Pavlov, E., Gilbert, J. A., Quinn, J. P., & McGrath, J. W. (2011). Direct quantification of inorganic polyphosphate in microbial cells using 4′-6-diamidino-2-phenylindole (DAPI). *Environmental Science & Technology, 45*, 7799−7803.

Kulakovskaya, T. V., Andreeva, N. A., Trilisenko, L. V., Vagabov, V. M., & Kulaev, I. S. (2004). Two exopolyphosphatases in *Saccharomyces cerevisiae* cytosol at different culture conditions. *Process Biochemistry, 39*, 1625−1630.

Kulakovskaya, T. V., Vagabov, V. M., & Kulaev, I. S. (2012). Inorganic polyphosphate in industry, agriculture and medicine: Modern state and outlook. *Process Biochemistry, 47*, 1−10.

Kuroda, A., & Kornberg, A. (1997). Polyphosphate kinase as a nucleoside diphosphate kinase in *Escherichia coli* and *Pseudomonas aeruginosa*. *Proceedings of the National Academy of Sciences, 94*, 439−442.

Li, Y., Rahman, S. M., Li, G., Fowle, W., Nielsen, P. H., & Gu, A. Z. (2018). The composition and implications of polyphosphate-metal in enhanced biological phosphorus removal systems. *Environmental Science & Technology, 53*, 1536−1544.

Lichko, L. P., Kulakovskaya, T. V., & Kulaev, I. S. (2010). Properties of partially purified endopolyphosphatase of the yeast *Saccharomyces cerevisiae*. *Biochemistry, 75*, 1404−1407.

Liu, J., Huang, W., Pang, Y., & Yan, D. (2015). Hyperbranched polyphosphates: Synthesis, functionalization and biomedical applications. *Chemical Society Reviews, 44*, 3942−3953.

Liu, J., Huang, W., Pang, Y., Zhu, X., Zhou, Y., & Yan, D. (2010). Hyperbranched polyphosphates for drug delivery application: Design, synthesis, and in vitro evaluation. *Biomacromolecules, 11*, 1564−1570.

Lui, E. L. H., Ao, C. K. L., Li, L., Khong, M. L., & Tanner, J. A. (2016). Inorganic polyphosphate triggers upregulation of interleukin 11 in human osteoblast-like SaOS-2 cells. *Biochemical and Biophysical Research Communications, 479*, 766−771.

Marchesini, N., Ruiz, F. A., Vieira, M., & Docampo, R. (2002). Acidocalcisomes are functionally linked to the contractile vacuole of *Dictyostelium discoideum*. *Journal of Biological Chemistry*, 277, 8146–8153.

McGrath, J. W., & Quinn, J. P. (2003). Microbial phosphate removal and polyphosphate production from wastewaters. *Advances in Applied Microbiology*, 52, 75–100.

McGrath, J. W., Cleary, S., Mullan, A., & Quinn, J. P. (2001). Acid-stimulated phosphate uptake by activated sludge microorganisms under aerobic laboratory conditions. *Water Research*, 35, 4317–4322.

Mekmene, O., & Gaucheron, F. (2011). Determination of calcium-binding constants of caseins, phosphoserine, citrate and pyrophosphate: A modelling approach using free calcium measurement. *Food Chemistry*, 127, 676–682.

Montag, M., Clough, C. R., Müller, P., & Cummins, C. C. (2011). Cyclophosphates as ligands for cobalt(III) in water. *Chemical Communication*, 47, 662–664.

Moreno-Sanchez, D., Hernandez-Ruiz, L., Ruiz, F. A., & Docampo, R. (2012). Polyphosphate is a novel proinflammatory regulator of mast cells and is located in acidocalcisomes. *Journal of Biological Chemistry*, 287, 28435–28444.

Mullan, A., Quinn, J. P., & McGrath, J. W. (2002). A nonradioactive method for the assay of polyphosphate kinase activity and its application in the study of polyphosphate metabolism in *Burkholderia cepacia*. *Analytical Biochemistry*, 308, 294–299.

Muller, W. E., Tolba, E., Schröder, H. C., Wang, S., Glaber, G., Munoz-espi, R., ... Wang, X. (2015). A new polyphosphate calcium material with morphogenetic activity. *Materials Letters*, 148, 163–166.

Nishimura, S., Nakamura, M., Natsui, R., & Yamada, A. (2010). New lithium iron pyrophosphate as 3.5 V class cathode material for lithium ion battery. *Journal of American Chemical Society*, 132, 13596–13597.

Niyogi, S., Jimenez, V., Girard-Dias, W., de-souza, W., Miranda, K., & Docampo, R. (2015). Rab32 is essential for maintaining functional acidocalcisomes and for growth and infectivity of *Trypanosoma cruzi*. *Journal of Cell Science*, 128, 2363–2373.

Ogawa, N., DeRisi, J., & Brown, P. O. (2000). New components of a system for phosphate accumulation and polyphosphate metabolism in *Saccharomyces cerevisiae* revealed by genomic expression analysis. *Molecular Biology of the Cell*, 11, 12.

Ohtake, H., Kato, J., Kuroda, A., Wu, H., & Ikeda, T. (1998). Regulation of bacterial phosphate taxis and polyphosphate accumulation in response to phosphate starvation stress. *Journal of Biosciences*, 23, 491–499.

Pavlov, E., Aschar-Sobbi, R., Campanella, M., Tumer, R. J., Gomez-Garcia, M. R., & Abramov, A. Y. (2010). Inorganic polyphosphate and energy metabolism in mammalian cells. *Journal of Biological Chemistry*, 285, 9420–9428.

Ramos, I., Gomes, F., Koeller, C. M., Saito, K., Heise, N., Masuda, H., ... Miranda, K. (2011). Acidocalcisomes as calcium- and polyphosphate storage compartments during embryogenesis of the insect *Rhodnius prolixus*. *PLoS One*, 6, e27276.

Rao, N. N., & Kornberg, A. (2008). Inorganic polyphosphate supports resistance and survival of stationary-phase *Escherichia coli*. *Journal of Bacteriology*, 178, 1394–1400.

Rao, N. N., Gomez-Garcia, M. R., & Kornberg, A. (2009). Inorganic polyphosphate: Essential for growth and survival. *Annual Review of Biochemistry*, 78, 605–647.

Rao, S., Nanbu, N., & Juneja, N. (2004). Iron absorption and bioavailability in rats of micronized dispersible ferric pyrophosphate. *International Journal for Vitamin and Nutrition Research*, 74, 3–9.

Rashchi, F., & Finch, J. A. (2000). Polyphosphates: A review their chemistry and application with particular reference to mineral processing. *Minerals Engineering*, 13, 1019–1035.

Rauchfuss, T. B. (1990). *Phosphorus: An outline of its chemistry, biochemistry, and technology*. Elsevier.

Rohloff, P., Miranda, K., Rodrigues, J. C., Fang, J., Galizzi, M., Plattner, H., ... Moreno, S. N. J. (2011). Calcium uptake and proton transport by acidocalcisomes of *Toxoplasma gondii*. *PLoS One*, 6, e18390.

Rothschild, L. J., & Mancinelli, R. L. (2001). Life in extreme environments. *Nature*, 409, 1092–1101.

Schelske, C. L., Aldridge, F. J., & Kenney, W. F. (1999). Assessing nutrient limitation and trophic state in Florida lakes. *Phosphorus biogeochemistry in subtropical ecosystems*. Boca Raton, FL: Lewis Publishers, (pp. 321–342).

Seufferheld, M., Vieira, M. C., Ruiz, F. A., Rodrigues, C. O., Moreno, S. N. J., & Docampo, R. (2003). Identification of organelles in bacteria similar to acidocalcisomes of unicellular eukaryotes. *Journal of Biological Chemistry*, 278, 29971–29978.

Seufferheld, M. J., Alvarez, H. M., & Farias, M. E. (2008). Role of polyphosphates in microbial adaptation to extreme environments. *Applied and Environmental Microbiology, 74*, 5867−5874.

Sidat, M., Bux, F., & Kasan, H. C. (1999). Polyphosphate accumulation by bacteria isolated from activated sludge. *Water SA, 25*, 0378−4738.

Stotz, S. C., Scott, C., Drummond-Main, C., Avchalumov, Y., Girotto, F., Davidsen, J., ... Colicos, M. A. (2014). Inorganic polyphosphate regulates neuronal excitability through modulation of voltage-gated channels. *Molecular Brain, 7*, 42.

Sun, Z. Y., & Zhao, L. (2002). Feasibility of calcium polyphosphate fiber as scaffold materials for tendon tissue engineering in vitro. *Chinese Journal of Reparative and Reconstructive Surgery, 16*, 426−428.

Szymona, M., & Ostrowski, W. (1964). Inorganic Polyphosphate Glucokinase of *Mycobacterium Phlei*. *Biochimica et Biophysica Acta, 85*, 283−295.

Tarayre, C., Nguyen, H. T., Brognaux, A., Delepierre, A., Cleroq, L. D., Charlier, R., ... Delvigne, F. (2016). Characterisation of phosphate accumulating organisms and techniques for polyphosphate detection: A review. *Sensors, 16*, 797.

Vagabov, V. M., Trilisenko, L. V., Kulakovskaya, T. V., & Kulaev, I. S. (2008). Effect of carbon source on polyphosphate accumulation in *Saccharomyces cerevisiae*. *FEMS Yeast Research, 8*, 877−882.

Varas, M. A., Riquelme-Barrios, S., Valenzuela, C., Marcoleta, A. E., Berrios-Pasten, C., Santiviago, C. A., & Chavez, F. P. (2018). Inorganic polyphosphate is essential for *Salmonella typhimurium* virulence and survival in *Dictyostelium discoideum*. *Frontiers in Cellular and Infection Microbiology, 8*, 8.

Wang, X., Schroder, H. C., & Muller, W. E. G. (2014). Enzymatically synthesized inorganic polymers as morphogenetically active bone scaffolds: Application in regenerative medicine. *International Review of Cell and Molecular Biology, 313*, 27−77.

Wang, X., Schroder, H. C., Schlossmacher, U., & Muller, W. E. G. (2013). *Inorganic polyphosphates: Biologically active biopolymers for biomedical applications*. In Biomedical inorganic polymers, Heidelberg, Berlin, Germany: Springer, (pp. 261−294).

Wang, Y., Li, M., Li, P., Teng, H., Fan, D., Du, W., & Guo, Z. (2019). Progress and applications of polyphosphate in bone and cartilage regeneration. *BioMed Research International, 5141204*, 2019.

Wolska-Mitaszko, B. (1997). Trehalases from spores and vegetative cells of yeast *Saccharomyces cerevisiae*. *Journal of Basic Microbiology, 37*, 295−303.

Wood, H. G., & Clark, J. E. (1988). Biological aspects of inorganic polyphosphates. *Annual Review of Biochemistry, 57*, 235−260.

Wurst, H., & Kornberg, A. (1994). A soluble exopolyphosphatase of *Saccharomyces cerevisiae*. *Journal of Biological Chemistry, 269*, 10996−11001.

Yan, D. (2016). Hyperbranched polyphosphates and their biomedical applications. *Nanomedicine: Nanotechnology, Biology and Medicine, 2*, 466.

Zakharian, E., Thyagarajan, B., French, R. J., Pavlov, E., & Rohacs, T. (2009). Inorganic polyphosphate modulates TRPM8 channels. *PLoS One, 4*, e5404.

Zhang, F., Blasiak, L. C., Karolin, J. O., Powell, R. J., Geddes, C. D., & Hill, R. T. (2015). Phosphorus sequestration in the form of polyphosphate by microbial symbionts in marine sponges. *Proceedings of the National Academy of Sciences, 112*, 4381−4386.

Zhang, H., Ishige, K., & Kornberg, A. (2010). A polyphosphate kinase (PPK2) widely conserved in bacteria. *Proceedings of the National Academy of Sciences, 99*, 16678−16683.

CHAPTER 12

Production and applications of polylactic acid

Ashutosh Kumar Pandey[1], Ranjna Sirohi[2], Sudha Upadhyay[3], Mitali Mishra[4], Virendra Kumar[4], Lalit Kumar Singh[5] and Ashok Pandey[6]

[1]Centre for Energy and Environmental Sustainability, Lucknow, India, [2]Department of Post-Harvest Process and Food Engineering, G.B. Pant University of Agriculture and Technology, Pantnagar, India, [3]Department of Chemical Engineering, Indian Institute of Technology, Guwahati, India, [4]Department of Biotechnology, Dr. Ambedkar Institute of Technology for Handicapped, Kanpur, India, [5]Department of Biochemical Engineering, Harcourt Butler Technical University, Kanpur, India, [6]Centre for Innovation and Translational Research, CSIR—Indian Institute of Toxicology Research (CSIR—IITR), Lucknow, India

12.1 Introduction

Polylactic acid (PLA) is a biodegradable polymer formed by lactic acid (LA) monomers. LA is an organic acid that is naturally present in many foods or forms as a result of microbial fermentation in food like yogurt. The commercial demand for PLA has increased over the years insisting on its improved and more economic production. Most of the LA being produced today comes from fermentation using sustainable and renewable sources (Abdel-Rahman, Tashiro, & Sonomoto, 2011). The production through fermentation provides the advantages of low production temperatures, low energy consumption, and production of optically pure D- or L-LA when proper microorganisms and substrate are used (Abdel-Rahman et al., 2011). As the monomer for PLA, the fermentative production of LA needs to be engineered and improved in terms of process efficiency as well as costs associated with both the raw materials and process control. Genetic engineering approaches have been used to improve microbial strains to increase the yield and optical purity of obtained LA. LA can be commercially produced by both chemical and fermentative methods wherein a racemic mixture of D, L-LA is produced by chemical synthesis while the fermentation process can result in L-stereoisomer of LA (Joglekar, Rahman, Babu, Kulkarni, & Joshi, 2006). The LA in fermented broth is mostly present as calcium salt precipitate which is crystallized and filtered to remove impurities. However, impurities still retained with the salts are further removed by methods like reactive extraction, membrane filtration, and electrodialysis. The purified LA is polymerized into PLA by methods such as polycondensation, ring-opening polymerization, and by direct methods like azeotropic dehydration and enzymatic polymerization (Auras, Lim, Selke, & Tsuji, 2010). PLA with high L-isomer content appears in crystalline form with high optical purity, whereas D-isomer is amorphous with comparatively less optical purity. PLA has a wide range of applications in the textile industry, plastic and packaging, pharmaceutical and biomedical, and bioremediation.

PLA was discovered by Carothers (DuPont) in 1932 when he obtained a low molecular weight product when LA was heated under vacuum. PLA refers to a family of polymers with pure poly-L-lactic acid (PLLA), pure poly-D-lactic acid (PDLA), and poly-D, L-lactic acid (PDLLA). It is a biodegradable and biocompatible polymer with applications in areas like medicine, waste treatment, packaging, and others (Griffith, 2000). PLA has helical conformation with chiral carbon atoms. Naturally occurring PLA is generally the L-isomer since most of the LA obtained from renewable sources exist in the L-form. A left-handed helical α-form which was a pseudoorthorhombic crystal structure of PLLA was reported (Marega, et al., 1992). The US Food and Drug Administration (FDA) and European regulatory authorities have approved PLA use in food industries and in drug-delivery systems (Lampe, Namba, Silverman, Bjugstad, & Mahoney, 2009). PLLA predominantly occurs in crystalline form which along with its extended degradation time can cause inflammatory reactions in the body. A combination of L-LA and D, L-LA monomers can be beneficial in overcoming the degradation issue since D, L-LA degrades rapidly without forming crystals during the process (Fukushima & Kimura, 2008). Isotactic and syndiotactic are the two structural forms present in the stereocenters of the repeating monomer units. Isotactic and syndiotactic polymers are the ones that contain sequential stereocenters with the same and opposite relative configuration, respectively (Auras et al., 2010). PLLA and PDLA are isotactic, optically active, and crystalline while PDLLA is a relatively atactic, optically inactive amorphous structure (Bouapao, Tsuji, Tashiro, Zhang, & Hanesaka, 2009).

12.2 Substrate

Plenty of materials of natural origin have been regarded as suitable substrates that can be utilized in the production of LA. To make the overall process cost-effective, it is very crucial to minimize cost procured with basic starting material in the production process. Unutilized biomass such as starch, lignocellulose, and algal biomass along with by-products of agriculture and food industries, whey and glycerol are feasible substrates obtained from a renewable resource. Sugars from molasses (glucose, sucrose), dextrose from hydrolyzed starch, and whey (lactose and maltose) are utilized for commercial LA fermentation. However, molasses contains impurities and requires proper pretreatment to decrease effects on downstream processes. Spent pulping liquor, obtained as a by-product in the manufacturing of wood pulp, as well as Sun root are important substrates for the production of LA. The economic feasibility of LA production significantly lies in pretreatment and saccharification of substrates by physicochemical and enzymatic methods (Abdel-Rahman et al., 2011). Starchy biomass is being used commercially as a substrate and few works regarding direct fermentation of starchy biomass are present but the commercial feasibility of such a process is still questionable (Reddy, Altaf, Naveena, Venkateshwar, & Kumar, 2008). Also, multiple and time-consuming pretreatments mediated by physicochemical processes and cascade of enzymatic processes are required for the degradation of cellulose (Okano, Tanaka, Ogino, Fukuda, & Kondo, 2010). Materials belonging to lignocellulosic polymers such as postharvest material of corn including leaves, cobs, sugarcane bagasse, and waste obtained from the processing of woods are utilized for the production of LA after cascades of various types of processing (Cui, Li, & Wan, 2011; Laopaiboon, Thani, Leelavatcharamas, & Laopaiboon, 2010). Various types of organic wastes used as a raw material in LA production are described in Table 12.1. Different types of substrates and fermentation processes utilized for the production of LA which has been reported in different studies are depicted in Tables 12.2 and 12.3.

Whey is a by-product of dairy industries, discharged in wastewater after milk processing, causing ample pollution. Since it is enriched in different types of proteins, sugars, fats, vitamins, salts, and other important nutrients, it can be employed as a potent raw material in the production of LA (Panesar, Kennedy, Gandhi, & Bunko, 2007; Kim, Wee, Kim, Yun, & Ryu, 2006; Li, Shahbazi, Coulibaly, & Mims, 2007). The synthetic pathway involved in LA production via lactose metabolism has been shown in Fig. 12.1. This pathway is mediated by many species of *Lactobacillus*, including *Lactobacillus delbrueckii subsp. bulgaricus*, *Lactobacillus casei, Lactobacillus acidophilus* as well as by *Kluyveromyces marxianus and Lactococcus lactis* utilizing whey as raw material

12.3 Microbial production

LA is being produced nowadays using many classes of microbes, each offering broad substrate ranges, increased yield of product, less critical nutrient concentration as well as better quality and purity of the product. Furthermore, use of mixed strains also facilitates the use of more readily available and complex substrates and further enhances the production (Cui et al., 2011). Genetically engineered strains are now being utilized in LA production (Okano et al., 2010). Microorganisms used in fermentation can be bacteria, fungi, cyanobacteria, or algae. The choice of microorganism solely lies in the type of substrate used for fermentation (Lunelli, 2010). A wide genus of bacteria produces LA, the most common being cocci. Lactic acid bacteria (LAB) are mostly facultative anaerobes, Gram-negative, nonspore-forming, and are devoid of cytochromes. It is due to this reason that they are unable to synthesize ATP from respiration. LAB can be categorized as homofermentative or heterofermentative strains.

Table 12.1: Lactic acid production from different microbes and substrates (Sin & Tueen, 2019; Narayanan, Roychoudhury, & Srivastava, 2004).

Organism	Substrate	Yield (g g^{-1})	Reference
Enterococcus faecalis RKY1	Cornstarch	1.04	Wee, Reddy, and Ryu (2008)
	Tapioca starch	1.01	
	Potato starch	0.99	
	Wheat starch	0.99	
Lactobacillus rhamnosus strain CASL	Cassava powder	0.71	Shi et al. (2012)
Lactobacillus pentosus	Trimming vine shoots	0.76	Moldes, Torrado, Converti, and Dominguez (2006)
Bacillus coagulan strains 36D1	Paper sludge	0.77	Budhavaram and Fan (2009)
Lactobacillus delbrueckii IFO 3202	Rice bran	0.78	Tanaka et al. (2006)
Lactobacillus delbrueckii mutant Uc-3	Molasses	0.87	Dumbrepatil et al. (2008)
Lactobacillus rhamnosus ATCC 7469		0.97	Marques et al. (2008)
Lactobacillus delbrueckii Uc-3	Cellobiose and cellotriose	0.9	Adsul et al. (2007)
Lactobacillus sp. RKY2	Lignocellulosic hydrolysates	0.9	Wee and Ryu (2009)
Lactococcus lactis IO-1	Sugarcane bagasse	0.36	Laopaiboon et al. (2010)
Lactobacillus rhamnosus CECT-288	Apple pomace	0.88	Gullón, Yáñez, Alonso, and Parajó (2008)
Lactobacillus bifermentans	Wheat bran hydrolysate	0.83	Givry, Prevot, and Duchiron (2008)
Bacillus sp. strain	Corncob molasses	0.5	Guo, Yan, Jiang, Teng, and Wang (2010)
Bacillus coagulans DSM 2314	Lime-treated wheat straw	0.43	Maas et al. (2008)
Lactobacillus rhamnosus CECT-288	Cellulosic biosludges	0.38	Romaní et al. (2008)
Lactobacillus delbrueckii	Sugarcane juice	0.95	Calabia and Tokiwa (2007)
Sporolactobacillus sp. CASD	Peanut meal, glucose	0.93	Wang et al. (2011)
Lb. delbreuckii	Alfalfa fibers	0.35	Sreenath et al. (2001)
Lb. plantarum	Alfalfa fibers	0.46	Sreenath et al. (2001)
Lb. rhamnosus ATCC 9595	Apple pomace	0.88	Gullón et al. (2008)
Lb. casei	Banana wastes	0.1	Chan-Blanco, Bonilla-Leiva, and Velázquez (2003)
Lb. delbrueckii NCIM 2025	Cassava bagasse	0.94	John, Nampoothiri, and Pandey (2006)
B. coagulans 36D1	Cellulose	0.8	Ou, Ingram, and Shanmugam (2011)
Lb. rhamnosus CECT-288	Cellulosic biosludge	0.38	Romaní et al. (2008)
E. coli AC-521	Glycerol	56.8 (g/L)	Hong et al. (2009)
Lb. delbrueckii subsp. delbrueckii IFO 3202	Defatted rice bran	0.28	Tanaka et al. (2006)
Lb. manihotivorans LMG18011	Food wastes	0.1	Ohkouchi and Inoue (2006)
Lactobacillus amylovorous ATCC-33620	Corn, rice, wheat starches	0.70	Sin and Tueen (2019)
Lactic acid bacteria and Clostridium sp.	Kitchen wastes	0.62	Zhang, He, Ye, and Shao (2008)
Lb. plantarum A6	Mussel-processing wastes	0.98	Pintado, Guyot, and Raimbault (1999)

(Continued)

Table 12.1: (Continued)

Organism	Substrate	Yield (g g^{-1})	Reference
B. coagulans strains 36D1	Paper sludge	0.77	Budhavaram and Fan (2009)
B. coagulans strains P4−102B	Paper sludge	0.78	Budhavaram and Fan (2009)
Lb. rhamnosus ATCC 7469	Paper sludge	0.97	Marques et al. (2008)
Lb. casei ATCC 10863	Ram horn hydrolysate	0.44	Kurbanoglu and Kurbanoglu (2003)
Lc. lactis IO-1	Sugarcane baggage	0.36	Laopaiboon et al. (2010)
Lb. pentosus ATCC 8041	Vine-trimming wastes	0.77	Bustos, Moldes, Cruz, and Domínguez (2004)
Lb. coryniformis ssp. torquens ATCC 25600	Waste cardboard	0.51	Yáñez, Alonso, and Parajó (2005)
Lb. delbrueckii mutant Uc-3	Waste sugarcane bagasse	0.83	Adsul et al. (2007)
Lb. paracasei strain LA1	Wastewater sludge	0.72	Nakasaki and Adachi (2003)
Lb. brevis CHCC 2097 and Lb. pentosus CHCC 2355	Wheat straw	0.95	Garde, Jonsson, Schmidt, and Ahring (2002)
K. marxianus	Cheese whey	0.24	Plessas et al. (2008)
Lb. helveticus		0.23	
Lb. bulgaricus		0.3	
Lb. helveticus and K. marxianus (mixed culture)		0.45	
Lb. bulgaricus and K. marxianus (mixed culture)		0.41	
Lb. helveticus and Lb. bulgaricus (mixed culture)		0.35	
Lb. helveticus and Lb. bulgaricus and K. marxianus (mixed culture)		0.47	
Lb. casei NRRL B-441	Cheese whey	0.93	Büyükkileci and Harsa (2004)
Lactococcus lactis ssp. ATCC 19435	Wheat starch	0.77−1	Sin and Tueen (2019)
Lactobacillus sp.	Wheat and rice bran	129 (g L^{-1})	Sin and Tueen (2019)
Rhizopus sp. MK.96-1196	Corncob	90 (g L^{-1})	Sin and Tueen (2019)
Lb. casei SU No. 22 and Lb. lactis WS 1042 (mixed culture)	Cheese whey	0.48	Roukas and Kotzekidou (1998)
B. coagulans LA204	Corn stover	97.59 (g L^{-1})	Hu et al. (2015)
Lb. bulgaricus ATCC 8001, PTCC 1332	Cheese whey	0.81	Fakhravar, Najafpour, Heris, Izadi, and Fakhravar (2012)
Bacillus sp. NL01	Lignocellulosic hydrolyzates	75.0 (g L^{-1})	Ouyang et al. (2013)
L. delbrueckii NCIM 2025, L. casei	Cassava bagasse	0.9−0.98	Sin and Tueen (2019)
L. amylovorous NRRL B-4542	Cornstarch	0.935 g g^{-1}	Sin and Tueen (2019)

Homofermentative LAB converts one molecule of glucose into two molecules of LA with the synthesis of two molecules of ATP. Heterofermentative LAB works under the phosphor ketolase pathway. They transform one molecule of xylose to one molecule of LA as well as synthesizes acetic acid from ethanol (Castillo Martinez et al., 2013).

Table 12.2: Batch fermentation of Lactic acid with different substrates and microbes (Abdel-Rahman et al., 2013).

Organism	Substrate	Fermentation mode	Yield (g/g)	Reference
Lb. delbrueckii	Broken rice	Batch	0.81	Nakano et al. (2012)
Lb. plantarum NCIMB 8826 (engineered)	Cornstarch	Batch	0.85	Okano et al. (2009)
Lb. rhamnosus and Lb. brevis (mixed culture)	Corn stover	Batch	0.7	Cui et al. (2011)
Lb. paracasei subsp. paracasei CHB2121	Glucose	Batch	0.96	Moon et al. (2012)
Bacillus sp. Na-2	Glucose	Batch	0.94	Qin et al. (2010)
Rhizopus oryzae GY18	Glucose	Batch	0.81	Guo et al. (2010)
Lb. paracasei KCTC13169	Jerusalem artichoke tuber extract	Batch	0.98	Choi et al. (2012)
Lb. rhamnosus ATCC 7469	Liquid distillery stillage	Batch	0.73	Djukić-Vuković et al. (2012)
H. halophilus JCM 21694	Sucrose	Batch	0.83	Calabia et al. (2011)
Escherichia coli (engineered)	Sucrose	Batch	0.85	Wang et al. (2012)
Rhizopus oryzae GY18	Sucrose	Batch	0.89	Guo et al. (2010)
Rhizopus oryzae GY18	Xylose	Batch	0.85	Guo et al. (2010)
Candida utilis (engineered)	Xylose	Batch	0.91	Saito et al. (2012)
Lb. Rhamnosus LA-04-1	White rice bran hydrolysate	Batch	0.81	Li et al. (2012)

Table 12.3: Fed-batch fermentation of different substrates using various microbes for lactic acid production (Abdel-Rahman et al., 2013).

Organism	Substrate	Fermentation mode	Yield (g g^{-1})	References
Lb. lactis BME5-18M	Continuous feeding, glucose	Fed-batch	0.97	Bai et al. (2003, 2004)
Lb. casei LA-04-1	Constant feed rate, glucose	Fed-batch	0.88	Ding and Tan (2006)
Lb. lactis-11	pH feedback-controlled substrate feeding, glucose	Fed-batch	0.99	Zhang et al. (2010)
Corynebacterium glutamicum	Pulse feeding, glucose	Fed-batch	0.87	Okino et al. (2008)
Rhizopus oryzae NRRL 395	Pulse feeding, glucose	Fed-batch	0.6	Liu et al. (2006)
Lb. casei LA-04-1	Pulse feeding, glucose	Fed-batch	0.89	Ding and Tan (2006)

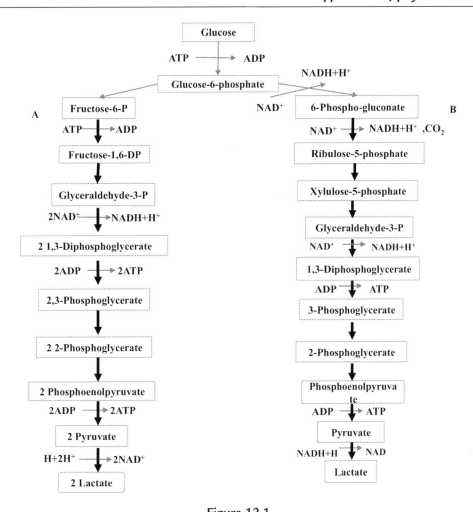

Figure 12.1
Metabolic pathways for lactic acid production. (A) Embden−Meyerhof−Parnas; (B) 6-phosphogluconate/phosphoketolase (Saad, 2006; Lasprilla et al., 2012).

Homofermentative LAB strains produce pure isomers of LA which is the reason why they are industrially more attractive than the heterofermentative strains of LAB (Reddy et al., 2008; Abdel-Rahman, Tashiro, & Sonomoto, 2013).

Filamentous fungi like *Rhizopus* species; *Rhizopus oryzae* can generate LA along with ethanol, LA is the main product (Meussen, De Graaff, Sanders, & Weusthuis, 2012). The yield of ethanol produced can be controlled with the addition of alcohol dehydrogenase inhibitors. Filamentous fungi show tremendous growth in nitrogen-rich media which often leads to the formation of chitin instead of LA. To overcome this, fungal pellets are used instead of spores, and growth is carried out in the presence of high carbon content instead of high nitrogen content (Ge, Qian, & Zhang, 2009).

12.3.1 Use of bacterial strains

Various classes of bacteria are being employed in the production of LA, which can broadly be classified into, LAB, *Bacilli, Corynebacterium glutamicum*, and *Escherichia coli*. To counter the various issues encountered in the LA production using bacteria (Budhavaram & Fan, 2009; Litchfield, 2009), removal of the gene associated with D- or L-lactate dehydrogenase to increase optical purity (Kylä-Nikkilä, Hujanen, Leisola, & Palva, 2000) and pyruvate formate lyase, alcohol dehydrogenase, acetate kinase to reduce the by-products formed (Zhou, Shanmugam, & Ingram, 2003), development of new strains (Zhou, Causey, Hasona, Shanmugam, & Ingram, 2003), and strain improvements for increased resistance against bacteriophages (Sreenath, Moldes, Koegel, & Straub, 2001) have been carried out.

Homofermentative LAB generally possesses aldolases, yielding LA as the main product, and therefore are of great importance in commercial LA production. Heterofermentative LAB, on the other hand, produces LA along with other by-products through phosphoketolase pathways. Some species like *Enterococcus mundtii* and genetically engineered *Lactobacillus plantarum* are known to show homofermentative production of LA (Abdel-Rahman et al., 2011; Abdel-Rahman et al., 2011). LAB are additionally advantageous as they do not cause any health hazards for the workers as well as consumers. Some strains are also known for their high acid tolerance, and can be easily engineered to produce D- or L-forms selectively (Kylä-Nikkilä et al., 2000; Lapierre, Mollet, & Germond, 2002). However, the necessity of complex nutrients in the fermentation broth increases the production cost (Litchfield, 2009; Hofvendahl & Hahn-Hägerdal, 2000). Also, low fermentation temperatures increase the risk of contamination. To counter this issue, alkaliphilic or thermotolerant strains like *Halolactibacillus halophilus* come up as promising alternatives (Calabia, Tokiwa, & Aiba, 2011).

In comparison to LAB, various strains of *Bacillus* like *Bacillus coagulans, Bacillus stearothermophilus, Bacillus licheniformis*, and *Bacillus subtilis* produce higher concentrations of product in mineral salt containing media with few N-sources (Wang, Zhao, Chamu, & Shanmugam, 2011). Alkaliphilic strains are exceptionally high yielding, simultaneously minimizing the risk of contamination (Meng, Xue, Yu, Gao, & Ma, 2012), and some are even known to undergo thermal fermentation. These properties are very economic, as coolant costs are reduced, and solid-state fermentation of lignocellulosic biomass can be easily carried out at optimal temperatures (Budhavaram & Fan, 2009; Maas et al., 2008; Qin et al., 2009). This also enables open fermentation at high temperatures (Qin et al., 2009) using various strains like *B. coagulans* (Ou, Mohammed, Ingram, & Shanmugam, 2009; Patel et al., 2006; Rosenberg, Rebroš, Krištofíková, & Malátová, 2005), *B. licheniformis* (Wang et al., 2011; Sakai & Yamanami, 2006) among others.

Since the screening of commercially successful LA-producing species is a rather tedious process, *E. coli* has been the subject of various studies about genetic manipulation of the bacterium to produce the maximum yield with the most minimal nutritional requirements. Strain improvements have been conducted on *E. coli* (Chang, Jung, Rhee, & Pan, 1999; Kochhar et al., 1992; Taguchi & Ohta, 1991) for the generation of LA from glucose, xylose, sucrose, and glycerol (Zhou et al., 2003; Zhou et al., 2003; Chang et al., 1999; Dien, Nichols, & Bothast, 2001; Zhou, Yomano, Shanmugam, & Ingram, 2005; Zhu, Eiteman, DeWitt, & Altman, 2007; Wang et al., 2012; Mazumdar, Clomburg, & Gonzalez, 2010). However, tolerance of the genetically modified species is lowered, when compared to other LAB and *Bacillus* species (Portnoy, Herrgård, & Palsson, 2008). Direct production of PLA from glucose using engineered *E. coli* through a single-step fermentation has been observed (Yang et al., 2010).

Several engineered *C. glutamicum* strains produce a combination of organic acids, including LA under oxygen-deficient conditions (Kawaguchi, Vertes, Okino, Inui, & Yukawa, 2006; Kawaguchi, Sasaki, Vertès, Inui, & Yukawa, 2008; Sasaki, Jojima, Inui, & Yukawa, 2008). Genetically engineered strains containing genes from species like *L. bulgaricus* produce 99.9% optically pure D-LA (Jia, Liu, Li, Li, & Wen, 2011). Oxygen deficiency is generally created to channel the nutrients into the production of LA and not on undue microbial growth (Inui et al., 2004), as a proportional relationship between microbial growth and LA production is observed (Okino, Inui, & Yukawa, 2005). Thus *C. glutamicum* can be optimally used to mass-produce LA without the use of complex and expensive media (Song et al., 2012).

12.3.2 Use of fungi and yeast

R. oryzae, among other species of the genus *Rhizopus*, has been utilized for the production of high quality, pure LA (Bai, Li, Liu, & Cui, 2008; Taskin, Esim, & Ortucu, 2012; Wu et al., 2011) because of the various advantages like the ability to metabolize starchy biomass without preprocessing (Jin, Huang, & Lant, 2003), and low nutritional needs (Bulut, Elibol, & Ozer, 2004; Marták, Schlosser, Sabolová, Krištofíková, & Rosenberg, 2003; Oda, Yajima, Kinoshita, & Ohnishi, 2003). Also, the filamentous nature of the pellet makes it easy to segregate from the fermentation broth, hence reducing the cost and tedium of downstream processing (Zhang, Jin, & Kelly, 2007). The filamentous morphology also affects the aeration and rheology of the culture, hence enabling repetitive and prolonged use of the same culture (Maneeboon, Vanichsriratana, Pomchaitaward, & Kitpreechavanich, 2010). However, the use of *Rhizopus* has its challenges, which include the production of by-products as well as difficulty in bulk mixing due to the filamentous nature of the mycelium (Litchfield, 2009; Liu et al., 2006; Liu, Liao, Liu, & Chen, 2006).

Yeasts have been extensively studied due to their ability to grow in mineral media (Dequin & Barre, 1994) and tolerance toward acidic pH, which reduces the cost of neutralizing agents as well as reduces the chances of contamination (Praphailong & Fleet, 1997). Using genetic modification, basic metabolic pathways of yeast can be altered and employed for increased LA production (Bianchi et al., 2001). *Saccharomyces cerevisiae* has been genetically altered to express beta-galactosidase from *Aspergillus aculeatus*, which enables it to utilize cellobiose as a raw material to generate LA and produce a fairly high concentration of the product (Tokuhiro, Ishida, Kondo, & Takahashi, 2008).

12.3.3 Use of cyanobacteria

Due to the recent awareness about global warming, any processes that utilize greenhouse gases are being channeled and incorporated into industrial procedures to minimize greenhouse gases. Microbes like blue-green algae are being used as alternative LA producers not only to minimize the carbohydrate feedstock cost but also to make the entire endeavor environment-friendly. Photosynthetic cyanobacteria species have been known for producing LA under anaerobic conditions from the starch produced by aerobic photosynthetic processes (Hirayama & Ueda, 2004; van der Oost, Bulthuis, Feitz, Krab, & Kraayenhof, 1989). Microfloral strains like *Scenedesmus obliquus* (Hirt, Tanner, & Kandler, 1971) and *Nannochlorum* sp. Hirayama and Ueda (2004) are suitable examples. Since cyanobacteria have almost minimal nutritional requirements as well as are photosynthetic, in addition to being easily engineered, they can be used for the production in a carbon-neutral and environment-friendly manner (Ducat, Way, & Silver, 2011). Genetically engineered *Synechocystis* species have been known to produce significant amounts of LA in this manner (Angermayr, Paszota, & Hellingwerf, 2012).

12.4 Strain improvement

LA is one of the significant substrates of PLA production and other important products. Microbes such as bacteria, fungi, and yeast synthesize PLA naturally by the means of fermentative processes utilizing C5 and C6 sugars. Enzymes like xylose isomerase, L and D lactate dehydrogenase, phosphoketolase, and transaldolase affect the overall productivity of LA. The efficiency of microbes to utilize unusual C sources can be improved by the means of metabolic engineering. The increasing demand for thermostable PLA gains the attention of many researchers to initiate the production of D-LA. The commercial production of D-LA is still limited. The homofermentative production of L-LA is mediated by *B. coagulans* using lignocellulosic sugars as substrates. Advancements in lab technologies including automation and high throughput screening make the process of specific strain collection less tedious (Holzapfel, 2014). Various wild-type microbial strains have crucial properties and are industrially important, however, certain modifications are required to make the most of their potentiality. Sometimes it might be required to eradicate their unnecessary properties. Genetic engineering practices would be considered as perfect techniques for carrying out favorable modifications in microbes of industrial importance but ethical issues associated with these r-DNA technologies restrict their applicability. Hence strain improvement methodologies adopt natural approaches like random mutagenesis, dominant selection, and directed evolution or adaptive evolution.

Random mutagenesis is a typical strain improvement method reported in many studies related to the food industry (Margolles & Sánchez, 2012; Saarela et al., 2011). Important steps of this methodology are alteration of the genomic constitution of microbes via introducing random mutations; characterization of variants; screening of variants with desired properties. Directed evolution (or adaptive evolution) reflects an application parameter (Lapierre et al., 2002; MacLean, Torres-Barceló, & Moxon, 2013). In this classical approach, the microbial population is augmented for strains with the required properties but, in this case also, there is a possibility of accretion of accidental mutations (Barrick & Lenski, 2013). In dominant selection, only the strains having a unique property is allowed to grow in certain media (Kibenich, Sørensen, & Johansen, 2012). This approach requires a significant understanding of microbial physiology.

12.5 Commercial strains

Over the years, a diverse range of microorganisms have been identified which are economically feasible for the mass production of this polymer. Many microbial isolates have been segregated not only based on their efficiency in converting pentose and hexose residues into LA but also based on the substrates they consume. These microbes can produce LA from various commonly available sources like soil, milk, fruits and vegetables, and animal excreta (Abdel-Rahman et al., 2013; Ye, Bin Hudari, Li, & Wu, 2014). Some soil-inhabiting thermophiles with the ability to metabolize C5 residues and many such similar microbial isolates are identified (Hong et al., 2009; Mohd Adnan & Tan, 2007). The prime concern in the industrial production of the monomer is the cost-effective procurement of high concentrations of quality sugar. Agricultural biomass discarded as refuse can be very easily used for conversion into LA after a thorough pretreatment procedure (Juturu & Wu, 2016).

Mostly, LA-producing microbes are Gram-negative bacteria, but some Gram-positive species have also been identified. These microbes do not produce spores, and fermentation is their chief energy production pathway since a well-defined electron transport system has not yet been identified. This also means that LA-producing bacteria undergo energy scarcity, and proliferate very slowly as a result, when compared to LA Bacilli (Reddy et al., 2008). One of the first microbial species employed for the commercial production of LA was *Ralstonia eutropha*, wherein two acetyl-CoA molecules undergo coupling by a

3-ketothiolase to generate acetoacetyl-CoA. *C. glutamicum* has evolved as the most widely used microbe in this respect (Nduko & Taguchi, 2019). Lignocellulosic residues are being commonly used for the homofermentative production of LA, as they can be commonly sourced from agricultural as well as vegetative refuse. A thermophilic bacterium *B. coagulans* produce L-LA from lignocellulosic residues homofermentatively under unsterilized conditions, but the industrial applicability of this microbe is still unexplored because of lack of adequate genetic tools (Juturu & Wu, 2016). Since a vast variety of microbes are readily available and can produce LA residues in adequate quantities, the economic and cost-efficient production of LA is not that difficult. Various genera of bacteria that are easily isolated are *Lactobacillus, Pediococcus, Aerococcus, Carnobacterium, Enterococcus, Tetragenococcus, Vagococcus, Leuconostoc, Oenococcus, Weissella, Streptococcus*, and *Lactococcus*. The most predominantly used species are *Lactobacilli* and *cocci*, and such species have been provided a "Generally Marked As Safe" or GRAS status (Reddy et al., 2008). Most commercial high-yielding strains include *Lactobacillus maninotivorans* and *Lactobacillus lactis*. Obligate heterolactic LAB includes *Lactobacillus bifermentans, Leuconostoc lactis, Lactobacillus sanfrancis*, etc. Facultative heterolactic LAB includes species like *Lactobacillus pentosus and L. plantarum*. Homolactic LAB like *Lactobacillus helveticus, Lactobacillus rhamnose*, and *Lactobacillus acidophilus* are also used, along with other bacteria like *Streptococcus* and *Lactococcus* species (Cubas-Cano, González-Fernández, Ballesteros, & Tomás-Pejó, 2018). *Lactobacillus paracasei* has been known to show productivity of about 1.03 g L^{-1} h^{-1} from 80 g biomass (Nguyen et al., 2012).

In consortium with these bacteria, many other classes of microbes have also been used to supplement the synthesis of LA monomeric residues. *R. oryzae* is a filamentous fungus that produces LA. *Aspergillus niger* has been used in consortium with *Lactobacillus* species to enhance LA production (Ge et al., 2009). Yeast species naturally do not possess the native pathways for the generation of LA. However, strains have been produced using metabolic engineering techniques to synthesize LA (Sauer, Porro, Mattanovich, & Branduardi, 2010). *Candida sonorensis* species is methylotrophic yeast that can transform pentoses and hexoses into LA. Also, certain algae and cyanobacteria species have been identified for enhancing LA production. Algae like *Hydrodictyon reticulum* contain polysaccharides (47.5%) which can be transformed into glucose and mannose and in turn into LA.

12.6 Fermentation modes and bioreactors

12.6.1 Batch fermentation

Currently, most of the LA fermentation is done in batch mode. As the operation occurs in closed state, the chances of contamination are less as compared to other modes and high

LA concentrations are obtained (Hofvendahl & Hahn-Hägerdal, 2000). Being a closed system, batch fermentations face the drawback of low cell concentration due to the limited amount of nutrients supplied in the feed which affects the productivity of LA due to substrate or product inhibition. LA production using sugarcane bagasse cellulose in batch fermentation provided 67 g L^{-1} LA with a yield of 83% but low productivity 0.93 g L^{-1} h^{-1} (Adsul, Varma, & Gokhale, 2007). The highest L-LA concentration (192 g L^{-1}) in batch fermentation is reported from 200 g L^{-1} glucose using *L. paracasei* subsp. *paracasei* CB2121 (Moon, Wee, & Choi, 2012). Kadam et al. (2006) compared the production of LA by wild and mutant strains and the highest production was obtained in batch fermentation giving 135 g L^{-1} LA with 150 g L^{-1} cane sugar. Both wild type and mutants utilized glucose more readily than other carbon sources; an increase in initial glucose concentration increased the LA concentration. Different methods for LA production are reported in batch state fermentation. Simultaneous saccharification and fermentation (SSF) (Cui et al., 2011; Nakano et al., 2012), separate hydrolysis and fermentation (SHF) (Li et al., 2012), or use of mixed culture (Cui et al., 2011) have been used to enhance the production of LA with different substrates. The use of single reaction vessels, reduction of end-product, inhibition of hydrolysis, and rapid processing time with high productivity are some of the major advantages of using SSF over SHF (Abdel-Rahman et al., 2011). The separate enzymatic hydrolysis step before fermentation in SHF reduces the overall productivity as reported in the work of Marques, Santos, Gírio, and Roseiro (2008) who obtained a lower LA yield in SHF (0.81 g g^{-1}) than SSF (0.97 g g^{-1}) from recycled paper sludge with the use of *Lactobacillus rhamnosus* ATCC 7469. The improved LA yield of 0.70 g g^{-1} by SSF was obtained in the presence of mixed strains of *L. rhamnosus* and *Lactobacillus Brevis* using sodium hydroxide (NaOH) treated corn stover (Cui et al., 2011). Similarly mixed strains of *L. casei subsp. rhamnosus* NRRL-B445 and *L. lactis subsp. lactis* ATCC 19435 provided the LA concentration of 60.3 g L^{-1}, higher than that obtained separately by single strains (Nancib et al., 2009).

Aeration has been reported to show effects on LA production in strains like *Bacillus* and *E. coli* (Zhu et al., 2007; Qin et al., 2010). A decrease in LA production is shown by *Bacillus* species in the absence of aeration due to low biomass while waste by-products are formed when too much aeration is provided (Qin et al., 2010). Neutralizing agents like sodium hydroxide, calcium carbonate, or ammonium solution, if added during fermentation, capture the undissociated LA as lactate salts, reducing the inhibition of microbes which would occur otherwise if the undissociated LA concentration keeps increasing in the fermentation broth (Qin et al., 2010; Tashiro et al., 2011). Tashiro et al. (2011) obtained higher productivity of D-LA by *L. delbrueckii* subsp. *lactis* QU41 using 1.67 g L^{-1} h^{-1} NH_4OH rather than using 1.3 g L^{-1} h^{-1} NaOH. LA concentration of 106 g L^{-1} was obtained by *Bacillus subtilis* CH 1 at pH 7.0, maintained by using potassium hydroxide, with glucose as a substrate (Romero-Garcia et al., 2009). The addition of low-cost media supplements including nitrogen sources led to the increased and economic production of LA. Molasses can be utilized as a suitable carbon as well as a nitrogen source (in place of yeast extract) for LA production (Dumbrepatil, Adsul, Chaudhari, Khire, & Gokhale, 2008). Yu et al. (2008) substituted corn steep liquor in place of yeast extract as a nitrogen source to achieve L-LA concentration of 115.1 g L^{-1}, yield 0.96 g g^{-1}, and productivity of 4.58 g L^{-1} h^{-1} in batch fermentation (Yu et al., 2008). LA production from molasses using chicken feather hydrolysate as nitrogen source provided a higher LA concentration of 38.5 g L^{-1} than yeast extract (33.2 g L^{-1}) and ammonium sulfate (28.5 g L^{-1}) (Taskin et al., 2012). Production of LA by the use of immobilized cells has been reported in batch fermentation systems. Polyurethane foam (PUF) was used for the immobilization of the co-culture of *L. paracasei* subsp. *paracasei* and *L. delbrueckii subsp. delbrueckii* mutant, further coated with cross-linked alginate for reducing cell leakage (John & Madhavan, 2011). More than 38 g L^{-1} LA yield was attained using PUF biofilms in 24 h while the productivity slightly decreased due to the alginate coating (John & Madhavan, 2011). LA production with immobilized *R. oryzae* on the cotton matrix was 37.83 g L^{-1} from 70 g L^{-1} glucose with productivity of 2.09 g L^{-1} h^{-1} (Chotisubha-Anandha et al., 2011).

12.6.2 Fed-batch fermentation

This mode of fermentation is especially desirable in cases where the change in nutrient concentration affects the biomass and productivity of the product (Lee et al., 1999). Fed-batch systems are considered better than batch and continuous because of their capability to reduce substrate inhibition, possibly due to the continuous or intermittent feeding of the substrate (Abdel-Rahman et al., 2011). Therefore the time and amount of substrate fed is an important factor to be considered while dealing with the fed-batch process. These systems are however comparatively less explored for the production of LA and few studies report its use for LA production. There are different methods of feeding in a fed-batch system such as intermittent, constant, and exponential which affect the productivity of LA. Bai et al. (2003) obtained the highest concentration of LA reported among batch or fed-batch with LAB by using low levels of initial glucose and continuous feeding. The inhibitory effects of glucose on LA production were reduced, as a result, obtaining 210 g L^{-1} LA (Bai et al., 2003). Pulse feeding methods (single and multi), peanut meal as a nitrogen source, and NaOH as a neutralizing agent were used for the production of LA using *Bacillus* species WL-S20 (Meng et al., 2012). In single pulse fed-batch fermentation, 180 g L^{-1} L-LA with 98.6% yield was obtained while 225 g L^{-1} L-LA was obtained by multiphase fed-batch fermentation with a yield of 99.3% (Meng et al., 2012). Ding and Tan (2006) studied the effect of different feeding strategies in the fed-batch system (pulse fed-batch, constant feed rate fed-batch, constant residual glucose concentration fed-batch, and exponential fed-batch) on LA production by *L. casei*. Exponential feeding of glucose (850 g L^{-1}) and yeast extract (1%) resulted in a maximum L-LA concentration of 180 g L^{-1} with yield 90.3% and productivity 2.14 g L^{-1} h^{-1} (Ding & Tan, 2006). As high as 226 g L^{-1} LA has been achieved by fed-batch fermentation (Wang et al., 2011) which makes it comparatively the preferred mode for industrial LA production, however, the limitation is the low productivity (Romaní, Yáñez, Garrote, & Alonso, 2008). Immobilization of cells in various bioreactors operating in the fed-batch system is still less studied. Immobilized *L. lactis* cells were used for LA production by hydrolysate of Jerusalem artichoke in a fibrous bed bioreactor by Shi et al. (2012) which resulted in a maximum LA concentration of 142 g L^{-1} in the fed-batch system. Fed-batch fermentation of immobilized *R. oryzae* NBRC 5384 was performed by Yamane and Tanaka (2013). Addition of calcium carbonate for pH control resulted in production of 280 g L^{-1} calcium lactate salt (initial glucose: 180 g L^{-1} and intermittent addition: 100 g L^{-1}) with yield being 92.5% and productivity 1.83 g L^{-1} h^{-1} (Yamane & Tanaka, 2013).

12.6.3 Continuous fermentation

Batch and fed-batch fermentation processes suffer from the effects of end-product inhibition. Processes that occur in continuous mode can overcome end-product inhibition due to the provision of continuous medium feeding which dilutes the product in the fermentation broth (Amrane & Prigent, 1996). Continuous fermentation systems occur in two modes: chemostat and turbid state. Turbidostatic systems control the changing turbidity of the fermentation broth due to varying cell concentration by controlling the dilution rates (Zhao et al., 2010). In chemostat, the rate of feeding of a medium into the fermenter and the rate of broth removal are the same which allows for the stable maintenance of concentration of cells, products, and medium in the reaction vessel. At steady-state conditions, the specific growth rate becomes equal to the dilution rate (Bustos et al., 2007). Batch fermentations give higher LA concentration but the operation is time-consuming because the vessel is emptied, cleaned, sterilized, and refilled after each batch while this is not the case in continuous processes and the productivity is comparatively high too (Dumbrepatil et al., 2008; Shibata et al., 2007). Although the productivity is high, certain drawbacks like the removal of residual substrate and cells as effluent along with the reduction in LA concentration with increasing dilution rate impart problems during LA production (Zhang et al., 2011). Continuous culture system with amylolytic LAB *Enterococcus faecium* No. 78 was used for high optical purity LA production obtaining productivity of 1.56 g L^{-1} h^{-1} from sago starch (Shibata et al., 2007). Tashiro et al. (2011) employed a conventional continuous system for D-LA production from glucose by *L. delbrueckii subsp. lactis* QU41 achieving 2.07–3.55 g L^{-1} h^{-1} productivity, higher than batch culture (1.67 g L^{-1} h^{-1}) (Tashiro et al., 2011). The productivity of LA from cassava starch increased with increasing dilution rate in a conventional continuous system with *L. plantarum* SW14 achieving productivity of 4.53 g L^{-1} h^{-1} at a dilution rate of 0.4 h^{-1} which was 4.7 times higher than batch fermentation (Bomrungnok et al., 2012).

Immobilization of cells is also used in continuous fermentation for LA production. Immobilized cells eliminate the chances of cell washout and hence can be operated even at high dilution rates. However, the long continuous operation lead to slow leakage of cells which can be avoided by using double layer beads (Tanaka et al., 1989). Pimtong et al. (2017) used immobilized *R. oryzae* cells for LA production from glucose in continuous operation (215 h) at an appropriate dilution rate obtaining LA concentration as high as 72.32 g L^{-1}. LA production in a bioreactor packed with a biofilm of *L. delbrueckii* immobilized on PUF resulted in productivity of 5 g L^{-1} h^{-1} (Rangaswamy & Ramakrishna, 2008). *L. delbrueckii* ZU-S2 cells immobilized in calcium alginate gel beads were used by Shen and Xia (2006) for the production of LA. At a dilution rate of 0.13 h^{-1}, the yield of LA was 92.4% with the productivity of 5.746 g L^{-1} h^{-1} (Shen & Xia, 2006).

12.7 Type of reactors used for production

12.7.1 Continuous stirred tank reactor

A stirred tank bioreactor is used for submerged fermentation mainly because of the provision of perfect mixing of fermentation broth making it ideal for commercial LA production. However, in the case of filamentous fungal fermentation, the formation of mycelial clumps causes operational problems due to insufficient mixing, low mass transfer, high shear rate, and increased power requirement (Chotisubha-Anandha et al., 2011; Jeong et al., 2001). To overcome such issues, immobilization has become a preferred option to avoid the problems associated with both fungal cells and losses in the case of bacterial cells. An acid-adapted preculture approach was used by Zhang et al. (2008) to control the morphology of *Rhizopus arrhizus*. Large clumps coalesced loose small pellets and freely dispersed small pellets were maintained using inoculum of acid-adapted precultures achieving a LA concentration and yield of 85.7 g L^{-1} and 86%, respectively (Zhang et al., 2008). LA production by cells of *L. casei* SU No. 22 and *L. lactis* WS 1042 coimmobilized in calcium alginate, K carrageenan, agar, and polyacrylamide gels was investigated from deproteinized whey (Roukas & Kotzekidou, 1991). The productivity of LA in continuous stirred tank reactor (CSTR) with repeated fermentation was 0.72–0.86 g L^{-1} h^{-1}, highest when entrapped in calcium alginate (Roukas & Kotzekidou, 1991). Norton et al. (1994) reported LA productivity of 28.5 g L^{-1} h^{-1} at a dilution rate of 1.21 h^{-1} from whey to permeate in CSTR with cells of *L. helveticus* L89 immobilized in K-carrageenan-locust bean gum gel beads (Norton et al., 1994). Another study of LA production in CSTR using whey permeate. Krischke et al. (1991) used cells of *L. casei* subsp. *casei* immobilized on to porous sintered glass beads obtaining productivity of 5.5 g L^{-1} h^{-1} at a dilution rate of 0.22 h^{-1}. During continuous fermentation, the mechanical stability of the beads and diffusional limitations of the substrate as well as the product within the gel bead matrix are the major problems that hinder effective production (Norton et al., 1994). Therefore dual reactor systems integrating CSTR and packed-bed reactors (PBRs) are to overcome the reduction in LA yield and productivities due to loss of support material (Zhang et al., 2011; Rangaswamy & Ramakrishna, 2008). The cells of *L. delbrueckii* NCIM2365 were immobilized on PUF for LA production in a dual reactor system in continuous fermentation operated for 1000 h achieving LA productivity of 5 g L^{-1} h^{-1}, higher than that of batch fermentation (2.0–2.5 g L^{-1} h^{-1}) and continuous fermentation with free cells (0.65 g L^{-1} h^{-1}) (Rangaswamy & Ramakrishna, 2008). In the dual reactor system, an increase in LA concentration by 16.6% and productivity by 12.5% was obtained with *L. lactis*-11 cells immobilized on ceramic beads as compared to free-cell systems in continuous fermentation (Zhang et al., 2011).

12.7.2 Packed-bed reactor

In PBRs, microorganisms are packed in a reactor column as spheres, chips, disks, beads, or pellets. The advantages of using PBR in continuous mode include the simultaneous removal of product and high productivities. However, nutrient depletion along the reactor column length leading to deviation in the viable cell population, large pH gradients, and elevated residual sugar concentrations are the major disadvantages of the system (Kosseva et al., 2009; Senthuran et al., 1999). Coimmobilized cells of *L. casei* and *L. lactis* on calcium alginate beads were used for LA production from deproteinized whey in a PBR in continuous fermentation resulting in maximum LA productivity of 7 g L^{-1} h^{-1} at dilution rate 0.4 h^{-1} and yield of 70% utilizing 50% lactose (Roukas & Kotzekidou, 1996). Sirisansaneeyakul et al. (2007) investigated the production of LA in a PBR in continuous, batch, and repeated-batch fermentation with both microencapsulations in the alginate membrane and entrapment in alginate beads of *L. lactis* IO-1 cells. The productivity obtained was 4.5 g L^{-1} h^{-1} in continuous fermentation at a dilution rate of 0.5 h^{-1}—which was higher than 2.16 g L^{-1} h^{-1} in batch, and 2.16–2.47 g L^{-1} h^{-1} in repeated-batch fermentation from glucose (Sirisansaneeyakul et al., 2007). LA production with immobilization of *L. helveticus* cells on alumina beads in PBR resulted in the production of 75.6 g L^{-1} LA from whey with the productivity of 3.90 g L^{-1} h^{-1} after a hydraulic retention time of 18 h (Tango & Ghaly, 2002).

12.7.3 Fluidized-bed reactor

In fluidized-bed reactors (FBRs), the combined upward and downward movement of particles leads to fluidized bed formation which also provides mixing of components. These reactors are comprised of a solid biocatalyst, the liquid containing substrate and/or product, and a gas phase. In comparison to PBR, these reactors expectedly provide higher mass transfer and heat transfer rates. The advantages of FBR include the approximate plug flow behavior of the liquid, no washout of cells unless at a very high flow rate, and volume of gas bubbles and growing biomass can be accommodated by the expandable fluidized beds preventing clogging, however, there is a need of regeneration of biocatalyst at regular intervals (Patel et al., 2008). In a study for LA production from whey permeate in CSTR and FBR, cells of *L. casei subsp. casei* DSM20244 were immobilized on sintered glass beads. Higher productivities of 10 g L^{-1} h^{-1} at dilution rate 0.4 h^{-1} and 13.5 g L^{-1} h^{-1} at dilution rate 1.0 h^{-1} were achieved in FBR as compared to CSTR (5.5 g L^{-1} h^{-1} at dilution rate of 0.22 h^{-1}) (Krischke et al., 1991). Fluidization of immobilized *Lactobacillus delbreuckii* in alginate beads was done by upward flowing liquid media in a tubular reactor in a study of Davison and Thomson (1992) followed by the addition of adsorbent polyvinyl pyridine resin Reillex 425 in the column which traveled through the biocatalyst bed. Adsorption of LA, pH moderation, and about fourfold increase in LA production (compared to FBR without resin addition) were observed after the addition of adsorbent resin (Davison & Thompson, 1992). A 12-fold increase in LA productivity along with the removal of inhibitory products was achieved by Kaufman et al. (1995) using particle FBR with *L. delbreuckii* NRRL B445 cells entrapped in gelatin beads and IRA-35 resin as adsorbent (Kaufman et al., 1995). An extractive fermentation process utilizing trialkyl phosphine oxide as an extractant for L-LA in three-phase FBRs and immobilized *R. oryzae* NRRL 395 was proposed by Lin et al. (2007) which resulted in high LA productivity (11 g L^{-1} h^{-1}) with relieving product inhibition. Liquid−solid circulating fluidized bed system is an alternative for continuous particle regeneration problems and achieves both the simultaneous reaction as well as regeneration of the bioparticles and/or chromatography particles (Zhu et al., 2000). Patel et al. (2008) achieved simultaneous production and recovery of LA by constructing a dual-particle liquid−solid circulating fluidized bed with immobilized *L. bulgaricus* ATCC 11842 and ion-exchange resins. A total of 240 g LA was produced with 42 g retention on resin, 124 g in the downer broth, and 74 g recovered in the riser outlet stream (Patel et al., 2008).

12.7.4 Airlift bioreactors

These are gas−liquid bioreactors in which compressed air is used for aeration and agitation and can be regarded as a special variation of FBRs (Rawat et al., 2019; Barragán et al., 2016). The absence of moving parts in tower reactors like airlift bestows them with the advantage of reduced shear stress which makes them appropriate to use with fungal cells. However, the scale-up of airlift bioreactors for LA production has not been studied much and requires more research to meet commercial-scale demands. Yin et al. (1998) employed an airlift bioreactor for L-LA production using *R. oryzae* achieving productivity of 1.07 g L^{-1} h^{-1} after 48-h culture. Induction of mycelial flocs morphology in *R. oryzae* with the addition of mineral support and poly(ethylene oxide) enhanced the L-LA production in an airlift bioreactor achieving a concentration of 104.6 g L^{-1} (from 43.2 g L^{-1}, without any addition) and yield of 0.87 using 120 g L^{-1} glucose as substrate (Park et al., 1998).

Sun, Li, and Bai (1999) immobilized *R. oryzae* cells on polyurethane foam cubes while studying the continuous production of L-LA in airlift bioreactor and concluded an increase in the productivity of LA with dilution rate or input glucose concentration during the pseudosteady state. Large-scale LA production by *Rhizopus sp.* MK-96-1196 with scale-up of airlift bioreactor (from 0.003 to 5 m^3) has also been studied (Liu et al., 2006). LA concentration higher than 90 g L^{-1} with a yield of about 80% was obtained at the oxygen transfer rate higher than 0.28 (g-O$_2$ L^{-1} h^{-1}) (Liu et al., 2006).

12.7.5 Fibrous-bed reactors

Fibrous-bed bioreactors use fibrous matrices like cotton cloths packed in a column in which cells are immobilized. The fibrous bed provides areas for cell immobilization causing high cell density and stability. Since their first use, packing designs have been greatly improved with uniform structures allowing for low diffusional limitations and enhanced productivity (Vijayakumar et al., 2008). Shi et al. (2012) reported the production of LA in a fibrous-bed reactor in batch mode and fed-batch mode with immobilized and free cells of *L. lactis* ATCC 15009 from Jerusalem artichoke hydrolysate. Fed-batch production with immobilized cells achieved 142 g L^{-1} LA which was 27.92% higher than that from fed-batch with free cells (Shi et al., 2012). Moreover, retention for a long time (approximately 780 h) by repeated fermentation in a fibrous-bed reactor resulted in increment of L-LA yields from 0.84 to 1.01 g g^{-1} and productivities ranged from 0.71 to 2.85 g L^{-1} h^{-1} (Shi et al., 2012). A rotating fibrous-bed bioreactor for L-LA production from glucose and cornstarch with immobilized mycelia of *R. oryzae* NRRL 395 was developed (Tay & Yang, 2002). LA concentration, yield, and productivity obtained from glucose were 126 g L^{-1}, 0.89 **g g^{-1}**, and 2.5 g L^{-1} h^{-1}, respectively, and that from cornstarch were 127 g L^{-1}, 0.90 g g^{-1}, and 1.65 g L^{-1} h^{-1}, respectively, in fed-batch fermentation (Tay & Yang, 2002). Mass transport was the rate-controlling step in LA fermentation in a static fibrous-bed reactor with immobilized *R. oryzae* NRRL 395; utilization of 70 g L^{-1} glucose resulted in a maximum LA concentration of 37.8 g L^{-1} with yield and productivity of 0.62 g g^{-1} and 2.09 g L^{-1} h^{-1}, respectively (Chotisubha-Anandha et al., 2011). Performance of fibrous-bed bioreactors in terms of kinetics and long-term stability for continuous LA production from unsupplemented acid whey [3.7% (w/v) lactose and 0.8% (w/v) LA] using immobilized *L. helveticus* ATCC 15009 cells was investigated by Silva and Yang (1995). Based on dilution rate and LA concentration, productivities in the range of 2.6–7 g L^{-1} h^{-1} were achieved, 10 times higher than that in batch fermentation with free cells (Silva & Yang, 1995).

12.8 Isolation, analysis, and determination technique and process

Conventional fermentation methods of LA production create economic hurdles because pH neutralization during the process leads to the formation of lactate salt whose conversion to

LA forms about 50% of the production cost (Eyal & Bressler, 1993). Precipitation and reacidification (using mineral acids) of the salt is required to be done to obtain LA. During the fermentative production of LA, impurities such as carbohydrates, metal ions, and proteins salts are produced along with the D- and L-LA enantiomer. The removal of these impurities is important because they might affect lactide production, quality and subsequently, the PLA formed. Evaporators can be employed to remove the residual water content leaving concentrated LA for polymerization. Recycling the water separated from crude LA can prevent the loss of feed material. The lactide production stream also contains LA, water, and LA oligomer, besides lactide enantiomers, which are separated by using a distillation column. The overhead stream contains low boiling point materials such as water and LA with some other low molecular weight by-products whereas the bottom stream contains LA oligomers which have less volatility than lactide. Lactide is present in the sidestream needed to be withdrawn with high purity because the quality of the polymer to be formed depends on the quality of the lactide obtained. Several other techniques for recovery of LA from fermentation broth include reactive extraction, ultrafiltration, reverse osmosis, electrodialysis, and adsorption.

12.8.1 Diffusion dialysis

The separation of molecules based on size using an ion-exchange membrane is called diffusion dialysis, first performed in the 1950s, and is being used commercially as a membrane separation process for the separation and purification of several products (Fu & Xu, 2008). The separation occurs due to the concentration gradient following the Donnan criteria across the ion-exchange membranes which can either be cation-exchange membranes (CEMs) or anion-exchange membranes (AEMs) (Stancheva, 2008). Recovery of inorganic acids, organic acids, and bases have been performed by diffusion dialysis using a plate or hollow fiber type CEMs or AEMs (Xu et al., 2008; Kiyono et al., 2004). The separation of weak organic acids from their salt depends upon their sorption and diffusion properties. In contrast to strong acids that have high diffusivity and low sorption, weak acids have low diffusivity and significant sorption causes their equal or high concentration in the membrane than in the surrounding solution as observed in a system of LA and sodium salt of LA (Narębska & Staniszewski, 2008). The high sorption could be attributed to the binding of acid molecules to water leading to separation from its salt (Narębska & Staniszewski, 1997, 2008). The separation performance for weak acids depends on the diffusivity, sorption characteristics, membrane permeability, membrane mass transfer coefficient, and dialysis coefficient. Sorption depends on molar mass, dissociation constant, and interaction with the membrane matrix whereby the acid concentration in the membrane increases with concentration to an extent after which an increase in acid concentration decreases the concentration in the membrane (Palatý et al., 2009). Permeability and/or diffusivity affect the membrane mass transfer coefficient. The permeability of the membrane depends on the acid concentration and its diffusivity in the membrane. Diffusivity (D) in the case of acetic acid (AA), LA, oxalic acid (OA), and tartaric acid (TA) increases in the order $D_{TA} < D_{LA} < D_{AA} < D_{OA}$ with increasing concentration in the membrane (Palatý et al., 2009). Permeability values increase with increasing temperature due to the increase in diffusivity with temperature, whereas the values decrease with increasing acid concentration (Palatý et al., 2007, 2009).

12.8.2 Membrane filtration

The separation of LA can be performed by the use of membrane separation techniques such as microfiltration, nanofiltration, ultrafiltration, reverse osmosis, and electrodialysis. The major advantage of using membrane separation techniques is that no expensive extractant or any other chemical is required for the separation. The increasing concentration of LA in the fermentation broth increases the acidity of the medium, thus inhibiting the microbial activity. To ensure high productivity by fermenter, it is important to control the concentration of acid inside by proper removal from the fermenter, maintaining the pH around 6, and to recycle the cells to avoid their loss. One way is to integrate membrane separation with a fermenter to ensure continuous separation of LA with cell recycle (Taniguchi et al., 1987). Giorno et al. (2002) studied the conversion of glucose to LA using a cross-flow membrane module which was equipped with microfiltration or ultrafiltration membranes achieving an average LA yield of 62%. For cell separation and recycling from the broth, polysulfone ultrafiltration capillary membranes with a molecular weight cutoff diameter 100 kDa and polyamide ultrafiltration membranes of the molecular weight cutoff value 50 kDa were used along with a microfiltration membrane of pore size 0.1 μm
(Giorno et al., 2002). Ceramic ultrafiltration membranes are easier to disinfect than polymer membranes and have been used for cells and protein separation and recycling with simultaneous removal of LA. LA separation and concentration from cheese whey fermentation broth was performed by Li et al. (2008) using combined nanofiltration and reverse osmosis membranes where the nanofiltration membrane (GE Osmonics HL membrane) retained 97% of lactose acquiring permeate with LA and water. Further, the separation with reverse osmosis (GE Osmonics) ADF membrane ensured 100% LA separation leaving water in the permeate (Li et al., 2008). Fouling due to the accumulation of microbes and impurities on the membrane surface is the major hindrance in the effective use of membrane technology as it greatly affects the filtration and separation characteristics of the membrane. Reverse osmosis and nanofiltration membranes are the most prone to rapid fouling (Timmer et al., 1994). During LA production and purification from glucose using a flat sheet cross-flow module of nanofiltration membrane integrated with a fermenter, the presence of sodium lactate salt decreased glucose retention (Bouchoux et al., 2005). This problem occurs due to the charge density of the nanofiltration membrane. Higher the charge density of the membrane (sodium lactate), the lower the retention of a neutral solute (glucose). The use of integrated membrane

systems could reduce the adverse effects of fouling and increase the membrane performance. Gonzalez et al. (2008) did ultrafiltration of whey before recovering LA by nanofiltration membranes. Hollow fiber ultrafiltration membrane for purification and cell recycling integrated with reverse osmosis for preconcentration of whey permeate was used by Tejayadi and Cheryan (1995) attaining a lactate concentration of 89 g L^{-1} in 150 h.

12.8.3 Electrodialysis

Conventional electrodialysis for separation and concentration of ions used ion-exchange membranes for ion transport under the influence of voltage. In the case of LA recovery, electrodialysis is considered a promising technique for effective separation where an in situ approach can separate LA with the lowering of product inhibition. The technique is environmentally suitable, fast, effectively concentrates LA, and removes nonionic molecules (Boniardi et al., 1997). The use of electrodialysis units with cell-free solutions is a feasible means to avoid the problem of membrane fouling due to bacterial accumulation. The problem with conventional in situ technique is the damage that occurs to the microbial cells due to the insertion of the cathode in the fermentation broth. This is overcome by applying ultrafiltration to separate the cells before electrodialysis (Nomura et al., 1991). One advancement in the conventional electrodialysis is the integration of a bipolar membrane which causes separation and concentration of salts, further converting them into acid and base (recycled for neutralization) with the minimum generation of salt waste as effluent (Timbuntam et al., 2008). This technique is also known as water-splitting electrodialysis as it contains a water-splitting stack with the bipolar membrane that splits the water molecule into its constituent H^+ and OH^- ions. The bipolar membrane has laminated cation and anion membrane layers. Electrodialysis with the bipolar membrane is used to recover LA suitable in quality for polymer formation. Thang et al. (2005) used electrodialysis for the separation and concentration of LA wherein first step lactate was separated from sugar and nonionic amino acids at pH 6.7 and then at pH 2, LA was recovered from inorganic salts in the second step which was more efficient in terms of energy consumption and operation time (Thang et al., 2005). Monopolar and bipolar electrodialysis for respective purification of sodium lactate and recovery of LA was performed (Madzingaidzo et al., 2002). A feed concentration of 125 g L^{-1} of sodium lactate was concentrated to 150 g L^{-1} with very few impurities present while the next step of bipolar electrodialysis provided 160 g L^{-1} of LA with negligible impurities (Madzingaidzo et al., 2002).

Current efficiency is calculated as the ratio of theoretical (I_T) and practical currents (I_P),

$$\eta = \frac{I_T}{I_P} \tag{12.1}$$

$$I_T = FJA_{eff}$$

where F is Faraday constant, A_{eff} is the effective membrane area, and J is flux density. The total current efficiency is calculated as

$$\eta T = \frac{I_{T1}t_1 + I_{T2}t_2}{I_{P1}t_1 + I_{P2}t_2} \quad (12.2)$$

Two-stage electrodialysis techniques have been assessed as appropriate and efficient for the recovery of lactate ions and their subsequent conversion to LA (Hbová et al., 2004).

12.8.4 Reactive extraction

The separation of LA from aqueous solution is a tedious task due to its hydrophilic nature. Reactive extraction uses an extractant with a high partition coefficient which reacts with LA in the aqueous phase forming a reaction mixture in the organic phase, followed by removing the acid from the organic layer. Reactive liquid-liquid extraction presents the advantage of easy removal of LA from fermentation broth reducing the effects of pH decrease. In cases where the extractant and diluent are nontoxic, these can be recycled back into the fermentation process after the removal of LA as critically reviewed by Wasewar et al. (2004) through the glucose fermentation process to LA followed by reactive extraction. The reactive extraction system of LA should have a high partition coefficient and high selectivity for LA. For efficient extraction to occur, the extractant should be less soluble in water with the distribution coefficient higher for LA than other impurities present (Götz, 1995). The distribution or partition coefficient is the ratio of the concentration of LA in the solvent phase (C_s) to the concentration in the aqueous phase (C_w) at equilibrium as denoted in Eq. (12.3)

$$K_d = \frac{C_s}{C_w} \quad (12.3)$$

Some other properties that an extractant(s) should possess include low viscosity, thermal stability, low heat of evaporation, melting point, and toxicity along with environmental sustainability and economic feasibility (Jung et al., 2000). Aliphatic amines and organophosphates have been widely used as extractants for carboxylic acids. Aliphatic amines have a higher distribution coefficient and are less expensive than organophosphates. Among the aliphatic amines, tertiary amines mixed with diluents are considered as the best suitable extractant for LA (Wasewar et al., 2002).

12.8.5 Adsorption

The use of solid sorbents specific to LA for its separation and recovery is a promising technique. For the separation to be efficient, the extractant or solid sorbent should be highly selective for the acid, biocompatible with the process microbes, and regenerable to overcome issues of excess cost. Most fermentative production of carboxylic acids occurs at pH above the pK_a of the acid, for example, LA is produced at pH 5–6 and it has pK_a 3.86 (Dai & King, 1996). Hence the recovery of LA can be done by using a basic substance that shows substantial uptake in a range of pH 5–6 (Tung & King, 1994). LA recovery from LA–glucose solution by anion-exchange method in an ion-exchange membrane-based extractive fermentation system was analyzed (Zihao & Kefeng, 1995). Monteagudo and Aldavero (1999) used Amberlite IRA-420, an anion-exchange resin for adsorption of lactate from the broth in the carbonate form. Percolation of ammonium carbonate solution through the resin led to the formation of ammonium lactate which was further converted to LA by cation-exchange resin Amberlite IR-120 in hydrogen form (Monteagudo & Aldavero, 1999). In a comparative study, activated carbon was found to be more effective for adsorption of LA and lactate than PVP since each cycle of adsorption and regeneration led to about 14% loss of PVP's adsorption capacity, restricting its practical application (Chen, 1998). Tong et al. (2004) used IRA-92, a faintly basic exchanger for the separation of LA in the fermentation broth, and achieved high purity (96.2%), yield (82.6%), and productivity (1.16 g LA g^{-1} resin day) from the supernatant of the fermentation broth using a single-step chromatographic procedure (Tong et al., 2004). Amberlite IRA-400 resin is an efficient adsorbent for LA with appropriate pore size and displays high adsorption capacity in a wide range of pH. The binding of lactate in only one step was achieved by refilling a fluidized bed column by Amberlite IRA-400, a strong anionic-exchange resin obtaining 0.126 g LA g^{-1} resin (Córdoba et al., 1996).

12.9 Synthesis and structure of polymers

12.9.1 Polylactic acid synthesis

PLA can be prepared by using LA as a monomer through different polymerization processes such as polycondensation, ring-opening polymerization, and direct polymerization through azeotropic dehydration and enzymatic polymerization (Auras et al., 2010; Garlotta, 2001). We have already discussed the microbes involved in the biosynthesis of LA and the biosynthesis pathways. Solution polycondensation and melts polycondensation are the most cost-effective polycondensation methods for PLA formation compared to others. In the case of polycondensation, different monomers are combined to form low- to intermediate-weight polymer by removal of water produced in condensation through the use of solvents and/or catalysts under high vacuum and temperature (Achmad et al., 2009). The polymer formed can be used as such or increased in molecular weight by coupling with isocyanates, epoxides, or peroxide (Gupta et al., 2007). Direct polymerization for PLA synthesis was reported by Achmad et al. (2009). No catalysts, solvents, and initiators were involved during synthesis which proceeded through variation of temperature from 150°C to 250°C and pressure from atmospheric pressure to vacuum for 96 h (Achmad et al., 2009).

Side reactions like transesterification can occur during polycondensation causing the formation of lactide-like ring structures that can alter the properties of the final polymer (Auras et al., 2010). The use of different catalysts and agents along with changing the polymerization conditions, however, can control the formation of such subproducts (Mehta et al., 2005). Direct condensation of LA involves removal of the free water followed by oligomer polycondensation and melt condensation of high molecular weight PLA. Removal of water is the rate-determining step for the first and third steps while for the second step, the rate-determining step is the catalyst-dependent chemical reaction (Auras et al., 2010). The fourth step involves particle formation and crystallization once the melt-polycondensation PLA is cooled below its melting temperature (Auras et al., 2010; Fukushima & Kimura, 2008). The low to intermediate molecular weight LA prepolymer can be linked to high molecular weight polymer for chain extension by treatment with chain extenders. A chain-extension process at 180°C in the presence of 1,6-hexamethylene diisocyanate as a chain extender was performed for obtaining a polymer with 27,500 g mol^{-1} molecular weight after 40 min (Gu et al., 2008). High molecular weight polymer can also be obtained by ring-opening polymerization in which the lactide ring gets opened in the presence of a catalyst (Auras et al., 2010) resulting in PLA with controlled molecular weight (Kim et al., 2009). The ratio and sequence of appearance of D- and L-LA units in the formed polymer depend upon the residence time, temperature, catalyst type, and concentration (Gupta et al., 2007).

Azeotropic dehydration is a direct method with comparatively easier removal of water formed due to reaction and for attaining higher molecular weight PLA (Garlotta, 2001). Azeotropic dehydration at 138°C for 48–72 h was used to form PLA with molecular weight 33,000 by Kim and Woo (2002) using a molecular sieve for drying and m-xylene as a solvent.

Enzymatic polymerization is an efficient, environmentally positive but the least studied approach for polymerization process control. Liquid ionic (1-hexyl-3 ethyl) imidazolium hexafluorophosphate [HMIM][PF6] for enzymatic synthesis of poly-L-lactide, mediated by lipase B from *Candida antarctica* (Novozyme 435), resulted in the highest PLA yield of 63% with a molecular weight of 37.89103 g mol^{-1} at 90°C (Chanfreau et al., 2010).

12.9.2 Structure of polymer

LA is the basic building monomer for PLA which itself is produced by either fermentation or chemical synthesis. The ability to economically produce high molecular weight PLA is the major reason behind its expanded demand and use. L-lactides and D, L-lactides are the building blocks for PLLA and PDLLA, respectively, which are further used to form PLA (Auras et al., 2004). The physical and chemical properties of the polymer are affected by the optical purity of PLA (Urayama et al., 2003; Tsuji et al., 2006; Sarasua et al., 2005).

LA obtained from biological sources exists mainly in L-form as a result of which PLA formed by such monomers has L-isomers as the main component. Thermal properties of PLA like melting temperature (T_m), glass transition temperature (T_g), as well as its crystallinity, decreases with a reduction in L-isomer content (Urayama et al., 2003; Tsuji et al., 2006). The water vapor transmission rate (WVTR) of the polymer remains unaffected when the optical purity of PLLA films ranges between 0% and 50%, whereas, with increasing crystallinity (0%–20%), WVTR values decrease (Tsuji et al., 2006). PLA resins with high L-isomer content can be injection molded for good heat-resistant properties. Higher D-isomer content (4%–8%) suggests low crystallinity and such resins are suitable for thermoformed, extruded, and blow-molded products (Drumright et al., 2000). Exposure of PLA to high temperatures results in thermal degradation to form lactide monomers which undergo racemization to form mesolactide and oligomers and this conversion is significant when the temperature goes above 200°C and for oligomers above 230°C (Tsukegi et al., 2007). Calcium oxide reduces the temperature required for pyrolysis and leads to L, L-lactide formation, hence, its addition to poly(L-lactide) can control racemization at 250°C–300°C (Fan et al., 2003).

12.9.3 Process flow diagram for production

Most of the LA produced today depends on anaerobic fermentation of carbohydrates near-neutral pH. Fig. 12.2 depicts the conventional process for LA production and recovery. The pH is maintained around 6.0 by the addition of calcium carbonate and hydroxide in excess

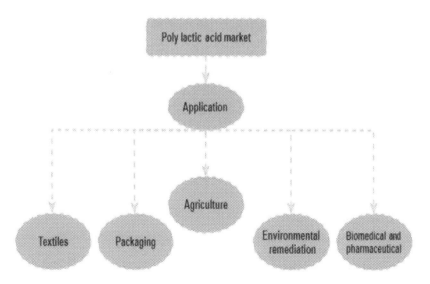

Figure 12.2
Conventional process for lactic acid manufacture from carbohydrate (Datta and Henry, 2006).

which in turn results in the formation of calcium salt of LA. Filtration of the broth containing calcium lactate is done to remove any cells or other insoluble followed by evaporation. Once the evaporation is over, sulfuric acid is employed for acidification converting the salt into LA and the insoluble sulfate salt of calcium is removed by filtration. Technical grade LA can be obtained by achieving further purification of the filtrate by ion-exchange resins and evaporation. However, for applications like polymer formation, a high purity product is desired which can be obtained by further processing of the technical-grade LA. This processing involves the esterification of obtained technical-grade products using ethanol or methanol followed by the recovery of the formed ester by distillation. The ester is hydrolyzed with water before evaporation and finally, residual ethanol is recycled resulting in highly pure heat-stable LA, suitable for use in polymer formation.

12.9.4 Properties of PLA polymers

PLA with more than 90% L-isomer content tends to be crystalline whereas, with a decrease in optical purity, amorphous nature predominates while T_m and T_g decrease with decreasing poly(L-lactide) (Auras et al., 2010). T_g affects chain mobility and is important to be considered for amorphous PLA while both T_g and T_m are important for predicting the behavior of semicrystalline PLA (Bouapao et al., 2009; Yamane & Sasai, 2003). The density of amorphous PLLA is reportedly 1.248 g mL^{-1}, crystalline PLLA is 1.290 g mL^{-1} and 1.36 and 1.33 g cm^{-3} for L-lactide and mesolactide, respectively, while for crystalline polylactide and amorphous polylactide the density is 1.36 and 1.25 g cm^{-3}, respectively (Auras et al., 2004). The melting enthalpy reported in most literature for an enantiopure PLA ($\Delta H°m$) is 93 J g^{-1}, however, higher values up to 148 J g^{-1} also have been reported (Södergård & Stolt, 2002). Solvents like dioxane, acetonitrile, chloroform, methylene chloride, 1,1,2-trichloroethane, and dichloroacetic acid dissolve PLA products. Polylactides are soluble in ethylbenzene, toluene, acetone, and tetrahydrofuran at boiling temperatures whereas crystalline poly(L-lactide) is not soluble in acetone, ethyl acetate, or tetrahydrofuran (Madhavan Nampoothiri et al., 2010). Copolymerization of lactide with other lactone type monomers, polyethylene glycol, or monomers with amino and carboxylic like functional groups can be used to induce alterations in PLA (Cheng et al., 2009). PLA can be altered through its blending with other materials where the change in PLA properties will depend upon the mechanical properties and microstructure of the blend and interfacial properties of the phases (Broz et al., 2003). Polylactide degradation is a two-step process where nonenzymatic chain scission of the ester groups cause molecular weight reduction until the LA and low molecular weight oligomers are naturally metabolized by microorganisms to yield carbon dioxide and water (Auras et al., 2004; Oyama et al., 2009). The rate of polymer degradation is influenced by particle size and shape, temperature, moisture, crystallinity, % isomer, residual LA concentration, molecular weight, water diffusion, and metal impurities from the catalyst (Auras et al., 2004; Drumright et al., 2000; Bleach et al., 2001; Cha & Pitt, 1990).

12.10 Commercialization and application

Due to the competitive cost, industrial applications and production of PLA have risen sharply. PLA's industrial applications are divided into categories of consumer enduring goods and consumer nonenduring goods. As per market perspective, merchandise products such as appliances, car, and medical products which have more than 3 years life span are consumer enduring goods while products such as packaging, medical items, and service ware which have a life span of up to 3 years are consumer nonenduring products. PLA industrial applications (Fig. 12.3) are categorized in biomedical and pharmaceutical, textiles, bioremediation, packaging of products, agriculture, and others.

12.10.1 Textiles

By spinning the PLA is processed into fibers and can absorb organic compounds which makes it suitable for textiles applications. PLA polymer is polar and can absorb moisture, hence, suitable for wipes. PLA antimicrobial wipes were developed by Biovation. Water filters made up of PLA mix fibers are produced by Fraunhofer UMSICHT and FKuR, that

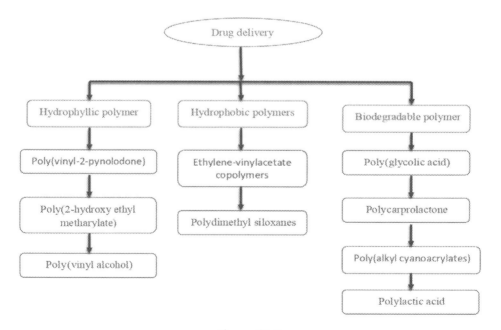

Figure 12.3
Application of polylactic acid in different fields.

is, Bio-Flex S 9533; the mixture of fibers contains coconut shell absorbent carbon (Bio-based News, 2020). The absorbing properties of PLA are excellent and so it can also act as a disposable product. For example, Bioarmour launched by Biovation is a single-use antimicrobial blood pressure cuff shield made up of 74 wt.% PLA and due to its breathable property, this cuff is directly used on the patient's skin and provides easiness to the patients (MarketWatch, 2015).

In the automotive industry, PLA fibers play an important role as a plastic substitute. Different companies including Ford Motor are finding environment-friendly polymers for interior parts of a car such as floor mats and trim parts. "bio-CAR" was a bio-based materials conference for automotive applications held in Germany in 2015 in which it was concluded that before PLA is purely used in automotive applications, it is necessary to solve some hurdles associated with it such as odors released during heating PLA polymer at high temperature, its moisture effect and degradation time (RP news wires, Noria Corporation). A comparative study to investigate automotive properties such as snagging and resistance to abrasion in the case of polyethylene terephthalate (PET) and PLA fiber-based products found that PLA met nearly all needs and performed equally as of PET but failed in some tests (Ghosh & Krishnan, 2007). PLA needs some improvements before it can replace synthetic polymers, for example, nylon and PET.

PLA textiles are also increased in the textile industry to make apparel, homeware, etc. Brands like Nike, Gap, and Under Armour are already using a combination of PET cotton fibers in the garment industries, although, the absorbing and breathability property of PLA textiles fibers is an alternative for PET cotton blends making PLA a cozy material for garment manufacturing. The research of Hohenstein Research Institute proved that the insulating and buffering capacity of PLA to sweat makes it fit for use in sports apparel. According to ATCC (Association of Textile Chemists and Colorists), PLA can resist multiple washing in laundry services and give high flexibility when used in jacket making, but some issues such as usage of high temperature in ironing and pressing limit the applicability of PLA textiles. PLA blends due to their good retention and pleat properties are used in knitted and embroidered fabrics, curtains, bedsheets, rugs, etc. (Blackburn, 2005).

12.10.2 Biomedical and pharmaceutical applications

In medical and pharmaceutical applications, biodegradable polymers are used over nondegradable polymers due to their lasting advantage of biocompatibility. Poly (α-hydroxy acids) like PLA, polyglycolic acid, and polydioxanone are the most prevalent synthetic biodegradable polymers used in medical and pharmaceutical applications (Middleton & Tipton, 2000). For use in the medical textile industry, surgical materials, drug-releasing systems as well as ecological materials, PLA shows a distinctive feature of perishability, biocompatibility, and eco-friendliness (Gupta et al., 2007).

A single polymer proved itself in the diversification of applications including biodegradable sutures, nanoparticles, porous scaffolds, and microparticle-based drug-delivery systems by implementing basic changes in its chemical structure. It attains the desired performance by the mixing or copolymerization with other polymeric and nonpolymeric components. In the application of biomaterials, the surface properties of PLA play an important role in determining its biocompatibility. Different methods like physical, chemical, and radiation have been used to establish surface-modification properties of PLA biomaterials. The main advantage of a biopolymer is that secondary operation is not necessary to remove the implants after repairing the defect site as the biopolymer disappears from the body itself with time due to its biodegradable nature (Dürselen et al., 2001).

12.10.3 Tissue engineering

Tissue engineering started in 1988 as a technique to reconstruct living tissues or replace them with biomaterials using 3D tissue scaffold under specific physiological conditions. This technique attracts huge attention in the medical field and science as the replacement of tissue and organ substitute (Liu et al., 2004). Due to the nondegradability of some metals in the environment, they are not good for scaffold applications despite their good mechanical properties. In the last two decades, people have seen a startling change in the organ reconstruction of human beings based on tissue engineering. In past days,
biostable substances were used to harvest the tissues from the cell culture. In the whole process, especially for the biocompatibility of biomaterials, the important role is played by the surface properties of PLA (Brigham & Sinskey, 2012). To get the desired properties of the surface, different methods with some alteration induced with chemical, radiation, and physical is used for PLA biomaterials. The principal central point is to study tissue engineering with materials possessing high-quality osteoconductive properties like inorganic material such as calcium compounds or hydroxyapatite which can be processed into permeable material and the good flexibility of design is attained by manipulation of structure and composition (Langer & Vacanti, 1993). Table 12.4 mentions some companies that have stepped in to manufacture orthopedic devices using PLA as the major material.

Table 12.4: An overview of a trading market available for bone-fixation devices made up of PLA used worldwide.

Company name	Material	Outcome	Country
Centerpulse orthopedics	PDLLA	Interference screw	USA
Biomet orthopedics	Drawn PLLA	Mini screw	USA
Geistlich biomaterial	P(LLA/DLLA)	Fixation pins for GTR and GBR membranes	Switzerland
Phusis	P(LLA/DLLA)	Interference screw	France
Take iron	Drawn PLLA	Pin, Screw, Miniplate, Rod	Japan

PDLA, poly-D-lactic acid; PLLA, poly-L-lactic acid; poly-D, PDLLA, L-lactic acid.

To get all the functions in one material is very difficult and the sole characteristics of PLA are not enough for application in tissue engineering. However, by minor modifications in its structural properties, that is, physical and chemical structures, PLA can be used in different applications. In urological surgery applications, PLLA material is preferred for recreation of the tendon; ligament and stunts having longer retention time. PLA has been used for orthopedic, cardiovascular, and neurological therapy through the culturing of different types of cells. In tissue engineering, an ideal tissue scaffold must possess the properties of biocompatibility, porosity, mechanical strength, and biodegradability (Behonick et al., 2007).

12.10.4 Drug-delivery system

For the maximization and minimization of therapeutic perspective and side effects to specific body parts, there is a need for the targeted release of bioactive materials. For medical use, the releasing tools used are different types of particles like liposomes, biodegradable polymers, and lipid nanoparticles (Roney et al., 2005). The release of the polymeric drug occurs by one of the ways such as diffusion, swelling, and erosion. When the ester bonds of monomeric units of polymer breakdown by the hydrolytic cleavage, the abrasion of the device occur and erosion starts after the infiltration of hydrophilic molecules. Bulk or homogeneous and surface abrasion are the two methods of erosion of biodegradable polymers (Kumari et al., 2010). Various polymers have been utilized in drug-delivery systems based on their feasible properties (Fig. 12.4).

In the form of nanoparticles, PLA and its copolymers have been utilized in the encapsulation process of different medicines, hormones, dermatological and physiological proteins. For obtaining polymeric nanoparticles different methods are used such as evaporation, displacement, diffusion of solvent, and salting-out (Rancan et al., 2009).

12.10.5 Packaging and service wares

In the past 5 years, the use of PLA in packaging has increased to a large extent. To fulfill consumer demands and to flourish the green packaging market, degradable packing materials made up of renewable resources are being developed by research through the collaboration of both academia and industry (Table 12.5).

Because of its low strength performance, PLA faces numerous challenges in packaging applications. By amending some processes such as blending or adding the polymer with other compounds like antioxidants and nucleating agents, the performance of PLA packaging has been improved to encounter the needs of industries (Auras et al., 2010). For example, by altering the processing of PLA, the production of oriented and nonoriented PLA occurs. In comparison with nonoriented PLA, oriented PLA has a substantial amount of thermal resistance. However, the loud noise produced by polymers due to its infirmity is

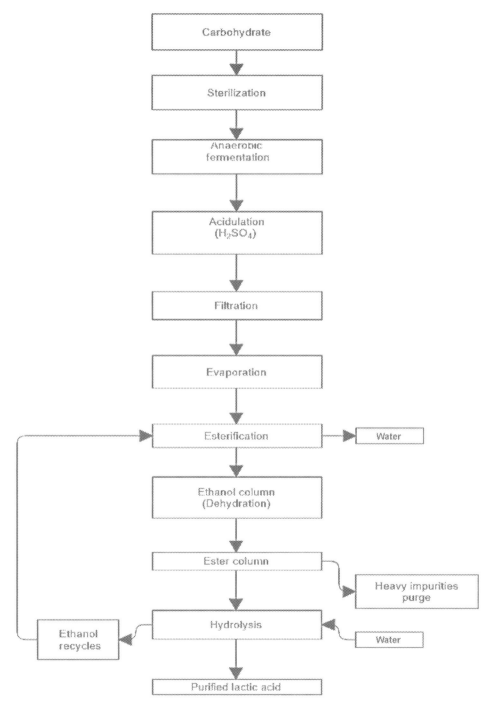

Figure 12.4
Flowchart of different drug-delivery systems of various polymers.

Table 12.5: Application of PLA in packaging and service ware and production origins.

Market brand	Active year	Applications	Country	Illustrations
Tenova	2003–current	Shopping bags	Sweden	Bags composed of 45 wt.% PLA and 55 wt.% Ecoflex
Wal-Mart	2005–current	Strawberries, Brussels sprouts	USA	Advertised as biodegradable clamshells among the first company to use commercialized PLA
Hypermarket chain Auchan	2005–current	Fresh vegetables	France	Advertised as biodegradable containers
Huhtamaki	2006–current	Dessert molds	Finland	Advertised as biodegradable containers
Shiseido-Urara	2009–current	Shampoo container	China	Favorable reception in the Chinese market as an environmentally friendly option Bottles are 50 wt.% PLA and 50 wt.% HDPE (High-density polyethylene)
Polenghi LAS	2010–current	Lemon juice container	Italy	Claimed to be the first blown extrusion PLA
PURALACT	2013–current	Serviceware	Netherlands	Safe food contact application and the containers are microwavable

PLA, Polylactic acid

still a matter of concern despite its desirable characteristics. The compostable PLA bag produced by Frito Lay went for big-time public perusal owing to the loud sound of crinkle on handling the bags of their Sunchips brand in 2010 and it was secluded from the market completely (BioCycle, 2020).

In the packaging of bakery and gifting cards, the oriented PLA are used while in the packaging of fresh products and the product with low shelf life, thermoformed clamshells nonoriented PLA are used. Some companies have declared that the mean life of packaged fruits is approximately 15% higher in PLA containers (NatureWorks, 2020).

As PLA is receptive to heat damage, it is very challenging to produce PLA containers for microwave and disposable service wares. To maintain the 3D shape of the polymer, heat deflection temperature of approximately 55°C–65°C is required to arrest the liquid polymer in the desired shape into the mold, however, this temperature is very low for the production of heat-stable PLA containers (Mohanty et al., 2002). The table shows examples of thermoformed PLA products used by different companies as PLA containers (Mohan, 2020).

12.10.6 Plasticulture/agriculture

The use of plastic in horticulture applications is known as plasticulture. It is used in agricultural applications such as the use of mulch films for the protection of soil and plants from erosion, birds, weeds, etc.; the process is known as mulching, as a filter spraying breeches and to hide the subway of greenhouses. In the 1950s the use of plastics started for horticulture purposes to upgrade and enhance the production and cultivation of agriculture products (Kasirajan & Ngouajio, 2012). The use of conventional nondegradable plastics raises concerns regarding pollution and waste management. It is economically unfeasible because of the associated labor costs, required to handle the cleaning of plastics, and transportation costs. As it is not an environmentally friendly option, the consumers prefer feasible biodegradable alternatives. Therefore PLAs, polyhydroxyalkanoates, and polybutylene adipate-co-terephthalate are the best options for biodegradable plastics (Kijchavengkul, 2010).

The use of plastic culture in horticulture is still at the initial research level in laboratories due to the high price of the polymers. However, the outcome is expected to be the most encouraging because of its biodegradability after usage. Due to their fragility, mulch films made up of PLA are not so successful to guard plants and soils from the invasion of microorganisms. Commercialized mulch films based on PLA are being used for mulching with a mixture of other biodegradable polymers (Hayes et al., 2012).

Mulching of crops enhances the growth and development of plants and allows the soil to warm over and is affected by the climate of the tunnel. Advances in agriculture crop production done by the help of mulching include low consumption of water, no herbicides, advanced reaping, and high-quality crop production (Tarara, 2000). The disappearance of sheets is affected by the microorganisms due to the moisture in the soil. Bulk erosion is the phenomenon in which the microorganisms have high accessibility to allow diffusion of water in the polymer and enhance hydrolysis, accelerating the biodegradability of the PLA mulching films. Surface erosion is also involved in this biodegradation (Marín et al., 2005).

12.10.7 Environmental remediation

Environmental remediation or bioremediation means polluting substance excision from the environment. Remediation is implemented by absorption and denitrification mechanisms for treating water and wastewater. Vaguely we can postulate that the regulation of mechanisms depends on the integration of van der Waals and the affinity of an electron in the middle of sorption and contaminants (Matsuzawa et al., 2010). Adsorbents such as charcoal, zeolite, and polymers compose the sorption media. Biodegradable polymers such as polycaprolactone (PCL), polybutylene succinate, and poly3-hydroxybutyrate-co-3-hydroxy valerate are used to lower the mechanisms. To facilitate the denitrification system, these polymers behave as an absorbing agent by absorbing contaminants from other source media or by giving energy and carbon chain supply to the microbes. Research is done on PLA for workable usage in bioremediation and availability as a raw material at lesser cost. For example, Regenesis Bioremediation Products (San Clemente, CA) produced commercially utilized PLA for bioremediation as the hydrogen release compound (HRC). It is produced in a liquid and a gel-like form for the controlled-release of LA for certain time-interval (Koenigsberg, 1999). The characterization of PLA and its research for bioremediation applications are limited. The environmental applications of biodegradable polymers depend upon their glass transition temperature (T_g). Accountably, PLA adsorbs contaminants up to 60°C and is reportedly more resistant to microbial flora as compared to the other biodegradable polymers like PCL (Caruso, 2015).

Commercial PLA ($\geq 2.0 \times 10^5$ Da) has a high molecular weight due to which they show resistance toward the microbial activity. So microorganisms consume PLA as food; hydrolysis reduces the molecular weight due to microbial consumption. For the denitrification mechanism, lower molecular weight PLA is likable. Practically, PLAs with a molecular weight of about 1.0×10^4 Da have a significant rate of N_2 removal than PLAs with higher molecular weight (Auras et al., 2010).

12.10.8 Other applications

Filters, space exploration parts, and 3D printing are some other applications of PLA. Conjoint work of Altran Italia, Thales Alenia Space, and the Italian Institute of Technology contributed to a 3D portable onboard printer for space; it was made by approximately 5.5 kg of PLA. Characteristics like glossiness and multicolor appearance make PLA suitable for 3D printing. PLA is preferred over commonly used printing filaments like acrylonitrile butadiene styrene (Chilson, 2013). For determining the high accuracy of dimensional parts, D.M. Enterprises Pvt. Limited (Hong Kong) used PLA tow blends instead of using cellulose acetate tow for cigarette filters as the former shows less deformation. Cellulose acetate is found naturally but its degradation rate is enhanced by concentrated acid addition making it unsafe to use (Robertson et al., 2012).

Fujitsu Laboratories Ltd. developed water-based paint incorporating PLA with an approach to eliminate the use of volatile organic compounds present in the solvent-based paints. The stability of the PLA emulsion and water-based paint applicants is upgraded by the isocyanate reactant (Hamakawa et al., 2017). Fujitsu collaborated with Toray Industries, Inc. in 2005, to generate PLA alloy which is thermally durable and inflammable for mass production of the Fujitsu's FMVBIBLO notebook models (Fujitsu Global, 2020). More knowledge in PLA properties including physiochemical, mechanical, stereochemical, and morphological helps to develop more advanced applicants of PLA by new technology.

12.11 Conclusions and perspectives

The initial stage in the production of PLA is the fermentative production of its monomer LA. Renewable substrates and LAB are the most widely used for commercial LA production. For further enhancement in production efficiency, strain improvement methods like natural mutagenesis and rDNA technology are still under research. The conventional fermentation methods and recovery techniques aided with approaches like immobilization have proven to be economic and efficient. The various isomers of LA are used to form PLA either by direct or ring-opening polymerization. The ring-opening polymerization method is the most prominently used because of associated higher yield and low toxicity effects. The selection of proper production and recovery methods for LA and PLA is important to ensure less waste generation and economic processing. Furthermore, the increasing commercial applications of PLA also demand improvement in the existing production methods, and hence more investigation in the current approaches is required to establish long-term feasibility of PLA.

References

Abdel-Rahman, M. A., Tashiro, Y., & Sonomoto, K. (2011). Lactic acid production from lignocellulose-derived sugars using lactic acid bacteria: Overview and limits. *Journal of Biotechnology*, *156*(4), 286–310.

Abdel-Rahman, M. A., Tashiro, Y., & Sonomoto, K. (2013). Recent advances in lactic acid production by microbial fermentation processes. *Biotechnology Advances*, *31*(6), 877–902.

Abdel-Rahman, M. A., Tashiro, Y., Zendo, T., Hanada, K., Shibata, K., & Sonomoto, K. (2011). Efficient homofermentative L-(+)-lactic acid production from xylose by a novel lactic acid bacterium, *Enterococcus mundtii* QU 25. *Applied and Environmental Microbiology*, *77*(5), 1892–1895.

Achmad, F., Yamane, K., Quan, S., & Kokugan, T. (2009). Synthesis of polylactic acid by direct polycondensation under vacuum without catalysts, solvents and initiators. *Chemical Engineering Journal*, *151*(1–3), 342–350.

Adsul, M. G., Varma, A. J., & Gokhale, D. V. (2007). Lactic acid production from waste sugarcane bagasse derived cellulose. *Green Chemistry*, *9*, 58–62.

Amrane, A., & Prigent, Y. (1996). A novel concept of bioreactor: Specialized function two-stage continuous reactor, and its application to lactose conversion into lactic acid. *Journal of Biotechnology*, *45*(3), 195–203.

Angermayr, S. A., Paszota, M., & Hellingwerf, K. J. (2012). Engineering a cyanobacterial cell factory for production of lactic acid. *Applied and Environmental Microbiology*, *78*(19), 7098–7106.

Auras, R., Harte, B., & Selke, S. (2004). An overview of polylactides as packaging materials. *Macromolecular Bioscience*, *4*(9), 835–864.

Auras, R., Lim, L. T., Selke, S. E. M., & Tsuji, H. (2010). *Poly(lactic acid): Synthesis, structures, properties, processing, and applications*. New York: John Wiley & Sons.

Bai, D.-M., Li, S.-Z., Liu, Z. L., & Cui, Z.-F. (2008). Enhanced l-(+)-lactic acid production by an adapted strain of *Rhizopus oryzae* using corncob hydrolysate. *Applied Biochemistry and Biotechnology*, *144*(1), 79–85.

Bai, D. M., Wei, Q., Yan, Z. H., Zhao, X. M., Li, X. G., & Xu, S. M. (2003). Fed-batch fermentation of *Lactobacillus lactis* for hyper-production of L-lactic acid. *Biotechnology Letters*, *25*, 1833–1835.

Bai, D. M., Yan, Z. H., Wei, Q., Zhao, X. M., Li, X. G., & Xu, S. M. (2004). Ammonium lactate production by *Lactobacillus lactis* BME5-18M in pH-controlled fed-batch fermentations. *Biochemical Engineering Journal*, *19*(1), 47–51.

Barragán, L. A. P., Figueroa, J. J. B., Durán, L. V. R., González, C. N. A., & Hennigs, C. (2016). *Fermentative production methods. Biotransformation of agricultural waste and by-products* (pp. 189–217). Elsevier.

Barrick, J. E., & Lenski, R. E. (2013). Genome dynamics during experimental evolution. *Nature Reviews Genetics*, *14*, 827–839.

Behonick, D. J., Xing, Z., Lieu, S., Buckley, J. M., Lotz, J. C., Marcucio, R. S., ... Colnot, C. (2007). Role of matrix metalloproteinase 13 in both endochondral and intramembranous ossification during skeletal regeneration. *PLoS One, 2*(11), e1150.

Bianchi, M. M., Brambilla, L., Protani, F., Liu, C. L., Lievense, J., & Porro, D. (2001). Efficient homolactic fermentation by *Kluyveromyces lactis* strains defective in pyruvate utilization and transformed with the heterologous LDH gene. *Applied and Environmental Microbiology, 67*(12), 5621−5625.

Bio-based News. (2020). Award-winning water filter made from bioplastic. Available from http://news.bio-based.eu/award-winning-water-filter-made-from-bioplastic/.

BioCycle. (2020). Chipmaker's All-Or-Nothing Claim Sets The Bar In Big And Bold. [Online]. Available from <https://www.biocycle.net/2010/08/17/chipmakers-all-or-nothing-claim-sets-the-bar-in-big-and-bold/> Accessed 05.01.20.

Blackburn, R. (2005). *Biodegradable and sustainable fibres* (47). Taylor & Francis US.

Bleach, N. C., Tanner, K. E., Kellomäki, M., & Törmälä, P. (2001). Effect of filler type on the mechanical properties of self-reinforced polylactide-calcium phosphate composites. *Journal of Materials Science: Materials in Medicine, 12*, 911−915.

Bomrungnok, W., Sonomoto, K., Pinitglang, S., & Wongwicharn, A. (2012). Single step lactic acid production from Cassava starch by *Laactobacillus plantarum* SW14 in conventional continuous and continuous with high cell density. *APCBEE Procedia, 2*, 97−103.

Boniardi, N., Rota, R., Nano, G., & Mazza, B. (1997). Lactic acid production by electrodialysis part I: Experimental tests. *Journal of Applied Electrochemistry, 27*, 125−133.

Bouapao, L., Tsuji, H., Tashiro, K., Zhang, J., & Hanesaka, M. (2009). Crystallization, spherulite growth, and structure of blends of crystalline and amorphous poly(lactide)s. *Polymer, 50*(16), 4007−4017.

Bouchoux, A., Roux-De Balmann, H., & Lutin, F. (2005). Nanofiltration of glucose and sodium lactate solutions: Variations of retention between single- and mixed-solute solutions. *Journal of Membrane Science, 258*(1−2), 123−132.

Brigham, C. J., & Sinskey, A. J. (2012). Applications of polyhydroxyalkanoates in the medical industry. *International Journal of Biotechnology for Wellness Industries, 1*(1), 52−60.

Broz, M. E., VanderHart, D. L., & Washburn, N. R. (2003). Structure and mechanical properties of poly (D, L-lactic acid)/poly (ε-caprolactone) blends. *Biomaterials, 24*(23), 4181−4190.

Budhavaram, N. K., & Fan, Z. (2009). Production of lactic acid from paper sludge using acid-tolerant, thermophilic *Bacillus coagulan* strains. *Bioresource Technology, 100*(23), 5966−5972.

Bulut, S., Elibol, M., & Ozer, D. (2004). Effect of different carbon sources on L (+)-lactic acid production by *Rhizopus oryzae*. *Biochemical Engineering Journal, 21*(1), 33−37.

Bustos, G., Moldes, A. B., Cruz, J. M., & Domínguez, J. M. (2004). Production of fermentable media from vine-trimming wastes and bioconversion into lactic acid by *Lactobacillus pentosus*. *Journal of the Science of Food and Agriculture, 84*(15), 2105−2112.

Bustos, G., de la Torre, N., Moldes, A. B., Cruz, J. M., & Domínguez, J. M. (2007). Revalorization of hemicellulosic trimming vine shoots hydrolyzates trough continuous production of lactic acid and biosurfactants by *L. pentosus*. *Journal of Food Engineering, 78*(2), 405−412.

Büyükkileci, A. O., & Harsa, S. (2004). Batch production of L (+) lactic acid from whey by *Lactobacillus casei* (NRRL B-441). *Journal of Chemical Technology and Biotechnology, 79*(9), 1036−1040.

Calabia, B. P., & Tokiwa, Y. (2007). Production of D-lactic acid from sugarcane molasses, sugarcane juice and sugar beet juice by *Lactobacillus delbrueckii*. *Biotechnology Letters, 29*(9), 1329−1332.

Calabia, B. P., Tokiwa, Y., & Aiba, S. (2011). Fermentative production of L-(+)-lactic acid by an alkaliphilic marine microorganism. *Biotechnology Letters, 33*(7), 1429−1433.

Caruso, G. (2015). Plastic degrading microorganisms as a tool for bioremediation of plastic contamination in aquatic environments. *Journal of Pollution Effect & Control, 3*(3).

Castillo Martinez, F. A., Balciunas, E. M., Salgado, J. M., Domínguez González, J. M., Converti, A., & de S Oliveira, R. P. (2013). Lactic acid properties, applications and production: A review. *Trends in Food Science and Technology, 30*(1), 70−83.

Cha, Y., & Pitt, C. G. (1990). The biodegradability of polyester blends. *Biomaterials, 11*(2), 108–112.

Chan-Blanco, Y., Bonilla-Leiva, A. R., & Velázquez, A. C. (2003). Using banana to generate lactic acid through batch process fermentation. *Applied Microbiology and Biotechnology, 63*(2), 147–152.

Chanfreau, S., Mena, M., Porras-Domínguez, J. R., Ramírez-Gilly, M., Gimeno, M., Roquero, P., . . . Bárzana, E. (2010). Enzymatic synthesis of poly-L-lactide and poly-L-lactide-co-glycolide in an ionic liquid. *Bioprocess and Biosystems Engineering, 33*, 629–638.

Chang, D. E., Jung, H. C., Rhee, J. S., & Pan, J. G. (1999). Homofermentative production of D- or L-lactate in metabolically engineered *Escherichia coli* RR1. *Applied and Environmental Microbiology, 65*(4), 1384–1389.

Chen, C.-C. (1998). Adsorption characteristics of polyvinylpyridine and activated carbon for lactic acid recovery from fermentation of *Lactobacillus delbrueckii*. *Separation Science and Technology, 33*(10), 1423–1437.

Cheng, Y., Deng, S., Chen, P., & Ruan, R. (2009). Polylactic acid (PLA) synthesis and modifications: A review. *Frontiers of Chemical in China, 4*, 259–264.

Chilson, L. (2013). The Difference between ABS and PLA for 3D printing. ProtoParadigm.

Choi, H. Y., Ryu, H.-K., Park, K.-M., Lee, E. G., Lee, H., Kim, S.-W., & Choi, E.-S. (2012). Direct lactic acid fermentation of Jerusalem artichoke tuber extract using *Lactobacillus paracasei* without acidic or enzymatic inulin hydrolysis. *Bioresource Technology, 114*, 745–747.

Chotisubha-Anandha, N., Thitiprasert, S., Tolieng, V., & Thongchul, N. (2011). Improved oxygen transfer and increased l-lactic acid production by morphology control of *Rhizopus oryzae* in a static bed bioreactor. *Bioprocess and Biosystems Engineering, 34*, 163–172.

Córdoba, P. R., Ragout, A. L., Siñeriz, F., & Perotti, N. I. (1996). Lactate from cultures of *Lactobacillus casei* recovered in a fluidized bed column using ion exchange resin. *Biotechnology Techniques, 10*, 629–634.

Cubas-Cano, E., González-Fernández, C., Ballesteros, M., & Tomás-Pejó, E. (2018). Biotechnological advances in lactic acid production by lactic acid bacteria: Lignocellulose as novel substrate. *Biofuels, Bioproducts and Biorefining, 12*(2), 290–303.

Cui, F., Li, Y., & Wan, C. (2011). Lactic acid production from corn stover using mixed cultures of *Lactobacillus rhamnosus* and *Lactobacillus brevis*. *Bioresource Technology, 102*(2), 1831–1836.

Dai, Y., & King, C. J. (1996). Selectivity between lactic acid and glucose during recovery of lactic acid with basic extractants and polymeric sorbents. *Industrial & Engineering Chemistry Research, 35*(4), 1215–1224.

Datta, R., & Henry, M. (2006). Lactic acid: Recent advances in products, processes and technologies—A review. *Journal of Chemical Technology and Biotechnology, 81*(7), 1119–1129.

Davison, B. H., & Thompson, J. E. (1992). Simultaneous fermentation and separation of lactic acid in a biparticle fluidized-bed bioreactor. *Applied Biochemistry and Biotechnology, 34*, 431–439.

Dequin, S., & Barre, P. (1994). Mixed lactic acid–alcoholic fermentation by *Saccharomyes cerevisiae* expressing the *Lactobacillus casei* L (+)–LDH. *Bio/Technology, 12*(2), 173.

Dien, B. S., Nichols, N. N., & Bothast, R. J. (2001). Recombinant *Escherichia coli* engineered for production of L-lactic acid from hexose and pentose sugars. *Journal of Industrial Microbiology & Biotechnology, 27*, 259–264.

Ding, S., & Tan, T. (2006). L-lactic acid production by *Lactobacillus casei* fermentation using different fed-batch feeding strategies. *Process Biochemistry, 41*(6), 1451–1454.

Djukić-Vuković, A. P., Vukašinović-Sekulić, M. S., Rakin, M. B., Nikolić, S. B., Pejin, J. D., & Bulatović, M. L. (2012). Effect of different fermentation parameters on L-lactic acid production from liquid distillery stillage. *Food Chemistry, 134*(2), 1038–1043.

Drumright, R. E., Gruber, P. R., & Henton, D. E. (2000). Polylactic acid technology. *Advanced Materials, 12*(23), 1841–1846.

Ducat, D. C., Way, J. C., & Silver, P. A. (2011). Engineering cyanobacteria to generate high-value products. *Trends in Biotechnology, 29*(2), 95–103.

Dumbrepatil, A., Adsul, M., Chaudhari, S., Khire, J., & Gokhale, D. (2008). Utilization of molasses sugar for lactic acid production by *Lactobacillus delbrueckii* subsp. *delbrueckii* mutant Uc-3 in batch fermentation. *Applied and Environmental Microbiology, 74*(1), 333−335.

Dürselen, L., Dauner, M., Hierlemann, H., Planck, H., Claes, L. E., & Ignatius, A. (2001). Resorbable polymer fibers for ligament augmentation. *Journal of Biomedical Materials Research, 58*(6), 666−672.

Eyal, A. M., & Bressler, E. (1993). Industrial separation of carboxylic and amino acids by liquid membranes: Applicability, process considerations, and potential advantage. *Biotechnology and Bioengineering, 41*(3), 287−295.

Fakhravar, S., Najafpour, G., Heris, S. Z., Izadi, M., & Fakhravar, A. (2012). Fermentative lactic acid from deproteinized whey using *Lactobacillus bulgaricus* in batch culture. *World Applied Sciences Journal, 17*(9), 1083−1086.

Fan, Y., Nishida, H., Shirai, Y., & Endo, T. (2003). Control of racemization for feedstock recycling of PLLA. *Green Chemistry, 5*, 575−579.

Fu, D., & Xu, J. (2008). *Diffusion dialysis for acid recovery and its development*, . Pollution control technologies (1). Oxford, UK: Eolss Publishers.

Fujitsu Global. (2020). Fujitsu and Toray Develop World's First Environmentally-Friendly Large-Size Plastic Housing for Notebook PCs. Available from https://www.fujitsu.com/global/about/resources/news/press-releases/2005/0113-01.html.

Fukushima, K., & Kimura, Y. (2008). An efficient solid-state polycondensation method for synthesizing stereocomplexed poly(lactic acid)s with high molecular weight. *Journal of Polymer Science Part A-Polymer Chemistry, 46*(11), 3714−3722.

Garde, A., Jonsson, G., Schmidt, A. S., & Ahring, B. K. (2002). Lactic acid production from wheat straw hemicellulose hydrolysate by *Lactobacillus pentosus* and *Lactobacillus brevis*. *Bioresource Technology, 81*(3), 217−223.

Garlotta, D. (2001). A literature review of poly(lactic acid). *Journal of Polymers and the Environment, 9*, 63−84.

Ge, X. Y., Qian, H., & Zhang, W. G. (2009). Improvement of l-lactic acid production from Jerusalem artichoke tubers by mixed culture of *Aspergillus niger* and *Lactobacillus* sp. *Bioresource Technology, 11*(5), 1872−1874.

Ghosh, S., & Krishnan, S. (2007). Application of poly (lactic acid) fibres in automotive interior. *Indian Journal of Fibre & Textile Research, 32*(1), 119−121.

Giorno, L., Chojnacka, K., Donato, L., & Drioli, E. (2002). Study of a cell-recycle membrane fermentor for the production of lactic acid by *Lactobacillus bulgaricus*. *Industrial & Engineering Chemistry Research, 41*(3), 433−440.

Givry, S., Prevot, V., & Duchiron, F. (2008). Lactic acid production from hemicellulosic hydrolyzate by cells of *Lactobacillus bifermentans* immobilized in Ca-alginate using response surface methodology. *World Journal of Microbiology and Biotechnology, 24*, 745−752.

González, M. I., Alvarez, S., Riera, F. A., & Álvarez, R. (2008). Lactic acid recovery from whey ultrafiltrate fermentation broths and artificial solutions by nanofiltration. *Desalination, 228*(1−3), 84−96.

Götz, P. (1995). Solvent extraction in biotechnology: Recovery of primary and secondary metabolites. *Chemie Ingenieur Technik, 67*(8), 1021−1022.

Griffith, L. G. (2000). Polymeric biomaterials. *Acta Materialia, 48*(1), 263−277.

Gu, S. Y., Yang, M., Yu, T., Bin Ren, T., & Ren, J. (2008). Synthesis and characterization of biodegradable lactic acid-based polymers by chain extension. *Polymer International, 57*(8), 982−986.

Gullón, B., Yáñez, R., Alonso, J. L., & Parajó, J. C. (2008). L-lactic acid production from apple pomace by sequential hydrolysis and fermentation. *Bioresource Technology, 99*(2), 308−319.

Guo, Y., Yan, Q., Jiang, Z., Teng, C., & Wang, X. (2010). Efficient production of lactic acid from sucrose and corncob hydrolysate by a newly isolated *Rhizopus oryzae* GY18. *Journal of Industrial Microbiology & Biotechnology, 37*(11), 1137−1143.

Gupta, B., Revagade, N., & Hilborn, J. (2007). Poly (lactic acid) fiber: An overview. *Progress in Polymer Science, 32*(4), 455−482.

Hamakawa, M., Ebisu, K., Kimura, K., Ishikawa, T., & Shinomura, Y. (2017). Greening throughout product life cycle. *Fujitsu Science & Technical Journal, 53*(6), 29–39.

Hayes, D. G., Dharmalingam, S., Wadsworth, L. C., Leonas, K. K., Miles, C., & Inglis, D. A. (2012). *Biodegradable agricultural mulches derived from biopolymers. Degradable polymers and materials: Principles and practice* (2nd ed., pp. 201–223). ACS Publications.

Hbová, V., Melzoch, K., Rychtera, M., & Sekavová, B. (2004). Electrodialysis as a useful technique for lactic acid separation from a model solution and a fermentation broth. *Desalination, 162*, 361–372.

Hirayama, S., & Ueda, R. (2004). Production of optically pure D-lactic acid by *Nannochlorum* sp. 26A4. *Applied Biochemistry and Biotechnology, 119*, 71–77.

Hirt, G., Tanner, W., & Kandler, O. (1971). Effect of light on the rate of glycolysis in *Scenedesmus obliquus*. *Plant Physiology, 47*(6), 841–843.

Hofvendahl, K., & Hahn-Hägerdal, B. (2000). Factors affecting the fermentative lactic acid production from renewable resources. *Enzyme and Microbial Technology, 26*(2–4), 87–107.

Holzapfel, W. H. (2014). Advances in fermented foods and beverages: Improving quality, technologies and health benefits. Woodhead Publishing.

Hong, A. A., Cheng, K.-K., Peng, F., Zhou, S., Sun, Y., Liu, C.-M., & Liu, D.-H. (2009). Strain isolation and optimization of process parameters for bioconversion of glycerol to lactic acid. *Journal of Chemical Technology and Biotechnology, 84*(10), 1576–1581.

Hu, J., Zhang, Z., Lin, Y., Zhao, S., Mei, Y., Liang, Y., & Peng, N. (2015). High-titer lactic acid production from NaOH-pretreated corn stover by *Bacillus coagulans* LA204 using fed-batch simultaneous saccharification and fermentation under non-sterile condition. *Bioresource Technology, 182*, 251–257.

Inui, M., Murakami, S., Okino, S., Kawaguchi, H., Vertès, A. A., & Yukawa, H. (2004). Metabolic analysis of *Corynebacterium glutamicum* during lactate and succinate productions under oxygen deprivation conditions. *Journal of Molecular Microbiology and Biotechnology, 7*(4), 182–196.

Jeong, J. C., Lee, J., & Park, Y. H. (2001). A unique pattern of mycelial elongation of *Blakeslea trispora* and its effect on morphological characteristics and β-carotene synthesis. *Current Microbiology, 42*(3), 225–228.

Jia, X., Liu, P., Li, S., Li, S., & Wen, J. (2011). D-lactic acid production by a genetically engineered strain *Corynebacterium glutamicum*. *World Journal of Microbiology and Biotechnology., 78*(3), 449–454.

Jin, B., Huang, L. P., & Lant, P. (2003). *Rhizopus arrhizus*—A producer for simultaneous saccharification and fermentation of starch waste materials to L (+)-lactic acid. *Biotechnology Letters, 25*(23), 1983–1987.

Joglekar, H. G., Rahman, I., Babu, S., Kulkarni, B. D., & Joshi, A. (2006). Comparative assessment of downstream processing options for lactic acid. *Separation and Purification Technology, 52*(1), 1–17.

John, R. P., & Madhavan Nampoothiri, K. (2011). Co-culturing of *Lactobacillus paracasei* subsp. *paracasei* with a *Lactobacillus delbrueckii* subsp. *delbrueckii* mutant to make high cell density for increased lactate productivity from cassava bagasse hydrolysate. *Current Microbiology, 62*(3), 790–794.

John, R. P., Nampoothiri, K. M., & Pandey, A. (2006). Simultaneous saccharification and fermentation of cassava bagasse for L-(+)-lactic acid production using *Lactobacilli*. *Applied Biochemistry and Biotechnology, 134*(3), 263–272.

Jung, M., Schierbaum, B., & Vogel, H. (2000). Extraction of carboxylic acids from aqueous solutions with the extractant system alcohol/tri-n-alkylamines. *Chemical Engineering & Technology, 23*(1), 70–74.

Juturu, V., & Wu, J. C. (2016). Microbial production of lactic acid: The latest development. *Critical Reviews in Biotechnology, 36*(6), 967–977.

Kadam, S. R., Patil, S. S., Bastawde, K. B., Khire, J. M., & Gokhale, D. V. (2006). Strain improvement of *Lactobacillus delbrueckii* NCIM 2365 for lactic acid production. *Process Biochemistry, 41*(1), 120–126.

Kasirajan, S., & Ngouajio, M. (2012). Polyethylene and biodegradable mulches for agricultural applications: A review. *Agronomy for Sustainable Development, 32*, 501–529.

Kaufman, E. N., Cooper, S. P., Clement, S. L., & Little, M. H. (1995). Use of a biparticle fluidized-bed bioreactor for the continuous and simultaneous fermentation and purification of lactic acid. *Applied Biochemistry and Biotechnology, 51*, 605.

Kawaguchi, H., Sasaki, M., Vertès, A. A., Inui, M., & Yukawa, H. (2008). Engineering of an L-arabinose metabolic pathway in *Corynebacterium glutamicum*. *Applied Microbiology and Biotechnology, 77*(5), 1053–1062.

Kawaguchi, H., Vertes, A. A., Okino, S., Inui, M., & Yukawa, H. (2006). Engineering of a xylose metabolic pathway in *Corynebacterium glutamicum*. *Applied and Environmental Microbiology, 72*(5), 3418–3428.

Kibenich, A., Sørensen, K., & Johansen, E. (2012). Texturizing lactic acid bacteria strains. *International Patent Application No. WO/2012/052557*.

Kijchavengkul, T. (2010). *Design of biodegradable aliphatic aromatic polyester films for agricultural applications using response surface methodology*. (PhD thesis, Michigan State University).

Kim, E., Shin, E. W., Yoo, I. K., & Chung, J. S. (2009). Characteristics of heterogeneous titanium alkoxide catalysts for ring-opening polymerization of lactide to produce polylactide. *Journal of Molecular Catalysis A: Chemical, 298*(1–2), 36–39.

Kim, H. O. K., Wee, Y. J., Kim, J. N., Yun, J. S., & Ryu, H. W. (2006). Production of lactic acid from cheese whey by batch and repeated batch cultures of *Lactobacillus* sp. RKY2. *Applied Biochemistry and Biotechnology, 131*(1–3), 694–704.

Kiyono, R., Koops, G. H., Wessling, M., & Strathmann, H. (2004). Mixed matrix microporous hollow fibers with ion-exchange functionality. *Journal of Membrane Science, 231*(1–2), 109–115.

Kochhar, S., Hottinger, H., Chuard, N., Taylor, P. G., Atkinson, T., Scawen, M. D., & Nicholls, D. J. (1992). Cloning and overexpression of *Lactobacillus helveticus* d-lactate dehydrogenase gene in *Escherichia coli*. *European Journal of Biochemistry, 208*(3), 799–805.

Koenigsberg, S.S. (1999). Hydrogen Release Compound (HRC): A novel technology for the bioremediation of chlorinated hydrocarbons. In *Proceedings of the 1999 conference on hazardous waste research* (p. 14), St Louis Missouri, May 24–27.

Kosseva, M. R., Panesar, P. S., Kaur, G., & Kennedy, J. F. (2009). Use of immobilised biocatalysts in the processing of cheese whey. *International Journal of Biological Macromolecules, 45*(5), 437–447.

Krischke, W., Schröder, M., & Trösch, W. (1991). Continuous production of l-lactic acid from whey permeate by immobilized *Lactobacillus casei* subsp. *casei*. *Applied Microbiology and Biotechnology, 34*, 573–578.

Kumari, A., Yadav, S. K., & Yadav, S. C. (2010). Biodegradable polymeric nanoparticles based drug delivery systems. *Colloids Surfaces B: Biointerfaces, 75*(1), 1–18.

Kurbanoglu, E. B., & Kurbanoglu, N. I. (2003). Utilization for lactic acid production with a new acid hydrolysis of ram horn waste. *FEMS Microbiology Letters, 225*(1), 29–34.

Kylä-Nikkilä, K., Hujanen, M., Leisola, M., & Palva, A. (2000). Metabolic engineering of *Lactobacillus helveticus* CNRZ32 for production of pure L-(+)-lactic acid. *Applied and Environmental Microbiology, 66*(9), 3835–3841.

Lampe, K. J., Namba, R. M., Silverman, T. R., Bjugstad, K. B., & Mahoney, M. J. (2009). Impact of lactic acid on cell proliferation and free radical-induced cell death in monolayer cultures of neural precursor cells. *Biotechnology and Bioengineering, 103*(6), 1214–1223.

Langer, R., & Vacanti, J. (1993). Tissue engineering. *Science, 260*, 920–926.

Laopaiboon, P., Thani, A., Leelavatcharamas, V., & Laopaiboon, L. (2010). Acid hydrolysis of sugarcane bagasse for lactic acid production. *Bioresource Technology, 101*(3), 1036–1043.

Lapierre, L., Mollet, B., & Germond, J. E. (2002). Regulation and adaptive evolution of lactose operon expression in *Lactobacillus delbrueckii*. *Journal of Bacteriology, 184*, 928–935.

Lasprilla, A. J. R., Martinez, G. A. R., Lunelli, B. H., Jardini, A. L., & Filho, R. M. (2012). Poly-lactic acid synthesis for application in biomedical devices—A review. *Biotechnology Advances, 30*(1), 321–328.

Lee, J., Lee, S. Y., Park, S., & Middelberg, A. P. J. (1999). Control of fed-batch fermentations. *Biotechnology Advances, 17*(1), 29–48.

Li, Y., Shahbazi, A., Coulibaly, S., & Mims, M. M. (2007). Semicontinuous production of lactic acid from cheese whey using integrated membrane reactor. *Applied Biochemistry and Biotechnology, 137*, 897–907.

Li, Y., Shahbazi, A., Williams, K., & Wan, C. (2008). Separate and concentrate lactic acid using combination of nanofiltration and reverse osmosis membranes. *Applied Biochemistry and Biotechnology, 147*(1–3), 1–9.

Li, Z., Lu, J. K., Yang, Z. X., Han, L., & Tan, T. (2012). Utilization of white rice bran for production of l-lactic acid. *Biomass and Bioenergy, 39*, 53–58.

Lin, J., Zhou, M., Zhao, X., Luo, S., & Lu, Y. (2007). Extractive fermentation of l-lactic acid with immobilized *Rhizopus oryzae* in a three-phase fluidized bed. *Chemical Engineering and Processing: Process Intensification*, 46(5), 369−374.

Litchfield, J. H. (2009). *Lactic acid, microbially produced. Encyclopedia of microbiology.* Oxford, UK: Elsevier.

Liu, C., Cui, N., Brown, N. M. D., & Meenan, B. J. (2004). Effects of DBD plasma operating parameters on the polymer surface modification. *Surface and Coatings Technology*, 185(2−3), 311−320.

Liu, T., Miura, S., Yaguchi, M., Arimura, T., Park, E. Y., & Okabe, M. (2006). Scale-up of L-lactic acid production by mutant strain *Rhizopus* sp. MK-96-1196 from 0.003 m^3 to 5 m^3 in airlift bioreactors. *Journal of Bioscience and Bioengineering*, 101(1), 9−12.

Liu, Y., Liao, W., Liu, C., & Chen, S. (2006). Optimization of L-(+)-lactic acid production using pelletized filamentous *Rhizopus oryzae* NRRL 395. In *Twenty-seventh symposium on biotechnology for fuels and chemicals* (pp. 844−853). New York: Springer.

Lunelli, B.H. (2010). *Produção e controle da síntese do éster de ácido acrílico através da fermentação do ácido láctico*. PhD thesis. Universidade Estadual De Campinas Faculdade De Engenharia Química.

Maas, R. H. W., Bakker, R. R., Jansen, M. L. A., Visser, D., de Jong, E., Eggink, G., & Weusthuis, R. A. (2008). Lactic acid production from lime-treated wheat straw by *Bacillus coagulans*: Neutralization of acid by fed-batch addition of alkaline substrate. *Applied Microbiology and Biotechnology*, 78(5), 751−758.

MacLean, R. C., Torres-Barceló, C., & Moxon, R. (2013). Evaluating evolutionary models of stress-induced mutagenesis in bacteria. *Nature Reviews Genetics*, 14, 221−227.

Madhavan Nampoothiri, K., Nair, N. R., & John, R. P. (2010). An overview of the recent developments in polylactide (PLA) research. *Bioresource Technology*, 101(22), 8493−8501.

Madzingaidzo, L., Danner, H., & Braun, R. (2002). Process development and optimisation of lactic acid purification using electrodialysis. *Journal of Biotechnology*, 96(3), 223−239.

Maneeboon, T., Vanichsriratana, W., Pomchaitaward, C., & Kitpreechavanich, V. (2010). Optimization of lactic acid production by pellet-form rhizopus oryzae in 3-l airlift bioreactor using response surface methodology. *Applied Biochemistry and Biotechnology*, 161(1−8), 137−146.

Marega, C., Marigo, A., Di Noto, V., Zannetti, R., Martorana, A., & Paganetto, G. (1992). Structure and Crystallization Kinetics of Poly (L-lactic acid). *Die Makromolekulare Chemie*, 193(7), 1599−1606. Available from https://doi.org/10.1002/macp.1992.021930704.

Margolles and Sanchez, 2012: Marega, C., Marigo, A., Di Noto, V., Zannetti, R., Martorana, A., & Paganetto, G. (1992). Structure and crystallization kinetics of poly (L-lactic acid). Die Makromolekulare Chemie, 193(7), 1599−1606.

Margolles, A., & Sánchez, B. (2012). Selection of a *Bifidobacterium animalis* subsp. *lactis* strain with a decreased ability to produce acetic acid. *Applied and Environmental Microbiology*, 78(9), 3338−3342.

Marín, J. L., Benavente-García, A. G., Arias, S. B., Franco, J. A., & López, F. C. (2005). *Materiales de acolchado biodegradables como alternativa al polietileno lineal de baja densidad*. V Congresso ibérico de ciências hortícolas: IV Congresso iberoamericano de ciências hortícolas:[comunicaçoes] (pp. 344−351). Associaçâo Portuguesa de Horticultura.

Marques, S., Santos, J. A. L., Gírio, F. M., & Roseiro, J. C. (2008). Lactic acid production from recycled paper sludge by simultaneous saccharification and fermentation. *Biochemical Engineering Journal*, 41(3), 210−216.

MarketWatch. (2015). Biovation Launches BioArmour(TM) Blood Pressure Cuff Shield: An infectious disease barrier product for hospitals and healthcare facilities. [Online]. Available from https://www.marketwatch.com/press-release/biovation-launches-bioarmourtm-blood-pressure-cuff-shield-an-infectious-disease-barrier-product-for-hospitals-and-healthcare-facilities.

Marták, J., Schlosser, Š., Sabolová, E., Krištofíková, L., & Rosenberg, M. (2003). Fermentation of lactic acid with *Rhizopus arrhizus* in a stirred tank reactor with a periodical bleed and feed operation. *Process Biochemistry*, 38(11), 1573−1583.

Matsuzawa, Y., Kimura, Z.-I., Nishimura, Y., Shibayama, M., & Hiraishi, A. (2010). Removal of hydrophobic organic contaminants from aqueous solutions by sorption onto biodegradable polyesters. *Journal of Water Resource and Protection*, 2(03), 214.

Mazumdar, S., Clomburg, J. M., & Gonzalez, R. (2010). *Escherichia coli* strains engineered for homofermentative production of D-lactic acid from glycerol. *Applied and Environmental Microbiology, 76* (13), 4327–4336.

Mehta, R., Kumar, V., Bhunia, H., & Upadhyay, S. N. (2005). Synthesis of poly(lactic acid): A review. *Journal of Macromolecular Science—Polymer Reviews, 45*(4), 325–349.

Meng, Y., Xue, Y., Yu, B., Gao, C., & Ma, Y. (2012). Efficient production of l-lactic acid with high optical purity by alkaliphilic *Bacillus* sp. WL-S20. *Bioresource Technology, 116*, 334–339.

Meussen, B. J., De Graaff, L. H., Sanders, J. P. M., & Weusthuis, R. A. (2012). Metabolic engineering of *Rhizopus oryzae* for the production of platform chemicals. *Applied Microbiology and Biotechnology, 94*(4), 875–886.

Middleton, J. C., & Tipton, A. J. (2000). Synthetic biodegradable polymers as orthopedic devices. *Biomaterials, 21*(23), 2335–2346.

Mohan, A. M. (2020). Danone first to switch to PLA for yogurt cup in Germany. [Online]. Available from <https://www.greenerpackage.com/bioplastics/danone_first_switch_pla_yogurt_cup_germany> Accessed 05.01.20.

Mohanty, A. K., Misra, M., & Drzal, L. T. (2002). Sustainable bio-composites from renewable resources: Opportunities and challenges in the green materials world. *Journal of Polymer and the Environment, 10* (1–2), 19–26.

Mohd Adnan, A. F., & Tan, I. K. P. (2007). Isolation of lactic acid bacteria from Malaysian foods and assessment of the isolates for industrial potential. *Bioresource Technology, 98*(7), 1380–1385.

Moldes, A. B., Torrado, A., Converti, A., & Dominguez, J. M. (2006). Complete bioconversion of hemicellulosic sugars from agricultural residues into lactic acid by *Lactobacillus pentosus*. *Applied Biochemistry and Biotechnology, 135*(3), 219–227.

Monteagudo, J. M., & Aldavero, M. (1999). Production of L-lactic acid by *Lactobacillus delbrueckii* in chemostat culture using an ion exchange resins system. *Journal of Chemical Technology and Biotechnology, 74*(7), 627–634.

Moon, S. K., Wee, Y. J., & Choi, G. W. (2012). A novel lactic acid bacterium for the production of high purity l-lactic acid, *Lactobacillus paracasei* subsp. *paracasei* CHB2121. *Journal of Bioscience and Bioengineering, 114*(2), 155–159.

Nakano, S., Ugwu, C. U., & Tokiwa, Y. (2012). Efficient production of d-(-)-lactic acid from broken rice by *Lactobacillus delbrueckii* using $Ca(OH)_2$ as a neutralizing agent. *Bioresource Technology, 104*, 791–794.

Nakasaki, K., & Adachi, T. (2003). Effects of intermittent addition of cellulase for production of L-lactic acid from wastewater sludge by simultaneous saccharification and fermentation. *Biotechnology and Bioengineering, 82*(3), 263–270.

Nancib, A., Nancib, N., & Boudrant, J. (2009). Production of lactic acid from date juice extract with free cells of single and mixed cultures of *Lactobacillus casei* and *Lactococcus lactis*. *World Journal of Microbiology and Biotechnology, 25*, 1423–1429.

Narayanan, N., Roychoudhury, P. K., & Srivastava, A. (2004). L (+) lactic acid fermentation and its product polymerization. *Electronic Journal of Biotechnology, 7*(2), 167–178.

Narebska, A., & Staniszewski, M. (1997). Separation of fermentation products by membrane techniques. I. Separation of lactic acid/lactates by diffusion dialysis. *Separation Science and Technology, 32*(10), 1669–1682.

Narębska, A., & Staniszewski, M. (2008). Separation of carboxylic acids from carboxylates by diffusion dialysis. *Separation Science and Technology, 43*(3), 490–501.

NatureWorks. (2020). SPAR Austria Enhances "Freshness" of Produce with NatureWorks PLA. [Online]. Available https://www.pressreleasefinder.com/NatureWorks/NWPR015/en/. [Accessed 05.01.2020].

Nduko, J. M., & Taguchi, S. (2019). *Microbial production and properties of LA-based polymers and oligomers from renewable feedstock. Production of materials from sustainable biomass resources* (pp. 361–390). Springer.

Nguyen, C. M., Kim, J.-S., Hwang, H. J., Park, M. S., Choi, G. J., Choi, Y. H., . . . Kim, J.-C. (2012). Production of l-lactic acid from a green microalga, *Hydrodictyon reticulum*, by *Lactobacillus paracasei* LA104 isolated from the traditional Korean food, makgeolli. *Bioresource Technology, 110*, 552–559.

Nomura, Y., Yamamoto, K., & Ishizaki, A. (1991). Factors affecting lactic acid production rate in the built-in electrodialysis fermentation, an approach to high speed batch culture. *Journal of Fermentation and Bioengineering, 71*(6), 450–452.

Norton, S., Lacroix, C., & Vuillemard, J. C. (1994). Kinetic study of continuous whey permeate fermentation by immobilized *Lactobacillus helveticus* for lactic acid production. *Enzyme and Microbial Technology, 16*(6), 457–466.

Oda, Y., Yajima, Y., Kinoshita, M., & Ohnishi, M. (2003). Differences of *Rhizopus oryzae* strains in organic acid synthesis and fatty acid composition. *Food Microbiology, 20*(3), 371–375.

Ohkouchi, Y., & Inoue, Y. (2006). Direct production of L (+)-lactic acid from starch and food wastes using *Lactobacillus manihotivorans* LMG18011. *Bioresource Technology, 97*(13), 1554–1562.

Okano, K., Zhang, Q., Shinkawa, S., Yoshida, S., Tanaka, T., Fukuda, H., & Kondo, A. (2009). Efficient production of optically pure D-lactic acid from raw corn starch by using a genetically modified L-lactate dehydrogenase gene-deficient and α-amylase-secreting *Lactobacillus plantarum* strain. *Applied and Environmental Microbiology, 75*(2), 462–467.

Okano, K., Tanaka, T., Ogino, C., Fukuda, H., & Kondo, A. (2010). Biotechnological production of enantiomeric pure lactic acid from renewable resources: Recent achievements, perspectives, and limits. *Applied Microbiology and Biotechnology, 85*(3), 413–423.

Okino, S., Inui, M., & Yukawa, H. (2005). Production of organic acids by *Corynebacterium glutamicum* under oxygen deprivation. *Applied Microbiology and Biotechnology, 68*(4), 475–480.

Okino, S., Suda, M., Fujikura, K., Inui, M., & Yukawa, H. (2008). Production of D-lactic acid by *Corynebacterium glutamicum* under oxygen deprivation. *Applied Microbiology and Biotechnology, 78*(3), 449–454.

Oost van der, J., Bulthuis, B. A., Feitz, S., Krab, K., & Kraayenhof, R. (1989). Fermentation metabolism of the unicellular cyanobacterium *Cyanothece* PCC 7822. *Archives of Microbiology, 152*, 415–419.

Ou, M. S., Ingram, L. O., & Shanmugam, K. T. (2011). L (+)-Lactic acid production from non-food carbohydrates by thermotolerant *Bacillus coagulans*. *Journal of Industrial Microbiology & Biotechnology, 38*(5), 599–605.

Ou, M. S., Mohammed, N., Ingram, L. O., & Shanmugam, K. T. (2009). Thermophilic *Bacillus coagulans* requires less cellulases for simultaneous saccharification and fermentation of cellulose to products than mesophilic microbial biocatalysts. *Applied Biochemistry and Biotechnology, 55*(1–3), 379–385.

Ouyang, J., Ma, R., Zheng, Z., Cai, C., Zhang, M., & Jiang, T. (2013). Open fermentative production of L-lactic acid by *Bacillus* sp. strain NL01 using lignocellulosic hydrolyzates as low-cost raw material. *Bioresource Technology, 135*, 475–480.

Oyama, H. T., Tanaka, Y., & Kadosaka, A. (2009). Rapid controlled hydrolytic degradation of poly(l-lactic acid) by blending with poly(aspartic acid-co-l-lactide). *Polymer Degradation and Stability, 94*(9), 1419–1426.

Palatý, Z., Stoček, P., Bendová, H., & Prchal, P. (2009). Continuous dialysis of carboxylic acids: Solubility and diffusivity in Neosepta-AMH membranes. *Desalination, 243*(1–3), 65–73.

Palatý, Z., Stoček, P., Žáková, A., & Bendová, H. (2007). Transport characteristics of some carboxylic acids in the polymeric anion-exchange membrane neosepta-AMH: Batch experiments. *Journal of Applied Polymer Science, 106*(2), 909–916.

Panesar, P. S., Kennedy, J. F., Gandhi, D. N., & Bunko, K. (2007). Bioutilisation of whey for lactic acid production. *Food Chemistry, 105*(1), 1–14.

Park, E. Y., Kosakai, Y., & Okabe, M. (1998). Efficient production of l-(+)-lactic acid using mycelial cotton-like flocs of *Rhizopusoryzae* in an air-lift bioreactor. *Biotechnology Progress, 14*(5), 699–704.

Patel, M., Bassi, A. S., Zhu, J. J. X., & Gomaa, H. (2008). Investigation of a dual-particle liquid-solid circulating fluidized bed bioreactor for extractive fermentation of lactic acid. *Biotechnology Progress, 24*(4), 821–831.

Patel, M. A., Ou, M. S., Harbrucker, R., Aldrich, H. C., Buszko, M. L., Ingram, L. O., & Shanmugam, K. T. (2006). Isolation and characterization of acid-tolerant, thermophilic bacteria for effective fermentation of biomass-derived sugars to lactic acid. *Applied and Environmental Microbiology*, 72(5), 3228−3235.

Pimtong, V., Ounaeba, S., Thitiprasert, S., Tolieng, V., Sooksai, S., Boonsombat, R., ... Thongchul, N. (2017). Enhanced effectiveness of *Rhizopus oryzae* by immobilization in a static bed fermentor for L-lactic acid production. *Process Biochemistry*, 52, 44−52.

Pintado, J., Guyot, J. P., & Raimbault, M. (1999). Lactic acid production from mussel processing wastes with an amylolytic bacterial strain. *Enzyme and Microbial Technology*, 24(8−9), 590−598.

Plessas, S., Bosnea, L., Psarianos, C., Koutinas, A. A., Marchant, R., & Banat, I. M. (2008). Lactic acid production by mixed cultures of *Kluyveromyces marxianus, Lactobacillus delbrueckii* ssp. *bulgaricus* and *Lactobacillus helveticus*. *Bioresource Technology*, 99(13), 5951−5955.

Portnoy, V. A., Herrgård, M. J., & Palsson, B. (2008). Aerobic fermentation of D-glucose by an evolved cytochrome oxidase-deficient *Escherichia coli* strain. *Applied and Environmental Microbiology*, 74(24), 7561−7569.

Praphailong, W., & Fleet, G. H. (1997). The effect of pH, sodium chloride, sucrose, sorbate and benzoate on the growth of food spoilage yeasts. *Food Microbiology*, 14(5), 459−468.

Qin, J., Zhao, B., Wang, X., Wang, L., Yu, B., Ma, Y., ... Xu, P. (2009). Non-sterilized fermentative production of polymer-grade L-lactic acid by a newly isolated thermophilic strain bacillus sp. 2-6. *PLoS One*, 4(2), e4359.

Qin, J., Wang, X., Zheng, Z., Ma, C., Tang, H., & Xu, P. (2010). Production of L-lactic acid by a thermophilic *Bacillus* mutant using sodium hydroxide as neutralizing agent. *Bioresource Technology*, 101(19), 7570−7576.

Rancan, F., Papakostas, D., Hadam, S., Hackbarth, S., Delair, T., Primard, C., ... Vogt, A. (2009). Investigation of polylactic acid (PLA) nanoparticles as drug delivery systems for local dermatotherapy. *Pharmaceutical Research*, 26(8), 2027−2036.

Rangaswamy, V., & Ramakrishna, S. V. (2008). Lactic acid production by *Lactobacillus delbrueckii* in a dual reactor system using packed bed biofilm reactor. *Letters in Applied Microbiology*, 46(6), 661−666.

Rawat, J. M., Bhandari, A., Raturi, M., & Rawat, B. (2019). Agrobacterium rhizogenes *mediated hairy root cultures: A promising approach for production of useful metabolites*. New and future developments in microbial biotechnology and bioengineering. Elsevier.

Reddy, G., Altaf, M., Naveena, B. J., Venkateshwar, M., & Kumar, E. V. (2008). Amylolytic bacterial lactic acid fermentation—A review. *Biotechnology Advances*, 26(1), 22−34.

Robertson, R. M., Thomas, W. C., Suthar, J. N., & Brown, D. M. (2012). Accelerated degradation of cellulose acetate cigarette filters using controlled-release acid catalysis. *Green Chemistry*, 14(8), 2266−2272.

Romaní, A., Yáñez, R., Garrote, G., & Alonso, J. L. (2008). SSF production of lactic acid from cellulosic biosludges. *Bioresource Technology*, 99(10), 4247−4254.

Romero-Garcia, S., Hernández-Bustos, C., Merino, E., Gosset, G., & Martinez, A. (2009). Homolactic fermentation from glucose and cellobiose using *Bacillus subtilis*. *Microbial Cell Factories*, 8, 23.

Roney, C., Kulkarni, P., Arora, V., Antich, P., Bonte, F., Wu, A., ... Aminabhavi, T. M. (2005). Targeted nanoparticles for drug delivery through the blood−brain barrier for Alzheimer's disease. *Journal of Controlled Release*, 108(2−3), 193−214.

Rosenberg, M., Rebroš, M., Krištofíková, L., & Malátová, K. (2005). High temperature lactic acid production by *Bacillus coagulans* immobilized in LentiKats. *Biotechnology Letters*, 27(23−24), 1943−1947.

Roukas, T., & Kotzekidou, P. (1991). Production of lactic acid from deproteinized whey by coimmobilized *Lactobacillus casei* and *Lactococcus lactis* cells. *Enzyme and Microbial Technology*, 22(3), 199−204.

Roukas, T., & Kotzekidou, P. (1996). Continuous production of lactic acid from deproteinized whey by coimmobilized *Lactobacillus casei* and *Lactococcus lactis* cells in a packed-bed reactor. *Food Biotechnology*, 10(3), 231−242.

Roukas, T., & Kotzekidou, P. (1998). Lactic acid production from deproteinized whey by mixed cultures of free and coimmobilized *Lactobacillus casei* and *Lactococcus lactis* cells using fedbatch culture. *Enzyme and Microbial Technology*, 22(3), 199–204.

RP news wires, Noria Corporation. Ford researchers aim to create greener, lighter plastics. [Online]. Available: https://www.reliableplant.com/Read/20034/ford-researchers-aim-to-create-greener,-lighter-plastics.

Saad, S. M. I. (2006). Lactic acid bacteria: Microbiological and functional aspects. *Revista Brasileira De Ciências Farmacêuticas*, 42(3), 473.

Saarela, M., Alakomi, H. L., Mättö, J., Ahonen, A. M., Puhakka, A., & Tynkkynen, S. (2011). Improving the storage stability of *Bifidobacterium breve* in low pH fruit juice. *International Journal of Food Microbiology*, 149(1), 106–110.

Saito, K., Hasa, Y., & Abe, H. (2012). Production of lactic acid from xylose and wheat straw by *Rhizopus oryzae*. *Journal of Bioscience and Bioengineering*, 114(2), 166–169.

Sakai, K., & Yamanami, T. (2006). Thermotolerant *Bacillus licheniformis* TY7 produces optically active l-lactic acid from kitchen refuse under open condition. *Journal of Bioscience and Bioengineering*, 102(2), 132–134.

Sarasua, J. R., Arraiza, A. L., Balerdi, P., & Maiza, I. (2005). Crystallinity and mechanical properties of optically pure polylactides and their blends. *Polymer Engineering and Science*, 45(5), 745–753.

Sasaki, M., Jojima, T., Inui, M., & Yukawa, H. (2008). Simultaneous utilization of d-cellobiose, d-glucose, and d-xylose by recombinant *Corynebacterium glutamicum* under oxygen-deprived conditions. *Applied Microbiology and Biotechnology*, 81(4), 691–699.

Sauer, M., Porro, D., Mattanovich, D., & Branduardi, P. (2010). 16 Years research on lactic acid production with yeast—ready for the market? *Biotechnology & Genetic Engineering Reviews*, 27(1), 229–256.

Senthuran, A., Senthuran, V., Hatti-Kaul, R., & Mattiasson, B. (1999). Lactic acid production by immobilized *Lactobacillus casei* in recycle batch reactor: A step towards optimization. *Journal of Biotechnology*, 73(1), 61–70.

Shen, X., & Xia, L. (2006). Lactic acid production from cellulosic waste by immobilized cells of *Lactobacillus delbrueckii*. *World Journal of Microbiology and Biotechnology*, 22, 1109–1114.

Shi, Z., Wei, P., Zhu, X., Cai, J., Huang, L., & Xu, Z. (2012). Efficient production of l-lactic acid from hydrolysate of Jerusalem artichoke with immobilized cells of *Lactococcus lactis* in fibrous bed bioreactors. *Enzyme and Microbial Technology*, 51(5), 263–268.

Shibata, K., Flores, D. M., Kobayashi, G., & Sonomoto, K. (2007). Direct l-lactic acid fermentation with sago starch by a novel amylolytic lactic acid bacterium, *Enterococcus faecium*. *Enzyme and Microbial Technology*, 41(1–2), 149–155.

Silva, E. M., & Yang, S. T. (1995). Kinetics and stability of a fibrous-bed bioreactor for continuous production of lactic acid from unsupplemented acid whey. *Journal of Biotechnology*, 41(1), 59–70.

Sin, L. T., & Tueen, B. S. (2019). *Polylactic acid: A practical guide for the processing. Manufacturing, and applications of PLA*. William Andrew.

Sirisansaneeyakul, S., Luangpipat, T., Vanichsriratana, W., Srinophakun, T., Chen, H. H. H., & Chisti, Y. (2007). Optimization of lactic acid production by immobilized *Lactococcus lactis* IO-1. *Journal of Industrial Microbiology & Biotechnology*, 34(5), 381–391.

Södergård, A., & Stolt, M. (2002). Properties of lactic acid based polymers and their correlation with composition. *Progress in Polymer Science*, 27(6), 1123–1163.

Song, Y., Matsumoto, K., Yamada, M., Gohda, A., Brigham, C. J., Sinskey, A. J., & Taguchi, S. (2012). Engineered *Corynebacterium glutamicum* as an endotoxin-free platform strain for lactate-based polyester production. *Applied Microbiology and Biotechnology*, 93(5), 1917–1925.

Sreenath, H. K., Moldes, A. B., Koegel, R. G., & Straub, R. J. (2001). Lactic acid production by simultaneous saccharification and fermentation of alfalfa fiber. *Journal of Bioscience and Bioengineering*, 92(6), 518–523.

Stancheva, K. A. (2008). Applications of dialysis. *Oxidation Communication*, 31(4), 758–775.

Sun, Y., Li, Y.-L., & Bai, S. (1999). Modeling of continuous L (+)-lactic acid production with immobilized R. *oryzae* in an airlift bioreactor. *Biochemical Engineering Journal*, 3(1), 87–90.

Taguchi, H., & Ohta, T. (1991). D-lactate dehydrogenase is a member of the D-isomer-specific 2-hydroxyacid dehydrogenase family: Cloning, sequencing, and expression in *Escherichia coli* of the D-lactate dehydrogenase gene of *Lactobacillus plantarum*. *The Journal of Biological Chemistry*, 266(19), 12588−12594.

Tanaka, H., Irie, S., & Ochi, H. (1989). A novel immobilization method for prevention of cell leakage from the Gel matrix. *Journal of Fermentation and Bioengineering*, 68(3), 216−219.

Tanaka, T., Hoshina, M., Tanabe, S., Sakai, K., Ohtsubo, S., & Taniguchi, M. (2006). Production of D-lactic acid from defatted rice bran by simultaneous saccharification and fermentation. *Bioresource Technology*, 97(2), 211−217.

Tango, M. S. A., & Ghaly, A. E. (2002). A continuous lactic acid production system using an immobilized packed bed of *Lactobacillus helveticus*. *Applied Microbiology and Biotechnology*, 58(6), 712−720.

Taniguchi, M., Kotani, N., & Kobayashi, T. (1987). High-concentration cultivation of lactic acid bacteria in fermentor with cross-flow filtration. *Journal of Fermentation Technology*, 65(2), 179−184.

Tarara, J. M. (2000). Microclimate modification with plastic mulch. *HortScience.*, 35(2), 169−180.

Tashiro, Y., Kaneko, W., Sun, Y., Shibata, K., Inokuma, K., Zendo, T., & Sonomoto, K. (2011). Continuous D-lactic acid production by a novelthermotolerant *Lactobacillus delbrueckii* subsp. *lactis* QU 41. *Applied Microbiology and Biotechnology*, 89(6), 1741−1750.

Taskin, M., Esim, N., & Ortucu, S. (2012). Efficient production of l-lactic acid from chicken feather protein hydrolysate and sugar beet molasses by the newly isolated *Rhizopus oryzae* TS-61. *Food and Bioproducts Processing*, 90(4), 773−779.

Tay, A., & Yang, S. T. (2002). Production of L(+)-lactic acid from glucose and starch by immobilized cells of *Rhizopus oryzae* in a rotating fibrous bed bioreactor. *Biotechnology and Bioengineering*, 80(1), 1−12.

Tejayadi, S., & Cheryan, M. (1995). Lactic acid from cheese whey permeate. Productivity and economics of a continuous membrane bioreactor. *Applied Microbiology and Biotechnology*, 43, 242−248.

Thang, V. H., Koschuh, W., Kulbe, K. D., & Novalin, S. (2005). Detailed investigation of an electrodialytic process during the separation of lactic acid from a complex mixture. *Journal of Membrane Science*, 249(1−2), 173−182.

Timbuntam, W., Sriroth, K., Piyachomkwan, K., & Tokiwa, Y. (2008). Application of bipolar electrodialysis on recovery of free lactic acid after simultaneous saccharification and fermentation of cassava starch. *Biotechnology Letters*, 30(10), 1747−1752.

Timmer, J. M. K., Kromkamp, J., & Robbertsen, T. (1994). Lactic acid separation from fermentation broths by reverse osmosis and nanofiltration. *Journal of Membrane Science*, 92(2), 185−197.

Tokuhiro, K., Ishida, N., Kondo, A., & Takahashi, H. (2008). Lactic fermentation of cellobiose by a yeast strain displaying β-glucosidase on the cell surface. *Applied Microbiology and Biotechnology*, 79(3), 481−488.

Tong, W.-Y., Fu, X.-Y., Lee, S.-M., Yu, J., Liu, J.-W., Wei, D.-Z., & Koo, Y. M. (2004). Purification of L (+)-lactic acid from fermentation broth with paper sludge as a cellulosic feedstock using weak anion exchanger Amberlite IRA-92. *Biochemical Engineering Journal*, 18(2), 89−96.

Tsuji, H., Okino, R., Daimon, H., & Fujie, K. (2006). Water vapor permeability of poly(lactide)s: Effects of molecular characteristics and crystallinity. *Journal of Applied Polymer Science*, 99(5), 2245−2252.

Tsukegi, T., Motoyama, T., Shirai, Y., Nishida, H., & Endo, T. (2007). Racemization behavior of l,l-lactide during heating. *Polymer Degradation and Stability*, 92(4), 552−559.

Tung, L. A., & King, C. J. (1994). Sorption and extraction of lactic and succinic acids at pH> pKa1. I. Factors governing equilibria. *Industrial & Engineering Chemistry Research*, 33(12), 3217−3223.

Urayama, H., Moon, S. Il, & Kimura, Y. (2003). Microstructure and thermal properties of polylactides with different L- and D-unit sequences: Importance of the helical nature of the L-sequenced segments. *Macromolecular Materials and Engineering*, 288(2), 137−143.

Vijayakumar, J., Aravindan, R., & Viruthagiri, T. (2008). Recent trends in the production, purification and application of lactic acid. *Chemical and Biochemical Engineering Quarterly*, 22(2), 245−264.

Wang, L., Zhao, B., Li, F., Xu, K., Ma, C., Tao, F., ... Xu, P. (2011). Highly efficient production of d-lactate by *Sporolactobacillus* sp. CASD with simultaneous enzymatic hydrolysis of peanut meal. *Applied Microbiology and Biotechnology*, *89*(4), 1009−1017.

Wang, Q., Zhao, X., Chamu, J., & Shanmugam, K. T. (2011). Isolation, characterization and evolution of a new thermophilic *Bacillus licheniformis* for lactic acid production in mineral salts medium. *Bioresource Technology*, *102*(17), 8152−8158.

Wang, Y., Tian, T., Zhao, J., Wang, J., Yan, T., Xu, L., ... Zhou, S. (2012). Homofermentative production of d-lactic acid from sucrose by a metabolically engineered *Escherichia coli*. *Biotechnology Letters*, *34*(11), 2069−2075.

Wasewar, K. L., Bert, A., Heesink, M., Versteeg, G. F., & Pangarkar, V. G. (2002). Equilibria and kinetics for reactive extraction of lactic acid using Alamine 336 in decanol. *Journal of Chemical Technology and Biotechnology*, *77*(9), 1068−1075.

Wasewar, K. L., Yawalkar, A. A., Moulijn, J. A., & Pangarkar, V. G. (2004). Fermentation of glucose to lactic acid coupled with reactive extraction: A review. *Industrial and Engineering Chemistry Research*, *43*(19), 5969−5982.

Wee, Y., Reddy, L. V. A., & Ryu, H. (2008). Fermentative production of L (+)-lactic acid from starch hydrolyzate and corn steep liquor as inexpensive nutrients by batch culture of *Enterococcus faecalis* RKY1. *Journal of Chemical Technology and Biotechnology*, *83*(10), 1387−1393.

Wee, Y. J., & Ryu, H. W. (2009). Lactic acid production by *Lactobacillus* sp. RKY2 in a cell-recycle continuous fermentation using lignocellulosic hydrolyzates as inexpensive raw materials. *Bioresource Technology*, *100*(18), 4262−4270.

Woong Kim, K., & Woo, S. I. (2002). Synthesis of high-molecular-weight poly(L-lactic acid) by direct polycondensation. *Macromolecular Chemistry and Physics*, *203*(15), 2245−2250.

Wu, X., Jiang, S., Liu, M., Pan, L., Zheng, Z., & Luo, S. (2011). Production of L-lactic acid by *Rhizopus oryzae* using semicontinuous fermentation in bioreactor. *Journal of Industrial Microbiology & Biotechnology*, *38*(4), 565−571.

Xu, T., Liu, Z., Huang, C., Wu, Y., Wu, L., & Yang, W. (2008). Preparation of a novel hollow-fiber anion-exchange membrane and its preliminary performance in diffusion dialysis. *Industrial & Engineering Chemistry Research*, *47*(16), 6204−6210.

Yamane, H., & Sasai, K. (2003). Effect of the addition of poly(D-lactic acid) on the thermal property of poly(L-lactic acid). *Polymer*, *44*(8), 2569−2575.

Yamane, T., & Tanaka, R. (2013). Highly accumulative production of L (+)-lactate from glucose by crystallization fermentation with immobilized *Rhizopus oryzae*. *Journal of Bioscience and Bioengineering*, *115*(1), 90−95.

Yáñez, R., Alonso, J. L., & Parajó, J. C. (2005). D-lactic acid production from waste cardboard. *Journal of Chemical Technology and Biotechnology*, *80*(1), 76−84.

Yang, T. H., Kim, T. W., Kang, H. O., Lee, S.-H., Lee, E. J., Lim, S.-C., ... Lee, S. Y. (2010). Biosynthesis of polylactic acid and its copolymers using evolved propionate CoA transferase and PHA synthase. *Biotechnology and Bioengineering*, *105*(1), 150−160.

Ye, L., Bin Hudari, M. S., Li, Z., & Wu, J. C. (2014). Simultaneous detoxification, saccharification and co-fermentation of oil palm empty fruit bunch hydrolysate for l-lactic acid production by *Bacillus coagulans* JI12. *Biochemical Engineering Journal*, *83*, 16−21.

Yin, P., Yahiro, K., Ishigaki, T., Park, Y., & Okabe, M. (1998). L (+)-Lactic acid production by repeated batch culture of *Rhizopus oryzae* in air-lift bioreactor. *Journal of Fermentation and Bioengineering*, *85*(1), 96−100.

Yu, L., Lei, T., Ren, X., Pei, X., & Feng, Y. (2008). Response surface optimization of l-(+)-lactic acid production using corn steep liquor as an alternative nitrogen source by Lactobacillus rhamnosus CGMCC 1466. *Biochemical Engineering Journal*, *39*(3), 496−502.

Zhang, B., He, Pj, Ye, Nf, & Shao, Lm (2008). Enhanced isomer purity of lactic acid from the non-sterile fermentation of kitchen wastes. *Bioresource Technology*, *99*(4), 855−862.

Zhang, Y., Cong, W., & Shi, S. (2010). Application of a pH feedback-controlled substrate feeding method in lactic acid production. *Applied Biochemistry and Biotechnology*, *162*(8), 2149−2156.

Zhang, Y., Cong, W., & Shi, S. Y. (2011). Repeated fed-batch lactic acid production in a packed bed-stirred fermentor system using a pH feedback feeding method. *Bioprocess and Biosystems Engineering*, *34*(1), 67−73.

Zhang, Z. Y., Jin, B., & Kelly, J. M. (2007). Production of lactic acid from renewable materials by *Rhizopus* fungi. *Biochemical Engineering Journal, 35*(3), 251–263.

Zhang, Z. Y., Jin, B., & Kelly, J. M. (2008). Production of L (+)-lactic acid using acid-adapted precultures of *Rhizopus arrhizus* in a stirred tank reactor. *Applied Biochemistry and Biotechnology, 149*(3), 265–276.

Zhao, Z., Wang, T., & Chen, L. (2010). Dynamic analysis of a turbidostat model with the feedback control. *Communications in Nonlinear Science and Numerical Simulation, 15*(4), 1028–1035.

Zhou, S., Causey, T. B., Hasona, A., Shanmugam, K. T., & Ingram, L. O. (2003). Production of optically pure D-lactic acid in mineral salts medium by metabolically engineered *Escherichia coli* W3110. *Applied and Environmental Microbiology, 69*(1), 399–407.

Zhou, S., Shanmugam, K. T., & Ingram, L. O. (2003). Functional replacement of the *Escherichia coli* D-(−)-lactate dehydrogenase gene (ldhA) with the L-(+)-lactate dehydrogenase gene (ldhL) from *Pediococcus acidilactici*. *Applied and Environmental Microbiology, 69*(4), 2237–2244.

Zhou, S., Yomano, L. P., Shanmugam, K. T., & Ingram, L. O. (2005). Fermentation of 10%(w/v) sugar to D (−)-lactate by engineered *Escherichia coli* B. *Biotechnology Letters, 27*(23–24), 1891–1896.

Zhu, J. X., Zheng, Y., Karamanev, D. G., & Bassi, A. S. (2000). (Gas-)liquid-solid circulating fluidized beds and their potential applications to bioreactor engineering. *Canadian Journal of Chemical Engineering, 78*(1), 82–94.

Zhu, Y., Eiteman, M. A., DeWitt, K., & Altman, E. (2007). Homolactate fermentation by metabolically engineered *Escherichia coli* strains. *Applied and Environmental Microbiology, 73*(2), 456–464.

Zihao, W., & Kefeng, Z. (1995). Kinetics and mass transfer for lactic acid recovered with anion exchange method in fermentation solution. *Biotechnology and Bioengineering, 47*(1), 1–7.

CHAPTER 13

Production and applications of bacterial cellulose

Fazli Wahid[1,2] and Cheng Zhong[1,2]

[1]State Key Laboratory of Food Nutrition & Safety, Tianjin University of Science & Technology, Tianjin, P.R. China, [2]Key Laboratory of Industrial Fermentation Microbiology (Ministry of Education), Tianjin University of Science & Technology, Tianjin, P.R. China

13.1 Introduction

Cellulose is the largest biopolymer available on the Earth and has widely contributed to human society in the form of paper, food additives, fabric, and biofuels (Ullah, Ul-Islam, Khana, Kim, & Park, 2015; Wang, Tavakoli, & Tang, 2019). It is mainly produced by vascular plants. However, the microorganism can also synthesize cellulose that is comparatively different in properties and applications from plant cellulose. Many bacterial genera, including *Rhizobium*, *Gluconacetobacter*, *Sarcina*, and *Rhodobacter* have been reported to produce cellulose (Azeredo, Barud, Farinas, Vasconcellos, & Claro, 2019; Lin et al., 2013). However, *Gluconoacetobacter xylinus* (known previously as *Acetobacter xylinum* and then classified as *Komagataeibacter xylinus*) has the capability to produce bacterial cellulose (BC) on the industrial scale. BC produced by this strain has been used in many industries, such as cosmetics, food, packing, and in a variety of advanced biomedical applications (Liu et al., 2018; Wang et al., 2017). It has a good reputation because of its high purity and good mechanical properties (Wang et al., 2019). BC is a β-glucan, composed of β-1,4-glucopyranosyl units with a high degree of polymerization ranging up to several million. In comparison to plant cellulose, BC exhibits superior properties because of the well-arranged 3D network structure (Fig. 13.1A). It comprises of randomly assembled ribbon-like fibrils (< 100 nm wide), which are further composed of nanofibrils (7–8 nm wide) accumulated in bundles (Fig. 13.1B) (Cacicedo et al., 2016). Because of this structure, BC possesses a combination of unique properties, including hydrophilicity, crystallinity, flexibility, high water-holding capability, moldability, and high purity (free of lignin and pectin, etc.) (Picheth et al., 2017). These features enable BC as an exciting class of nanomaterials and since its discovery; it has attracted remarkable interest for a variety of applications such as wound dressings, tissue engineering, artificial blood vessels, hemostatic material, drug delivery, cosmetics, food industry, water purification, paper making, electronics, and so on (de Oliveira Barud et al., 2016). Some of the BC-based technologies have

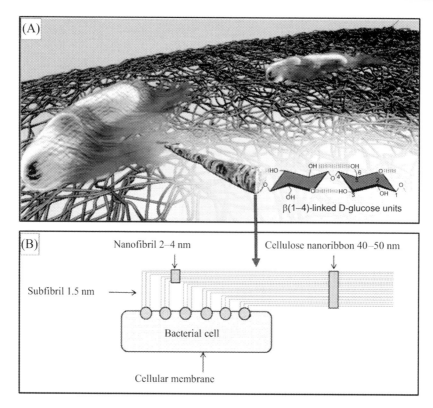

Figure 13.1
(A) Schematic representation of the 3D network structure of BC. (B) BC nanoribbon configuration. Source: *Reproduced with permission from Cacicedo, M. L., Castro, M. C., Servetas, I., Bosnea, L., Boura, K., Tsafrakidou, P., ... Castro, G. R. (2016). Progress in bacterial cellulose matrices for biotechnological applications. Bioresource Technology, 213, 172−180; reproduced with permission from de Oliveira Barud, H. G., da Silva, R. R., Barud, H. D. S., Tercjak, A., Gutierrez, J., Lustri, W. R., ... Ribeiro, S. J. L. (2016). A multipurpose natural and renewable polymer in medical applications: Bacterial cellulose. Carbohydrate Polymers, 153, 406−420.*

been already commercialized such as Bioprogress, Biofill, and Gegiflex (Ullah, Wahid, Santos, & Khan, 2016). This chapter discusses the production of BC, properties, and applications in various fields. It also highlights future insights and research about BC.

13.2 A brief history of bacterial cellulose

Extracellular cellulose was described as a kind of moist skin and slippery gelatinous swollen material by Louis Pasteur. This material had long been used as a traditional dessert by the people of the Philippine called Natta-de-coco (1-cm-thick gel sheets produced by the fermentation of coconut water) (Foresti, Vázquez, & Boury, 2017). Meanwhile, BC was scientifically reported by Adrian Brown while working on *Bacterium aceti* in 1886 (Brown,

1887). A pellicle was formed on the surface of the fermented medium and rapidly grown until the whole surface of the liquid was covered by a gelatinous membrane. This solid pellicle was named as "vinegar plant" and was commonly used for home-made vinegar preparation. The constituents of the pellicle were identified as cellulose and the microorganism responsible for its synthesis was named *Acetobacterium xylinum* (Lin et al., 2013). Since its discovery, many names were assigned to this bacterium but *Acetobacter xylinum* became its official name according to the International Code of Nomenclature of Bacteria. Later on, it was called *G. xylinus* which is a subspecies of *Acetobacter aceti* (Ross, Mayer, & Benziman, 1991), but nowadays it is classified as *K. xylinus*. Since that time, a momentous research has been devoted to the synthesis, modification, and novel applications of BC. Today, BC has been used for various applications, including food, drug delivery, wound dressing, and cosmetics (de Oliveira Barud et al., 2016; Ullah, Santos, & Khan, 2016; Blanco Parte et al., 2020).

13.3 Bacterial cellulose production

BC can be produced by both static and agitated cultivation conditions. For large-scale and commercial production, semicontinuous or continuous fermentation techniques are suitable to meet the requirements. However, the maximum and cost-effective production of BC is always the main objective. Therefore the choice of bacterial strain and the conditions for bioprocess are of prime importance.

13.3.1 Selection of bacterial strain

Cellulose can be produced by the species of *Aerobacter, Achromobacter, Agrobacterium, Alcaligenes, Komagataeibacter, Rhizobium, Pseudomonas, Dickeya, Rhodobacter*, and *Azotobacter*. Some examples are shown in Fig. 13.2. However, only *Komagataeibacter* is used for the commercial production of BC due to its capability to metabolize a large volume of nitrogen and carbon sources (Azeredo et al., 2019). The *Komagataeibacter* species are Gram-negative, aerobic, and live mainly in vegetables and fruits in the decomposition process. They have the capability to convert common carbon sources such as glycerol, fructose, sucrose, glucose, and others at a temperature ranging from 25°C to 30°C (Cacicedo et al., 2016). Moreover, the development of genetically modified strains is also an important approach that has been investigated for improved production, yield, and structural features of BC. For example, a strain of *Komagataeibacter rhaeticus* which can be cultivated in a low nitrogen condition was isolated, its genome was sequenced and a toolkit was developed for genetic engineering study. The toolkit provided the data required to engineer *K. rhaeticus* iGEM, which assisted to design a system that allowed to control the production of natural cellulose and to produce novel pattern of functionalized cellulose-based biomaterials (Florea et al., 2016). The genetically engineered strains can also produce BC in a low oxygen environment that can assist the production of BC in static conditions (Liu et al., 2018).

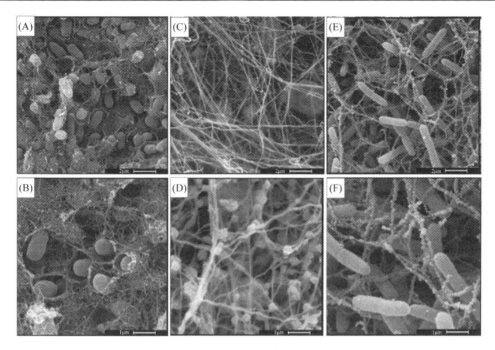

Figure 13.2
The micrographs of different bacteria producing BC [(A) and (B)], *S. enterica*, [(C) and (D)] *G. xylinus*, and [(E) and (F)] *D. dadantii*. Scale bars: [(A), (C), and (E)] 2 μm and [(B), (D), and (F)] 1 μm. Source: *Reproduced with permission from Foresti, M. L., Vázquez, A., & Boury, B. (2017). Applications of bacterial cellulose as precursor of carbon and composites with metal oxide, metal sulfide and metal nanoparticles: A review of recent advances.* Carbohydrate Polymers, *157, 447−467*.

13.3.2 Culture medium

The production of BC requires a culture medium rich in glucose with other nutrients, which leads to a high-cost production and limits the potential applications of BC (Azeredo et al., 2019). Generally, Hestrin and Schramm (HS) culture medium is used for the production of BC, which contains glucose, yeast extracts, and peptone as carbon and nitrogen sources (Lin et al., 2013). However, changing the carbon or nitrogen sources can change the productivity of BC. Moreover, alternative culture media such as fruits and fruit juice, sugarcane molasses, corn steep liquors, and many others have also been investigated for the production of BC. It is noticeable that fruits with high sugar content have been frequently used as a medium for BC production. These include pineapple, orange, apple, and grapes juices, persimmon vinegar, litchi extracts, and coconut milk. The molasses from sugar industries are suitable and a cheaper carbon source for the production of BC with similar characteristics to those obtained from expensive carbon sources such as fructose or glucose (Tyagi & Suresh, 2016).

Some industrial and agricultural wastes can also be used for the production of BC. The use of these wastes for BC production is not only a cheap source for the production of BC but also environment friendly. For example, citrus peel and pomace from beverage industrial wastes showed potential for the production of BC. In addition, the use of these wastes as a medium resulted in 5.7 g L^{-1} production of BC, which was comparatively higher than normal HS medium (3.9 g L^{-1}) (Fan et al., 2016).

13.3.3 Cultivation methods

As aforementioned, currently *Komagataeibacter* is used for the synthesis of BC due to its economical and high production rate. This is a Gram-negative bacteria, which is aerobic in nature, and produce BC extracellularly at the air/liquid interface, normally at $25°C < T < 30°C$, and $4 < pH < 7$ (Ul-Islam, Khan, Ullah, & Park, 2015). Bacteria produce BC as a primary metabolite, this mechanism helps aerobic bacteria to move to the oxygen-rich surface. Furthermore, the formed BC film protects bacteria from UV-light and holds moisture (Klemm, Schumann, Udhardt, & Marsch, 2001). The formation of BC can be divided into three phases. In the first phase, several molecules of glucose are polymerized (forming β-1,4-glucosidic linkages) to synthesize a cellulosic chain. Several cellulosic chains are combined to form 1.5 nm (in diameter) subfibrils (protofibril). In the second phase, numerous subfibrils are combined together to form 2–4 nm (in diameter) nanofibrils. In the third phase, different nanofibrils are joined together to form 20- to 100-nm-wide fibers (Fig. 13.1). The interwoven matrix of these fibers is called BC pellicle (Castro et al., 2011; Klemm et al., 2001). The production of BC is a slow and continuous process, the formed fibers join the pellicle and hence increase its thickness. BC can be produced mainly by static, agitating/shaking culture conditions, and bioreactor cultures (Wang et al., 2019). The produced BC is quite different in different aspects such as macroscopic morphology, microstructure, and properties. A gelatinous film is formed on the surface of the medium by a static method. The agitating/shaking method produces spherical or irregular masses. Choosing a method depends on the physical, mechanical, and morphological requirements as well as the final applications of BC (Wang et al., 2019). In this chapter, we will mainly discuss static and shaking/agitating cultivation methods for BC production.

13.3.3.1 Static cultivation method for bacterial cellulose production

The static cultivation method is a simple approach and extensively used for the production of BC. In this approach, the nutrient-filled containers are incubated at a suitable temperature and pH for 1–14 days to produce BC. The resultant BC is in the form of hydrogel sheets with excellent structural properties. The freshly harvested BC membranes are yellowish in color, after treating with an alkaline solution and washing with water they turn into white color (Fig. 13.3A; Wang et al., 2019). As BC membranes are formed on the surface of the nutrient media, therefore, the BC production depends on the

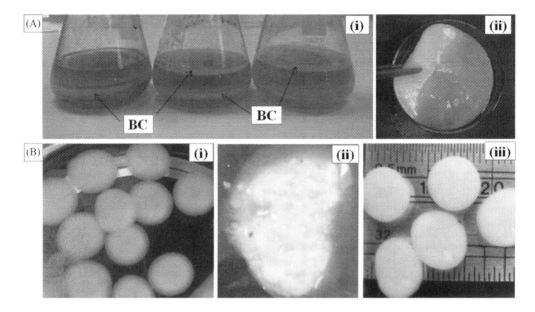

Figure 13.3
Production of BC: (A) static condition: (a) the formation of BC pellicles in culture media and (b) BC pellicle after purification. (B) Agitated/shaking method: (a) the formed spherical, (b) fibrous, and (c) purified BC spheres. Source: *Reproduced with permission from Hu, Y., Catchmark, J. M., & Vogler, E. A. (2013). Factors impacting the formation of sphere-like bacterial cellulose particles and their biocompatibility for human osteoblast growth.* Biomacromolecules, 14, 3444–3452; *reproduced with permission from Wang, J., Tavakoli, J., & Tang, Y. H. (2019). Bacterial cellulose production, properties and applications with different culture methods—A review.* Carbohydrate Polymers, 219, 63–76.

surface area of the air/liquid interface. Moreover, the thickness of the membrane increases with the increase of the fermentation time (Lin et al., 2013).

In a traditional static cultivation method, containers of different sizes and shapes can be used; therefore the produced membranes acquire the shape of the container. This is a benefit when BC with predefined shapes are needed, which have been used widely in regenerative medicines (Azeredo et al., 2019). On the other hand, the conventional cultivation method requires a long time with low productivity, which may limit its wide-range application (Lin et al., 2013).

13.3.3.2 Agitated/shaking cultivation method

In this cultivation method, the containers having culture media are shaken at a certain speed. BC can be formed throughout the medium because of the continuous mixing of oxygen in it. Nevertheless, the excessive oxygen supply also resulting in a decrease in the production of BC (Ul-Islam et al., 2015). Therefore the main goal of the agitated/shaking cultivation approach is to provide the optimum amount of oxygen to the medium. The BC produced in this method is comparatively higher than the static cultivation method, which contributes to the cost-effective yield. This method leads to the production of BC in different forms, from fibers suspension to spheres and pellets (Fig. 13.3B), the size and shape of the BC depend on the applied rotational speed (Ul-Islam et al., 2015).

Although the BC production by agitated/shaking method is higher than static cultivation, the major limitation of this method is the appearance of cellulose-negative mutants and the genetic instability of bacteria under agitating conditions (Kim, Kim, Wee, Park, & Ryu, 2007). Therefore various reactors have been designed to improve the productivity of BC and inhibiting the formation of cellulose-negative strains. Among these reactors, stirred-tank reactors have been often used for fibrous BC production. However, the crystallinity, elasticity, and the degree of polymerization of BC produced by this method are lower than those pellicles produced by the static cultivation method (Azeredo et al., 2019). In these reactors, the produced fibrous BC suspension possesses high cell density and high viscosity, which restricts the supply of oxygen. In this situation, a higher agitation power is needed, resulting in higher energy consumption (Shoda & Sugano, 2005).

Another type of bioreactor is the airlift reactor, in which the oxygen can be continuously supplied from the bottom of the reactor to the fermentation medium. The production process requires less energy and involving less shear stress in comparison to the stirred-tank reactor. This reactor was first developed for BC production in 1997, since then several modified airlift reactors have been proposed (Ul-Islam et al., 2015). For example, Wu and Li (2015) developed an airlift reactor with a series of net plates, to produce BC in pellicle form. The produced pellicles showed a higher water-holding capacity than the membrane produced by the static cultivation method. The elastic modulus could be manipulated by changing the number of net plates.

13.4 Structural and functional features of bacterial cellulose

13.4.1 Mechanical properties

BC exhibits significant mechanical properties because of its reticulated structure consisting of ultrafine fibers (Hu, Chen, Yang, Li, & Wang, 2014). The mechanical properties of BC are higher than plant cellulose and many other synthetic fibers. The high mechanical properties make BC an attractive material for the regeneration of many kinds of tissue such as blood vessels and meniscus (Rajwade, Paknikar, & Kumbhar, 2015). The mechanical properties of BC are generally investigated in its sheet form. Among the earliest investigations, Yamanaka and coworkers found the tensile strength of the BC film to be ~ 260 MPa, while Young's modulus was found to be as high as ~ 18 GPa (Yamanaka et al., 1989). The high mechanical properties can be attributed to the hydrogen bonding in the fibrils and large contact area due to the ultrafine nature of these fibrils. The Young's modulus of BC films can be improved by further purification with oxidative or alkaline solution (Rajwade et al., 2015). The mechanical properties of BC fibers have also been investigated. For example, the mechanical properties of a single nanofiber with a diameter ranging from 70 to 90 nm were measured by atomic force microscopy. The Young's modulus of the nanofibers was recorded to be ∼78 GPa (Guhados, Wan, & Hutter, 2005).

BC can be produced in different forms and shapes to meet the requirements for the desired applications. Therefore the mechanical properties vary according to the requirements. For example, BC was produced in the tube forms for blood vessel applications with Young's modulus of 3.5 MPa, which could sustain a blood pressure of 250 mmHg (Bodin et al., 2007). BC in a crescent shape was developed to mimic meniscus with Young's modulus of 1 MPa, which was higher than collagen (0.01 MPa). Moreover, BC with effective cellulose content of 13.7% was found to have an equilibrium modulus of 2.4 MPa, which was comparable to the native ear cartilage (3.3 MPa) (Bodin, Concaro, Brittberg, & Gatenholm, 2007).

13.4.2 Water holding/release capacity

BC has a complex polymeric structure, where the water molecules are normally attached through hydrogen bonding to its molecular structure. Nevertheless, some of the water (unbonded) that can penetrate and exit the BC structure is responsible for maintaining its hydration level, which is crucial for wound-dressing applications (Portela, Leal, Almeida, & Sobral, 2019). The water-holding capacity of BC is in a range of 60–700 times of its dry weight, which depends on the preparation conditions of BC. Typically, a BC pellicle produced by the static cultivation method possesses approximately 1% of the total weight, while the rest 99% is water. The high hydrophilicity of BC could be due to the formation of cellulose ribbons in which the liquid medium and numerous formed micelles are trapping a huge amount of water molecules (Portela et al., 2019). In addition, the hydrophilicity of BC pellicles is due to the large internal surface area of the wet pellicles. However, by drying, BC possesses poor rehydration, its high crystallinity limits the water reabsorption capability (Huang, Chen, Lin, Hsu, & Chen, 2010).

13.4.3 Structure, pore size, and morphology

The chemical structure and the formation of BC fibrous network have been already discussed in the introduction part and the production of BC. In this section, we will talk about the arrangement of the fibrous structure and its impact on the related properties such as pore size, morphology, and the water-holding capacity of BC. The continuous spinning of fibrils leads to the formation of a 3D nanofibers structure stabilized by inter- and intrafibrillar hydrogen bonding. Because of this structure, BC possesses unique properties such as hydrophilicity, crystallinity, flexibility, high water-holding capability, and a high surface area (Sulaeva, Henniges, Rosenau, & Potthast, 2015). However, a denser arrangement of the BC fibers presenting a compact structure, resulting in reduced surface area, a smaller pore volume, and a decreased water-holding capacity. In such an arrangement of the fibers, the available space, the number of trapping sites for capturing water molecules is reduced (Wang, Wan, Luo, Gao, & Huang, 2012). However, this structure is associated with a lower water release rate. The compact structure retaining a larger amount of water in the system because of the formation of hydrogen bonding and a smaller amount of bulk water, preventing the evaporation of water (Gelin et al., 2007). Considering the counteracting effects related to the structure of BC on the capability of holding and releasing water, researchers are focusing on the adjustment of the structural parameters of BC during the biosynthesis and postsynthetic modifications. The alteration of fermentation systems, such as culture media components, growth condition, and the addition of specific additives during the synthesis process of BC, causing changes in its structural features (crystallinity, morphology, and porosity) (Sulaeva et al., 2015). For example, a compact structure of BC with reduced porosity was obtained by the addition of hydroxypropyl methylcellulose to the culture medium, the resultant BC exhibited a reduced water-holding capacity. On the other hand, BC structures with high porosity showed high water-holding capacity (Huang, Chen, Lin, & Chen, 2011). BC synthesized in the presence of carboxymethyl cellulose in the culture media resulted in a network with broader fibers due to the adhesion of carboxymethyl cellulose with the BC fibers. The resultant BC showed high water-holding capacity (Chen et al., 2016).

A higher porosity with increased water-holding capacity can also be obtained by the postsynthetic modification of BC. For instance, Paximada, Dimitrakopoulou, Tsouko, Koutinas, and Fasseas (2016) found that a short ultrasonic pretreatment of the BC suspension prompted the breakdown of the fibrils which resulted in an increase in the water-holding capacity.

13.4.4 Biodegradability

BC does not show degradation in the mammalian body because of its high crystallinity and the lack of enzymes in the mammalian body that can break the $\beta(1-4)$ glycosidic bond of the BC chain (Zaborowska et al., 2010). Therefore BC is considered a slow/nondegradable material, which is suitable to be used as a scaffold to provide long-term support. Many studies have been conducted to test the biodegradability of BC. For instance, BC was implanted subcutaneously in mice, which retained its shape and size even after 12 weeks (Avila et al., 2014). In another study, BC was implanted in the nasal dorsum of rabbits that exhibited minor fragmentation after 6 months, indicating the nonbiodegradability of BC (Amorim, Costa, Souza, Castro, & Silva, 2009).

Considering the slow/nonbiodegradation of BC, many efforts have been made to improve its biodegradability. In an attempt, the amorphous region of BC was oxidized with peroxidate to form a degradable 2,3-dialdehyde BC. The modified BC showed degradation in water, phosphate buffer solution, and in simulated body fluid (Li, Wan, Li, Liang, & Wang, 2009). In another study, BC membranes were irradiated with γ-radiations, which exhibited a rapid degradation in 2–4 weeks in vivo (using rabbits as a model animal) (Czaja, Kyryliouk, DePaula, & Buechter, 2014). The incorporation of enzyme cellulase into BC can also impart biodegradable characteristics to BC (Hu & Catchmark, 2011).

13.4.5 Biocompatibility

Biocompatibility is the ability of a material to show nontoxicity to a biological system, and elicit a suitable response from the host upon a particular application (Torres, Commeaux, & Troncoso, 2012). Therefore the biocompatibility is a combination of the complex interactions of an implant with the host tissues. It means that besides the biocompatibility, a biomaterial should have a suitable surface topography, low friction coefficient, and proper hydrophilicity. BC is biocompatible because of its structural similarity to the extracellular matrix, such as collagen. In addition, compared to proteins, BC is lesser or nonimmunogenic material due to its polysaccharide nature (Petersen & Gatenholm, 2011). The biocompatibility of BC is comparable with other biomaterials, such as polytetrafluoroethylene and polyglycolic acid, normally used for tissue engineering (Esguerra et al., 2010).

Various in vitro studies reported the biocompatibility of BC, which were mainly focusing on cell attachment and proliferation. For example, the human fibroblasts and osteoblasts cells were found to be proliferated very well on BC (Chen et al., 2009), while the mesenchymal stem cells cultured on BC exhibited more than 95% viability (Mendes et al., 2009). In another study, Schwann cells were grown on BC, the results revealed no considerable difference in the shape and functions of the cells, indicating its biocompatibility (Zhu, Li, Zhou, Lin, & Zhang, 2014). The effect of different morphologies of BC was also evaluated in different independent studies. BC was compressed to produce densified cellulose (17%), which exhibited no toxicity toward human fibroblast cells (Avila et al., 2014). The osteoblast cells were attached and spread well on the BC (large particles produced by agitated culture conditions) (Hu, Catchmark, & Vogler, 2013).

Although BC showed biocompatibility in vitro tests, the in vivo biocompatibility tests are also important for biomedical applications. Therefore a detailed and systematic study of the in vivo biocompatibility of BC was conducted by Helenius and coworkers (2006). BC was implanted in subcutaneous in Wistar rats, the shape of the implants were retained up to 12 weeks without any macroscopic inflammation. No expression of fibrosis, capsules formation, or giant cells was found around the implant. In addition, new blood vessels were formed around and growing to the implant, indicating the good biocompatibility of BC (Helenius et al., 2006). Bionext is a commercially available cellulose sponge that was investigated for the nasal reconstruction in rabbits, indicating good biocompatibility and good stability (Amorim et al., 2009).

In addition to biocompatibility, for a biomaterial used as a vascular implant, the hemocompatibility is another important feature to be investigated. The hemocompatibility of BC was compared with commercially available grafts of poly(ethyleneterephtalate) (PET) and expanded poly(tetrafluoroethylene) (ePTFE), the results revealed that the BC-based grafts did not cause plasma coagulation and showed better hemocompatibility (Fink et al., 2010). Moreover, the contact of biomaterial with blood results in the adsorption and denaturation of the protein. The adsorption of protein on BC was also observed but the hemolysis of the red blood cells was found to be less than 2%, thus indicating the nonhemolytic nature of BC (Andrade et al., 2011).

13.5 Applications of bacterial cellulose

Based on the distinctive structural properties of BC such as adjustable shape, high water-holding capability, high degree of crystallinity, high density, and high surface area, it has numerous applications in various fields such as food, wastewater treatment, electronics, and biomedical fields. The various applications of BC are shown in Fig. 13.4 and discussed in the following sections.

13.5.1 Biomedical applications

BC has attracted momentous research interest for biomedical applications. Today, the main utilization of BC is its use for tissue engineering and wound-dressing applications. It is because of the good biocompatibility and ability to provide a 3D substrate for cell

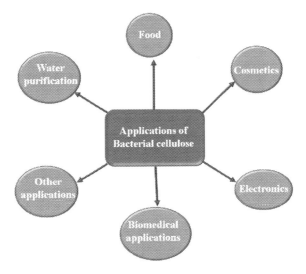

Figure 13.4

Summary of various applications of bacterial cellulose in different fields.

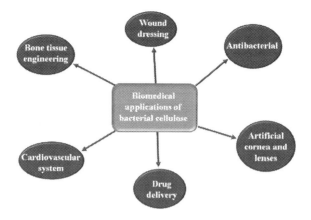

Figure 13.5
Biomedical applications of bacterial cellulose.

attachment. BC has several biomedical applications, which are summarized in Fig. 13.5 and discussed in the following sections.

13.5.1.1 Wound-healing applications

Wound healing is a complex process, involving an association of various kinds of cells and their products, such as extracellular matrix (mainly collagen) and secreted soluble compounds (such as growth factor) for cell proliferation (Eming, Martin, & Tomic-Canic, 2014). The modern medical practice demands wound healing beyond mechanical protection and moisturization. Therefore it is leading to develop new materials with improved properties, such as infection prevention, elimination of exudates, suitable diffusion of gases, painless removal, and cost-effectiveness. BC is used for wound-healing applications due to its suitable characteristics for wound healing, such as good biocompatibility, high water-holding ability, flexibility, high tensile strength, and permeability to gases and liquids (Sulaeva et al., 2015). BC can be used in its pure form or modified to achieve the characteristics of wound-healing applications (Blanco Parte et al., 2020).

Among the first proposed direct application of BC membranes for biomedical applications is related to wound healing. Fontana and coworkers (1990) were the innovators to describe the use of BC for replacing burn skin. Later on, a number of researchers reported the use of BC for wound-healing applications. The BC dressings are suggested as a temporary cover for wounds, such as diabetic wounds, skin tears, pressure sores, second-degree burns, biopsy sites, and skin graft donor sites by the manufacturers (de Oliveira Barud et al., 2016). Some BC-based wound dressings have been commercialized, such as BioFill, XCell, Bioprocess, and Gengiflex. The BioFill is the first commercial biomembrane that meets the main requirements of an ideal wound dressing, such as good adherence to the wound, barrier for bacteria, water and vapor permeability, flexibility, easy handling, and cost-effective. Moreover, by applying BioFill provides pain relief and accelerate the healing process (Czaja, Young, Kawecki, & Brown, 2007).

Wet BC can be used as a novel wound-dressing material for the treatment of burn wounds (Czaja et al., 2007). This type of wound dressing revealed excellent results because it provides a moist environment and rapid healing. Similarly, Portal, Clark, and Levinson (2009) used BC-based dressing for chronic wounds and found 75% epithelialization within 81 days, while the similar results were obtained in 315 days without the use of BC membranes. The great conformability of the BC material has been proven when applied to a large number of patients in clinical trials. The BC adhered well to the wound sites, and its elastic properties allowed to a significant molding to facial contours and other parts of the body as shown in Fig. 13.6. It is important to mention that BC has obviously reduced the healing time of the nonhealing ulcers in comparison to standard care products (Czaja, Krystynowicz, Bielecki, & Brown, 2006).

Besides the direct use of BC for wound healing, it can be easily modified to form its derivatives or composites with improved properties. Therefore numerous BC-based nanocomposites have been developed for wound dressing. For instance, BC/collagen (Moraes et al., 2016) and BC/gelatin (Wang et al., 2012) nanocomposite have shown improved thermal stability and mechanical properties. BC composites with kaolin (an agent used for blood clotting) were found to be suitable for wound-healing applications (Wanna, Alam, Toivola, & Alam, 2013). Napavichayanun and coworkers (Napavichayanun, Amornsudthiwat, Pienpinijtham, & Aramwit, 2015) designed silk sericin-BC that improved

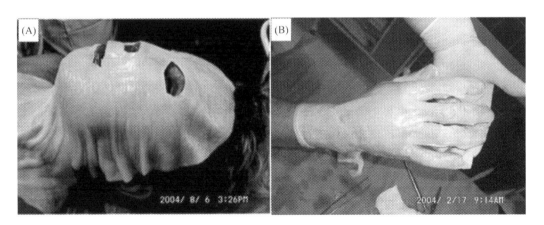

Figure 13.6
Wound dressing based on wet BC: (A) on wounded face and (B) on wounded hands. Source: *Reproduced with permission from Czaja, W., Krystynowicz, A., Bielecki, S., & Brown, R. M. (2006). Microbial cellulose—The natural power to heal wounds. Biomaterials, 27, 145–151. Reproduced with permission from Czaja, W. K., Young, D. J., Kawecki, M., & Brown, R. M. (2007). The future prospects of microbial cellulose in biomedical applications. Biomacromolecules, 8, 1–12.*

fibroblast proliferation and increased the production of the extracellular matrix that reduced the wound-healing time. Polyhexamethylene biguanide was also added as an antimicrobial material to sericin-BC composites, resulting in collagen production and facilitated cell migration. Furthermore, it prevented infection and caused rapid wound healing.

One of the important features of designing wound-dressing materials is the improvement of the water-retention ability of the final product for a long time. The BC/chitosan composites have been found to release water slowly; therefore they could be used for treating skin ulcers, chronic wounds, and wounds requiring frequent dressing change (Ul-Islam, Shah, Ha, & Park, 2011). Lin, Lien, Yeh, Yu, and Hsu (2013) demonstrated the wound-healing ability of BC/chitosan composite by using mice as a model animal. They found that the composites did not show any toxicity toward animal cells. Furthermore, wound treated with BC/chitosan composite showed faster wound healing in comparison to pure BC or commercially available dressings.

BC has several excellent properties as a wound-healing material, but it does not possess any antimicrobial activity. The antimicrobial activity is an important characteristic of a good wound-healing material. Metal and metal oxides possess a broad spectrum of antimicrobial activity. Therefore several metal/metal oxide nanoparticles, such as silver (Wu et al., 2014), copper (Araujo et al., 2018), zinc oxide (Khalid, Khan, Ul-Islam, Khan, & Wahid, 2017), and silver sulfadiazine (Berndt, Wesarg, Wiegand, Kralisch, & Mueller, 2013) have been incorporated to BC to improve the wound-healing characteristics of BC-based dressings.

13.5.1.2 Antimicrobial applications

One of the major problems in the biomedical field is the emerging of antibiotic-resistant microbes. Therefore the new antimicrobial materials should not be only based on antibiotics, but some other antimicrobial materials should also be designed. One of such approaches can be the use of BC-based materials (Ullah et al., 2016). Nevertheless, BC does not possess any antimicrobial activity; therefore various methods have been used to introduce the antimicrobial activity to BC (Wahid et al., 2019; 2019). Generally, the incorporation of antimicrobial agents into BC is not a complicated process. The presence of hydroxyl groups and the large surface area make BC an attractive material for adsorption of micro- and macromolecules such as quaternary compound, nanoparticles, polysaccharides, and proteins (Sulaeva et al., 2015).

BC/polymer composites generally improve the physical characteristics of BC, such as improving the water-retention capacity and high tensile strength at the same time, as well as introducing antibacterial properties. BC/chitosan antibacterial composites were prepared by adding chitosan nanocrystals into the culture medium. The composites were also prepared through the ex situ method by blending chitosan nanocrystals with BC. The composite showed antibacterial activity against tested bacterial strains (Butchosa et al., 2013). In another similar approach, the BC membranes were modified by in situ method with chitosan, the resulted BC/chitosan composite showed several valuable properties, including good mechanical properties, moisture-retention properties, and good antimicrobial properties (Ciechańska, 2004). BC was modified with polihexanide (PHMB), the resultant BC/PHMB composite revealed antibacterial activity against *Staphylococcus aureus* *(S. aureus)* and completely inhibited the growth of bacteria (Wiegand et al., 2015).

BC has also been modified with various metal/metal oxide nanoparticles to introduce antimicrobial properties. BC/Ag-NPs-based nanoparticles have been prepared by various physical or chemical processes (Elayaraja, Zagorsek, Li, & Xiang, 2017; Pourali et al., 2014). These nanocomposites have been proven to be antimicrobial against various microorganisms. BC has been fabricated with Fe_3O_4, which presented bactericidal activity against *S. aureus*, *Pseudomonas aeruginosa* (*P. aeruginosa*), and *Staphylococcus epidermidis* (Moniri et al., 2018). We have recently incorporated ZnO nanoparticles into BC, the resultant BC/ZnO nanocomposite exhibited antibacterial activity against *S. aureus* and *Bacillus substilis* (Gram-positive) and *Escherichia coli* and *P. aeruginosa* (Gram-negative) bacterial strains (Wahid et al., 2019). Moreover, several other metal/metal oxide nanoparticles have been incorporated into BC such as copper/copper oxide (Almasi, Jafarzadeh, & Mehryar, 2018; Xie et al., 2020), TiO_2 (Liu et al., 2017), and MgO (Mirtalebi, Almasi, & Khaledabad, 2019) for introducing antibacterial properties.

13.5.1.3 Drug-delivery application

Besides the favorable properties for wound dressing, closed contact of the BC-based materials with the lesion site makes it a promising platform for transdermal drug delivery. The common way of the drug loading to the BC membrane is by the immersion of membrane into the drug suspension. Furthermore, chemical modification or BC-based composites have been designed for control release. The simplest preparation method, easy application, and comparable drug release demonstrate the promising potentialities of BC for drug-delivery applications. Generally, antimicrobial and nonsteroidal anti-inflammatory drugs are loaded to BC (Abeer, Amin, & Martin, 2014). For example, the lyophilized BC membrane was immersed in the benzalkonium chloride solution (as a model drug) and lyophilized again. The drug-loading capacity of the membrane was 0.116 mg cm^{-2} and exhibited a prolonged (24 h) antibacterial effect against *S. aureus* and *B. substilis* (Wei, Yang, & Hong, 2011). Silva and coworkers (2014) explored BC as a transdermal delivery membrane for diclofenac sodium (a model drug) and glycerol as a plasticizer. The 3D network structure of BC provided a sustainable drug release. This feature can also be combined with the good absorption and biocompatibility of BC. A comparative study of the lidocaine delivery to the human epidermis of the three different systems (BC membrane, a gel, and an aqueous solution) was investigated in vitro. It was observed that the permeation rate of the drug into BC was lower in comparison to the other two systems (aqueous solution and gel) (Trovatti et al., 2011).

The rate of drug release can also be customized by controlling the porosity of BC by any chemical or physical method. For instance, de Olyveira, Manzine Costa, and Basmaji (2013) used gamma-irradiated and nonirradiated BC in vitro drug release. It was revealed that the irradiated BC has greater pore density in comparison to nonirradiated samples. Thus the irradiated membrane showed a slower diffusion rate than the nonirradiated membrane. Similarly, Stoica-Guzun, Stroescu, Tache, Zaharescu, and Grosu (2007) investigated the effect of electron beam irradiations on the transdermal release of antibiotics (tetracycline) from BC. The study revealed that the irradiated BC significantly reduced the in vitro diffusion of tetracycline. These results suggested the potential of BC in the form of transdermal patches.

13.5.1.4 Cardiovascular grafts

Typically blood circulates in the body through blood vessels under normal physiological conditions. Nevertheless, these blood vessels are injured or blocked in pathological circumstances, which in some cases need to be substituted by another blood vessel collected from a donor or patient's own body, or can be replaced by artificial blood vessels (Adams, Xiao, & Xu, 2007). There are several serious concerns about the patient's own blood vessels, and donor strategies (Parveen, Krishnakumar, & Sahoo, 2006). Therefore researchers are working to design artificial blood vessels. The distinctive characteristics of BC, such as biocompatibility, ultrafine fibrous structure, moldability, and foldability make it a suitable candidate for artificial blood vessels (Zang et al., 2015). The designing of BASYC a synthetic blood vessel based on BC, has made a breakthrough in the biomedical field. BASYC was used to substitute carotid arteries in mice, which revealed patency for 1 year. Similarly, BASYC showed patency for 90 days in pig prostheses (Schumann et al., 2009). The prostheses developed three distinct layered structures by epithelialization, which was similar to natural blood vessels (Schumann et al., 2009). The BC-based graft showed advantages over conventional vascular grafts (PET and ePTFE), in which the BC showed slower thrombin generation on its surface (Fink et al., 2010). Bodin et al. (2007) deposited BC in the tubular form on the top of the silicone tubes by fermenting *K. xylinus*. The formed tubes had a uniform inner side and a porous outer surface. The tubes could be prepared in different diameters and can also be produced in branched form. These tubes showed the potential to be used as vascular grafts.

In addition to blood vessels, BC-based materials have also been investigated for designing artificial heart valves (Andrade et al., 2011; Fink et al., 2010). Particularly, BC/PVA composites have been extensively investigated as blood-contacting materials. They have shown superior platelet adhesion and activation profile as compared to conventional ePTFE. Moreover, BC/PVA composites showed similar properties to heart valves and exhibited hemocompatibility, and hence have promising applications in cardiac prosthesis (Mohammadi, 2011). Furthermore, BC/PVA possesses mechanical properties almost similar to the heart valve originated from porcine. Therefore they have been used to make leaflets of heart valves (de Oliveira Barud et al., 2016).

13.5.1.5 Artificial cornea and lenses

Corneal disease is the foremost reason for blindness, and approximately 10 million people around the globe have lost their eyesight due to this disease. Therefore to deal with this problem, a wide variety of biomaterials have been investigated for the cornea bioengineering (Rajwade et al., 2015). A fascinating and innovative utilization of BC-based materials involves its capability to adhere and improve the proliferation of the keratinocytes and retinal pigment epithelium. In turn, BC-based grafts may reduce the rejection rate of the transplanted cornea and enhance the treatment of eye diseases (Picheth et al., 2017). Thus several BC-based biocomposites have been designed to completely adapt the properties suitable for eye therapeutics. For instance, Wang, Gao, Zhang, and Wan (2010), synthesized a BC/PVA hydrogel that exhibited water content and light transmittance similar to the natural cornea. The OH groups in both polymers provide the interfacial interactions in the biocomposite. The water-holding capability, suitable mechanical properties, and light-transmitting properties make it a promising optically functional material.

Poor eyesight is another major visionary problem, more than 2.3 billion people worldwide have been suffered from visionary defects because of the refractive error (RE) (Naidoo & Jaggernath, 2012). RE can be measured and corrected very simply by refractive surgical strategies or devices, such as contact lenses and spectacles. The contact lenses are available in the market in both soft and hard forms. Soft contact lenses are made of soft polymer, while hard contact lenses are made of glass or stiff polymers, including poly(hydroxyethyl methacrylate) (PHEMA), poly(methyl methacrylate), and polycarbonates (Levinson, 2010). BC-based contact lenses were prepared from a viscous solution of BC in 1-butyl-3-methylimidazolium chloride. The solution was poured into a mold and precipitated by treating with isopropanol and then with water. The prepared lenses were naturally separated from the mold surfaces. The hydrated contact lenses exhibited stability for more than 8 weeks, retaining its transparency and shapes (Levinson, 2010). Another type of contact lenses was prepared by combining water-soluble polymer PHEMA and water-insoluble BC nanofibers. The resultant composite material held about 40% water content and exhibited good mechanical integrity. The designed BC/PHEMA composites showed suitability to be used as contact lenses (Li, Wan, & Panchal, 2015).

13.5.1.6 Bone tissue engineering

Nanocellulose and its composites are promising materials for culturing various cells, including chondroblast and osteoblast cells, indicating their potential for bone tissue regeneration. For example, a membrane based on BC and hydroxyapatite (Hap) was designed for bone tissue regeneration. The composite membrane showed the growth of osteoblast cells and a good bone module formation. Moreover, BC/Hap revealed osteoblast adhesion, proliferation, and mineralization. Therefore the composite could be promising to facilitate quick bone regeneration. A similar study showed the suitability of the BC/Hap scaffold for the regeneration of connective tissues and bones (Grande, Torres, & Gomez, 2009).

The loading of bone marrow mesenchymal stem cells into BC is another strategy to prepare a scaffold for evaluating the BC-based material for bone repair. For instance, the horse mesenchymal stem cells were cultivated on BC scaffold, the porous structure of BC acted as a support for cell adhesion, growth, and migration. The results indicated the differentiation of the cells into chondrocytes and bone cells on the BC scaffold (Favi et al., 2013). BC has a low degradation rate which makes it an ideal material for the slow healing process because a durable scaffold is needed for a long time regenerating tissues (Pang et al., 2020). Noh and coworkers (2019) cultivated human umbilical cord blood mesenchymal stem cells on the BC/collagen scaffold. It was found that by increasing the content of BC, the scaffold showed higher strength and the deposition of calcium was also higher, indicating the effectiveness of the scaffold with a higher content of BC for osteogenic differentiation of mesenchymal stem cells. In addition, the cells loaded BC/collagen (5:1) scaffold was subcutaneously transplanted in mice, the results exhibited more red blood cells and angiogenic markers, indicating neovascularization. It could be due to the maintenance of a stable porous microenvironment, allowing better oxygen and nutrient diffusion in the initial stages of bone generation.

13.5.2 Applications of bacterial cellulose in food

BC is classified as "generally recognized as safe" material by the Food and Drug Administration in 1992. BC can be utilized in food in different forms, such as fats replacer, meat analog (artificial meat), and rheological modifier (Fig. 13.7). One of the first applications of BC as the raw material for a chewy and juicy Filipino traditional dessert "nata-de-coco" with a smooth mouthfeel. It is basically produced by fermenting the coconut water to synthesize BC, then BC is chopped into small cubes, and immersed in the sugar syrup. Nata-de-coco is a well-liked food and spreading around the globe in the form of sweet candies (Blanco Parte et al., 2020).

The main advantage of BC as a food ingredient is its appeal for dietetic foods because it is indigestible by humans. Therefore fat-free or low-calorie foods can be prepared by using BC (Lin et al., 2013). However, the BC gel is very difficult for biting, but it may become edible by treating with calcium chloride and alginate. The addition of these ingredients into BC make it edible by holding the water in the gelatinous form, thus making it easier to cut with teeth. These facts make BC as a novel material for processed food such as low-calorie desserts and salads (Ullah et al., 2016).

Figure 13.7
Main applications of BC in food. Source: *Reproduced from with permission from Azeredo, H. M. C., Barud, H., Farinas, C. S., Vasconcellos, V. M., & Claro, A. M. (2019). Bacterial cellulose as a raw material for food and food packaging applications.* Frontiers in Sustainable Food Systems, 3, 2019.

The combination of BC with *Monascus* extracts (obtained from a natural red pigment mold) can be used as a meat analog. The composite is highly stable against changing its morphology and color, and its flavor is much similar to meat. The produced meat analog may have a number of market applications, including BC as a dietary fiber and the low-cholesterol effect from *Monascus*, which may make the product a good alternative to meat products for diet-restricted consumers (Azeredo et al., 2019).

BC also exhibits promising applications as an additive for thickening, gelling, and stabilizing agents in the food industry. It is used as a filler for enhancing the mechanical properties of fragile food hydrogels and thus improving the worth of the pasty food. A 0.2%–0.3% BC considerably increased the strength of Tofu (a food made of soy milk), providing a better texture and firmness (Shi, Zhang, Phillips, & Yang, 2014). BC has also endowed Kamaboko (a Japanese processed sea food) with improved brittleness and stiffness, reducing its springiness. The addition of BC into creamy condiments improving its stickiness which could be easily served quantitatively by using a spoon. Likewise, the BC-containing food items can maintain their humidity for a longer storage period. The contour of ice cream having BC showed stability for a minimum of 1 h after removal from the freezer, which would melt at the same time without using BC. Thus it is an evidence that BC can be extensively used in processed food to enhance their stability, quality, and storage conditions (Shi et al., 2014).

BC can also be used as food-packaging material to increase shelf-life and increase food safety. Antimicrobial agents, moisture removal, and ethylene and oxygen scavengers are all used in the BC-based food-packaging system (Tomé et al., 2010).

13.5.3 Applications in cosmetics

Cosmetics are the materials that are utilized to enhance certain organoleptic characteristics of the human body (Norhasliza Hasan & Kamarudin, 2012). It includes the substances that are applied to the human body to alter its appearance and to beautify the organs without disturbing their normal structures and functions (Norhasliza Hasan & Kamarudin, 2012). However, most of the consumers are using cosmetics for beauty without consideration of their ill effects on the body. To avoid the harmful effects of cosmetics on the human body, natural skin-care products are suggested that consisting of natural or herbal ingredients. In this perspective, cellulose fibers are added into cosmetics for the stabilization of the oil-in-water emulsion. Such a formulation may not irritate sensitive skin due to the absence of surfactants (Norhasliza Hasan & Kamarudin, 2012).

BC is an excellent nonallergenic polymer used in cosmetics in different forms. For instance, facial masks of BC are used as cosmetic materials for the treatment of dry skin due to its biocompatibility and the ability to hydrate the skin (Amnuaikit, Chusuit, Raknam, & Boonme, 2011). Therefore these masks are applied for improving the moisture content of the skin. BC facial masks with holes for eyes, nose, and mouth can meet the requirements for a prolonged and repeated use for skin beauty, moisturizing, nutrition, and cosmetic effects (Ullah et al., 2016). The BC-based gel is biocompatible and less adhesive (removed without pain) as compared to a paper mask. These masks showed the potentials to be used in medicated cosmetics, antiaging, and moisturizing facial masks (Ullah et al., 2016). BC is a suitable carrier for cosmetically active agents, such as whitening materials (ursolic acid or kojic acid), moisturizers (hyaluronic acid or salicylic acid), antiwrinkling agents (polypeptides), enzymes, growth factors, or a combination of them. For example, BC with glycerin gave a considerably high skin-moisturizing effect, suggesting its potential as a moisturizing facial mask (Almeida et al., 2014). Likewise, facial masks consisting of BC membranes containing ginseng extracts have revealed favorable results in terms of moist feel, skin elasticity, and overall satisfaction in women over 30 years of age (Ullah et al., 2016). Similarly, BC gels with control release of silk sericin were prepared with enhanced moisture-retention properties as compared to commercially existing paper masks (Aramwit & Bang, 2014). In addition, the BC-based materials may find potential applications in preparing skin, lips, and nails care products and long-lasting perfumes (Ullah et al., 2016).

13.5.4 Electronics

Though BC has been investigated mainly for biomedical and pharmaceutical applications, it has also found applications in the electronics field because of its fibrous network structure, light mass, flexibility, high mechanical properties, and printability. Moreover, BC can be easily modified to find its applications in electronics such as conductive membranes, biosensors, and batteries (Jang, Hwang, Kim, Ryu, & Lee, 2017; Jiang, Yin, Yu, Zhong, & Zhang, 2015). For instance, functionalized BC with polyaniline and silicon nanoparticles is capable to maintain its flexibility and shows potential applications as anodic material for Li-ion batteries (Park, Lee, Shin, Kim, & Hyun, 2016). In another similar approach, the BC membrane incorporated with silver nanoparticles has been used as an electrode for the reduction of oxygen in a fuel cell (Zhang et al., 2015). BC has been used as a template to construct cobalt ferrate nanotubes which may be used as nanowires for several electronics (Menchaca-Nal et al., 2016). Likewise, BC/ZnS/epoxy resin nanocomposite with good flexibility and transparency was developed for optoelectronic biomaterials (Guan, Chen, Yao, Zheng, & Wang, 2016). Yoon, Jin, Kook, and Pyun (2006) incorporated multiwalled carbon nanotubes to the BC membrane to develop a high electrical conductive composite membrane.

BC can be applied as a battery separator due to its high mechanical properties, high crystallinity, and good thermostability. The battery separators should be chemically and physically stable to prevent direct contact between electrodes but should allow the transport of ions. Its structure and properties can highly affect the performance of the battery. Since BC exhibits a three-dimensional porous structure; therefore it is considered as a promising material for next-generation battery separators (Jiang et al., 2015).

BC is considered a favorable material for sensors. For instance, BC incorporated with silver nanoparticles was used for the detection of cysteine enantiomers (Zor, 2018). Other applications of BC for electronics include the surface-enhanced Raman spectroscopy (SERS), an analytical technique used widely for the detection of an analyte in a low concentration. BC with gold and silver nanoparticles has been proven to exhibit SERS enhancement (Wei & Vikesland, 2015; Wei, Rodriguez, Renneckar, Leng, & Vikesland, 2015).

13.5.5 Water purification

The worldwide growing industrialization brings toxic contamination into the natural water in the form of heavy metals, dyes, and other pollutants. To encounter these impending challenges, several water remediation techniques based on membrane filtration, adsorption, chemical precipitation, and biological treatment have been used to clean the pollutant water (Ma et al., 2017; Zou et al., 2016).

BC is considered an eco-friendly filter material, different materials have been incorporated to BC to improve its filtration activity. A BC-based composite filtration membrane was developed for water purification. BC severed as a porous template, where graphene oxide (GO) was loaded for improving the filtration properties. The resulting BC-GO membrane exhibited advanced properties of high mechanical strength, water stability, and selective ions permeation (Fang, Zhou, Deng, Zheng, & Liu, 2016). In a recent study, palladium nanoparticles were incorporated into the BC-GO membrane which showed remarkable efficacy in the removal of organic dyes from water, eliminating up to 93% methylene orange and other organic pollutants such as 4-nitrophenol and methylene blue (Xu et al., 2018). Furthermore, BC-chitosan composite has shown potential for the removal of heavy metal from water, achieving up to 50% removal of copper from copper-containing water (Urbina et al., 2018). Recently, BC was fabricated with polydopamine to develop a filtration membrane. Heavy metals such as cadmium and lead, along with organic pollutants such as methyl orange, methylene blue, and rhodamine 6G or their combination were successfully removed by the filtration process (Gholami Derami et al., 2019).

Adsorption is another technique used for removal of heavy metals and dyes from water. Since BC fibers exhibit high surface area which can have maximum contact with pollutants and thus possess good adsorption capability. Moreover, BC can be chemically modified by different chemical reactions to improve its adsorption characteristics (Cheng et al., 2019). For instance, polyethyleneimine was grafted into BC for copper and lead ions adsorptions from aqueous solutions (Jin, Xiang, Liu, Chen, & Lu, 2017). Furthermore, other groups, including phosphate, carboxyl, amidoxime, and diethylenetriamine have been grafted into BC for heavy metal ions adsorption (Cheng et al., 2019).

BC acts as a template for photocatalyst to degrade dyes in polluted water. For instance, BC was fabricated with TiO_2 nanoparticles for the degradation of organic dyes in water (Sun, Yang, & Wang, 2010). Li et al. (2017) fabricated oxidized BC with TiO_2, which exhibited good photodegradation of organic dyes under UV-irradiation (Li et al., 2017). BC fabricated with ZnO nanoparticles showed 91% degradation of methylene blue within 2 h under UV light (Wahid et al., 2019). Moreover, several other BC-based nanocomposites have been reported for the degradation of organic dyes, such as BC/polydopamine/TiO_2 (Yang et al., 2020), BC/GO/TiO_2 (Liu et al., 2017), and BC/CuO (Phutanon, Motina, Chang, & Ummartyotin, 2019).

13.5.6 Other applications

The distinctive characteristics of BC make it an appropriate material for various other applications than those discussed before. The paper manufacturing industry is evaluating the use of BC as an additive in high-quality papers, which has improved the tensile index with reduced elongation and porosity. Moreover, the addition of BC is effective to increase the folding endurance of the paper. All these characteristics are currently demanding in the market (Fillat et al., 2018; Tabarsa, Sheykhnazari, Ashori, Mashkour, & Khazaeian, 2017).

The textile industry also uses BC because of its gelatinous film by interlacing microfibrils and leather-like appearance. For advanced applications of BC as a fabric material, several studies are focusing on dyeing, bleaching, and other methods to improve BC properties (Chan, Shin, & Jiang, 2018). A British fashion designer developed BC textile directly from a cultivation container and created BC gloves and jackets by using conventional garment-designing techniques (Wood, 2019).

Recently, biopolymers have become a fundamental matrix for the immobilization of enzymes. Therefore BC has also been explored as supporting material in this field. The 3D network structure of BC is expected to entrap enzyme molecules because of its large surface area and high protein-holding capability (Mohite & Patil, 2014). Several enzymes, such as lipases (Kim, Jin, Kan, Kim, & Lee, 2017), laccase (Sampaio et al., 2016), and glucoamylase (Wu & Lia, 2008) have been immobilized on BC-based materials for various applications in medical, environmental, and textile industries.

13.6 Commercialization of BC-based products

The distinctive properties such as 3D porous structure, high mechanical properties, and good hydrophilicity make BC as an important material for numerous applications. Though BC is synthesized in laboratories for research purposes on a small scale, there are several commercial channels of BC. At the industrial scale, BC is mainly prepared in the membrane forms by the static culturing process. However, various reactors, including silicone tubes, revolving wires, and packed bead reactors have also been used for desired applications. The Philippine dessert, nata-de-coco has been produced on a commercial scale in Philippine for the past several decades. The export of nata-de-coco from the Philippines to Japan has a great impact on the prospect of expanding BC production (Esa, Tasirin, & Rahman, 2014). A German company Fzmb GmbH is one of the largest companies to produce BC for cosmetic and biomedical applications. A kind of BC Prima cell™ utilized as a wound dressing for treating ulcers produced by Xylos Co., USA. Other BC-based brands for wound dressings include Biofill, Bioprocess, Dermafill, Xcell, and Gengiplex (Sulaeva et al., 2015). BASYC has been used as artificial blood vessels and cuff for nerve suturing. In addition to the medical utilization of BC products, there are several other commercial products of BC. For instance, Sony Corporation in Japan with association with other corporations develop a BC-based diaphragm for audio speakers (Chawla, Bajaj, Survase, & Singhal, 2009). Ajinomoto Co. in association with Mitsubishi Paper Mills in Japan is working on the development of BC-based paper products (Esa et al., 2014).

13.7 Conclusions and perspectives

BC is mostly synthesized by the bacteria *K. xylinus*, which can produce BC on the industrial scale. In the appropriate cultural conditions, bacteria synthesize fibrils which are free from impurities such as hemicellulose and lignin. These fibrils are organized in a 3D hierarchical network. BC exhibits distinctive characteristics, including high purity, high water-holding capability, high crystallinity, and excellent mechanical properties. Despite holding high water content, BC exhibits good mechanical properties and it can be synthesized in different shapes due to its moldable characteristics. BC is an emerging nontoxic biomaterial and it shows remarkable potential as a biopolymer for various applications. Therefore this chapter has summarized the production of BC, its properties, and various applications.

Despite the remarkable properties and promising applications, the industrial production of BC is facing several problems. Particularly, the expensive production techniques have prevented their broad-range applications. Numerous efforts have been made to improve various features of BC, such as reducing the production costs and improving its mechanical and biological properties. Several attempts have been made to reduce the production cost of BC, such as the development of innovative production approaches, designing new bioreactors, using cheaper carbon sources, and innovation of new microbial strains and genetic modification of the existing microbial species. However, additional research is necessary to further expand the applications in many fields, especially in biomedical fields. Therefore future research should be focused on the designing of cost-effective, renewable media, and the discovery of new microbes for the production of BC.

Another problem with BC is its poor biodegradation, the biodegradability of BC should be improved by physical and chemical modifications. Therefore more research is needed for improving the biodegradability of BC. Moreover, the applications of BC in food, cosmetics, and other biomedical fields need further research to further widen the scope of BC in various fields. Finally, the synthesis and applications of BC are likely to be increased significantly, and BC may replace different traditional materials someday.

13.8 Acknowledgments

The authors are thankful for the financial support from the National Natural Science Foundation of China, and the National Natural Science Foundation of Tianjin.

References

Abeer, M. M., Amin, M. C. I. M., & Martin, C. (2014). A review of bacterial cellulose-based drug delivery systems: their biochemistry, current approaches and future prospects. *Journal of Pharmacy and Pharmacology*, 66, 1047–1061.

Adams, B., Xiao, Q., & Xu, Q. (2007). Stem cell therapy for vascular disease. *Trends in Cardiovascular Medicine, 17*, 246−251.

Almasi, H., Jafarzadeh, P., & Mehryar, L. (2018). Fabrication of novel nanohybrids by impregnation of CuO nanoparticles into bacterial cellulose and chitosan nanofibers: Characterization, antimicrobial and release properties. *Carbohydrate Polymers, 186*, 273−281.

Almeida, I. F., Pereira, T., Silva, N. H. C. S., Gomes, F. P., Silvestre, A. J. D., Freire, C. S. R., . . . Costa, P. C. (2014). Bacterial cellulose membranes as drug delivery systems: An in vivo skin compatibility study. *European Journal of Pharmaceutics and Biopharmaceutics, 86*, 332−336.

Amnuaikit, T., Chusuit, T., Raknam, P., & Boonme, P. (2011). Effects of a cellulose mask synthesized by a bacterium on facial skin characteristics and user satisfaction. *Medical Devices (Auckland, N.Z.), 4*, 77−81.

Amorim, W. L., Costa, H. O., Souza, F. Cd, Castro, M. Gd, & Silva, Ld (2009). Experimental study of the tissue reaction caused by the presence of cellulose produced by *Acetobacter xylinum* in the nasal dorsum of rabbits. *Brazilian Journal of Otorhinolaryngology, 75*, 200−207.

Andrade, F. K., Silva, J. P., Carvalho, M., Castanheira, E. M. S., Soares, R., & Gama, M. (2011). Studies on the hemocompatibility of bacterial cellulose. *Journal of Biomedical Materials Research Part A, 98A*, 554−566.

Aramwit, P., & Bang, N. (2014). The characteristics of bacterial nanocellulose gel releasing silk sericin for facial treatment. *BMC Biotechnology, 14*.

Araujo, I. M. S., Silva, R. R., Pacheco, G., Lustri, W. R., Tercjak, A., Gutierrez, J., . . . Barud, H. S. (2018). Hydrothermal synthesis of bacterial cellulose-copper oxide nanocomposites and evaluation of their antimicrobial activity. *Carbohydrate Polymers, 179*, 341−349.

Avila, H. M., Schwarz, S., Feldmann, E.-M., Mantas, A., von Bomhard, A., Gatenholm, P., & Rotter, N. (2014). Biocompatibility evaluation of densified bacterial nanocellulose hydrogel as an implant material for auricular cartilage regeneration. *Applied Microbiology and Biotechnology, 98*, 7423−7435.

Azeredo, H. M. C., Barud, H., Farinas, C. S., Vasconcellos, V. M., & Claro, A. M. (2019). Bacterial cellulose as a raw material for food and food packaging applications. *Frontiers in Sustainable Food Systems, 3*.

Berndt, S., Wesarg, F., Wiegand, C., Kralisch, D., & Mueller, F. A. (2013). Antimicrobial porous hybrids consisting of bacterial nanocellulose and silver nanoparticles. *Cellulose, 20*, 771−783.

Blanco Parte, F. G., Santoso, S. P., Chou, C.-C., Verma, V., Wang, H.-T., Ismadji, S., & Cheng, K.-C. (2020). Current progress on the production, modification, and applications of bacterial cellulose. *Critical Reviews in Biotechnology*.

Bodin, A., Backdahl, H., Fink, H., Gustafsson, L., Risberg, B., & Gatenholm, P. (2007). Influence of cultivation conditions on mechanical and morphological properties of bacterial cellulose tubes. *Biotechnology and Bioengineering, 97*, 425−434.

Bodin, A., Concaro, S., Brittberg, M., & Gatenholm, P. (2007). Bacterial cellulose as a potential meniscus implant. *Journal of Tissue Engineering and Regenerative Medicine, 1*, 406−408.

Brown, A. J. (1887). LXII.—Further notes on the chemical action of *Bacterium aceti*. *Journal of the Chemical Society, Transactions, 51*, 638−643.

Butchosa, N., Brown, C., Larsson, P. T., Berglund, L. A., Bulone, V., & Zhou, Q. (2013). Nanocomposites of bacterial cellulose nanofibers and chitin nanocrystals: Fabrication, characterization and bactericidal activity. *Green Chemistry, 15*, 3404−3413.

Cacicedo, M. L., Castro, M. C., Servetas, I., Bosnea, L., Boura, K., Tsafrakidou, P., . . . Castro, G. R. (2016). Progress in bacterial cellulose matrices for biotechnological applications. *Bioresource Technology, 213*, 172−180.

Castro, C., Zuluaga, R., Putaux, J.-L., Caro, G., Mondragon, I., & Ganan, P. (2011). Structural characterization of bacterial cellulose produced by *Gluconacetobacter swingsii* sp. from Colombian agroindustrial wastes. *Carbohydrate Polymers, 84*, 96−102.

Chan, C. K., Shin, J., & Jiang, S. X. K. (2018). Development of tailor-shaped bacterial cellulose textile cultivation techniques for zero-waste design. *Clothing and Textiles Research Journal, 36*, 33−44.

Chawla, P. R., Bajaj, I. B., Survase, S. A., & Singhal, R. S. (2009). Microbial cellulose: Fermentative production and applications. *Food Technology and Biotechnology, 47*, 107−124.

Cheng, R., Kang, M., Zhuang, S., Shi, L., Zheng, X., & Wang, J. (2019). Adsorption of Sr(II) from water by mercerized bacterial cellulose membrane modified with EDTA. *Journal of Hazardous Materials, 364*, 645−653.

Chen, Y. M., Xi, T., Zheng, Y., Guo, T., Hou, J., Wan, Y., & Gao, C. (2009). In vitro cytotoxicity of bacterial cellulose scaffolds used for tissue-engineered bone. *Journal of Bioactive and Compatible Polymers, 24*, 137−145.

Chen, C., Zhang, T., Dai, B., Zhang, H., Chen, X., Yang, J., . . . Sun, D. (2016). Rapid fabrication of composite hydrogel microfibers for weavable and sustainable antibacterial applications. *ACS Sustainable Chemistry & Engineering, 4*, 6534−6542.

Ciechańska, D. (2004). Multifunctional bacterial cellulose/chitosan composite materials for medical applications. *Fibres & Textiles in Eastern Europe, 4*(12), 69−72.

Czaja, W., Krystynowicz, A., Bielecki, S., & Brown, R. M. (2006). Microbial cellulose—The natural power to heal wounds. *Biomaterials, 27*, 145−151.

Czaja, W., Kyryliouk, D., DePaula, C. A., & Buechter, D. D. (2014). Oxidation of gamma-irradiated microbial cellulose results in bioresorbable, highly conformable biomaterial. *Journal of Applied Polymer Science, 131*.

Czaja, W. K., Young, D. J., Kawecki, M., & Brown, R. M. (2007). The future prospects of microbial cellulose in biomedical applications. *Biomacromolecules, 8*, 1−12.

de Oliveira Barud, H. G., da Silva, R. R., Barud, H. D. S., Tercjak, A., Gutierrez, J., Lustri, W. R., . . . Ribeiro, S. J. L. (2016). A multipurpose natural and renewable polymer in medical applications: Bacterial cellulose. *Carbohydrate Polymers, 153*, 406−420.

de Olyveira, G. M., Manzine Costa, L. M., & Basmaji, P. (2013). Physically modified bacterial cellulose as alternative routes for transdermal drug delivery. *Journal of Biomaterials and Tissue Engineering, 3*, 227−232.

Elayaraja, S., Zagorsek, K., Li, F., & Xiang, J. (2017). In situ synthesis of silver nanoparticles into TEMPO-mediated oxidized bacterial cellulose and their antivibriocidal activity against shrimp pathogens. *Carbohydrate Polymers, 166*, 329−337.

Eming, S. A., Martin, P., & Tomic-Canic, M. (2014). Wound repair and regeneration: Mechanisms, signaling, and translation. *Science Translational Medicine, 6*.

Esa, F., Tasirin, S. M., & Rahman, N. A. (2014). Overview of bacterial cellulose production and application. *Agriculture and Agricultural Science Procedia, 2*, 113−119.

Esguerra, M., Fink, H., Laschke, M. W., Jeppsson, A., Delbro, D., Gatenholm, P., . . . Risberg, B. (2010). Intravital fluorescent microscopic evaluation of bacterial cellulose as scaffold for vascular grafts. *Journal of Biomedical Materials Research Part A, 93A*, 140−149.

Fang, Q., Zhou, X., Deng, W., Zheng, Z., & Liu, Z. (2016). Freestanding bacterial cellulose-graphene oxide composite membranes with high mechanical strength for selective ion permeation. *Scientific Reports, 6*.

Fan, X., Gao, Y., He, W., Hu, H., Tian, M., Wang, K., & Pan, S. (2016). Production of nano bacterial cellulose from beverage industrial waste of citrus peel and pomace using *Komagataeibacter xylinus*. *Carbohydrate Polymers, 151*, 1068−1072.

Favi, P. M., Benson, R. S., Neilsen, N. R., Hammonds, R. L., Bates, C. C., Stephens, C. P., & Dhar, M. S. (2013). Cell proliferation, viability, and in vitro differentiation of equine mesenchymal stem cells seeded on bacterial cellulose hydrogel scaffolds. *Materials Science & Engineering C-Materials for Biological Applications, 33*, 1935−1944.

Fillat, A., Martinez, J., Valls, C., Cusola, O., Blanca Roncero, M., Vidal, T., . . . Javier Pastor, F. I. (2018). Bacterial cellulose for increasing barrier properties of paper products. *Cellulose, 25*, 6093−6105.

Fink, H., Faxalv, L., Molnar, G. F., Drotz, K., Risberg, B., Lindahl, T. L., & Sellborn, A. (2010). Real-time measurements of coagulation on bacterial cellulose and conventional vascular graft materials. *Acta Biomaterialia, 6*, 1125−1130.

Fink, H., Faxälv, L., Molnár, G. F., Drotz, K., Risberg, B., Lindahl, T. L., & Sellborn, A. (2010). Real-time measurements of coagulation on bacterial cellulose and conventional vascular graft materials. *Acta Biomaterialia, 6*, 1125−1130.

Florea, M., Hagemann, H., Santosa, G., Abbott, J., Micklem, C. N., Spencer-Milnes, X., ... Ellis, T. (2016). Engineering control of bacterial cellulose production using a genetic toolkit and a new cellulose-producing strain. *Proceedings of the National Academy of Sciences of the United States of America, 113*, E3431−E3440.

Fontana, J. D., de Souza, A. M., Fontana, C. K., Torriani, I. L., Moreschi, J. C., Gallotti, B. J., ... Farah, L. F. (1990). Acetobacter cellulose pellicle as a temporary skin substitute. *Applied Biochemistry and Biotechnology, 24-25*, 253−264.

Foresti, M. L., Vázquez, A., & Boury, B. (2017). Applications of bacterial cellulose as precursor of carbon and composites with metal oxide, metal sulfide and metal nanoparticles: A review of recent advances. *Carbohydrate Polymers, 157*, 447−467.

Gelin, K., Bodin, A., Gatenholm, P., Mihranyan, A., Edwards, K., & Stromme, M. (2007). Characterization of water in bacterial cellulose using dielectric spectroscopy and electron microscopy. *Polymer, 48*, 7623−7631.

Gholami Derami, H., Jiang, Q., Ghim, D., Cao, S., Chandar, Y. J., Morrissey, J. J., ... Singamaneni, S. (2019). A robust and scalable polydopamine/bacterial nanocellulose hybrid membrane for efficient wastewater treatment. *ACS Applied Nano Materials, 2*, 1092−1101.

Grande, C. J., Torres, F. G., Gomez, C. M., & Carmen Bano, M. (2009). Nanocomposites of bacterial cellulose/hydroxyapatite for biomedical applications. *Acta Biomaterialia, 5*, 1605−1615.

Guan, F., Chen, S., Yao, J., Zheng, W., & Wang, H. (2016). ZnS/Bacterial cellulose/epoxy resin (ZnS/BC/E56) nanocomposites with good transparency and flexibility. *Journal of Materials Science & Technology, 32*, 153−157.

Guhados, G., Wan, W. K., & Hutter, J. L. (2005). Measurement of the elastic modulus of single bacterial cellulose fibers using atomic force microscopy. *Langmuir, 21*, 6642−6646.

Helenius, G., Backdahl, H., Bodin, A., Nannmark, U., Gatenholm, P., & Risberg, B. (2006). In vivo biocompatibility of bacterial cellulose. *Journal of Biomedical Materials Research Part A, 76A*, 431−438.

Huang, H.-C., Chen, L.-C., Lin, S.-B., & Chen, H.-H. (2011). Nano-biomaterials application in situ modification of bacterial cellulose structure by adding HPMC during fermentation. *Carbohydrate Polymers, 83*, 979−987.

Huang, H.-C., Chen, L.-C., Lin, S.-B., Hsu, C.-P., & Chen, H.-H. (2010). In situ modification of bacterial cellulose network structure by adding interfering substances during fermentation. *Bioresource Technology, 101*, 6084−6091.

Hu, Y., & Catchmark, J. M. (2011). In vitro biodegradability and mechanical properties of bioabsorbable bacterial cellulose incorporating cellulases. *Acta Biomaterialia, 7*, 2835−2845.

Hu, Y., Catchmark, J. M., & Vogler, E. A. (2013). Factors impacting the formation of sphere-like bacterial cellulose particles and their biocompatibility for human osteoblast growth. *Biomacromolecules, 14*, 3444−3452.

Hu, W., Chen, S., Yang, J., Li, Z., & Wang, H. (2014). Functionalized bacterial cellulose derivatives and nanocomposites. *Carbohydrate Polymers, 101*, 1043−1060.

Jang, W. D., Hwang, J. H., Kim, H. U., Ryu, J. Y., & Lee, S. Y. (2017). Bacterial cellulose as an example product for sustainable production and consumption. *Microbial Biotechnology, 10*, 1181−1185.

Jiang, F. J., Yin, L., Yu, Q. C., Zhong, C. Y., & Zhang, J. L. (2015). Bacterial cellulose nanofibrous membrane as thermal stable separator for lithium-ion batteries. *Journal of Power Sources, 279*, 21−27.

Jin, X., Xiang, Z., Liu, Q., Chen, Y., & Lu, F. (2017). Polyethyleneimine-bacterial cellulose bioadsorbent for effective removal of copper and lead ions from aqueous solution. *Bioresource Technology, 244*, 844−849.

Khalid, A., Khan, R., Ul-Islam, M., Khan, T., & Wahid, F. (2017). Bacterial cellulose-zinc oxide nanocomposites as a novel dressing system for burn wounds. *Carbohydrate Polymers, 164*, 214−221.

Kim, H. J., Jin, J. N., Kan, E., Kim, K. J., & Lee, S. H. (2017). Bacterial cellulose-chitosan composite hydrogel beads for enzyme immobilization. *Biotechnology and Bioprocess Engineering, 22*, 89−94.

Kim, Y.-J., Kim, J.-N., Wee, Y.-J., Park, D.-H., & Ryu, H.-W. (2007). Bacterial cellulose production by *Gluconacetobacter* sp RKY5 in a rotary biofilm contactor. *Applied Biochemistry and Biotechnology, 137*, 529–537.

Klemm, D., Schumann, D., Udhardt, U., & Marsch, S. (2001). Bacterial synthesized cellulose—Artificial blood vessels for microsurgery. *Progress in Polymer Science, 26*, 1561–1603.

Levinson, D. J., & Glonek, T. (2010). *Microbial cellulose contact lens* (U.S. Patent Application No. US7832857 B2).

Lin, W.-C., Lien, C.-C., Yeh, H.-J., Yu, C.-M., & Hsu, S.-h (2013). Bacterial cellulose and bacterial cellulose–chitosan membranes for wound dressing applications. *Carbohydrate Polymers, 94*, 603–611.

Lin, S.-P., Loira Calvar, I., Catchmark, J. M., Liu, J.-R., Demirci, A., & Cheng, K.-C. (2013). Biosynthesis, production and applications of bacterial cellulose. *Cellulose, 20*, 2191–2219.

Liu, M., Liu, L., Jia, S., Li, S., Zou, Y., & Zhong, C. (2018). Complete genome analysis of *Gluconacetobacter xylinus* CGMCC 2955 for elucidating bacterial cellulose biosynthesis metabolic regulation. *Scientific Reports, 8*.

Liu, M., Li, S., Xie, Y., Jia, S., Hou, Y., Zou, Y., & Zhong, C. (2018). Enhanced bacterial cellulose production by *Gluconacetobacter xylinus* via expression of *Vitreoscilla* hemoglobin and oxygen tension regulation. *Applied Microbiology and Biotechnology, 102*, 1155–1165.

Liu, L.-P., Yang, X.-N., Ye, L., Xue, D.-D., Liu, M., Jia, S.-R., ... Zhong, C. (2017). Preparation and characterization of a photocatalytic antibacterial material: Graphene oxide/TiO$_2$/bacterial cellulose nanocomposite. *Carbohydrate Polymers, 174*, 1078–1086.

Li, G., Nandgaonkar, A. G., Wang, Q., Zhang, J., Krause, W. E., Wei, Q., & Lucia, L. A. (2017). Laccase-immobilized bacterial cellulose/TiO$_2$ functionalized composite membranes: Evaluation for photo- and bio-catalytic dye degradation. *Journal of Membrane Science, 525*, 89–98.

Li, J., Wan, Y. Z., Li, L. F., Liang, H., & Wang, J. H. (2009). Preparation and characterization of 2,3-dialdehyde bacterial cellulose for potential biodegradable tissue engineering scaffolds. *Materials Science & Engineering C-Biomimetic and Supramolecular Systems, 29*, 1635–1642.

Lim, X., Wan, W., & Panchal, C. J. (2015). *Transparent bacterial cellulose nanocomposite hydrogels* (U.S. Patent No. US8940337 B2).

Ma, Q. L., Yu, Y. F., Sindoro, M., Fane, A. G., Wang, R., & Zhang, H. (2017). Carbon-based functional materials derived from waste for water remediation and energy storage. *Advanced Materials, 29*.

Menchaca-Nal, S., Londono-Calderon, C. L., Cerrutti, P., Foresti, M. L., Pampillo, L., Bilovol, V., ... Martinez-Garcia, R. (2016). Facile synthesis of cobalt ferrite nanotubes using bacterial nanocellulose as template. *Carbohydrate Polymers, 137*, 726–731.

Mendes, P. N., Rahal, S. C., Marques Pereira-Junior, O. C., Fabris, V. E., Rahal Lenharo, S. L., de Lima-Neto, J. F., & Landim-Alvarenga, Fd. C. (2009). In vivo and in vitro evaluation of an *Acetobacter xylinum* synthesized microbial cellulose membrane intended for guided tissue repair. *Acta Veterinaria Scandinavica, 51*.

Mirtalebi, S. S., Almasi, H., & Khaledabad, M. A. (2019). Physical, morphological, antimicrobial and release properties of novel MgO-bacterial cellulose nanohybrids prepared by in-situ and ex-situ methods. *International Journal of Biological Macromolecules, 128*, 848–857.

Mohammadi, H. (2011). Nanocomposite biomaterial mimicking aortic heart valve leaflet mechanical behaviour. *Proceedings of the Institution of Mechanical Engineers, Part H: Journal of Engineering in Medicine, 225*, 718–722.

Mohite, B. V., & Patil, S. V. (2014). A novel biomaterial: Bacterial cellulose and its new era applications. *Biotechnology and Applied Biochemistry, 61*, 101–110.

Moniri, M., Moghaddam, A. B., Azizi, S., Rahim, R. A., Saad, W. Z., Navaderi, M., ... Mohamad, R. (2018). Molecular study of wound healing after using biosynthesized BNC/Fe$_3$O$_4$ nanocomposites assisted with a bioinformatics approach. *International Journal of Nanomedicine, 13*, 2955–2971.

Moraes, P., Saska, S., Barud, H., de Lima, L. R., Martins, V. D. A., Plepis, A. M. D., ... Gaspar, A. M. M. (2016). Bacterial cellulose/collagen hydrogel for wound healing. *Materials Research-Ibero-American Journal of Materials, 19*, 106–116.

Naidoo, K. S., & Jaggernath, J. (2012). Uncorrected refractive errors. *Indian Journal of Ophthalmology, 60*, 432–437.

Napavichayanun, S., Amornsudthiwat, P., Pienpinijtham, P., & Aramwit, P. (2015). Interaction and effectiveness of antimicrobials along with healing-promoting agents in a novel biocellulose wound dressing. *Materials Science & Engineering C-Materials for Biological Applications*, *55*, 95–104.

Noh, Y. K., Da Costa, A. D. S., Park, Y. S., Du, P., Kim, I.-H., & Park, K. (2019). Fabrication of bacterial cellulose-collagen composite scaffolds and their osteogenic effect on human mesenchymal stem cells. *Carbohydrate Polymers*, *219*, 210–218.

Norhasliza Hasan, D. R. A. B., & Kamarudin, S. (2012). Application of bacterial cellulose (BC) in natural facial scrub. *International Journal on Advanced Science Engineering Information Technology*, *4*.

Pang, M., Huang, Y., Meng, F., Zhuang, Y., Liu, H., Du, M., . . . Cai, Y. (2020). Application of bacterial cellulose in skin and bone tissue engineering. *European Polymer Journal*, *122*, 109365.

Park, M., Lee, D., Shin, S., Kim, H.-J., & Hyun, J. (2016). Flexible conductive nanocellulose combined with silicon nanoparticles and polyaniline. *Carbohydrate Polymers*, *140*, 43–50.

Parveen, S., Krishnakumar, K., & Sahoo, S. (2006). New era in health care: Tissue engineering. *Journal of Stem Cells & Regenerative Medicine*, *1*, 8–24.

Paximada, P., Dimitrakopoulou, E. A., Tsouko, E., Koutinas, A. A., Fasseas, C., & Mandala, I. G. (2016). Structural modification of bacterial cellulose fibrils under ultrasonic irradiation. *Carbohydrate Polymers*, *150*, 5–12.

Petersen, N., & Gatenholm, P. (2011). Bacterial cellulose-based materials and medical devices: Current state and perspectives. *Applied Microbiology and Biotechnology*, *91*, 1277–1286.

Phutanon, N., Motina, K., Chang, Y. H., & Ummartyotin, S. (2019). Development of CuO particles onto bacterial cellulose sheets by forced hydrolysis: A synergistic approach for generating sheets with photocatalytic and antibiofouling properties. *International Journal of Biological Macromolecules*, *136*, 1142–1152.

Picheth, G. F., Pirich, C. L., Sierakowski, M. R., Woehl, M. A., Sakakibara, C. N., de Souza, C. F., . . . de Freitas, R. A. (2017). Bacterial cellulose in biomedical applications: A review. *International Journal of Biological Macromolecules*, *104*, 97–106.

Portal, O., Clark, W. A., & Levinson, D. J. (2009). Microbial cellulose wound dressing in the treatment of nonhealing lower extremity ulcers. *Wounds*, *21*, 1–3.

Portela, R., Leal, C. R., Almeida, P. L., & Sobral, R. G. (2019). Bacterial cellulose: A versatile biopolymer for wound dressing applications. *Microbial Biotechnology*, *12*, 586–610.

Pourali, P., Yahyaei, B., Ajoudanifar, H., Taheri, R., Alavi, H., & Hoseini, A. (2014). Impregnation of the bacterial cellulose membrane with biologically produced silver nanoparticles. *Current Microbiology*, *69*, 785–793.

Rajwade, J. M., Paknikar, K. M., & Kumbhar, J. V. (2015). Applications of bacterial cellulose and its composites in biomedicine. *Applied Microbiology and Biotechnology*, *99*, 2491–2511.

Ross, P., Mayer, R., & Benziman, M. (1991). Cellulose biosynthesis and function in bacteria. *Microbiological Reviews*, *55*, 35–58.

Sampaio, L. M. P., Padrao, J., Faria, J., Silva, J. P., Silva, C. J., Dourado, F., & Zille, A. (2016). Laccase immobilization on bacterial nanocellulose membranes: Antimicrobial, kinetic and stability properties. *Carbohydrate Polymers*, *145*, 1–12.

Schumann, D. A., Wippermann, J., Klemm, D. O., Kramer, F., Koth, D., Kosmehl, H., . . . Salehi-Gelani, S. (2009). Artificial vascular implants from bacterial cellulose: Preliminary results of small arterial substitutes. *Cellulose*, *16*, 877–885.

Shi, Z., Zhang, Y., Phillips, G. O., & Yang, G. (2014). Utilization of bacterial cellulose in food. *Food Hydrocolloids*, *35*, 539–545.

Shoda, M., & Sugano, Y. (2005). Recent advances in bacterial cellulose production. *Biotechnology and Bioprocess Engineering*, *10*, 1–8.

Silva, N. H. C. S., Rodrigues, A. F., Almeida, I. F., Costa, P. C., Rosado, C., Neto, C. P., . . . Freire, C. S. R. (2014). Bacterial cellulose membranes as transdermal delivery systems for diclofenac: In vitro dissolution and permeation studies. *Carbohydrate Polymers*, *106*, 264–269.

Stoica-Guzun, A., Stroescu, M., Tache, F., Zaharescu, T., & Grosu, E. (2007). Effect of electron beam irradiation on bacterial cellulose membranes used as transdermal drug delivery systems. *Nuclear Instruments & Methods in Physics Research Section B-Beam Interactions with Materials and Atoms, 265*, 434−438.

Sulaeva, I., Henniges, U., Rosenau, T., & Potthast, A. (2015). Bacterial cellulose as a material for wound treatment: Properties and modifications. A review. *Biotechnology Advances, 33*, 1547−1571.

Sun, D., Yang, J., & Wang, X. (2010). Bacterial cellulose/TiO_2 hybrid nanofibers prepared by the surface hydrolysis method with molecular precision. *Nanoscale, 2*, 287−292.

Tabarsa, T., Sheykhnazari, S., Ashori, A., Mashkour, M., & Khazaeian, A. (2017). Preparation and characterization of reinforced papers using nano bacterial cellulose. *International Journal of Biological Macromolecules, 101*, 334−340.

Tomé, L. C., Brandão, L., Mendes, A. M., Silvestre, A. J. D., Neto, C. P., Gandini, A., ... Marrucho, I. M. (2010). Preparation and characterization of bacterial cellulose membranes with tailored surface and barrier properties. *Cellulose, 17*, 1203−1211.

Torres, F. G., Commeaux, S., & Troncoso, O. P. (2012). Biocompatibility of bacterial cellulose based biomaterials. *Journal of Functional Biomaterials, 3*, 864−878.

Trovatti, E., Silva, N. H. C. S., Duarte, I. F., Rosado, C. F., Almeida, I. F., Costa, P., ... Neto, C. P. (2011). Biocellulose membranes as supports for dermal release of lidocaine. *Biomacromolecules, 12*, 4162−4168.

Tyagi, N., & Suresh, S. (2016). Production of cellulose from sugarcane molasses using Gluconacetobacter intermedius SNT-1: Optimization & characterization. *Journal of Cleaner Production, 112*, 71−80.

Ul-Islam, M., Khan, S., Ullah, M. W., & Park, J. K. (2015). Bacterial cellulose composites: Synthetic strategies and multiple applications in bio-medical and electro-conductive fields. *Biotechnology Journal, 10*, 1847−1861.

Ul-Islam, M., Shah, N., Ha, J. H., & Park, J. K. (2011). Effect of chitosan penetration on physico-chemical and mechanical properties of bacterial cellulose. *Korean Journal of Chemical Engineering, 28*, 1736.

Ullah, H., Santos, H. A., & Khan, T. (2016). Applications of bacterial cellulose in food, cosmetics and drug delivery. *Cellulose, 23*, 2291−2314.

Ullah, M. W., Ul-Islam, M., Khana, S., Kim, Y., & Park, J. K. (2015). Innovative production of bio-cellulose using a cell-free system derived from a single cell line. *Carbohydrate Polymers, 132*, 286−294.

Ullah, H., Wahid, F., Santos, H. A., & Khan, T. (2016). Advances in biomedical and pharmaceutical applications of functional bacterial cellulose-based nanocomposites. *Carbohydrate Polymers, 150*, 330−352.

Urbina, L., Guaresti, O., Requies, J., Gabilondo, N., Eceiza, A., Corcuera, M. A., & Retegi, A. (2018). Design of reusable novel membranes based on bacterial cellulose and chitosan for the filtration of copper in wastewaters. *Carbohydrate Polymers, 193*, 362−372.

Wahid, F., Bai, H., Wang, F.-P., Xie, Y.-Y., Zhang, Y.-W., Chu, L.-Q., ... Zhong, C. (2019). Facile synthesis of bacterial cellulose and polyethyleneimine based hybrid hydrogels for antibacterial applications. *Cellulose, 27*, 369−383.

Wahid, F., Duan, Y.-X., Hu, X.-H., Chu, L.-Q., Jia, S.-R., Cui, J.-D., & Zhong, C. (2019). A facile construction of bacterial cellulose/ZnO nanocomposite films and their photocatalytic and antibacterial properties. *International Journal of Biological Macromolecules, 132*, 692−700.

Wahid, F., Hu, X.-H., Chu, L.-Q., Jia, S.-R., Xie, Y.-Y., & Zhong, C. (2019). Development of bacterial cellulose/chitosan based semi-interpenetrating hydrogels with improved mechanical and antibacterial properties. *International Journal of Biological Macromolecules, 122*, 380−387.

Wang, J., Gao, C., Zhang, Y., & Wan, Y. (2010). Preparation and in vitro characterization of BC/PVA hydrogel composite for its potential use as artificial cornea biomaterial. *Materials Science & Engineering C-Materials for Biological Applications, 30*, 214−218.

Wang, S.-S., Han, Y.-H., Ye, Y.-X., Shi, X.-X., Xiang, P., Chen, D.-L., & Li, M. (2017). Physicochemical characterization of high-quality bacterial cellulose produced by *Komagataeibacter* sp strain W1 and identification of the associated genes in bacterial cellulose production. *RSC Advances, 7*, 45145−45155.

Wang, J., Tavakoli, J., & Tang, Y. H. (2019). Bacterial cellulose production, properties and applications with different culture methods—A review. *Carbohydrate Polymers, 219*, 63−76.

Wang, J., Wan, Y. Z., Luo, H. L., Gao, C., & Huang, Y. (2012). Immobilization of gelatin on bacterial cellulose nanofibers surface via crosslinking technique. *Materials Science & Engineering C-Materials for Biological Applications, 32*, 536–541.

Wanna, D., Alam, C., Toivola, D. M., & Alam, P. (2013). Bacterial cellulose-kaolin nanocomposites for application as biomedical wound healing materials. *Advances in Natural Sciences-Nanoscience and Nanotechnology, 4*.

Wei, H., Rodriguez, K., Renneckar, S., Leng, W., & Vikesland, P. J. (2015). Preparation and evaluation of nanocellulose-gold nanoparticle nanocomposites for SERS applications. *Analyst, 140*, 5640–5649.

Wei, H., & Vikesland, P. J. (2015). pH-Triggered molecular alignment for reproducible SERS detection via an AuNP/nanocellulose platform. *Scientific Reports, 5*.

Wei, B., Yang, G., & Hong, F. (2011). Preparation and evaluation of a kind of bacterial cellulose dry films with antibacterial properties. *Carbohydrate Polymers, 84*, 533–538.

Wiegand, C., Moritz, S., Hessler, N., Kralisch, D., Wesarg, F., Mueller, F. A., . . . Hipler, U.-C. (2015). Antimicrobial functionalization of bacterial nanocellulose by loading with polihexanide and povidone-iodine. *Journal of Materials Science-Materials in Medicine, 26*.

Wood, J. (2019). Bioinspiration in fashion—A review. *Biomimetics (Basel, Switzerland), 4*.

Wu, S.-C., & Li, M.-H. (2015). Production of bacterial cellulose membranes in a modified airlift bioreactor by *Gluconacetobacter xylinus*. *Journal of Bioscience and Bioengineering, 120*, 444–449.

Wu, S.-C., & Lia, Y.-K. (2008). Application of bacterial cellulose pellets in enzyme immobilization. *Journal of Molecular Catalysis B-Enzymatic, 54*, 103–108.

Wu, J., Zheng, Y., Song, W., Luan, J., Wen, X., Wu, Z., . . . Guo, S. (2014). In situ synthesis of silver-nanoparticles/bacterial cellulose composites for slow-released antimicrobial wound dressing. *Carbohydrate Polymers, 102*, 762–771.

Xie, Y.-Y., Hu, X.-H., Zhang, Y.-W., Wahid, F., Chu, L.-Q., Jia, S.-R., & Zhong, C. (2020). Development and antibacterial activities of bacterial cellulose/graphene oxide-CuO nanocomposite films. *Carbohydrate Polymers, 229*.

Xu, T., Jiang, Q., Ghim, D., Liu, K.-K., Sun, H., Derami, H. G., . . . Singamaneni, S. (2018). Catalytically active bacterial nanocellulose-based ultrafiltration membrane. *Small, 14*.

Yamanaka, S., Watanabe, K., Kitamura, N., Iguchi, M., Mitsuhashi, S., Nishi, Y., & Uryu, M. (1989). The structure and mechanical properties of sheets prepared from bacterial cellulose. *Journal of Materials Science, 24*, 3141–3145.

Yang, L., Chen, C., Hu, Y., Wei, F., Cui, J., Zhao, Y., . . . Sun, D. (2020). Three-dimensional bacterial cellulose/polydopamine/TiO$_2$ nanocomposite membrane with enhanced adsorption and photocatalytic degradation for dyes under ultraviolet-visible irradiation. *Journal of Colloid and Interface Science, 562*, 21–28.

Yoon, S. H., Jin, H. J., Kook, M. C., & Pyun, Y. R. (2006). Electrically conductive bacterial cellulose by incorporation of carbon nanotubes. *Biomacromolecules, 7*, 1280–1284.

Zaborowska, M., Bodin, A., Backdahl, H., Popp, J., Goldstein, A., & Gatenholm, P. (2010). Microporous bacterial cellulose as a potential scaffold for bone regeneration. *Acta Biomaterialia, 6*, 2540–2547.

Zang, S., Zhang, R., Chen, H., Lu, Y., Zhou, J., Chang, X., . . . Yang, G. (2015). Investigation on artificial blood vessels prepared from bacterial cellulose. *Materials Science and Engineering: C, 46*, 111–117.

Zhang, T., Zheng, Y., Liu, S., Yue, L., Gao, Y., & Yao, Y. (2015). Bacterial cellulose membrane supported three-dimensionally dispersed silver nanoparticles used as membrane electrode for oxygen reduction reaction in phosphate buffered saline. *Journal of Electroanalytical Chemistry, 750*, 43–48.

Zhu, C., Li, F., Zhou, X., Lin, L., & Zhang, T. (2014). Kombucha-synthesized bacterial cellulose: Preparation, characterization, and biocompatibility evaluation. *Journal of Biomedical Materials Research Part A, 102*, 1548–1557.

Zor, E. (2018). Silver nanoparticles-embedded nanopaper as a colorimetric chiral sensing platform. *Talanta, 184*, 149–155.

Zou, Y. D., Wang, X. X., Khan, A., Wang, P. Y., Liu, Y. H., Alsaedi, A., ... Wang, X. K. (2016). Environmental remediation and application of nanoscale zero-valent iron and its composites for the removal of heavy metal ions: A review. *Environmental Science & Technology, 50*, 7290–7304.

PART IV
Biopolymer Composites

CHAPTER 14

Biodegradable polymer composites

R. Reshmy[1], Eapen Philip[1], P.H. Vaisakh[1], Raveendran Sindhu[2], Parameswaran Binod[2], Aravind Madhavan[3], Ashok Pandey[4], Ranjna Sirohi[5] and Ayon Tarafdar[6]

[1]Post-Graduate and Research Department of Chemistry, Bishop Moore College, Mavelikara, India, [2]Microbial Processes and Technology Division, CSIR—National Institute for Interdisciplinary Science and Technology (CSIR—NIIST), Thiruvananthapuram, India, [3]Rajiv Gandhi Center for Biotechnology, Thiruvananthapuram, India, [4]Centre for Innovation and Translational Research, CSIR—Indian Institute for Toxicology Research (CSIR—IITR), Lucknow, India, [5]Department of Post-Harvest Process and Food Engineering, G.B. Pant University of Agriculture and Technology, Pantnagar, India, [6]Department of Food Engineering, National Institute of Food Technology, Entrepreneurship and Management, Sonipat, India

14.1 Introduction

The utilization of waste and contaminants from agricultural processes for the developments of valuable biodegradable composites is still remain unexplored. Especially during this decade, the possibilities of using organic wastes and biomass residues as additives or reinforcements in polymer composites have attracted considerable interest. Thermoplastic matrix composites are widely used in industry mainly due to their quick processability, low absorption of moisture, abrasion and chemical resistance, low cost, shelf life, and high resistance to impact and delamination. Additionally, recycling capability of thermoplastic matrix composites is one of the major advantages over thermosetting. For example, polypropylene (PP) is an amazing category of thermoplastics that have immense commercialization potentials due to its characteristics such as low density, good processability, and environmental barriers. The pressure and temperature play a vital role during the processing of the thermoplastic matrix composites to facilitate the impregnation and reduction of void fractions of the fibers. The applied pressure includes transversal flux of the polymer melt that impregnates the fibers into the matrix-forming condensed composites. The global crisis associated with petroleum resources offers options for production of novel green end materials that are eco-friendly and can preserve the sustainability of environment. The production of biodegradable polymer composites reinforced by natural fiber facilitates the use of environmental-friendly materials. The utilization of green materials provides an efficient route to solve crop residue issues.

Residues of crops, such as sugar cane, banana, oil palm, and pineapple leaf are produced in billions of tons worldwide (Deepa et al., 2015). These can be obtained in abundance, at minimal cost, and they are also products of sustainable biomass. Normally a small portion of the residues are only utilized as fertilizers or household fuels and the rest of the residues have been burned on the ground. Consequently air pollution has an adverse environmental effect. A feasible solution to solve this problem is to utilize these crop residues as reinforcement in the production of viable polymer composites for different applications (Taj, Munawar, & Khan, 2007).

14.1.1 Polymer composites

Polymer composites can be defined as certain high performance, flexible material consisting of a mixture of different phases of the material, at least one of which is usually a polymer. Matrix and reinforcement are two main phases required for producing polymer composites (Taj et al., 2007). Usually, organic polymers are used as matrix and fiber as reinforcement (Wang, Zheng, & Zheng, 2011a). Such polymer composites are widely applicable for products with specifications such as moderate stiffness, excellent specific strength, and low density. In the case of products with optimal friction and wear, the design of polymer composites with specific fillers and reinforcements is of particular relevance. Polymer composites have been found to meet up the requirements of designing for various applications due to their versatility in microstructure, thermal, mechanical, and physical properties (Pie & Friedrich, 2016). Polymer composites are manufactured based on the starting materials. Nanocomposites are prepared using different techniques such as solution intercalation, in situ intercalative polymerization, and melt intercalation. Methods like solution casting, twin-screw extrusion, and melt mixing are used to manufacture composites of natural fiber. These techniques have the ability to generate well-mixed polymeric composites with effective filler dispersion that enhances the fiber–matrix interaction. In recent years, researchers showed considerable interest in revealing the potentials of plant fibers by concerning environmental parameters to develop biocomposites with low density and good mechanical properties. The major benefits of the biodegradable polymer composites are their normal soil degradation without the release of harmful materials. Some surface modifications are to be carried out to make these composites more biodegradable and water soluble. The biodegradable polymer composites have immense commercialization potentials in the fields like disposable carry bags, sanitary diapers, and tissue engineering applications (Li, Zhang, & Zhang, 2017).

14.1.2 Advantages of biodegradable polymer composites

Polymers composites have many benefits such as lightweight, good resistance to corrosion, low cost, and easy to process. Natural fibers have recently attracted researcher's attention as a supplement, due to their advantages over other existing materials and their production requires less energy. These are eco-friendly, readily available, entirely biodegradable, low density, sustainable, and cheap that fulfill economic interest in the industry. A number of polymers such as graphene, aluminum oxide, diamond, carbon nanotubes (CNTs), and boron nitride (BN) show high thermal conductivity and can be introduced as fillers to improve their thermal conductivity (Li et al., 2017). When natural fiber-reinforced plastics are subjected to the combustion or landfill process at the end of their life-cycle, the amount of CO_2 produced by the fibers during their production is neutral in relation to the acculturated volume. The abrasive quality of fiber is much smaller, resulting in overall benefits over the technical process and recycling of the composite materials. Natural fiber-reinforced plastics are the most environmentally friendly materials that can be composed at the end of their life-cycle by using biodegradable polymers as matrixes (Taj et al., 2007).

Natural fiber composites help to maintain nonrenewable resources that are currently used as the main source for production of packaging and other products. They also reduce many of the environmental degradation issues, which indicate that the removal of natural fiber composites compared to advanced fiber composites is easier, safer, and less expensive. During processing, either thermosetting or thermoplastic, the fibers are treated very often on the surface to prepare adhesion with the polymer matrix. Therefore fibers with their high strength and rigidity must be strongly bonded to the matrix if they are to be imparted to the composite. The action of fractures also depends on the strength of the interface. A poor interface leads to low strength and stiffness but high fracture resistance, whereas a strong interface provides high strength and stiffness but often low brittle resistance. The characteristics of the interface also affect other composite properties, such as fatigue, environmental degradation, and resistance to creep. The choice of the correct fiber is highly dependent on the requirement. With military aircraft it is important to have both a high modulus and strength. In contrast, satellite applications benefit from the use of high fiber modules to improve the stiffness and stability of reflector plates, antennas, and supporting structures. The advantage of these materials as reinforcements are due to their property to drape or conform to curved surfaces without wrinkling (Balakrishnan, John, Pothen, Sreekala, & Thomas, 2016).

14.1.3 General commercialization processes

In 1914 the first commercial polymer structure containing two distinct polymers was discovered by Jonas Aylsworth. The substance was essentially a continuous interpenetrating network consisting of phenol-formaldehyde resins (phenolics) and sulfur-crossed natural rubber (Kim et al., 2017). Composites are the combination of at least two materials that exhibit superior quality than the properties of the individual base materials. Biodegradable polymer composite processing consists of two types of processes, one step and two step. In one-step process, the composites are formed by direct combination of soaked fiber and resin with simultaneous curing and molding.

In the early years, one-step process was generally adopted for the production of composites. The benefit of the one-step process is that technology and equipment are easily available, even though it has certain fatal drawbacks. It is difficult to remove the volatiles including water and solvent, which are quickly mixed up with the substance and create void holes. The uneven spreading of resin also led to the formation of poor and rich glue regions. In several cases, it forms gray thread due to incomplete fibers impregnation. Two-step method is invented to rectify the above-mentioned limitations of one-step process. In two-step process, first mixing and wetting of fiber and resin to produce a semifinished product and then convert into a composite product. In this process, an advanced dipping process controls the uneven resin distribution problems. Consequently the production process removes water, solvents, semiproducts, and other low molecular weight materials and thereby reduces the voids. The reliability of the composite material depends upon the quality of the semifinished product. During the manufacturing of semifinished products, fiber and resin are mixed with other fillers which later converted by compression molding or injection into composite products. The quality of the final composites also depends on the resin-fiber ratio, molding conditions, and manufacturing processes (Wang, Zheng, & Zheng, 2011b).

The technologies using in composite processing are the fundamental and prerequisite for the advancement of the composite industry (Verbeek & Van Den Berg, 2009). A new manufacturing technique has been implemented in recent years, in which the reinforcing materials and thermoplastic resin are first converted into semifinished board and then cut the semifinished board into a flan which is used to shape items through compression molding or stamping. Introduction of advanced technologies makes the availability of quality products in the market with reasonable price. Nowadays polymer-blend technology is widely using in the manufacturing of rubber-toughened plastics. The commercialization methods currently using in the market are based on preforming and molding methods (Fig. 14.1).

14.2 Types of biodegradable polymer composites

Based on the raw material resources and their production methods, biodegradable polymeric composite can be classified into various categories. These are natural fiber composites, double-layer polymer composites, CNT-reinforced composites, and petrochemical-based composites. The chemical structures of different biopolymers are depicted in Table 14.1.

14.2.1 Natural fiber composites

The matrixes in the polymer composites have many advantages, like low density, less abrasive, and low cost. Research will concentrate on designing these materials with a combination of structure and properties of composites with reduced cost and improved

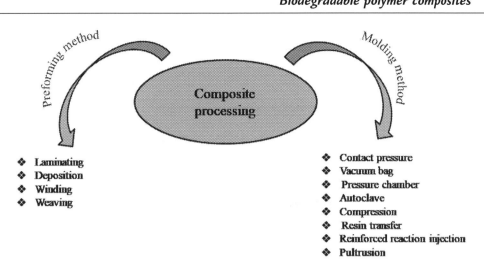

Figure 14.1
Different commercialization methods of composite processing.

processibility. There are mainly matrixes or reinforcement-based composites derived from agricultural or food waste resources to achieve biodegradable composites.

14.2.1.1 Matrix-based composites

The developments of newer technologies for improving the properties of matrix-based composites are of great important. Many researches are going on in this field by using fiber as matrix to produce new composites for various applications. Over the last centuries, the traditional materials including metals, ceramics, clay, etc. are replaced by polymers in different applications. The polymer matrix composites (PMCs) are made up of different kind of organic fibers (short or continuous fibers) combined with a number of reinforcing materials that are responsible for the properties like high strength, toughness, and rigidity. The PMC is designed in such a way that the fibers have the capacity to withstand the mechanical loads applied. The PMCs have numerous advantages namely, manufacturability, reliability, and easy handling compared to ceramics or metal matrix composites (Wang et al., 2011a). The application of thermoplastic polymers for engineering field is popular because of their excellent chemical resistance, improved mechanical properties, and lower cost. The only disadvantage of these polymers is that after end-use they are nonbiodegradable which can be overcome to some extent by creating composite polymer materials containing natural fibers.

Polypropylene exhibits reasonable mechanical and thermal characteristics; therefore it is utilized as one of the most commonly used thermoplastic polymers. Hence it is a palpable choice as matrix materials in the processing of natural fiber-reinforced composites. The composites based on polypropylene have greater flexural resistance relative to

Table 14.1: Chemical structure of different biodegradable polymers.

Name	Structure
Polyhydroxyalkanoates	
Polyvinyl alcohol	
Polyurethane	
Polycaprolactone	
Polyaniline	
Polypyrrole	
Polyester amide	
Graphene	

polypropylene. Such materials are of medium dimensional stability, thermal deformation at high temperatures, and burn resistance. It has been documented that there is a direct relationship between fiber load and mechanical characteristics like tensile strength, tensile module, and unidirectional composite impact strength of the polypropylene-reinforced coconut fibers. Coir fiber chemical treatment showed better mechanical properties with lower propensity of absorption of water in composite materials due to greater adhesion among coconut fiber and matrix content of polypropylene. In general the sodium hydroxide treatments improve the adhesion between thermoplastic polymers and plant fibers. In order to improve the adhesion between hydrophobic polypropyl matrixes with cellulosic fibers, the natural fibers were chemically modified with vinyltriethoxysilane, maleic anhydride, and its copolymers.

The incorporation of maize fibers into high-density polyethylene (HDPE) matrices decreases the thermal conductivity and diffusivity of fiber content. The mechanical properties of hybridized pineapple leaf fiber (PALF) and kenaf have also improved by reinforcing with HDPE. Due to higher cellulose content, which improved tensile strength and Young's modules, the PALF's traction strength was found to be superior to that of kenaf. The effect of the ratio of thermoplastic/bagasse fiber on the mechanical properties of composites made from LDPE and HDPE was investigated. Mechanical and physical properties have been found to deteriorate significantly when the composite bagasse fiber LDPE and HDPE exceeded 50%. The PALF has been prepared with various lengths and fiber content (0%−25%). This study showed that long fibers are stronger than short fibers, and also improved tensile strength as the fiber content improved. The composites are laminated using different injection process steps, such as drying, cutting, mixing, compounding, grinding, and molding. Such composite specimens are chosen for the study of their thermal and mechanical properties with a weight percentage of 10%−40% of the fibers. HDPE-reinforced in coconut fiber demonstrated higher mechanical properties compared to HDPE matrix without any compatibilizer and fiber alterations.

Mechanical properties of palm fiber-reinforced polystyrene were improved by melting and hot pressing using graft copolymerization of methyl methacrylate on the fiber surface. Polycarbonate (PC) is a class of thermoplastic polymers comprising of carbonate groups in its structure. PC is used with added transparency in construction applications that require strength. They can be conveniently molded and thermoformed. The tamarind fruit fibers were chemically treated followed by the surface modification using PC was found to enhance the tensile and thermal stability with lesser moisture absorption for the composites. Such PC matrix materials are significant since it surpasses the nylon due to the excellent strength, high thermal resistance, glass-like transparency, and compatible processability.

Wood plastic composite materials are widely used in the construction industry. Polyvinyl chloride (PVC) has turn out to be very suitable for building and other construction purposes because of its compatibility with natural fibers, lower cost, durability, and chemical and flame resistance. The coconut fiber-PVC/acrylonitrile composites have better thermal properties and impact strength than the PVC/wood composites when there is no variation in fiber content.

Most structural composites use polymers of the thermoset as material of the matrix. The thermoset polymers have very low density and viscosity, so that they can be integrated into fibers at low pressure. For the manufacturing of thermosets, basic manufacturing methods such as hand laying and spraying, compression, transfer, transfer of resin, injection, vacuum injection, and molding of pressure bags are used (Kim et al., 2017; Verbeek & Van Den Berg, 2009). The use of many other approaches such as cold-press molding, filament winding, pultrusion, reinforced reaction injection molding, and vacuum forming is seldom mentioned in the case of composites. Natural fiber-reinforced composite materials that use coir, coconut fiber, jute, pineapple leaf, banana, kenaf, sisal, and flax fibers have demonstrated superior mechanical characteristics with thermoset matrices, allowing them desirable for low-load applications like furniture and food packaging. The mechanical properties of the short randomly oriented sisal and banana fiber with hybrid-reinforced polyester were tested at different fiber loads. Tensile and flexural properties showed a favorable hybrid effect while impact efficiency showed a bad effect on the hybrid. The findings suggested that the mixture of sisal-jute-glass fiber can improve the properties which could be used as a substitute source for polymer composites reinforced with pure fiberglass. A great deal of work has been made to identify biodegradable composites having similar characteristics to conventional materials. The date palm fibers were also used as reinforcements and polyester as resin for preparation of the composites. These composites have been tested for different physical and mechanical properties where the material has been found to be acceptable (Wang et al., 2011b).

14.2.1.2 Reinforcement-based composites

Natural fiber polymer composites (NFPCs) have advantageous characteristics such as low density, less costly, and less rigid compared to traditional composite materials, providing benefits in automobile, building, and design applications in industry. Use of natural fibers in matrices as reinforcement has a beneficial impact on polymer composite mechanical behavior. The physical and mechanical characteristics of these NFPCs can be further improved by chemical treatment, while the absorption of moisture by NFPC can be reduced by surface modification of fibers such as alkalization and binding agents. Use of natural fiber for reinforcing composites has gained growing attention from both educational and industrial sectors. Many varieties of natural fibers are currently under investigation for use in plastics, including sugar cane, bamboo, flax, cotton, rice husk, wheat, barley, corn, kenaf, ramie, sisal, coir, hyacinth, kapok, PALF, raphia, banana fiber, and papyrus. Because of their many advantages, thermoplastics filled with special wood fillers enjoy rapid growth; lightweight, rational strength, and rigidity (Taj et al., 2007). Some plant proteins, due to their thermoplastic properties, are interesting renewable materials. Hibiscus family members are both an environmentally friendly and biodegradable crop. For composite and other consumer uses, it was found to be a major source of fiber. Kenaf fiber polyester resin composites have mechanical properties that blast to a height of at least 10 m. Hemp has been used traditionally to create ropes, but today the fiber is used to produce items like clothes, toys, and shoes. The fiber is absolutely biodegradable, recyclable, and nontoxic. Flax fibers in thermoplastic biocomposites can be outstanding fillers. These biocomposites may have tremendous potential to reduce the usage of petroleum-based plastics. The automobile, construction, and appliance industries are growing the usage of flax fibers every day owing to cost reductions, nonabrasive, and green movement. Biocomposites which include thermoplastics and modified flax fiber have similar mechanical properties to those of glass fiber-based composites.

The influence of compatible agents has been documented on the mechanical properties and morphology of thermoplastic polymer composites packed with rice-husk flours. When the filler load rises, the composites produced without any compatible agent demonstrate reduced tensile strength and improved fragility but significantly enhanced mechanical properties by adding the compatible agent. The poor interface connection between the filler and also the polymer matrix causes the composites to decrease the tensile strength but raises the tensile strength and modulus with the addition of the compatibilizing agents such as ethylene/propylene copolymers (Gaylord, 1989). Wheat straw has been used to create composites, anion exchangers, and panel boards where the straw acts as matrix material instead of fibrous. Jute is also among the most popular agrofibers used only as a part of thermoplastics and thermosets. Bamboo is also an available natural source throughout Asia and South America and was used to produce bamboo-reinforced thermosetting plastic and processed bamboo fiber ecocomposites utilizing the traditional hot-press process. The high weight component of bamboo fiber allowed bamboo composites to improve their capacity by manipulating the bamboo fiber most effectively into a cotton form (Wang, Zheng, & Zheng, 2011c).

14.2.2 Double-layer polymer composites

Double-layer polymer composites is an arrangement of layers of fibrous composite materials that can be joined to provide the required technological properties including in-plane rigidity, strength, thermal expansion coefficient, and bending rigidity. The individual layers in a polymeric, ceramic, or metallic matrix material consist of high-modulus, high-strength fibers. Typical fibers used include cellulose, graphene, chitin, and certain matrix materials are epoxy, polyimides, polyhydroxyalkanoates (PHAs), and polyhydroxybutyrate. Some examples of double-layer composites are polyhydroxybutyrate-cellulose (Cyras, Commisso, Mauri, & Vazquez, 2007), graphene-epoxy (Li et al., 2017), and chitin-calcium carbonate (Kato, 2000).

14.2.2.1 Polyhydroxyalkanoates composites

PHAs are a multistructure group of polyhydroxyesters. The composition of these polyesters may include monomers of 3-, 4-, 5-, and 6-hydroxyalkanoic acid that are accumulated under conditions deprived of nutrients by different bacterial species-native or mutant but abundant sources of carbon. A wide range of carbon sources may be used as a substratum for microbial PHA accumulation, but the use of renewable resources such as fatty acids is very beneficial in terms of mass production. The accumulated PHAs serve as a carbon and energy reserve within the microorganisms. A number of PHAs composed of both copolymers and block copolymers were generated utilizing different methods, including fermentation and enzymatic catalysis, resulting in an aggregate PHA content of up to 90% of the dry microbial mass. Furthermore, alternate production schemes focused on genetically modified plants and leaves are gaining traction and may become the chosen way to generate PHA. Unlike petrochemical plastics, PHAs are biodegradable, biocompatible with excellent gas barrier properties. These are almost identical to those of PVC and polyethylene terephthalate. These combinations of excellent physicochemical properties warrant the increasing commercial use of biopolymers in various niche applications ranging from biomedical, packaging, automotive, infrastructure, and aerospace to military applications. Unfortunately, despite their promising commercial potential, most of the developed PHAs, especially those with a higher monomeric composition of 3-hydroxybutyric acid, have been reported to exhibit fragility, low heat distortion temperature, and poor gas barrier properties, limiting processing malevolence and ductility. This in turn explained the current concern about the materials durability based on their cost of energy, shelf life, replacement, and maintenance. The nanoreinforcement of these types of biopolymers into nanocomposites in order to improve their quality and properties has already been shown to extend their applications in aggressive environments where neat polymers may fail. Therefore nanoreinforcements and the resulting nanocomposites is currently the subject of intense work to improve the performance of PHAs. By modifying the biopolymers using nanofillers such as layered silicates or phyllosilicates through nanoreinforcements, is obviously expanding their applications by significantly improving the polymer's performance. Nanofillers most widely used include silylated kaolinite, multiwall carbon nanotubes (MWCNTs), bioactive glass, nanoclay, cellulose nanocrystals, double-layered hydroxides, double-layered cobalt-aluminum hydroxides, etc.

In the processing of nanocomposite materials, a variety of approaches have often been used. These include intercalation of the solution, intercalation of the melt, polymerization, sol-gel, deposition, sputtering of the magnetron, laser, ultrasound, supercritical fluid, etc. The most commonly studied techniques are the methods of solution intercalation and melt intercalation in the manufacture of PHA nanocomposite. Nevertheless, it is shown that using in situ intercalative polymerization, supercritical fluids, and electrospinning are promising and evolving techniques. A nanocomposite's efficiency and consistency depends on how well the nanofillers scatter or combine in the matrix. These methods are therefore different strategies for improving the thermomechanical and physicochemical properties of the composites by enhancing effective interactions between the nanofiller and the polymer matrices. The type of polymeric matrix often dictates the type of surfactant and/or modifier to be used for better composition. For example, in a nanocomposite using a hydrophobic polymer matrix and layered hydrophilic silicate nanofillers, an alkyl ammonium surfactant has been found to be the best modifier. It is because the alkyl ammonium cations undergoing an ion-exchange reaction that turns the hydrophilic silicate surface into a stable hydrophobic sheet, while at the same time supplying a functional group that can communicate with the polymer matrix.

In this process, nanofillers are usually allowed to swell in the solvent by incubation for a specified period of time. The solution intercalation (solution casting) method is based on solving the polymer matrix into the solvent (e.g., chloroform, toluene, and dichloromethane) and nanofillers swelling within the solvent. After homogeneous mixing at a specific temperature, the polymer is applied to solubilize and intercalate with the coated nanofillers. The resulting mixture may be permitted to stand for some time depending on the experimental design, usually 4–5 days under daily periodic shaking before the solvent evaporates. Nevertheless, the solvent is evaporated directly after mixing in another method. Upon solvent evaporation, the intercalated polymeric materials give a nanocomposite that is normally dried in vacuum. The melt intercalation method gained momentum in the early 1990s as an alternative to avoiding excessive use of organic solvents during the solution intercalation process. This method involves mixing the polymer powder with the nanofiller, then pressing the mixture into pellets, followed by heating in a thermal chamber at a specific temperature under continuous mixing. The production of an intercalated PHBHV-clay nanocomposite using melt extrusion in a Brabender mixer at 165°C, with an agitation rate of 50 rpm for 15 min, was first described using ammonium-modified montmorillonite clay (Ohashi, Drumond, & Zane, 2009). Using two various types of naturally modified nanoclays (montmorillonite ion—with dimethyl—and synthetic fluoromica ion—with dimethyl diammonium), the researchers noted an inverse proportional relationship between the content of nanocomposite-clay and the characterization of d-spacing of the sample on X-diffraction (Gumel & Annuar, 2015). Nanofillers are allowed to swell in a liquid monomer solution, followed by activation of polymerization using an appropriate initiator such as water, heat, illumination, and catalyst, resulting in linear or cross-linked polymer matrix nanocomposites. Copolymers with hydroxyvalerate (HV) have been developed to reduce the thermal volatility and fragility of PHB. Cellulose has an impact on T_g (glass-transition temperature) in improving crystallization of the PHB polymer composites, and thermal activity has been established. Reinforced by jute measurements enhancing tensile strength by 50%–150%, PHB/HV reported bending strength by 30%–50% and impact strength by 90%. Regarding short abaca fibers/PHB composites, the influence of fiber thickness, surface alteration on tensile and flexural properties has been examined and an increased flexural intensity is observed on rising fiber content.

14.2.2.2 Graphene composites

Graphene, one in the allotropes of the carbon family (diamond, graphite, fullerenes, and CNTs), is a planar monolayer of carbon sp^2 hybrid atoms arranged in a 2D lattice. Like many other dimensional design products, this became known as the fundamental building stone. Because of the low-cost and high-yield performance, graphene oxide (GO) and reduced graphene oxide (RGO) as well as their derivatives as polymer composite fillers have shown tremendous potential for numerous essential applications, but in many instances they have lower physical properties than ideal single-layer graphene. Of the polymer composites loaded with GO and RGO close to CNT-based polymer composites, positive attempts have been made over the past decade. Based on the variation of beginning graphene materials (GO, thermally reduced graphene oxide, and chemically reduced graphene oxide) there are three major routes of polymer composites. Use a GO sheet as a composite starting filler, oxygen groups comprising carboxylic, hydroxyl, and epoxy groups on basal planes and edges of the sheet for extremely hydrophilic and quite stable aqueous dispersions (Tang, Zhao, & Guan, 2017). The inclusion of graphene will increase the conductivity and strength of bulk materials, and help build composites of superior quality. Graphene may also be added to plastics, polymers, and ceramics to create heat-resistant, conductive composites under friction. Graphene composites have many potential applications, producing unique and innovative materials with a lot of research going on. The uses appear infinite, as one graphene-polymer proves to be strong, versatile, and an outstanding electrical conductor. In contrast, another dioxide-graphene composite has been found to be of impressive photocatalytic efficiencies. Graphene has a wide surface area and robust interactions with the polymer matrix, which ensures that even limited quantities of graphene will improve the output significantly (Liu, Ullah, Kuo, & Cai, 2019).

14.2.3 Carbon nanotube-reinforced composites

CNTs were discovered by Iijima in 1991 and the first CNT-incorporated polymer nanocomposites were documented in 1994. Then several nanocomposites are reported with nanoscale fillers namely, carbon blacks, nanoclays, silicas, and carbon nanofibers (CNFs) to boost mechanical, electrical, and thermal polymers. CNT have high thermal conductivity, low density of mass, high flexibility, and a high aspect ratio. It has a unique blend of electrical, thermal, and mechanical properties that make nanotubes excellent candidates in the manufacture of multifunctional polymer nanocomposites to replace or supplement conventional nanofillers. Most nanotubes are heavier than titanium, lighter than steel, and more conductive than copper. There are two major types of CNT: single-wall carbon nanotubes (SWCNTs) and multiwall carbon nanotubes (MWCNTs). CNTs come in various types of mixture based on the methods of synthesis and vary greatly (Moniruzzaman & Winey, 2006). For many potential applications such materials show great promise: with example, aerospace, nanoelectronics, and sports goods.

14.2.3.1 Polyaniline composites

Due to its low monomer quality, environmental resilience, wide conductivity spectrum, particular redox states, and ease of synthesis, the preparation of conductive polymer nanocomposites using CNTs has evolved enormously in recent years. Polymer-nanotube interaction is one of the most important characteristics to be controlled in CNT/polyaniline (PANI) nanocomposites. The synergistic properties of the nanocomposites are essential for the existence of chemical and electronic interactions between CNT and PANI. Specific characterization methods such as FT-IR, UV-vis, Raman spectroscopy, thermal examination, and cyclic voltammetry analyzed the essence of the interactions between both the CNTs and the PANI. For the processing of CNT/PANI nanocomposites, some synthetic methods have been published, such as the solid-state blending of both powder components, the combination of dispersions of each part, and aniline electrochemical polymerization over a CNT-based electrode. The most effective route for CNT/PANI synthesis is aniline in situ chemical polymerization with CNTs that requires certain chemical pretreatments to ensure dispersion stability (Wu et al., 2017). Such CNT pretreatments have an important function in the morphology and conductance of the obtained nanocomposites. Two main approaches for dispersing CNTs include the chemical functionalization of nanotube walls and noncovalent adhesion of surfactant molecules. Some CNT/PANI nanocomposites are prepared utilizing silver nanoparticles via interfacial polymerization, resulting in a product with high transparency, stability, and optical properties. Several potential applications are known for these materials as supercapacitors, biosensors, actuators, biomedical devices, transistors, battery electrodes, and transparent thin-film conductors (Salvatierra, Oliveira, & Zarbin, 2010).

14.2.3.2 Polypyrrole composites

Polypyrrole (PPy) is one of the most common conducting polymers. It has important applications in the manufacture of polymer/CNT composites, based on its environmental resilience and excellent electrical conductivity. PPy can be prepared by chemical or electrochemical oxidation of pyrrole in different organic solvents and aqueous media. Wang et al. documented the synthesis and electrochemical capacitance characteristics of SWCNT/PPy composites and found that; these composites have very low load transfer resistance, high specific flexibility, and very good load/discharge efficiency (Wang, Xu, Chen, & Sun, 2007). Another method for CNT/PPy synthesis was by thermal deposition of hydrocarbons through in situ pyrrole polymerization on CNT (Fan et al., 1999). A novel composite CNT/PPy was developed using an electrochemical route where CNTs act as a conductive and powerful dopant in the formation of a conducting polymer (Chen et al., 2000). The synthesis of MWCNTs/PPy composite materials by in situ chemical oxidation polymerization using ammonium peroxodisulfate and poly(styrenesulfonate) combinations was demonstrated by Wu et al. The presence of cationic electrolyte improved the conductivity and solubility of the fabricated PPy/MWCNT composites (Wu, Chang, & Lin, 2009). Some possibilities for the synthesis of new CNTs/PPy composites utilized incorporation of CNTs as a conductive filler in polymer matrix and as reinforcement in structural materials. PPy has dominance for commercial applications among conducting polymers due to its unique characteristics, such as ease of preparation, high conductivity, and air stability. The development of nanosized ultrahigh catalysts, field emitters, probe tips, nanoscale template, electrooptical devices, and nanoelectronic components was another potential application of CNTs/PPy.

14.2.3.3 Polyvinyl alcohol composites

Polyvinyl alcohol (PVA) is a biopolymer with very high moisture-absorbing efficiency, fiber formability, biocompatibility, chemical tolerance, biodegradability, and swelling properties, suitable for different applications. The SWCNT/PVA composites developed lightweight polymer composites with enhanced thermal, electrical, and mechanical properties. Naebe, Lin, Staiger, Dai, and Wang (2008) have demonstrated the production of MWNT/PVA composites using electrospinning that could lead to an increased PVA crystallinity for the electrospun nanofiber. The MWNT/PVA sensor's humidity switching properties were investigated and showed better nonlinear response, good switching characteristics, and high sensitivity. Such sensors are better suited for monitoring and regulating other moist atmospheres than normal humidity sensors (Fei, Jiang, Jiang, Mu, & Zhang, 2014). Another work on a novel CNTs/PVA-AgNP composite was developed by electrospinning and their application in wound healing was investigated by Jatoi et al. The antibacterial studies have confirmed prolonged antibacterial effect which will help to avoid repeated dressing removal (Jatoi, Ogasawara, Kim, & Ni, 2019). The wide range applications of CNT/PVA composites are due to their properties such as actuators, sensors, electronic devices, and biomedical field.

14.2.4 Petrochemical-based biocomposites

Biodegradable composites can be made not just from bio-based feed stock, but from petrochemical raw materials as well. The eco-friendly polymer composite consists mainly of renewable natural polymers and synthetic degradable polymers. The biodegradability of biocomposites together with rigidity of petroleum-based raw materials made the composite suitable for specific applications. In recent years the development of new bio-based or partially bio-based versions has increased. These are also biodegradable and exhibit high elasticity, durability, and resistance to fracture, hence being used as a preferred alternative for use in products like containers, wraps, and other packaging materials. Some common examples of petrochemical-based biocomposites are polycaprolactone (PCL), polyester amide (PEA), and polyurethane (PU) composites.

14.2.4.1 Polycaprolactone composites

PCL is a petroleum-based linear polymer which is recognized as one of the few fully biodegradable and biocompatible synthetic polymers. It is aliphatic polyester with a high tolerance to liquids, solvents, and gasoline, synthesized through caprolactone ring-opening polymerization. PCL's exceptional mechanical, chemical, and outstanding bioresorbable properties have led to its complete commercial production for biomedical applications (Sayyar et al., 2013). To make PCL useful, its temperature profile must be adjusted to allow it to be manufactured into films, sheets, or commodities molded with injection. Mixing, adding mineral fibers, animal fibers, and natural fibers as reinforcements make PCL composites more cost-effective. Wu et al. reported mechanical and thermal properties blends of maleated PCL and starch composites (PCL-g-MAH/starch). Because of the formation of an ester carbonyl group, the greater compatibility of PCL-g-MAH with starch resulted in much better dispersion and homogeneity of starch in the PCL-g-MAH composite and consequently improved the properties (Wu, 2003).

Production of surface-modified biocomposite jute fiber hybrid and differing polylactide and PCL fractions has been recorded with improved rigidity and durability. This biocomposites fabricated by hot pressing of solvent-impregnated prepregs process. The addition of PCL led to an improvement in the biodegradation rate of biocomposites, thus making them more environmentally friendly (Goriparthi, Suman, & Nalluri, 2012). Sayyar et al. introduce two different routes for the production of graphene/PCL biocomposites using simple mixing and esterification methods (Sayyar et al., 2013). The esterification reaction resulted in the improvement of covalent linkage between graphene chains and PCL leads to better composite with excellent processability for tissue engineering. Biomedical and tissue engineering PCL has significant benefits and future applications in the area of food packaging.

14.2.4.2 Polyester amide composites

PEA composites are a developing group of biodegradable polymers that are suitable for the production of specialty products. A considerable endeavor has been made in producing PEAs since the preparation of the first derivatives in the 1970s (Wu, 2003). Such polymers have groups of ester and amide that are degradable and offer strong mechanical and thermal properties. Because of the blocky, irregular, and organized microstructures, the properties like hydrophobic/hydrophilic ratio and biodegradability can be effectively tuned. PEA was developed commercially from polyamide statistical copolycondensation (PA 6 or PA 6-6) monomers and adipic acid (Bordes, Pollet, & Avérous, 2009). For the biodegradable PEA composite the influence of chemical surface modifications of the jute fibers has been studied. The mechanical properties of composites such as tensile and bending capabilities improve as a consequence of the surface adjustment (Mohanty, Khan, & Hinrichsen, 2000). Mohanty et al. (2005) have reported utilizing twin-screw extrusion and injection molding processes to produce soy-based bioplastic composite with PEA. These green composites enhanced mechanical properties including impact resistance, tensile, and flexural properties. The film-stacking process was used by Jiang et al. to hot press biodegradable composites consisting of flax and cotton fiber and PEA (Jiang & Hinrichsen, 1999). However, considerable efforts to validate the highly promising properties of new PEAs composites and improve their performance properties still seem necessary. The potential benefits of PEA composites include disposable pots for plants, disposables for consumers, packaging, etc.

14.2.4.3 Polyurethane composites

PUs are among the most versatile materials with desirable properties, such as elasticity, flexibility, tear strength, high resistance to abrasion, and excellent shock absorption and biocompatibility. It has the advantages of relatively low price, quick reaction time, low viscosity, and excellent matrix binding without special fiber sizing. These composites have extensive use in many civilian and military sectors. The PU composites can be recycled according to the nature of waste materials and are generally converted to more usable forms like pellets, flakes, or powder.

Husić, Javni, and Petrović (2005) reported glass reinforced soy-based PU composites as a viable alternative to the conventional petroleum-based nondegradable plastics. The thermal and mechanical properties of PU resin was improved by incorporating soy because of better oxidative, thermal, and hydrolytic stability of these novel composites compared to other petrochemical composites. Starch PU films were prepared as the main polyol component and their mechanical properties were investigated by Kim, Kwon, Yang, and Park (2007). Due to the PU melting behavior, films could only be prepared at a suitable temperature by hot pressing. In addition, the strain rate had a significant effect on mechanical properties in both the tensile and bending tests. Kurańska, Aleksander, Mikelis, & Ugis (2013) have shown that rapeseed oil-based polyols are a raw material that can be used efficiently in the production of porous PU composites. Partial replacement of petrochemical polyol with biopolyols greatly reduces water absorption in modified PUR content. The incorporation of natural fillers in the formulation of PU enables the properties of biocomponent. All of these attempts to improve production technologies together with the introduction of new design principles will surely result in the development of newer PU composites in the nearby future.

14.3 Potentials and applications

The biodegradable polymer composites have acquired a critical role in modern applications as materials in almost every manufacturing sector, including aerospace, infrastructure, military, sports equipments, consumer products, and automotive and biomedical devices. It has numerous advantages namely, tunable properties, low cost, high strength, easy processability, chemical and physical degradation resistance, biocompatibility, and bioresorability. The advantages and potential applications of various biodegradable composites are depicted in Table 14.2.

Composite materials have been used in the transport sector in vehicles, trains, ships, and other transport devices that increase consumption of these goods year after year, with the highest percentage being the amount inside the shipping sector. The composites are primarily used in automobile industry for a wide range of body parts like engine cover, door, seat, fire engine, and other tanker transport. In the ship industry, composite materials are used to produce a variety of work vessels, fishing boats, shipping boats, motor boats, lifeboats, cruises, military minesweeper, and submarines. The glass fiber-reinforced plastics that are highly resistant to corrosive chemicals are widely used in anticorrosion devices. Environmentally friendly polymer composites are currently used for the production of

Table 14.2: Some biodegradable polymer composites, their advantages, and applications.

Matrix	Reinforcement	Advantages	Potential applications
Polyhydroxyalkanoates	Cellulose/Starch/Clay/Layered hydrophilic silicate/hydroxyvalerate	Biodegradability, biocompatible, improved barrier properties	Food packagings, biomedical devices
Polyvinyl alcohol	Carbon nanotubes/Silver nanoparticle	Biocompatibility, chemical tolerance, biodegradability, and swelling properties	Sensors, electronic devices, wound healing
Polyurethane	Glass/Soy/Starch	Elasticity, flexibility, tear strength, low cost, quick reaction time, low viscosity, and excellent matrix bonding	Military equipments, packagings
Polycaprolactone	Starch/Natural fibers/Graphene	Bioresorbable, biodegradable, excellent processibility	Tissue engineering, biomedical packagings, food packagings
Polyaniline	Carbon nanotubes/Silver nanoparticles	High transparency, stability, and optical properties	Supercapacitors, biosensors, actuators, biomedical devices, transistors, battery electrodes, transparent thin-film conductors
Polypyrrole	Carbon nanotubes	Low load transfer resistance, high specific flexibility and very good load/discharge efficiency, improved conductivity	Ultrahigh catalysts, field emitters, probe tips, nanoscale template, electrooptical devices, and nanoelectronic components
Polyester amide	Adipic acid/Jute fibers/Soy/Flax/Cotton	Tunable hydrophobicity/hydrophilicity impact resistance, tensile, and flexural properties	Disposable garden pots, disposable cutleries, packagings
Graphene composites	Plastics/Polymers/Ceramics	Increased conductivity and strength of bulk materials, high heat resistance, impressive photocatalytic efficiencies	Supercapacitors, sensors, biomedical devices

laminate, copper-coated laminate, insulating pipe, electrical retaining plate, pin, isolator, street lamps, and telegraph pole and live electrical and computer-operating equipment. Recently, biodegradable composites have grabbed attentions in the field of construction, maritime manufacturing, civil engineering components, and waterproof coatings. Such composites are very compactable with strong mechanical efficiency. Other fields that utilize biodegradable composites are biomedical, packaging, aerospace, biomass energy production, etc. (Asim, Saba, Jawaid, and Nasir, 2018). Products design using biocomposites is extremely important that guarantee material sustainability. A number of commercially available polymer composites are depicted in Fig. 14.2.

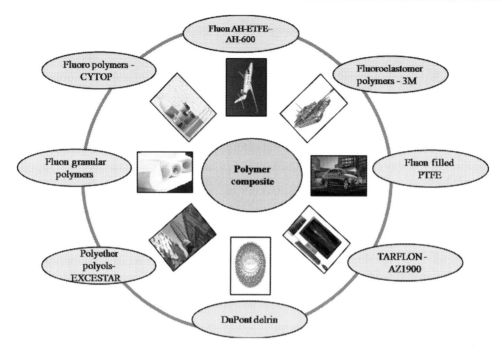

Figure 14.2
Commercially available polymer composite in various fields.

14.4 Conclusions and perspectives

The developments of environmental-friendly biodegradable composites from waste materials are of considerable interest over the last decade. A large number of companies are now working in this sector for manufacturing a wide array of innovative products. These advances have explored bio-based polymers and composites from a limited field into widespread applications. Nonetheless, the obstacles that need to be addressed in the coming years are the lower production performance of certain bio-based polymers, lifetime stability during in-service existence regardless of their biodegradation, their comparatively high manufacturing and processing costs, and the need to reduce the use of agricultural land and forests. The diverse possibilities to improve their processability and performance put together biodegradable composites the most promising material in packaging, food, aerospace, and biomedical applications with multitudes of market scopes. The biodegradable polymer composite could mostly be modified and functionalized in a variety of ways, showing many new possibilities to extend the function and applications. So, the implementation of more stable and effective polymer bionanocomposites for innovative applications will be needed in the coming future.

Acknowledgments

Reshmy R. and Raveendran Sindhu acknowledge Department of Science and Technology for sanctioning projects under DST WOS-B scheme. Ranjna Sirohi acknowledges CSIR for providing fellowship under direct SRF scheme.

References

Asim, M., Saba, N., Jawaid, M., & Nasir, M. (2018). *Potential of natural fiber/biomass filler-reinforced polymer composites in aerospace applications*. Elsevier Ltd.

Balakrishnan, P., M.J. John, L. Pothen, M.S. Sreekala, & S. Thomas (2016). Natural fibre composites and their applications in aerospace engineering. In *Advanced composite materials for aerospace engineering: Processing, properties, and applications*. Elsevier Science.

Bordes, P., Pollet, E., & Avérous, L. (2009). Nano-biocomposites: Biodegradable polyester/nanoclay systems. *Progress in Polymer Science, 34*(2), 125–155.

Chen, G. Z., Shaffer, M., Coleby, D., Dixon, G., Zhou, W., Frey, D. J., & Windle, A. H. (2000). Carbon nanotube and polypyrrole composites: Coating and doping. *Advanced Materials, 12*(7), 522–526.

Cyras, V. P., Commisso, M. S., Mauri, A. N., & Vazquez, A. (2007). Biodegradable double-layer films based on biological resources: Polyhydroxybutyrate and cellulose. *Journal of Applied Polymer Science, 106*, 749–756.

Deepa, B., Abraham, E., Cordeiro, N., Mozetic, M., Mathew, A. P., Oksman, K., . . . Pothan, L. A. (2015). Utilization of various lignocellulosic biomass for the production of nanocellulose: A comparative study. *Cellulose, 22*(2), 1075–1090.

Fan, J., Wan, M., Zhu, D., Chang, B., Pan, Z., & Xie, S. (1999). Synthesis and properties of carbon nanotube-polypyrrole composites. *Synthetic Metals, 102*(1–3), 1266–1267.

Fei, T., Jiang, K., Jiang, F., Mu, R., & Zhang, T. (2014). Humidity switching properties of sensors based on multiwalled carbon nanotubes/polyvinyl alcohol composite films. *Journal of Applied Polymer Science, 131*(1), 1–7.

Gaylord, N. G. (1989). Compatibilizing agents: Structure and function in polyblends. *Journal of Macromolecular Science: Part A—Chemistry, 26*(8), 1211–1229.

Goriparthi, B. K., Suman, K. N. S., & Nalluri, M. R. (2012). Processing and characterization of jute fiber reinforced hybrid biocomposites based on polylactide/polycaprolactone blends. *Polymer Composites, 33*(2), 237–244.

Gumel, A. M., & Annuar, M. S. M. (2015). Nanocomposites of polyhydroxyalkanoates (PHAs). *RSC Green Chemistry, 2015*(30), 98–118.

Husić, S., Javni, I., & Petrović, Z. S. (2005). Thermal and mechanical properties of glass reinforced soy-based polyurethane composites. *Composites Science and Technology, 65*(1), 19–25.

Jatoi, A. W., Ogasawara, H., Kim, I. S., & Ni, Q. Q. (2019). Polyvinyl alcohol nanofiber based three phase wound dressings for sustained wound healing applications. *Materials Letters, 241*, 168–171.

Jiang, L., & Hinrichsen, G. (1999). Flax and cotton fiber reinforced biodegradable polyester amide composites, 1. Manufacture of composites and characterization of their mechanical properties. *Angewandte Makromolekulare Chemie, 268*(4649), 13–17.

Kato, T. (2000). Polymer/calcium carbonate layered thin-film composites. *Advanced Materials, 12*(20), 1543–1546.

Kim, D. H., Kwon, O. J., Yang, S. R., & Park, J. S. (2007). Preparation of starch-based polyurethane films and their mechanical properties. *Fibers and Polymers, 8*(3), 249–256.

Kim, D.-Y., Kadam, A., Shinde, S., Saratale, R. G., Patra, J., & Ghodake, G. (2017). Recent developments in nanotechnology transforming the agricultural sector: A transition replete with opportunities. *Journal of the Science of Food and Agriculture, 98*(3), 849–864.

Kurańska, M., Aleksander, P., Mikelis, K., & Ugis, C. (2013). Porous polyurethane composites based on biocomponents. *Composites Science and Technology*, 75, 70–76.

Li, A., Zhang, C., & Zhang, Y. F. (2017). Thermal conductivity of graphene-polymer composites: Mechanisms, properties, and applications. *Polymers*, 9(9), 1–17.

Liu, W., Ullah, B., Kuo, C.-C., & Cai, X. (2019). Two-dimensional nanomaterials-based polymer composites: Fabrication and energy storage applications. *Advances in Polymer Technology*, 2019, 1–15.

Mohanty, A. K., Khan, M. A., & Hinrichsen, G. (2000). Influence of chemical surface modification on the properties of biodegradable jute fabrics—Polyester amide composites. *Composites Part A: Applied Science and Manufacturing*, 31(2), 143–150.

Mohanty, A. K., Tummala, P., Liu, W., Misra, M., Mulukutla, P. V., & Drzal, L. T. (2005). Injection molded biocomposites from soy protein based bioplastic and short industrial hemp fiber. *Journal of Polymers and the Environment*, 13(3), 279–285.

Moniruzzaman, M., & Winey, K. I. (2006). Polymer nanocomposites containing carbon nanotubes. *Macromolecules*, 39(16), 5194–5205.

Naebe, M., Lin, T., Staiger, M. P., Dai, L., & Wang, X. (2008). Electrospun single-walled carbon nanotube/polyvinyl alcohol composite nanofibers: Structure-property relationships. *Nanotechnology*, 19(30).

Ohashi, E., Drumond, W. S., & Zane, N. P. (2009). Biodegradable poly (3-hydroxybutyrate) nanocomposite. *Macromolecular Symposia*, 279, 138–144.

Pie, X., & Friedrich, K. (2016). *Friction and wear of polymers and composites*. IntechOpen.

Salvatierra, R. V., Oliveira, M. M., & Zarbin, A. J. G. (2010). One-pot synthesis and processing of transparent, conducting, and freestanding carbon nanotubes/polyaniline composite films. *Chemistry of Materials*, 22(18), 5222–5234.

Sayyar, S., Murray, E., Thompson, B. C., Gambhir, S., Officer, D. L., & Wallace, G. G. (2013). Covalently linked biocompatible graphene/polycaprolactone composites for tissue engineering. *Carbon*, 52, 296–304.

Taj, S., Munawar, M. A., & Khan, S. (2007). Natural fiber-reinforced polymer composites. *Proceedings of the Pakistan Academy of Sciences*, 44(2), 129–144.

Tang, L., Zhao, L., & Guan, L. , (2017). *Graphene/Polymer composite materials: Processing, properties and applications. Advanced composite materials: Properties and applications*. Warsaw, Berlin: De Gruyter Open Poland.

Verbeek, C. J. R., & Van Den Berg, L. E. (2009). Recent developments in thermo-mechanical processing of proteinous recent developments in thermo-mechanical processing of proteinous bioplastics. *Recent Patents on Materials Science*, 2, 171–189, November.

Wang, J., Xu, Y., Chen, X., & Sun, X. (2007). Capacitance properties of single wall carbon nanotube/polypyrrole composite films. *Composites Science and Technology*, 67(14), 2981–2985.

Wang, R.-M., Zheng, S.-R., & Zheng, Y.-P. (2011a). *Introduction to polymer matrix composites. Polymer matrix composites and technology*. Woodhead Publishing.

Wang, R.-M., Zheng, S.-R., & Zheng, Y.-P. (2011b). *Matrix materials. Polymer matrix composites and technology*. Woodhead Publishing.

Wang, R.-M., Zheng, S.-R., & Zheng, Y.-P. (2011c). *Reinforced materials. Polymer matrix composites and technology*. Woodhead Publishing. (Chap. 4).

Wu, C. S. (2003). Physical properties and biodegradability of maleated-polycaprolactone/starch composite. *Polymer Degradation and Stability*, 80(1), 127–134.

Wu, G., Tan, P., Wang, D., Li, Z., Peng, L., Hu, Y., ... Chen, S. (2017). High-performance supercapacitors based on electrochemical-induced vertical-aligned carbon nanotubes and polyaniline nanocomposite electrodes. *Scientific Reports*, 7, 1–8.

Wu, T. M., Chang, H. L., & Lin, Y. W. (2009). Synthesis and characterization of conductive polypyrrole/multi-walled carbon nanotubes composites with improved solubility and conductivity. *Composites Science and Technology*, 69(5), 639–644.

CHAPTER 15

Thermal/rheological behavior and functional properties of biopolymers and biopolymer composites

Prachi Gaur[1], Vivek Kumar Gaur[2], Poonam Sharma[3] and Ashok Pandey[4,5]

[1]Institute of Information Management and Technology, Aligarh, India, [2]Amity Institute of Biotechnology, Amity University, Lucknow, India, [3]Department of Bioengineering, Integral University, Lucknow, India, [4]Centre for Innovation and Translational Research, CSIR—Indian Institute of Toxicology Research (CSIR—IITR), Lucknow, India, [5]Centre for Energy and Environmental Sustainability, Lucknow, India

15.1 Introduction

The alarming environmental conditions caused due to increasing use of petroleum-derived plastic material and the increasing awareness of replenishing nonrenewable fossil fuels reserves have knocked the attention of scientific community toward exploring sustainable bio-based renewable alternatives, which can replace traditional fossil-based plastics with commercial, recyclable, and biodegradable materials, suitable as per environmental legislation, the REACH Act (registration, evaluation, authorization, and restriction of chemical substances) (Bharti & Swetha, 2016; Mngomezulu, John, Jacobs, & Luyt, 2014; Pattanashetti, Heggannavar, & Kariduraganavar, 2017; Soroudi & Jakubowicz, 2013; Tanase-Opedal, Espinosa, Rodríguez, & Chinga-Carrasco, 2019). Conventional plastic marks its versatile presence in almost every segment due to its mechanical, thermal, and functional properties. It is quite challenging to treasure a gem-like plastic material but the shortcomings associated with the long-term usage of plastic material urges serious demand toward searching for a biogenic polymer. The word polymer is derived from the Greek word "poly" meaning many, and the word "meros" stands for particle. Thus polymer is a series of repeating units of monomer, such as polylactide, poly(butylene succinate), polyhydroxyalkanoates (PHAs), poly(p-dioxanone), and poly(ε-caprolactone) (Nair, Sekhar, Nampoothiri, & Pandey, 2017). Among all of these, PHAs and poly(lactic acids) (PLAs) were superior biopolymers in terms of properties and feasibility (Bharti & Swetha, 2016).

Biopolymers are smart polymers of new generation and are harvested from plants or microbial cells or it could be synthesized from a basic biological system, offering a viable solution to the economy and environmental issues (Rebelo, Fernandes, & Fangueiro, 2017; Younes, 2017). Biopolymers are also termed as bioplastic and are defined as biomass polymers mainly constituted with carbon-based compounds (Bharti & Swetha, 2016). Microbially synthesized PHA, its copolymers, and bio-derived PLA proposed strong candidature as biodegradable polymer but their inherent shortcomings such as lower functional properties, weak mechanical properties, narrow processing window, and low electric conduction further stipulate upgradation in their qualities (González-López, Robledo-Ortíz, Manríquez-González, Silva-Guzmán, & Pérez-Fonseca, 2018; Singh, Kumar, Ray, & Kalia, 2015; Sun et al., 2018). Tailoring a unique polymer with a perspective of sustainable development or eco-designing with natural material components such as natural fibers (jute, cotton, hemp, sisal, elephant grass, bamboo, kenaf, and flax), natural resins, matrix materials (natural rubber), and biodegradable polymers (cellulose, lignin, and pectin) stood up as an interesting and eco-friendly alternatives in new biodegradable "green composite" or biocomposite fabrication (Frone, Berlioz, Chailan, & Panaitescu, 2013; Gordobil, Delucis, Egüés, & Labidi, 2015; Mngomezulu et al., 2014; Mukherjee & Kao, 2011; Roy, Shit, Sengupta, & Shukla, 2014; Yıldızhan, Çalık, Özcanlı, & Serin, 2018).

Synthesis of biocomposite promotes the biopolymer functionality and its application in multiple areas, concurrently reducing the cost of manufacturing, imposing economic and environmental benefits (Boufarguine, Guinault, Miquelard-Garnier, & Sollogoub, 2013). Introduction of natural fiber induce desired characteristics such as good mechanical strength, nontoxicity, light weight, and low cost to fiber-reinforced biocomposites (Tanase-Opedal et al., 2019). The natural polymers obtained from a plant source or animal source are used for incorporation as filler or reinforcement material. Owing to the highly elastic, nonreactive, antimicrobial, and antiinflammatory properties of biopolymer, they are suitable for medical applications (Suarato, Bertorelli, & Athanassiou, 2018). Two (or more) distinct phases or constituents participate in an amalgamation to produce biocomposite material with a unique set of properties differing from the parent components. In fiber-reinforced composite material, natural fibers contribute toward stiffness and strength whereas biopolymer matrix entraps these fibers in a continuous phase to grant suitable structure (Fig. 15.1; Roy et al., 2014).

Reinforcement of nanomaterials such as nanoclay, nanocellulose, and carbon nanotubes has a synergistic effect in enhancing the properties of biocomposites (Sun et al., 2018). A variety of PLA and poly([R]-3-hydroxybutyrate) (PHB) biocomposites harnessed with different reinforced natural materials have varied potential and are used in different application due to their increased modulus, processing window extension, improved tensile strength, toughness, crystallinity, stiffness, and heat distortion temperature (Gurunathan, Mohanty, & Nayak, 2015; Anderson, Zhang, & Wolcott, 2013; Peelman et al., 2015).

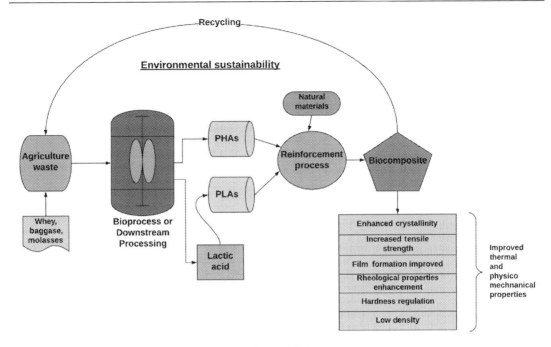

Figure 15.1
Scheme of biocomposite formulation.

Biocomposites have many commercial applications in various fields such as in medical, dental, automotive industry, food additives, food-packaging industry, plastic bottles, cosmetics, textile, water treatment chemicals, absorbents, electronics and electric devices, data-storage elements, and biosensors (Fig. 15.2; Ju et al., 2016; Kale et al., 2018; Rebelo et al., 2017; Soroudi & Jakubowicz, 2013; Sun et al., 2018; Tanase-Opedal et al., 2019; Xian, Wang, Zhu, Guo, & Tian, 2018; Xie, Cao, Rodríguez-Lozano, Luong-Van, & Rosa, 2017). In medical and dentistry, the composites are widely used as restorative materials, in tissue engineering; as drug carriers, prosthetic parts manufacturing like bone scaffolds, tissue graft, skin patches, and others including sutures (Liu et al., 2019; Rogina, 2014). Both PHA- and PLA-based composites extend great interest as food-packaging material with excellent gas and moisture barrier properties (Gupta & Katiyar, 2017; Kovalcik, Machovsky, Kozakova, & Koller, 2015; Marra, Silvestre, Duraccio, & Cimmino, 2016). In the automotive industry, thermosetting composites are replaced with thermoplastic fiber-reinforced biocomposites. This transformation offers several advantages owing to the light weight, corrosion resistance, attractive design flexibility, and cost-effectiveness of the biocomposites. These biocomposite-derived thermoplastic is used in parts of interior panel, shelves, parcels, floor mats, interior door trims, and spare tire covers. Automobile manufacturing companies, including Mercedes, Ford, and Toyota are using bio-based components for interior designing of automobiles (Roy et al., 2014; Soroudi & Jakubowicz, 2013; Wei & McDonald, 2016). As per a directive

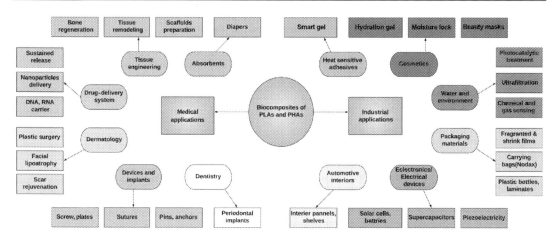

Figure 15.2
Multifarious application of polyhydroxyalkanoates (PHA) and polylactic acid (PLA) biocomposites.

initiated by Japanese government, 20% of entire plastic material used in Japan is substituted with the bio-derived plastics or biopolymers by the year 2020. As a result of encouragement, "hybrid bioplastic" materials are developed in Japan by incorporating bio-based resins, alloys, or blends of bio-based and petrochemical-based materials. Commercially available hybrid plastics include poly(ethylene terephthalate) (PET), acrylonitrile–butadiene–styrene, polycarbonate, high-impact polystyrene, polypropylene, and poly(methyl methacrylate) (PMMA). These polymers are economically viable commodities that have already been employed in common-use articles and packaging materials (Soroudi & Jakubowicz, 2013). This study focuses on recent research works related to biocomposite formulation, natural reinforcement materials, and thermal and functional properties of biocomposites making them a versatile candidate replacing conventional fossil-based plastic. The effect of various constituents as reinforcement agent and the processing techniques dealing with modification of thermal properties and functional characteristics are discussed in detail.

15.2 Biocomposites derived from polylactic acid

Increasing awareness of environmental concerns and constant reduction of limited petroleum resources have emerged as a crucial force behind the search of renewable substitutes to commute traditional fossil-based plastics material (Tanase-Opedal et al., 2019). PLA is an excellent thermoplastic polymer with the ability to supersede many petroleum-based polymers, yet its brittleness, poor impact resistance, slow crystallization rate, low melt strength, low heat deflection temperature, and relatively extravagant nature have led to further exploration of natural sources, to cast PLA-integrated biocomposites, which tailors the properties of final product, concurrently reduces the cost of biocomposites

(Nagarajan, Zhang, Misra, & Mohanty, 2015). Biocomposites are a special class of composite material or multicomponent system that is obtained by blending of at least one component of biotic genesis or biodegradable polymer, such as potato pulp powder and wood fiber (Awal, Rana, & Sain, 2015; Righetti et al., 2019). PLA biocomposites are gaining progressive popularity as an ecological and economic alternative to conventional petroleum-derived materials, and are utilized for a wide variety of applications (Righetti et al., 2019).

Reinforcement with natural filler materials like lignocellulosic material, eggshell powder, nutshell powder, holocellulose, and natural fibers derived from animal and plant sources, including wool, silk, horns, wood fiber/flour, flex, hemp, and sisal, bring forth considerable advantages, such as pervasive availability at an economical price, modest abrasiveness, industrial wide acceptability, low density, and biorenewability (Agustin-Salazar et al., 2018; Aranberri, Montes, Azcune, Rekondo, & Grande, 2017; Awal et al., 2015; Beigbeder, Soccalingame, Perrin, Bénézet, & Bergeret, 2019).

Natural fibers marked as filler are hydrophilic while biopolymers are hydrophobic or less hydrophilic matrices, which lack binding compatibility among themselves. It personifies greater complications for the realization of the biocomposites with revised properties. Thus surface properties of natural fiber are modified to improve this scarce compatibility. Stretching- or calendaring-type physical treatment can enhance the interfacial interaction of natural fiber. Corona/plasma treatment gives rise to the formation of free radicals and surface cross-linking for inducing compatibility amid hydrophilic natural fibers and hydrophobic polymer matrix. The chemical treatment method employs coupling agents such as bioadimide, lignin, titanates, maleic anhydride, silanes, triazine, and zirconates (Awal et al., 2015; Righetti et al., 2019). Bioadimide is a novel additive advancing the adhesion qualities and processability of composites by increasing the adhesion between fibers and matrix (Awal et al., 2015). This adhesion vigorously influences tensile strength and stiffness of the biocomposite (Righetti et al., 2019). Besides being a natural filler material lignin performs exceptionally by enhancing the compatibility between matrix and biopolymer components, thus strengthening the biopolymer interface besides improving its mechanical properties (Tanase-Opedal et al., 2019).

The substitution of multivariant natural compounds as filler material and their method of incorporation greatly improve the characteristics of PLA-based biocomposites. In the same refinement process, several research endeavors have ended up with an array of biocomposites holding varying properties. PLA-based biocomposite filament carrying 20% and 40% lignin shows reduced tensile strength, elasticity modulus, and higher crystallinity due to the nucleating-agent-like activity of lignin (Tanase-Opedal et al., 2019).

Valorization of agro-food by-products like wood fiber, eggshell powder, chicken feathers, potato pulp powder, or pecan (*Carya illinoinensis*) nutshell powder reduces the cost of final product and simultaneously promotes economy circulation (Agustin-Salazar et al., 2018; Awal et al., 2015; Kong et al., 2018; Righetti et al., 2019). Agro-food waste is 100% biodegradable so they emerge as a material of choice for incorporation in PLA-based biopolymer as a filler material. Biocomposite of PLA and wood fiber/chicken feathers waste, along with bioadditive is adequate for use in automotive panel components due to refined mechanical properties (Aranberri et al., 2017; Awal et al., 2015). Chicken feather waste carrying PLA biocomposites is used in flooring and building material in place of wooden and plastic material (Aranberri et al., 2017). Bio-$CaCO_3$ obtained from eggshell had successfully substituted inorganic $CaCO_3$ and is being used as biofiller material in biocomposites (Kong et al., 2018). PLA biocomposites carrying potato pulp powder as filler material show a progressive decline in elastic modulus, tensile strength, and elongation so they are technically suitable for the production of rigid packaging item (Righetti et al., 2019). Surface treatment of potato pulp powder with petroleum-based or biogenic waxes contributes toward improving mechanical properties (Righetti et al., 2019).

PLA biomatrix composed of *Pennisetum purpureum* fiber or Napier grass, as a reinforcement filler shows commending mechanical properties and controlled biodegradability. Silkworm silk fiber and PLA composite material exhibits superior mechanical and thermal properties thus making it a suitable material for forming an ideal scaffold (Revati et al., 2017). UV-resistant lignin-PLA biocomposites are produced via lignin-lactide grafting along with methanol and dioxane/water. Owing to the UV-inhibiting effect of lignin, this biocomposite shows effective UV-blocking ability as compared to other composites (Park, Kim, Youn, & Choi, 2019). Nanocomposite material is fabricated from PLA polymer and natural nanocellulose obtained from different bast fiber sources like hemp, flax, jute, kenaf, ramie, and roselle. The nanocomposite exhibits higher firmness, sustainability, and biodegradability (Kian, Saba, Jawaid, & Sultan, 2019).

Magnesium (Mg)-filled PLA biocomposites as filaments are convenient feedstock material for extrusion-based additive manufacturing technology, known as rapid prototyping or 3D printing for the development of versatile range of building materials (Antoniac et al., 2019). Biocomposites of PLA and hydroxyapatite overcome challenges associated with pure PLA and show high mechanical strength, osteoconductivity, osteoinductivity, and biodegradability (Tajbakhsh & Hajiali, 2017).

15.3 Biocomposites derived from polyhydroxyalkanoate

Fabrication of biocomposites from PHA polyesters is a remedy to many of the deformities encountered with the use of PHA. Although PHA offers a variety of assistance including biocompatibility, nontoxicity, biodegradability, renewability, and an amalgamation of excellent mechanical, physical, and functional properties yet its high processing cost, brittleness, and stability under narrow range of temperature have opened a new horizon of biocomposite formulation prior to commercialization (Vandi et al., 2018). PHB and its copolyester [R]-3-hydroxyvalerate [(poly(3-hydroxybutyrate-co-3-hydroxyvalerate)) (PHBV)] are the most common types of PHA. PHB and its copolymer are an excellent renewable resource, which possess comparatively high crystallization rate, hydrophobicity, biocompatibility, and biodegradability (García-Quiles, Fernández Cuello, & Castell, 2019; Ten, Jiang, Zhang, & Wolcott, 2015). PHAs are more suitable for food packaging but they are much costlier (€7–12 kg^{-1}) as compared to other biopolymers such as PLA (€2.5–3 kg^{-1}) and this has restricted their application for expensive goods in the medical and pharmaceutical domain (Cinelli, Seggiani, Mallegni, Gigante, & Lazzeri, 2019).

Blending of PHA with natural biodegradable polymer or low-cost lignocellulosic fiber significantly reduces the cost of production and concurrently improves the properties of PHA-based biocomposites (Cinelli et al., 2019; Ten et al., 2015). Natural polymers like corn gluten, soy protein isolates, starch, collagen, bacterial cellulose (BC), dried distillers grains with solubles, and poly (butylene adipate-co-terephthalate) (PBAT) were successfully incorporated in biocomposites to showcase varying physical and morphological characteristics (Chiulan et al., 2016; Ten et al., 2015; Wu, 2015). Lignocellulosic waste fibers are a by-product of agro-food processing obtained from agricultural and industrial crops. They are incorporated as fillers and reinforcement material in the fabrication process of PHA-based composites (Cinelli et al., 2019). Reinforcement of natural fibers obtained as waste like rice straw, wheat paddy, flax fiber, wood fiber, wood flour, jute fiber, bamboo fiber, yarn fiber, pineapple waste, kenaf, and yarn with PHB and PHBV by melt compounding method enhances flexural properties, mechanical properties, and degradation rate of the final biocomposite (Cinelli et al., 2019; Ten et al., 2015). The reinforcement of PHBV with natural fibers significantly improves brittleness and low impact resistance characteristics in the biocomposite material (Ten et al., 2015). Biocomposites produced by melt extrusion of PHBV and waste wood sawdust fibers (a by-product of the wood processing industry) have improved morphology, processability, mechanical properties, and thermal stability (Cinelli et al., 2019). Wood fiber-PHA composites offer similar properties to commercially available polypropylene and polyethylene-based wood plastic composites. Both the reinforcement material and the matrix ingredients are entirely biogenic and biodegradable in wood fiber-PHA composite. It has extended engineering applications at reduced cost of as little as 37% of neat PHA (Vandi et al., 2018).

Incorporation of BC (biopolymer synthesized by *Acetobacter xylinum*) in PHB as filler, fine-tunes the mechanical properties by proportionately reducing the crystallization temperature, and extending the processing window of PHA. PHA biocomposite carrying BC shows better cytocompatibility, controlled cell attachment, and surface properties, which makes this material most relevant for biomedical applications (Chiulan et al., 2016). Collagen is a biocompatible, biodegradable, nontoxic, and low allergic reactivity imposing biopolymer, extracted from extracellular matrix. Collagen type I is a structural protein, which provides elasticity and mechanical strength to tissues and ambient environment for cell growth. Therefore fabrication of composite material from collagen type I and PHA produces a biocomposite material "collagen/maleic anhydride-grafted PHA" with biomedical applications. Maleic anhydride was grafted in between the PHA matrix and collagen to reduce surface incompatibility by enhancing the interface compatibility in the matrix (Wu, 2015). The blending of PHBV with PBAT improves the impact strength and toughness of PHBV biocomposite by two orders of magnitude. Poly(propylene carbonate) (PPC) is a biodegradable aliphatic polycarbonate synthesized by an amalgamation of propylene oxide and carbon dioxide. It is also found to be a suitable filler material for biocomposite production. The addition of 40 wt.% PPC to PHBV can significantly improve impact strength and almost doubles the strain failure of pure PHBV.

Nanomaterials hold expertise in expanding surface to volume ratio and high aspect ratio due to nonspherical nature. Natural nanomaterials, including sepiolite nanoclay fibers and polysaccharide nanocrystals of cellulose, chitosan, chitin, and cellulose nanocrystals (CNCs)/nanowhiskers are encouraging claimant for biocomposite formulation (García-Quiles et al., 2019; Ten et al., 2015). Reinforcement of sepiolite nanoclay fibers contributes magnificent sorption capacity and improves the physical and chemical properties of resultant biocomposite by cross-linking (García-Quiles et al., 2019). Cellulose nanowhiskers/nanocrystals like wood, sugar beet, ramie, hemp, sisal, cotton, and tunicin breaks the economical barrier associated with synthetic polymer. Chitin nanofibrils association with PHBV significantly improve Young's modulus and tensile strength of the PHBV matrix (Ten et al., 2015). Introduction of fibers or nanoclay as filler material in biocomposite formulation is an opportunistic approach to enhance the suitability of biopolymers while reducing the demand of petroleum-based plastic polymer (García-Quiles et al., 2019).

15.4 Thermal and rheological properties

15.4.1 Polylactides and its biocomposites

Low toughness, poor thermal property, and no antimicrobial activity are the major drawbacks that restrict the application of polylactides (PLA) in many fields (Hamad, Kaseem, Ayyoob, Joo, & Deri, 2018). Different composites have been formulated with improved properties to overcome these barriers. Addition of fillers and fibers offers a way to improve the thermal and mechanical properties of PLA (Table 15.1; Cheng et al., 2009; Huda, Drzal, Mohanty, & Misra, 2006; Mohanty, Misra, & Drzal, 2002; Oksman & Selin, 2004). The thermal property is of utmost importance for identifying the possible applications of PLA composite (Kaseem, Hamad, & Ur Rehman, 2019). The addition of TiO_2 nanoparticle in PLA/TiO_2 composite alters its thermal property (Mallick et al., 2018).

Table 15.1: Modification in different properties by reinforcement of polylactic acid (PLA) and polyhydroxyalkanoates (PHA).

Biopolymer type	Reinforcement material/Biofiller	Binder/Solvent method	Property enhanced	Reference
PLA Lignin composite	Lignin (20%, 40%)	--	Reduces tensile strength, elasticity modulus, higher crystallinity	Tanase-Opedal et al. (2019)
Lignin-g-poly(lactic acid)	Lignin	Triazabicyclodecene	UV absorption, dispersion modified, reduced brittleness	Chung et al. (2013)
PLA lignin-polypropylene composites	40% Lignin	Polypropylene	Reduced tensile strength, poor adhesion, reduced Young's modulus	Gao (2017)
Lignin PLA biocomposite	Lignin-lactide grafting	Methanol and dioxane/water	Effective UV blocking ability	Park et al. (2019)
PLA chicken feather composite	Chicken feather powder	--	Improved composite strength, refined mechanical properties, used in flooring, building material	Aranberri et al. (2017)
PLA chicken feather fiber	5% Chicken feather fiber	--	Increased thermal stability	Cheng et al. (2009)
PLA wood flour composite	Wood flour	--	Refined mechanical properties, used in aircraft or automobile interior	Awal et al. (2015)
Wood-fiber-reinforced poly (lactic acid) composites	30% Wood fiber	Polypropylene	High mechanical properties, stiffness, used in automotive and packaging industries	Huda, Drzal, Misra, and Mohanty (2006)
MAH-g-PLA composite	30% Maleic anhydride (MAH)	Dicumyl peroxide (DCP)	High crystallinity, improved toughness	Zhang et al. (2017)
PLA eggshell powder composite	Bio $CaCO_3$ obtained from eggshell (4%)	Chloroform	Higher tensile strength, thermal stability, and modulus	Ashok et al. (2014); Kong et al. (2018)
PLA PPP composite	Potato pulp powder	Bio-based waxes	Improved mechanical properties, reduced elastic modulus, tensile strength, and elongation makes suitable for rigid packaging items	Righetti et al. (2019)
Elephant grass/PLA composite	Elephant grass fiber	--	Increased water absorption rate, tensile strength, impact strength	Gunti, Ratna Prasad, & Gupta (2018)

(Continued)

Table 15.1: (Continued)

Biopolymer type	Reinforcement material/Biofiller	Binder/Solvent method	Property enhanced	Reference
PLA/TiO$_2$ composite	TiO$_2$ nanoparticle		Reduced melting point	Mallick et al. (2018)
PLA/TiO$_2$ZnO composite	ZnO and TiO$_2$ nanoparticles		Reduced activation energy	Wang et al. (2019)
SSF PLA composite	Silkworm silk fiber		Superior mechanical and thermal properties, suitable material for light weight scaffold	Revati et al. (2017)
Magnesium (Mg) filled PLA biocomposites	Magnesium	Extrusion-based additive manufacturing technology	Rapid prototyping or 3D printing for building material manufacturing	Antoniac et al. (2019)
PLA/kenaf composites	10% (wt./wt.) Kenaf	--	Reduced glass transition temperature and cold crystallization temperature	Tawakkal et al. (2014)
PLA/kenaf composites	40% (wt./wt.) Kenaf	--	Reduced cold crystallization temperature	Tawakkal et al. (2014)
PLA CF composite	Short coir fibers	Sodium hydroxide and hydrogen peroxide solution 1.5 % (wt./wt.)	Reduced thermal stability and manufacturing cost	Sun et al. (2017)
PLA WF composite	10% (wt./wt.) Wood flour	Triclosan	Melting peak at lower temperature	Prapruddivongs and Sombatsompop (2012)
PLA SF composite	Short *Spartium junceum* L. fibers	Montmorillonite nanoclay	Increased fire resistance, enhanced thermal properties	Kovacevic et al. (2015)
PLA/nano-TiO$_2$/nano-Ag composite	nano-TiO$_2$ and nano-Ag	Solvent volatilization method	Increases elongation at break, decreased transparency, tensile strength (TS), and elastic modulus (EM)	Li et al. (2017)
PHB HV SDF biocomposite	Wood sawdust fibers (SDF)	Melt extrusion	Processability, mechanical, and thermal stability	Cinelli et al. (2019)
Wood fiber-PHA composites	Wood fiber	Polypropylene	Cost reduction and mechanical stability	Vandi et al. (2018)

Composite	Component	Additive	Properties	Reference
Collagen/maleic anhydride-grafted PHA (PHA-g-MA) composite	Collagen type I	Maleic anhydride	Biocompatible, biodegradable, nontoxic, and low allergic material suitable for biomedical applications	Wu (2015)
Poly(hydroxyalkanoate) (PHA-g-AA)	Acrylic acid		Decreased melting temperature	Wu (2014)
Poly(3-hydroxybutyrate-co-4-hydroxybutyrate) (P3HB4HB)/(MicroPCMs) composite	Microencapsulated phase change materials (MicroPCMs) and poly(3-hydroxybutyrate-co-4-hydroxybutyrate) (P3HB4HB)		Improved thermal stability, decreased melting and freezing point	Chen et al. (2014)
PHA chestnut shell fiber glycidyl methacrylate (PHA-g-GMA/CSF) composite	Chestnut shell fiber	Glycidyl methacrylate and epoxidized hydroxyl group (eOH)	Increased melting point, glass transition temperature, low crystallinity	Wu and Liao (2014)
PHA CNC eCO composite	5 wt.% Cellulose nanocrystals (CNCs) from sugarcane waste, and 10 wt.% epoxidized canola oil (eCO)	Zinc acetate	Improved tensile strength, enhanced stain at break	Lopera-Valle et al. (2019)
(P(3HB-co- 4HB)) poly (3-hydroxybutyrate-co-4-hydroxybutyrate) PHBV composite	3-Hydroxyvalerate (3HV) or 4-hydroxybutyrate (4HB) into poly(3-hydroxybutyrateco-3-hydroxyvalerate) ( ) or		Improved handling profile for film formation, melt-spinning, and cold-drawing	Cherpinski et al. (2017)
Poly(3-hydroxybutyrate), P (3HB) and medium-chain length PHA (mcl-PHA) composite	48 wt.% P(3HB) and 52 wt.% mcl-PHA	--	High tensile strength, Young's modulus	Rebocho et al. (2020)
PPC PHBV (β-hydroxybutyrate-co-3-hydroxyvalerate) composite	40 wt.% Poly(propylene carbonate) (PPC)	Propylene oxide and carbon dioxide	Improved impact strength, thermal stability, and stain at failure	Tao et al. (2009)

It causes a reduction in the melting temperature of PLA/TiO$_2$ composite as compared to the neat PLA. This change in the property was attributed to the symmetrical disturbance in the PLA chain structure caused by TiO$_2$ particles, this leads to increasing PLA chain distance. Owing to this, the glass transition temperature and the crystallization temperature exhibited by neat PLA in curves of differential scanning calorimetry are not observed in the curves PLA/TiO$_2$ composite. The introduction of ZnO and TiO$_2$ nanoparticles affects the thermal stability of PLA (Wang et al., 2019). The addition of these nanoparticles into the PLA matrix effectively reduces the PLA activation energy required for the pyrolysis process and increases the degradation rate constant. The addition of H$_2$O$_2$ and TiO$_2$ into the polymer matrix decreases the maximum and the onset of the thermal decomposition temperature. PLA composites formulated by the addition of chicken feather fiber showed increased thermal stability as compared to pure PLA. The thermomechanical analysis revealed that chicken feather fiber at a concentration of 5 wt.% in the composite exhibited the best thermal stability (Cheng et al., 2009).

Thermal properties of PLA are modified by addition of thymol and kenaf fiber. The addition of kenaf at 10% (wt./wt.) led to a slight change in the glass transition temperature, whereas the addition of kenaf at a concentration of 40% (wt./wt.) in PLA matrix exhibits no measurable change in the glass transition temperature of the composite (Tawakkal, Cran, & Bigger, 2014). Kenaf at a concentration of 40% (wt./wt.) poses reduction in cold crystallization temperature (101.7°C) as compared to pure PLA, owing to its nucleating-agent-like properties. The addition of thymol at a concentration of 10% (wt./wt.) to PLA/kenaf composites further decreases cold crystallization temperature and glass transition temperature. The reason for the reduction in glass transition temperature is attributed to the plasticizing property of thymol. The glass transition temperature of pure PLA and PLA/wood flour 10% (wt./wt.) reduces by the addition of triclosan at 1.5% (wt./wt.) (Prapruddivongs & Sombatsompop, 2012). With the increase in fiber loading concentration, the melting peaks of PLA composite shifts to a lower temperature, and a secondary small melting peak appears (Tawakkal et al., 2014). The appearance of a second minor peak is due to the melting of transcrystalline zone or difference in the PLA crystalline phase formed in the presence of fibers (Prapruddivongs & Sombatsompop, 2012). The biocomposite formed by reinforcing PLA with short coir fibers shows affected thermal property (Sun, Zhang, Liang, Xiao, & Lin, 2017). Adhesion between PLA matrix and fibers is improved after the treatment of the short coir fibers with sodium hydroxide and hydrogen peroxide solution. The addition of treated coir fiber led to a reduction in thermal stability. Reinforcement of chemically modified corn fiber with PLA yields PLA/corn fiber composite with improved thermal properties (Luo, Zhang, Xiong, & Wan, 2016). The glass transition temperature for PLA/untreated corn fiber composite and unreinforced PLA is 61.18°C and 61.7°C, respectively, thus indicating that untreated corn fiber does not exhibit a significant effect on the glass transition temperature. Interestingly, the reinforcement with the treated fiber enhances the glass transition temperature by up to 28°C in comparison to unreinforced PLA and untreated corn fiber/PLA composite. The surface-treated fiber exhibits hindrance effect owing to which temperature of the composite increases (Idicula, Malhotra, Joseph, & Thomas, 2005). Short *Spartium junceum* L. fibers amalgamate in random orientation with PLA molecules to produce PLA biocomposite (Kovacevic, Bischof, & Fan, 2015). The fire resistance of the composite is increased by treating the fibers with montmorillonite nanoclay. The biocomposite thus formed exhibits enhanced thermal properties. These montmorillonite nanoclay-treated fiber composites have improved thermal stability at extreme of temperature (300°C) (Kovacevic et al., 2015).

15.4.2 Polyhydroxyalkanoate and its biocomposites

The thermomechanical properties of PHA mainly arise because of the involvement of monomers and their structural arrangement including the distance between the ester linkage, length of groups extending from the polymer backbone, and nature of these groups. Thermal properties include glass transition temperature, flexibility, toughness of the polymer, and heat enthalpy (Roy & Visakh, 2014). Attention has been inclined toward the production of low-cost, bio-based composites by improving the thermal properties of PHAs. Polymeric composites extend major application to the environment, where the parent polymer fails (Gumel & Annuar, 2014). To obtain new property in PHA, a number of multiphase material are mixed with macrofillers, plasticizers, or other polymers. Biocomposites are generally obtained by the addition of fillers in a biopolymer matrix (Roy & Visakh, 2014). Rice husk is an effective wetting and uniform dispersing substance in both the matrix of composites, containing acrylic acid-grafted poly(hydroxyalkanoate) (PHA-g-AA) and PHA implemented with rice husk (PHA/RH) (Wu, 2014). Thermal properties of PHA-g-AA are superior as compared to PHA/RH. The melting temperature, glass-transition temperature, and heat of fusion as evaluated by differential scanning calorimetry show a decreased value with the increasing content of rice husk. The incorporation of rice husk in both composites enables the polymer to open and expand thereby causing a reduction in melting temperature. On the other hand, the value of glass transition temperature increases with increasing content of rice husk.

Polyhydroxybutyrate production by using crude glycerol and improved thermal property of extracted PHB was performed by Sindhu et al. (2011). PHB used for the preparation of polymer blends such as PHB-PLA, PHB-thermoplastic starch (TS), PHB-PLA-TS, and PHB-polyethyl glycol-TS (PHB-PEG-TS). The melting temperature of standard PHB, extracted PHB, and PHB blended with thermoplastic starch is nearly the same whereas it is slightly higher for PHB blended with PLA (Sindhu et al., 2011).

Sridhar, Lee, Chun, and Park (2013) successfully prepared graphene-reinforced poly (3-hydroxybutyrate-co-4-hydroxybutyrate) (P3HB4HB) nanocomposites by solution casting method. The endothermic bimodal melting peak suggests the formation of two different crystalline phases with different sizes and thicknessess. The incorporation of graphene increases the crystallinity of nanocomposite and the increasing graphene concentration reduces the melting temperature of the composite. Two melting peaks of heterogenous and homogenous nucleation of PHBV occurred due to the chain aggregation in which the homogenous nucleation exhibited a higher melting temperature.

In nanocomposites, an increase in glass transition temperature is observed with the increase in graphene concentration. Lu et al. (2014) quantified the effect on thermal property by adding distillers dried grains with solubles (DDGS) to the PHA matrix (Lu et al., 2014). Differential scanning calorimetry parameters revealed the differences in the thermal properties of PHA/DDGS blends in different ratios. The inclusion of DDGS to PHA decreases the melting enthalpy. This behavior is attributed to the reduction in crystallinity or decrease in the melt viscosity by DDGS. Similarly, the crystallization enthalpy exhibits a reduction with the increasing concentration of DDGS in prepared composites. The rheological properties of composites are measured by rheometer. Viscoelastic property of the blends is affected by the weight percentage ratio of DDGS. With an increase in the ratio of DDGS to PHA, the viscosity increases at low angular frequency. PHA with a 30% ratio of DDGS shows seven times higher viscosity as compared to PHA containing 10% DDGS. A bio-based composite, microencapsulated phase change materials (MicroPCMs), and P3HB4HB were fabricated in a study performed by Chen, Chen, Ouyang, Zuo, and Ye (2014). The addition of MicroPCMs has improved the thermal stability of the composite. Thermograms of P3HB4HB and the composites at cooling and heating phase had various contents of MicroPCMs.

No formation of endo- and exothermic peaks during the phase transition shows that thermal effect of the composite is associated to the MicroPCMs only. The integration of MicroPCMs to the composite led to a decrease in melting and freezing temperature. The increase in the content of MicroPCMs in composite causes elevation in the heat-storage capability of the composite. After the incorporation of 40 wt.% of the MicroPCM in the composites, melting enthalpy arrives at 23.28 J g^{-1}, compared to the MicroPCMs with 10 of wt.%. The PHA composites synthesized with chestnut shell fiber (PHA/CSF) and glycidyl methacrylate-grafted PHA (PHA-g-GMA) with CSF (PHA-g-GMA/CSF) exhibit altered thermal properties such as heat of fusion, glass transition temperature, and melting point temperature (Wu & Liao, 2014). Thermal properties of the PHA/CSF and PHA-g-GMA/CSF composites change with the different ratio of CSF used. The glass transition temperature and melting point increases with increase in CSF content in PHA/CSF and PHA-g-GMA/CSF composites. The condensation reaction occurring between epoxidized hydroxyl group (eOH) of CSF and glycidylmethacrylate group of GMA in PHA-g-GMA/CSF composite enhances the value for glass transition temperature. The heat enthalpy in PHA/CSF increases in significantly indicating low crystallinity. Advancement in the development of nanocomposites by the incorporation of nanoscaled fillers to the polymer matrix is a leading technology.

Improvement in properties can be obtained by the addition of clay to polymers (Bordes, Pollet, & Avérous, 2009). A study performed by D'Amico, Manfredi, and Cyras (2012) aimed to improve the thermal property of PHB by the addition of two different types of clay Cloisite Na$^+$ (CNa$^+$) and Cloisite15A (C15A) (D'Amico et al., 2012). The addition of this clay to the PHB enhances the thermal stability by acting as a heat barrier. Despite this, in PHB, the percentage of crystallinity decreases after the addition of clay. However, clay exhibits no influence on the melting temperature of the nanocomposite. The glass transition temperature rises with an increase in the interlayer space between clay. The possible reason could be the interference of motion in the molecular string of PHB caused by clay.

Microfibrils from raw bamboo is extracted and employed to prepare a novel PHA-based biocomposite with different loadings of microfibrils (Jain, Kumar, & Jindal, 1992). The thermal property of the composite is improved as compared to neat PHB. Differential scanning calorimetry scan of PHB composite suggests the thermal phase variation in composites loaded with microfibrils. The melting temperature for all the sample is approximately same including PHB as standard. The temperature of crystallinity for pure PHB was 59.27°C whereas, the value decreased by the addition of 5% of microfibrils (Krishnaprasad et al., 2009).

15.5 Functional properties of biopolymers and biocomposites

15.5.1 Tensile strength of biopolymer

The two major obstacles in the commercial utilization of biopolymers such as PHA and PLA are their low tensile strength ($<$10 MPa) and elongation at break ($\sim$5%) properties. These properties are strongly warranted for replacing plastic-like polymer, so modification of biopolymer by blending with biogenic copolymer as additives is the best solution to the existing situation (Lopera-Valle et al., 2019). Reinforcement of 5 wt.% CNCs from sugarcane waste, 10 wt.% epoxidized canola oil (eCO) as a plasticizing agent, and zinc acetate as blending agent into 25 wt.% PLA in microextruder results in the composite formulation. Molding of composite into a 2-mm sheet using a hot press reveals that incorporation of CNCs in lesser volume improves the tensile strength (from 47 to 56 MPa), and Young's modulus of the composite (from 1200 to 1500 MPa). The addition of eCO enhances the strain at break property (Lopera-Valle et al., 2019).

Escherichia coli LSBJ is a living biocatalyst capable of improving the strength of copolymer by cross-linking and controlling the composition of PHA repeating units with terminal alkenes in the side chain, along with incorporating a specific ratio of unsaturated repeating units into PHA polymer. Incorporation of polar functional groups by cross-linking improves the tensile strength, Young's modulus, and hydrophilic characteristics of the copolymer (Levine, Heberlig, & Nomura, 2016). A natural blend of PHB, and medium-chain length

PHA (mcl-PHA), was synthesized by a coculture biomass of *Pseudomonas citronellolis* NRRL B-2504 and *Cupriavidus necator* DSM 428 by feeding upon apple pulp waste. This natural blend comprises of 48 wt.% PHB and 52 wt.% mcl-PHA, and a molecular weight of 4.3×10^5 Da with a polydispersity index of 2.2, the tensile strength of the film is around 1.47 ± 0.07 MPa, and Young's modulus of 5.42 ± 1.02 MPa (Rebocho et al., 2020).

In a low-cost desktop 3D printer, commercial PLA (PLA 3D850), PLA-graphene, and graphene nanoplatelet PLA is used to concoct different samples of PLA graphene-based nanocomposite material by the fused filament fabrication technique. PLA graphene-based nanocomposite sample shows significant improvement in their properties such as tensile strength, flexural stress, surface texture, and upright orientation while flat printing. With increasing concentration of graphene nanoplatelet, the impact strength of the PLA-graphene-based nanocomposite reduces up to 1.2–1.3 times (Caminero et al., 2019). PLA membranes of different pore sizes (large pore—479 μm, small pore—273 μm, and no pore) are 3D printed using 3D membrane printer, and the conventional solvent casting method. The comparative study reflects that 3D-printed membranes are superior in mechanical properties and does not impose effects on cell growth. Therefore the 3D-printing method is accepted for the fabrication of customized barrier property bearing membranes (Zhang et al., 2019).

Zhao et al. (2017) show that incorporation of 0.4% ZnO to PHA enhances the tensile strength, elongation to break, and elastic module by 55%, 41.2%, and 62.2%, respectively as compared to neat PHA, but it lacks thermal stability (Zhao et al., 2017). A blend of thermodynamically immiscible PLA and poly(e-caprolactone) 7 wt.%, with compatibilizer (3% tributyl citrate) is prepared. The increase in the concentration of PLA from 93 to 95 wt.% improves the tensile strength significantly (42.9 ± 3.5 MPa) to 54.1 ± 3.4 MPa and the values of elongation at break of the material have reached from $10.3\% \pm 2.7\%$ to $8.8\% \pm 1.8\%$ (Jeong et al., 2018).

Composite material fabricated by melt-spinning of PHA copolyester PHBV with PLA molecule hold very similar properties to that of human bone, such as high tensile strength, elasticity modulus, and tensile strain, therefore it offers promising candidature as an implant material (Koller, 2018). Polyurethanes films are synthesized by the amalgamation of PEG, L-lysine di-isocyanate, and PHA as a stretchable, biodegradable, biocompatible, and nontoxic film. Tensile strength, Young's modulus, and elongation at break properties of polyurethanes film falls in the range of 1.01–9.49 MPa, 3.07–25.61 MPa, and 102%–998%, respectively (Wang, Xie, Xiao, Chen, & Wang, 2019).

PLA-based biocomposites such as PLA–starch, PLA–chitin, and PLA–chitin–starch with increasing concentration of PLA, namely, 92%, 96%, and 98% (wt./wt.), respectively, have been formulated. In PLA–starch composite, the addition of alcohol as a plasticizer improves miscibility, but adversely affects the tensile strength (53.5 MPa), yield strength (60 MPa), and Young's elastic modulus (3500 MPa). Whereas, in the PLA-chitin composite the tensile strength decreases with reducing chitin concentration. This shows that the dispersion property of chitin in PLA is superlative to starch molecules (Olaiya et al., 2019).

Low-cost PLA/wood flour composite is prepared with the composition of 80 wt.% PLA, 20 wt.% wood flour, and PMMA. The composite shows increased interface compatibility, the tensile strength and bending strength by 4.60% and 26.54%, respectively (Wan & Zhang, 2018). The incorporation of natural polymer lignin (7%) as filler in PLA biopolymer contributes toward the reduction of the tensile strength and water sorption capacity of the PLA matrix and reaches below the values of neat PLA molecule. Furthermore, with a gradual increase in lignin concentration from 7 to 15 wt.%, the tensile strength of the matrix increases along with a significant decline in the water sorption capacity. It shows that lignin content holds control over PLA-lignin biocomposite properties. The tensile strength increment is an attribute of the plasticizing nature of lignin augmented toward the mechanical properties of the biocomposite (Spiridon & Tanase, 2018).

15.5.2 Crystallinity of biocomposites

Crystallinity of biopolymer material like PLA and PHA is a crucial property that decides versatility of its application. Crystallization is a fundamental property of polymers that influences the thermomechanical properties, fine structure, processing technique, and functional properties of the polymer. The degree of crystallinity is the ratio of amorphous region to crystalline region in a polymeric material. PHAs are semicrystalline, its degree of crystallization is adjudged by the structure of its carbon backbone and side chains (Gopi, Kontopoulou, Ramsay, & Ramsay, 2018). The presence of both the amorphous and crystalline regions in PHA has led to the increase in crystallization temperature (Arrieta, Samper, Aldas, & López, 2017; Zhila & Shishatskaya, 2018).

The main chain of medium-chain length PHAs (mcl-PHAs) adopts a similar helical conformation, like short-chain length PHAs, whereas the side chains are responsible for zig-zag conformation. Crystallization of mcl-PHA with C5-C7 carbon side chains creates a 21-helix in an orthorhombic lattice with two molecules per unit cell. The number of carbons in the side chain greatly influences the thermal properties, like crystallization and melting point, glass transition temperature, and heat of fusion of mcl-PHA. Long side chain holds a crystallizing tendency, thus controls thermal properties of mcl-PHAs (Gopi et al., 2018).

The slow crystallization rate of PLA is one of the shortcomings limiting its practical applications. Therefore during fabrication of biocomposites, the enrichment of dominant features like crystallinity is of prime importance. Fabrication techniques like mechanical ball milling influence the crystallinity of biocomposite material formed by mixing of pulp cellulose fibers and PLA. The crystallinity of cellulose fibers is 78.5%, while after 10 min treatment of ball milling the crystallinity of the composite material reduces to 47.3%.

By increasing the treatment time to 30 min the cellulose fibers became amorphous and the crystallinity declines progressively namely, 46.9% (10 min), 21.1% (30 min), and 14.2% (60 min) (Qiang, Wang, & Wolcott, 2018).

PHB bears a melting temperature similar to PLA molecule but has high crystallinity; hence PHB represents strong candidature as blending material with PLA. Melt blending of PLA with PHB matrix significantly enhances the PLA crystallinity along with the regulation of its physical properties. Solvent casting technique and compression molding at 200°C show that melt-compressed samples of PLA carrying low molecular weight PHB (9400 g mol^{-1}) is 50 wt.% miscible in the melt within the PHB. This shows that PHB addition facilitates the crystallization of PLA. Thermally degraded lignomers of PHB namely, hydroxybutyrate (OHB) (molecular weight between 4000 and 83,000 g mol^{-1}) when combined with low molecular weight PLAs (110,000 and 253,000 g mol^{-1}) increase the miscibility of both of the polymers. High-molecular weight OHB (4000 g mol^{-1}) enhances the crystallization of PLA by forming small spherulitic crystals, which further act as nucleating agent for PLA (Arrieta et al., 2017; Sedničková et al., 2018).

An investigation conducted on pure PLA, PLA with triacetine (TAC), a plasticizer, a mixture PLA/PHB, and TAC shows that the amorphous region degrades substantially faster as compared to the crystalline part (Sedničková et al., 2018). XRD measurements of 3D-printed PLA confirm that there are α and δ crystalline forms of PLA which occur due to the orientation of the filament deposition and their formation. At lower temperatures, the polymer fully crystallize in α form and partially crystallize in δ form after 3D-printing heat treatment, because the cold crystallization temperature of PLA is approximately 100°C (Liao et al., 2019). The cold crystallization of PLA takes place upon melting of the semantic structure formed due to main chain and side chain (Gopi et al., 2018).

Homopolymeric poly-3-hydroxydecanoate (PHD) composite was prepared by annealing mcl-PHA samples with different percentage of hydroxydecanoate (HD) at 50°C. It was analyzed that the melting point and crystallinity increase with increasing concentration of HD. In the first crystallization cycle when the concentration of HD increased from 43 to 97 mol%, the crystallinity temperature also increases from 44°C to 72.2°C. While, in the second cycle of crystallization, the temperature reduces between 40°C and 60°C because of the annealing pattern. The increased crystallinity stipulates toward higher tensile strength and Young's modulus of HD materials creating crystalline cross-linking (Gao, Leng, Zhang, Wei, & Li, 2020). Wide-angle X-ray diffraction patterns of PHD molecules show that crystallization of side chain occurs faster as compared to the main chain (Gopi et al., 2018).

15.5.3 Biopolymer film formation

PHA and PLA reflect excellent film-forming properties, capable to replace conventional plastic-based polymers. Rigorous research for biodegradable polymer finds PHA and

PLA as one of the most suitable renewable materials carrying moisture resistance, oxygen barrier, and thermoplastic properties (Cheng, Khan, Khan, & Rabnawaz, 2018; Cherpinski, Torres-Giner, Cabedo, & Lagaron, 2017). There are various methods employed for the fabrication of biopolymer-based films, including the electrospinning technique, ring-opening polymerization, and solvent volatilization method (Cherpinski et al., 2017; Cheng et al., 2018; Li et al., 2017).

PHA films are made by electrospinning technique having unique barrier layer design, adhesive interlayers, coatings for casting fiber, and bioplastic-based food-packaging material (Cherpinski et al., 2017). PHB is the most capacitive member of PHA family. It is suitable for a series of biomedical applications such as drug-delivery systems, implant generation, stents manufacturing, and surgical plaster material. PHB lacks degree of crystallinity due to its low nucleation density this aids in copolymerization of 3-hydroxybutyrate with 3-hydroxyvalerate or 4-hydroxybutyrate into PHBV or P(3HB-co-4HB), thereby improves the handling profile of the PHB for film preparation. Other techniques like melt-spinning and cold-drawing processing are being used for improving the mechanical properties of PHB films (Cherpinski et al., 2017).

Solvent volatilization method is employed for the formation of nanoblend film of polylactide (PLA)/nano-TiO_2 and PLA/nano-TiO_2/nano-Ag. The nanoblend films exhibits low water vapor permeability and poor transparency as compared to pure PLA film. Increasing concentration of nanoparticles in the PLA nanoblend film increases the elongation at break property while transparency, tensile strength, and elastic modulus are decreased (Li et al., 2017).

A film prepared from lactide of PLA molecules by ring-opening polymerization has shown a moderate water vapor permeation and oxygen permeability, that is, 168 g-mil/(m^2-day-atm) and 1528 cc-mil/(m^2-day-atm), respectively. As a packaging material, these films can extend the shelf-life of perishable food items after improving its barrier properties (Cheng et al., 2018). The morphology of PHA-based films extruded from PHA and organomodified montmorillonite or Cloisite 30B contents is highly dependent on PHBV and P3HB4HB matrix, the nanoclay content, the matrix/nanoclay interfacial regions, and the polymer crystalline phase fraction. Nanoclay increases the water solubility nanocomposite films by sorption and tortuosity effect phenomenon (Follain, Crétois, Lebrun, & Marais, 2016).

Fabricating a porous material from amalgamation of poly(butylene succinate) (PBS) and PLA in a concentration of 40 wt.% to 50 wt.%, respectively, led to significant increase in the mean pore diameter from 6.91 to 120 μm, the porosity from 81.52% to 96.90%, and the contact angle decreases from 81.08 degrees to 46.56 degrees. A study on the rate of degradation of this PBS/PLA blend polymer shows that the tensile strength of 50:50 PBS/PLA is 8.72 MPa after 16 d degradation by proteinase K, and it improves with the rising concentration of PLA. PBS/PLA also showed good corrosion resistance and did not cause an inflammatory response when used as a subcutaneous implant and it further reduces the elongation at break property (Shi, Ma, Su, & Wang, 2020). PET grafts used in prosthetic valve endocarditis or prosthetic vascular graft infection are highly susceptible to bacterial infection and biofilm formation. Therefore coating of the crimped grafts with a film of di-block copolymer of polyhexylene adipate-b-methoxy polyethylene oxide by selective cleavage of the ester bond is highly advantageous to maintain the essential graft's folding ability and producing microbially inert graft with local antibacterial effect and biocompatibility (Al Meslmani, Mahmoud, Sommer, Lohoff, & Bakowsky, 2015). PHA-grafted functionalized cellulose films with ^{177}Lu and other β-emitting radioisotopes such as ^{90}Y and ^{166}Ho, as skin patches are fabricated under the optimized conditions for contact brachytherapy and are useful in tumor responses (Saxena et al., 2017). To determine the miscibility of PHA and PLA in a composite matrix-mercaptopropionic acid and 2-aminoethanethiol are incorporated into the vinyl-bearing PHA membrane which is functionalized by UV-initiated thiol-ene click reaction and chemical modification using fluorescein or GRGDS (a fibronectin active fragment). The PHA derivative thus produced emanates fluorescence under UV irradiation (Tajima et al., 2016).

15.6 Conclusions and perspectives

Biocomposites are biogenic alternative to petrochemical-based plastic material offering freedom from a channel of environmental concerns and economic issues. Natural fiber-reinforced composite material is a valorization of agrowaste to derive composite formulation aiding in strengthening and modifying the thermal and functional properties of biopolymer. In biocomposite, natural fibers are embedded in a continuous polymer matrix to enhance structural properties. Functional properties such as film formation, crystallinity, tensile strength, and flexural properties of PLA and PHA polymer are modified to exponential level using different concentration of natural filler material. The enhancement in the functional and thermal behavior of these biocomposites opens the doorway toward their increased application in the medical and industrial sectors.

Acknowledgment

Vivek Kumar Gaur acknowledges Council of Scientific and Industrial Research (CSIR), New Delhi for Senior Research Fellowship.

References

Agustin-Salazar, S., Cerruti, P., Medina-Juárez, L. Á., Scarinzi, G., Malinconico, M., Soto-Valdez, H., & Gamez-Meza, N. (2018). Lignin and holocellulose from pecan nutshell as reinforcing fillers in poly (lactic acid) biocomposites. *International Journal of Biological Macromolecules*, *115*, 727−736.

Al Meslmani, B. M., Mahmoud, G. F., Sommer, F. O., Lohoff, M. D., & Bakowsky, U. (2015). Multifunctional network-structured film coating for woven and knitted polyethylene terephthalate against cardiovascular graft-associated infections. *International Journal of Pharmaceutics*, *485*(1−2), 270−276.

Anderson, S., Zhang, J., & Wolcott, M. P. (2013). Effect of interfacial modifiers on mechanical and physical properties of the PHB composite with high wood flour content. *Journal of Polymers and the Environment*, *21*(3), 631−639.

Antoniac, I., Popescu, D., Zapciu, A., Antoniac, A., Miculescu, F., & Moldovan, H. (2019). Magnesium filled polylactic acid (PLA) material for filament based 3D printing. *Materials*, *12*(5), 719.

Aranberri, I., Montes, S., Azcune, I., Rekondo, A., & Grande, H. J. (2017). Fully biodegradable biocomposites with high chicken feather content. *Polymers*, *9*(11), 593.

Arrieta, M. P., Samper, M. D., Aldas, M., & López, J. (2017). On the use of PLA-PHB blends for sustainable food packaging applications. *Materials*, *10*(9), 1008.

Ashok, B., Naresh, S., Reddy, K. O., Madhukar, K., Cai, J., Zhang, L., & Rajulu, A. V. (2014). Tensile and thermal properties of poly (lactic acid)/eggshell powder composite films. *International Journal of Polymer Analysis and Characterization*, *19*(3), 245−255.

Awal, A., Rana, M., & Sain, M. (2015). Thermorheological and mechanical properties of cellulose reinforced PLA biocomposites. *Mechanics of Materials*, *80*, 87−95.

Beigbeder, J., Soccalingame, L., Perrin, D., Bénézet, J. C., & Bergeret, A. (2019). How to manage biocomposites wastes end of life? A life cycle assessment approach (LCA) focused on polypropylene (PP)/wood flour and polylactic acid (PLA)/flax fibres biocomposites. *Waste Management*, *83*, 184−193.

Bharti, S. N., & Swetha, G. (2016). Need for bioplastics and role of biopolymer PHB: A short review. *Journal of Petroleum and Environmental Biotechnology*, *7*(272), 2.

Bordes, P., Pollet, E., & Avérous, L. (2009). *Potential use of polyhydroxyalkanoate (PHA) for biocomposite development*. Nano- and biocomposites (pp. 193−225). Boca Raton, FL, USA: CRC Press.

Boufarguine, M., Guinault, A., Miquelard-Garnier, G., & Sollogoub, C. (2013). PLA/PHBV films with improved mechanical and gas barrier properties. *Macromolecular Materials and Engineering*, *298*(10), 1065−1073.

Caminero, M. Á., Chacón, J. M., García-Plaza, E., Núñez, P. J., Reverte, J. M., & Becar, J. P. (2019). Additive manufacturing of PLA-based composites using fused filament fabrication: Effect of graphene nanoplatelet reinforcement on mechanical properties, dimensional accuracy and texture. *Polymers*, *11*(5), 799.

Chen, D., Chen, Y., Ouyang, X., Zuo, J., & Ye, X. (2014). Influence of MicroPCMs on thermal and dynamic mechanical properties of a biodegradable P3HB4HB composite. *Composites Part B: Engineering*, *56*, 245−248.

Cheng, S., Khan, B., Khan, F., & Rabnawaz, M. (2018). Synthesis of high molecular weight polyester using in situ drying method and assessment of water vapor and oxygen barrier properties. *Polymers*, *10*, 1113.

Cheng, S., Lau, K. T., Liu, T., Zhao, Y., Lam, P. M., & Yin, Y. (2009). Mechanical and thermal properties of chicken feather fiber/PLA green composites. *Composites Part B: Engineering*, *40*(7), 650−654.

Cherpinski, A., Torres-Giner, S., Cabedo, L., & Lagaron, J. M. (2017). Post-processing optimization of electrospun submicron poly (3-hydroxybutyrate) fibers to obtain continuous films of interest in food packaging applications. *Food Additives & Contaminants: Part A*, *34*(10), 1817−1830.

Chiulan, I., Mihaela Panaitescu, D., Nicoleta Frone, A., Teodorescu, M., Andi Nicolae, C., Căşărică, A., ... Sălăgeanu, A. (2016). Biocompatible polyhydroxyalkanoates/bacterial cellulose composites: Preparation, characterization, and in vitro evaluation. *Journal of Biomedical Materials Research. Part A*, *104*(10), 2576−2584.

Chung, Y. L., Olsson, J. V., Li, R. J., Frank, C. W., Waymouth, R. M., Billington, S. L., & Sattely, E. S. (2013). A renewable lignin−lactide copolymer and application in biobased composites. *ACS Sustainable Chemistry & Engineering*, *1*(10), 1231−1238.

Cinelli, P., Seggiani, M., Mallegni, N., Gigante, V., & Lazzeri, A. (2019). Processability and degradability of PHA-based composites in terrestrial environments. *International Journal of Molecular Sciences*, *20*(2), 284.

D'Amico, D. A., Manfredi, L. B., & Cyras, V. P. (2012). Relationship between thermal properties, morphology, and crystallinity of nanocomposites based on polyhydroxybutyrate. *Journal of Applied Polymer Science*, *123*(1), 200−208.

Follain, N., Crétois, R., Lebrun, L., & Marais, S. (2016). Water sorption behaviour of two series of PHA/montmorillonite films and determination of the mean water cluster size. *Physical Chemistry Chemical Physics*, *18*(30), 20345–20356.

Frone, A. N., Berlioz, S., Chailan, J. F., & Panaitescu, D. M. (2013). Morphology and thermal properties of PLA–cellulose nanofibers composites. *Carbohydrate Polymers*, *91*(1), 377–384.

Gao, Y. (2017). *Biorefinery lignin as filler material in polylactic acid composite* (Master of Science thesis). Iowa State University.

Gao, M., Leng, X., Zhang, W., Wei, Z., & Li, Y. (2020). A biobased aliphatic polyester derived from 10-hydroxydecanoic acid: Molecular weight dependence of physical properties. *Polymer Testing*, *82*, 106–295.

García-Quiles, L., Fernández Cuello, Á., & Castell, P. (2019). Sustainable materials with enhanced mechanical properties based on industrial polyhydroxyalkanoates reinforced with organomodifiedsepiolite and montmorillonite. *Polymers*, *11*(4), 696.

González-López, M. E., Robledo-Ortíz, J. R., Manríquez-González, R., Silva-Guzmán, J. A., & Pérez-Fonseca, A. A. (2018). Polylactic acid functionalization with maleic anhydride and its use as coupling agent in natural fiberbiocomposites: A review. *Composite Interfaces*, *25*(5–7), 515–538.

Gopi, S., Kontopoulou, M., Ramsay, B. A., & Ramsay, J. A. (2018). Manipulating the structure of medium-chain-length polyhydroxyalkanoate (MCL-PHA) to enhance thermal properties and crystallization kinetics. *International Journal of Biological Macromolecules*, *119*, 1248–1255.

Gordobil, O., Delucis, R., Egüés, I., & Labidi, J. (2015). Kraft lignin as filler in PLA to improve ductility and thermal properties. *Industrial Crops and Products*, *72*, 46–53.

Gumel, A. M., & Annuar, M. S. M. (2014). *Nanocomposites of polyhydroxyalkanoates (PHAs). Polyhydroxyalkanoate (PHA) based blends, composites and nanocomposites* (pp. 98–118). Royal Society of Chemistry.

Gunti, R., Ratna Prasad, A. V., & Gupta, A. V. S. S. K. S. (2018). Mechanical and degradation properties of natural fiber-reinforced PLA composites: Jute, sisal, and elephant grass. *Polymer Composites*, *39*(4), 1125–1136.

Gupta, A., & Katiyar, V. (2017). Cellulose functionalized high molecular weight stereocomplex polylactic acid biocomposite films with improved gas barrier, thermomechanical properties. *ACS Sustainable Chemistry & Engineering*, *5*(8), 6835–6844.

Gurunathan, T., Mohanty, S., & Nayak, S. K. (2015). A review of the recent developments in biocomposites based on natural fibres and their application perspectives. *Composites Part A: Applied Science and Manufacturing*, *77*, 1–25.

Hamad, K., Kaseem, M., Ayyoob, M., Joo, J., & Deri, F. (2018). Polylactic acid blends: The future of green, light and tough. *Progress in Polymer Science*, *85*, 83–127.

Huda, M. S., Drzal, L. T., Misra, M., & Mohanty, A. K. (2006). Wood-fiber-reinforced poly (lactic acid) composites: Evaluation of the physicomechanical and morphological properties. *Journal of Applied Polymer Science*, *102*(5), 4856–4869.

Huda, M. S., Drzal, L. T., Mohanty, A. K., & Misra, M. (2006). Chopped glass and recycled newspaper as reinforcement fibers in injection molded poly (lactic acid)(PLA) composites: A comparative study. *Composites Science and Technology*, *66*(11–12), 1813–1824.

Idicula, M., Malhotra, S. K., Joseph, K., & Thomas, S. (2005). Dynamic mechanical analysis of randomly oriented intimately mixed short banana/sisal hybrid fibre reinforced polyester composites. *Composites Science and Technology*, *65*(7–8), 1077–1087.

Jain, S., Kumar, R., & Jindal, U. C. (1992). Mechanical behaviour of bamboo and bamboo composite. *Journal of Materials Science*, *27*(17), 4598–4604.

Jeong, H., Rho, J., Shin, J. Y., Lee, D. Y., Hwang, T., & Kim, K. J. (2018). Mechanical properties and cytotoxicity of PLA/PCL films. *Biomedical Engineering Letters*, *8*(3), 267–272.

Ju, Y., Liao, F., Dai, X., Cao, Y., Li, J., & Wang, X. (2016). Flame-retarded biocomposites of poly (lactic acid), distiller's dried grains with solubles and resorcinol di (phenyl phosphate). *Composites Part A: Applied Science and Manufacturing*, *81*, 52–60.

Kale, R. D., Gorade, V. G., Madye, N., Chaudhary, B., Bangde, P. S., & Dandekar, P. P. (2018). Preparation and characterization of biocomposite packaging film from poly (lactic acid) and acylated microcrystalline cellulose using rice bran oil. *International Journal of Biological Macromolecules, 118*, 1090−1102.

Kaseem, M., Hamad, K., & Ur Rehman, Z. (2019). Review of recent advances in polylactic acid/TiO_2 composites. *Materials, 12*(22), 3659.

Kian, L. K., Saba, N., Jawaid, M., & Sultan, M. T. H. (2019). A review on processing techniques of bast fibers nanocellulose and its polylactic acid (PLA) nanocomposites. *International Journal of Biological Macromolecules, 121*, 1314−1328.

Koller, M. (2018). Biodegradable and biocompatible polyhydroxy-alkanoates (PHA): Auspicious microbial macromolecules for pharmaceutical and therapeutic applications. *Molecules, 23*(2), 362.

Kong, J., Li, Y., Bai, Y., Li, Z., Cao, Z., Yu, Y., ... Dong, L. (2018). High-performance biodegradable polylactide composites fabricated using a novel plasticizer and functionalized eggshell powder. *International Journal of Biological Macromolecules, 112*, 46−53.

Kovacevic, Z., Bischof, S., & Fan, M. (2015). The influence of *Spartium junceum* L. fibres modified with montmorrilonitenanoclay on the thermal properties of PLA biocomposites. *Composites Part B: Engineering, 78*, 122−130.

Kovalcik, A., Machovsky, M., Kozakova, Z., & Koller, M. (2015). Designing packaging materials with viscoelastic and gas barrier properties by optimized processing of poly (3-hydroxybutyrate-co-3-hydroxyvalerate) with lignin. *Reactive and Functional Polymers, 94*, 25−34.

Krishnaprasad, R., Veena, N. R., Maria, H. J., Rajan, R., Skrifvars, M., & Joseph, K. (2009). Mechanical and thermal properties of bamboo microfibril reinforced polyhydroxybutyratebiocomposites. *Journal of Polymers and the Environment, 17*(2), 109.

Levine, A. C., Heberlig, G. W., & Nomura, C. T. (2016). Use of thiol-ene click chemistry to modify mechanical and thermal properties of polyhydroxyalkanoates (PHAs). *International Journal of Biological Macromolecules, 83*, 358−365.

Li, W., Zhang, C., Chi, H., Li, L., Lan, T., Han, P., ... Qin, Y. (2017). Development of antimicrobial packaging film made from poly (lactic acid) incorporating titanium dioxide and silver nanoparticles. *Molecules (Basel, Switzerland), 22*(7), 1170.

Liao, Y., Liu, C., Coppola, B., Barra, G., Di Maio, L., Incarnato, L., & Lafdi, K. (2019). Effect of porosity and crystallinity on 3D printed PLA properties. *Polymers, 11*(9), 1487.

Liu, S., Wu, G., Chen, X., Zhang, X., Yu, J., Liu, M., ... Wang, P. (2019). Degradation behavior in vitro of carbon nanotubes (CNTs)/poly (lactic acid)(PLA) composite suture. *Polymers, 11*(6), 1015.

Lopera-Valle, A., Caputo, J. V., Leão, R., Sauvageau, D., Luz, S. M., & Elias, A. (2019). Influence of epoxidized canola oil (eCO) and cellulose nanocrystals (CNCs) on the mechanical and thermal properties of polyhydroxybutyrate (PHB)—Poly (lactic acid) (PLA) blends. *Polymers, 11*(6), 933.

Lu, H., Madbouly, S. A., Schrader, J. A., Kessler, M. R., Grewell, D., & Graves, W. R. (2014). Novel bio-based composites of polyhydroxyalkanoate (PHA)/distillers dried grains with solubles (DDGS). *RSC Advances, 4*(75), 39802−39808.

Luo, H., Zhang, C., Xiong, G., & Wan, Y. (2016). Effects of alkali and alkali/silane treatments of corn fibers on mechanical and thermal properties of its composites with polylactic acid. *Polymer Composites, 37*(12), 3499−3507.

Mallick, S., Ahmad, Z., Touati, F., Bhadra, J., Shakoor, R. A., & Al-Thani, N. J. (2018). PLA-TiO_2 nanocomposites: Thermal, morphological, structural, and humidity sensing properties. *Ceramics International, 44*(14), 16507−16513.

Marra, A., Silvestre, C., Duraccio, D., & Cimmino, S. (2016). Polylactic acid/zinc oxide biocomposite films for food packaging application. *International Journal of Biological Macromolecules, 88*, 254−262.

Mngomezulu, M. E., John, M. J., Jacobs, V., & Luyt, A. S. (2014). Review on flammability of biofibres and biocomposites. *Carbohydrate Polymers, 111*, 149−182.

Mohanty, A. K., Misra, M., & Drzal, L. T. (2002). Sustainable biocomposites from renewable resources: opportunities and challenges in the green materials world. *Journal of Polymers and the Environment, 10*(1-2), 19−26.

Mukherjee, T., & Kao, N. (2011). PLA based biopolymer reinforced with natural fibre: A review. *Journal of Polymers and the Environment, 19*(3), 714.

Nagarajan, V., Zhang, K., Misra, M., & Mohanty, A. K. (2015). Overcoming the fundamental challenges in improving the impact strength and crystallinity of PLA biocomposites: Influence of nucleating agent and mold temperature. *ACS Applied Materials & Interfaces, 7*(21), 11203−11214.

Nair, N. R., Sekhar, V. C., Nampoothiri, K. M., & Pandey, A. (2017). *Biodegradation of biopolymers. Current developments in biotechnology and bioengineering* (pp. 739−755). Elsevier.

Oksman, K., & Selin, J. F. (2004). Plastics and composites from polylactic acid. In F. T. Wallenberger, & N. E. Weston (Eds.), *Natural fibers, plastics and composites* (Vol. 1). Boston, MA: Springer.

Olaiya, N. G., Surya, I., Oke, P. K., Rizal, S., Sadiku, E. R., Ray, S. S., ... Abdul Khalil, H. P. S. (2019). Properties and characterization of a PLA−chitin−starch biodegradable polymer composite. *Polymers, 11*(10), 1656.

Park, S. Y., Kim, J. Y., Youn, H. J., & Choi, J. W. (2019). Utilization of lignin fractions in UV resistant lignin-PLA biocomposites via lignin-lactide grafting. *International Journal of Biological Macromolecules, 138*, 1029−1034.

Pattanashetti, N. A., Heggannavar, G. B., & Kariduraganavar, M. Y. (2017). Smart biopolymers and their biomedical applications. *Procedia Manufacturing, 12*, 263−279.

Peelman, N., Ragaert, P., Ragaert, K., De Meulenaer, B., Devlieghere, F., & Cardon, L. (2015). Heat resistance of new biobased polymeric materials, focusing on starch, cellulose, PLA, and PHA. *Journal of Applied Polymer Science, 132*(48).

Prapruddivongs, C., & Sombatsompop, N. (2012). Roles and evidence of wood flour as an antibacterial promoter for triclosan-filled poly (lactic acid). *Composites Part B: Engineering, 43*(7), 2730−2737.

Qiang, T., Wang, J., & Wolcott, M. P. (2018). Facile fabrication of 100% bio-based and degradable ternary cellulose/PHBV/PLA composites. *Materials, 11*(2), 330.

Rebelo, R., Fernandes, M., & Fangueiro, R. (2017). Biopolymers in medical implants: A brief review. *Procedia Engineering, 200*, 236−243.

Rebocho, A. T., Pereira, J. R., Neves, L. A., Alves, V. D., Sevrin, C., Grandfils, C., ... Reis, M. A. (2020). Preparation and characterization of films based on a natural P (3HB)/mcl-PHA blend obtained through the co-culture of *Cupriavidus necator* and *Pseudomonas citronellolis* in apple pulp waste. *Bioengineering, 7*(2), 34.

Revati, R., Majid, M. A., Ridzuan, M. J. M., Normahira, M., Nasir, N. M., & Gibson, A. G. (2017). Mechanical, thermal and morphological characterisation of 3D porous *Pennisetum purpureum*/PLA biocomposites scaffold. *Materials Science and Engineering: C, 75*, 752−759.

Righetti, M. C., Cinelli, P., Mallegni, N., Massa, C. A., Aliotta, L., & Lazzeri, A. (2019). Thermal, mechanical, viscoelastic and morphological properties of poly (lactic acid) based biocomposites with potato pulp powder treated with waxes. *Materials, 12*(6), 990.

Righetti, M. C., Cinelli, P., Mallegni, N., Massa, C. A., Bronco, S., Stäbler, A., & Lazzeri, A. (2019). Thermal, mechanical, and rheological properties of biocomposites made of poly (lactic acid) and potato pulp powder. *International Journal of Molecular Sciences, 20*(3), 675.

Rogina, A. (2014). Electrospinning process: Versatile preparation method for biodegradable and natural polymers and biocomposite systems applied in tissue engineering and drug delivery. *Applied Surface Science, 296*, 221−230.

Roy, I., & Visakh, P. M. (2014). *Polyhydroxyalkanoate (PHA) based blends, composites and nanocomposites* (Vol. 30). Royal Society of Chemistry.

Roy, S. B., Shit, S. C., Sengupta, R. A., & Shukla, P. R. (2014). A review on biocomposites: Fabrication, properties and applications. *International Journal of Innovative Research in Science, Engineering and Technology, 3*(10), 16814−16824.

Saxena, S. K., Kumar, Y., Shaikh, S. H., Pandey, U., Kumar, S. A., & Dash, A. (2017). Preparation of radioactive skin patches using polyhydroxamic acid-grafted cellulose films toward applications in treatment of superficial tumors. *Cancer Biotherapy & Radiopharmaceuticals, 32*(10), 364−370.

Sedničková, M., Pekařová, S., Kucharczyk, P., Bočkaj, J., Janigová, I., Kleinová, A., ... Sedlařík, V. (2018). Changes of physical properties of PLA-based blends during early stage of biodegradation in compost. *International Journal of Biological Macromolecules, 113*, 434–442.

Shi, K., Ma, Q., Su, T., & Wang, Z. (2020). Preparation of porous materials by selective enzymatic degradation: Effect of in vitro degradation and in vivo compatibility. *Scientific Reports, 10*(1), 1–9.

Sindhu, R., Ammu, B., Binod, P., Deepthi, S. K., Ramachandran, K. B., Soccol, C. R., & Pandey, A. (2011). Production and characterization of poly-3-hydroxybutyrate from crude glycerol by *Bacillus sphaericus* NII 0838 and improving its thermal properties by blending with other polymers. *Brazilian Archives of Biology and Technology, 54*(4), 783–794.

Singh, M., Kumar, P., Ray, S., & Kalia, V. C. (2015). Challenges and opportunities for customizing polyhydroxyalkanoates. *Indian Journal of Microbiology, 55*(3), 235–249.

Soroudi, A., & Jakubowicz, I. (2013). Recycling of bioplastics, their blends and biocomposites: A review. *European Polymer Journal, 49*(10), 2839–2858.

Spiridon, I., & Tanase, C. E. (2018). Design, characterization and preliminary biological evaluation of new lignin-PLA biocomposites. *International Journal of Biological Macromolecules, 114*, 855–863.

Sridhar, V., Lee, I., Chun, H. H., & Park, H. (2013). Graphene reinforced biodegradable poly (3-hydroxybutyrate-co-4-hydroxybutyrate) nano-composites. *Express Polymer Letters, 7*(4).

Suarato, G., Bertorelli, R., & Athanassiou, A. (2018). Borrowing from nature: Biopolymers and biocomposites as smart wound care materials. *Frontiers in Bioengineering and Biotechnology, 6*, 137.

Sun, J., Shen, J., Chen, S., Cooper, M. A., Fu, H., Wu, D., & Yang, Z. (2018). Nanofiller reinforced biodegradable PLA/PHA composites: Current status and future trends. *Polymers, 10*(5), 505.

Sun, Z., Zhang, L., Liang, D., Xiao, W., & Lin, J. (2017). Mechanical and thermal properties of PLA biocomposites reinforced by coir fibers. *International Journal of Polymer Science*.

Tajbakhsh, S., & Hajiali, F. (2017). A comprehensive study on the fabrication and properties of biocomposites of poly (lactic acid)/ceramics for bone tissue engineering. *Materials Science and Engineering: C, 70*, 897–912.

Tajima, K., Iwamoto, K., Satoh, Y., Sakai, R., Satoh, T., & Dairi, T. (2016). Advanced functionalization of polyhydroxyalkanoate via the UV-initiated thiol-ene click reaction. *Applied Microbiology and Biotechnology, 100*(10), 4375–4383.

Tanase-Opedal, M., Espinosa, E., Rodríguez, A., & Chinga-Carrasco, G. (2019). Lignin: A biopolymer from forestry biomass for biocomposites and 3D printing. *Materials, 12*(18), 3006.

Tao, J., Song, C., Cao, M., Hu, D., Liu, L., Liu, N., & Wang, S. (2009). Thermal properties and degradability of poly (propylene carbonate)/poly (β-hydroxybutyrate-co-β-hydroxyvalerate)(PPC/PHBV) blends. *Polymer Degradation and Stability, 94*(4), 575–583.

Tawakkal, I. S., Cran, M. J., & Bigger, S. W. (2014). Effect of kenaf fibre loading and thymol concentration on the mechanical and thermal properties of PLA/kenaf/thymol composites. *Industrial Crops and Products, 61*, 74–83.

Ten, E., Jiang, L., Zhang, J., & Wolcott, M. P. (2015). *Mechanical performance of polyhydroxyalkanoate (PHA)-based biocomposites. Biocomposites* (pp. 39–52). Elsevier Inc.

Vandi, L. J., Chan, C. M., Werker, A., Richardson, D., Laycock, B., & Pratt, S. (2018). Wood-PHA composites: Mapping opportunities. *Polymers, 10*(7), 751.

Wan, L., & Zhang, Y. (2018). Jointly modified mechanical properties and accelerated hydrolytic degradation of PLA by interface reinforcement of PLA-WF. *Journal of the Mechanical Behavior of Biomedical Materials, 88*, 223–230.

Wang, C., Xie, J., Xiao, X., Chen, S., & Wang, Y. (2019). Development of nontoxic biodegradable polyurethanes based on polyhydroxyalkanoate and L-lysine diisocyanate with improved mechanical properties as new elastomers scaffolds. *Polymers, 11*(12), 1927.

Wang, X. J., Huang, Z., Wei, M. Y., Lu, T., Nong, D. D., Zhao, J. X., ... Teng, L. J. (2019). Catalytic effect of nanosized ZnO and TiO_2 on thermal degradation of poly (lactic acid) and isoconversional kinetic analysis. *Thermochimica Acta, 672*, 14–24.

Wei, L., & McDonald, A. G. (2016). A review on grafting of biofibers for biocomposites. *Materials*, *9*(4), 303.

Wu, C. S., & Liao, H. T. (2014). The mechanical properties, biocompatibility and biodegradability of chestnut shell fibre and polyhydroxyalkanoate composites. *Polymer Degradation and Stability*, *99*, 274−282.

Wu, C. S. (2014). Preparation and characterization of polyhydroxyalkanoate bioplastic-based green renewable composites from rice husk. *Journal of Polymers and the Environment*, *22*(3), 384−392.

Wu, C. S. (2015). Influence of modified polyester on the material properties of collagen-based biocomposites and in vitro evaluation of cytocompatibility. *Materials Science and Engineering: C*, *48*, 310−319.

Xian, X., Wang, X., Zhu, Y., Guo, Y., & Tian, Y. (2018). Effects of MCC content on the structure and performance of PLA/MCC biocomposites. *Journal of Polymers and the Environment*, *26*(8), 3484−3492.

Xie, H., Cao, T., Rodríguez-Lozano, F. J., Luong-Van, E. K., & Rosa, V. (2017). Graphene for the development of the next-generation of biocomposites for dental and medical applications. *Dental Materials*, *33*(7), 765−774.

Yıldızhan, Ş., Çalık, A., Özcanlı, M., & Serin, H. (2018). Biocomposite materials: A short review of recent trends, mechanical and chemical properties, and applications. *European Mechanical Science*, *2*(3), 83−91.

Younes, B. (2017). Classification, characterization, and the production processes of biopolymers used in the textiles industry. *The Journal of the Textile Institute*, *108*(5), 674−682.

Zhang, H. Y., Jiang, H. B., Ryu, J. H., Kang, H., Kim, K. M., & Kwon, J. S. (2019). Comparing properties of variable pore-sized 3D-printed PLA membrane with conventional PLA membrane for guided bone/tissue regeneration. *Materials*, *12*(10), 1718.

Zhang, L., Lv, S., Sun, C., Wan, L., Tan, H., & Zhang, Y. (2017). Effect of MAH-g-PLA on the properties of wood fiber/polylactic acid composites. *Polymers*, *9*(11), 591.

Zhao, Y., Liu, Z., Cao, C., Wang, C., Fang, Y., Huang, Y., ... Tang, C. (2017). Self-sacrificed template synthesis of ribbon-like hexagonal boron nitride nano-architectures and their improvement on mechanical and thermal properties of PHA polymer. *Scientific Reports*, *7*(1), 1−6.

Zhila, N., & Shishatskaya, E. (2018). Properties of PHA bi-, ter-, and quarter-polymers containing 4-hydroxybutyrate monomer units. *International Journal of Biological Macromolecules*, *111*, 1019−1026.

CHAPTER 16

Synthesis and applications of chitosan and its composites

Thana Saffar[1], Narisetty Vivek[2], Sara Magdouli[1], Joseph Amruthraj Nagoth[3], Maria Sindhura John[3], Raveendran Sindhu[4], Parameswaran Binod[4] and Ashok Pandey[5]

[1]Centre Technologique des Résidus Industriels, University of Quebec in Abitibi Témiscamingue, Quebec, Canada, [2]Department of Biochemical Engineering and Biotechnology, Indian Institute of Technology Delhi, New Delhi, India, [3]School of Biosciences and Veterinary Medicine, University of Camerino, Camerino, Italy, [4]Microbial Processes and Technology Division, CSIR—National Institute for Interdisciplinary Science and Technology (CSIR—NIIST), Thiruvananthapuram, India, [5]Centre for Innovation and Translational Research, CSIR—Indian Institute of Toxicology Research (CSIR—IITR), Lucknow, India

16.1 Introduction

Chitosan, a derivative of chitin, is a linear polysaccharide (1–4)-linked 2-acetoamido-2-deoxy-β-D-glucopyranose units. The polymer has immense potential, with ease of incorporating required chemical modifications to attain important structural, physical, and functional properties to be a suitable biopolymer either in native form or as a composite (Cheaburu-Yilmaz, Yilmaz, & Vasile, 2015; Lebreton & Andrady, 2019; van den Broek, Knoop, Kappen, & Boeriu, 2015). Though cellulose-producing organisms do not produce the polymer, it is considered as the derivative of cellulose, and are structurally similar except acetamide (-NHCOCH$_3$) group at second carbon (Fig. 16.1) (Blind, Petersen, & Riillo, 2017; Prateepchanachai, Thakhiew, Devahastin, & Soponronnarit, 2019; Zargar, Asghari, & Dashti, 2015).

The presence of NH$_2$ groups in chitosan makes the industrial demand of this polymer higher compared to chitin. Chitosan, a pseudonatural cationic polymer, is mainly obtained by deacetylation of chitin under alkaline conditions. Chitin can be extracted from the shells of molluscs, crustaceans, and insects and produced by few fungi and algae. For instance, it has been proved that the use of starch in biofilm, coating and bending applications is of interest since it represents a good promise to reduce environmental footprint. Nonetheless, due to its moisture sensibility and poor mechanical properties, commercial applications of

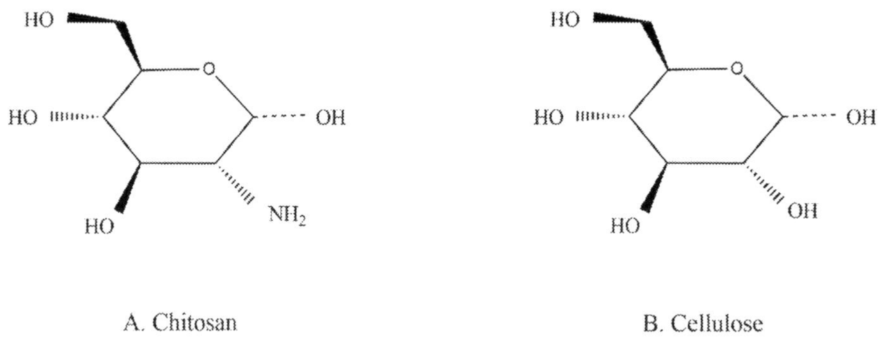

Figure 16.1
Structural illustration of monomeric units of chitosan and cellulose.

starch are limited (Jamróz, Kulawik, & Kopel, 2019). Chitosan is a nontoxic, biodegradable, and biocompatible natural polymer with antibacterial activity and applications in the areas of biomedicine, packaging, wastewater treatment, etc. The polymer has reactive hydroxyl and amino groups that can be used to impart modifications by chemical reactions according to the application or structures. Chitosan presents NH_2 group on the C-2 position of the D-glucosamine repeating unit (Fig. 16.2). Chitosan is synthesized by deacetylation of α-chitin in a 40%−50% aqueous alkali solution at 100°C−160°C for a few hours. The chitosan derived by this process has up to 0.95 degree of deacetylation. Further deacetylation can be obtained by repeating the alkaline treatment. In case of β-chitin, a much lower temperature near 80°C is enough for deacetylation and colorless chitosan products were obtained in comparison to α-chitin.

As ionic strength of the environment and acidic nature demonstrate the physiological behavior of the chitosan biopolymer and its composites, these natural materials can be used as vehicle for the controlled release of drugs or other site-specific active ingredients (Vargas, Chiralt, Albors, & González-Martínez, 2009) working on chitosan and its polymer composites.

The main purpose of this chapter is to present a comprehensive overview and state-of-art technological developments on the synthesis and applications of chitosan, highlighting its chemical structure and functionalities. Furthermore, extraction and chemical modification have also been elaborated. Finally, special emphasis has been attached on the versatile applications of chitosan such as engineering tissue scaffold, drug delivery, wastewater treatment, and bioremediation.

16.2 Biofunctionality of chitosan

Chitosan and its composite blends could be molded into edible films, coatings, and bioactive ingredients due to their thermochemical behavior, biological compatibility, and

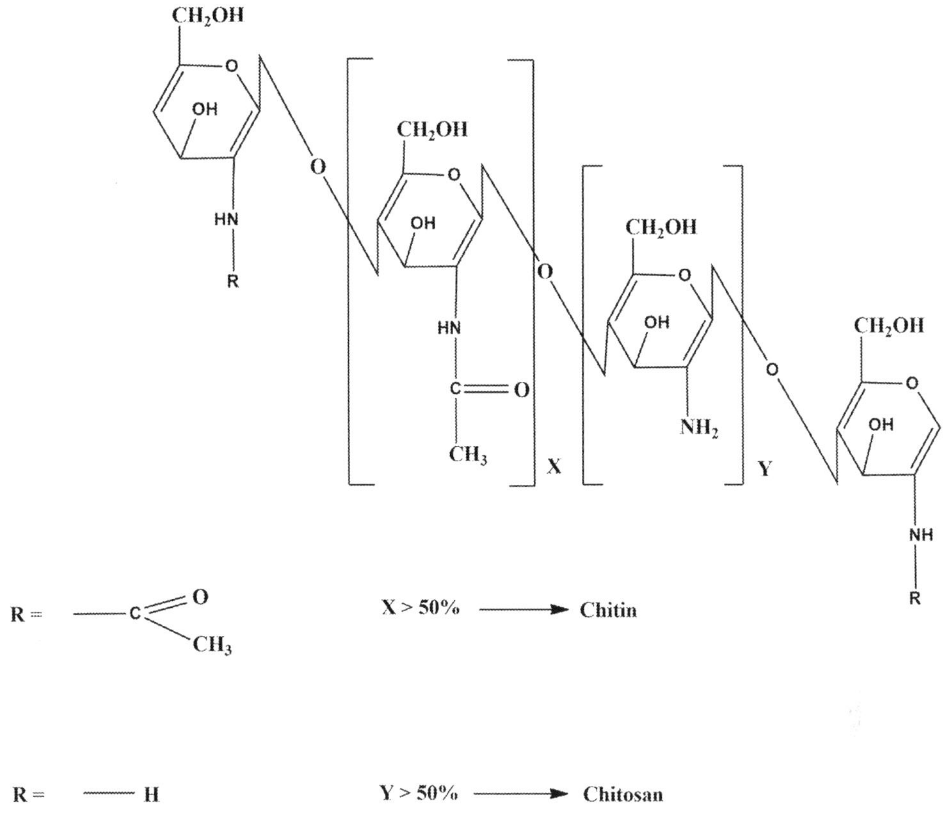

Figure 16.2
Structural illustration of chitin and chitosan.

nontoxic nature (Lazaridou & Biliaderis, 2002). Over the last two decades increase in concerns to the ecology and environment, tremendous research on chitosan and other biopolymers are found to be viable solutions to the waste disposal of food- and plastic-packaging materials (Cooper, Oldinski, Ma, Bryers, & Zhang, 2013; Nitschke, Altenbach, Malolepszy, & Mölleken, 2011). Furthermore, biopolymer films are the excellent matrix for incorporating a wide variety of functional additives, such as antioxidants, antifungal agents, antimicrobials (Zemljič, Tkavc, Vesel, & Šauperl, 2013), colors, and nutrients, and these active materials can improve storage life, shelf life, and keep the quality intact (Abdollahi, Rezaei, & Farzi, 2012a, 2012b). Such matrix biopolymers include starches, cellulose derivatives, chitosan/chitin, gums, proteins (animal or plant-based), and lipids. Chitin and chitosan molecular weight may range in several million grams per mole (g mol^{-1}), but the commercially available chitosan has a molecular weight ranging from 3800 to 5,00,000 g mol^{-1}, with 2%−40% degree of acetylation and 6.89% nitrogen. The chitosan is usually nontoxic with an approximate LD50 value of 16 g kg^{-1} of body weight of mice, the value corresponds to sugar and salt content. The reactive oxygen species, O_2^-, HO^*, and H_2O_2, are unstable molecules that readily react with the chemical groups and other substances in the human body resulting in the cell or tissue damage and diseases. The chemical groups are amino acids, DNA molecule, and membrane components, the chitosan as the scavenger has an activity of 12.35% against these unstable molecules. The -OH groups in the chitosan readily react with HO^*, and $-NH_2$ groups on reaction forms stable NH^3+ macromolecules (Kumirska et al., 2010; Nguyen, Nguyen, & Hsieh, 2013). Similarly, chitosan has various characteristics that impart biological function to the polymer, which is discussed in detail in the medical applications of chitosan in this chapter.

16.2.1 Extraction and chemical modification of chitosan for bio-based materials

Given the growing concern regarding petrochemical environmental footprint, alternatives derived from bio-based products has gained interest. Therefore renewable sources can provide a sustainable raw matter for the development of novel functional biomaterials (Thomas, Mishra, & Asiri, 2019). On account of the waste residues from crab and shrimp canning industries, chitosan is among the most abundant polymer in terms of availability. The advantage of chitosan as a promising biomaterial is attended (Rinaudo & Goycoolea, 2019). Because of its biocompatibility in several matrices and biological tissues, chitosan becomes a raw matter of interest in the biomedical and environmental fields (Thomas et al., 2019). Chitosan is distinguishable from other biopolymers by its unique cationic polysaccharide nature. It is suitable as a delivery vector in drug delivery, condensation of DNA and RNA, complexation with electron interactions, etc. (Samal, Dash, Dubruel, Müllen, & Rajadas, 2014). In the structural view, chitosan is a linear polysaccharide copolymer of N-acetyl-D-glucosamine and D-glucosamine linked with covalent β-(1—4)-glycosidic bonds (Fig. 16.2). It is produced from chitin (the main component of crustacean shells) via deacetylated reaction (2-acetamido-2-deoxy-b-D-glucopyranose). The first isolation-processing step of chitosan involves protein removal from crustacean shells where calcium carbohydrate is dissolved. After the removal of protein, the subsequent step is the deacetylation with caustic soda (NaOH) or enzymatic hydrolysis (deacetylase). Later, minerals were removed by acidic washing (HCl). Finally, water washing and dewatering neutralize pH (Zargar et al., 2015; Fig. 16.3).

Basic functionalities, their abundance, and accessibility on the chitosan backbone define physical, chemical, and biological properties. Besides the fact of the interesting features like biocompatibility and biodegradability, chitosan on degradation/depolymerization does not generate any toxic products. On an industrial scale, chitosan obtained via alkaline deacetylation of chitin. Nonetheless, the existence of acetylglucosamine and glucosamine groups may contribute to the degree of crystallinity. A high-purity fraction of chitosan is frequently required for medical applications and can come by further deacetylation steps that imply fragmentation and depolymerization.

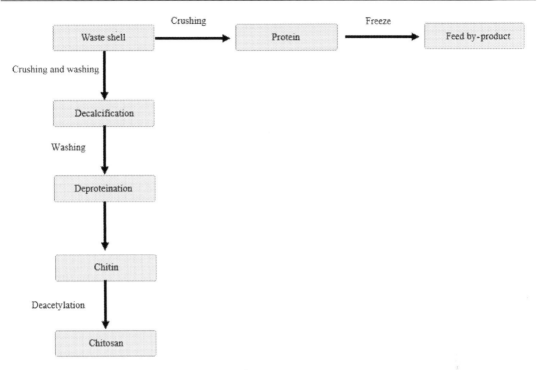

Figure 16.3
Illustration of a chemical-processing method for extraction of chitin and chitosan.

Although the chemical method is widely used commercially, there are few limitations like (1) the amount of energy consumed for the various steps, (2) the amount of concentrated acid and alkali required for the demineralization and deproteination, and (3) the broad range of chitin with different molecular weights, due to heterogeneity in the extent of deacetylation is observed. As biological methods are adding advantage in various fields of science and technology, being alternative to petrochemical substrates and methods, chitin and chitosan extraction from the waste shells through a biological process (Fig. 16.4). The biological method comprises of proteolytic and chitinolytic microorganisms or purified exo or endo enzymes. The biological reactions with enzymes are highly specific, resulting in oligomers with the optimal degree of polymerization and high molecular weight chitin polymers. The added advantages can be the replacement of hazardous chemicals and wastewater treatment methods after using chemicals.

In the biological approach, the proteolytic bacteria like *Lactobacillus sp.*, *Pseudomonas sp.*, and *Bacillus subtilis*, demineralize and deproteinate ground crustacean shells. The organic acids and exo-proteases released by these microorganisms cause demineralization and deproteination. Several enzymes like alcalse, pepsin, papain, pancreatin, devolvase, and trypsin, can demineralize and deproteinize the shells to extract the high molecular

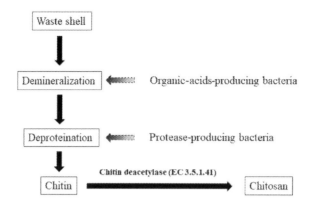

Figure 16.4
Biological process for extraction of chitin and chitosan.

weight chitin polymer (Bajaj, Freiberg, Winter, Xu, & Gallert, 2015; Xu, Gallert, & Winter, 2008). The differences between the chemical and biological methods are tabulated (Table 16.1).

It is worth mentioning that molecular weight and the degree of crystallinity are the most important parameters that characterize chitosan. Contrary to other polysaccharides (i.e., chitin, cellulose, agar, pectin, etc.), chitin is a highly basic biopolymer which made it very suitable for an improved solubility in water, organic solvent, and biological media.

16.3 Synthesis of composite blends of chitosan

The most relevant parameters that control chemical grafting are the degree of purity, acetylation of chitosan, the type and percentage of catalyst, temperature, time reaction, etc. For instance, in the regenerative medicine field, the development of high value-added materials leading to growth cells is of interest. Abbarategi et al. (2008) have tested chitosan as an engineering tissue for cell adhesion in vitro (Abarrategi et al., 2008). Multiwall carbon nanotubes and chitosan scaffolds were prepared via the ice-segregation using the self-assembly method. This assembly of porous carbon and chitosan scaffold successfully attributed to C_2C_{12} myoblast cell to assess the viability and proliferation.

Through several chemical modifications like alkylation, esterification, acylation, nitration, sulfonation, and phosphorylation, chitosan improved its properties. Chemical modification could bring especially wished or desired properties (solubility, antioxidant activity, new active sites, etc.) via different reactions. Chemical grafting usually refers to a small moieties insertion on chitosan to enhance or to create. In this way, chitosan derivatives can

Table 16.1: Comparison of chemical and biological extraction methods of chitin and chitosan from various sources (Lazaridou & Biliaderis, 2002; Nitschke et al., 2011; Rinaudo & Goycoolea, 2019; Xu et al., 2008)

Sl. no.	Process	Methodology	Chemical method	Biological method
1	Recovery	Demineralization	Acidic pretreatment (HCl, HNO_3, H_2SO_4, CH_3COOH, and HCOOH)	Acids produced by microorganisms (e.g., lactic acid by lactic acid bacterial strains)
		Deproteinization	Alkali pretreatment	Enzymatic pretreatment (valorization carried out by endo enzymes or exo enzymes released by microorganisms into surroundings, e.g., proteases by proteolytic microorganisms)
		Deacetylation	Thermal and alkaline pretreatment for removal of N-acetyl groups	Enzymatic pretreatment (Chitin deacetylases produced by bacteria and fungi)
2	Quality	Function of molecular mass and degree of acetylation	Acidic, and alkaline pretreatment leads to depolymerization along with demineralization and deacetylation resulting in different varieties of final product	Method maintains a homogenous and high-quality final product
3	Economic considerations		Sequential acidic, and alkaline pretreatment methods required neutralization, which adds extra unit procedures Effluent treatment after acidic and alkaline extraction may increase the economy of the whole process	The biological process is usually carried out in environmental friendly process conditions and the secondary metabolites produced and the microorganisms used can be added income to the processes

serve for the preparation of new biopolymers with improved or competitive properties. Hence these chemical processes are used for either controlling the reactivity of the polymer (improving the affinity or selectivity) or enhancing sorption kinetics (managing diffusion properties).

The functionalization of chitosan through graft polymerization is a better technique. For example, carboxyalkylation is among the most commonly reported methods for chemical modification. Carboxyalkylation occurs when an acidic group is introducing on chitosan backbone grafting. Skorik, Pestov, Kodess, and Yatluk (2012) tested carboxyalkylation on chitosan via two pathways: nucleic addition and nucleophilic substitution reactions using numerous acids. Skorik found that acrylic, crotonic, and halocarboxylic acids were the most advantageous. The chitosan carboxyalkyl derivative presented new desired properties, including high yield, a high degree of grafting, and a relatively low reaction time with no need to stir the reaction (Skorik Pestov, Kodess, & Yatluk, 2012).

Moreover, esterification has been performed between chitosan and maleic anhydride to obtain a water-soluble product (Fu & Xiao, 2017). However, the chitosan-esterified derivative has occasionally been insoluble. Fu and Xiao (2017) studied this change and attributed it to the chitosan crystallinity. They worked on chitosan water-soluble through the reaction of esterification through modulating the concentration of the solution. They obtained a successfully modified chitosan derivative that displayed a higher degree of crystallinity and, consequently, a water-soluble precipitate (Fu & Xiao, 2017).

Chitosan reactive amine group was also suitable for sulfonation to acquire a useful induced composite for the preparation of membrane ion exchange and water uptake. Chitosan was modified by phthaloylation and then blended with the sulfonated polyethersulfone to produce different blend membranes according to their composition. Characterization of the resulted materials investigated solubility on organic solvent, film formability, and ion conductivity (Muthumeenal, Neelakandan, Kanagaraj, & Nagendran, 2016). It has been found that the capacity of proton conductivity improved when the ratio of N-phthaloyl chitosan increase depends on sulfonated polyethersulfone due to the easy proton mobility implied by a higher degree of hydrophobicity.

More recently, graft copolymerization has been performed on chitosan as a pathway to chemical modification. Ammonium persulfate was used as a radical initiator, to create active sites onto chitosan. Because of this radical attack, labile hydrogens are very susceptible to be eliminated from hydroxyl and amino groups. The copolymerization of acrylamide with chitosan produced hydrogels with improved elastic properties and high grafting degree up to 90% (Mochalova et al., 2006).

Thiol-containing chitosan is obtained by the reaction between chitosan and thiolactic acid. Primary amino is bonded thiolactic acid to form the amide bonds. Jayakumar, Reis, and Mano (2007) showed that thiol-containing chitosan could be used in the drug-delivery system where the optimum release rate was around pH = 7.4, water-solubility, swelling capacity, and the ionization of thiol groups were pH sensitive. According to Jayakumar et al., thiol-containing chitosan was a good drug carrier in vivo (Jayakumar et al., 2007).

Furthermore, polystyrene graft on chitosan was investigated using $^{60}Co\ \gamma$ radiation. The effect of various parameters, such as adsorbed dose, solvent, and oxygen on grafting, was tested. Han, Wang, Sun, Zhao, and Zhang (2017) prepared chitosan-polystyrene microtiter plates with a modified surface propitious for higher protein adsorption (Han, Wang, Sun, Zhao, & Zhang, 2017). By chemical modification, the potential use of chitosan becomes more critical. A more extended range of applications can be considered for chitosan and chitosan derivatives for a high added-value application.

16.4 Applications

Numerous applications employed chitosan as a raw material in several fields namely, biomedicine (wound healing, inhibition of growth of microorganisms), pharmacy (drug release, pain alleviation), water treatment (heavy metal), food industry (bio-based film for packaging), and nanotechnology (carbon nanotubes) (Table 16.2; Badawy et al., 2004; Balakrishnan, Mohanty, Umashankar, & Jayakrishnan, 2005; Howling et al., 2001; Ribeiro et al., 2009; Rodríguez-Vázquez, Vega-Ruiz, Ramos-Zúñiga, Saldaña-Koppel, & Quiñones-Olvera, 2015).

Table 16.2: Applications of chitosan and its composite blends in various fields.

Chitosan/Chitosan derivatives/Modified chitosan in different fields	Examples	Reference
Nanomaterial for metals removal in wastewater treatment	Copper [Cu(II)] removal	(Ngah, Endud, & Mayanar, 2002)
	Zinc [Zn(II)] removal	(Kyzas, Siafaka, Pavlidou, Chrissafis, & Bikiaris, 2015)
	Nickel [Ni(II)] removal	(Eser, Tirtom, Aydemir, Becerik, & Dinçer, 2012)
	Lead [Pb(II)] removal	(Fan, Luo, Sun, Li, & Qiu, 2013)
	Cadmium [Cd(II)] removal	(Hydari, Sharififard, Nabavinia, & reza Parvizi, 2012)
	Mercury [Hg(II)] removal	(Kyzas and Deliyanni, 2013)
	Chrome [Cr(VI)] removal	(Bhatt, Sreedhar, & Padmaja, 2015)
Pharmaceutical and biomedical materials	Gel for wound healing	(Cardoso et al., 2019)
	Anti-microbial activity	(Anush, Vishalakshi, Kalluraya, & Manju, 2018)
	Drug carrier	(Ito, Takami, Uchida, & Murakami, 2018)
Tissue engineering	Bone tissue scaffold	(Balagangadharan, Dhivya, & Selvamurugan, 2017)
	Cell growth enhancer	(Sivashankari and Prabaharan, 2016)
	Cartilage repair	(Man et al., 2016)
Agriculture	Defensive activator in plant system	(Pichyangkura and Chadchawan, 2015)
	Seed fertilizer	(Aziz, Hasaneen, & Omer, 2016)
	Fungicide	(Pham et al., 2018)
Food additives and food packaging	Flavor in food	(Yu et al., 2018)
	Emulsifier	(Ho et al., 2016)
	Antioxidant and antimicrobial activity	(Yuan, Chen, & Li, 2016)
	Food packaging film	(Siripatrawan and Vitchayakitti, 2016)

16.4.1 Food and packaging

Chitosan has been effectively used in food packaging because of its ability to form semipermeable films (Dutta, Tripathi, Mehrotra, & Dutta, 2009; Elsabee & Abdou, 2013). Recent studies have reported the antibacterial activity of chitosan. They have revealed that this latter is useful for inhibiting the growth of bacteria from extending the duration of food conservation. The latter is considered as a promising natural solution to prevent food spoilage and pathogens. Generally, chitosan showed more substantial bactericidal effects for Gram-positive bacteria (No, Park, Lee, & Meyers, 2002; Zhong et al., 2007) in comparison with Gram-negative bacteria. Besides, growth inhibition and yeast inactivation appear to be dependent on chitosan concentration, pH, and temperature. Currently, the addition of chitosan to acidic foods is reported to enhance its effectiveness as a natural preservative (No et al., 2002). The antibacterial activity of chitosan on *Aeromonas hydrophila* has been notified, and only 0.04% chitosan was sufficient to prevent the growth and the production of hemolysin by *A. hydrophila* (Taha & Swailam, 2002) and this activity increased with increasing temperature and decreasing pH. According to the literature, antimicrobial properties of chitosan is influenced by intrinsic and extrinsic factors such as the type of chitosan, the degree of polymerization, the composition nutrient content, and substrate water activity (Fei Liu, Lin Guan, Zhi Yang, Li, & De Yao, 2001; Wang & Chen, 2005). Added to these factors, the composition of food components (starch, whey protein, and oil) influenced the antimicrobial activity of chitosan was investigated (Devlieghere, Vermeulen, & Debevere, 2004). Salt concentration between 10 and 25 mM may decrease the chitosan's activity. However, irradiation degrades chitosan and increases its antimicrobial activity. Besides, chitosan acetate, a mixture of chitosan with acetic acid, is reported to have antibacterial activity, especially at a concentration between 0.15% and 0.2% (w/v) (Li et al., 2007).

Recently, some blend films of starch and chitosan treated with irradiation show a great antibacterial activity (Zhai, Wei, Zeng, Gong, & Yin, 2006), and coating of chitosan solution on the starch films increased their mechanical and physical properties by a change in the tensile stress, decreasing % elongation at the break, water uptake rate, and water vapor permeability (Bangyekan, Aht-Ong, & Srikulkit, 2006). Other semipermeable films based on chitosan are hard, durable, and very difficult to tear. They present moderate values of water permeability, have low oxygen permeability, decrease of respiration, and delay the process of maturation due to the reduction of the evolution of ethylene and carbon dioxide that may inhibit fungal development. Chitosan is soluble in aqueous solutions of organic and mineral acids. Solubility allows the formation of membranes with excellent mechanical properties (Zhai et al., 2006). Chitosan is a naturally abundant and renewable polymer with unique features such as biodegradability, biocompatibility, nontoxicity, solubility in various media, viscosity, polyelectrolyte behavior, the formation of polyoxygel, the ability to form films, and the chelation of metals. These properties result from the protonation of NH_2 groups on the skeleton of chitosan (Ausar, Passalacqua, Castagna, Bianco, & Beltramo, 2002; Kanatt, Chander, & Sharma, 2008). The use of chitosan films led to an extension of the shelf life of foods and an improvement in their quality (Ausar et al., 2002; Meng, Li, Liu, & Tian, 2008). New types of chitosan derivatives have been recently developed with much higher antimicrobial activity (Zhong et al., 2007). The antimicrobial activity of chitosan and its derivatives against different groups of microorganisms such as bacteria, yeast, and fungi has been reviewed (Ausar et al., 2002; Kanatt et al., 2008). The antimicrobial activity of chitosan derivatives is usually dependent on the presence of acetyl, chloroacetyl, and benzoyl groups, which confer a higher antimicrobial activity compared to native chitosan. A new biopolymer has been investigated, which consists of blending chitosan with nylon-6. This combination works well antibacterial against both Gram-positive and Gram-negative bacteria. However, increasing the nylon-6 content in the blended membranes may decrease the antibacterial property (Ma, Zhou, & Zhao, 2008). Thiourea chitosan derivatives are distinguished with their higher fungicidal activity, which accounts for more than 60 times compared to pure chitosan (Eweis, Elkholy, & Elsabee, 2006).

C-irradiation depolymerization and deacetylation process are among standard procedures used to synthesize functional biopolymers and chitosan derivatives (Fei Liu et al., 2001). By controlling the deacetylation degree, numerous chitosan derivatives and different lipid emulsions containing chitosan can be obtained. Currently, Jumaa, Furkert, and Müller (2002) have investigated chitosan with different molecular weights, degree of deacetylation, and viscosity (Jumaa, Furkert, & Müller, 2002), and results showed that the polymer with a higher degree of deacetylation and lower viscosity displayed higher antimicrobial activity. Moreover, compared to native chitosan, chitosan-based oligomers presented more moderate antibacterial activities (No et al., 2002).

Complexing chitosan with some base metals has been carried out, and the efficacy of antimicrobial activity has been studied. For instance, complex chitosan with Zn offered a broad spectrum of active antimicrobial activities against microbial species [bacteria (Gram-positive and Gram-negative), fungi] (Wang, Du, & Liu, 2004). Further often, by grafting the sulfanilamide group onto the chitosan chain, various sulfanilamide derivatives based on chitosan can be formulated. The reaction consists of the 4-acetamidobenzenesulfonyl chloride solution with different types of chitosan (carboxymethyl chitosan, chitosan sulfates) (Zhong et al., 2007). These derivatives showed an excellent inhibiting effect on various fungi and increasing more molecular weight. The concentration of chitosan, as well as the degree of substitution, influences the antifungal activity positively.

Durango et al. have synthesized new antimicrobial films based on yam starch and chitosan with excellent flexibility and able to attack *S. enteritidis* (Durango et al., 2006). In the same context, a recent study has reported the efficacy of mixing mint extract and chitosan to preserve meat and meat products and to improve their shelf life and safety taking into consideration the antioxidant property of mint extract and the excellent antimicrobial properties of chitosan. Results indicated that a mixture of 0.05% chitosan and the mint extract is highly effective against Gram-positive bacteria (Kanatt, Chander, & Sharma, 2008).

Immobilized collagen/chitosan via grafting acrylic acid on polypropylene with various weight ratios of collagen/chitosan is again distinguished for its higher antibacterial activity (Wang & Chen, 2005). In another study, chitosan was complexed with glucose by heating chitosan with glucose (Kanatt et al., 2008). Compared to chitosan alone, this complex is much better with excellent antioxidant activity compared with chitosan or glucose alone. Recent work has reported a new process based on coating chitosan solution with a vinyl acetate-ethylene copolymer to produce antimicrobial paperboard highly recommended for universal antimicrobial packaging with a variety of foods (Lee, An, Park, & Lee, 2003).

16.4.2 Wastewater treatment

Currently, chitosan is used as an adsorbent in water and wastewater treatment industries due to its high contents of amino and hydroxyl functional groups (Hasan, Ghosh, Viswanath, & Boddu, 2008). Thus it can well adsorb dyes, metal ions, and proteins. Recent studies have been carried out to immobilize potential algal candidates such as *Scenedesmus* spp. on chitosan beads to remove phosphate and nitrate from water. Results were impressive in terms of phosphate and nitrate removal compared to the conventional free-cell system (Fierro, del Pilar Sánchez-Saavedra, & Copalcua, 2008). Besides, chitosan with various degrees of substitution reported efficient to adsorb anionic dye Congo red. The adsorption capacity was dependent on many factors such as temperature and pH (Wang & Wang, 2008). For instance, the adsorptions of the blue 19 dye onto cross-linked chitosan/oil palm ash composite blends are more efficient at pH 6.0 (Hasan et al., 2008). The formation of new adsorbents from chitosan via a cross-linking method investigated for the removal of Cd(II) and Zn(II). Herein, sulfoximine is used as a chelating agent on the surface of chitosan. The adsorption capacity was more relevant at pH 8 (Vitali, Laranjeira, Gonçalves, & Fávere, 2008). Some other nanocomposites of Cu_2O and chitosan with a mass ratio of 50% prepared by electrochemical deposition to treat drinking water (Chen, Chang, & Chen, 2008). Another study has reported that polyacrylamide-chitosan has higher adsorption capacity for Pb compared to chitosan alone (Akkaya & Ulusoy, 2008), suggesting that the use of chitosan with polyacrylamide can increase the efficiency of chitosan in separation and removal procedures of toxic metals. Recently, the carboxymethyl chitosan prepared via irradiation—cross-linked was investigated for the adsorption of humic acids in aqueous solutions under acidic pH conditions (pH 6.0) (Akkaya & Ulusoy, 2008).

Further, chitosan can play the role of resin during the ionic-exchange process to adsorb Hg with better efficiency than the commercially available resins (Donia, Atia, & Elwakeel, 2008). Other types of adhesives, such as glycine-modified cross-linked chitosan resin, were investigated for the removal of Au, Pt, and Pd ions (Ramesh, Hasegawa, Sugimoto, Maki,

& Ueda, 2008). Molecular imprinting technology and photocatalytic degradation reported decreasing environmental pollution. Experiments showed that these resins were effective in the removal of the mentioned ions from water and contaminated with Au. Other contaminants such as heavy metal ions and some organic compounds were easily removed by ion-imprinted chitosan-TiO_2 adsorbent (Li, Su, & Tan, 2008) or chitosan-ceramic alumina (Boddu, Abburi, Talbott, Smith, & Haasch, 2008). A composite chitosan flocculant synthesized from polyaluminium chloride, and silicate could remove aluminum and other suspended solids from the water and also made the process economical by reducing the cost of almost 34% (Zeng, Wu, & Kennedy, 2008). According to Sugunan, Thanachayanont, Dutta, and Hilborn (2005), the capability of electrostatic behavior and attraction between chitosan and gold nanoparticles can be used as an indicator of heavy metal pollution in water bodies (Sugunan, Thanachayanont, Dutta, & Hilborn, 2005).

16.4.3 Bioremediation

Functional properties of chitosan in environmental applications have been widely reported and reviewed for the adsorption of dye (DR80, DR81, and RR 180); ethylbenzene, gold (III), Au (III), and Pd (II); rhodamine B, Ni (II), and Cr (VI); phenolic pollutants, etc. (Zhao, Mai, Kang, & Zou, 2008; Wang, Zhang, & Wang, 2009; Chang, Chang, & Chen, 2006; Mahmoodi & Mokhtari-Shourijeh, 2015; Mohamed & Ouki, 2011). For a long, chitosan was investigated in the bioremediation field as a coating material and support for the immobilization of viable microorganisms or free extracellular enzymes because of its higher specificity and activity against a wide range of contaminants and under harsh environmental conditions (high temperatures, extreme pH, and other denaturing agents). Accordingly, chitosan-enzyme encapsulation presents an innovative for practical and industrial applications due to its biocompatibility, biodegradability, available functional groups, and adjustable structural characteristics (Croisier & Jérôme, 2013). Although its efficiency, this latter can fail, especially if selected microorganisms are chitosanase producers and produce chitosan-degrading enzymes (Somashekar & Joseph, 1996). In electrochemistry, clay, chitosan, and gold nanoparticles are used for horseradish peroxidase-based biosensor development (Zhao et al., 2008).

Secondly, chitosan derivatives were used as pesticides (Rabea et al., 2005). Among wide derivatives investigated, N-alkyl chitosan and N-benzyl chitosan were distinguished with their insecticidal and fungicidal activities. Chitin and chitosan have been widely reported to bind the different types of metals effectively.

16.4.4 Drug delivery

As a natural, biocompatible, and nontoxic biodegradable biopolymer, chitosan has been used and investigated in the development of drug-delivery vehicles and related formulations. Various chitosan derivatives are used for the controlled release of drugs.

Different methods investigated for the synthesis of these derivatives such as chitosan microspheres (Kulkarni, Kulkarni, & Keshavayya, 2007) and some amphiphilic derivatives of chitosan (Sui, Wang, Dong, & Chen, 2008).

The optimization of different methods and their various operating conditions has been well-reviewed. Among parameters to be highly controlled to optimize the process as well as the size distribution, microsphere size, degree of swelling, drug entrapment efficiency, the control of stirring rate, cross-linking agent concentration, and the drug-to-polymer ratio are very primordial.

New hydrogel-type ophthalmic drug release systems developed. These systems are prepared via the impregnation process of N-chitosan derivatives with flurbiprofen and timolol maleate (Braga et al., 2008). These derivatives can be easily handled and used for patients.

Rather than the impregnation process, a cross-linking method investigated for the development of self-aggregated nanoparticles from methoxypoly(ethylene glycol)-grafted chitosan (Yang et al., 2008). These nanoparticles can serve as a carrier for the poorly water-soluble drug to sustain its release, prolong its circulation time, increase its therapeutic index, and decrease its toxic effects.

Besides, in colonic drug delivery, the coating process is documented for its potential efficacy (Nunthanid et al., 2008). Recently, sonication was utilized to prepare chitosan-based microspheres. Results showed that these products presented numerous advantages (resistance against gastric pH, a potential of regeneration at neutral pH, etc.) (Oosegi, Onishi, & Machida, 2008). Further often, a casting solvent-evaporation method (Wang, Du, Luo, Lin, & Kennedy, 2007) has been applied to synthesize some nanocomposite films from chitosan/organic rectorite. These nanocomposites films showed a slower and more continuous release compared to chitosan film.

Functionalizing chitosan to produce chitosan derivatives or some peptide-chitosans with higher thermosensitivity, a large porosity, and higher water-holding capacity offered a new horizon for drug-delivery systems. Moreover, new dry chitosan powders were developed for drug release (Learoyd, Burrows, French, & Seville, 2008). Other cross-linked copolymer particles with a core-shell structure, have been synthesized as a drug-delivery system via soapless emulsion copolymerization of N-isopropylacrylamide and chitosan (Lee, Ha, Cho, Kim, & Lee, 2004); meanwhile, other carboxymethyl chitosan nanoparticles complexed with calcium ions were used as a delivery vehicle to carry doxorubicin drug (Shi, Du, Yang, Zhang, & Sun, 2006).

16.4.5 Medical

Chitosan and its derivatives have found their way in medical research, as means of artificial tissues, skin, bone, cartilage, and other parts of the human body, due to their irresistible and noncomparable physical properties like biocompatibility, biodegradability, and their antimicrobial activity (Kim et al., 2008).

A new study has investigated the efficacy of multiwall carbon nanotubes scaffolds composed with a minor fraction of chitosan and a porous structure for tissue engineering purposes. These compounds are investigated as support for culture growth. These scaffolds are distinguished by their cellular behavior; this fact makes them a potential alternative to be used for the construction of different tissues (Abarrategi et al., 2008). Mun et al. (2008) have reported that grafted polymers such as 2-hydroxyethylacrylate onto chitosan for medicine and pharmaceutical applications (Mun et al., 2008).

A polymer Ne-carbobenzyloxy-L-lysine synthesized via ring-opening polymerization has been investigated as a suitable material in cell culture and tissue engineering. Chitosan/collagen/polyethylene nanofibrous membranes obtained by electrospinning are advantageous applications in skin regeneration and wound healing (Baek, Kim, & Suh, 2008; Chi, Wang, & Liu, 2008). Similarly, the other derivatives like chitosan-silicate or PVA/bioactive glass-based composites synthesized by the freeze-drying method (Shirosaki et al., 2008), sol-gel, and foaming method have potential biomedical applications (Mansur & Costa, 2008).

16.5 Conclusions and perspectives

Chitin and chitosan are natural linear polysaccharides with exceptional properties such as the biocompatible, renewable, biodegradable chelating agent, and as a drug carrier. There are several sources of chitin such as crustaceans, molluscs, insects, fungi, and algae. Chitin and chitosan are insoluble in organic and aqueous solutions but can be chemically modified to be desirable for the application. These flexible characteristics of the chitosan enhance the physical and mechanical properties for the development of various functional biocomposites and blends. These blends are used in as biodegradable films in food packaging, wound dressings, and drug carriers in the pharmaceutical and medical fields, as porous membranes or coagulating agents in wastewater treatments, and as nitrogen source or encapsulating material in the agriculture and bioremediation applications. Currently the process of using chitosan and its blends is confined to laboratory studies, due to limitation of its availability. Hence further studies should be done on economical downstream processing of chitosan from its renewable sources. Later chemical modifications resulting in blends of chitosan can be effectively commercialized for different applications.

Acknowledgment

One of the authors Raveendran Sindhu acknowledges Department of Science and Technology for sanctioning a project under DST WOS-B scheme.

References

Abarrategi, A., Gutiérrez, M. C., Moreno-Vicente, C., Hortigüela, M. J., Ramos, V., López-Lacomba, J. L., Ferrer, M. L., & del Monte, F. (2008). Multiwall carbon nanotube scaffolds for tissue engineering purposes. *Biomaterials, 29*(1), 94–102.

Abdollahi, M., Rezaei, M., & Farzi, G. (2012a). A novel active bionanocomposite film incorporating rosemary essential oil and nanoclay into chitosan. *Journal of Food Engineering, 111*(2), 343–350.

Abdollahi, M., Rezaei, M., & Farzi, G. (2012b). Improvement of active chitosan film properties with rosemary essential oil for food packaging. *International Journal of Food Science and Technology, 47*(4), 847–853.

Akkaya, R., & Ulusoy, U. (2008). Adsorptive features of chitosan entrapped in polyacrylamide hydrogel for Pb^{2+}, UO_2^{2+}, and Th^{4+}. *Journal of Hazardous Materials, 151*(2–3), 380–388.

Anush, S., Vishalakshi, B., Kalluraya, B., & Manju, N. (2018). Synthesis of pyrazole-based Schiff bases of chitosan: Evaluation of antimicrobial activity. *International Journal of Biological Macromolecules, 119*, 446–452.

Ausar, S. F., Passalacqua, N., Castagna, L. F., Bianco, I. D., & Beltramo, D. M. (2002). Growth of milk fermentative bacteria in the presence of chitosan for potential use in cheese making. *International Dairy Journal, 12*(11), 899–906.

Aziz, H. M. A., Hasaneen, M. N., & Omer, A. M. (2016). Nano chitosan-NPK fertilizer enhances the growth and productivity of wheat plants grown in sandy soil. *Spanish Journal of Agricultural Research, 14*(1), 17.

Badawy, M. E., Rabea, E. I., Rogge, T. M., Stevens, C. V., Smagghe, G., Steurbaut, W., & Höfte, M. (2004). Synthesis and fungicidal activity of new N, O-acyl chitosan derivatives. *Biomacromolecules, 5*(2), 589–595.

Baek, S.-H., Kim, B., & Suh, K.-D. (2008). Chitosan particle/multiwall carbon nanotube composites by electrostatic interactions. *Colloids and Surfaces A: Physicochemical and Engineering, 316*(1–3), 292–296.

Bajaj, M., Freiberg, A., Winter, J., Xu, Y., & Gallert, C. (2015). Pilot-scale chitin extraction from shrimp shell waste by deproteination and decalcification with bacterial enrichment cultures. *Applied Microbiology and Biotechnology, 99*(22), 9835–9846.

Balagangadharan, K., Dhivya, S., & Selvamurugan, N. (2017). Chitosan based nanofibers in bone tissue engineering. *International Journal of Biological Macromolecules, 104*, 1372–1382.

Balakrishnan, B., Mohanty, M., Umashankar, P., & Jayakrishnan, A. (2005). Evaluation of an in situ forming hydrogel wound dressing based on oxidized alginate and gelatin. *Biomaterials, 26*(32), 6335–6342.

Bangyekan, C., Aht-Ong, D., & Srikulkit, K. (2006). Preparation and properties evaluation of chitosan-coated cassava starch films. *Carbohydrate Polymers, 63*(1), 61–71.

Bhatt, R., Sreedhar, B., & Padmaja, P. (2015). Adsorption of chromium from aqueous solutions using crosslinked chitosan–diethylenetriaminepentaacetic acid. *International Journal of Biological Macromolecules, 74*, 458–466.

Blind, K., Petersen, S. S., & Riillo, C. A. (2017). The impact of standards and regulation on innovation in uncertain markets. *Research Policy, 46*(1), 249–264.

Boddu, V. M., Abburi, K., Talbott, J. L., Smith, E. D., & Haasch, R. (2008). Removal of arsenic (III) and arsenic (V) from aqueous medium using chitosan-coated biosorbent. *Water Research, 42*(3), 633–642.

Braga, M. E., Vaz Pato, M. T., Costa Silva, H. S. R., Ferreira, E. I., Gil, M. H., Duarte, C. M. M., & de Sousa, H. C. (2008). Supercritical solvent impregnation of ophthalmic drugs on chitosan derivatives. *Journal of Supercritical Fluids, 44*(2), 245–257.

Cardoso, A. M., de Oliveira, E. G., Coradini, K., Bruinsmann, F. A., Aguirre, T., Lorenzoni, R., Barcelos, R. C. S., Roversi, K., Rossato, D. R., Pohlmann, A. R., Guterres, S. S., Burger, M. E., & Beck, R. C. R. (2019). Chitosan hydrogels containing nanoencapsulated phenytoin for cutaneous use: Skin permeation/penetration and efficacy in wound healing. *Material Science and Engineering C, 96*, 205–217.

Chang, Y.-C., Chang, S.-W., & Chen, D.-H. (2006). Magnetic chitosan nanoparticles: Studies on chitosan binding and adsorption of Co (II) ions. *Reactive & Functional Polymers, 66*(3), 335–341.

Cheaburu-Yilmaz, C. N., Yilmaz, O., & Vasile, C. (2015). Eco-friendly chitosan-based nanocomposites: Chemistry and applications. In *Eco-friendly polymer nanocomposites* (pp. 341–386). Springer.

Chen, J.-P., Chang, G.-Y., & Chen, J.-K. (2008). Electrospun collagen/chitosan nanofibrous membrane as wound dressing. *Colloids and Surfaces A: Physicochemical and Engineering, 313*, 183–188.

Chi, P., Wang, J., & Liu, C. (2008). Synthesis and characterization of polycationic chitosan-graft-poly (l-lysine). *Materials Letters, 62*(1), 147–150.

Cooper, A., Oldinski, R., Ma, H., Bryers, J. D., & Zhang, M. (2013). Chitosan-based nanofibrous membranes for antibacterial filter applications. *Carbohydrate Polymers, 92*(1), 254–259.

Croisier, F., & Jérôme, C. (2013). Chitosan-based biomaterials for tissue engineering. *European Polymer Journal, 49*(4), 780–792.

Devlieghere, F., Vermeulen, A., & Debevere, J. (2004). Chitosan: Antimicrobial activity, interactions with food components and applicability as a coating on fruit and vegetables. *Food Microbiology, 21*(6), 703–714.

Donia, A. M., Atia, A. A., & Elwakeel, K. Z. (2008). Selective separation of mercury (II) using magnetic chitosan resin modified with Schiff's base derived from thiourea and glutaraldehyde. *Journal of Hazardous Materials, 151*(2–3), 372–379.

Durango, A., Soares, N. F. F., Benevides, S., Teixeira, J., Carvalho, M., Wobeto, C., & Andrade, N. J. (2006). Development and evaluation of an edible antimicrobial film based on yam starch and chitosan. *Packaging Technology and Science: An International Journal, 19*(1), 55–59.

Dutta, P., Tripathi, S., Mehrotra, G., & Dutta, J. (2009). Perspectives for chitosan based antimicrobial films in food applications. *Food Chemistry, 114*(4), 1173–1182.

Elsabee, M. Z., & Abdou, E. S. (2013). Chitosan based edible films and coatings: A review. *Materials Science and Engineering C, 33*(4), 1819–1841.

Eser, A., Tirtom, V. N., Aydemir, T., Becerik, S., & Dinçer, A. (2012). Removal of nickel (II) ions by histidine modified chitosan beads. *Chemical Engineering Journal, 210*, 590–596.

Eweis, M., Elkholy, S., & Elsabee, M. (2006). Antifungal efficacy of chitosan and its thiourea derivatives upon the growth of some sugar-beet pathogens. *International Journal of Biological Macromolecules, 38*(1), 1–8.

Fan, L., Luo, C., Sun, M., Li, X., & Qiu, H. (2013). Highly selective adsorption of lead ions by water-dispersible magnetic chitosan/graphene oxide composites. *Colloids and Surfaces. B, Biointerfaces, 103*, 523–529.

Fei Liu, X., Lin Guan, Y., Zhi Yang, D., Li, Z., & De Yao, K. (2001). Antibacterial action of chitosan and carboxymethylated chitosan. *Journal of Applied Polymer Science, 79*(7), 1324–1335.

Fierro, S., del Pilar Sánchez-Saavedra, M., & Copalcua, C. (2008). Nitrate and phosphate removal by chitosan immobilized *Scenedesmus*. *Bioresource Technology, 99*(5), 1274–1279.

Fu, Y., & Xiao, C. (2017). A facile physical approach to make chitosan soluble in acid-free water. *International Journal of Biological Macromolecules, 103*, 575–580.

Han, S., Wang, H., Sun, Z., Zhao, H., & Zhang, P. (2017). Surface modification of PS microtiter plate with chitosan oligosaccharides by 60Co irradiation. *Carbohydrate Polymers, 176*, 135–139.

Hasan, S., Ghosh, T. K., Viswanath, D. S., & Boddu, V. M. (2008). Dispersion of chitosan on perlite for enhancement of copper (II) adsorption capacity. *Journal of Hazardous Materials, 152*(2), 826–837.

Howling, G. I., Dettmar, P. W., Goddard, P. A., Hampson, F. C., Dornish, M., & Wood, E. J. (2001). The effect of chitin and chitosan on the proliferation of human skin fibroblasts and keratinocytes in vitro. *Biomaterials, 22*(22), 2959–2966.

Ho, K. W., Ooi, C. W., Mwangi, W. W., Leong, W. F., Tey, B. T., & Chan, E.-S. (2016). Comparison of self-aggregated chitosan particles prepared with and without ultrasonication pretreatment as Pickering emulsifier. *Food Hydrocolloids, 52*, 827–837.

Hydari, S., Sharififard, H., Nabavinia, M., & reza Parvizi, M. (2012). A comparative investigation on removal performances of commercial activated carbon, chitosan biosorbent and chitosan/activated carbon composite for cadmium. *Chemical Engineering Journal, 193*, 276–282.

Ito, T., Takami, T., Uchida, Y., & Murakami, Y. (2018). Chitosan gel sheet containing drug carriers with controllable drug-release properties. *Colloids and Surfaces. B, Biointerfaces, 163*, 257–265.

Jamróz, E., Kulawik, P., & Kopel, P. (2019). The effect of nanofillers on the functional properties of biopolymer-based films: A review. *Polymers*, *11*(4), 675.

Jayakumar, R., Reis, R., & Mano, J. (2007). Synthesis and characterization of pH-sensitive thiol-containing chitosan beads for controlled drug delivery applications. *Drug Delivery*, *14*(1), 9−17.

Jumaa, M., Furkert, F. H., & Müller, B. W. (2002). A new lipid emulsion formulation with high antimicrobial efficacy using chitosan. *European Journal of Pharmaceutics and Biopharmaceutics*, *53*(1), 115−123.

Kanatt, S. R., Chander, R., & Sharma, A. (2008). Chitosan and mint mixture: A new preservative for meat and meat products. *Food Chemistry*, *107*(2), 845−852.

Kanatt, S. R., Chander, R., & Sharma, A. (2008). Chitosan glucose complex—A novel food preservative. *Food Chemistry*, *106*(2), 521−528.

Kim, I.-Y., Seo, S. J., Moon, H. S., Yoo, M. K., Park, I. Y., Kim, B. C., & Cho, C. S. (2008). Chitosan and its derivatives for tissue engineering applications. *Biotechnology Advances*, *26*(1), 1−21.

Kulkarni, V., Kulkarni, P., & Keshavayya, J. (2007). Glutaraldehyde-crosslinked chitosan beads for controlled release of diclofenac sodium. *Journal of Applied Polymer Science*, *103*(1), 211−217.

Kumirska, J., Czerwicka, M., Kaczyński, Z., Bychowska, A., Brzozowski, K., Thöming, J., & Stepnowski, P. (2010). Application of spectroscopic methods for structural analysis of chitin and chitosan. *Marine Drugs*, *8*(5), 1567−1636.

Kyzas, G. Z., & Deliyanni, E. A. (2013). Mercury (II) removal with modified magnetic chitosan adsorbents. *Molecules*, *18*(6), 6193−6214.

Kyzas, G. Z., Siafaka, P. I., Pavlidou, E. G., Chrissafis, K. J., & Bikiaris, D. N. (2015). Synthesis and adsorption application of succinyl-grafted chitosan for the simultaneous removal of zinc and cationic dye from binary hazardous mixtures. *Chemical Engineering Journal*, *259*, 438−448.

Lazaridou, A., & Biliaderis, C. G. (2002). Thermophysical properties of chitosan, chitosan−starch and chitosan−pullulan films near the glass transition. *Carbohydrate Polymers*, *48*(2), 179−190.

Learoyd, T. P., Burrows, J. L., French, E., & Seville, P. C. (2008). Chitosan-based spray-dried respirable powders for sustained delivery of terbutaline sulfate. *European Journal of Pharmaceutics and Biopharmaceutics*, *68*(2), 224−234.

Lebreton, L., & Andrady, A. (2019). Future scenarios of global plastic waste generation and disposal. *Palgrave Communications*, *5*(1), 1−11.

Lee, C. H., An, D. S., Park, H. J., & Lee, D. S. (2003). Wide-spectrum antimicrobial packaging materials incorporating nisin and chitosan in the coating. *Packaging Technology and Science: An International Journal*, *16*(3), 99−106.

Lee, S. B., Ha, D. I., Cho, S. K., Kim, S. J., & Lee, Y. M. (2004). Temperature/pH-sensitive comb-type graft hydrogels composed of chitosan and poly (N-isopropylacrylamide). *Journal of Applied Polymer Science*, *92*(4), 2612−2620.

Li, Q., Su, H., & Tan, T. (2008). Synthesis of ion-imprinted chitosan-TiO$_2$ adsorbent and its multi-functional performances. *Biochemical Engineering Journal*, *38*(2), 212−218.

Li, Y., Chen, X. G., Liu, N., Liu, C. S., Liu, C. G., Meng, X. H., Yu, L. J., & Kenedy, J. F. (2007). Physicochemical characterization and antibacterial property of chitosan acetates. *Carbohydrate Polymers*, *67*(2), 227−232.

Mahmoodi, N. M., & Mokhtari-Shourijeh, Z. (2015). Preparation of PVA-chitosan blend nanofiber and its dye removal ability from colored wastewater. *Fibers and Polymers*, *16*(9), 1861−1869.

Man, Z., Hu, X., Liu, Z., Huang, H., Meng, Q., Zhang, X., Dai, L., Zhang, J., Fu, X., Duan, X., Zhou, C., & Ao, Y. (2016). Transplantation of allogenic chondrocytes with chitosan hydrogel-demineralized bone matrix hybrid scaffold to repair rabbit cartilage injury. *Biomaterials*, *108*, 157−167.

Mansur, H. S., & Costa, H. S. (2008). Nanostructured poly (vinyl alcohol)/bioactive glass and poly (vinyl alcohol)/chitosan/bioactive glass hybrid scaffolds for biomedical applications. *Chemical Engineering Journal*, *137*(1), 72−83.

Ma, Y., Zhou, T., & Zhao, C. (2008). Preparation of chitosan−nylon-6 blended membranes containing silver ions as antibacterial materials. *Carbohydrate Research*, *343*(2), 230−237.

Meng, X., Li, B., Liu, J., & Tian, S. (2008). Physiological responses and quality attributes of table grape fruit to chitosan preharvest spray and postharvest coating during storage. *Food Chemistry*, *106*(2), 501−508.

Mochalova, A., Zaborshchikova, N. V., Knyazev, A. A., Smirnova, L. A., Izvozchikova, V. A., Medvedeva, V. V., & Semchikov, Y. D. (2006). Graft polymerization of acrylamide on chitosan: Copolymer structure and properties. *Polymer Science, Series A*, *48*(9), 918−923.

Mohamed, M., & Ouki, S. (2011). Removal mechanisms of toluene from aqueous solutions by chitin and chitosan. *Industrial & Engineering Chemistry Research*, *50*(16), 9557−9563.

Mun, G. A., Nurkeeva, Z. S., Dergunov, S. A., Nam, I. K., Maimakov, T. P., Shaikhutdinov, E. M., Lee, S. C., & Park, K. (2008). Studies on graft copolymerization of 2-hydroxyethyl acrylate onto chitosan. *Reactive & Functional Polymers*, *68*(1), 389−395.

Muthumeenal, A., Neelakandan, S., Kanagaraj, P., & Nagendran, A. (2016). Synthesis and properties of novel proton exchange membranes based on sulfonated polyethersulfone and N-phthaloyl chitosan blends for DMFC applications. *Renewable Energy*, *86*, 922−929.

Ngah, W. W., Endud, C., & Mayanar, R. (2002). Removal of copper (II) ions from aqueous solution onto chitosan and cross-linked chitosan beads. *Reactive & Functional Polymers*, *50*(2), 181−190.

Nguyen, V. C., Nguyen, V. B., & Hsieh, M.-F. (2013). Curcumin-loaded chitosan/gelatin composite sponge for wound healing application. *International Journal of Polymer Science*, *2013*.

Nitschke, J., Altenbach, H.-J., Malolepszy, T., & Mölleken, H. (2011). A new method for the quantification of chitin and chitosan in edible mushrooms. *Carbohydrate Research*, *346*(11), 1307−1310.

No, H. K., Park, N. Y., Lee, S. H., & Meyers, S. P. (2002). Antibacterial activity of chitosans and chitosan oligomers with different molecular weights. *International Journal of Food Microbiology*, *74*(1−2), 65−72.

Nunthanid, J., Huanbutta, K., Luangtana-anan, M., Sriamornsak, P., Limmatvapirat, S., & Puttipipatkhachorn, S. (2008). Development of time-, pH-, and enzyme-controlled colonic drug delivery using spray-dried chitosan acetate and hydroxypropyl methylcellulose. *European Journal of Pharmaceutics and Biopharmaceutics*, *68*(2), 253−259.

Oosegi, T., Onishi, H., & Machida, Y. (2008). Novel preparation of enteric-coated chitosan-prednisolone conjugate microspheres and in vitro evaluation of their potential as a colonic delivery system. *European Journal of Pharmaceutics and Biopharmaceutics*, *68*(2), 260−266.

Pham, D. C., Nguyen, T. H., Ngoc, U. T. P., Le, N. T. T., Tran, T. V., & Nguyen, D. H. (2018). Preparation, characterization and antifungal properties of chitosan-silver nanoparticles synergize fungicide against *Pyricularia oryzae*. *Journal of Nanoscience and Nanotechnology*, *18*(8), 5299−5305.

Pichyangkura, R., & Chadchawan, S. (2015). Biostimulant activity of chitosan in horticulture. *Scientia Horticulturae*, *196*, 49−65.

Prateepchanachai, S., Thakhiew, W., Devahastin, S., & Soponronnarit, S. (2019). Improvement of mechanical and heat sealing properties of chitosan films via the use of glycerol and gelatin blends in film-forming solution. *International Journal of Biological Macromolecules*, *131*, 589−600.

Rabea, E. I., Badawy, M. E. I., Rogge, T. M., Stevens, C. V., Höfte, M., Steurbaut, W., & Smagghe, G. (2005). Insecticidal and fungicidal activity of new synthesized chitosan derivatives. *Pest Management Science*, *61*(10), 951−960.

Ramesh, A., Hasegawa, H., Sugimoto, W., Maki, T., & Ueda, K. (2008). Adsorption of gold (III), platinum (IV) and palladium (II) onto glycine modified cross linked chitosan resin. *Bioresource Technology*, *99*(9), 3801−3809.

Ribeiro, M. P., Espiga, A., Silva, D., Baptista, P., Henriques, J., Ferreira, C., Silva, J. C., Borges, J. P., Pires, E., Chaves, P., & Correia, I. J. (2009). Development of a new chitosan hydrogel for wound dressing. *Wound Repair and Regeneration*, *17*(6), 817−824.

Rinaudo, M., & Goycoolea, F. M. (2019). *Advances in chitin/chitosan characterization and applications*. MDPI-Multidisciplinary Digital Publishing Institute.

Rodríguez-Vázquez, M., Vega-Ruiz, B., Ramos-Zúñiga, R., Saldaña-Koppel, D. A., & Quiñones-Olvera, L. F. (2015). Chitosan and its potential use as a scaffold for tissue engineering in regenerative medicine. *BioMed Reserach International*, *2015*.

Samal, S. K., Dash, M., Dubruel, P., Müllen, K., & Rajadas, J. (2014). Chapter 20. Cationic polymers as carriers through the blood–brain barrier. In S. Samal, & P. Dubruel (Eds.), *Polymer chemistry series* (pp. 539–556). Cambridge: Royal Society of Chemistry.

Shi, X., Du, Y., Yang, J., Zhang, B., & Sun, L. (2006). Effect of degree of substitution and molecular weight of carboxymethyl chitosan nanoparticles on doxorubicin delivery. *Journal of Applied Polymer Science*, *100*(6), 4689–4696.

Shirosaki, Y., Okayama, T., Tsuru, K., Hayakawa, S., & Osaka, A. (2008). Synthesis and cytocompatibility of porous chitosan–silicate hybrids for tissue engineering scaffold application. *Chemical Engineering Journal*, *137*(1), 122–128.

Siripatrawan, U., & Vitchayakitti, W. (2016). Improving functional properties of chitosan films as active food packaging by incorporating with propolis. *Food Hydrocolloids*, *61*, 695–702.

Sivashankari, P., & Prabaharan, M. (2016). Prospects of chitosan-based scaffolds for growth factor release in tissue engineering. *International Journal of Biological Macromolecules*, *93*, 1382–1389.

Skorik, Y. A., Pestov, A. V., Kodess, M. I., & Yatluk, Y. G. (2012). Carboxyalkylation of chitosan in the gel state. *Carbohydrate Polymers*, *90*(2), 1176–1181.

Somashekar, D., & Joseph, R. (1996). Chitosanases—properties and applications: A review. *Bioresource Technology*, *55*(1), 35–45.

Sugunan, A., Thanachayanont, C., Dutta, J., & Hilborn, J. (2005). Heavy-metal ion sensors using chitosan-capped gold nanoparticles. *Science and Technology of Advanced Materials*, *6*(3–4), 335.

Sui, W., Wang, Y., Dong, S., & Chen, Y. (2008). Preparation and properties of an amphiphilic derivative of succinyl-chitosan. *Colloids and Surfaces A: Physicochemical and Engineering*, *316*(1–3), 171–175.

Taha, S. M., & Swailam, H. M. (2002). Antibacterial activity of chitosan against *Aeromonas hydrophila*. *Food/Nahrung*, *46*(5), 337–340.

Thomas, S., Mishra, R. K., & Asiri, A. M. (2019). *Sustainable polymer composites and nanocomposites*. Springer.

van den Broek, L. A. M., Knoop, R. J. I., Kappen, F. H. J., & Boeriu, C. G. (2015). Chitosan films and blends for packaging material. *Carbohydrate Polymers*, *116*(Feb), 237–242. Available from https://doi.org/10.1016/j.carbpol.2014.07.039.

Vargas, M., Chiralt, A., Albors, A., & González-Martínez, C. (2009). Effect of chitosan-based edible coatings applied by vacuum impregnation on quality preservation of fresh-cut carrot. *Postharvest Biology and Technology*, *51*(2), 263–271.

Vitali, L., Laranjeira, M. C., Gonçalves, N. S., & Fávere, V. T. (2008). Spray-dried chitosan microspheres containing 8-hydroxyquinoline-5 sulphonic acid as a new adsorbent for Cd (II) and Zn (II) ions. *International Journal of Biological Macromolecules*, *42*(2), 152–157.

Wang, C., & Chen, C. (2005). Anti-bacterial and swelling properties of acrylic acid grafted and collagen/chitosan immobilized polypropylene non-woven fabrics. *Journal of Applied Polymer Science*, *98*(1), 391–400.

Wang, L., & Wang, A. (2008). Adsorption properties of Congo Red from aqueous solution onto surfactant-modified montmorillonite. *Journal of Hazardous Materials*, *160*(1), 173–180.

Wang, Q., Zhang, J., & Wang, A. (2009). Preparation and characterization of a novel pH-sensitive chitosan-g-poly (acrylic acid)/attapulgite/sodium alginate composite hydrogel bead for controlled release of diclofenac sodium. *Carbohydrate Polymers*, *78*(4), 731–737.

Wang, X., Du, Y., & Liu, H. (2004). Preparation, characterization and antimicrobial activity of chitosan–Zn complex. *Carbohydrate Polymer*, *56*(1), 21–26.

Wang, X., Du, Y., Luo, J., Lin, B., & Kennedy, J. F. (2007). Chitosan/organic rectorite nanocomposite films: Structure, characteristic and drug delivery behaviour. *Carbohydrate Polymers*, *69*(1), 41–49.

Xu, Y., Gallert, C., & Winter, J. (2008). Chitin purification from shrimp wastes by microbial deproteination and decalcification. *Applied Microbiology and Biotechnology*, *79*(4), 687–697.

Yang, X., Zhang, Q., Wang, Y., Chen, H., Zhang, H., Gao, F., & Liu, L. (2008). Self-aggregated nanoparticles from methoxy poly (ethylene glycol)-modified chitosan: Synthesis; characterization; aggregation and methotrexate release in vitro. *Colloids and Surfaces. B, Biointerfaces*, *61*(2), 125–131.

Yu, D., Regenstein, J. M., Xia, W., Yang, F., Jiang, Q., & Wang, B. (2018). The effects of edible chitosan-based coatings on flavor quality of raw grass carp (*Ctenopharyngodon idellus*) fillets during refrigerated storage. *Food Chemistry, 242*, 412–420.

Yuan, G., Chen, X., & Li, D. (2016). Chitosan films and coatings containing essential oils: The antioxidant and antimicrobial activity, and application in food systems. *Food Research International, 89*, 117–128.

Zargar, V., Asghari, M., & Dashti, A. (2015). A review on chitin and chitosan polymers: Structure, chemistry, solubility, derivatives, and applications. *ChemBioEng Reviews, 2*(3), 204–226.

Zemljič, L. F., Tkavc, T., Vesel, A., & Šauperl, O. (2013). Chitosan coatings onto polyethylene terephthalate for the development of potential active packaging material. *Applied Surface Science, 265*, 697–703.

Zeng, D., Wu, J., & Kennedy, J. F. (2008). Application of a chitosan flocculant to water treatment. *Carbohydrate Polymers, 71*(1), 135–139.

Zhai, X., Wei, W., Zeng, J., Gong, S., & Yin, J. (2006). Layer-by-layer assembled film based on chitosan/carbon nanotubes, and its application to electrocatalytic oxidation of NADH. *Microchimica Acta, 154*(3–4), 315–320.

Zhao, X., Mai, Z., Kang, X., & Zou, X. (2008). Direct electrochemistry and electrocatalysis of horseradish peroxidase based on clay–chitosan-gold nanoparticle nanocomposite. *Biosensors & Bioelectronics, 23*(7), 1032–1038.

Zhong, Z., Ji, X., Xing, R., Liu, S., Guo, Z., Chen, X., & Li, P. (2007). The preparation and antioxidant activity of the sulfanilamide derivatives of chitosan and chitosan sulfates. *Bioorganic & Medicinal Chemistry, 15*(11), 3775–3782.

CHAPTER 17

Nanocellulose-reinforced biocomposites

Sam Sung Ting[1,2], Gan Pei Gie[1], Mohd Firdaus Omar[2,3] and Muhammad Faiq Abdullah[1]

[1]School of Bioprocess Engineering, Universiti Malaysia Perlis, Arau, Malaysia, [2]Center of Excellence Geopolymer and Green Technology, Universiti Malaysia Perlis (UniMAP), Kangar, Perlis, Malaysia, [3]School of Material Engineering, Universiti Malaysia Perlis, Arau, Malaysia

17.1 Introduction

Over the last decades, the utilization of petroleum-based polymers has increased significantly due to their ease of processability, low cost, light weight, and outstanding barrier properties (Guidotti et al., 2017). However, the increased consumption of plastic products has led to the severe pollution in marine and terrestrial environments resulted by the accumulation of a great amount of plastic wastes in marine or landfills site. These plastic wastes are nonbiodegradable with consequence permanence in the environment for at least hundreds years (Miandad, Barakat, Aburiazaiza, Rehan, & Nizami, 2016). Therefore there is an increase in the public awareness on the negative effects of the use of plastic products and concern on environmental sustainability, which in turn to stimulate the production of degradable and environmental-friendly composites. The polymer composites also play an important role in expanding the circular bioeconomy. It can be designed to be fully biodegraded into carbon dioxide or contribute to the carbon capture and storage via the integration into nonbiodegradable infrastructure.

In order to meet the requirements of end users, the properties of polymers can be modified by the addition of reinforcing filler to suit the different applications. The exploration of filler such as nanocellulose as the reinforcement for polymer composites is found to be promising to enhance the properties of polymeric composites. Nanocellulose has garnered tremendous level of attention from the scientific community. It has been the topic of a wide range of research efforts that aimed at various application due to their high mechanical strength, large surface area, biodegradability, and sustainability (Gan, Sam, bin Abdullah, & Omar, 2019). Therefore this chapter discusses the preparation methods of nanocellulose from different sources. Concurrently, the processing methods, properties and application of nanocellulose-reinforced biocomposites are elucidated and elaborated as well.

17.2 Cellulose

Cellulose was first discovered in the plant cell wall by Anselme Payen in 1838 (Payen, 1838). Cellulose is one of the most abundant biopolymers in Earth, occurring in wood, cotton, and other plant-based materials. Its annual worldwide production is estimated to be 10^{10} and 10^{11} tons (Maaloul, Ben Arfi, Rendueles, Ghorbal, & Diaz, 2017). An elementary plant fiber is a single cell, typically a length from 1 to 50 mm and a diameter of around 10–50 μm (Chirayil, Mathew, & Thomas, 2014). In nature, cellulose does not occur as an isolated individual molecule, but it is found as assemblies of individual cellulose chain-forming fiber. Pristine fiber surface composed of bundles of microfibrils with deposited noncellulosic material such as lignin, hemicellulose, and wax is shown in Fig. 17.1. Plant fibers are like microscopic tubes and it consists of several cell walls. The cellulose and noncellulosic materials are located in secondary cell wall.

Cellulose is a linear homopolysaccharide which made up of repeated D-glucopyranose units linked together by β-1–4-linkages at C1 and C4 hydroxyl groups as shown in Fig. 17.2. Each monomer consists of three hydroxyl groups which has the ability to form the hydrogen bonds with other cellulose units. The hydroxyl groups and their ability to form hydrogen bonding allowing the crystalline packing of cellulose chains into highly compact system (Brinchi, Cotana, Fortunati, & Kenny, 2013).

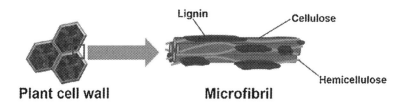

Figure 17.1
Structural organization of lignin, hemicellulose, and cellulose in cell wall. Source: *Adapted from Sharma, A., Thakur, M., Bhattacharya, M., Mandal, T., & Goswami, S. (2019). Commercial application of cellulose nano-composites—A review. Biotechnology Reports, 21.*

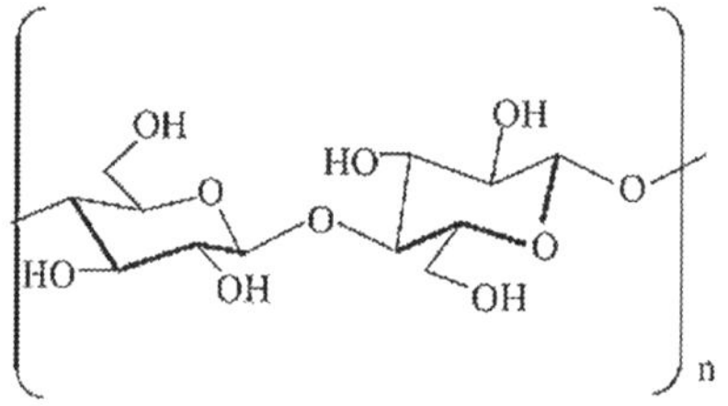

Figure 17.2
Structure of cellulose. Source: *Adapted from De Prez, J., Van Vuure, A. W., Ivens, J., Aerts, G., & Van de Voorde, I. (2018). Enzymatic treatment of flax for use in composites. Biotechnology Reports, 20, e00294.*

Table 17.1: Mechanical properties of natural fiber.

Fiber	Density (g cm^{-3})	Failure strain (%)	Tensile strength (MPa)	Young's modulus (GPa)
Ramie	1.5	2.0–3.8	400–938	44–128
Flax	1.5	1.2–3.2	345–1830	27–80
Hemp	1.5	1.6	550–1110	58–70
Jute	1.3–1.5	1.5–1.8	393–800	10–55
Sisal	1.3–1.5	2.0–2.5	507–855	9.4–28
Cotton	1.5–1.6	3.0–10	287–800	5.5–13
Empty fruit bunch	0.7–1.55	4–18	100–550	1.7–9

Sources: *Adapted from Pickering, K. L., Efendy, M. G. A., & Le, T. M. (2016). A review of recent developments in natural fibre composites and their mechanical performance. Composites Part A: Applied Science and Manufacturing, 83, 98–112; Mahjoub, R., Bin Mohamad Yatim, J., & Mohd Sam, A. R. (2013). A review of structural performance of oil palm empty fruit bunch fiber in polymer composites. Advances in Materials Science and Engineering, 2013.*

Cellulose has attracted significant attention due to its excellent properties such as biocompatibility, biodegradability, mechanical properties, and thermal and chemical stability (Chen et al., 2015). Its high strength and stiffness make it a suitable material as reinforcing filler in composites with structural requirements. Table 17.1 summarizes the comparison of the mechanical properties of natural fibers.

17.3 Nanocellulose

Nanocellulose can be categorized into different categories based on their source, dimension, shape, and preparation methods. Basically, nanocellulose is divided into two types which are cellulose nanocrystal (CNC) and cellulose nanofiber (CNF), respectively. It can be derived from various lignocellulosic sources such as tunicates, bacteria, agricultural and forestry wastes, and energy crops (Halib et al., 2017). Nanocellulose offers numerous attractive characteristics such as great surface area, high tensile strength, high modulus elasticity, and high crystallinity. Besides, nanocellulose is capable to attain a similar enhancement as cellulose which requires higher loading to improve the properties of composites (Li, Tian, Jin, & Li, 2018). Therefore nanocellulose can be considered as a significant reinforcing material for the development of high-performance composites.

17.3.1 Cellulose nanofiber

CNF, which also known as nanofibrillated cellulose or cellulose nanofibril, is a micrometer-long entangled cellulose with a diameter less than 100 nm and a length up to several micrometer as shown in Fig. 17.3 (Nair & Yan, 2015). It consists of both amorphous and crystalline parts. The CNF shows densely packed structure with small interstices between the cellulose. High-pressure homogenization is the most common technique for the isolation

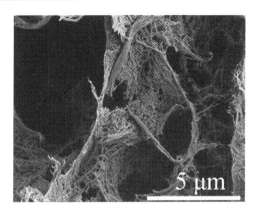

Figure 17.3
Representative micrograph of cellulose nanofiber. Source: *Adapted from Suzuki, A., Sasaki, C., Asada, C., & Nakamura, Y. (2018). Production of cellulose nanofibers from Aspen and Bode chopsticks using a high temperature and high pressure steam treatment combined with milling. Carbohydrate Polymers, 194, 303–310.*

of CNF. It involves passing the cellulose into a small nozzle at high pressure. During the high-pressure homogenizing, the high pressure and velocity in the fluid generate shear rates in the stream, the interfibrillar hydrogen bonds in cellulose is delaminated by mechanical force and the amorphous region of cellulose is break down, hence CNF can be isolated (Frone, Panaitescu, & Donescu, 2011). High-pressure homogenization has attracted the attention from researchers for the application of cellulose refining due to its high efficiency, simplicity, and do not require organic solvents for processing (Li et al., 2012). The first CNF was developed from wood pulps in 1983. It was prepared by treating the cellulose to the high-pressure homogenization without any pretreatment (Herrick, Casebier, Hamilton, & Sandberg, 1983). Subsequently, different raw materials have been applied in high-pressure homogenization for the isolation of CNF such as bleached wood pulp (Quiévy et al., 2010), cotton (Wang et al., 2013), sugarcane bagasse (Li et al., 2012), bamboo pulp (Zhang et al., 2012), oil palm empty fruit bunch (Fatah et al., 2014), coconut palm (Xu et al., 2015), and bleached softwood Kraft pulp (Niazi, Jahan, Berg, & Gregersen, 2017). However, problems such as clogging can occur due to the small nozzle size of homogenizer and lead to the immature termination of the homogenizing process. Therefore pretreatments are needed in order to reduce the size of cellulose and also prevent the agglomeration of cellulose during homogenizing (Wang, Zhang, Jiang, Li, & Yu, 2015). Tian et al. (2016) have produced the highly negative CNF via high-pressure homogenization by 20 times under a pressure of 600 bar. The homogenization process was preceded by strong acid hydrolysis pretreatment in order to reduce the energy required during homogenizing of cellulose (Chen et al., 2017). The application of pretreatments is very essential as it can decrease the number of pass to go through during homogenization and shorten the time required for homogenization.

Another method, microfluidization, is similar to high-pressure homogenization except that the intensifier pump and the interaction chamber are applied in microfluidization system. The pressure within the microfluidizer can be increased by the intensifier pump and lead to the defibrillation of cellulose inside the chamber (Ferrer, Filpponen, Rodríguez, Laine, & Rojas, 2012). CNF produced by microfluidization has a more uniform size compared to high-pressure homogenization. The effect of different passing number and chamber size of the microfluidizer on the morphological property of CNF was investigated. The results proven that the number of passes has a profound effect on the homogeneity of CNF and showed insignificant difference on the diameter of CNF. The diameter of the CNF was mainly affected by the size of chamber. As the chamber size reduced, the diameter of CNF decreased as well (Taheri & Samyn, 2016).

Besides, high-intensity ultrasonication can also be used to convert the cellulose to CNF. High-intensity ultrasonication is a simple technique to isolate the CNF by using hydrodynamics force of ultrasound. During the ultrasonication, the cavitation has led to formation of strong oscillating power which comprises of the formation, expansion, and implosion of microscopic gas bubbles when the molecules absorb the ultrasonic energy. The oscillating power breaks down the interfibrillar hydrogen bonding between the cellulose and defibrillate it into CNF (Chen et al., 2013).

The mechanical separation techniques such as high-pressure homogenization, microfluidization, and high-intensity ultrasonication are efficient methods for the isolation of CNF, however, these processes have two main problems which are high-energy consumption and clogging issues. Therefore pretreatments are required in order to reduce the size of cellulose and reduce the energy consumption as well as the clogging problems. Alkaline pretreatment and bleaching are the most common pretreatment applied for the dissolution of lignin and hemicellulose to increase the accessibility of cellulose for the isolation of CNF (Kouadri & Satha, 2018; Leite, Zanon, & Menegalli, 2017). Besides, 2,2,6,6-tetramethylpiperidine-1-oxyl radical (TEMPO) oxidation is another promising method for reducing the energy consumption. The increased aspect ratio and surface charge of CNF by TEMPO-mediated oxidation could generate the repulsive forces between fibrils, hence contributing to a faster and easier defibrillation during mechanical shearing of CNF (Cao, 2018).

17.3.2 Cellulose nanocrystal

The outstanding properties of CNC have attracted significant concern on the research of the isolation of CNC from lignocellulosic biomass, especially from agricultural residues. The main principle of the isolation of CNC involves the removal of disordered amorphous domains from cellulose chains and the release of crystalline regions. Acid hydrolysis can be considered as the most well-known chemical treatment that employs strong acids to hydrolyze the amorphous domains of cellulose (Gan, Sam, Abdullah, Omar, & Tan, 2020).

Initially, the acid diffuses into the amorphous regions of the cellulose and the glycosidic bonds are break down by hydrolysis. Afterward, the more easily accessible glycosidic bonds are hydrolyzed and lastly hydrolysis takes place at the reducing end group and at the surface of the CNC (Trache, Hussin, Haafiz, & Thakur, 2017). Various acids are used in the acid hydrolysis of cellulose such as hydrochloric acid (Rafieian, Shahedi, Keramat, & Simonsen, 2014), phosphoric acid (Espinosa, Kuhnt, Foster, & Weder, 2013), formic acid (Liu et al., 2016), and sulfuric acid (Martelli-Tosi et al., 2018). Among all, sulfuric acid hydrolysis is the most popular and commonly used technique. The isolation of CNC using sulfuric acid leads to a negatively charged crystallites surface due to the esterification by sulfate ions after acid hydrolysis. As a result, a more stable CNC dispersion can be formed using sulfuric acid compared to other acids (Yang et al., 2017).

The comparison among the effect of four different acids on the isolation of CNC from bamboo at 60°C for 2 h was determined. The acids were hydrochloric acid, phosphoric acid, sulfuric acid, and a mixture of acetic acid and nitric acid. CNC hydrolyzed by sulfuric acid presented the highest degree of crystallinity followed by phosphoric acid. Whereas, the CNC hydrolyzed by hydrochloric acid and the acetic acid and nitric acid mixture displayed the lowest crystallinity index. Hydrochloric acid and the acetic acid and nitric acid mixture could promote the swelling of cellulose, thereby facilitate the breakage of hydrogen bonding in cellulosic crystalline domains. The authors concluded that the CNC produced by sulfuric acid hydrolysis has the highest crystallinity and stability compared to other acids (Zhang et al., 2014).

The variation of hydrolysis parameters using sulfuric acid with different plant sources is summarized in Table 17.2. The properties of CNC are usually dependent not only on the type of treatment, time, acid concentration, and temperature of the reaction but also on the different sources of cellulose. The isolation of CNC from *Posidonia oceanica* leaves where the reaction time was in focus was reported (Luzi et al., 2016). *P. oceanica* leaves was hydrolyzed using a 64 wt.% sulfuric acid at 40°C with different reaction times. The CNC produced with 60 min reaction time had a smaller diameter and higher yield compared to other reaction time. The relationship between reaction time and thermal properties of CNC using a 64 wt.% sulfuric acid at 45°C for 30–90 min was also studied (Razali et al., 2017). The CNC produced with 60 min reaction time had the highest thermal stability and the lowest char residue. The introduction of sulfate groups to the surface of CNC could affect the thermal stability by the sulfation area and the sulfate group's flame-retardant activity.

While, the effect of hydrolysis time on the properties of CNC isolated from oil palm fronds was examined (Saurabh et al., 2016). Among all the CNC, CNC produced with 45-min hydrolysis was the best possible choice as a reinforcing agent due to its highest aspect ratio, crystallinity percentage, and moderate crystallinity size. The penetration of H^+ ions into the

Table 17.2: Variation of sulfuric acid hydrolysis parameters of various plant-based sources.

Raw material	Concentration of acid (wt.%)	Temperature (°C)	Time (min)	References
Alfa fibers	64	50	30	El Achaby, Kassab, Barakat, and Aboulkas (2018)
Kenaf core	60–64	45	60	Sabaruddin and Paridah (2018)
Empty fruit bunch	64	40	45	Shanmugarajah et al. (2015)
Cotton linter pulps	64	50	90	Mandal and Chakrabarty (2011)
Sugarcane bagasse	50	40	10	Wulandari, Rochliadi, and Arcana (2016)
Ushar seed fiber	64	50	75	Oun and Rhim (2016)
Potato peel	64	45	90	Chen, Lawton, Thompson, and Liu (2012)
Barley husk	65	50	60	Espino et al. (2014)
Rice straw	64	45	45	Lu and Hsieh (2012)
Almond shells	40	60	45	Maaloul et al. (2017)
Posidonia oceanica	64	45	30	Luzi et al. (2017)
Mandacaru	60	45	60–120	Nepomuceno, Santos, Oliveira, Glenn, and Medeiros (2017)
Oil palm mesocarp	65	45	45	Chieng, Lee, Ibrahim, Then, and Loo (2017)
Corncob	64	45	60	Liu et al. (2016)
Red algae	64	45	45	Chen, Lee, Juan, & Phang (2016)

crystallites and the actual cleavage of the glycosidic bond depended on the hydrolysis conditions, including the type of the acid used, duration and temperature of hydrolysis as well as acid concentration.

The probable mechanism of acid hydrolysis is represented in Fig. 17.4. Hydrolysis of cellulose with sulfuric acid involves protonation of glycosidic oxygen (path 1) or cyclic oxygen (path 2), followed by dissociation of glycosidic bonds induced by the addition of water (Lu & Lo, 2010). This hydrolysis process yields two fragments with shorter chains while preserving the nature of the chain polymer. Beside chain scission, sulfuric acid hydrolysis of cellulose also involves esterification of the hydroxyl groups.

Depending on reaction conditions, the appearance of the whiskers suspension can be white with some starting pulp particles (low yield), ivory white viscous suspension (optimal), yellowish or even black viscous suspension (overhydrolyzed) (Dong, Revol, & Gray, 1998). A temperature of 45°C yields an ivory-white suspension in 1 h is considered to be optimal for the production of CNC via sulfuric acid hydrolysis. Besides, in order to avoid the possible aggregation of CNC, ultrasonic treatment is carried out to prevent the desulfation of sulfate groups on CNC surface (Lu & Lo, 2010).

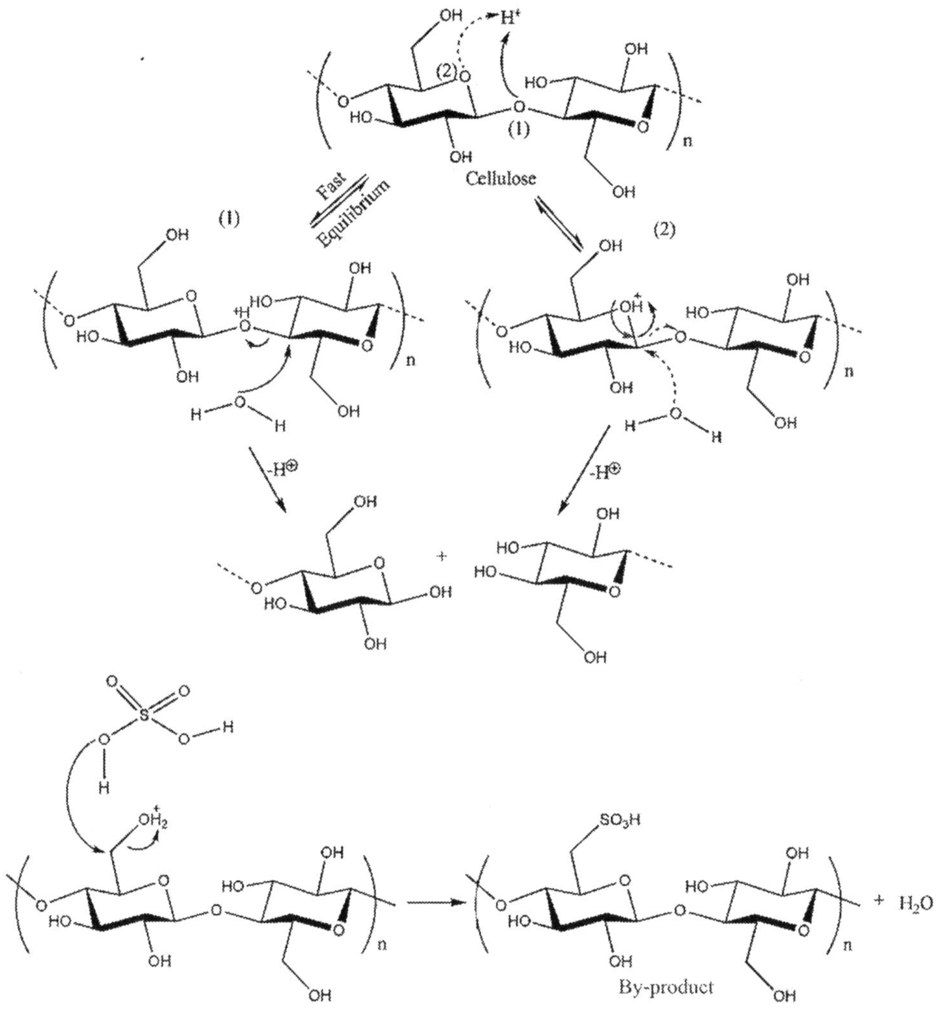

Figure 17.4
Mechanism of acid-catalyzed hydrolysis of cellulose. Source: *Adapted from Lu, P., & Lo Hsieh, Y. (2010). Preparation and properties of cellulose nanocrystals: Rods, spheres, and network. Carbohydrate Polymers, 82, 2, 329−336.*

17.4 Processing methods of nanocellulose-reinforced biocomposites

17.4.1 Solvent casting

Solvent casting has been widely employed in the preparation of polymeric biocomposites. It is a process which involves the mixing of polymeric matrix and reinforcing filler, followed by casting and drying steps (Benini, Cioffi, & Voorwald, 2017). The nanocellulose is dispersed in a solvent that is suitable to the polymer. After that, the nanocellulose and

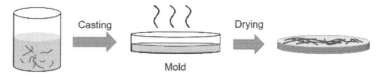

Figure 17.5
Fabrication of nanocellulose-reinforced composites by solvent casting. Source: *Adapted from Dai, Z., Ottesen, V., Deng, J., Helberg, R. M. L., & Deng, L. (2019). A brief review of nanocellulose based hybrid membranes for CO2 separation. Fibers, 7, 5, 1–18.*

polymeric matrix solution is casted onto a plate/dish and dried it by evaporation methods. Once the solvent is evaporated, the nanocellulose is entrapped with the polymeric chains and preserved the state of dispersion of nanocellulose in the solvent as shown in Fig. 17.5 (Mondragon, Peña-Rodriguez, González, Eceiza, & Arbelaiz, 2015). The advantage of solvent casting is that the fabrication of polymeric composites can be carried out without any need of specialized facility or equipment. Nevertheless, solvent-casting technique is suitable for industrial scale-up due to the large amount of organic solvent is required during casting of composites.

Different types of reinforcing fillers such as cellulose from cotton (Spagnol et al., 2018), wood (Butron, Llorente, Fernandez, Meaurio, & Sarasua, 2019), bark (Peng, Nair, Chen, Yan, & Cao, 2018), Luffa cylindrica (Follain et al., 2013), pineapple leaf (Cherian et al., 2011), and flax yarns (Liu, Sui, & Bhattacharyya, 2014) had been applied in the preparation of polymeric composites by the solvent-casting technique. The properties of polyvinyl alcohol (PVOH) composites with the incorporation of CNC from two different cellulose sources fabricated via solvent-casting method were determined (Fortunati et al., 2013). The field emission scanning electron micrograph showed that the preparation of CNC/PVOH composites by solvent casting exhibited a well-dispersed and homogeneous surface. The transparency test also showed that the incorporation of CNC did not affect the transparency of the PVOH composites. It implied that the CNC was dispersed homogeneously and no agglomeration of CNC during the processing of PVOH composites.

On the other hand, the disadvantages of solvent-casting method are the retention of toxic in polymer matrix and limited shapes of the polymer composites (Bhat, Dasan, Khan, Soleimani, & Usmani, 2017). Furthermore, due to the hydrophilic nature of nanocellulose, the choices of compatible solvents are limited to water-soluble polymers only. Therefore several studies had been carried out to overcome the incompatibility issue between hydrophilic nanocellulose and nonpolar matrices. Kasa et al. (2017) investigated the effect of surface modification of CNC on the compatibility between CNC and polylactic acid (PLA) matrix. Acetylation of CNC was performed in order to decrease the hydrophilicity and polarity of CNC. They observed that the addition of acetylated-CNC as reinforcing filler has a more even dispersion within PLA matrix compared to the unmodified-CNC/PLA composites. It was attributed to the enhanced interfacial adhesion between filler and matrix resulted by the declination in CNC surface polarity.

Another approach to resolve the incompatibility issue between CNF and poly (3-hydroxybutyrate-co-3-hydroxyvalerate) (PHBV) was attempted (Benini et al., 2017). The mixture of dimethylformamide and chloroform was used as the solvent to suspend CNF and dissolve PHBV. The CNF/PHBV composites that were prepared by solvent casting has an average thickness of 93.33 ± 5.77 μm. The dispersion of CNF in PHBV was observed by using scanning electron microscope (SEM). The SEM images exhibited a homogeneous surface without the presence of visible agglomerates were obtained by using solvent casting. The composite transparency was also not affected by the addition of CNC. Similar observation was also made by Gu and Catchmark (2013).

17.4.2 Melt processing

Melt processing is a technique which involves the mixing of nanocellulose and polymer matrix at melt condition by using extruder. It can be considered as an environmental-friendly method as no organic solvent is required during the processing of composites (Fornes & Paul, 2003). Besides, it is also more technoeconomically viable due to is compatibility with the industrial process as it offers a higher degree of freedoms in the products specifications (Qaiss, Bouhfid, & Essabir, 2015).

The properties of polymeric composites are mainly affected by the aspect ratio of nanocellulose, the dispersion of nanocellulose in the polymer matrix, and the interfacial adhesion between nanocellulose and polymer matrix (Ferreira, Pinheiro, de Souza, Mei, & Lona, 2019). Several studies have proven that the melt processing could produce composites with better dispersion of nanocellulose and less nanocellulose agglomerates. CNF/thermoplastic starch composites with varies CNF concentration (0, 5, 10, 15, and 20 wt.%) were melt compounded by using a twin screw extruder as displayed in Fig. 17.6 (Hietala, Mathew, & Oksman, 2013). The temperature profile of various zones in extruder was set as 80°C, 90°C, 90°C, 100°C, 100°C, 110°C, and 110°C. The extrusion process was conducted at a speed of 200 rpm. At a higher rotational speed, the excessive water could be evaporated by the vacuum ventilation and also decreased the residence time in extruder. The tensile strength of composites was significantly enhanced by the incorporation of CNF up to 10 wt.%. The further addition of CNF has no significant enhancement on the tensile strength of thermoplastic starch composites. The transparency of thermoplastic starch composites was observed by using UV-vis spectrometer. There was no significant reduction in clarity of the composites and no visible aggregation of nanocellulose was observed in the composites with the addition of up to 15 wt.% CNF.

Due to the hydrophilic nature of nanocellulose, it is always challenging to disperse the nanocellulose evenly in the hydrophobic matrix such as poly(ε-caprolactone).

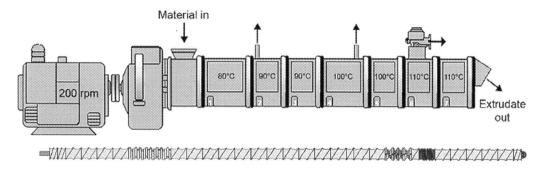

Figure 17.6

Fabrication of CNF/thermoplastic starch composites by extrusion method. (*CNF*, Cellulose nanofiber) Source: *Adapted from Hietala, M., Mathew, A. P., & Oksman, K. (2013). Bionanocomposites of thermoplastic starch and cellulose nanofibers manufactured using twin-screw extrusion. European Polymer Journal, 49, 4, 950–956.*

Agglomeration of nanocellulose can form with the hydrophobic polymer matrix resulted by the interaction between nanocellulose which lead to the formation of interhydrogen-bonding network. Wei et al. (2017) proposed that the surface modification of CNC could produce the homogeneous structure and enhance the cellulose-hydrophobic polymer matrix interface which in turn improve the properties of composites. Transesterification of CNC by using canola oil fatty acid methyl ester was performed to reduce the hydrophilicity of CNC. The transesterified CNC was compounded together with PLA by using twin screw compounder at a temperature of 185°C. The morphology of PLA composites was investigated by using SEM. The SEM micrographs showed that the neat PLA film exhibited a smooth and homogeneous surface. The PLA composites with the addition of unmodified CNC exhibited a brittle surface with the discrete cavities, implying the poor interfacial adhesion between hydrophilic CNC and hydrophobic PLA. After the transesterification of CNC, a smoother surface and no phase separation was observed. This indicated that the transesterification of CNC is an effective method to improve the compatibility between CNC and hydrophobic matrix prepared by melt compounding technique.

17.4.3 Electrospinning

Electrospinning, which is also known as electrostatic fiber spinning, is a simple yet effective method for the fabrication of polymeric nanofibers with diameter ranging from micro- to nanometer via the electrostatic forces (Ghelich, Keyanpour Rad, & Youzbashi, 2015). The polymer solution is contained in a syringe pump, which is connected to the positive electrode by the surface tension while the other electrode is connected to the metal plate. When the intensity of voltage increases, the electrostatic forces could overcome the surface tension of polymer solution and a jet is ejected from the surface to produce

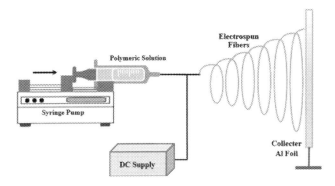

Figure 17.7

Schematic diagram of electrospinning technique. Source: *Adapted from Asmatulu, R. (2016). Highly hydrophilic electrospun polyacrylonitrile/polyvinypyrrolidone nanofibers incorporated with gentamicin as filter medium for dam water and wastewater treatment. Journal of Membrane and Separation Technology, 5, 2, 38–56.*

solvent-free nanofibers as illustrated in Fig. 17.7 (Lee, Jung, Lee, & Il Lee, 2019). The advantage of electrospinning is that the nanofibers produced have a relatively higher surface area to volume ratio. Besides, electrospinning also offers a high production rate, low processing cost, and consistent quality of nanofibers (Lee et al., 2018).

Electrospinning has gained a great level of interest from researchers in the preparation of polymer nanofibers and composites. It has lately been applied as an alternative technique for the processing of nanocellulose-reinforced composites. Fabrication of CNC-reinforced PVOH composite fibers via electrospinning was reported (Sutka et al., 2015). The incorporation of CNC reduced the diameter of composite nanofiber and enhanced the diameter distribution of composite nanofibers. The addition of CNC increased the conductivity of polymer solution and hence decreased the distance between the electrodes to eject the jet. Similar results were also observed by Sanders, Han, Rushing, and Gardner (2019), where the author reported the diameter of electrospun PVOH composite fibers was narrowed down by the incorporation of CNC.

The electrospun CNC-reinforced starch composite fiber mats by using cationic starch as the compatibilizer was prepared. The effect of varying cationic starch and CNC concentration and ratio of cationic starch to CNC on the properties of starch composite fibers was investigated. The results showed that a 2 wt.% of CNC-cationic starch concentration is the optimal concentration to improve the strength of composite fibers. The author also noticed that the ratio of cationic starch to CNC at 1:2 and 1:1 could enhance the reinforcing effect of CNC and improve the interfacial adhesion between filler and polymer matrix (Wang, Kong, & Ziegler, 2019).

17.4.4 Layer-by-layer

Layer-by-layer self-assembly (LBL) method is a versatile technique for the fabrication of thin membranes and composite film with controllable thickness down to nanoscale (Detzel, Larkin, & Rajagopalan, 2011). The application of LBL technique can maximize the interfacial adhesion between the nanofiller and polar polymeric composites. Besides, it also allows the preparation of biocomposites at high filler loading and at the same time provides a homogeneous dispersion of filler (Wei et al., 2015).

The multilayer film composite by alternately depositing CNF/PVOH on the CNF/polyethylene terephthalate (PET) substrate was prepared via LBL method (Hou, Wang, & Ragauskas, 2019). The results showed that the deposition of a single layer of PVOH on the CNF/PET substrate has increased the hydrophilicity of the composites. The thermal stability of the multilayer film composite was also found to be increased significantly due to the enhanced intermolecular bonding between CNF and PVOH.

The preparation of CNC/chitosan multilayer composites with highly deacetylated chitosan (cationic phase) and CNC (anionic phase) using LBL technique was also reported (de Mesquita, Donnici, & Pereira, 2010). The LBL technique was initiated by the driving forces of electrostatic interactions and hydrogen bonding between the sulfate groups of CNC and the amino groups of chitosan. The multilayer composite presented a smooth surface with the roughness of less than 11 nm. The 20-bilayer CNC/chitosan film composites obtained with a thickness of 140 nm displayed a highly uniform and dense packing of CNC without agglomeration which resulted by the good interaction between CNC and chitosan matrix.

In another study, by increasing the number of bilayers (chitosan multiwalled carbon nanotubes film and CNC multiwalled carbon nanotubes film) from 5 to 20, the electrochemical properties was decreased by two orders of magnitude resulted by the increase in the amount of multiwalled carbon nanotube as conductive material (Trigueiro, Silva, Pereira, & Lavall, 2014). The film composites showed a smooth surface with homogeneous dispersion of CNC and without the formation of CNC aggregation. Each layer with an average thickness of 10.0 nm was reported.

17.5 Properties of nanocellulose-reinforced biocomposites

17.5.1 Tensile properties

Tensile properties are the physical properties of polymeric composites indicate the deformation of composites under the application of force or load. Under the tensile testing, the composites are subjected to the tensile forces until failure takes place. The tensile properties of composites play an important role in the selection of suitable polymer for particular applications.

An improvement in tensile properties of kappa-carrageenan film composites after the addition of CNF was observed (Savadekar, Karande, Vigneshwaran, Bharimalla, & Mhaske, 2012). The kappa-carrageenan composites with different weight ratio of CNF (0.1, 0.2, 0.3, 0.4, 0.5, and 1 wt.% based on dry weight on kappa-carrageenan powder) were prepared. The incorporation of CNF enhanced the tensile strength significantly. The tensile strength of composites increased by 44% with the addition of 0.4 wt.% CNF compared to neat kappa-carrageenan film. This was attributed to the homogeneous distribution of CNF within the polymer matrix. However, the addition of CNF at higher concentration has led to the declination of tensile strength. It was ascribed to the addition of CNF exceeded the threshold limit, which resulted in the uneven distribution of CNF and the agglomeration of CNF within the kappa-carrageenan matrix. Other researches also show significant improvement in tensile strength for a great number of nanocellulose-reinforced polymer composites, such as PVOH (Cho & Park, 2011), soy protein isolate (Han, Yu, & Wang, 2018), sugar palm starch (Ilyas, Sapuan, Ishak, & Zainudin, 2018), hydroxypropyl methylcellulose (Bilbao-Sainz, Bras, Williams, Sénechal, & Orts, 2011), whey protein isolate (Qazanfarzadeh & Kadivar, 2016), thermoplastic starch (Savadekar & Mhaske, 2012), and unsaturated polyester (Jose, Mathew, Hassan, Mozetic, & Thomas, 2014).

Optimization of the tensile properties of CNC/chitosan composites was carried out by Dehnad, Emam-Djomeh, Mirzaei, Jafari, and Dadashi (2014). The influence of various concentration of chitosan, CNC, and glycerol was studied by using response surface method. With the addition of 1 wt./v% chitosan, 0.18 wt.% CNC, and 30 wt.% glycerol, the tensile strength and Young's modulus of 245.05 and 4430 MPa with an elongation at break at 46.80% could be achieved. The formation of strong intermolecular hydrogen bonding between CNC and chitosan resulted in sufficient stress transfer and hence improved the tensile properties of chitosan composites.

The difference in nanocellulose isolation methods could influence the quality of nanocellulose as well as its performance in reinforcing effect. The tensile properties of PVOH composites reinforced with nanocellulose isolated by acid hydrolysis, TEMPO-oxidation, and ultrasonication were determined (Zhou, Fu, Zheng, & Zhan, 2012). TEMPO-oxidated nanocellulose/PVOH composites exhibited the highest tensile strength and modulus compared to acid-hydrolyzed nanocellulose and ultrasonicated nanocellulose composites. The better tensile properties of TEMPO-oxidated nanocellulose/PVOH composites was attributed to the higher aspect ratio of nanocellulose which allowed a stronger interfacial adhesion between filler and matrix and effective stress transfer within the polymer composites.

The effect of the addition of nanocellulose on the tensile properties of hydrophobic matrix was examined (Cao et al., 2013). CNC isolated from nitrile rubber was used as the reinforcing phase in the nitrile rubber by extrusion. The addition of CNC showed significant improvement on the tensile strength of the composites. The tensile strength of composites increased doubled from 7.7 to 15.8 MPa with the inclusion of CNC from 0 to 20 phr. However, the addition of CNC also resulted in the declination in elongation at break of nitrile rubber composites. This could be attributed to the enhanced intermolecular reaction between filler and matrix, which diminished the ability to resist the deformation of rubber chains and restricted the chain mobility of polymer composites (Thomas et al., 2015).

Surface modification of nanocellulose can also lead to an improvement in tensile properties, especially with hydrophobic polymer matrices. The surface modification by acetylation of CNF could produce the PLA composites with improved tensile properties (Abdulkhani, Hosseinzadeh, Ashori, Dadashi, & Takzare, 2014). The tensile strength and Young's modulus of composites increased notably with the addition of acetylated CNF. This could be explained by the even dispersion of reinforcing CNC within the polymer matrix and the improved compatibility between CNF and PLA. On the other hand, the hydrophobicity of CNC can also be modified by esterification through the grafting of lauric acid (Trinh & Mekonnen, 2018). The esterified CNC exhibited significant reinforcing effect on the tensile strength and modulus of elasticity of the epoxy composites compared to unmodified-CNC/epoxy composites. It could be due to the improved interfacial adhesion between CNC and epoxy that led to an effective stress transfer.

17.5.2 Thermal properties

17.5.2.1 Thermal stability

Thermal property of the nanocellulose-reinforced polymer composite is gaining greater attention in the present scenario. It is one of the essential parameters in measuring the thermal property of polymer composites. The understanding of the thermal property plays a significant role in optimizing the processing and designing conditions of polymer composites as well as developing the composites with improved thermal properties in order to meet the requirements for different applications.

Thermal degradation of polymeric biocomposites generally comprises three major phases. The first stage involves the elimination of absorbed moisture or volatile matter; the second stage is the dehydration process, while the last stage is associated with the degradation of carbonaceous matter (Kiziltas et al., 2016). The understanding in the fundamental properties of nanocellulose such as morphology and dimensions is important in developing the structure-property relationship of composites for a great variety of applications (Sacui et al., 2014). The comparison between the effect of CNC or CNF incorporation on the thermal stability soy protein isolate film composites was performed (Martelli-Tosi et al., 2018). The thermal stability of CNC and CNF were found to be at approximately 220°C. Decomposition temperature of the composites increased remarkably regardless of the types of nanocellulose. However, the CNF-reinforced composites presented a higher decomposition temperature compared to CNC-reinforced soy protein isolate composites. This could be explained by the higher heat transfer rate resulted by high surface area of CNC, which in turn to reduce the thermal stability of composites (Ghaemi et al., 2018). Similar finding was also reported by Mondragon et al. (2015) in which the thermal stability of gelatin composites was enhanced notably by 7°C–9°C compared to the neat gelatin film, regardless CNC or CNF.

According to Yadav, Behera, Chang, and Chiu (2020), the inclusion of CNC enhanced the thermal stability of chitosan biocomposites by improving the interactions between CNC and chitosan matrix via hydrogen bonding. The enhanced intermolecular reaction between filler and matric increased the thermal energy required to break down the hydrogen-bonding network, hence lead to a higher decomposition temperature of composites. Similar observation was made by Ilyas et al. (2018). Besides, they also observed that amount of char residue of composites was increased with the concentration of CNC. The content of char residue increased from 4.91% (neat sugar palm starch film) to 5.25% (0.1 wt.% CNC/sugar palm starch composite). The authors deduced that the improvement in thermal stability could attributed to the large composition of carbonate in the CNC-reinforced sugar palm starch composite.

Nevertheless, a different behavior was observed, in which the thermal stability of starch composites declined upon the addition of CNC (Agustin, Ahmmad, Alonzo, & Patriana, 2014). The declination in the thermal decomposition temperature of CNC-reinforced starch composites was attributed to the inherent low thermal stability of CNC resulting from the sulfuric acid hydrolysis. Sulfation of CNC took place on the surface of CNC during acid hydrolysis and led to a lower thermal stability of CNC. Besides, the sulfation of CNC also increased the amount of charred residues due to the role of sulfate which acted as a flame retardant. The effect of sulfate group on the thermal stability of CNC was investigated (Lin & Dufresne, 2014). The experimental results proved that the sulfated CNC showed a lower heat resistance compared to the desulfated CNC. The substitution of hydroxyl group in CNC by sulfate group via sulfuric acid hydrolysis decreased the thermal stability of cellulose. This could be explained by the sulfate group that decomposed at a lower temperature (Nan et al., 2017). Moreover, the sulfate group also acted as the dehydrating catalyst in decomposing nanocellulose and thus led to a lower thermal stability (Sahlin et al., 2018). Similar findings were reported in polymer matrices such as chitosan (Tang, Zhang, Zhao, Guo, & Zhang, 2018) and PVOH (Li, Yue, & Liu, 2012), in which the inclusion of CNC showed declination in thermal stability of composites.

The hydrophilic characteristic of nanocellulose hinders its application as its dispersibility of nanocellulose in hydrophobic matrices is a big obstacle in developing the nanocomposites due to the incompatibility issue between filler and matrix. The effect of surface modification of nanocellulose on the thermal stability of PLA composites was determined (Kasa et al., 2017). The incorporation of unmodified CNC into the PLA matrix showed no significant improvement in the thermal stability of composites. However, with the incorporation of acetylated-CNC, the thermal decomposition temperature of PLA composites increased from 309°C for neat PLA film to 324°C for 1 wt.% acetylated-CNC/PLA composites. The significant increment in thermal stability was due to the homogeneous dispersion of acetylated-CNC within PLA composites which resulted in the good intermolecular reaction and led to a higher thermal stability of composites.

Recently, the effect of silane as a compatibilizer with various concentration on the thermal behavior of CNC/PLA composites was reported (Qian & Sheng, 2017). The experimental results showed that the CNC/PLA composites with the addition of 4 wt.% silane has the highest decomposition temperature compared to other concentration as presented in Fig. 17.8. The onset temperature and maximum decomposition temperature were 359.0°C and 383.0°C and increased to 361.2°C and 389.1°C, respectively. The improvement in thermal stability could be explained by the enhanced interfacial adhesion between CNC and PLA and, thus higher energy was required in order to disrupt the intermolecular hydrogen bonding of composites. In addition, the composites compatibilized with 16 wt.%

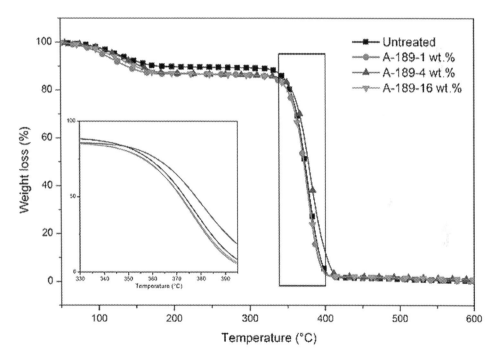

Figure 17.8

TGA curve of CNC/PLA composites compatibilized with silane at different concentrations. (*CNC*, Cellulose nanocrystal; *PLA*, polylactic acid) Source: *Adapted from Qian, S., & Sheng, K. (2017). PLA toughened by bamboo cellulose nanowhiskers: Role of silane compatibilization on the PLA bionanocomposite properties. Composites Science and Technology, 148, 59–69.*

silane showed a declination in thermal stability. This could be attributed to the self-condensation of silane which interrupts the intermolecular reaction between CNC and PLA.

17.5.2.2 Thermal behavior

Thermal behaviors such as glass-rubber transition temperature (T_g), crystallization temperature (T_c), melting temperature (T_m), and crystallinity index (X_c) can be investigated using differential scanning calorimetry (DSC). Thermal behavior of composites is an important parameter which may affect the properties of the final products. Its performance is greatly influence by the interfacial adhesion between filler and polymer matrix. The thermal characteristics of the composites reinforced with nanocellulose from different sources are tabulated in Table 17.3. The thermal properties of biocomposites are greatly affected by the incorporation of nanocellulose.

Nanocellulose was isolated by using different techniques (acid hydrolysis, TEMPO-oxidation, and ultrasonication) and investigated their effect on the thermal property of PVOH composites (Zhou et al., 2012). The inclusion of nanocellulose increased the T_g of composites from 70.3°C to 75.1°C, 76.9°C, and 71.2°C for acid hydrolyzed nanocellulose, TEMPO-oxidated nanocellulose, and ultrasonicated nanocellulose, respectively. This could be attributed to the strong intermolecular hydrogen bonding between filler and matrix which constrained the polymer-chain movement of composites. Besides, T_m and X_c of composites were also found to increased notably by the addition of nanocellulose. However, the results showed that the ultrasonicated-nanocellulose was not able to significantly improve the thermal property of composites. This could be due to the aggregation of ultrasonicated-nanocellulose with the polymer matrix which restricted the growth of crystalline domains in PVOH.

T_g can be considered as a significant parameter in thermal analysis. It is largely influenced by polymer-chain mobility, branching, interfacial adhesion, and degree of cross-linking. T_g of composites mostly shows no substantial improvement with the incorporation of nanocellulose, for examples, chitosan (Celebi & Kurt, 2015), natural rubber/butadiene rubber/styrene-butadiene rubber (Chen, Gu, & Xu, 2014), and wheat gluten (El-Wakil, Hassan, Abou-Zeid, & Dufresne, 2015). In addition, Luzi et al. (2017) observed a declination in T_g of PLA composites by the addition of surface-modified CNC. The surface modification of CNC was done by using a commercial surfactant, acid phosphate ester of ethoxylatednonyl phenol. T_g decreased from 56.2°C for neat PLA to 40.6°C for 1 wt.% CNC/PLA composite. The presence of surfactant could induce a plasticizing effect which led to an increase in the polymer-chain flexibility, hence reduced the T_g. Similar result was also reported by El-Hadi (2017). However, a few researches have reported an improvement in T_g of the nanocellulose-reinforced composites. A significant enhancement in T_g of CNC/sugar palm starch composites was reported (Cao, Huang, & Chen, 2018). The T_g of neat sugar palm starch film was 37.91°C and the inclusion of 1 wt.% CNC increased the T_g to a temperature of 49.15°C. This phenomenon could be explained by enhanced intermolecular reaction between filler and matrix, which restricted the polymer-chain mobility of composites and led to a higher T_g.

Table 17.3: Thermal characteristics of nanocellulose-reinforced composites.

Type of matrix	Source of nanocellulose	Thermal characteristics of neat matrix	Melting temperature, T_m (°C)	Glass-rubber transition temperature, T_g (°C)	Crystallization temperature, T_c (°C)	Crystallinity index, X_c (%)	References
PVA-borax	Bleached wood pulp	T_m: 108°C T_g: Not obvious	119–123	51–53	–	–	(Han et al., 2017)
PVA	Bleached eucalyptus Kraft pulp	T_m: 227.9°C X_c: 39%	221.2–226.8	–	–	40.4–42.3	(Peng, Ellingham, Sabo, Clemons, & Turng, 2015)
	Commercial cellulose microcystals	T_m: 178.9°C T_g: 33.7°C X_c: 20.9%	177.4–182.0	34.0–55.9	–	14.6–29.2	(Cataldi, Rigotti, Nguyen, & Pegoretti, 2018)
PVA/chitin	Oil palm empty fruit bunch	T_m: 236.0°C T_g: 59.8°C X_c: 7.1%	211.2–230.1	61.8–77.0	–	9.2–27.0	(Mok et al., 2020)
PLA-polyethylene glycol	Microcrystalline cellulose	T_m: 139°C T_g: 58°C T_c: 99°C X_c: 45%	132–137	59–61	98–99	39–41	(Çelebi & Kurt, 2014)
PLA	Microcrystalline cellulose	T_m: 149.9°C T_g: 58.5°C X_c: 30.9%	144.4–151.8	58.1–59.4	–	27.7–34.5	(Khoo, Ismail, & Chow, 2016)
Polyamide-6	Flax and microcrystalline cellulose	T_m: 216.0°C T_c: 185.1°C X_c: 46.8 %	217.9	–	188.1–189.2	47.5–49.3	(Qua & Hornsby, 2011)
Polycaprolactone	Filter paper	T_m: 56.7°C T_c: 32.2°C X_c: 49.1%	59.0–63.4	–	31.7–34.5	42.7–52.8	(Boujemaoui et al., 2017)

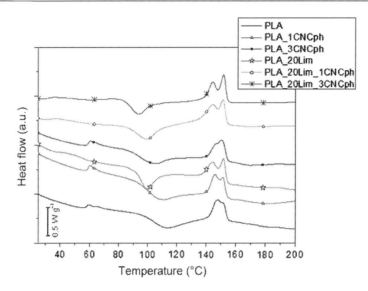

Figure 17.9
DSC Thermogram of CNC-reinforced PLA composites. (*CNC*, Cellulose nanocrystal; *DSC*, differential scanning calorimetry; *PLA*, polylactic acid) Source: *Adapted from Fortunati, E., Luzi, F., Puglia, D., Dominici, F., Santulli, C., Kenny, J. M., & Torre, L. (2014). Investigation of thermo-mechanical, chemical and degradative properties of PLA-limonene films reinforced with cellulose nanocrystals extracted from Phormium tenax leaves. European Polymer Journal, 56, 1, 77−91.*

The effect of limonene as a plasticizer on the thermal characteristics of CNC/PLA composites was examined (Fortunati et al., 2014). The incorporation of CNC showed no significant improvement on the T_g of PLA composites as illustrated in Fig. 17.9. The limonene acted as a plasticizer interrupted the intermolecular bonding of polymer chains and increased the polymer chain flexibility and mobility, reduced the T_g of composites. Besides, the inclusion of plasticizer also led to a lower crystallinity of the composites. Nevertheless, the incorporation of CNC favored the crystallization at a lower temperature, the crystallinity index of the composites increased notably especially for CNC/limonene/PLA composites resulted by nucleation effect of CNC. The addition of CNC increased the nucleation sites in amorphous regions of composites, which in turn to shorten the time for nucleation induction and lead to a greater crystallization kinetics.

17.5.3 Barrier property

The understanding in the barrier property of polymeric biocomposites is a prerequisite to fabricate versatile materials for high-performance applications. The barrier property of polymeric composites can be affected by numerous factors such as concentration of various additives, temperature, polymer structure, content of the filler, size and shape of filler particles as well as presence of voids within the matrix (Kalachandra & Turner, 1987). Barrier property can be considered as one of the most significant criteria that should be highlighted as it could influence the performance of the composites for the application, especially in water-sensitive products. Hence in order to enhance the reliability of composites, it is necessary to understand the mechanism of moisture diffusion.

Diffusion of water vapor across the polymeric composites is governed by three different mechanisms (Alomayri, Assaedi, Shaikh, & Low, 2014; Dhakal, Zhang, & Richardson, 2007). The first mechanism involves the water diffusion inside the microgaps between polymer chains. This is followed by the capillary transport into the voids and microcracks at interfaces between filler and matrix. The third mechanism involves swelling effect which leads to the debonding between filler and matrix and microcracks propagation in the matrix.

The moisture diffusion coefficient of composites was decreased remarkably with the increasing of CNF loading. The enhanced interfacial adhesion between CNF and thermoplastic starch as well as the formation of CNF network structure restrained the swelling of the starch matrix, which decreased the migration of water molecules through the film composites. Besides, CNF was less hygroscopic in comparison to starch due to its higher crystallinity, hence the moisture absorption reduced with the addition of CNF (Hietala et al., 2013). Similar result was also reported by Sun et al. (2018). However, they noticed that the addition of CNC at high concentration had a reverse effect on the water vapor permeability (WVP) of composites. This could be attributed to the aggregation of CNC within the matrix and led to the formation of more voids and increased the hydrophilicity of the composites. Fig. 17.10 depicts the mechanism of water vapor transmission in the composite with and without the incorporation of nanocellulose. The incorporation of nanocellulose has resulted in tortuous pathway and leaded to lower diffusion and thus lower the WVP. Homogeneous dispersion of nanocellulose into the matrix, which resulted in enhanced interaction between nanocellulose and matrix as well as between filler, has also leading to a better water vapor barrier.

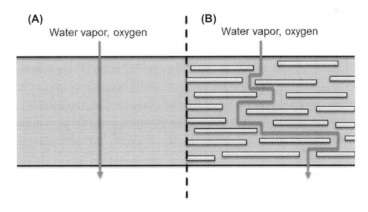

Figure 17.10
Schematic diagram of tortuous pathway in nanocellulose-reinforced composites. Source: *Adapted from Bugnicourt, E., Kehoe, T., Latorre, M., Serrano, C., Philippe, S., & Schmid, M. (2016). Recent prospects in the inline monitoring of nanocomposites and nanocoatings by optical technologies. Nanomaterials, 6, 8.*

On the other hand, the effect of CNC posttreatment on the barrier property of agar composites was investigated (Oun & Rhim, 2015). The CNC was neutralized by using either dialysis or sodium hydroxide while the CNC without any posttreatment was used as the control. The results showed that the sodium hydroxide neutralized-CNC-reinforced agar composites presented a better barrier property compared to other CNC. The presence of sulfate group on the surface of CNC could hinder the interfacial adhesion between CNC and agar matrix. Hence by using the sodium hydroxide posttreatment, the sulfate groups on the CNC could be reduced notably. Besides, the authors also found out that the WVP of composites increased significant by the addition of 10 wt.% CNC which could be resulted by the agglomeration of CNC at high filler concentration.

Aside from surface modification of nanocellulose, cross-linking can also be used as the alternative for improving the interaction between filler and polymer matrix. Owi et al. (2019) observed that the barrier property of CNC/tapioca starch composites was enhanced by the cross-linking reaction. Two different cross-linkers were employed in this study, which were citric acid and lime juice, respectively. The WVP of both composites cross-linked by citric acid and lime juice declined significantly by approximately 55% with the incorporation of 1 wt.% CNC as presented in Fig. 17.11. The enhancement in barrier property for cross-linked composites was due to the replacement of hydrophilic hydroxyl group by hydrophobic ester

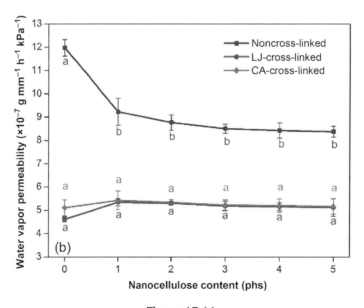

Figure 17.11

WVP of noncross-linked and cross-linked CNC/tapioca starch composites. (*CNC*, Cellulose nanocrystal; *WVP*, water vapor permeability) Source: *Adapted from Owi, W. T., Ong, H. L., Sam, S. T., Villagracia, A. R., kuo Tsai, C., & Akil, H. M. (2019). Unveiling the physicochemical properties of natural Citrus aurantifolia crosslinked tapioca starch/nanocellulose bionanocomposites. Industrial Crops and Products, 139, 111548.*

group through esterification which reduced the availability of hydroxyl group to react with water molecules. Furthermore, the addition of CNC created the tortuous pathway and increased the tortuosity for the water molecules diffused across the film composites.

The significant declination in WVP of hydrophobic PLA composites by increasing the hydrophobicity of CNF via grafting of hydrophobic monomers was observed (Song, Xiao, & Zhao, 2014). For the composites with the CNF of less than 5 wt.%, all composites showed better water vapor barrier property. This was attributed to the increased hydrophobicity of CNF enhanced its compatibility with PLA matrix which lead to a compact and homogeneous cellulose network formed within the hydrophobic polymer matrix. However, for the composites with the addition of CNF higher than 5 wt.%, the barrier property of composites decreased. Due to the increased hydrophobicity of CNF, the hydrogen bonding between the cellulose was deteriorated and resulted in the formation of porous structure. In addition to that, the higher number of hydroxyl groups was left after the modification as only partial hydroxyl groups were replaced by the grafting of hydrophobic monomers.

17.5.4 Biodegradation property

Biodegradation is the degradation of materials brought about by the action of living microbial organisms such as bacteria, fungi, and algae (Joutey, Bahafid, Sayel, & El Ghachtouli, 2013). A number of mechanisms have been proposed for the degradation of polymer materials by microorganisms. Shah and Pandey (2017) proposed that the presence of filler can accelerate the degradation rate of composites. Nanocellulose is a natural polymer and exhibits faster biodegradation compared to the other matrices. The nanocellulose component in the nanocomposites is consumed by the microorganisms faster than the matrix, leading to the formation of voids, increase in porosity, and decrease in the integrity of the matrix (Arjmandi, Hassan, Mohamad Haafiz, & Zakaria, 2016). The polymer matrix is broken down into smaller parts and led to the rapid degradation of the nanocomposites.

Nanocellulose is more susceptible to biodegradation in an outdoor environment depending on the environmental conditions such as moisture, temperature, ultraviolet (UV) radiation, and the microbial activity. The amorphous region of nanocellulose is degraded by the microorganisms more easily than the crystalline parts, which will lead to a high rate of biodegradation of the nanocomposites (Abraham et al., 2012). The biodegradability of nanocellulose composites is also affected by the crystallinity, hydration, and swelling of nanocellulose (Lin & Dufresne, 2014).

Microbial degradation of polymeric composites usually involves two types of enzymes, extracellular and intracellular depolymerases (Gu, 2003). During degradation, the exo-enzymes secreted from microorganisms will break down complex polymers into simpler molecules such as oligomers, dimers, and monomers. These molecules will be absorbed by the microorganisms and utilized as carbon and energy sources. The process of breaking down complex molecules into smaller molecules is known as depolymerization (Mohan, 2011). The polymers can be degraded by the microorganisms through aerobic or anaerobic conditions. CO_2 and H_2O are produced by microbial degradation in the presence of oxygen. Whereas, for anaerobic circumstances, polymeric composites are normally degraded into CO_2, CH_4, and H_2O under methanogenic conditions or into H_2S, CO_2, and H_2O under sulfidogenic conditions as their end product of biodegradation (Mohan, 2011). Numerous degrading-microorganisms for polymer composites were reported such as *Aspergillus nomius* (Munir, Harefa, Priyani, & Suryanto, 2018), *Pseudomonas fluorescens* (Hussein et al., 2015), *Aspergillus nidulans* (Usha, Sangeetha, & Palaniswamy, 2011), and *Phanerochaete chrysosporium* (Ali et al., 2014).

Polymers such as cellulose and chitin are biologically synthesized and can be completely and rapidly biodegraded by microorganisms in a wide range of natural environments. During degradation process, a small part of the polymer will be converted into the microbial biomass or stabilized in the soil humus and other natural products (Madras, 2005). This opens the possibility of using specific microorganisms to enhance the degradation process of polymeric material.

A number of factors determine the degradation rate of polymer composites, for example, the addition of cross-linkers, the modification of nanocellulose and environmental factors. Environmental factors such as the soil type, ultraviolet exposure, the amount and type of microorganisms present, their sensitivity to environmental parameters, and the adaptability of the microbiota are the important parameters affecting the biodegradation of nanocomposites (César, Mariani, Innocentini-Mei, & Cardoso, 2009). Therefore understanding the biodegradability of composites is important when a polymeric system is applied in daily life, as its weight loss rate has a direct impact on the environment. Lani, Ngadi, Johari, and Jusoh (2014) studied the influence of different CNC loading levels on the degradability of PVA/starch composites through soil burial tests. All PVA/starch composites with CNC exhibited a lower degradability compared to PVA/starch composites without CNC. The addition of nanocellulose reduced the water absorption by the film and delayed its degradation. Microorganisms used the polymer matrix as an energy source, whereas the water absorbed promotes the microbial growth on the film. Hence the higher the amount of water diffused through the films, the faster the composites were degraded.

A similar result was also observed, in which the CNF-reinforced PVA/starch film exhibited a lower degradation rate than the PVA/starch film (Heidarian, Behzad, & Sadeghi, 2017). After 70 days, the neat PVA/starch composites showed a 21% of residual weight compared to PVA/starch/5 wt.% CNF composites (44%). The addition of highly crystalline CNF reduced the diffusion of water and slowed microbial degradation. The ease of accessibility and high concentration of starch in PVA/starch composites further promoted the biodegradability of polymer composites. This is because starch could be biodegraded by α- and β-amylase enzymes, which could be found in most of the microbes. Strong interfacial adhesion between nanocellulose and the matrix and good dispersion of nanofillers in the polymer matrix were identified as factors contributing to the slow degradation rate of polymer composites (Deepa et al., 2016)

The biodegradability of hydrophobic composites exhibited a different trend compared to hydrophilic composites. The addition of CNC significantly accelerated the biodegradability of the polyvinylchloride composites. The hydrophobic nature of the matrix reduces its ability to absorb water and thus slows the biodegradation rate of the composites. However, nanocellulose has a higher degradation rate and will be consumed by microorganisms faster than the matrix. The addition of CNC resulted in the formation of pores and voids and the loss of the integrity of the matrix, resulting in a faster degradation rate in nanocellulose-reinforced composites (Kadry, El-Hakim, & Abd El-Hakim, 2015). The nanocomposites will be broken down into small fragments. Therefore the biodegradation rate of nanocellulose composites will be higher than that of nonreinforced composites as proven by these researchers (Bras et al., 2010; Kadry et al., 2015).

Surface modification such as acetylation was performed in order to increase the hydrophobicity of CNF and the effect of acetylation on the biodegradability of CNF/low-density polyethylene (LDPE)/starch composites was studied (Ahmadi, Behzad, Bagheri, & Heidarian, 2018). The biodegradation rate of composites increased with the CNF concentration and the composites with 5 wt.% of CNF recorded a highest biodegradation rate. Whereas, the acetylated CNF/LDPE/starch composites showed a slower biodegradation compared to unmodified CNF/LDPE/starch composites. The acetylation of CNF improved the compatibility between CNF and LDPE and reduced the formation of voids and cracks between the interface of nanocomposites. The biodegradation started with nanocellulose and could leave holes and voids, which subsequently increase the surface area for the biodegradation of the remaining nanocomposites. Each cellulose has three free hydroxyl groups which are the active sites for the microorganism attack and it is very susceptible to the hydrolytic cleavage by water molecules (Abraham et al., 2012). Therefore the replacement of hydroxyl group by acetyl group via acetylation increased the composites resistance against microorganism's attack.

17.6 Conclusions and perspectives

The potential of nanocellulose-reinforced composites widely extended in various applications. However, the use of nanocellulose offers several challenges for researchers as the differences in polarity between the cellulose and matrix have to be taken into account. In this chapter, different processing techniques of the nanocellulose-reinforced biocomposites were discussed in brief. Significant attention was placed on the solvent casting and extrusion methods as these are the most applied techniques in both laboratory and industrial-scale process.

Owing to its high abundancy, excellent tensile strength and modulus, good biodegradability and low weight, nanocellulose materials could act as a promising reinforcing filler for the fabrication of composites. The review on tensile, thermal, barrier, and biodegradation properties of the nanocellulose-reinforced composites has been presented. There is an agreement in the literature regarding the enhancement in tensile, thermal, barrier, and biodegradation properties of polymeric composites by the inclusion of nanocellulose, limited to the cases in which the nanocellulose are evenly dispersed and compatible with the polymer matrices. The interfacial adhesion between nanocellulose and polymeric matrix plays an essential role in improving the properties of composites, especially in hydrophobic matrices. The influence of nanocellulose concentration, types of nanocellulose (CNF or CNC), surface modification of both nanocellulose and polymer matrix, and the application of cross-linkers and compatibilizers on the tensile, thermal, barrier, and biodegradation properties of polymeric biocomposites were described.

Nanocellulose has outstanding characteristics, however, to translate it from the laboratory scale to the industrial scale is quite complicated due to the high manufacturing cost and energy for the isolation of nanocellulose. The isolation of cellulose into nanocellulose and its compatibility problems with hydrophobic polymeric materials are still the major issues facing now. Continued explorations are required in order to improve the performance of the nanocellulose-reinforced composites for sustainable developments and advancements.

References

Abdulkhani, A., Hosseinzadeh, J., Ashori, A., Dadashi, S., & Takzare, Z. (2014). Preparation and characterization of modified cellulose nanofibers reinforced polylactic acid nanocomposite. *Polymer Testing, 35*, 73–79.

Abraham, E., Elbi, P. A., Deepa, B., Jyotishkumar, P., Pothen, L. A., Narine, S. S., & Thomas, S. (2012). X-ray diffraction and biodegradation analysis of green composites of natural rubber/nanocellulose. *Polymer Degradation and Stability, 97*(11), 2378–2387.

Agustin, M. B., Ahmmad, B., Alonzo, S. M. M., & Patriana, F. M. (2014). Bioplastic based on starch and cellulose nanocrystals from rice straw. *Journal of Reinforced Plastics and Composites, 33*(24), 2205–2213.

Ahmadi, M., Behzad, T., Bagheri, R., & Heidarian, P. (2018). Effect of cellulose nanofibers and acetylated cellulose nanofibers on the properties of low-density polyethylene/thermoplastic starch blends. *Polymer International, 67*(8), 993–1002.

Ali, M. I., Ahmed, S., Robson, G., Javed, I., Ali, N., Atiq, N., & Hameed, A. (2014). Isolation and molecular characterization of polyvinyl chloride (PVC) plastic degrading fungal isolates. *Journal of Basic Microbiology, 54*(1), 18–27.

Alomayri, T., Assaedi, H., Shaikh, F. U. A., & Low, I. M. (2014). Effect of water absorption on the mechanical properties of cotton fabric-reinforced geopolymer composites. *Journal of Asian Ceramic Societies, 2*(3), 223–230.

Arjmandi, R., Hassan, A., Mohamad Haafiz, M. K., & Zakaria, Z. (2016). Biodegradability and thermal properties of hybrid montmorillonite/microcrystalline cellulose filled polylactic acid composites: Effect of filler ratio. *Polymers and. Polymer Composites, 24*(9), 741–746.

Asmatulu, R. (2016). Highly hydrophilic electrospun polyacrylonitrile/polyvinypyrrolidone nanofibers incorporated with gentamicin as filter medium for dam water and wastewater treatment. *Journal of Membrane and Separation Technology, 5*(2), 38–56.

de Carvalho Benini, K. C. C., Cioffi, M. O. H., & Voorwald, H. J. C. (2017). PHBV/cellulose nanofibrils composites obtained by solution casting and electrospinning process. *Matéria (Rio de Janeiro)*, *22*(2).

Bhat, A. H., Dasan, Y. K., Khan, I., Soleimani, H., & Usmani, A. (2017). *Application of nanocrystalline cellulose: Processing and biomedical applications. Cellulose-reinforced nanofibre composites: Production, properties and applications* (1st ed., pp. 215−240). Cambridge, UK: Woodhead Publishing.

Bilbao-Sainz, C., Bras, J., Williams, T., Sénechal, T., & Orts, W. (2011). HPMC reinforced with different cellulose nano-particles. *Carbohydrate Polymers*, *86*(4), 1549−1557.

Boujemaoui, A., Sanchez, C. C., Engström, J., Bruce, C., Fogelström, L., Carlmark, A., & Malmström, E. (2017). Polycaprolactone nanocomposites reinforced with cellulose nanocrystals surface-modified via covalent grafting or physisorption: A comparative study. *ACS Applied Materials & Interfaces*, *9*(40), 35305−35318.

Bras, J., Hassan, M. L., Bruzesse, C., Hassan, E. A., El-Wakil, N. A., & Dufresne, A. (2010). Mechanical, barrier, and biodegradability properties of bagasse cellulose whiskers reinforced natural rubber nanocomposites. *Industrial Crops and Products*, *32*(3), 627−633.

Brinchi, L., Cotana, F., Fortunati, E., & Kenny, J. M. (2013). Production of nanocrystalline cellulose from lignocellulosic biomass: Technology and applications. *Carbohydrate Polymers*, *94*(1), 154−169.

Bugnicourt, E., Kehoe, T., Latorre, M., Serrano, C., Philippe, S., & Schmid, M. (2016). Recent prospects in the inline monitoring of nanocomposites and nanocoatings by optical technologies. *Nanomaterials*, *6*(8).

Butron, A., Llorente, O., Fernandez, J., Meaurio, E., & Sarasua, J. R. (2019). Morphology and mechanical properties of poly(ethylene brassylate)/cellulose nanocrystal composites. *Carbohydrate Polymers*, *221*, 137−145.

Cao, Y. (2018). Applications of cellulose nanomaterials in pharmaceutical science and pharmacology. *Express Polymer Letters*, *12*(9), 768−780.

Cao, L., Huang, J., & Chen, Y. (2018). Dual cross-linked epoxidized natural rubber reinforced by tunicate cellulose nanocrystals with improved strength and extensibility. *ACS Sustainable Chemistry & Engineerng*, *6*(11), 14802−14811.

Cao, X., Xu, C., Wang, Y., Liu, Y., Liu, Y., & Chen, Y. (2013). New nanocomposite materials reinforced with cellulose nanocrystals in nitrile rubber. *Polymer Testing*, *32*(5), 819−826.

Cataldi, A., Rigotti, D., Nguyen, V. D. H., & Pegoretti, A. (2018). Polyvinyl alcohol reinforced with crystalline nanocellulose for 3D printing application. *Materials Today Communications*, *15*, 236−244, January.

Çelebi, H., & Kurt, A. (2014). Thermal and mechanical properties of PLA/PEG blend and its nanocomposites. In 16th European conference on composite materials, ECCM 2014 (pp. 22−26).

Celebi, H., & Kurt, A. (2015). Effects of processing on the properties of chitosan/cellulose nanocrystal films. *Carbohydrate Polymers*, *133*, 284−293.

César, M. E. F., Mariani, P. D. S. C., Innocentini-Mei, L. H., & Cardoso, E. J. B. N. (2009). Particle size and concentration of poly(ε-caprolactone) and adipate modified starch blend on mineralization in soils with differing textures. *Polymer Testings*, *28*(7), 680−687.

Chen, J., Guan, Y., Wang, K., Zhang, X., Xu, F., & Sun, R. (2015). Combined effects of raw materials and solvent systems on the preparation and properties of regenerated cellulose fibers. *Carbohydrate Polymers*, *128*, 147−153.

Chen, W. J., Gu, J., & Xu, S. H. (2014). Exploring nanocrystalline cellulose as a green alternative of carbon black in natural rubber/butadiene rubber/styrene-butadiene rubber blends. *Express Polymer Letters*, *8*(9), 659−668.

Chen, Y., He, Y., Fan, D., Han, Y., Li, G., & Wang, S. (2017). An efficient method for cellulose nanofibrils length shearing via environmentally friendly mixed cellulase pretreatment. *Journal of Nanomaterials*, *2017*.

Chen, D., Lawton, D., Thompson, M. R., & Liu, Q. (2012). Biocomposites reinforced with cellulose nanocrystals derived from potato peel waste. *Carbohydrate Polymers*, *90*(1), 709−716.

Chen, Y. W., Lee, H. V., Juan, J. C., & Phang, S. M. (2016). Production of new cellulose nanomaterial from red algae marine biomass *Gelidium elegans*. *Carbohydrate Polymers*, *151*, 1210−1219.

Chen, P., Yu, H., Liu, Y., Chen, W., Wang, X., & Ouyang, M. (2013). Concentration effects on the isolation and dynamic rheological behavior of cellulose nanofibers via ultrasonic processing. *Cellulose*, *20*(1), 149−157.

Cherian, B. M., Leão, A. L., Souza, S. F., Costa, L. M. M., Olyveira, G. M., Kottaisamy, M., Nagarajan, E. R., & Thomas, S. (2011). Cellulose nanocomposites with nanofibres isolated from pineapple leaf fibers for medical applications. *Carbohydrate Polymers*, *86*(4), 1790–1798.

Chieng, B. W., Lee, S. H., Ibrahim, N. A., Then, Y. Y., & Loo, Y. Y. (2017). Isolation and characterization of cellulose nanocrystals from oil palm mesocarp fiber. *Polymers (Basel)*, *9*(8), 355.

Chirayil, C. J., Mathew, L., & Thomas, S. (2014). Review of recent research in nano cellulose preparation from different lignocellulosic fibers. *Review on Advanced Materials Science*, *37*(1–2), 20–28.

Cho, M., & Park, B. (2011). Tensile and thermal properties of nanocellulose-reinforced poly (vinyl alcohol) nanocomposites. *Journal of Industrial and Engineering Chemistry*, *17*(1), 36–40.

Dai, Z., Ottesen, V., Deng, J., Helberg, R. M. L., & Deng, L. (2019). A brief review of nanocellulose based hybrid membranes for CO_2 separation. *Fibers*, *7*(5), 1–18.

Deepa, B., Abraham, E., Pothan, L. A., Cordeiro, N., Faria, M., & Thomas, S. (2016). Biodegradable nanocomposite films based on sodium alginate and cellulose nanofibrils. *Materials (Basel).*, *9*(1), 1–11.

Dehnad, D., Emam-Djomeh, Z., Mirzaei, H., Jafari, S. M., & Dadashi, S. (2014). Optimization of physical and mechanical properties for chitosan-nanocellulose biocomposites. *Carbohydrate Polymers*, *105*(1), 222–228.

Detzel, C. J., Larkin, A. L., & Rajagopalan, P. (2011). Polyelectrolyte multilayers in tissue engineering. *Tissue Engineering—Part B: Reviews*, *17*(2), 101–113.

Dhakal, H. N., Zhang, Z. Y., & Richardson, M. O. W. (2007). Effect of water absorption on the mechanical properties of hemp fibre reinforced unsaturated polyester composites. *Composites Science and Technology*, *67*(7–8), 1674–1683.

Dong, X. M., Revol, J.-F., & Gray, D. G. (1998). Effect of microcrystallite preparation conditions on the formation of colloid crystals of cellulose. *Cellulose*, *5*, 19–32.

El-Hadi, A. M. (2017). Increase the elongation at break of poly (lactic acid) composites for use in food packaging films. *Scientific Reports*, *7*, 1–14, March.

El-Wakil, N. A., Hassan, E. A., Abou-Zeid, R. E., & Dufresne, A. (2015). Development of wheat gluten/nanocellulose/titanium dioxide nanocomposites for active food packaging. *Carbohydrate Polymers*, *124*, 337–346.

El Achaby, M., Kassab, Z., Barakat, A., & Aboulkas, A. (2018). Alfa fibers as viable sustainable source for cellulose nanocrystals extraction: Application for improving the tensile properties of biopolymer nanocomposite films. *Industrial Crops and Products*, *112*, 499–510, November 2017.

Espinosa, S. C., Kuhnt, T., Foster, E. J., & Weder, C. (2013). Isolation of thermally stable cellulose nanocrystals by phosphoric acid hydrolysis. *Biomacromolecules*, *14*(4), 1223–1230.

Espino, E., Cakir, M., Domenek, S., Roman-Gutierrez, A. D., Belgacem, M. N., & Bras, J. (2014). *Isolation and characterization of cellulose nanocrystals from industrial by-products of* Agave tequilana *and barley*. *Industrial Crops and Products* (62, pp. 552–559).

Fatah, I. Y. A., Abdul Khalil, H. P. S., Hossain, M. S., Aziz, A. A., Davoudpour, Y., Dungani, R., & Bhat, A. (2014). Exploration of a chemo-mechanical technique for the isolation of nanofibrillated cellulosic fiber from oil palm empty fruit bunch as a reinforcing agent in composites materials. *Polymers (Basel)*, *6*(10), 2611–2624.

Ferreira, F., Pinheiro, I., de Souza, S., Mei, L., & Lona, L. (2019). Polymer composites reinforced with natural fibers and nanocellulose in the automotive industry: A short review. *Journal of Composites Science*, *3*(2), 51.

Ferrer, A., Filpponen, I., Rodríguez, A., Laine, J., & Rojas, O. J. (2012). Valorization of residual empty palm fruit bunch fibers (EPFBF) by microfluidization: Production of nanofibrillated cellulose and EPFBF nanopaper. *Bioresource Technology*, *125*, 249–255.

Siqueira, G., Bras, J., Follain, N., Belbekhouche, S., Marais, S., & Dufresne, A. (2013). Thermal and mechanical properties of bio-nanocomposites reinforced by *Luffa cylindrica* cellulose nanocrystals. *Carbohydrate Polymers*, *91*(2), 711–717.

Fornes, T. D., & Paul, D. R. (2003). Formation and properties of nylon 6 nanocomposites. *Polímeros*, *13*(4), 212–217.

Fortunati, E., Puglia, D., Luzi, F., Santulli, C., Kenny, J. M., & Torre, L. (2013). Binary PVA bio-nanocomposites containing cellulose nanocrystals extracted from different natural sources: Part I. *Carbohydrate Polymers*, *97*(2), 825−836.

Fortunati, E., Luzi, F., Puglia, D., Dominici, F., Santulli, C., Kenny, J. M., & Torre, L. (2014). Investigation of thermo-mechanical, chemical and degradative properties of PLA-limonene films reinforced with cellulose nanocrystals extracted from *Phormium tenax* leaves. *European Polymer Journal*, *56*(1), 77−91.

Frone, A. N., Panaitescu, D. M., & Donescu, D. (2011). Some aspects concerning the isolation of cellulose micro- and nano-fibers. *UPB Scientific Bulletin, Series B: Chemistry and Materials Science*, *73*(2), 133−152.

Gan, P. G., Sam, S. T., Abdullah, M. F., Omar, M. F., & Tan, L. S. (2020). An alkaline deep eutectic solvent based on potassium carbonate and glycerol as pretreatment for the isolation of cellulose nanocrystals from empty fruit bunch. *BioResources*, *15*(1), 1154−1170.

Gan, P. G., Sam, S. T., bin Abdullah, M. F., & Omar, M. F. (2019). Thermal properties of nanocellulose-reinforced composites: A review. *Journal of Applied Polymer Science*, *48544*, 1−14.

Ghaemi, F., Abdullah, L. C., Kargarzadeh, H., Abdi, M. M., Azli, N. F. W. M., & Abbasian, M. (2018). Comparative study of the electrochemical, biomedical, and thermal properties of natural and synthetic nanomaterials. *Nanoscale Research Letters*, *13*(1).

Ghelich, R., Keyanpour Rad, M., & Youzbashi, A. A. (2015). Study on morphology and size distribution of electrospun NiO-GDC composite nanofibers. *Journal of Engineered Fibers and Fabrics*, *10*(1), 12−19.

Guidotti, G., Soccio, M., Siracusa, V., Gazzano, M., Salatelli, E., Munari, A., & Lotti, N. (2017). Novel random PBS-based copolymers containing aliphatic side chains for sustainable flexible food packaging. *Polymers (Basel)*, *9*(12), 1−16.

Gu, J., & Catchmark, J. M. (2013). Polylactic acid composites incorporating casein functionalized cellulose nanowhiskers. *Journal of Biological Engineering*, *7*(1), 1−10.

Gu, J. D. (2003). Microbiological deterioration and degradation of synthetic polymeric materials: Recent research advances. *International Biodeterioration & Biodegradation*, *52*(2), 69−91.

Halib, N., Perrone, F., Cemazar, M., Dapas, B., Farra, R., Abrami, M., Chiarappa, G., Forte, G., Zanconati, F., Pozzato, G., Murena, L., Fiotti, N., Lapasin, R., Cansolino, L., Grassi, G., & Grassi, M. (2017). Potential applications of nanocellulose-containing materials in the biomedical field. *Materials (Basel)*, *10*(8), 1−31.

Han, Y., Yu, M., & Wang, L. (2018). Soy protein isolate nanocomposites reinforced with nanocellulose isolated from licorice residue: Water sensitivity and mechanical strength. *Industrial Crops and Products*, *117*, 252−259, September 2017.

Han, J., Yue, Y., Wu, Q., Huang, C., Pan, H., Zhan, X., Mei, C., & Xu, X. (2017). Effects of nanocellulose on the structure and properties of poly(vinyl alcohol)-borax hybrid foams. *Cellulose*, *24*(10), 4433−4448.

Heidarian, P., Behzad, T., & Sadeghi, M. (2017). Investigation of cross-linked PVA/starch biocomposites reinforced by cellulose nanofibrils isolated from aspen wood sawdust. *Cellulose*, *24*(8), 3323−3339.

Herrick, F. W., Casebier, R. L., Hamilton, J. K., & Sandberg, K. R. (1983). Microfibrillated cellulose: Morphology and accessibility. *Journal of Applied Polymer Science, Applied Polymer Symposium*, *37*, 797−813.

Hietala, M., Mathew, A. P., & Oksman, K. (2013). Bionanocomposites of thermoplastic starch and cellulose nanofibers manufactured using twin-screw extrusion. *European Polymer Journal*, *49*(4), 950−956.

Hou, Q., Wang, X., & Ragauskas, A. J. (2019). Preparation and characterization of nanocellulose−polyvinyl alcohol multilayer film by layer-by-layer method. *Cellulose*, *26*(8), 4787−4798.

Hussein, A. A., Al-Mayaly, I. K., Khudeir, S. H., Hussein, A. A., Al-Mayaly, I. K., & Kudier, S. H. (2015). Isolation, screening and identification of low density polyethylene (LDPE) degrading bacteria from contaminated soil with plastic wastes isolation, screening and identification of low density polyethylene (LDPE) degrading. *Mesopotomia Environmental Journal*, *1*(4), 1−14.

Ilyas, R. A., Sapuan, S. M., Ishak, M. R., & Zainudin, E. S. (2018). Development and characterization of sugar palm nanocrystalline cellulose reinforced sugar palm starch bionanocomposites. *Carbohydrate Polymers*, *202*, 186−202.

Jose, C., Mathew, L., Hassan, P. A., Mozetic, M., & Thomas, S. (2014). Rheological behaviour of nanocellulose reinforced unsaturated polyester nanocomposites. *International Journal of Biological Macromolecules, 69*, 274–281.

Joutey, N. T., Bahafid, W., Sayel, H., & El Ghachtouli, N. (2013). *Biodegradation: Involved microorganisms and genetically engineered microorganisms. Biodegradation—Life of Science*. IntechOpen.

Kadry, G., El-Hakim, A. E. F. A., & Abd El-Hakim, A. E. F. (2015). Effect of nanocellulose on the biodegradation, morphology and mechanical properties of polyvinylchloride/nanocellulose nanocomposites. *Research Journal of Pharmaceutical, Biological and Chemical Sciences, 6*(6), 659–666.

Kalachandra, S., & Turner, D. T. (1987). Water sorption of polymethacrylate networks: Bis-GMA/TEGDM copolymers. *Journal of Biomedical Materials Research, 21*(3), 329–338.

Kasa, S. N., Omar, M. F., Al Bakri Abdullah, M. M., Ismail, I. N., Ting, S. S., Vac, S. C., & Vizureanu, P. (2017). Effect of unmodified and modified nanocrystalline cellulose reinforced polylactic acid (PLA) polymer prepared by solvent casting method: Morphology, mechanical and thermal properties. *Materiale Plastice, 54*(1), 91–97.

Khoo, R. Z., Ismail, H., & Chow, W. S. (2016). Thermal and morphological properties of poly (lactic acid)/nanocellulose nanocomposites. *Procedia Chemistry, 19*, 788–794.

Kiziltas, A., Nazari, B., Kiziltas, E. E., Gardner, D. J. S., Han, Y., & Rushing, T. S. (2016). Cellulose nanofiber-polyethylene nanocomposites modified by polyvinyl alcohol. *Journal of Applied Polymer Science, 133*(6), 1–8.

Kouadri, I., & Satha, H. (2018). Extraction and characterization of cellulose and cellulose nanofibers from *Citrullus colocynthis* seeds. *Industrial Crops and Products, 124*, 787–796, April.

Lani, N. S., Ngadi, N., Johari, A., & Jusoh, M. (2014). Isolation, characterization, and application of nanocellulose from oil palm empty fruit bunch fiber as nanocomposites. *Journal of Nanomaterials, 2014*.

Lee, D., Jung, J., Lee, G. H., & Il Lee, W. (2019). Electrospun nanofiber composites with micro-/nano-particles for thermal insulation. *Advanced Composites Materials, 28*(2), 193–202.

Lee, J. K. Y., Chen, N., Peng, S., Li, L., Tian, L., Thakor, N., & Ramkrishna, S. (2018). Polymer-based composites by electrospinning: Preparation and functionalization with nanocarbons. *Progress in Polymer Science, 86*, 40–84.

Leite, A. L. M. P., Zanon, C. D., & Menegalli, F. C. (2017). Isolation and characterization of cellulose nanofibers from cassava root bagasse and peelings. *Carbohydrate Polymers, 157*, 962–970.

Lin, N., & Dufresne, A. (2014). Surface chemistry, morphological analysis and properties of cellulose nanocrystals with gradiented sulfation degrees. *Nanoscale, 6*(10), 5384–5393.

Lin, N., & Dufresne, A. (2014). Nanocellulose in biomedicine: Current status and future prospect. *European Polymer Journal, 59*, 302–325.

Liu, C., Li, B., Du, H., Lv, D., Zhang, Y., Yu, G., Mu, X., & Peng, H. (2016). Properties of nanocellulose isolated from corncob residue using sulfuric acid, formic acid, oxidative and mechanical methods. *Carbohydrate Polymers, 151*, 716–724.

Liu, D. Y., Sui, G. X., & Bhattacharyya, D. (2014). Synthesis and characterisation of nanocellulose-based polyaniline conducting films. *Composites Science and Technology, 99*, 31–36.

Li, M., Tian, X., Jin, R., & Li, D. (2018). Preparation and characterization of nanocomposite films containing starch and cellulose nanofibers. *Industrial Crops and Products, 123*, 654–660.

Li, J., Wei, X., Wang, Q., Chen, J., Chang, G., Kong, L., Su, J., & Liu, Y. (2012). Homogeneous isolation of nanocellulose from sugarcane bagasse by high pressure homogenization. *Carbohydrate Polymers, 90*(4), 1609–1613.

Li, W., Yue, J., & Liu, S. (2012). Preparation of nanocrystalline cellulose via ultrasound and its reinforcement capability for poly(vinyl alcohol) composites. *Ultrasonics Sonochemistry, 19*(3), 479–485.

Luzi, F., Fortunati, E., Jiménez, A., Puglia, D., Chiralt, A., & Torre, L. (2017). PLA nanocomposites reinforced with cellulose nanocrystals from *Posidonia oceanica* and ZnO nanoparticles for packaging application. *Journal of Renewable Materials, 5*(2), 103–115.

Luzi, F., Fortunati, E., Puglia, D., Petrucci, R., Kenny, J. M., & Torre, L. (2016). Modulation of acid hydrolysis reaction time for the extraction of cellulose nanocrystals from *Posidonia oceanica* leaves. *Journal of Renewable Materials, 4*(3), 190–198.

Lu, P., & Hsieh, Y.-L. (2012). Preparation and characterization of cellulose nanocrystals from rice straw. *Carbohydrate Polymers, 87*(1), 564–573.

Lu, P., & Lo, Y. (2010). Hsieh, Preparation and properties of cellulose nanocrystals: Rods, spheres, and network. *Carbohydrate Polymers, 82*(2), 329–336.

Maaloul, N., Ben Arfi, R., Rendueles, M., Ghorbal, A., & Diaz, M. (2017). Dialysis-free extraction and characterization of cellulose crystals from almond (*Prunus dulcis*) shells. *Journal of Materials and Environmental Science, 8*(11), 4171–4181.

Madras, G. (2005). *Chapter 15: Enzymatic degradation of polymers. Biodegradable Polymers for Industrial Applications* (pp. 411–440). Woodhead Publishing Limited.

Mahjoub, R., Bin Mohamad Yatim, J., & Mohd Sam, A. R. (2013). A review of structural performance of oil palm empty fruit bunch fiber in polymer composites. *Advances in Materials Science and Engineering, 2013*.

Mandal, A., & Chakrabarty, D. (2011). Isolation of nanocellulose from waste sugarcane bagasse (SCB) and its characterization. *Carbohydrate Polymers, 86*(3), 1291–1299.

Martelli-Tosi, M., Masson, M. M., Silva, N. C., Esposto, B. S., Barros, T. T., Assis, O. B. G., & Tapia-Blácido, D. R. (2018). Soybean straw nanocellulose produced by enzymatic or acid treatment as a reinforcing filler in soy protein isolate films. *Carbohydrate Polymers, 198*, 61–68.

de Mesquita, J. P., Donnici, C. L., & Pereira, F. V. (2010). Biobased nanocomposites from layer-by-layer assembly of cellulose nanowhiskers with chitosan. *Biomacromolecules, 11*(2), 473–480.

Miandad, R., Barakat, M. A., Aburiazaiza, A. S., Rehan, M., & Nizami, A. S. (2016). Catalytic pyrolysis of plastic waste: A review. *Process Safety and Environmental Protection, 102*, 822–838.

Mohan, K., & Srivastava, T. (2011). Microbial deterioration and degradation of polymeric materials. *Journal of Biochemical Technology, 2*(4), 210–215.

Mok, C. F., Ching, Y. C., Abu Osman, N. A., Muhamad, F., Mohd Junaidi, M. U., & Choo, J. H. (2020). Preparation and characterization study on maleic acid cross-linked poly(vinyl alcohol)/chitin/nanocellulose composites. *Journal of Applied Polymer Science*, 1–16, January.

Mondragon, G., Peña-Rodriguez, C., González, A., Eceiza, A., & Arbelaiz, A. (2015). Bionanocomposites based on gelatin matrix and nanocellulose. *European Polymer Journal, 62*, 1–9.

Munir, E., Harefa, R. S. M., Priyani, N., & Suryanto, D. (2018). Plastic degrading fungi *Trichoderma viride* and *Aspergillus nomius* isolated from local landfill soil in Medan. *IOP Conference Series: Earth and Environmental Science, 126*(1).

Nair, S. S., & Yan, N. (2015). Bark derived submicron-sized and nano-sized cellulose fibers: From industrial waste to high performance materials. *Carbohydrate Polymers, 134*, 258–266.

Nan, F., Nagarajan, S., Chen, Y., Liu, P., Duan, Y., Men, Y., & Zhang, J. (2017). Enhanced toughness and thermal stability of cellulose nanocrystal iridescent films by alkali treatment. *ACS Sustainable Chemistry & Engineering, 5*(10), 8951–8958.

Nepomuceno, N. C., Santos, A. S. F., Oliveira, J. E., Glenn, G. M., & Medeiros, E. S. (2017). Extraction and characterization of cellulose nanowhiskers from Mandacaru (*Cereus jamacaru* DC.) spines. *Cellulose, 24*(1), 119–129.

Niazi, M. B. K., Jahan, Z., Berg, S. S., & Gregersen, Ø. W. (2017). Mechanical, thermal and swelling properties of phosphorylated nanocellulose fibrils/PVA nanocomposite membranes. *Carbohydrate Polymers, 177*, 258–268, August.

Oun, A. A., & Rhim, J. W. (2016). Characterization of nanocelluloses isolated from Ushar (*Calotropis procera*) seed fiber: Effect of isolation method. *Materials Letters, 168*, 146–150.

Oun, A. A., & Rhim, J. W. (2015). Effect of post-treatments and concentration of cotton linter cellulose nanocrystals on the properties of agar-based nanocomposite films. *Carbohydrate Polymers, 134*, 20–29.

Owi, W. T., Ong, H. L., Sam, S. T., Villagracia, A. R., kuo Tsai, C., & Akil, H. M. (2019). Unveiling the physicochemical properties of natural *Citrus aurantifolia* crosslinked tapioca starch/nanocellulose bionanocomposites. *Industrial Crops and Products, 139*, 111548, October 2018.

Payen, A. (1838). Mémoire sur la composition du tissu propre des plantes et du ligneux. *Comptes Rendus, 7*, 1052–1056.

Peng, J., Ellingham, T., Sabo, R., Clemons, C. M., & Turng, L. S. (2015). Oriented polyvinyl alcohol films using short cellulose nanofibrils as a reinforcement. *Journal of Applied Polymer Science, 132*(48).

Peng, Y., Nair, S. S., Chen, H., Yan, N., & Cao, J. (2018). Effects of lignin content on mechanical and thermal properties of polypropylene composites reinforced with micro particles of spray dried cellulose nanofibrils. *ACS Sustainable Chemistry & Engineering, 6*(8), 11078–11086.

Pickering, K. L., Efendy, M. G. A., & Le, T. M. (2016). A review of recent developments in natural fibre composites and their mechanical performance. *Composites Part A: Applied Science and Manufacturing, 83*, 98–112.

De Prez, J., Van Vuure, A. W., Ivens, J., Aerts, G., & Van de Voorde, I. (2018). Enzymatic treatment of flax for use in composites. *Biotechnology Reports, 20*, e00294.

Qaiss, A., Bouhfid, R., & Essabir, H. (2015). *Effect of processing conditions on the mechanical and morphological properties of vomposites reinforced by natural fibres. Manufacturing of natural fibre reinforced polymer composites* (pp. 177–197). Switzerland: Springer International Publishing.

Qazanfarzadeh, Z., & Kadivar, M. (2016). Properties of whey protein isolate nanocomposite films reinforced with nanocellulose isolated from oat husk. *International Journal of Biological Macromolecules, 91*, 1134–1140.

Qian, S., & Sheng, K. (2017). PLA toughened by bamboo cellulose nanowhiskers: Role of silane compatibilization on the PLA bionanocomposite properties. *Composites Science and Technology, 148*, 59–69.

Qua, E. H., & Hornsby, P. R. (2011). Preparation and characterisation of nanocellulose reinforced polyamide-6. *Plastics Rubber and Composites, 40*(6–7), 300–306.

Quiévy, N., Jacquet, N., Sclavons, M., Deroanne, C., Paquot, M., & Devaux, J. (2010). Influence of homogenization and drying on the thermal stability of microfibrillated cellulose. *Polymer Degradation and Stability, 95*(3), 306–314.

Rafieian, F., Shahedi, M., Keramat, J., & Simonsen, J. (2014). Mechanical, thermal and barrier properties of nano-biocomposite based on gluten and carboxylated cellulose nanocrystals. *Industrial Crops and Products, 53*, 282–288.

Razali, N., Hossain, M. S., Taiwo, O. A., Ibrahim, M., Nadzri, N. W. M., Razak, N., Rawi, N. F. M., Mahadar, M. M., & Mohamad Kassim, M. H. (2017). Influence of acid hydrolysis reaction time on the isolation of cellulose nanowhiskers from oil palm empty fruit bunch microcrystalline cellulose. *BioResources, 12*(3), 6773–6788.

Sabaruddin, F. A., & Paridah, M. T. (2018). Effect of lignin on the thermal properties of nanocrystalline prepared from kenaf core. *IOP Conference Series: Materials Science and Engineering, 368*(1).

Sacui, I. A., Nieuwendaal, R. C., Burnett, D. J., Stranick, S. J., Jorfi, M., Weder, C., Foster, E. J., Olsson, R. T., & Gilman, J. W. (2014). Comparison of the properties of cellulose nanocrystals and cellulose nanofibrils isolated from bacteria, tunicate, and wood processed using acid, enzymatic, mechanical, and oxidative methods. *ACS Applied Materials & Interfaces, 6*(9), 6127–6138.

Sahlin, K., Forsgren, L., Moberg, T., Bernin, D., Rigdahl, M., & Westman, G. (2018). Surface treatment of cellulose nanocrystals (CNC): Effects on dispersion rheology. *Cellulose, 25*(1), 331–345.

Sanders, J. E., Han, Y., Rushing, T. S., & Gardner, D. J. (2019). Electrospinning of cellulose nanocrystal-filled poly (vinyl alcohol) solutions: Material property assessment. *Nanomaterials, 9*(5).

Saurabh, C. K., Dungani, R., Owolabi, A. F., Atiqah, N. S., Zaidon, A., Aprilia, N. A. S., Sarker, Z. M., & Khalil, H. P. S. A. (2016). Effect of hydrolysis treatment on cellulose nanowhiskers from oil palm (*Elaeis guineesis*) fronds: Morphology, chemical, crystallinity, and thermal characteristics. *BioResources, 11*(3), 6742–6755.

Savadekar, N. R., Karande, V. S., Vigneshwaran, N., Bharimalla, A. K., & Mhaske, S. T. (2012). Preparation of nano cellulose fibers and its application in kappa-carrageenan based film. *International Journal of Biological Macromolecules, 51*(5), 1008–1013.

Savadekar, N. R., & Mhaske, S. T. (2012). Synthesis of nano cellulose fibers and effect on thermoplastics starch based films. *Carbohydrate Polymers*, *89*(1), 146–151.

Shah, P., & Pandey, K. (2017). Advancement in packaging film using microcrystalline cellulose and TiO_2. *American Journal of Polymer Science and Technology*, *3*(6), 97–102.

Shanmugarajah, B., Loo, P., Mei, I., Chew, L., Yaw, S., & Tan, K. W. (2015). Isolation of nanocrystalline cellulose (NCC) from palm oil empty fruit bunch (EFB): Preliminary result on FTIR and DLS analysis. *Chemical Engineering Transactions*, *45*, 1705–1710, March.

Sharma, A., Thakur, M., Bhattacharya, M., Mandal, T., & Goswami, S. (2019). Commercial application of cellulose nano-composites—A review. *Biotechnology Reports*, *21*.

Song, Z., Xiao, H., & Zhao, Y. (2014). Hydrophobic-modified nano-cellulose fiber/PLA biodegradable composites for lowering water vapor transmission rate (WVTR) of paper. *Carbohydrate Polymers*, *111*, 442–448.

Spagnol, C., Fragal, E. H., Witt, M. A., Follmann, H. D. M., Silva, R., & Rubira, A. F. (2018). Mechanically improved polyvinyl alcohol-composite films using modified cellulose nanowhiskers as nano-reinforcement. *Carbohydrate Polymers*, *191*, 25–34.

Sun, Q., Zhao, X., Wang, D., Dong, J., She, D., & Peng, P. (2018). Preparation and characterization of nanocrystalline cellulose/*Eucommia ulmoides* gum nanocomposite film. *Carbohydrate Polymers*, *181*, 825–832.

Sutka, A., Sutka, A., Gaidukov, S., Timusk, M., Gravitis, J., & Kukle, S. (2015). Enhanced stability of PVA electrospun fibers in water by adding cellulose nanocrystals. *Holzforschung*, *69*(6), 737–743.

Suzuki, A., Sasaki, C., Asada, C., & Nakamura, Y. (2018). Production of cellulose nanofibers from Aspen and Bode chopsticks using a high temperature and high pressure steam treatment combined with milling. *Carbohydrate Polymers*, *194*, 303–310, April.

Taheri, H., & Samyn, P. (2016). Effect of homogenization (microfluidization) process parameters in mechanical production of micro- and nanofibrillated cellulose on its rheological and morphological properties. *Cellulose*, *23*(2), 1221–1238.

Tang, Y., Zhang, X., Zhao, R., Guo, D., & Zhang, J. (2018). Preparation and properties of chitosan/guar gum/nanocrystalline cellulose nanocomposite films. *Carbohydrate Polymers*, *197*, 128–136.

Thomas, M. G., Abraham, E., Jyotishkumar, P., Maria, H. J., Pothen, L. A., & Thomas, S. (2015). Nanocelluloses from jute fibers and their nanocomposites with natural rubber: Preparation and characterization. *International Journal of Biological Macromolecules*, *81*, 768–777.

Tian, C., Yi, J., Wu, Y., Wu, Q., Qing, Y., & Wang, L. (2016). Preparation of highly charged cellulose nanofibrils using high-pressure homogenization coupled with strong acid hydrolysis pretreatments. *Carbohydrate Polymers*, *136*, 485–492.

Trache, D., Hussin, M. H., Haafiz, M. K. M., & Thakur, V. K. (2017). Recent progress in cellulose nanocrystals: Sources and production. *Nanoscale*, *9*(5), 1763–1786.

Trigueiro, J. P. C., Silva, G. G., Pereira, F. V., & Lavall, R. L. (2014). Layer-by-layer assembled films of multi-walled carbon nanotubes with chitosan and cellulose nanocrystals. *Journal of Colloid and Interface Science*, *432*, 214–220.

Trinh, B. M., & Mekonnen, T. (2018). Hydrophobic esterification of cellulose nanocrystals for epoxy reinforcement. *Polymer (Guildf)*, *155*, 64–74, May.

Usha, R., Sangeetha, T., & Palaniswamy, M. (2011). Screening of polyethylene degrading microorganisms from garbage soil. *Libyan Agriculture Research Center Journal International*, *2*(4), 200–204.

Wang, H., Kong, L., & Ziegler, G. R. (2019). *Fabrication of starch—Nanocellulose composite fibers by electrospinning*, . Food Hydrocolloids (90, pp. 90–98).

Wang, Y., Wei, X., Li, J., Wang, Q., Wang, F., & Kong, L. (2013). Homogeneous isolation of nanocellulose from cotton cellulose by high pressure homogenization. *Journal of Materials Science and Chemical Engineering*, *01*(05), 49–52.

Wang, H., Zhang, X., Jiang, Z., Li, W., & Yu, Y. (2015). A comparison study on the preparation of nanocellulose fibrils from fibers and parenchymal cells in bamboo (*Phyllostachys pubescens*). *Industrial Crops and Products*, *71*, 80–88.

Wei, L., Luo, S., McDonald, A. G., Agarwal, U. P., Hirth, K. C., Matuana, L. M., Sabo, R. C., & Stark, N. M. (2017). Preparation and characterization of the nanocomposites from chemically modified nanocellulose and poly(lactic acid). *Journal of Renewable Materials*, *5*(5), 410−422.

Wei, C., Zeng, S., Tan, Y., Wang, W., Lv, J., & Liu, H. (2015). Impact of layer-by-layer self-assembly clay-based nanocoating on flame retardant properties of sisal fiber cellulose microcrystals. *Advances in Materials Science and Engineering*, *2015*, 1−7.

Wulandari, W. T., Rochliadi, A., & Arcana, I. M. (2016). Nanocellulose prepared by acid hydrolysis of isolated cellulose from sugarcane bagasse. *IOP Conference Series: Materials Science and Engineering*, *107*(1).

Xu, C., Zhu, S., Xing, C., Li, D., Zhu, N., & Zhou, H. (2015). Isolation and properties of cellulose nanofibrils from coconut palm petioles by different mechanical process. *PLoS One*, *10*(4), 1−11.

Yadav, M., Behera, K., Chang, Y. H., & Chiu, F. C. (2020). Cellulose nanocrystal reinforced chitosan based UV barrier composite films for sustainable packaging. *Polymers (Basel)*, *12*(1).

Yang, X., Han, F., Xu, C., Jiang, S., Huang, L., Liu, L., & Xia, Z. (2017). Effects of preparation methods on the morphology and properties of nanocellulose (NC) extracted from corn husk. *Industrial Crops and Products*, *109*, 241−247.

Zhang, J., Song, H., Lin, L., Zhuang, J., Pang, C., & Liu, S. (2012). Microfibrillated cellulose from bamboo pulp and its properties. *Biomass and Bioenergy*, *39*, 78−83.

Zhang, P. P., Tong, D. S., Lin, C. X., Yang, H. M., Zhong, Z. K., Yu, W. H., Wang, H., & Zhou, C. H. (2014). Effects of acid treatments on bamboo cellulose nanocrystals. *Asia-Pacific Journal of Chemical Engineering*, *9*(5), 686−695.

Zhou, Y. M., Fu, S. Y., Zheng, L. M., & Zhan, H. Y. (2012). Effect of nanocellulose isolation techniques on the formation of reinforced poly(vinyl alcohol) nanocomposite films. *Express Polymer Letters*, *6*(10), 794−804.

CHAPTER 18

Biomedical applications of microbial polyhydroxyalkanoates

Aravind Madhavan[1], K.B. Arun[1], Raveendran Sindhu[2], Parameswaran Binod[2], Ashok Pandey[3], Ranjna Sirohi[4], Ayon Tarafdar[5] and R. Reshmy[6]

[1]Rajiv Gandhi Centre for Biotechnology, Thiruvananthapuram, India, [2]Microbial Processes and Technology Division, CSIR—National Institute for Interdisciplinary Science and Technology (CSIR—NIIST), Thiruvananthapuram, India, [3]Center for Innovation and Translational Research, CSIR—Indian Institute of Toxicology Research (CSIR—IITR), Lucknow, India, [4]Department of Post-Harvest Process and Food Engineering, G. B. Pant University of Agriculture and Technology, Pantnagar, India, [5]Department of Food Engineering, National Institute of Food Technology, Entrepreneurship and Management, Sonipat, India, [6]Post-Graduate and Research Department of Chemistry, Bishop Moore College, Mavelikara, India

18.1 Introduction

Polyhydroxyalkanoates (PHAs) are a group of natural polyesters produced by the microorganisms under different stress environment, which possess elastomeric and thermoplastic characteristics (Zhang, Shishatskaya, Volova, da Silva, & Chen, 2018). The basic characteristics of PHA-based bioplastics include:

- Hydrolytic degradation resistant and insoluble.
- High resistance to ultraviolet rays but susceptible to acids and bases.
- Chloroform soluble and also soluble various other solvents.
- Highly biocompatible and hence suitable for medical applications.
- Sinks in water and this makes easy anaerobic biodegradation in sediments.
- Nontoxic.
- Not sticky compared to other polymers while melting.

PHAs (and bioplastics in general) are very interesting biomaterials for three basic reasons: produced using renewable microbial sources, are biodegradable, and biocompatible. Generally, PHAs are produced by the bacteria and are potential candidates to replace the petroleum-derived plastics, which although have found tremendous useful applications in every walks of our life, are considered as serious environmental concerns (Rodriguez-Contreras, 2019).

PHAs are biopolymers synthesized by the microorganisms as energy-storage complex. PHAs are produced from sugars and lipids with the help of different enzymes. They are nontoxic, biodegradable, insoluble in water (hydrophobic), biocompatible, and thermoplastic, which makes PHAs a good candidate for the packaging industry, pharmacy, agriculture, etc. (Laycock, Halley, Pratt, Werker, & Lant, 2014). They can be converted into nanoparticles, microspheres, films, microcapsules, and porous matrices and thus drugs can easily be encapsulated in PHAs copolymer or homopolymer. PHAs might be a promising material in treating extremely resistant infections, due to the capability of delivery and maintenance of sufficient concentrate of antibiotics for a prolonged period (Zhang et al., 2018).

18.2 Types of polyhydroxyalkaonates for biomedical application

Monomeric composition architecture at supramolecular level determines the structural and chemical properties of PHA. Depending up on the monomeric composition, PHAs can be classified into scl-PHA (short-chain-length PHA) containing monomers of three to five carbon atoms, and mcl-PHA (medium-chain-length PHA) with six and more carbon atoms per monomer. Scl-PHA possesses thermoplastic properties and mcl-PHA exhibits elastomeric and latex-like properties. Based on the biological production and applications, PHAs are classified into homopolyesters, which contain only one kind of monomer or heteropolyesters, which contain copolyesters or monomers of diverse side chains or backbones and terpolyesters with different side chains and backbones (Koller, Maršálek, Miranda de Sousa Dias, & Braunegg, 2017; Kourmentza et al., 2017; Tan et al., 2014). To manufacture the heteropolyesters, PHA needs addition of structurally similar substrates to the basic building blocks (Koller et al., 2008; Koller et al., 2014; Koller et al., 2015; Koller, Hesse, Fasl, Stelzer, & Braunegg, 2017; Lefebvre, Rocher, & Braunegg, 1997; Miranda de Sousa Dias et al., 2017; Raposo, de Almeida, da Fonseca, & Cesário, 2017).

Poly-3-hydroxybutyrate (PHB), copolymers of 3-hydroxybutyrate and 3-hydroxyvalerate (PHBV), poly-4-hydroxybutyrate (P4HB), copolymers of 3-hydroxybutyrate and 3-hydroxyhexanoate (PHBHHx), poly-3-hydroxyoctanoate (PHO), and their composites have been applied in many areas of biomedicine such as sutures development, patches repair, cardiovascular stents, orthopedic field, guided tissue engineering devices, bone and nerve tissue engineering, and wound dressings (Hazer et al., 2012; Li, Thouas, & Chen, 2016; Shishatskaya, Nikolaeva, Vinogradova, & Volova, 2016). The strategies for the production and applications of PHAs in the medical field is outlined in Fig. 18.1.

18.3 Genetic-engineered strains for the production of polyhydroxyalkaonates

PHAs can be categorized into several kinds based on the structure of the repeating units in the polymers and its composition mainly depends on the enzyme PHA synthase, the metabolic pathway involved, the substrate, that is, carbon source, and the cultivation

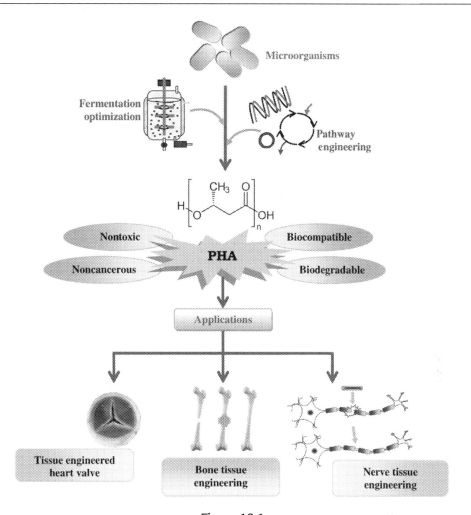

Figure 18.1
Schematic representation for the production of polyhydroxyalkanoates (PHAs) and its applications in biomedicine.

conditions (Rehm & Steinbüchel, 1999). Currently, their use is limited due to the high cost of the production. To limit the cost of production, various genetically modified bacterial strains that can use low cost substrate as carbon sources have been developed (Lee, 1996).

18.3.1 Rational strategies for cost-effective, good-quality large-scale production of PHAs

Following are considered as the main strategies for the cost-effective production of PHAs.

1. Searching new natural strains that efficiently convert the selected feedstock (substrate) into PHAs.

2. Genetic engineering of microorganisms that can make PHAs production highly efficient.
3. Creating recombinant bacterial strain able to produce PHAs that can naturally utilize wide variety simple and complex carbon source (Favaro, Basaglia, & Casella, 2019).

A number of genetic engineering approaches can be used to improve the production of bacterial PHAs. The main aim of genetic engineering approaches is to produce recombinant strains that can produce PHAs using different waste streams and that can resist contamination.

The biosynthesis pathway for PHAs production is well known and understood. Several strategies can be used for achieving improved efficiency of PHAs production such as changing the growth pattern and growth rate, increasing the cell size for more accumulation, altering the PHA synthesis pathway using DNA reprogramming technology like optimizing ribosome-binding site or promoter of PHA operon, overexpression of PHA synthesis operon, chromosomal overexpression of PHA synthesis, and removing other competing pathways like beta-oxidation and CRISPR/Cas9 tools (Chen & Jiang, 2018). The type and availability of the substrate are main factors for the efficient production of PHAs. Fatty acids are the main substrates for PHAs production. The composition of the PHAs can vary according to the type of fatty acid used. But usually fatty acids are converted into acetyl-coA through beta-oxidation for the use in many other biosynthesis pathways such as Kreb's cycle and cholesterol biosynthesis rather than PHAs synthesis.

Several bacteria such as *Bacillus* sp., *Pseudomonas* sp., *Alkaligenes lactus*, and *Cupriavidus necator* can produce PHAs via beta-oxidation. However, generally the synthesis of PHAs is very low, which eventually results in high cost of PHA production. The two enzymes, fadA (ketothiolase) and fadB (hydroxyacyl-coA dehydrogenase) catalyze the last two steps of the fatty acid degradation (Ouyang et al., 2007). In *Pseudomonas entomophila* or *Pseudomonas putida*, the gene which codes for these two enzymes in beta-oxidation pathway can be deleted using gene-knockout method. Deletion of these enzymes results in the conversion of fatty acid substrates into a different product, that is, hydroxyacyl-CoA, which forms PHAs rather than forming acetyl-coA. Deletion of the beta-oxidation pathway in different species of *Pseudomonas* eventually could produce different types copolymers containing 3-hydroxydecanoate, 3-hydroxyhexanoate (3HHx), 3-hydroxyoctanoate (3HO), and 3-hydroxydodecanoate (Zhang, Lin, Wu, Wang, & Chen, 2019; Zou et al., 2017).

PhaC, which produces PHA synthase, one of the key enzymes involved in PHA biosynthesis is clustered together along with the other genes and it functions by polymerizing the monomeric hydroxyalkanoate substrate (Fonseca & Antonio, 2006). PHA synthase genes from *Pseudomonas aeruginosa* and *Ralstonia eutropha* were cloned into two different strains of *E. coli* (DH10B and JM101) and cultured in the medium containing soybean oil and corn starch as carbons source and acrylic acid as fatty acid beta-oxidation inhibitor, the recombinant DH10B strain produced 5.9%

of PHAs, having a total cell dry weight of 0.92 g L^{-1} (Tan, Xue, Aibaidula, & Chen, 2011). *Halomonas* TD01, a halophile, grew at high salt concentration and accumulated high amount of PHA when grown in salt medium supplemented with glucose (Tan, Wu, Chen, & Chen, 2014). This strain, however, can be engineered by partial inactivation of the DNA restriction/methylation system to increase the stability. Also, those genes that are involved in the multiple pathways are expressed with the help of an inducible high-copy number plasmid named pSEVA341 with a containing a LacIq-Ptrc system. The enzymes, PHA depolymerases and 2-methylcitrate synthase can be removed within the genome for the overexpression of threonine dehydrogenase, resulting in the construction of the modified *Halomonas* TD08 strain. The overexpression strain uses various carbohydrates as carbohydrate source and resulted in the production of PHBV containing of 4–6 mol% or 3-hydroxyvalerate (3HV) from various carbohydrates as the sole carbon source. The inducible overexpression of MinCD can inhibits the formation of FtsZ rings in the *Halomonas* TD08 during the stationary phase. This generates larger cells or filamentous cells that are 1.4-fold times longer than its original size and can accumulate about 69%–82% (wt./wt.) of PHAs (Schrader et al., 2009).

Methanol can be used as an important feedstock in the chemical and biotechnological industries due to its sustainability and low price (Schrader et al., 2009). Methylotrophic bacteria can use methylamine and methanol as energy and carbon sources. They use two types of pathways, the ribulose monophosphate (RuMP) pathway and the serine cycle for assimilating C1 compounds. Methylotrophic bacteria that use serine cycle pathway can accumulate poly-3-hydroxybutyrate, which is the best distinctive natural PHA that could be obtained using methanol. *Methylobacterium extorquens* AM1 produce PHA composed solely of 3-hydroxybutyrate (3HB) through methylotrophic growth. *M. extorquens* AM1 is genetically engineered in a way that the original gene *phaCMe* coding for PHA synthase is replaced with *phaCAc* of *Aeromonas caviae*. They can produce a PHA composed of different polymers such as 3HV, 3HB and a C6-monomer, (R)-3-hydroxyhexanoate from methanol. PHA accumulation in the recombinant strain AM1CAc is higher when compared to PHA produced in the native strain (Orita, Nishikawa, Nakamura, & Fukui, 2014).

Recombinant strain with an accelerated cell growth can also be used for the increased PHA production. For example, bacteria usually divide by binary fission, but deleting certain fission genes such as *minC* and *minD* together can cause multiple fission. Deletion of these genes leads to more than one fission ring formation, called z-ring (multiple fission rings) in different positions of an enlarged cell and forms more number of daughter cells at the same time. The overexpression of cell shape controlling gene *mreB* along with other genes such as *ftsL*, *ftsQ*, *ftsN*, *ftsW*, and *ftsZ* in *E. coli* JM109 ΔminCD improved the cell growth and eventually PHA production with more cell dry weight and about 80% PHB accumulation (Chen & Jiang, 2017; Wu, Fan, Jiang, Chen, & Chen, 2016). Engineered bacterial strains for the production of PHAs are listed in Table 18.1 (Ahn, Park, & Lee, 2001; Arikawa, Matsumoto, & Fujiki, 2017; Nikel, Pettinari, Galvagno, & Méndez, 2006; Pais et al., 2014;

Table 18.1: Polyhydroxyalkanoates (PHAs) production by different recombinant strains.

Recombinant strain	Organisms from which gene is taken	Type of PHA produced	Substrate used	Total PHA concentration (g L^{-1})	Total PHA content (CDM %)	Reference
C. necator mREPT	lacz, laci&laco genes from E. coli	3HB	Whey permeate	1.4	22	Povolo et al. (2010)
E. coli CGSC 4401 (pJC4)	PHA synthesis genes from Alcaligenes latus	3HB	Processed bovine whey powder solution	168	87	Ahn et al. (2001)
C. necator 142SR	csc genes from E. coli	3HB-co-3HHx	sucrose	113	81	Arikawa et al. (2017)
E. coli P8-X8	PHB synthesis genes from Cupriavidus necator	3HB	Cheese whey	19	39	Pais et al. (2014)
C. necator NCIMB11599	MBEL55E sacC gene from Mannheimia succiniciproducens	3HB	Sucrose	2.0	73	Park et al. (2015)
E. coli GCSC6576 (pSYL107)	PHA synthesis genes from Ralstonia eutropha, and ftsZ gene from E. coli	3HB	Bovine whey powder	69.0	87	Wong et al. (1998)
Klebsiella aerogenes 2688 (pJM9131)	PHB synthesis genes from Alcaligene seutrophus	3HB	Sucrose	3.0	50	Zhang et al., 1994
E. coli ΔarcA	Azotobacter sp. strain FA8	3HB	Whey	51.1	73	Nikel et al. (2006)

Park et al., 2015; Povolo, Toffano, Basaglia, & Casella, 2010; Wong, Lee, Wong, & Lee, 1998; Zhang, Obias, Gonyer, & Dennis, 1994).

18.4 Polyhydroxyalkanoates for drug delivery

Drug-delivery systems (DDSs) are engineered technology used for the targeted delivery of a pharmaceutical compound. In conventional drug treatment, typically the bioavailability of the drugs at target is not maintained for an extended period. The major goal of DDS is controlled delivery of a drug to the target at the therapeutically optimal dose and rate. Polymers have been used as the carriers of drug as they can effectively provide the drug to a target site, and thus enhance the therapeutic advantage while minimizing side effects (Barouti, Jaffredo, & Guillaume, 2017). The most commonly developed polymers are polyester-based DDSs. Polyester-based DDSs can be prepared from the naturally producing biopolymers such as polyhydroxyalkanotes (Li, Yang, & Loh, 2016) or chitosan (Bugnicourt & Ladavière, 2016) or artificial macromolecules such as polyglycolic acid (PGA), polylactic acid (PLA), poly-ε-caprolactone (PCL), or polyethylene glycol (PEG).

PHB-derived self-assembly systems have been established from both the artificial and natural PHB materials. This involves nanoparticles, microparticles, micelles, and gels (Barouti et al., 2017). Nanoparticles can be derived from PHB homopolymer as well as from the copolymers by using a higher stirring solvent evaporation method. The characteristics and size of nanoparticles are determined by the nature of the surfactant, the choice of solvent, the volume ratio of organic/aqueous solvent, and the rate of mixing. Common methods of the synthesis of nanoparticles are nanoprecipitation, emulsion techniques, or electrospraying (Kamaly et al., 2016; Pouton & Akhtar, 1996). PHB homopolymer-derived nanoparticles have been used to deliver drugs with hydrophobicity. This system shows excellent drug-loading efficiency (DLE) because of the lack of a hydrophilic group in the PHB backbone, which is hydrophobic in nature. One example is FOL-PHB (a nanoparticles conjugate) used to target colorectal cancer, which is encapsulated with arsenic trioxide. In this case, high DLE is achieved because of the interaction between the ester carbonyl group electronegative oxygen of PHB with the arsenic metal ion as well as the PHB and arsenic trioxide hydrophobic interactions. Fourteen percent of arsenic trioxide was released within 30 min of administration due to burst effect and then 40% of release in the next 48 h, which resulted in the sustained release of the drug (Althuri et al., 2013).

PHB-based microparticles are prepared using the similar polymers like the polymers used for the preparation of nanoparticles but the method used is a slow stirring rate solvent evaporation method. These are used to encapsulate either hydrophobic or hydrophilic molecules dependent on the formulation method and the physicochemical characteristics of the copolymer. Microparticles are more appropriate for the hydrophilic drugs than the hydrophobic drugs. For example, PHB-co-CL-based microparticles, which are encapsulated with calcein (hydrophilic in nature) show a DLE of 100%, but 1-(2,5-dimethyl-4-(2,5-dimethyl phenyl) phenyl diazenyl) azonapthalen-2-ol (OilRed O), which has encapsulated a hydrophobic molecule, show lower DLE (Pignatello et al., 2009).

PHB-based micelles are prepared from the PHB/PEG-based, PHB-based stimuli-responsive, and PHB-based copolymers by the aggregation of amphiphilic block copolymers in a medium, which is aqueous by the aggregation of their hydrophobic blocks (Nicolai, Colombani, & Chassenieux, 2010). The main disadvantage of these systems is that the hydrophobic block length determines the critical micelle concentration of the copolymer. Also, the number of aggregation and the molecular weight vary with the temperature, which makes these systems less stable (Mao, Sukumaran, Beaucage, Saboungi, & Thiyagarajan, 2001). These systems are commonly used for the hydrophobic drugs. DLE of these systems is determined by the parameters such as partition equilibrium coefficient (K), crystallinity of the core, copolymer architecture, and other conditions surrounding it, for example, temperature and pH. PEG helps in the increase in vivo circulation of the drug because of the steric inherence and hydration effect it helps to prevent the adhesion to opsonins. For example, PDMAEMA-b-PHB-b-PDMAEMA micelle based is used to encapsulate doxorubicin (Kim, Mount, Gombotz, & Pun, 2010) and PEEP-b-PHB-b-PEEP, a triblock copolymer is used to encapsulate Paclitaxel (Cheng & Wang, 2009).

PHB-based hydrogels are the less used systems compared to others. Hydrogels are 3D cross-linked polymer chains, which are hydrophilic and can hold substantial quantity of water without dissolution. Hydrogel response is mostly determined by the kind of cross-linking, the composition, and cross-linking degree. These systems are used to encapsulate the stem cells, and thus used in the stem cell therapy (Vashist, Vashist, Gupta, & Ahmad, 2014).

Different strategies are used to target the PHB system to the site of action. One strategy is to conjugate a target ligand to the polymeric carrier. For example, A33scFv-GFP-PHB-based nanoparticles have been used in colon cancer in which A33scFV is the antibody against A33 antigen and GFP is used for imaging. Another approach is the conjugation of folates (FOLs), a compound in nucleotide synthesis to the polymer. Folates are highly required in the dividing cells. For example, for the L929 fibrosarcoma targeting, FOL-PHB conjugates are used (Sudimack & Lee, 2000). Yet another approach is the conjugation of particular peptide to the polymer to produce a polymer-protein hybrid particle. For example, RGD4C-PHB nanoparticle PHB conjugates are effective regarding MDA-MB 231 human breast cancer (Lee et al., 2011). Furthermore, target HT-29 human colon cancer cells has also been used to conjugate the polymer with the bile salt (DOCA-PHB-b-PEG conjugate), which disrupts the tight junction and facilitates the cellular uptake (Chaturvedi et al., 2013). In all the cases, a fluorescent protein, such as Nile Red can be conjugated to monitor the cellular uptake.

18.5 Polyhydroxyalkanoates in tissue engineering

Tissue engineering is a rapidly emerging area that has received extensive responsiveness in medical science. Due to the amphiphilic behavior, higher biocompatibility, more supportive in proliferation, and its enormous antimicrobial properties, microbial PHAs are extensively used in bone, cartilage, cardiac, blood, and other tissue engineering. More specifically PHB, PHBV, 3HB, P4HB, PHO, PHBHHx, and their composites are broadly used in tissue engineering. For this purpose, engineered scaffolds are made with an extremely porous biodegradable PHA and incubated with suitable cell type for the seeding, to finally allow to tissue growth (ex vivo). Then the bioactive PHA scaffold is grafted as a support for the development of new tissues (Elmowafy et al., 2019; Masood, Yasin, & Hameed, 2015).

18.5.1 Tissue engineering—bone

PHA alone or in a mixture has been used in bone tissue engineering such as partial bone marrow replacement, repair, reinforcement, and regeneration. Hence they are considered as promising orthobiomaterials. P3HB, P3HB-co-3HHx, and P3HB-co-3HV are studied widely in this area. However, scl-PHA attained the main emphasis because of its greater mechanical strength. Many investigations have been conducted with PHAs as a bone fixation device. In the 1980s studies have used T-plates made of P(3HB) and supported with fibers of carbon (7%) in the rabbits for fixing tibial diaphysis osteotomies (Vainionpää et al., 1986). The microporous piezoelectric P(3HB-co-3HV)-glass fibers composites also accelerated native bone formation. Subcutaneous use of with phosphate glass induced new bone generation in the rats (Knowles & Hastings, 1993). The newly formed bone at the interphase between P(3HB-co-3HV)/HAp scaffold implant and rabbit tibia was morphologically, chemically, and biologically active. The bioactive nature of HA facilitated the direct attachment of bones with the implant (Wang, Zhou, Xia, Dai, & Liu, 2013).

High-porosity PHBV scaffold integrated with Ca_2SiO_4 nanoparticles mimic identical to the extracellular matrix and provide good proliferation, differentiation, and adhesion of human MG-63 such as osteoblast (Wang et al., 2005). PHBHHx and PHB films with different concentrations of HHx (hydroxyhexanoate) accelerate the growth of osteoblasts (Luklinska & Schluckwerder, 2003). A scaffold of P(HB-HHx)—mesoporous bioactive glass (MBG) composite was used for studying its osteoblasts generation on the implantation of rat calvarial bone defects. Incorporation of MBG improved osteogenic and bioactive properties of PHBHHx polymer (Zhao et al., 2014). P3HB-ZrO_2 and Herafill (varying concentrations) composite were studied in infant rat femora. Herafill composed of calcium carbonate ($CaCO_3$), calcium sulfate ($CaSO_4$), and glycerol tripalmitate. The study reported PHB-ZrO_2 with 30% Herafill induced higher accumulation bone around the implant (Meischel et al., 2016).

18.5.2 Tissue engineering—cartilage

Cartilage is a critical tissue having chondrocytes in bone terminus regularly exposed to large mechanical loads. The volumetric density of cartilage is small in the human body and the main function is to provide a lubricant and smooth surface to enable diffusion of mechanical load with a short frictional coefficient. Hence they are easily prone to injuries. Damage gives rise to osteoarthritis and loss of function of joints. Therefore the healing is indispensable. Due to avascular nature, cartilage regeneration is highly challenging, although, regeneration using PHB, PHBHHx, PHBV, and its composites has been reported. The PHA chondrocytes showed better adhesion, enhanced proliferation, and differentiation on PHA scaffolds (Sun, Wu, Li, & Chang, 2005). A three-dimensional PHBHHx scaffold was evaluated for articular cartilage repair in a rabbit. After 16 weeks, full thickened cartilage was obtained (Wang, Bian, Wu, & Chen, 2008). Incorporation of poly(L-lactide-co-ε-caprolactone) (PLCL) into PHBV microspheres improves the mechanical properties. The chondrogenesis studies confirmed good adhesion and increased type II collagen contents during the proliferation and cartilage development (Li et al., 2013). Medium-chain length PHA combined with a PHA-binding protein revealed improved multiplication of the cell lines. Surface modification of PHBHHx scaffolds with the help of a PHA binding protein bonded with arginyl-glycyl-aspartic acid promoted the proliferation of chondrogenic umbilical cord blood-derived mesenchymal stem cells in humans (Li et al., 2015).

18.5.3 Tissue engineering—nerve

Nerve tissue engineering primarily emphasizes on neurodegenerative diseases and injuries of the peripheral nervous system. PHAs such as P(3HB-co-4HB), P(3HB), P(3HBV), and P(3HB-co-3HHx) have shown good effects on neural stem cell (NSC) multiplication, viability, and growth. They also increased the survival and axon-dendrite segregation of NSC (Chen & Tong, 2012). For cervical spinal cord injury, an alginate hydrogel plus fibronectin-coated PHB fiber graft was implanted. After 8 weeks of incubation, 45% of injured neuron reduction was noticed in the untreated model, whereas the PHB-treated model reduced the neuron loss by 50%. The addition of the neonatal Schwann cells showed the regenerating axons passing through the PHB graft and spread along its whole length. It confirmed that PHB scaffolded with the Schwann cells fibronectin and alginate hydrogel, enhancing persistence and regeneration of neurons after spinal cord damage (Novikov et al., 2002).

PHBHHx nerve channels are ultimate for the renewal of Schwann cells. The effective regeneration of the long nerve gap injuries requires an efficient microarchitecture, which imitated the native nerve tissues materials (Lizarraga-Valderrama et al., 2015). Another study revealed that the treatment with alkali like NaOH on P(3HB-co-3HHx) films enhanced the hydrophilicity, led to increasing the attachment to NSCs/neural progenitor cell

(NPCs) with low serum (Lu et al., 2014). Thus owing to greater biocompatibility with NSCs, PHAs are excellent options for the central nervous system regeneration when compared with other biomaterials.

18.5.4 Tissue engineering—peridontal

Currently, only a few reports are available with PHAs applications in periodontal tissue engineering. In 2012 Wang et al. (2012) reported the adeptness of PHBV/Ecoflex composite at different ratios for the periodontal tissue engineering. Ecoflex is faster degrading biodegradable polyester having greater elasticity and hydrophilicity. In this study, PHBV was dissolved in Ecoflex with different ratios (30:70, 70:30, 50:50, and 100:0 of PHBV:Ecoflex). The Ecoflex improved the hydrophilicity and flexibility of PHBV and enhanced the mechanical, biological, and chemical properties of PHBV. Hence in comparison with the pure PHBV, the PHBV/Ecoflex scaffold enhanced adhesion and multiplication of periosteum-derived stem cells and periodontal ligament stem cells. The findings concluded that surface modifications could be one of the critical factors for stimulating a signaling cascade in tissue regeneration (Geiger, Spatz, & Bershadsky, 2009). Accordingly, modified PHAs can be considered as a powerful extracellular matrix.

18.5.5 Tissue engineering—cardiovascular

The properties of PHAs such as nontoxicity, long durability, lack of immunogenicity, and resistance to infection make their use in cardiovascular engineering interesting for vascular grafts, artery augmentation, heart valves, cardiologic stents, and pericardial patches. Vascular grafting is the repairing or replacing broken blood vessels in artery or veins systems that might be caused due to damage or by diseases such as atherosclerosis and traumatic injury. In such cases, synthetic grafting, autologous grafting, xenografting, etc., have many drawbacks such as poor stability, lack of immune response, and risk of infection. According to Marois et al. (1999), the utilization of P(3HHx-co-3HO) in the rats as an impregnation substrate proved very deliberate dilapidation of the polymer. A complex of poly(3-hydroxybutyrate-co-3-hydroxyhexanoate) (PHBHHx) and poly(propylene carbonate) (PPC) at varying ratios have also been used for blood vessel tissue engineering. Here, the addition of PPC improved the mechanical properties of PHBHHx (Zhang et al., 2007).

Presently, inborn cardiovascular defects are incurable birth defects. Highly porous patches of P(4HB) scaffold have been used for manufacturing autologous cardiovascular tissue with good success. In a sheep model, these scaffolds were seeded with autologous smooth muscle, fibroblast, and endothelial cells to expand the pulmonary artery. Tissue regeneration without the issues of thrombus, dilation, or stenosis in the patch was perceived. It exhibited a self-sealing capacity when related with a polytetrafluoroethylene, left a hole for blood leakage (Stock et al., 2000).

Heart valve scaffolds developed by the tissue engineering are good substitutes to heart valve replacement surgery. High rigidity of chemical-synthesized polyesters makes them inefficient to function as bendable leaflets of the pulmonary valve. Interestingly, a more porous and stretchable P(3HHx-co-3HO)-PGA mesh was more fit in a study conducted with lamb models. P(3HHx-co-3HO) scaffolds seeded with the vascular cells showed no thrombus formation, but mild stenosis was observed after 17 weeks of implantation (Sodian et al., 2000). Hoerstrup et al. (2000) established a trileaflet heart valve using porous PHA scaffold [P(4HB) with a nonwoven PGA meshcoat]. The implanted valve properly functioned without any thrombus, stenosis, and aneurysm. It was one of the best results of cardiac tissue engineering. Castellano et al. (2014) proved the efficiency of PHB as implants on the epicardial surface of rats, which promoted angiogenesis. Tables 18.2 and 18.3 show the biocompatibility studies and medical applications of PHAs (Basnett et al., 2013; Bian, Wang, Aibaidoula, Chen, & Wu, 2009; Chaturvedi, Kulkarni, & Aminabhavi, 2011; Deng et al., 2003; Gursel, Yagmurlu, Korkusuz, & Hasirci, 2002; Hufenus, Reifler, Maniura-Weber, Spierings, & Zinn, 2012; Koller, Bona, Chiellini, & Braunegg, 2013; Korsatko et al., 1984; Lizarraga-Valderrama et al., 2015; Masood et al., 2013; Naveen et al., 2010; Puppi et al., 2017; Puppi, Morelli, & Chiellini, 2017; Rezaie Shirmard et al., 2017; Shishatskaya, Volova, Puzyr, Mogilnaya, & Efremov, 2004; Volova, Shishatskaya, Sevastianov, Efremov, & Mogilnaya, 2003; Wang et al., 2010; 2010; Wang, Wu, & Chen, 2005; Wu et al., 2014; Zhao et al., 2014).

Table 18.2: Polyhydroxyalkanoates (PHAs) and biocompatibility studies.

PHA	Study	Reference
Poly(3-hydroxybutyrate-co-3-hydroxyvalerate)	Compatibility studies of PHBHV/PLA fibers Evaluation of biochemical and physiological parameters of rat implanted with biopolymer sutures	Hufenus et al. (2012); Shishatskaya et al. (2004)
Poly(3-hydroxybutyrate) (PHB)	Biocompatibility studies of PHB tablets in mice fibroblast Physiological studies of rats implanted with PHB sutures	Korsatko et al. (1984); Volova et al. (2003)
Poly(3-hydroxybutyrate-co-3-hydroxyvalerate-co-4-hydroxybutyrate)	Production of pure polymer	Koller et al. (2013)
Poly(3-hydroxybutyrate-co-3-hydroxyhexanoate)	Biocompatibility of mouse osteoblast cells	Wang et al. (2005)
Poly(3-hydroxyoctanoate) Homopolyester; scl-PHA	Biocompatibility studies neuronal cells	Lizarraga-Valderrama et al. (2015)
Poly(3-hydroxyhexanoate-co-3-hydroxyoctanoate) (PHHxHO) (copolyester; mcl-PHA)	Production of pure polymer	Wang et al. (2010)

Table 18.3: Application of polyhydroxyalkanoate (PHA) in medicine.

PHA	Application	Reference
Poly(3-hydroxybutyrate-co-3-hydroxyvalerate) (PHBHV)	Immobilization of kanamycin Immobilization of sulbactam:cefoperazone and gentamicin Immobilization of drug 5-fluorouracil for colon cancer treatment Immobilization of fingolimod Immobilization of ellipticine for cancer treatment	Naveen et al. (2010); Chaturvedi et al. (2011); Gursel et al. (2002); Masood et al. (2013); Rezaie Shirmard et al. (2017)
Poly(3-hydroxybutyrate-co-3-hydroxyhexanoate)	Immobilization of rhodamine B	Wu et al. (2014)
Poly(3-hydroxybutyrate-co-3-hydroxyhexanoate)	Production of blood vessel stents Scaffolds for chondrocytes proliferation Scaffold for osteoblast proliferation Nerve tissue engineering Scaffolds for bone marrow mesenchymal stem cells Scaffolds for bone regeneration	Puppi et al. (2017); Bian et al. (2009); Deng et al. (2003); Puppi et al. (2017); Wang et al. (2010); Zhao et al. (2014)
Poly(3-hydroxyoctanoate)	For blood vessel stents	Basnett et al. (2013)

18.6 Conclusions and perspectives

PHAs possess distinct positions in the medical field involving their usage in different medical implants. Research is currently being carried out to elucidate the capability of PHAs for expanding biomedical applications. The main advantage of PHA is its ecofriendly characteristics such as elasticity, less cytotoxicity, biodegradability, nontoxicity, biocompatibility, capability for surface modification, and changeable physical and chemical properties. Since most of PHAs are from the microbes, especially bacteria, it also opens the possibility of producing PHAs with unique monomer composition by exploiting the tools and techniques of metabolic engineering. Some novel bacteria with versatile production capability also produce PHAs with unique properties. Nonetheless, the immune rejection reaction of human body to these implants should be studied extensively. It is expected that the use of more PHA-based implants in the medical field would emerge as a major application in future with the production of biocompatible PHAs.

Acknowledgments

Raveendran Sindhu and Reshmy R acknowledge Department of Science and Technology, New Delhi for the support under DST WOS-B scheme. Ranjna Sirohi acknowledges CSIR, New Delhi for providing fellowship under direct SRF scheme bearing the grant no. 09/171(0136)/19.

References

Ahn, W. S., Park, S. J., & Lee, S. Y. (2001). Production of poly (3-hydroxybutyrate) from whey by cell recycle fed-batch culture of recombinant Escherichia coli. *Biotechnology Letters*, *23*(3), 235–240.

Althuri, A., Mathew, J., Sindhu, R., Banerjee, R., Pandey, A., & Binod, P. (2013). Microbial synthesis of poly-3-hydroxybutyrate and its application as targeted drug delivery vehicle. *Bioresource Technology*, *145*, 290–296.

Arikawa, H., Matsumoto, K., & Fujiki, T. (2017). Polyhydroxyalkanoate production from sucrose by *Cupriavidus necator* strains harboring csc genes from *Escherichia coli* W. *Applied Microbiology and Biotechnology*, *101*(20), 7497–7507.

Barouti, G., Jaffredo, C. G., & Guillaume, S. M. (2017). Advances in drug delivery systems based on synthetic poly (hydroxybutyrate)(co) polymers. *Progress in Polymer Science*, *73*, 1–31.

Basnett, P., Ching, K. Y., Stolz, M., Knowles, J. C., Boccaccini, A. R., Smith, C., ... Roy, I. (2013). Novel poly (3-hydroxyoctanoate)/poly (3-hydroxybutyrate) blends for medical applications. *Reactive and Functional Polymers*, *73*(10), 1340–1348.

Bian, Y. Z., Wang, Y., Aibaidoula, G., Chen, G. Q., & Wu, Q. (2009). Evaluation of poly (3-hydroxybutyrate-co-3-hydroxyhexanoate) conduits for peripheral nerve regeneration. *Biomaterials*, *30*(2), 217–225.

Bugnicourt, L., & Ladavière, C. (2016). Interests of chitosan nanoparticles ionically cross-linked with tripolyphosphate for biomedical applications. *Progress in Polymer Science*, *60*, 1–7.

Castellano, D., Blanes, M., Marco, B., Cerrada, I., Ruiz-Saurí, A., Pelacho, B., ... Sepúlveda, P. (2014). A comparison of electrospun polymers reveals poly (3-hydroxybutyrate) fiber as a superior scaffold for cardiac repair. *Stem Cells and Development*, *23*(13), 1479–1490.

Chaturvedi, K., Ganguly, K., Kulkarni, A. R., Nadagouda, M. N., Stowbridge, J., Rudzinski, W. E., & Aminabhavi, T. M. (2013). Ultra-small fluorescent bile acid conjugated PHB–PEG block copolymeric nanoparticles: Synthesis, characterization and cellular uptake. *RSC Advances*, *3*(19), 7064–7070.

Chaturvedi, K., Kulkarni, A. R., & Aminabhavi, T. M. (2011). Blend microspheres of poly (3-hydroxybutyrate) and cellulose acetate phthalate for colon delivery of 5-fluorouracil. *Industrial & Engineering Chemistry Research*, *50*(18), 10414–10423.

Chen, G. Q., & Jiang, X. R. (2017). Engineering bacteria for enhanced polyhydroxyalkanoates (PHA) biosynthesis. *Synthetic and Systems Biotechnology*, *2*(3), 192–197.

Chen, G. Q., & Jiang, X. R. (2018). Engineering microorganisms for improving polyhydroxyalkanoate biosynthesis. *Current Opinion in Biotechnology*, *53*, 20–25.

Chen, W., & Tong, Y. W. (2012). PHBV microspheres as neural tissue engineering scaffold support neuronal cell growth and axon–dendrite polarization. *Acta Biomaterialia*, *8*(2), 540–548.

Cheng, J., & Wang, J. (2009). Syntheses of amphiphilic biodegradable copolymers of poly (ethyl ethylene phosphate) and poly (3-hydroxybutyrate) for drug delivery. *Science in China Series B: Chemistry*, *52*(7), 961–968.

Deng, Y., Lin, X. S., Zheng, Z., Deng, J. G., Chen, J. C., Ma, H., & Chen, G. Q. (2003). Poly (hydroxybutyrate-co-hydroxyhexanoate) promoted production of extracellular matrix of articular cartilage chondrocytes in vitro. *Biomaterials*, *24*(23), 4273–4281.

Elmowafy, E., Abdal-Hay, A., Skouras, A., Tiboni, M., Casettari, L., & Guarino, V. (2019). Polyhydroxyalkanoate (PHA): Applications in drug delivery and tissue engineering. *Expert Review of Medical Devices*, *16*(6), 467–482.

Favaro, L., Basaglia, M., & Casella, S. (2019). Improving polyhydroxyalkanoate production from inexpensive carbon sources by genetic approaches: A review. *Biofuels, Bioproducts and Biorefining*, *13*(1), 208–227.

Fonseca, G. G., & Antonio, R. V. (2006). Polyhydroxyalkanoates production by recombinant *Escherichia coli* harboring the structural genes of the polyhydroxyalkanoate synthases of *Ralstonia eutropha* and *Pseudomonas aeruginosa* using low cost substrate. *Journal of Applied Science*, *6*, 1745–1750.

Geiger, B., Spatz, J. P., & Bershadsky, A. D. (2009). Environmental sensing through focal adhesions. *Nature Reviews Molecular Cell Biology*, *10*(1), 21–33.

Gursel, I., Yagmurlu, F., Korkusuz, F. E., & Hasirci, V. A. (2002). In vitro antibiotic release from poly (3-hydroxybutyrate-co-3-hydroxyvalerate) rods. *Journal of Microencapsulation, 19*(2), 153−164.

Hazer, D. B., Burcu, D., Mut, M., Dincer, N., Saribas, Z., Hazer, B., & Ozgen, T. (2012). The efficacy of silver embedded polypropylene-grafted polyethylene glycolcoated ventricular catheters on prevention of shunt catheter infection in rats. *Child's Nervous System, 28*, 839−846.

Hoerstrup, S. P., Sodian, R., Daebritz, S., Wang, J., Bacha, E. A., Martin, D. P., ... Vacanti, J. P. (2000). Functional living trileaflet heart valves grown in vitro. *Circulation, 102*(suppl_3), Iii-44.

Hufenus, R., Reifler, F. A., Maniura-Weber, K., Spierings, A., & Zinn, M. (2012). Biodegradable bicomponent fibers from renewable sources: Melt-spinning of poly (lactic acid) and poly [(3hydroxybutyrate)-co-(3-hydroxyvalerate)]. *Macromolecular Materials and Engineering, 297*(1), 75−84.

Kamaly, N., Yameen, B., Wu, J., & Farokhzad, O. C. (2016). Degradable controlled-release polymers and polymeric nanoparticles: Mechanisms of controlling drug release. *Chemical Reviews, 116*(4), 2602−2663. Available from https://doi.org/10.1021/acs.chemrev.5b00346.

Kim, T. H., Mount, C. W., Gombotz, W. R., & Pun, S. H. (2010). The delivery of doxorubicin to 3-D multicellular spheroids and tumors in a murine xenograft model using tumor-penetrating triblock polymeric micelles. *Biomaterials, 31*(28), 7386−7397.

Knowles, J. C., & Hastings, G. W. (1993). In vitro and in vivo investigation of a range of phosphate glass-reinforced polyhydroxybutyrate-based degradable composites. *Journal of Materials Science: Materials in Medicine, 4*(2), 102−106.

Koller, M., Bona, R., Chiellini, E., & Braunegg, G. (2013). Extraction of short-chain-length poly-[(R)-hydroxyalkanoates] (scl-PHA) by the "anti-solvent" acetone under elevated temperature and pressure. *Biotechnology Letters, 35*(7), 1023−1028.

Koller, M., Bona, R., Chiellini, E., Fernandes, E. G., Horvat, P., Kutschera, C., ... Braunegg, G. (2008). Polyhydroxyalkanoate production from whey by *Pseudomonas hydrogenovora*. *Bioresource Technology, 99*, 4854−4863.

Koller, M., Hesse, P., Fasl, H., Stelzer, F., & Braunegg, G. (2017). Study on the effect of levulinic acid on whey-based biosynthesis of poly (3-hydroxybutyrate-co-3-hydroxyvalerate) by *Hydrogenophaga pseudoflava*. *Applied Food Biotechnology, 4*, 65−78.

Koller, M., Maršálek, L., Miranda de Sousa Dias, M., & Braunegg, G. (2017). Producing microbial polyhydroxyalkanoate (PHA) biopolyesters in a sustainable manner. *New Biotechnology, 37*, 24−38.

Koller, M., Miranda de Sousa Dias, M., Rodríguez-Contreras, A., Kunaver, M., Žagar, E., Kržan, A., & Braunegg, G. (2015). Liquefied wood as inexpensive precursor-feed stock for bio-mediated incorporation of (R)-3-hydroxyvalerate into polyhydroxyalkanoates. *Materials., 8*, 6543−6557.

Koller, M., Salerno, A., Strohmeier, K., Schober, S., Mittelbach, M., Illieva, V., ... Braunegg, G. (2014). Novel precursors for production of 3-hydroxyvalerate-containing poly [(R)-hydroxyalkanoate]s. *Biocatalysis and Biotransformation, 32*, 161−167.

Korsatko, W., Wabnegg, B., Tillian, H. M., Egger, G., Pfrager, R., & Walser, V. P. (1984). 3-hydroxybutyric acid—A biodegradable carrier for long term medication dosagies. Studies on compatibility of Poly-D (-)-3-hydroxybutyric acid implantation tablets in tissue culture and animals. *Die Pharmazeutische Industrie, 46*, 952−954.

Kourmentza, C., Plácido, J., Venetsaneas, N., Burniol-Figols, A., Varrone, C., Gavala, H., & Reis, M. A. (2017). Recent advances and challenges towards sustainable polyhydroxyalkanoate (PHA) production. *Bioengineering., 4*(2), 55.

Laycock, B., Halley, P., Pratt, S., Werker, A., & Lant, P. (2014). The chemomechanical properties of microbial polyhydroxyalkanoates. *Progress in Polymer Science, 39*, 397.

Lee, S. Y. (1996). Plastic bacteria? Progress and prospects for polyhydroxyalkanoate production in bacteria. *Trends in Biotechnology, 14*(11), 431−438.

Lee, J., Jung, S. G., Park, C. S., Kim, H. Y., Batt, C. A., & Kim, Y. R. (2011). Tumor-specific hybrid polyhydroxybutyrate nanoparticle: surface modification of nanoparticle by enzymatically synthesized functional block copolymer. *Bioorganic & Medicinal Chemistry Letters, 21*(10), 2941−2944.

Lefebvre, G., Rocher, M., & Braunegg, G. (1997). Effects of low dissolved-oxygen concentrations on poly-(3-hydroxybutyrate-co-3-hydroxyvalerate) production by *Alcaligenes eutrophus*. *Applied and Environmental Microbiology, 63*, 827−833.

Li, C., Zhang, J., Li, Y., Moran, S., Khang, G., & Ge, Z. (2013). Poly (l-lactide-co-caprolactone) scaffolds enhanced with poly (β-hydroxybutyrate-co-β-hydroxyvalerate) microspheres for cartilage regeneration. *Biomedical Materials, 8*(2), 025005.

Li, X., Chang, H., Luo, H., Wang, Z., Zheng, G., Lu, X., ... Xu, M. (2015). Poly (3-hydroxybutyrate-co-3-hydroxyhexanoate) scaffolds coated with PhaP-RGD fusion protein promotes the proliferation and chondrogenic differentiation of human umbilical cord mesenchymal stem cells in vitro. *Journal of Biomedical Materials Research. Part A, 103*(3), 1169−1175.

Li, Y., Thouas, G. A., & Chen, Q. Z. (2012). Biodegradable soft elastomers: Synthesis/properties of materials and fabrication of scaffolds. *RSC Advances, 2*, 8229−8242.

Li, Z., Yang, J., & Loh, X. J. (2016). Polyhydroxyalkanoates: Opening doors for a sustainable future. *NPG Asia Materials.*, e265. (4).

Lizarraga-Valderrama, L. R., Nigmatullin, R., Taylor, C., Haycock, J. W., Claeyssens, F., Knowles, J. C., & Roy, I. (2015). Nerve tissue engineering using blends of poly (3-hydroxyalkanoates) for peripheral nerve regeneration. *Engineering in Life Sciences, 15*(6), 612−621.

Lu, H. X., Yang, Z. Q., Jiao, Q., Wang, Y. Y., Wang, L., Yang, P. B., ... Lu, X. Y. (2014). Low concentration of serum helps to maintain the characteristics of NSCs/NPCs on alkali-treated PHBHHx film in vitro. *Neurological Research, 36*(3), 207−214.

Luklinska, Z. B., & Schluckwerder, H. (2003). In vivo response to HA-polyhydroxybutyrate/polyhydroxyvalerate composite. *Journal of Microscopy, 211*(2), 121−129.

Mao, G., Sukumaran, S., Beaucage, G., Saboungi, M. L., & Thiyagarajan, P. (2001). PEO − PPO − PEO block copolymer micelles in aqueous electrolyte solutions: Effect of carbonate anions and temperature on the micellar structure and interaction. *Macromolecules, 34*(3), 552−558.

Marois, Y., Zhang, Z., Vert, M., Beaulieu, L., Lenz, R. W., & Guidoin, R. (1999). In vivo biocompatibility and degradation studies of polyhydroxyoctanoate in the rat: A new sealant for the polyester arterial prosthesis. *Tissue Engineering, 5*(4), 369−386.

Masood, F., Chen, P., Yasin, T., Fatima, N., Hasan, F., & Hameed, A. (2013). Encapsulation of Ellipticine in poly-(3-hydroxybutyrate-co-3-hydroxyvalerate) based nanoparticles and its in vitro application. *Materials Science and Engineering: C, 33*(3), 1054−1060.

Masood, F., Yasin, T., & Hameed, A. (2015). Polyhydroxyalkanoates—What are the uses? Current challenges and perspectives. *Critical Reviews in Biotechnology, 35*(4), 514−521.

Meischel, M., Eichler, J., Martinelli, E., Karr, U., Weigel, J., Schmöller, G., ... Stanzl-Tschegg, S. E. (2016). Adhesive strength of bone-implant interfaces and in-vivo degradation of PHB composites for load-bearing applications. *Journal of the Mechanical Behavior of Biomedical Materials, 53*, 104−118.

Miranda de Sousa Dias, M., Koller, M., Puppi, D., Morelli, A., Chiellini, F., & Braunegg, G. (2017). Fed-batch synthesis of poly(3-hydroxybutyrate) and poly(3-hydroxybutyrate-co-4-hydroxybutyrate) from sucrose and 4-hydroxybutyrate precursors by *Burkholderia sacchari* strain DSM 17165. *Bioengineering., 4*, 36.

Naveen, N., Kumar, R., Balaji, S., Uma, T. S., Natrajan, T. S., & Sehgal, P. K. (2010). Synthesis of nonwoven nanofibers by electrospinning—A promising biomaterial for tissue engineering and drug delivery. *Advanced Engineering Materials, 12*(8), B380−B387.

Nicolai, T., Colombani, O., & Chassenieux, C. (2010). Dynamic polymeric micelles versus frozen nanoparticles formed by block copolymers. *Soft Matter, 6*(14), 3111−3118.

Nikel, P. I., Pettinari, M. J., Galvagno, M. A., & Méndez, B. S. (2006). Poly (3-hydroxybutyrate) synthesis by recombinant *Escherichia coli* arcA mutants in microaerobiosis. *Applied and Environmental Microbiology, 72*(4), 2614−2620.

Novikov, L. N., Novikova, L. N., Mosahebi, A., Wiberg, M., Terenghi, G., & Kellerth, J. O. (2002). A novel biodegradable implant for neuronal rescue and regeneration after spinal cord injury. *Biomaterials, 23*(16), 3369−3376.

Orita, I., Nishikawa, K., Nakamura, S., & Fukui, T. (2014). Biosynthesis of polyhydroxyalkanoate copolymers from methanol by *Methylobacterium extorquens* AM1 and the engineered strains under cobalt-deficient conditions. *Applied Microbiology and Biotechnology, 98*(8), 3715−3725.

Ouyang, S. P., Luo, R. C., Chen, S. S., Liu, Q., Chung, A., Wu, Q., & Chen, G. Q. (2007). Production of polyhydroxyalkanoates with high 3-hydroxydodecanoate monomer content by fadB and fadA knockout mutant of *Pseudomonas putida* KT2442. *Biomacromolecules, 8*(8), 2504−2511.

Pais, J., Farinha, I., Freitas, F., Serafim, L. S., Martínez, V., Martínez, J. C., ... Reis, M. A. (2014). Improvement on the yield of polyhydroxyalkanotes production from cheese whey by a recombinant *Escherichia coli* strain using the proton suicide methodology. *Enzyme and Microbial Technology, 55*, 151−158.

Park, S. J., Jang, Y. A., Noh, W., Oh, Y. H., Lee, H., David, Y., ... Lee, S. H. (2015). Metabolic engineering of *Ralstonia eutropha* for the production of polyhydroxyalkanoates from sucrose. *Biotechnology and Bioengineering, 112*(3), 638−643.

Pignatello, R., Musumeci, T., Impallomeni, G., Carnemolla, G. M., Puglisi, G., & Ballistreri, A. (2009). Poly (3-hydroxybutyrate-co-ε-caprolactone) copolymers and poly (3-hydroxybutyrate-co-3-hydroxyvalerate-co-ε-caprolactone) terpolymers as novel materials for colloidal drug delivery systems. *European Journal of Pharmaceutical Sciences, 37*(3-4), 451−462.

Pouton, C. W., & Akhtar, S. (1996). Biosynthetic polyhydroxyalkanoates and their potential in drug delivery. *Advanced Drug Delivery Reviews, 18*(2), 133−162.

Povolo, S., Toffano, P., Basaglia, M., & Casella, S. (2010). Polyhydroxyalkanoates production by engineered *Cupriavidus necator* from waste material containing lactose. *Bioresource Technology, 101*(20), 7902−7907.

Puppi, D., Morelli, A., & Chiellini, F. (2017). Additive manufacturing of poly (3-hydroxybutyrate-co-3-hydroxyhexanoate)/poly (ε-caprolactone) blend scaffolds for tissue engineering. *Bioengineering., 4*(2), 49.

Puppi, D., Pirosa, A., Lupi, G., Erba, P. A., Giachi, G., & Chiellini, F. (2017). Design and fabrication of novel polymeric biodegradable stents for small caliber blood vessels by computer-aided wet-spinning. *Biomedical Materials, 12*(3), 035011.

Raposo, R. S., de Almeida, M. C. M., da Fonseca, M. M. R., & Cesário, M. T. (2017). Feeding strategies for tuning poly(3-hydroxybutyrate-co-4-hydroxybutyrate) monomeric composition and productivity using *Burkholderia sacchari*. *International Journal of Biological Macromolecules, 105*, 825−833.

Rehm, B. H., & Steinbüchel, A. (1999). Biochemical and genetic analysis of PHA synthases and other proteins required for PHA synthesis. *International Journal of Biological Macromolecules, 25*(1-3), 3−19, 1.

Rezaie Shirmard, L., Bahari Javan, N., Khoshayand, M. R., Kebriaee-zadeh, A., Dinarvand, R., & Dorkoosh, F. A. (2017). Nanoparticulate fingolimod delivery system based on biodegradable poly (3-hydroxybutyrate-co-3-hydroxyvalerate) (PHBV): Design, optimization, characterization and in-vitro evaluation. *Pharmaceutical Development and Technology, 22*(7), 860−870.

Rodriguez-Contreras, A. (2019). Recent advances in the use of polyhydroyalkanoates in biomedicine. *Bioengineering (Basel), 6*(3), E82.

Schrader, J., Schilling, M., Holtmann, D., Sell, D., Villela Filho, M., Marx, A., & Vorholt, J. A. (2009). Methanol-based industrial biotechnology: Current status and future perspectives of methylotrophic bacteria. *Trends in Biotechnology, 27*(2), 107−115.

Shishatskaya, E. I., Nikolaeva, E. D., Vinogradova, O. N., & Volova, T. G. (2016). Experimental wound dressings of degradable PHA for skin defect repair. *Journal of Materials Science: Materials in Medicine, 27*, 165.

Shishatskaya, E. I., Volova, T. G., Puzyr, A. P., Mogilnaya, O. A., & Efremov, S. N. (2004). Tissue response to the implantation of biodegradable polyhydroxyalkanoate sutures. *Journal of Materials Science: Materials in Medicine, 15*(6), 719−728.

Sodian, R., Hoerstrup, S. P., Sperling, J. S., Martin, D. P., Daebritz, S., Mayer, J. E., Jr, & Vacanti, J. P. (2000). Evaluation of biodegradable, three-dimensional matrices for tissue engineering of heart valves. *ASAIO Journal, 46*(1), 107−110.

Stock, U. A., Sakamoto, T., Hatsuoka, S., Martin, D. P., Nagashima, M., Moran, A. M., ... Mayer, J. E., Jr (2000). Patch augmentation of the pulmonary artery with bioabsorbable polymers and autologous cell seeding. *The Journal of Thoracic and Cardiovascular Surgery*, *120*(6), 1158−1167.

Sudimack, J., & Lee, R. J. (2000). Targeted drug delivery via the folate receptor. *Advanced Drug Delivery Reviews*, *41*(2), 147−162.

Sun, J., Wu, J., Li, H., & Chang, J. (2005). Macroporous poly (3-hydroxybutyrate-co-3-hydroxyvalerate) matrices for cartilage tissue engineering. *European Polymer Journal*, *41*(10), 2443−2449.

Tan, D., Wu, Q., Chen, J. C., & Chen, G. Q. (2014). Engineering *Halomonas* TD01 for the low-cost production of polyhydroxyalkanoates. *Metabolic Engineering*, *26*, 34−47.

Tan, D., Xue, Y. S., Aibaidula, G., & Chen, G. Q. (2011). Unsterile and continuous production of polyhydroxybutyrate by *Halomonas* TD01. *Bioresource Technology*, *102*(17), 8130−8136.

Tan, G. Y. A., Chen, C. L., Li, L., Ge, L., Wang, L., Razaad, I. M. N., ... Wang, J. Y. (2014). Start a research on biopolymer polyhydroxyalkanoate (PHA): A review. *Polymers.*, *6*, 706−754.

Vainionpáá, S., Vihtonen, K. M., Mero, M., Pátiálá, H., Rokkanen, P., Kilpikari, J., & Törmálá, P. (1986). Biodegradable fixation of rabbit osteotomies. *Acta Orthopaedica Scandinavica*, *57*(3), 237−239.

Vashist, A., Vashist, A., Gupta, Y. K., & Ahmad, S. (2014). Recent advances in hydrogel based drug delivery systems for the human body. *Journal of Materials Chemistry B*, *2*(2), 147−166.

Volova, T., Shishatskaya, E., Sevastianov, V., Efremov, S., & Mogilnaya, O. (2003). Results of biomedical investigations of PHB and PHB/PHV fibers. *Biochemical Engineering Journal*, *16*(2), 125−133.

Wang, A., Gan, Y., Qu, J., Cheng, B., Wang, F., & Yu, H. Application of electrospun poly [(R)-3-hydroxybutyrate-co-(R)-3-hydroxyvalerate]-Ecoflex mats in periodontal regeneration: A primary study. In *2012 International Conference on Biomedical Engineering and Biotechnology*. pp. 972−975. IEEE. 2012.

Wang, L., Wang, Z. H., Shen, C. Y., You, M. L., Xiao, J. F., & Chen, G. Q. (2010). Differentiation of human bone marrow mesenchymal stem cells grown in terpolyesters of 3-hydroxyalkanoates scaffolds into nerve cells. *Biomaterials*, *31*(7), 1691−1698.

Wang, N., Zhou, Z., Xia, L., Dai, Y., & Liu, H. (2013). Fabrication and characterization of bioactive β-Ca2SiO4/PHBV composite scaffolds. *Materials Science and Engineering: C*, *33*(4), 2294−2301.

Wang, Y., Bian, Y. Z., Wu, Q., & Chen, G. Q. (2008). Evaluation of three-dimensional scaffolds prepared from poly (3-hydroxybutyrate-co-3-hydroxyhexanoate) for growth of allogeneic chondrocytes for cartilage repair in rabbits. *Biomaterials*, *29*(19), 2858−2868.

Wang, Y. W., Wu, Q., & Chen, G. Q. (2005). Gelatin blending improves the performance of poly (3-hydroxybutyrate-co-3-hydroxyhexanoate) films for biomedical application. *Biomacromolecules*, *6*(2), 566−571.

Wang, Y. W., Yang, F., Wu, Q., Cheng, Y. C., Peter, H. F., Chen, J., & Chen, G. Q. (2005). Effect of composition of poly (3-hydroxybutyrate-co-3-hydroxyhexanoate) on growth of fibroblast and osteoblast. *Biomaterials*, *26*(7), 755−761.

Wong, H. H., Lee, S. Y., Wong, H. H., & Lee, S. Y. (1998). Poly-(3-hydroxybutyrate) production from whey by high-density cultivation of recombinant *Escherichia coli*. *Applied Microbiology and Biotechnology*, *50*(1), 30−33.

Wu, H., Fan, Z., Jiang, X., Chen, J., & Chen, G. Q. (2016). Enhanced production of polyhydroxybutyrate by multiple dividing *E. coli*. *Microbial Cell Factories*, *15*(1), 1−3.

Wu, L. P., Wang, D., Parhamifar, L., Hall, A., Chen, G. Q., & Moghimi, S. M. (2014). Poly (3-hydroxybutyrate-co-R-3-hydroxyhexanoate) nanoparticles with polyethylenimine coat as simple, safe, and versatile vehicles for cell targeting: Population characteristics, cell uptake, and intracellular trafficking. *Advanced Healthcare Materials*, *3*(6), 817−824.

Zhang, H., Obias, V., Gonyer, K., & Dennis, D. (1994). Production of polyhydroxyalkanoates in sucrose-utilizing recombinant *Escherichia coli* and *Klebsiella* strains. *Applied and Environmental Microbiology*, *60*(4), 1198−1205.

Zhang, J., Shishatskaya, E. I., Volova, T. G., da Silva, L. F., & Chen, G. Q. (2018). Polyhydroxyalkanoates (PHA) for therapeutic applications. *Materials Science and Engineering: C.*, *86*, 144−150.

Zhang, L., Zheng, Z., Xi, J., Gao, Y., Ao, Q., Gong, Y., ... Zhang, X. (2007). Improved mechanical property and biocompatibility of poly (3-hydroxybutyrate-co-3-hydroxyhexanoate) for blood vessel tissue engineering by blending with poly (propylene carbonate). *European Polymer Journal*, *43*(7), 2975−2986.

Zhang, X., Lin, Y., Wu, Q., Wang, Y., & Chen, G. Q. (2019). Synthetic biology and genome-editing tools for improving pha metabolic engineering. *Trends in Biotechnology*, *38*(7), P689−P700.

Zhao, S., Zhu, M., Zhang, J., Zhang, Y., Liu, Z., Zhu, Y., & Zhang, C. (2014). Three dimensionally printed mesoporous bioactive glass and poly (3-hydroxybutyrate-co-3-hydroxyhexanoate) composite scaffolds for bone regeneration. *Journal of Materials Chemistry*, *B2*(36), 6106−6118.

Zou, H., Shi, M., Zhang, T., Li, L., Li, L., & Xian, M. (2017). Natural and engineered polyhydroxyalkanoate (PHA) synthase: Key enzyme in biopolyester production. *Applied Microbiology and Biotechnology*, *101*(20), 7417−7426.

PART V

Process engineering and commercialization

CHAPTER 19

Process engineering and commercialization of polyhydroxyalkanoates

Lalit R. Kumar[1], Bhoomika Yadav[1], Rajwinder Kaur[1], Sravan Kumar Yellapu[1], Sameer Pokhrel[1], Aishwarya Pandey[1], Bhagyashree Tiwari[1] and R.D. Tyagi[2]

[1]Centre Eau Terre Environnement, Institut National de la Recherche Scientifique, Québec City, QC, Canada, [2]BOSK Bioproducts, Québec City, QC, Canada

19.1 Introduction

Modernization, urbanization along with industrialization, is responsible for the rapid fossil energy sources depletion. A wide-array usage of nonrenewable means augments the release of various greenhouse gases (GHGs), which in turn leads in the direction of worsening the Earth's atmosphere. This originates a prerequisite for neoteric renewable means for chemicals and energy as substitutions (Fernández-Dacosta, 2018). There are three substitutes to petroleum-based plastics, which are assumed to be capable of replacing them are (1) photodegradable plastics, (2) partially degradable plastics, and (3) bioplastics that are completely biodegradable. Accordingly, there is abundant keenness around the embellishment of bioplastics, the plastic that has green and biodegradable constituents wherever either of the chief components is an animal, plant, or microbial by-product (Kalia, Raizada, & Sonakya, 2000).

The challenge for bioplastics is to maintain the physiochemical features of the conventional plastics but allow biodegradability and, thus biocompatibility too (Ryberg, Hauschild, Wang, Averous-Monnery, & Laurent, 2019). Polyhydroxyalkanoates (PHAs) are the utmost versatile, completely biodegradable bioplastics that exhibit properties equivalent to traditional petroleum-derived plastics (Kourmentza et al., 2017). Other polymers that are biodegradable, for example, chemically synthesized plastics and starch-based plastics, also contribute to the scene, but they face scarcity of alterability in structure and widespread material properties (Kumar, Kumar, & Singh, 2020). However, the production of this material is still limited because of two major reasons: an expensive production and several optimizations required with the currently available bioprocessing fermentations. Also, this fermentation is discontinuous most of the time, which makes it yet more time consuming. Therefore an economical, extensive, and stable PHA process development procedure is required concerning the commercialization of PHA-related usage and acceptance in the market (Zheng, Chen, Ma, & Chen, 2020).

19.2 Types and properties of polyhydroxyalkanoates

PHAs remain diverse in conditions of their mechanical attributes and thus remain sorted in the following two subgroups: (1) PHA_{SCL} that are short-chain in length with monomeric units that have five atoms of carbon (Anastas & Kirchhoff, 2002; Madkour, Heinrich, Alghamdi, Shabbaj, & Steinbüchel, 2013) and (2) PHA_{MCL} which have been regarded as medium chain in length whose monomeric part consists of carbon atoms higher than five. The PHAs are nebulous macromolecules that have a glass transition temperature, which declines alongside the intensification within the length of the side chain (Zhila & Shishatskaya, 2018). The main observable characteristics concerning to PHAs include water insolubility, no hydrolytic degradation at all, and the propensity of defying ultraviolet nonetheless incompetent to tolerate both acids and bases, are suitable for medical usage, dissipate effortlessly in chloroform and other chlorinated hydrocarbons, accelerate the anaerobic biodegradation within the residues by allowing its breakdown in water, encompass zero toxicity, and show nonsticky attributes once melted, dissimilar to the conventional polymers (Aznar et al., 2019).

19.3 Applications of polyhydroxyalkanoates

Nontoxic, hydrophobic, biocompatible, elastomeric, biodegradable, piezoelectric, having the maximum degree of polymerization, water insolubility, and optically active are eye-catching characteristics of PHAs that make them extremely reasonable with traditional plastics for a range of industrial applications (Amelia, Govindasamy, Tamothran, Vigneswari, & Bhubalan, 2019). Plastics replaced by PHA polymers have a range of important features with real marketable usages. Some of the promising applications related to PHA have been packaging, utilization for disposable materials, and various utilities within the field of medical science (Koller, Maršálek, de Sousa Dias, & Braunegg, 2017). These PHA polymers have also been successful in conquering fields like tissue engineering, agriculture, nanocomposites, and polymer blends (Philip, Keshavarz, & Roy, 2007). These polymers can also be used as stereoregular compounds that may be able to serve as precursors that have chirality helping the chemical synthesis of optically active compounds (Kourmentza et al., 2017; Wang, Yin, & Chen, 2014).

19.4 Process development at lab scale

The process flow diagram for the production of PHA is highlighted in Fig. 19.1. This section will discuss important parameters to be considered in PHA upstream as well as downstream processing during the development phase.

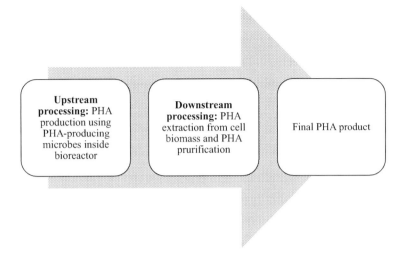

Figure 19.1
Process flow diagram for polyhydroxyalkanoate (PHA) production.

19.4.1 Upstream processing

19.4.1.1 Selection and type of microbes for production of polyhydroxyalkanoates

PHA can be produced by wild-type strains, mesophiles, extremophiles, or using genetically modified organisms. Investigations of mixed microbial (MMC) for PHA accumulation can be well defined as dynamic feast-and-famine regimes, enrichment of PHA accumulating strains alternate with the phase of feeding substrate pulses. PHA-producing microbes are present in various ranges of ecological places such as marine, groundwater sediments, estuarine sediments, rhizosphere, as well as anthropogenic-originated ecosystems such as wastewater treatment plants (Koller et al., 2017). Therefore tracing and isolation of PHA-producing microbes in the ecosystem is important to discover consortium.

Polyhydroxyalkanoate production by pure cultures

Depending on the feedstock, a PHA producer microbe has to be selected. For example, for renewable feedstocks that have high carbon and nutrients, growth associated with PHA producer, *Alcaligenes latus* can be chosen. Whereas feedstock lacking essential growth nutrients like nitrogen or phosphorus, nongrowth-associated bacteria are mostly chosen, for example, *Cupriavidus necator* (Kourmentza et al., 2017). To select promising PHA producers, they should have features like utilizing the renewable feedstock, potential for contamination-free bioprocess, production under nonsterile conditions, as well as high PHA yield and purity. It has been observed that the extremophiles such as microbes from hypersaline environments are promising microbes for PHA due to their potential to reduce PHA-production cost (sterilization and downstream steps). Microbes using lignocellulosic materials for PHA production have also been studied (Govil et al., 2020). *Saccharophagus degradans* are capable of degrading complex polysaccharide, for example, algal origin cellulose, other plant materials (Kourmentza et al., 2017).

Certain bacteria belonging to the genus of *Bacillus, Sphingobacterium*, and *Pseudomonas* have been used to combine two important processes, that is, bioremediation and production of value-added products like PHA. These strains have been characterized to utilize and degrade several carbon sources, for example, certain *Pseudomonas* strains from hydrocarbon-contaminated soils have been reported to use hydrocarbons for PHA production (Kourmentza et al., 2017).

Cyanobacteria are also investigated as PHA-producing oxygenic photoautotrophs that can reduce C-source and oxygen supply for the process. Efforts devoted to design optimized photobioreactors and genetically engineered cyanobacterial strains are required considerably in this field. Such a process can be planned to simultaneously decrease effluent gases like CO_2 and organic materials from poultry wastes as shown by *Nostoc muscorum* Agardh (nitrogen-fixating cyanobacterium) (Bhati & Mallick, 2016). Cyanobacteria such as purple nonsulfur bacteria are also studied as PHA producers. For example, *Rhodospirillum rubrum* has been studied to convert syngas to PHA (Do et al., 2007).

Extremophiles such as halophiles can be encouraging and exhibit a remarkable potential for PHA production because of the following advantages: low contamination risk as other microbes cannot grow under high salt concentrations of media, less energy requirements under nonsterile conditions, minimal water (fresh) consumption, and operating in more efficient contamination-free continuous process of fermentation. A very well-known example of PHA halophile is *Haloferax mediterranei* that has shown the accumulation of high PHA contents using waste feedstocks. One study showed the PHA content of 70% and 66% by *H. mediterranei* using pretreated vinasse (25% and 50% vol./vol., respectively) (Bhattacharyya et al., 2012). *Bacillus megaterium* isolated from salterns has shown up to 39% PHA with or without 5% w/v NaCl using glucose (Kourmentza et al., 2017).

Polyhydroxyalkanoate production by a MMC

In addition to the pure cultures, efforts for the application of MMC are often made for the PHA-production process (Huang, Chen, Wen, & Lee, 2017). In MMC, selective pressure is exerted for the desired metabolism on the MMC by choosing appropriate conditions in the bioreactor. Cyclic feast-famine conditions are used as an effective selective pressure for microbes accumulating PHA. Using MMC is attractive because it can accumulate PHA at a low cost. MMC does not have sterility demands and has fewer requirements for control devices and equipment when compared to monoseptic cultures (Mannina, Presti, Montiel-Jarillo, Carrera, & Suárez-Ojeda, 2020).

Unlike pure cultures, the MMC process uses volatile fatty acids (VFAs) as precursors for PHA production. Substrates like glycerol, carbohydrates in MMCs, and other complex substrates like olive mill wastewater, cheese whey, and food wastes form glycogen apart from PHA, and therefore a step where they could be converted into VFA in continuous stirred tank reactor (CSTR) is included. They can be adapted easily to inexpensive and complex feedstocks like agroindustrial streams and domestic wastewaters. The process consists of two steps, in the first step, sequential batch reactors (SBRs) are used for selecting enriching a microbial population that can produce PHA using various transient conditions. In the next step, the microbes from SBR are given conditions to maximize PHA accumulation, after which the PHA-accumulated cells are harvested for extraction (Kourmentza et al., 2017). Researchers have observed PHA production using MMC with substrates such as acetate, domestic waste streams, sugarcane molasses, food wastes, and oil mill effluents (Koller et al., 2017). It has been observed that VFA can affect PHA composition. VFA with even carbon atoms tends to produce polyhydroxybutyrate (PHB), whereas that with odd numbers produces poly(3-hydroxybutyrate-co-3-hydroxyvalerate) (PHBV) polymers containing different % hydroxyvalerate (HV) molar fractions (Kourmentza et al., 2017).

Low biomass output is generally encountered in MMC that leads to low PHA production. Huang et al. (2017) researched to enhance PHA production using MMC with an extended cultivation strategy. They found the overall PHA storage yield of 0.49 g chemical oxygen demand (COD) PHA g^{-1} COD VFA and volumetric productivity of 1.21 g PHA $L^{-1} d^{-1}$ that can serve as a favorable approach to use in industrial-scale practice. Also, the MMC consists of various PHA-producing microbes, and each of them accumulates PHA with different molar mass, molar mass distribution, the composition of monomers, and crystallinity. Therefore MMC approaches cannot be selected as a method of choice if highly uniform PHA is required. However, one needs to keep in mind that the PHA composition is also related to the feedstocks than just the biodiversity of MMC (Sabapathy et al., 2020).

Polyhydroxyalkanoate production by genetically modified microbial strains

Genetic engineering can serve as an effective and interesting means for optimizing the biosynthetic processes of microbes to transfer the genetic information from PHA producers to microbes that are nonPHA producers (Gahlawat, Kumari, & Bhagat, 2020). Gene manipulation related to the PHA biosynthetic mechanisms, oxygen uptake, and quorum sensing may improve PHA production (Kourmentza et al., 2017). For example, this approach is shown when whey was used for PHB production using *Escherichia coli* (recombinant). *E. coli* can convert the whey lactose and is engineered with *PhaC*-PHA syntheses genes from *C. necator*, which lacks β-galactosidase activity and cannot utilize lactose as a carbon source (Koller et al., 2017). PHB pathway optimization in *E. coli* has also been investigated for other genes involved in the pathway, such as *phb*C, *phb*A, and *phb*B. Besides, the vice versa condition is also available where *C. necator* is engineered to convert additional substrates such as whey lactose to PHB (Koller et al., 2017).

Genetic engineering could also increase the molar fraction of 3-hydroxyvalerate (3HV) in copolyesters. The gene encoding for membrane-bound transhydrogenase *Pnt*AB (*E. coli*) and nicotinamide adenine dinucleotide phosphate (NADPH)-dependent acetoacetyl-CoA reductase *Pha*B1 (*C. necator*) has been engineered in *R. rubrum* cells, and these cells showed 13-folds improved 3HV molar fraction (Heinrich, Raberg, & Steinbüchel, 2015). Another example of genetic manipulation is halophilic bacteria, recombinant *Halomonas campaniensis* produced around 70% PHB on kitchen wastes in the presence of NaCl (26.7 g L^{-1}) (Yue et al., 2014). PHA diversity can be altered by engineering PHA biosynthesis pathways such as acetoacetyl-CoA pathway, β-oxidation, and in situ fatty acid synthesis and by the specificity of PHA synthase. Genetic-engineered microbes can enhance substrate conversion, optimize the composition of polyesters, optimize the oxygen availability to enhance biomass growth, increase the PHA productivity under oxygen limitations, and lower the viscosity of fermentation broth (facilitating the recovery of PHA granules) (Koller et al., 2017; Koller, 2019).

Economically, the development of genetically engineered microbes can use inexpensive substrates such as waste streams and convert them into PHA (Gahlawat et al., 2020). The current R&D needs to exploit the wild-type or genetically engineered microbes that can convert such inexpensive carbon sources to PHA without the use of any expensive precursors. Besides, the yield needs to be enhanced by optimizing the cultivation conditions, or by genetic engineering to change the metabolic carbon flux for PHA synthesis instead of side products such as organic acids or CO_2 (Koller, 2019).

19.4.1.2 Choice of substrate in the fermentation

Pure carbon substrates

Pure substrates such as glucose, fructose, sucrose, and plant oils can be easily metabolized by microbes. A study reported *H. mediterranei* utilize pure glycerol and produced 75% (wt./wt.) of biomass PHA content with the productivity of 0.12 g L^{-1} h^{-1} (Hermann-Krauss et al., 2013). Another study used glucose in fed-batch strategy as the substrate for *Ralstonia eutropha* and produced 76% (wt./wt.) biomass PHA content with 2.42 g L^{-1} h^{-1} productivity (Kim et al., 1994).

Raw materials (carbon sources) cost is the most important factor that affects the economics of the PHAs-production process, specifically for large scale, without taking account of the downstream and recovery process (Amache, Sukan, Safari, Roy, & Keshavarz, 2013). Commonly used substrates such as sugars, fatty acids, and plant oils can cover up to 50% of total production cost. Therefore the selection of carbon source is a major factor for determining the performance of the bioprocess and final product cost. The simplest method that can be used is choosing inexpensive, renewable, and readily available carbon sources such as waste substrates.

Complex substrates

Microbes are capable of producing PHA from several carbon sources such as plant oils, fatty acids, alkanes, and simple carbohydrates. Every year, thousands to millions of tons of wastes are discharged from various processing industries. Utilizing these wastes not only decreases the substrate cost but also saves the waste disposal cost (Gahlawat et al., 2020; Koller et al., 2012; Ujang, Salim, Din, & Ahmad, 2007). Various strains have been studied that can produce PHAs from agricultural waste substrates such as beet molasses and sugarcane. However, mostly a pretreatment method is necessary before such waste streams could be exploited by microbial PHA producers.

However, using inexpensive substrates can provide lower PHA productivity when compared to pure substrates because of the low concentration of convertible carbon in the waste stream. Another reason could be the presence of certain substances that inhibit the growth and PHA accumulation in the system. Such low overall yield of the product can further counteract the decrease of production costs as it would require more efficient bioconversion, extended process timing, and more quantity of waste substrate to generate a similar amount of product like that with pure substrates (Koller et al., 2017).

One of the classic examples of utilizing waste substrates is the feedstock whey that contains carbon substrate concentration (lactose) around 4–5 wt.%, which can be further ultrafiltered and concentrated to around 21 wt.%. Whey from the dairy sector constitutes wastes and surplus material in major parts of the world. It is less-expensive raw material and causes a disposal problem for the dairy industry due to its high COD and biochemical oxygen demand (BOD) (Kim, Wee, Kim, Yun, & Ryu, 2006).

Crude glycerol is produced in large quantities as a coproduct stream of the biodiesel industry and thus can be of great potential to be used as a carbon source (Ashby, Solaiman, & Foglia, 2005; Vasudevan & Briggs, 2008). EU-FP7 ANIMPOL project used lipid-rich streams from slaughterhouses and rendering industry where these materials get transformed into biodiesel and crude glycerol phase that can be converted to PHA (Koller & Braunegg, 2015). Waste lipids such as waste cooking oil (WCO), greasy oil, and animal waste constitute inexpensive sources of triacylglycerides and can be further applied as substrates for PHA production.

Lignocellulosic is one of the richest organic materials with total global quantities to be estimated around 80,000 Mt (Koller et al., 2017). Industries producing the major lignocellulosic wastes are the agroindustry, the paper-pulp industry, and the wood-processing industry. The conversion rates of pentoses and hexoses by a microbe determines the selection of production strains for PHA biosynthesis from lignocellulose-derived substrates (Kumar, Singh, & Singh, 2008). *Burkholderia cepacia* and *B. sacchari* were used for PHA production using hydrolysate of spruce sawdust. The hydrolysis is important to pretreat waste and solubilize it (Govil et al., 2020). Here, sawdust hydrolysis was done by using acids and enzymes to hydrolyze hemicellulose and cellulose fractions that formed a considerable amount of fermentable sugars, which were converted to biomass and PHA (Koller, 2017). The waste streams from the sugar industry and bioethanol industry are also used for PHA production; thus making the process economically competitive (Gahlawat et al., 2020; Koller et al., 2008).

19.4.1.3 Identification of key-process parameters in upstream processing

Type of substrate and concentration

The type of substrate as a carbon source directly influences its uptake rate and PHA synthesis. The nature of the substrate also determines its metabolic route in the cell and the final polymeric composition. The % of PHB and % polyhydroxyvalerate (PHV) is primarily affected by the nature and concentration of the substrate used in the production media (Chaitanya, Mahmood, Kausar, & Sunilkumar, 2014). The recombinant *E. coli* produced 2.5% (wt./wt.) of PHV with the addition of 1 mM valine to 1% (w/v) glucose medium, while up to 4% (wt./wt.) PHV was produced when 1 mM threonine was added. The appropriate concentration of carbon is necessary to obtain maximum possible biomass concentration and PHA production. The low concentration of carbon substrate can limit the microbial growth due to lack of enough substrate, while a high concentration of reducing sugars or carbon can cause the substrate inhibition. In a study by Sangkharak and Prasertsan (2008), the effect of carbon substrates such as glucose, fructose, acetate, and their concentration was studied on cell growth and PHA accumulation using *R. sphaeroides* N20. The strain was able to utilize all these carbon sources, and maximum biomass concentration and PHB content of 8.95 g L^{-1} and 72.9% (wt./wt.), respectively, were obtained after 60 h using 4 g L^{-1} acetate as a carbon substrate. In another study by Babruwad, Prabhu, Upadhyaya, and Hungund (2016), the effect of glucose, fructose, sucrose, and galactose on PHB production was studied and PHB yield of 0.325, 0.317, 0.298, and 0.264 g L^{-1}, respectively, were obtained. Thus PHB yield is affected while shifting from one carbon source to another.

pH

The pH of media has a huge impact on PHA productivity and monomer composition. Moreover, metabolic routes are highly prone to even small pH change (Wei et al., 2011). The pH is very much specific for every microorganism for growth as well as metabolite production. According to the recent study by Wagle, Dixit, and Vakil (2017), isolates *Bacillus flexus* WY2, *Bacillus subtilis* 6833, *Brucella melitensis* AUH2, and *Pseudomonas aeruginosa* S164S produced maximum PHA content of 30, 27, 32, and 35% (wt./wt.), respectively, at pH 7.0, while the isolates *P. aeruginosa* VSS (25% wt./

wt.) and *Microbacterium aurum* TPL18 (24% wt./wt.) produced the highest amount of polymer at pH 6.0 under similar conditions of fermentation. Panigrahi and Badveli (2013) reported that a pH of 6.0–7.0 supports better polymer yield. During PHA production with oil palm frond juice as carbon substrate using *C. necator*, the pH was varied from 6.0 to 8.0, and biomass concentration (6.42–8.57 g L^{-1}) and poly(3-hydroxybutyrate) content [20%–34% (wt./wt.)] were increased with increase in medium pH from 6.0 to 7.0. However, pH above 7.0 decreased biomass concentration and PHB content as well (Zahari et al., 2012).

Temperature

Temperature is a fundamental factor that affects the growth of microorganisms. It also impacts enzymatically catalyzed reaction rates and diffusion of the substrate into the microbial cells (Grady, 1999). However, the influence of temperature on PHA synthesis varies among genera to genera. According to Wei et al. (2011), PHB production by *Cupriavidus taiwanensis* was observed at 30°C–37°C and the highest biomass concentration and PHB production were obtained at 30°C. However, in another study, the optimal PHA production (0.36 g L^{-1}) was obtained at 37°C in *V. azureus* BTKB33. Whereas 30°C and 45°C were found to be optimal for PHA production in *C. taiwanensis* and *H. mediterranei*, respectively (Wei et al., 2011). In another study by Karbasi et al. (2012), the investigation of different fermentation conditions (temperature was varied from 20°C to 35°C) for PHA production was done by using *C. necator*, and maximum biomass concentration (10.2 g L^{-1}) and PHA concentration (5.8 g L^{-1}) was obtained using fructose at 30°C temperature.

Dissolved oxygen, aeration, and agitation

Agitation provides dispersion of homogeneous heat and proper mixing in the fermentation broth. Besides, agitation also enhances the oxygen transfer rate for microbial cells. Generally, low agitation speed can lead to microbial cell aggregation, and thus make the culture medium heterogeneous, which further influence microbial growth and also affects PHA production. Also, the shear forces caused by higher agitation speed may have a negative effect on microbial growth and PHA accumulation (Serafim, Lemos, Albuquerque, & Reis, 2008). During PHA production with oil palm frond juice using *C. necator*, the agitation speed was varied from 180 to 260 rpm, and maximum biomass concentration and PHA content of 9.42 g L^{-1} and 40% (wt./wt.) were obtained at 220 rpm (Zahari et al., 2012).

Restricted availability of nitrogen, oxygen, or phosphate initiates PHA storage in most of the microorganisms. The role of dissolved oxygen (DO) concentration in PHA synthesis and accumulation while employing different microbes have been well demonstrated. During the scale-up of PHB production (30 L fermenter) using *Methylobacterium* sp. under limiting DO conditions, it resulted in a 4.58 times increase in PHA production (Nath, Dixit, Bandiya, Chavda, & Desai, 2008). According to a reported study, PHA content was increased using oxygen limitation, that is, 1%−4% of air saturation conditions for *Halomonas boliviensis* (Babruwad et al., 2016). While for *Bacillus mycoides*, the limitation of oxygen inhibited both microbial growth and PHB accumulation (Borah, Thakur, & Nigam, 2002). Controlling the DO level in different phases of growth and PHA accumulation phase reduces the cost of aeration and, consequently, the overall cost of the PHA production. Zahari et al. (2012) reported the highest production of PHA at 20%−30% (wt./wt.) DO level with oil palm frond juice using *C. necator*. These results suggested that PHA accumulation (44%−46% wt./wt.) was three times more at lower DO concentration (20%−30%) as compared to the higher DO (50% DO), where biomass PHA content of only 15% (wt./wt.) was obtained. The insufficient oxygen supply to the bacteria may decrease nicotinamide adenine dinucleotide (NADH) oxidation and lead to PHA biosynthesis.

Nutrient limitation

Nutrient limitation in many pure cultures is essential for PHA accumulation. Low concentration of nitrogen and phosphorus favors PHB accumulation. Optimized limiting concentrations of phosphorus can increase the PHB accumulation by ninefold (Lee et al., 2000). It has been reported that a lower concentration of nitrogen and phosphorus are essential for PHA accumulation, but a minimal level of these nutrients are required for PHA accumulation. The nitrogen limitation raises the intracellular concentration of both NADH and NADPH, which leads to an inhibition of the tricarboxylic acid cycle, and consequently, increasing the amount of acetyl-CoA in the cell. Hence when cells consume more carbon than what they can use for growth, cells metabolize it all to acetyl-CoA, then the excess is used for producing PHA (Silva, 2015). In addition, the nutrients have their effect on various enzymes involved in the PHA-accumulation process like dehydrogenases and phosphatases, which results in variation in accumulation potential of the culture. Nitrogen or phosphorus limitation induces higher dehydrogenase activity. Low phosphorus concentration leads to higher phosphatase activity, whereas higher concentration lowers down the phosphatase activity.

The effect of different $(NH_4)_2SO_4$ concentrations on the PHB production by *C. necator* using oil palm fond juice was investigated (Zahari et al., 2012). The cell dry weight increased from 5.25 to 10.15 g L^{-1} with an increased concentration of $(NH_4)_2SO_4$ from 0 to 2.0 g L^{-1}. While PHA synthesis decreased with the increase of $(NH_4)_2SO_4$ concentration and highest biomass PHA content of 42−44 (%wt./wt.) was achieved at $(NH_4)_2SO_4$ concentration of 0−0.5 g L^{-1}, which indicates the significance of imbalanced growth conditions in PHA-accumulation process.

C/N ratio

Bacterial PHA synthesis is largely affected by the carbon-to-nitrogen (C/N) ratio. The C:N ratio is a very crucial factor for PHA production and, thus needs to be optimized because it changes with the type of microbe used. During the growth of *A. latus ATCC 29713* for PHA accumulation, it was reported that the highest PHA accumulation occurs at C:N ratio of 28.3, which increased the PHA accumulation by eightfold (Grothe, Moo-Young, & Chisti, 1999). The PHB production of 70% (wt./wt.) was obtained with *B. megaterium* at C:N ratio of 15:1 using sucrose as a carbon substrate and ammonium sulfate as a nitrogen source (Faccin et al., 2009). However, similar PHA production of 70% (wt./wt.) with *R. eutropha* was obtained at the C:N ratio of 80:1 using propionic acid, acetic acid, or butyric acid as a carbon substrate and ammonium chloride as a source of nitrogen (Yang et al., 2009). The effect of C:N ratio (glucose as a carbon substrate and ammonium chloride as a source of nitrogen) was investigated on PHA accumulation by *C. necator* (Ahn et al., 2015). The increase in C/N ratio (3/1–360/1) increased PHA concentration (0.15–0.72 g L^{-1}); however, PHA concentration was decreased with a further increase in the C/N ratio. Thus nitrogen-deficient conditions should be preferred over the nitrogen-free condition for PHA accumulation.

Feeding strategy

The feeding strategy used for PHA production is a critical factor for process optimization as well as bioprocess improvement. The fed-batch strategy is considered to be more effective to have higher biomass concentration, PHA yield, and PHA productivity than batch mode (Cruz, Gouveia, Dionísio, Freitas, & Reis, 2019). However, the selection of an appropriate substrate feeding strategy is required to control the carbon concentration throughout the fermentation process. Several strategies, such as pulse feeding, exponential feeding, pH control, DO stat mode, and continuous feeding have been tested for PHA production (Kaur & Roy, 2015; Peña, Castillo, García, Millán, & Segura, 2014). To understand the effect of feeding regime on PHA production, Kedia, Passanha, Dinsdale, Guwy, and Esteves (2014) studied batch and continuous feed strategy during PHA production using *C. necator*. It was reported that pH-based (whenever the pH of fermentation media goes up by 0.1–0.2, the volatile acids are automatically fed into the fermenter) continuous volatile acid feed [VFAs concentration—2% (vol./vol.) in the medium] was better than one single batch mode of feeding. Acid concentrations beyond 2% (vol./vol.) in the medium were reportedly inhibitory. Acid concentration higher than tolerable concentration causes to increase in the undissociated form of acid entering the cell by diffusion and they lose the proton to decrease the intracellular pH causing inhibition.

19.4.1.4 Application of modeling and simulations in upstream processing

For high PHA accumulation, most essential fermentation conditions require maintenance of appropriate conditions of carbon (in excess) and nitrogen (in limitation) during cultivation. In some cases, inhibition of bacterial growth can be due to the high initial concentrations of the substrate, such as VFAs, crude glycerol that can affect the growth and rates of product formation. Therefore it is important to understand the process kinetics to design fermentation strategies for maximum productivity (Kaur & Roy, 2015). Mathematical models facilitate the optimization of the process settings in a limited period without any trial or error. These models can simplify the actual phenomenon, and based on the measurements, different models can be developed. Models are used to define the biomass growth, substrate consumption, and kinetics of product formation. The number of feeding strategies has also been designed for optimizing the process for the highest production of PHA that could be further implemented. Further, they are also used to start or stop the nutrient feeding, feeding profiles, and substrate concentration in the feed (Kaur & Roy, 2015).

The improvement in the genetic properties of PHA-producing strains and large-scale productivity also requires mathematical modeling in addition to the process and strain optimization procedures. The models present currently for dealing with PHAs, can be divided into low structured, formal kinetic, dynamic, metabolic (high structured), cybernetic, hybrid, and neural network models. No single mathematical model exists that expresses all the characteristics of strains and also the characteristics of industrial-scale plants with different requirements for PHA production. Therefore the various modeling approach is required to fine-tune the actual steps of the process. For example, for standard microbial cultivations, formal kinetic models and low-structured models are mostly preferred due to their simplicity, user-friendly behavior, and low computational demand (Novak, Koller, Braunegg, & Horvat, 2015).

Various modeling approaches like unstructured and low-structured models, dynamic models by pure and mixed cultures, metabolic models, cybernetic models, neural networks, and hybrid models have been applied for the modeling of PHA biosynthesis and are reviewed in detail elsewhere (Novak et al., 2015). A single type of model does not generally state problems in PHA production related to bioreactor performance, metabolism, kinetics, and cultivation conditions. For example, the mathematical models are used for modeling processes in mixing, bioreactors, metabolic pathways, genetic manipulations, cell cycles, and metabolic engineering. Whereas the low-structured models are more accurate and used for more simple cases. Computational fluid dynamics (CFD) offers tools for the computational characterization of fluid streaming, reactant concentration, temperature, and momentum. Combining these models to form a hybrid model consisting of kinetic or genome-scale metabolic models with CFD modeling to solve complex situations of biotechnology and bioprocess. Fig. 19.2 schematically represents the various steps involved in model development for process optimization.

Interesting studies and results have been attained by using the model-based production process optimization of PHA (Mamat et al., 2018; Wang, Carvalho, Reis, & Oehmen, 2018). Though, there is still a big room of improvement and development in the field of modeling dealing with the variety of bioprocess technologies and the steps involved in the process.

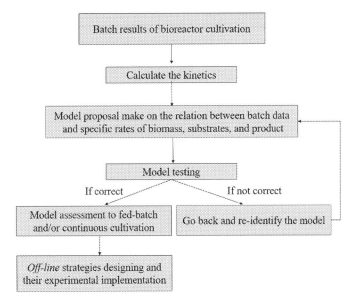

Figure 19.2
Steps involved in model development for a process optimization.

19.4.2 Downstream processing

The bioplastics should compete with the conventional petrochemical plastics not only in cost but also concerning the overall environmental impact. There are two methods to recover PHAs after fermentation: one way to separate the PHA granules is by dissolving the biomass (using strong oxidants) and the second way is by extracting PHA directly from biomass (with the help of solvents) (Yadav, Pandey, Kumar, & Tyagi, 2019). To obtain the desired PHA yield and purity, various optimization in the downstream processing is required. If the oxidant concentration in the recovery step is not optimized and controlled, the PHA dissolution with the nonPHA cell mass (NPCM) can occur, causing a low PHA recovery. The downstream processing of PHA is an important cost factor in the process and can have a noteworthy effect on the process's ecological footprint. Halogenated solvents used for PHA extraction, for example, chloroform and dichloromethane, have high extraction yields of PHA and good product purity. However, they have high costs and have a high risk to the environment and people working with them. The recovery steps are chosen by several factors, for example, microbial strain, required product purity, molecular mass requirement, and availability of separation techniques (Chee et al., 2010).

19.4.2.1 Solvent extraction

Using organic solvents for extracting PHA from cells is one of the oldest and widely used methods of PHA recovery. PHA is insoluble in water and is soluble in a limited number of solvents, for example, chloroform, methylene chloride, propylene, 1,2-dichloroethane, ethylene carbonates, and cyclic carbonic esters (Madkour et al., 2013). In the solution of PHA in the organic solvent, the solvent should be removed next. It is done by solvent evaporation; thus parting the PHA in a solvent-free state. Another method to separate PHA from solvent is PHA precipitation using a solvent where the PHA is insoluble. Such solvents are called antisolvents. The solution with PHA is slowly poured into antisolvent where precipitation occurs, and the precipitated PHA is separated by centrifugation or filtration. However, this method is not generally applied at an industrial scale due to economic constraints (Madkour et al., 2013). Methylene chloride is shown to give the purity more than 98% of PHBV from *R. eutropha*. Solvents, for example, 1,2-propanediol, diethyl succinate, and butyrolactone show recovery were ranging from 70% to 90% and purity between 99% and 100% when the temperature of $110^{\circ}C-140^{\circ}C$ was used. Other solvents such as acetic anhydride tetrahydrofuran, methyl cyanide, and tetrahydrofuran ethyl cyanide yielded the much lower recovery of PHA (Madkour et al., 2013). Solvent extraction may be expensive, and the separation process is tedious, affecting humans and the environment. For most of the methods describes, optimized protocols have to be established to be used at a large-scale process. The molecular weight and its distribution remain almost constant when organic solvents are utilized for PHA extraction.

19.4.2.2 Digestion methods

The principle of digestion methods is the solubilization of cellular materials that surround the PHA granules. These methods could be broadly categorizing into chemical digestion and enzymatic digestion. These methods are based on the solubilization of NPCM, mainly by sodium hypochlorite surfactants (Chee et al., 2010). Surfactants such as sodium dodecyl sulfate (SDS), betaine, Triton X-100, and palmitoyl have been used for PHA recovery. Isolation of PHA by surfactant digestion gave a lower degree of purity but higher molecular weight than sodium hypochlorite, whereas PHA of higher purity was obtained using sodium hypochlorite but with molecular weight degradation up to 50%. Therefore a combination of sodium hypochlorite and surfactant (sequential treatment) yielded high and rapid recovery with a 50% overall cost reduction as compared to solvent extraction (Kunasundari & Sudesh, 2011). However, there are certain problems caused by surfactants in wastewater treatment, and this method could be costly due to the high cost associated with SDS and sodium hypochlorite. One of the economical and green methods of disruption is the dissolution of NPCM by acidic and alkaline methods. They are said to decrease the recovery costs by 90% (at large scale) and give a high PHA yield and purity (Madkour et al., 2013; Table 19.1).

Another recovery process is by using enzymatic digestion that involves solubilization of cell components by heat treatment, enzymatic hydrolysis, and surfactant washing. This

Table 19.1: Advantages and disadvantages of different polyhydroxyalkanoate (PHA)-recovery methods.

Method	Advantages	Disadvantages
Solvent extraction	• High purity • Less PHA degradation • Higher molecular weight	• Not eco-friendly • Toxic and volatile solvents involved in high volumes • High capital and operation cost • Harmful for humans • Most suitable at lab scale
Sodium hypochlorite	• High purity • No drying required • Can be applied to large scale	• Higher reduction of molecular weight
Surfactants	• Can be recovered directly from broth • Degradation of PHA is limited	• High costs • Low purity • SDS difficult to remove from polymer • Wastewater treatment requires removal of surfactant
Surfactant-chelate	• Convenient • High-purity product	• Large volumes of wastewater produced • Degradation of polymer may occur
Dissolution of NPCM by acids	• Inexpensive • Eco-friendly • Good yield and high product purity	• Reduced molecular weight if process parameters are not controlled strictly
Enzymatic digestion	• Mild conditions • Good recovery • High purity	• Complex process • High cost of enzymes
Bead mill disruption	• Chemicals not required • Less contamination • No need for micronization of PHA granules	• Difficult to scale-up • Several passes required for reasonable recovery • Long processing time
High pressure homogenization	• No chemicals required • Less contamination	• Severe micronization of granules • Thermal degradation may occur • Formation of fine cellular debris • Poor disruption at low biomass concentration
Ultrasonication disruption	• Combination with other methods brings high purity of product	• Large-scale application difficult
Supercritical fluid	• Simple • Moderate operating conditions • Inexpensive • Rapid and eco-friendly	• Depend on process parameters • Strict parameters to be followed • Further chemicals required for high degree of disruption

technique is attractive due to its mild operation conditions and the good quality of recovered PHA. However, the high cost of enzymes and various complexities associated with the steps of the process limits its industrial-scale use (Kunasundari & Sudesh, 2011).

19.4.3 Other methods

Mechanical disruption to recover PHA from microbial cells has also been used. Various techniques used in this process involve bead mill, high-pressure homogenization, sonication, and SDS-high-pressure homogenization. However, most of these methods require high capital costs and long processing time (Kunasundari & Sudesh, 2011). Other methods for PHA recovery involve supercritical fluid (SCF), cell fragility, aqueous two-phase system (ATPS), flotation, gamma irradiation, air classification, and spontaneous liberation (Gahlawat et al., 2020).

The final intended application of PHA will determine the purity of PHA based on which the extraction method of PHA from cell biomass has to be carried out. The recovery of PHA significantly affects the overall economics, and developing a clean, simple, cost-effective, and efficient process for extraction with good quality and purity of PHA is required. Some of the efficient techniques at the laboratory level have no opportunity for commercialization due to factors like high-energy requirements, equipment's high cost, and low accessibility. Therefore significant parameters that can affect the performance of the recovery processes have to be determined and optimized at an industrial scale.

19.5 Scale-up from lab scale to pilot scale

19.5.1 Scale-up parameters for equipment

For a successful scale-up of any technology, the scale-up of each equipment must be carefully designed such that conditions at a larger scale should be similar to that of conditions on a smaller scale. Each equipment has certain parameters that need to be kept in mind during scale-up.

19.5.1.1 Bioreactor

While designing a bioreactor for a bigger scale, the most important parameter is vessel geometry (height to diameter ratio) (Doran, 1995). The height to diameter ratio should remain the same for laboratory and large-scale bioreactors. In bioreactors, besides reactor geometry, other factors need to be considered. As mixing is a crucial parameter in the bioreactor, it is important to keep mixing time constant in scale-up (Doran, 1995). For bioreactors with longer mixing times, mixing can be improved by installing baffles, which produce greater turbulence. The stirrer functions are sensitive to various aspects of tank geometry, the base profile of the tank, including impeller off-bottom clearance, the distance between sparger and impeller, and type of sparger. Optimizing these features results in improved mixing without requiring large power inputs (Doran, 1995). During bioprocess, the DO concentration in the medium affects cell growth and metabolite production. It has been reported that the volumetric mass (oxygen) transfer coefficient (K_{La}) could be used as a parameter for scale-up for aerobic fermentation (Ndao, Sellamuthu, Gnepe, Tyagi, & Valero, 2017). K_{La} can be calculated using dynamic gassing-out method (Tourlousse & Ahmad, 2007).

19.5.1.2 Centrifuge

At a laboratory scale, the batch centrifuge is used for harvesting of fermented broth. However, at industrial scale, continuous centrifugation is preferred instead of the batch centrifuge as batch centrifuge might prove time-consuming, and the batch process will require holding tanks, which furthers increases equipment cost. However, large-scale continuous centrifuge can be designed using the database of the laboratory-scale centrifuge (Brar, Verma, Tyagi, Valéro, & Surampalli, 2006). The performance (Q) of a large-scale centrifuge is product of settling velocity (Vg) and sigma factor ($\sum$). The settling velocity (Vg) of the particle present in the broth is a function of broth (fluid) characteristics. The sigma factor ($\sum$) corresponds to the design characteristic and is independent of broth and particle characteristics and is the quantitative parameter for the scale-up of centrifuge (Brar et al., 2006). This $\sum$ ultimately decides the best design criteria and sets the specifications of an industrial-scale centrifuge. The required value $\sum$ for a centrifuge can be estimated from Vg of broth and process requirements of Q (Brar et al., 2006).

19.5.2 Process validation

Process validation is defined as the raw data collection, evaluation at the process design step, and continuing it through the pilot-scale testing and commercialization phase. Process validation ensures that the process, including equipment, buildings, and materials, achieves the projected results on a constant basis (Vancov, Palmer, & Keen, 2019). The process validation will assure the quality of the product without any compromise. Product-quality testing and process control need to be incorporated at each step of the manufacturing process for effective process validation.

Process validation is a critical part of good manufacturing practices (GMPs), and it should be conducted with predefined protocols. Technical reports should be prepared, including summarized results and conclusions. For process validation to be successful, the qualified rooms, facility including equipment, and the relevant instructions, such as maintenance and cleaning schedules, are required. The specifications of the process and product must be clearly defined before validation begins. Periodic revalidation should be performed to ensure that processes and procedures remain capable of achieving the proposed results. The process validation plan will change based upon the product and its quality (Vancov et al., 2019).

Process validation must consider all critical parameters which can influence product quality. Determining critical process parameters is essential during development and improvement phases of the process. The data extracted here can be incorporated in the process validation

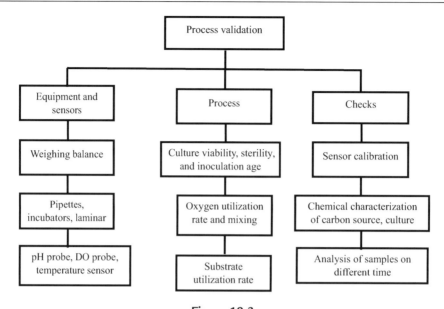

Figure 19.3
General process validation plan for polyhydroxyalkanoate (PHA) production.

and is then also an object of the regulatory inspection. For this reason, during the development and optimization phase, all changes implemented during these phases need to be carefully documented, so that process design is traceable. Process validation of PHA production is subjected to inspection of all necessary items to validate the intended results and quality of the product (Fig. 19.3).

19.5.3 Good manufacturing practices

Generally, manufacturing standards are followed by pharmaceutical companies. But similar rules can be followed for other commercial products like enzymes, microbial protein, and bioplastics to maintain manufacturing standards and product quality (Maithani, Grover, Raturi, Gupta, & Bansal, 2019). Quality control (QC) testing is an integral part of GMP. This testing is carried out in the QC laboratory located within the plant premises while some of the activity may be conducted outside plant premises. Both are under the control of quality systems deployed by the firm. Hence QC testing conducted within the framework of GMP or its requirements is Good Quality Control Laboratory Practices, that is, GQCLPs (Fabio et al., 2016). In industries and research laboratories, GMP rules will be followed to maintain the quality of the materials in different stages such as raw materials, intermediates, bulk, and finished products and packaging materials. These are referred to as Good Quality Control Laboratory Practice, or GQCLP, standards. Different authorities from the regulatory board inspect QC laboratories for amenability with these standards (Fabio et al., 2016). QC labs work according to GQCLP, and this is the term used to describe the GMP principles for QC laboratory work (Maithani et al., 2019).

19.5.4 Problems and challenges encountered in scale-up

Scale-up from lab scale to pilot scale is important to determine the feasibility of the process at a commercial level. The data generated at pilot scale like power consumption (for the agitator), chilling, and steam requirements can be used to design commercial-scale facilities. The pilot-scale trials should generate a technical and batch record and data that will allow efficient scale-up to industrial scale.

19.5.4.1 Upstream processing

There are several steps that need to be considered in scale-up of upstream processing, such as strain development, feed preparation, and fermentation process.

19.5.4.1.1 Inoculum preparation

In the PHA-production process, the first stage is the preparation of inoculum using the lyophilized culture or culture plate or glycerol stock. Stability of the culture and sterility of the medium need to be considered during inoculum preparation. Inoculum size and age need to be considered while transferring culture during multistage seed preparation. Depending on the strain, inoculum age and inoculum volume can vary between 12 and 24 h and 1% and 10% vol./vol., respectively. For inoculating large-scale production fermenter (e.g., 100,000 L fermenter), multistage seed preparation is required, where increases the risk of contamination. Hence sterility needs to be checked and maintained at every stage to avoid contamination.

19.5.4.1.2 Feed preparation

The feed preparation includes the preparation of different salts and their sterilization. During scale-up, feed preparation volume increases with the working volume of the production medium. Salt saturation and solubility, salt precipitation during sterilization need to be considered during scale-up. The major challenge in scale-up of feed preparation is the difference in sterilization method between laboratory scale (lab-scale autoclave) and pilot scale (sterilization in place), which increases thermal exposure at large scale due to both increased sterilization duration and higher temperatures. Increased exposure to thermal stress increases the chances of medium compound getting degraded, leading to the destruction of nutritive material or generation of inhibitory compounds (Wynn, Hanchar, Kleff, Senyk, & Tiedje, 2016). The solution to this problem is the use of continuous sterilizers.

19.5.4.1.3 Fermentation

In the fermentation scale-up process, the main challenge is mixing. In bench-scale fermenters, the oxygen mass transfer is high due to homogenous mixing with a high mixing rate. As scale increases, mixing (agitation) becomes a complicated issue. In stirred tank fermenters, oxygen transfer rates decrease from 400 mmol $L^{-1} h^{-1}$ in the bench scale to around 200 mmol $L^{-1} h^{-1}$ in the pilot scale (Wynn et al., 2016). In small-scale fermenters, concentration gradient (soluble nutrients) is readily dispersed due to efficient mixing, but in large-scale fermenters, there are regions of concentration gradient due to which cells experience nutrient limitation even in the presence of abundant nutrients. The mixing time in the bench-scale fermenter is around 1 s while it is around 30 s in pilot-scale fermenter (Wynn et al., 2016). Hence the required time period can be increased to stabilize pH at a higher scale.

19.5.4.2 Biomass harvesting

In lab scale, a batch centrifuge with high g force is used for biomass harvesting, which operates at an efficiency of 95%. However, in the pilot and commercial scale, the continuous centrifuge is used with lower g force as that of the batch centrifuge. As a result, the centrifuge efficiency is lower in pilot- or commercial-scale centrifuge as compared to the lab scale. This problem can be avoided by maintaining consistent back pressure (through control valves) on the supernatant line of the centrifuge. This will help in the separation of the heavy/light phase and may give good separation efficiency.

19.5.4.3 Downstream processing

Downstream processing is one of the most neglected aspect in the scale-up process. The main challenges in going from lab scale to pilot scale and then the pilot scale to the commercial scale are explained below.

19.5.4.3.1 Nonlinear scale-up

It is the first and foremost challenge faced during the scale-up of downstream processing. A chemical process cannot be simply picked from a laboratory scale and applied on a larger scale such as a pilot or industrial scale with a significant increase in the amount of chemicals in a proportional pattern. With the increased size and volume to process, a number of different properties related to the size of the system can be altered, for instance, change in turbulent and laminar flow occurs thereby causing the change in reaction kinetics of the process, change in fluid mechanics as well as thermodynamics in a nonlinear pattern have to be taken into consideration (Donati & Paludetto, 1997).

For instance—if chemical 'A' is used in downstream processing of PHA where a chemical reaction is made to occur in a certain size vessel. The reaction leads to the formation of the product as well as a certain amount of heat is generated in the process. On linear scale-up of this process, if the vessel size is increased to 100 times and the amount of chemical 'A' is increased by 100 times, then 100 times the amount of final product is expected with 100 times more heat released in the process. But in a practical case, the result will be different due to the involvement of several factors that could affect the process. Firstly, a 100 times larger vessel than the lab scale behave differently with a large amount of the chemicals in it. Of course, the lab-scale vessel will have relatively larger contact with the broth inside in comparison to a large-sized tank where a major portion of the chemicals in the tank will not be in contact with the wall of the large-sized vessel. Thus the reaction rate and kinetics of a reaction are affected by the surface area to mass relationship. Hence the aforementioned factors should be considered for the scale-up of the downstream process.

19.5.4.3.2 Reaction kinetics

In any chemical reaction, the molecules involved in the reaction must mix well to attain the equilibrium state. Since mixing in large-scale mixing will not be too effective than on the smaller scale, the state of equilibrium reaches after a long time thereby affecting the reaction rates, kinetics, and the amount of product produced (Gabelle, Augier, Carvalho, Rousset, & Morchain, 2011).

19.5.4.3.3 Chemical equilibrium

During the scale-up process, the time to reach the chemical equilibrium increases due to a higher amount of chemicals being used (Wynn et al., 2016). Consequently, batch scheduling will alter from the one that was initially projected.

19.5.4.3.4 Properties of materials

Physical as well as the chemical characteristics of the materials being used during the downstream process have an important role in the process economy in the long run as conditions such as early corrosion in the equipment add unnecessary cost for a process (Wynn et al., 2016). Hence in the scale-up process, material selection is very crucial too. Normally, in small-scale operations such as a laboratory scale, this type of problem gets unnoticed as a small amount of chemicals is used.

19.5.4.3.5 Fluid dynamics

With the increase in the size of the vessel, the fluid dynamics gets altered in a nonlinear pattern. Reynold's number is vital for heat transfer and the extent of mixing. Because of the nonlinear pattern of fluid dynamics on a large scale, variation among laminar and turbulent flow is difficult to predict (Doran, 1995). Computational fluid dynamics can be used for estimating flow patterns in commercial scale using the pilot-scale data.

19.5.4.3.6 Thermodynamics

In detail thermodynamic analysis of the process is required for the successful scale-up of a process since the chemical reactions and products may be temperature sensitive (Donati & Paludetto, 1997). Normally, the heat generated in small-scale operations gets unnoticed, but the same in large-scale operation may be problematic to manage due to the generation of a higher amount of heat in the process. A proper cooling system should be arranged to manage it.

19.5.4.3.7 Equipment size

Incorrectly sized equipment during scale-up can make the reaction difficult to control, which affects the thermodynamics, fluid dynamics, and reaction efficiency (Robert, Don, & James, 1984). Downstream efficiency is largely dependent on the correct equipment size selection.

19.5.4.3.8 Agitation

Agitation is an important parameter in the scale-up of an upstream and downstream process. Optimal agitation in the system is most critical for generating a good reaction kinetics in the downstream processing as optimal agitation ensures homogeneity in the system. In order to address the problem with agitation, angled agitators and baffles are being effectively used (Robert et al., 1984).

For effective downstream processing to achieve the desired purity and higher recovery, the number of processing steps should be as low as possible since each unit operation adds cost and lower the recovery yield (Mannina et al., 2020). There is a benefit of testing recovery and product purity at the pilot scale and the data obtained from the pilot-scale largely resemble the commercial-scale operation and it also serves as a guide for the design equipment design as well. Also, mass balances, energy balances, heating, and cooling requirements can be developed from the pilot-scale operation, which will be helpful for scale-up to go to commercial scale.

19.6 Commercialization of polyhydroxyalkanoates

19.6.1 Social factors affecting polyhydroxyalkanoate commercialization

19.6.1.1 Market demand

Market demand is an important social factor for the commercialization of any technology. The annual market demand decides to what scale the product is to be commercialized. If the annual scale of production is larger than market demand, then some of the products may remain unsold, leading to a loss in revenue for the manufacturer. If the annual scale of production is less than market demand, then it leads to increased unit production costs for the manufacturer. Market demand decides the annual plant capacity, which has an impact on PHA unit production cost. In one of the studies, unit production cost decreased from

$4.29 to $2.71 kg^{-1} PHA on increasing plant capacity from 2000 tonnes/annum to 10,000 tonnes/annum (Pavan et al., 2019). Hence the manufacturer should have an idea about product market demand, area (province, country) where he/she intends to target and annual plant capacity besides the process parameters.

19.6.1.2 Location of industry

The location where the industry needs to be constructed is also an important social factor for the commercialization of any type of technology. Since there is the generation of wastewater and GHG emission from the industry, the industry should be constructed away from the residential areas and aquatic bodies as it may pollute the aquatic bodies and may affect the health of residents living nearby. If downstream processing of PHA involves the use of toxic solvents (like chloroform, methanol), the manufacturing plant should be constructed away from residential areas. Due to a surge in housing prices, people prefer residential areas, away from the industrial area.

19.6.2 Technoeconomic studies

Technoeconomic studies are important for the commercialization of any technology. It reveals the industrial feasibility of any technology to the investors and policymakers. For the commercialization of any technology, economic parameters like capital investments, annual operating cost, unit production cost, gross and net profits, payback period, and return on investment need to be considered (Brar, Verma, Tyagi, Valéro, & Surampalli, 2007; Chen et al., 2018; Kumar, Ndao, Valéro, & Tyagi, 2019; Ram, Kumar, Tyagi, & Drogui, 2018). Technoeconomic studies also reveal bottlenecks of a process and guide the researchers for developing a cost-effective process during the development phase. Several technoeconomic studies have been reported for PHAs.

In one of the studies, PHB was produced using citric molasses (waste product of orange juice industry) as fermentation carbon substrate and *C. necator* as the microorganism (Pavan et al., 2019). During 42 h of microbial cultivation, 61.6 g L^{-1} biomass with 68.8% PHB content was observed. The cost analysis revealed that if PHB concentration is enhanced from 42.5 to 96.6 g L^{-1}, the upstream processing and total production cost will decrease by 42.6% and 18.22%, respectively, for the annual plant capacity of 2000 tonnes.

Four different carbon sources—WCO, soybean oil, crude, and refined glycerol were tested for PHA production using *C. necator* (Leong et al., 2016). PHA concentration obtained for different substrates at 72 h was 11.05, 20.73, 25.01, and 31.07 g L^{-1}, respectively, while chloroform and ethanol were used for PHA extraction. The unit PHA production cost was least for crude glycerol ($0.36 kg^{-1}) and highest for soybean oil ($1.63 kg^{-1}). The study concluded that the cost of carbon substrate and PHA concentration during fermentation impacts PHA production cost. PHB was produced using mixed microbial culture from industrial wastewater, while surfactant-hypochlorite-based chemical treatment was used for PHB extraction (Dacosta, Posada, & Ramirez, 2015). The cost assessment revealed that the unit production cost was $1.72 kg^{-1} PHB, which was lower than obtained with commercial substrates. Sterilization of fermenter is avoided when microbial culture is employed for PHA production, which saves the utility cost.

PHB was produced using mixed methanotrophic culture, while the acetone−water solvent mixture was used for PHB extraction (Levett et al., 2016). Through technoeconomic analysis, PHB unit production cost was calculated to be $4.1−$6.8 kg^{-1} PHB, which was lower than obtained with commercial substrates. In annual operating cost, raw material cost contribution decreased, proving methane as a low-cost substrate (Levett et al., 2016). Further reduction (21.95%) in PHB-production cost is possible by isolating and using PHB-producing thermophilic methanotrophs.

H. mediterranei has been used for PHA production using the stillage from the ethanol industry (Bhattacharyya et al., 2015). During 135 h of fermentation, 13.12 g L^{-1} PHA with 63% (wt./wt.) PHA content was observed. The salts recovered during the desalination of spent stillage were used for PHA production. For an annual production of 1890 tonnes PHA, the unit production cost of PHA was calculated to be $2.05 kg^{-1}. Desalination was a major cost impacting factor in PHA-production cost. Based on the reported studies, the major cost impacting parameters in upstream processing are PHA productivity, PHA content, and carbon substrate cost. However, PHA content is an essential parameter as it impacts the PHA extraction cost.

PHB was produced using *C. necator*, while citric molasses (waste product of orange juice industry) were used as a carbon substrate for fermentation (Pavan et al., 2019). Four biomass pretreatment methods were investigated before solvent-based PHA extraction: thermal pretreatment (95°C for 45 min), ultrasonication (10 kHz), high-pressure pretreatment (90 MPa), and no pretreatment. The unit production cost of PHB for different pretreatment methods was $4.28, $4.46, $4.28, and $4.72 kg^{-1}, respectively. Thermal and high-pressure pretreatment methods were most cost-effective in the downstream processing of PHB. PHB was produced using mixed microbial culture (Fernández-Dacosta, Posada, Kleerebezem, Cuellar, & Ramirez, 2015). For downstream processing, three routes were evaluated: (a) chemical treatment with solvent dichloromethane, (b) chemical treatment with surfactant and NaOCl (sodium hypochlorite), and (c) chemical treatment of cell biomass with 0.2 M NaOH and 0.2% (w/v) surfactant. Through economic analysis, it was found that alkali-surfactant-based treatment was most cost-effective while solvent-based treatment was least cost-effective.

19.6.3 Environmental assessment

For the commercialization of any technology, it is important to estimate nonrenewable energy consumption and environmental impacts associated with it. Life-cycle assessment (LCA) is conducted by decision makers and policy makers for this purpose. LCA assesses land use, GHG emission and air pollution, nonrenewable energy use (NREU), water use and pollution, degradation of soil, forests, etc. associated with any manufacturing process. It also identifies the critical process steps that impart high GHG emission and is helpful for researchers and scientists to design eco-friendly processes during the development phase.

LCA of upstream processing (USP) of bioplastic production is highly dependent on a carbon substrate. In one of the studies, several raw materials were used for bioplastic production—corn starch, soybean oil, sucrose, biogas, and municipal organic wastes (Kookos, Koutinas, & Vlysidis, 2019). LCA revealed that corn starch had the lowest GHG emission and NREU, while biogas had the highest GHG emission and NREU. Moreover, corn starch resulted in the lowest eutrophication and acidification potential. Employing waste substrates for PHA production reduces the problem of their disposal and treatment, reducing GHG emissions. Moreover, waste substrates also prevent the use of energy-intensive and GHG-emitting commercial substrates.

PHB was produced using industrial wastewater containing mixed microbial communities (Dacosta et al., 2015). Through LCA, it was found that GHG emission and NREU contribution of USP was 40% and 28% of the total process, respectively. Total GHG emission of the process was half of that obtained from using commercial substrates.

LCA for different PHB extraction methods [dimethyl carbonate (DMC) and halogenated hydrocarbons] has been performed (Righi et al., 2017). It was revealed that GHG emission of DMC-based extraction is lower than those with halogenated hydrocarbons. Four scenarios were further evaluated using DMC protocol: PHB extraction from wet or dried microbial biomass, and recovery by polymer precipitation or solvent evaporation. It was found that PHB extraction from dried biomass followed by polymer precipitation was most promising.

Alkali-based, acid-based, and bleaching-based PHA extraction has been reported (Dietrich, Dumont, Del Rio, & Orsat, 2017). LCA of several extraction methods revealed that sodium hydroxide treatment had the lowest GHG emission (4.08 kg CO_2 eq h^{-1}) when compared to sulfuric acid treatment (6.27 kg CO_2 eq h^{-1}) and sodium hypochlorite digestion (29.46 kg CO_2 eq h^{-1}). A reduction in polymers' molecular weight was observed with all treatments when compared to traditional chloroform extraction. Among all recovery methods, sulfuric acid treatment was the most promising one with low GHG emission, high purity (98%), and recovery (79%) (Dietrich et al., 2017).

In one of the studies, three PHB recovery routes were evaluated: (a) chemical treatment with sodium hypochlorite and surfactant, (b) chemical treatment of cell biomass with 0.2 M NaOH and 0.2% (w/v) surfactant, and (c) chemical treatment with dichloromethane (solvent) (Fernández-Dacosta et al., 2015). LCA revealed that alkali-based treatment had the lowest GHG emission and NREU while solvent-based treatment had the highest GHG emission and NREU.

19.6.4 Commercial production of polyhydroxyalkanoates

There are several companies that produce PHAs at the commercial level. The companies producing PHA with plant capacity, microbial strain, substrate, product type, and applications have been tabulated in Table 19.2.

Table 19.2: Worldwide polyhydroxyalkanoate (PHA)-producing companies.

Company's name	Microbial strain	Substrate and plant capacity	Product type	Application(s)	References
Biomatera, Canada	Soil isolated bacteria	Renewable substrates	PHA	Resins	Kourmentza et al. (2017)
Biomer, Germany	*Alcaligenes latus*	Sucrose 300 t/a	P3HB	Extrusion and injection molding	Chen (2009)
Bio-On Srl., Italy	*C. necator*	Sugar beets 10,000 t/a	PHA	Cosmetic industry	Tan et al. (2014)
BluePHA, China	Strain developed by synthetic biology	—	P3HP, P4HB	—	Kourmentza et al. (2017)
Danimer Scientific, USA	—	Canola Oil	PHB	Packaging, laminates, coatings, nonwoven fibers	Amelia et al. (2019)
Kaneka Corporation, Japan	—	Plant oils 3500 t/a	PHBH	Bags, automotive and electrical components	Wang, Modjinou, and Mangeon (2016)
PHB Industrial S.A., Brazil	*Alcaligenes* sp.	Saccharose 3000 t/a	PHA	Resins	Kourmentza et al. (2017)
Polyferm, Canada	*Aeromonas hydrophila*	Sugars or vegetable oil	PHA	Biomedical and plastic components	Tan et al. (2014)
Shenzhen Ecoman Biotech. Co. Ltd., China	—	Glucose 5000 t/a	PHA	Resins and microbeads	Kourmentza et al. (2017)
SIRIM Bioplastics Pilot Plant, Malaysia	—	Crude Palm oil 2000 t/a	PHA	—	Kourmentza et al. (2017)
Tepha Inc., USA	*Escherichia coli*	—	Copolymer	Medical devices	Suriyamongkol, Weselake, Narine, Moloney, and Shah (2007)
TianAn Biologic Materials Co., China	*R. eutropha*	Dextrose 10,000 t/a	PHBV	Injection molding, extrusion thermoforming, blown films	Verlinden, Hill, Kenward, Williams, and Radecka (2007)
Tianjin GreenBio Materials Co., China	*Escherichia coli*	Glucose 10,000 t/a	Copolymer	Resins	Rivero et al. (2017)

P3HB, Poly(3-hydroxybutyrate); *P4HB*, poly-4-hydroxybutyrate; *PHB*, polyhydroxybutyrate; *PHBH*, polyhydroxybutyrate-hexanoate; *PHBV*, poly(3-hydroxybutyrate-co-3-hydroxyvalerate)

19.6.5 Challenges in the commercialization of polyhydroxyalkanoates

In recent years, bioplastic companies have introduced packaging materials and bottles. These products are being sold as offering significant environmental advantages over conventional plastics. But there has been a difficulty for bioplastics to achieve a good market. According to one study, less than 0.5% of the world's plastic consumption is contributed by bioplastics (Iles & Martin, 2013). Petroleum-based (conventional) plastics continue to dominate the market due to its lower market prices ($1–$2 kg^{-1}). During 2000–2005, there was no significant increase in petroleum prices, which led to the shutdown of many PHA-related projects in some of the plastic-producing companies (Chen, 2009). Petroleum-based bioplastics benefit from large-scale economies and developed technologies at an industrial scale. There are also difficulties in substitution of conventional plastics with new, untried bioplastics and scale-up of small-scale PHA-based technologies to industrial levels (Iles & Martin, 2013). It is still unclear whether the bioplastics market will be profitable at an industrial scale despite the high costs incurred in R&D activities and creating new infrastructure. Besides, there are some sustainability issues related to bioplastics that remain unresolved, raising the question of the bioplastic value proposition (Yadav et al., 2019).

19.7 Conclusions and perspectives

1. Research on the production and application of PHA requires interdisciplinary knowledge. Successful commercialization of PHAs requires joint efforts by biotechnologists, microbiologists, polymer scientists, geneticists, design engineers, chemists, botanists, medical scientists, venture capitalists, and government agencies.
2. Current chemical companies should start thinking for bioplastics production as there is a market for bioplastics. Merely reducing costs and increasing yields will not guarantee bioplastics survival in the market. Rather, innovative business models should be developed to create a sustainable value proposition.
3. To lower PHA-production costs, high PHA concentration and PHA content need to be achieved in short fermentation time. Another way of reducing PHA production cost is by employing waste substrates during fermentation. High PHA-producing strains that can utilize complex waste substrates activated sludge needs to be isolated and tested in the laboratory.
4. For successful commercialization, the scale-up of each equipment must be carefully designed and tested such that conditions at a larger scale should be similar to that of conditions on a smaller scale. During the scale-up of downstream processing equipment, fluid dynamics, material properties, and thermodynamics need to be taken into account.

5. Technoeconomic and LCA studies revealed that waste substrate during fermentation reduces the production cost and GHG emission from the process. PHA content in fermented biomass is an essential process parameter as it impacts the PHA-extraction cost. Among several PHA-extraction methods, NaOH and H_2SO_4 treatment are the most promising ones with high recovery and low GHG emission.
6. To increase low-cost PHA applications, PHA can be blended with low-cost materials like starch and cellulose in packaging materials. To increase high-cost PHA applications, PHA can be used as bioimplant materials or tissue engineering materials.

Conflicts of Interest

Authors declare no conflict of interest. No copyright material has been used.

Acknowledgment

Sincere thanks to the Natural Sciences and Engineering Research Council of Canada (Grant A 4984, Canada Research Chair) for the financial support.

References

Ahn, J., Jho, E. H., Nam, K., Ahn, J., Jho, E. H., & Nam, K. (2015). Effect of C/N ratio on polyhydroxyalkanoates (PHA) accumulation by *Cupriavidus necator* and its implication on the use of rice straw hydrolysates. *Environmental Engineering Research, 20*, 246–253.

Amache, R., Sukan, A., Safari, M., Roy, I., & Keshavarz, T. (2013). Advances in PHAs production. *Chemical Engineering Transactions, 32*, 931–936.

Amelia, T. S. M., Govindasamy, S., Tamothran, A. M., Vigneswari, S., & Bhubalan, K. (2019). *Applications of PHA in agriculture. Biotechnological applications of polyhydroxyalkanoates* (pp. 347–361). Springer.

Anastas, P. T., & Kirchhoff, M. M. (2002). Origins, current status, and future challenges of green chemistry. *Accounts of Chemical Research, 35*, 686–694.

Ashby, R. D., Solaiman, D. K., & Foglia, T. A. (2005). Synthesis of short-/medium-chain-length poly (hydroxyalkanoate) blends by mixed culture fermentation of glycerol. *Biomacromolecules, 6*, 2106–2112.

Aznar, A., De León, J., Popescu, M., Serra-Ricart, M., Short, P., Pravec, P., . . . Sota, A. (2019). Physical properties of PHA 2014 JO25 from a worldwide observational campaign. *Monthly Notices of the Royal Astronomical Society, 483*, 4820–4827.

Babruwad, P. R., Prabhu, S. U., Upadhyaya, K. P., & Hungund, B. S. (2016). Production and characterization of thermostable polyhydroxybutyrate from *Bacillus cereus* PW3A. *Journal of Biochemical Technology, 6*, 990–995.

Bhati, R., & Mallick, N. (2016). Carbon dioxide and poultry waste utilization for production of polyhydroxyalkanoate biopolymers by *Nostoc muscorum* Agardh: A sustainable approach. *Journal of Applied Phycology, 28*, 161–168.

Bhattacharyya, A., Jana, K., Haldar, S., Bhowmic, A., Mukhopadhyay, U. K., De, S., & Mukherjee, J. (2015). Integration of poly-3-(hydroxybutyrate-co-hydroxyvalerate) production by *Haloferax mediterranei* through utilization of stillage from rice-based ethanol manufacture in India and its techno-economic analysis. *World Journal of Microbiology and Biotechnology, 31*, 717–727.

Bhattacharyya, A., Pramanik, A., Maji, S. K., Haldar, S., Mukhopadhyay, U. K., & Mukherjee, J. (2012). Utilization of vinasse for production of poly-3-(hydroxybutyrate-co-hydroxyvalerate) by *Haloferax mediterranei*. *AMB Express, 2*, 34.

Borah, B., Thakur, P., & Nigam, J. (2002). The influence of nutritional and environmental conditions on the accumulation of poly-β-hydroxybutyrate in *Bacillus mycoides* RLJ B-017. *Journal of Applied Microbiology, 92*, 776−783.

Brar, S. K., Verma, M., Tyagi, R., Valéro, J., & Surampalli, R. (2006). Efficient centrifugal recovery of *Bacillus thuringiensis* biopesticides from fermented wastewater and wastewater sludge. *Water Research, 40*, 1310−1320.

Brar, S. K., Verma, M., Tyagi, R., Valéro, J., & Surampalli, R. (2007). Techno-economic analysis of Bacillus thuringiensis biopesticides production from wastewater and wastewater sludge. In R. J. LeBlanc, P. J. Laughton, & R. D. Tyagi (Eds.), *Moving forward wastewater biosolids sustainability: Technical, managerial, and public synergy* (pp. 731−738). New Brunswick: GMSC.

Chaitanya, K., Mahmood, S., Kausar, R., & Sunilkumar, N. (2014). Biotechnological production of polyhydroxyalkonates by various isolates: A review. *International Journal of Pharmaceutical Science Invention, 3*, 1−11.

Chee, J.-Y., Yoga, S.-S., Lau, N.-S., Ling, S.-C., Abed, R. M., & Sudesh, K. (2010). *Bacterially produced polyhydroxyalkanoate (PHA): Converting renewable resources into bioplastics, . Current research, technology and education topics in applied microbiology and microbial biotechnology* (2, pp. 1395−1404)). Spain: Formatex Research Center.

Chen, G.-Q. (2009). A microbial polyhydroxyalkanoates (PHA) based bio-and materials industry. *Chemical Society Reviews, 38*, 2434−2446.

Chen, J., Tyagi, R. D., Li, J., Zhang, X., Drogui, P., & Sun, F. (2018). Economic assessment of biodiesel production from wastewater sludge. *Bioresource Technology, 253*, 41−48.

Cruz, M. V., Gouveia, A. R., Dionísio, M., Freitas, F., & Reis, M. A. (2019). A process engineering approach to improve production of P (3HB) by *Cupriavidus necator* from used cooking oil. *International Journal of Polymer Science, 2019*.

Dacosta, C. F., Posada, J. A., & Ramirez, A. (2015). Large scale production of polyhydroxyalkanoates (PHAs) from wastewater: A study of techno-economics, energy use and greenhouse gas emissions. *International Journal of Environmental, Chemical, Ecological, Geological and Geophysical Engineering, 9*, 433−438.

Dietrich, K., Dumont, M.-J., Del Rio, L. F., & Orsat, V. (2017). Producing PHAs in the bioeconomy—Towards a sustainable bioplastic. *Sustainable Production and Consumption, 9*, 58−70.

Do, Y. S., Smeenk, J., Broer, K. M., Kisting, C. J., Brown, R., Heindel, T. J., ... DiSpirito, A. A. (2007). Growth of *Rhodospirillum rubrum* on synthesis gas: Conversion of CO to H2 and poly-β-hydroxyalkanoate. *Biotechnology and Bioengineering, 97*, 279−286.

Donati, G., & Paludetto, R. (1997). Scale up of chemical reactors. *Catalysis Today, 34*, 483−534.

Doran, P. M. (1995). *Bioprocess engineering principles*. Elsevier.

Fabio, P., Alessandro, C., Stefania, S., Iolanda, G., Elena, M., & Aldo, C. (2016). The requirements for manufacturing highly active or sensitising drugs comparing good manufacturing practices. *Acta Bio Medica: Atenei Parmensis, 90*, 288.

Faccin, D. J. L., Martins, I., Cardozo, N. S. M., Rech, R., Ayub, M. A. Z., Alves, T. L. M., ... Resende Secchi, A. (2009). Optimization of C:N ratio and minimal initial carbon source for poly (3-hydroxybutyrate) production by *Bacillus megaterium*. *Journal of Chemical Technology & Biotechnology, 84*, 1756−1761.

Fernández-Dacosta, C. (2018). Alternative sources to fossil carbon: Ex-ante assessment of novel technologies using waste as a source, Utrecht University.

Fernández-Dacosta, C., Posada, J. A., Kleerebezem, R., Cuellar, M. C., & Ramirez, A. (2015). Microbial community-based polyhydroxyalkanoates (PHAs) production from wastewater: Techno-economic analysis and ex-ante environmental assessment. *Bioresource Technology, 185*, 368−377.

Gabelle, J. C., Augier, F., Carvalho, A., Rousset, R., & Morchain, J. (2011). Effect of tank size on kLa and mixing time in aerated stirred reactors with non-Newtonian fluids. *The Canadian Journal of Chemical Engineering, 89*, 1139−1153.

Gahlawat, G., Kumari, P., & Bhagat, N. R. (2020). Technological advances in the production of polyhydroxyalkanoate biopolymers. *Current Sustainable/Renewable Energy Reports, 7*(3), 73−83.

Govil, T., Wang, J., Samanta, D., David, A., Tripathi, A., Rauniyar, S., ... Sani, R. K. (2020). Lignocellulosic feedstock: A review of a sustainable platform for cleaner production of nature's plastics. *Journal of Cleaner Production, 270*, 122521.

Grady, J. O. (1999). *System engineering deployment*. CRC Press.

Grothe, E., Moo-Young, M., & Chisti, Y. (1999). Fermentation optimization for the production of poly (β-hydroxybutyric acid) microbial thermoplastic. *Enzyme and Microbial Technology, 25*, 132−141.

Heinrich, D., Raberg, M., & Steinbüchel, A. (2015). Synthesis of poly (3-hydroxybutyrate-co-3-hydroxyvalerate) from unrelated carbon sources in engineered *Rhodospirillum rubrum*. *FEMS Microbiology Letters, 362*.

Hermann-Krauss, C., Koller, M., Muhr, A., Fasl, H., Stelzer, F., & Braunegg, G. (2013). Archaeal production of polyhydroxyalkanoate (PHA) co- and terpolyesters from biodiesel industry-derived by-products. *Archaea, 2013*.

Huang, L., Chen, Z., Wen, Q., & Lee, D.-J. (2017). Enhanced polyhydroxyalkanoate production by mixed microbial culture with extended cultivation strategy. *Bioresource Technology, 241*, 802−811.

Iles, A., & Martin, A. N. (2013). Expanding bioplastics production: Sustainable business innovation in the chemical industry. *Journal of Cleaner Production, 45*, 38−49.

Kalia, V., Raizada, N., & Sonakya, V. (2000). Bioplastics. *Journal of Scientific and Industrial Research, 59*, 433−445.

Karbasi, S., Mirr, C. R., Yarandi, P. G., Frazier, R. J., Koch, K. W., & Mafi, A. (2012). Observation of transverse Anderson localization in an optical fiber. *Optics Letters, 37*, 2304−2306.

Kaur, G., & Roy, I. (2015). Strategies for large-scale production of polyhydroxyalkanoates. *Chemical and Biochemical Engineering Quarterly, 29*, 157−172.

Kedia, G., Passanha, P., Dinsdale, R. M., Guwy, A. J., & Esteves, S. R. (2014). Evaluation of feeding regimes to enhance PHA production using acetic and butyric acids by a pure culture of *Cupriavidus necator*. *Biotechnology and Bioprocess Engineering, 19*, 989−995.

Kim, B. S., Lee, S. C., Lee, S. Y., Chang, H. N., Chang, Y. K., & Woo, S. I. (1994). Production of poly (3-hydroxybutyric acid) by fed-batch culture of *Alcaligenes eutrophus* with glucose concentration control. *Biotechnology and Bioengineering, 43*, 892−898.

Kim, H.-O., Wee, Y.-J., Kim, J.-N., Yun, J.-S., & Ryu, H.-W. (2006). *Production of lactic acid from cheese whey by batch and repeated batch cultures of* Lactobacillus *sp. RKY2. Twenty-seventh symposium on biotechnology for fuels and chemicals* (pp. 694−704). Springer.

Koller, M. (2017). *Advances in polyhydroxyalkanoate (PHA) production*. Multidisciplinary Digital Publishing Institute.

Koller, M. (2019). Switching from petro-plastics to microbial polyhydroxyalkanoates (PHA): The biotechnological escape route of choice out of the plastic predicament? *The EuroBiotech Journal, 3*, 32−44.

Koller, M., Salerno, A., Muhr, A., Reiterer, A., Chiellini, E., Casella, S., ... Braunegg, G. (2012). *Whey lactose as a raw material for microbial production of biodegradable polyesters, . Polyester* (Vol. 347, pp. 51−92). Rijeka: IntechOpen.

Koller, M., & Braunegg, G. (2015). Biomediated production of structurally diverse poly (hydroxyalkanoates) from surplus streams of the animal processing industry. *Polimery, 60*.

Koller, M., Maršálek, L., de Sousa Dias, M. M., & Braunegg, G. (2017). Producing microbial polyhydroxyalkanoate (PHA) biopolyesters in a sustainable manner. *New Biotechnology, 37*, 24−38.

Koller, M., Bona, R., Chiellini, E., Fernandes, E. G., Horvat, P., Kutschera, C., ... Braunegg, G. (2008). Polyhydroxyalkanoate production from whey by Pseudomonas hydrogenovora. *Bioresource Technology*, 99, 4854−4863.

Kookos, I. K., Koutinas, A., & Vlysidis, A. (2019). Life cycle assessment of bioprocessing schemes for poly (3-hydroxybutyrate) production using soybean oil and sucrose as carbon sources. *Resources, Conservation and Recycling*, 141, 317−328.

Kourmentza, C., Plácido, J., Venetsaneas, N., Burniol-Figols, A., Varrone, C., Gavala, H. N., & Reis, M. A. (2017). Recent advances and challenges towards sustainable polyhydroxyalkanoate (PHA) production. *Bioengineering*, 4, 55.

Kumar, L. R., Ndao, A., Valéro, J., & Tyagi, R. (2019). Production of *Bacillus thuringiensis* based biopesticide formulation using starch industry wastewater (SIW) as substrate: A techno-economic evaluation. *Bioresource Technology*, 294, 122144.

Kumar, R., Singh, S., & Singh, O. V. (2008). Bioconversion of lignocellulosic biomass: Biochemical and molecular perspectives. *Journal of Industrial Microbiology & Biotechnology*, 35, 377−391.

Kumar, V., Kumar, S., & Singh, D. (2020). Microbial polyhydroxyalkanoates from extreme niches: Bioprospection status, opportunities and challenges. *International Journal of Biological Macromolecules*, 147, 1255−1267.

Kunasundari, B., & Sudesh, K. (2011). Isolation and recovery of microbial polyhydroxyalkanoates. *Express Polymer Letters*, 5.

Lee, S. Y., Wong, H. H., Choi, Ji, Lee, S. H., Lee, S. C., & Han, C. S. (2000). Production of medium-chain-length polyhydroxyalkanoates by high-cell-density cultivation of *Pseudomonas putida* under phosphorus limitation. *Biotechnology and Bioengineering*, 68, 466−470.

Leong, Y. K., Show, P. L., Lin, H. C., Chang, C. K., Loh, H.-S., Lan, J. C.-W., & Ling, T. C. (2016). Preliminary integrated economic and environmental analysis of polyhydroxyalkanoates (PHAs) biosynthesis. *Bioresources and Bioprocessing*, 3, 41.

Levett, I., Birkett, G., Davies, N., Bell, A., Langford, A., Laycock, B., ... Pratt, S. (2016). Techno-economic assessment of poly-3-hydroxybutyrate (PHB) production from methane—The case for thermophilic bioprocessing. *Journal of Environmental Chemical Engineering*, 4, 3724−3733.

Madkour, M. H., Heinrich, D., Alghamdi, M. A., Shabbaj, I. I., & Steinbüchel, A. (2013). PHA recovery from biomass. *Biomacromolecules*, 14, 2963−2972.

Maithani, M., Grover, H., Raturi, R., Gupta, V., & Bansal, P. (2019). Ethanol content in traditionally fermented ayurvedic formulations: Compromised good manufacturing practice regulations—Compromised health. *The American Journal of Drug and Alcohol Abuse*, 45, 208−216.

Mamat, N.H., Noor, S.B.M., Soh, A.C., Rashid, A.H.A., Ahmad, N.L.J., & Yusoff, I.M. (2018). Optimization of neural network architecture using particle swarm algorithm for dissolved oxygen modelling in a 200L bioreactor PHA production. In *2018 IEEE student conference on eesearch and development (SCOReD)*. Selangor, Malaysia.

Mannina, G., Presti, D., Montiel-Jarillo, G., Carrera, J., & Suárez-Ojeda, M. E. (2020). Recovery of polyhydroxyalkanoates (PHAs) from wastewater: A review. *Bioresource Technology*, 297, 122478.

Nath, A., Dixit, M., Bandiya, A., Chavda, S., & Desai, A. (2008). Enhanced PHB production and scale up studies using cheese whey in fed batch culture of *Methylobacterium* sp. ZP24. *Bioresource Technology*, 99, 5749−5755.

Ndao, A., Sellamuthu, B., Gnepe, J. R., Tyagi, R. D., & Valero, J. R. (2017). Pilot-scale biopesticide production by *Bacillus thuringiensis* subsp. *kurstaki* using starch industry wastewater as raw material. *Journal of Environmental Science and Health, Part B*, 52, 623−630.

Novak, M., Koller, M., Braunegg, M., & Horvat, P. (2015). Mathematical modelling as a tool for optimized PHA production. *Chemical and Biochemical Engineering Quarterly*, 29, 183−220.

Panigrahi, S., & Badveli, U. (2013). Screening, isolation and quantification of PHB-producing soil bacteria. *International Journal of Science and Engineering Invention*, 2, 1−6.

Pavan, F. A., Junqueira, T. L., Watanabe, M. D., Bonomi, A., Quines, L. K., Schmidell, W., & de Aragao, G. M. (2019). Economic analysis of polyhydroxybutyrate production by *Cupriavidus necator* using different routes for product recovery. *Biochemical Engineering Journal*, 146, 97−104.

Peña, C., Castillo, T., García, A., Millán, M., & Segura, D. (2014). Biotechnological strategies to improve production of microbial poly-(3-hydroxybutyrate): A review of recent research work. *Microbial Biotechnology*, 7, 278–293.

Philip, S., Keshavarz, T., & Roy, I. (2007). Polyhydroxyalkanoates: Biodegradable polymers with a range of applications. *Journal of Chemical Technology & Biotechnology*, 82, 233–247.

Ram, S. K., Kumar, L., Tyagi, R. D., & Drogui, P. (2018). Techno-economic evaluation of simultaneous production of extra-cellular polymeric substance (EPS) and lipids by *Cloacibacterium normanense* NK6 using crude glycerol and sludge as substrate. *Water Science and Technology*, 77, 2228–2241.

Righi, S., Baioli, F., Samorì, C., Galletti, P., Tagliavini, E., Stramigioli, C., ... Fantke, P. (2017). A life cycle assessment of poly-hydroxybutyrate extraction from microbial biomass using dimethyl carbonate. *Journal of Cleaner Production*, 168, 692–707.

Rivero, C., Hu, Y., Kwan, T., Webb, C., Theodoropoulos, C., Daoud, W., & Lin, C. (2017). *Bioplastics from solid waste. Current developments in biotechnology and bioengineering* (pp. 1–26). Elsevier.

Robert, H. P., Don, W. G., & James, O. M. (1984). *Perry's chemical engineers' handbook*. Nova Iorque, New York: McGraw-Hill.

Ryberg, M. W., Hauschild, M. Z., Wang, F., Averous-Monnery, S., & Laurent, A. (2019). Global environmental losses of plastics across their value chains. *Resources, Conservation and Recycling*, 151, 104459.

Sabapathy, P. C., Devaraj, S., Meixner, K., Anburajan, P., Kathirvel, P., Ravikumar, Y., ... Qi, X. (2020). Recent developments in polyhydroxyalkanoates (PHAs) production in the past decade—A review. *Bioresource Technology*, 306, 123132.

Sangkharak, K., & Prasertsan, P. (2008). Nutrient optimization for production of polyhydroxybutyrate from halotolerant photosynthetic bacteria cultivated under aerobic-dark condition. *Electronic Journal of Biotechnology*, 11, 83–94.

Serafim, L. S., Lemos, P. C., Albuquerque, M. G., & Reis, M. A. (2008). Strategies for PHA production by mixed cultures and renewable waste materials. *Applied Microbiology and Biotechnology*, 81, 615–628.

Silva, F.R. (2015). Impact of carbon/nitrogen feeding strategy on polyhydroxyalkanoates production using mixed microbial cultures.

Suriyamongkol, P., Weselake, R., Narine, S., Moloney, M., & Shah, S. (2007). Biotechnological approaches for the production of polyhydroxyalkanoates in microorganisms and plants—A review. *Biotechnology Advances*, 25, 148–175.

Tan, G.-Y. A., Chen, C.-L., Li, L., Ge, L., Wang, L., Razaad, I. M. N., ... Wang, J.-Y. (2014). Start a research on biopolymer polyhydroxyalkanoate (PHA): A review. *Polymers*, 6, 706–754.

Tourlousse, D., & Ahmad, F. (2007). *Design of an experimental unit for the determination of oxygen gas-liquid volumetric mass transfer coefficients using the dynamic re-oxygenation method ENE806. Laboratory feasibility studies in environmental engineering*. East Lansing, MI: Michigan State University.

Ujang, Z., Salim, M. R., Din, M. M., & Ahmad, M. A. (2007). Intracellular biopolymer productions using mixed microbial cultures from fermented POME. *Water Science and Technology*, 56, 179–185.

Vancov, T., Palmer, J., & Keen, B. (2019). Two-stage pretreatment process validation for production of ethanol from cotton gin trash. *BioEnergy Research*, 12, 593–604.

Vasudevan, P. T., & Briggs, M. (2008). Biodiesel production—Current state of the art and challenges. *Journal of Industrial Microbiology & Biotechnology*, 35, 421.

Verlinden, R. A., Hill, D. J., Kenward, M., Williams, C. D., & Radecka, I. (2007). Bacterial synthesis of biodegradable polyhydroxyalkanoates. *Journal of Applied Microbiology*, 102, 1437–1449.

Wagle, A., Dixit, Y., & Vakil, B. (2017). Optimization of bacterial polyhydroxyalkanoate production using one-factor-at-a-time approach. *International Journal of Pharma and Bio Sciences*, 8, 339–348.

Wang, J., Modjinou, T., & Mangeon, C. (2016). AONILEX®, 3 Invitation 4 Committees 5 General Information 6 Plan and Exhibition area 7 Speakers, (6) 1.

Wang, X., Carvalho, G., Reis, M. A., & Oehmen, A. (2018). Metabolic modeling of the substrate competition among multiple VFAs for PHA production by mixed microbial cultures. *Journal of Biotechnology*, 280, 62–69.

Wang, Y., Yin, J., & Chen, G.-Q. (2014). Polyhydroxyalkanoates, challenges and opportunities. *Current Opinion in Biotechnology*, *30*, 59−65.

Wei, Y.-H., Chen, W.-C., Huang, C.-K., Wu, H.-S., Sun, Y.-M., Lo, C.-W., & Janarthanan, O.-M. (2011). Screening and evaluation of polyhydroxybutyrate-producing strains from indigenous isolate *Cupriavidus taiwanensis* strains. *International Journal of Molecular Sciences*, *12*, 252−265.

Wynn, J. P., Hanchar, R., Kleff, S., Senyk, D., & Tiedje, T. (2016). Biobased technology commercialization: The importance of lab to pilot scale-up. In S. Van Dien (Ed.), *Metabolic engineering for bioprocess commercialization* (pp. 101−119). Cham: Springer International Publishing.

Yadav, B., Pandey, A., Kumar, L. R., & Tyagi, R. D. (2020). Bioconversion of waste (water)/residues to bioplastics-A circular bioeconomy approach. *Bioresource Technology*, *298*, 122584.

Yang, Y.-H., Brigham, C. J., Budde, C. F., Boccazzi, P., Willis, L. B., Hassan, M. A., . . . Sinskey, A. J. (2009). Optimization of growth media components for polyhydroxyalkanoate (PHA) production from organic acids by *Ralstonia eutropha*. *Applied Microbiology and Biotechnology*, *87*, 2037−2045.

Yue, H., Ling, C., Yang, T., Chen, X., Chen, Y., Deng, H., . . . Chen, G.-Q. (2014). A seawater-based open and continuous process for polyhydroxyalkanoates production by recombinant *Halomonas campaniensis* LS21 grown in mixed substrates. *Biotechnology for Biofuels*, *7*, 108.

Zahari, M. A. K. M., Ariffin, H., Mokhtar, M. N., Salihon, J., Shirai, Y., & Hassan, M. A. (2012). Factors affecting poly (3-hydroxybutyrate) production from oil palm frond juice by Cupriavidus necator (C C U G 5 2 2 3 8 T). *Journal of Biomedicine and Biotechnology*, *201*, 125865.

Zheng, Y., Chen, J.-C., Ma, Y.-M., & Chen, G.-Q. (2020). Engineering biosynthesis of polyhydroxyalkanoates (PHA) for diversity and cost reduction. *Metabolic Engineering*, *58*, 82−93.

Zhila, N., & Shishatskaya, E. (2018). Properties of PHA bi-, ter-, and quarter-polymers containing 4-hydroxybutyrate monomer units. *International Journal of Biological Macromolecules*, *111*, 1019−1026.

CHAPTER 20

Lignin production in plants and pilot and commercial processes

Ayyoub Salaghi[1], Long Zhou[1], Preety Saini[1], Fangong Kong[2], Mohan Konduri[3] and Pedram Fatehi[1,2]

[1]Green Processes Research Centre and Chemical Engineering Department, Lakehead University, Thunder Bay, ON, Canada, [2]State Key Laboratory of Biobased Material and Green Papermaking, Qilu University of Technology, Jinan, P.R. China, [3]Bio-Economy Technology Centre, FPInnovations, Thunder Bay, ON, Canada

20.1 Introduction

Naturally occurring polymers have been adjudged as superior substitutes to replace unsustainable oil-based polymers. Several types of biopolymers exist in nature, which vary in their molecular properties and functional features (Shankar & Rhim, 2018). Among all, lignin, a major component of vascular plant tissues (xylem and phloem) (Lourenco et al., 2016), is one of the world's most abundant naturally occurring three-dimensional, nontoxic, heterogeneous, renewable, and aromatic organic polymer. It is perhaps second only to cellulose in terms of abundance (Kirk, Higuchi, & Chang, 1980a), and might well be the only richest source for bioenergy production (Laurichesse & Averous, 2014). It functions to store and release carbon dioxide, aids in carrying water and nutrients to critical areas of the plants, and provides strength to grow and fight against microbial degradation (Kirk, Higuchi, & Chang, 1980b). Lignin plays a key role in the earth's carbon cycle, and there is a constant exchange of carbon to the biosphere: the plants produce carbon dioxide through respiration, and the atmospheric carbon dioxide produces lignin that is degraded naturally by microorganisms via the sequence of chemical reactions to balance the atmospheric carbon content (Gougoulias, Clark, & Shaw, 2014). However, the importance of lignin has been overshadowed for quite some time now because of its structural complexity due to intricate bonding to other components present in biomass, namely, cellulose and hemicelluloses (Li, Pu, & Ragauskas, 2016). Depending on the lignocellulosic resources, lignin is further categorized into three major classes: (1) softwood lignin, (2) hardwood lignin, and (3) grass lignin, where each type differs significantly from one another (Holtzapple, 2003). Lignin, however, is also a major side-product in the pulp and paper industry, and bioethanol industry, although it is chiefly burned as a fuel for energy production (Branco, Sarafim, & Xavier, 2019) and recovery of pulping reagents (Phillips, Jameel, & Chanq, 2013). Hopefully, with a better understanding of the chemistry of lignin, its inefficient usage from mere burning to a more economical end usage can be more meaningfully recognized.

20.1.1 Occurrence and formation of lignin

Lignin was isolated for the first time by Anselme Payen in 1838 (Payane, 1838) when he extracted insoluble fibrous cellulose from wood, upon successive treatment with concentrated nitric acid and an alkaline solution, after leaving behind a dissolved incrusting material that was later named as lignin by Frank Schulze in 1865 (Alder, 1977). The structural similarity of lignosulfonates to coniferyl alcohol was reported firstly in 1897 by Peter Klason who later asserted that lignin is essentially a macromolecule in 1907. His further studies put forward the belief that the repeating units in lignin are held preferentially by ether bonds. In chemical terms, the main components of wood (cellulose, hemicellulose, and lignin) along with the minor fractions of inorganics and extractives are interconnected to form a three-dimensional network (Alder, 1977). However, unlike most biopolymers, lignin does not have a single type of monomers, for instance, cellulose is solely a polymer of glucose molecules (Sjostron, 1981), but lignin has rather an extended disposition of substituted phenyl propane units, which are commonly called monolignols. Today, the availability of advanced sophisticated characterization/analytical tools has largely resolved its chemical structure; however, many aspects in the chemistry of lignin remain obscure though (Glasser & Sarkanen, 1989).

Of the three broad classes of lignin, the primary monomers of softwood lignins (obtained from evergreen coniferous trees) are derived from *trans*-coniferyl alcohol **1** or alternatively known as guaiacylpropane (G-unit) (IUPAC name: 4-hydroxy-3-methoxyphenylpropane) (Scheme 20.1). In addition to the guaiacylpropane polymers, the hardwood lignins (obtained from deciduous trees that shed leaves annually) contains polymers of *trans*-sinapyl alcohol **2** or commonly known as syringylpropane (S-unit) (IUPAC name: 3,5-dimethoxy-4-hydroxyphenylpropane) monomers (Scheme 20.1). The grass or compression wood lignin contains polymers of **1** and **2** and *trans-p*-coumaryl alcohol **3** or better known as *p*-hydroxyphenylpropane (H-unit) (IUPAC name: 4-hydroxyphenylpropane) (Scheme 20.1).

The irregularity observed in the structure of lignin is a result of the cross-linking of these monolignols **1**–**3** during lignification through different kinds of chemical bonds (Table 20.1). Several oxidative enzymes are present in plant species that cause oxidation of hydroxyl groups present in lignin. In addition to hydroxyl functional groups, the aldehyde and carboxyl groups are noticed in the structure of lignin (Allan, 1971).

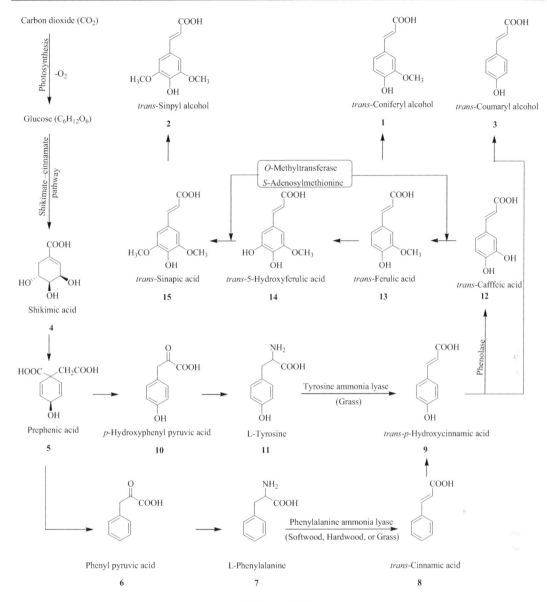

Scheme 20.1
Synthesis of monolignols **1–3**. Source: *Modified from Kirk, T. K., Higuchi, T., & Chang, H. (1980). Lignin biodegradation: Microbiology, chemistry, and potential application (Vol. 1, pp. 1–231). Boca Raton, FL: CRC Press Inc.*

The synthesis of lignin occurs via an array of interconnected chemical and biochemical reactions where the initially formed monolignol free radicals undergo enzyme-mediated oxidative coupling to afford the lignin production (Boerjan, Ralph, & Baucher, 2003;

Table 20.1: Predominant linkages and their percentage in different lignin types (Guadix-Montero and Sankar, 2018).

S. no.	Linkage type: common name	Chemical structure	Softwood lignin (%)	Hardwood lignin (%)	Grass lignin (%)
1.	β−O−4: β-aryl ether		45−50	60−62	74−84
2.	α−O−4: α-aryl ether		~8	~8	−
3.	β−5 or α−O−4: phenylcoumaran		9−12	3−11	5−11
4.	β−1: 1,2-Diarylpropane		~7	~7	−
5.	β−β or α−O−γ: resinol		2−6	3−16	1−7

Source: *From Guadix-Montero, S., & Sankar, M. (2018). Review on catalytic cleavage of C−C-interunit linkages in lignin model compounds: Towards lignin depolymerisation.* Topics in Catalysis, 61, 183−198.

Glasser, 1980; Sarkanen & Hergert, 1971; Terashima, Fukushima, He, & Takabe, 1993). Lignifying plants (softwood, hardwood ore grass) contain all kinds of enzymes that are necessary to carry out the full biochemical transformations as depicted in Scheme 20.1 (Kirk et al., 1980a). The biosynthesis of monolignols **1**−**3** starts in the differentiated wood cells of primary walls and the lignification is extended to their secondary cell walls (Saleh, Leney, & Sarkanen, 1967; Wardrop & Bland, 1959). The precursor glucose molecules are manufactured from atmospheric carbon dioxide by photosynthesis (Tanaka & Makino, 2009). The glucose molecules produce shikimite acid **4** via shikimate−cinnamate pathway (Pearl, 1967). The specific enzymes present in plants convert **4** into prephenic acid **5**. There are two different routes available to convert **5** into *p*-hydroxycinnamic acid **9**. In one sequence, prephenic acid **5** is transformed into phenyl pyruvic acid **6**, which is then converted to an essential amino acid L-phenylalanine **7**. This essential amino acid **7** is converted to *trans*-cinnamic acid **8** with phenylalanine ammonia lyase (Koukol & Conn, 1962), which exists among grasses, deciduous (hardwood) and coniferous (softwood) trees. Afterward, hydroxylation at *para*-position of **8** affords *trans-p*-hdroxycinnamic acid **9**. In another reaction, the sequence of converting **5** into *p*-hydroxycinnamic acid **9** occurs in grass species only because they have another enzyme called tyrosine ammonia lyase. In this transformation, first, there is the formation of *p*-hydroxyphenyl pyruvic acid **10** from **5**. Then, **10** is converted to L-tyrosine **11**, the next **11** is converted to **9** by tyrosine ammonia lyase (Neish, 1961). The further hydroxylation of **9** with oxidative phenolase enzyme makes *trans*-caffeic acid **12** (Russel, 1971; Vaughan & Butt, 1970). The enzyme-mediated transmethylation of **12** with *S*-adenosylmethinine in the presence of *O*-methyltransferases produces *trans*-ferulic acid **13** (Higuchi, Shimada, Nakatsubo, & Tanahashi, 1977). Again hydroxylation followed by another cycle of transmethylation is performed on **13** with *S*-adenosylmethinine and *O*-methyltransferases to furnish *trans*-sinapic acid **15** via *trans*-5-hydoxyferulic acid **14** (Shimada, Fushiki, & Higuchi, 1973). Lastly, the reduction of the carboxyl groups of cinnamic acid derivatives **9**, **13** and **15** to hydroxyl groups by some mechanisms to generate *trans-p*-coumaryl alcohol **3**, *trans*-coniferyl alcohol **1**, and *trans*-sinapyl alcohol **2**, respectively. Thereafter, these monolignols **1**−**3** participate in polymerization to afford final lignin molecules.

The polymerization starts with the radicalization of the monolignols, (**1**−**3**) followed by dimerization via nonenzymatic radical−radical coupling reactions. The dimers grow further to yield oligo- and polylignols, and ultimately constitute the final structure of lignin (Higuchi, 1997; Sarkanen & Hergert, 1971; Terashima, 2000).

The formation of different bonds has been discussed by taking the general structure of monolignols **1**−**3**, as described in Scheme 20.2. Different types of oxidases, such as peroxidase or laccase, are present in plant species (Janusz et al., 2017). Also, another type of oxidizing agent, hydrogen peroxide, is produced immensely during plant respiration and photosynthesis (Ismail, Khandaker, Mat, & Boyce, 2015). These enzymes and hydrogen peroxide, either

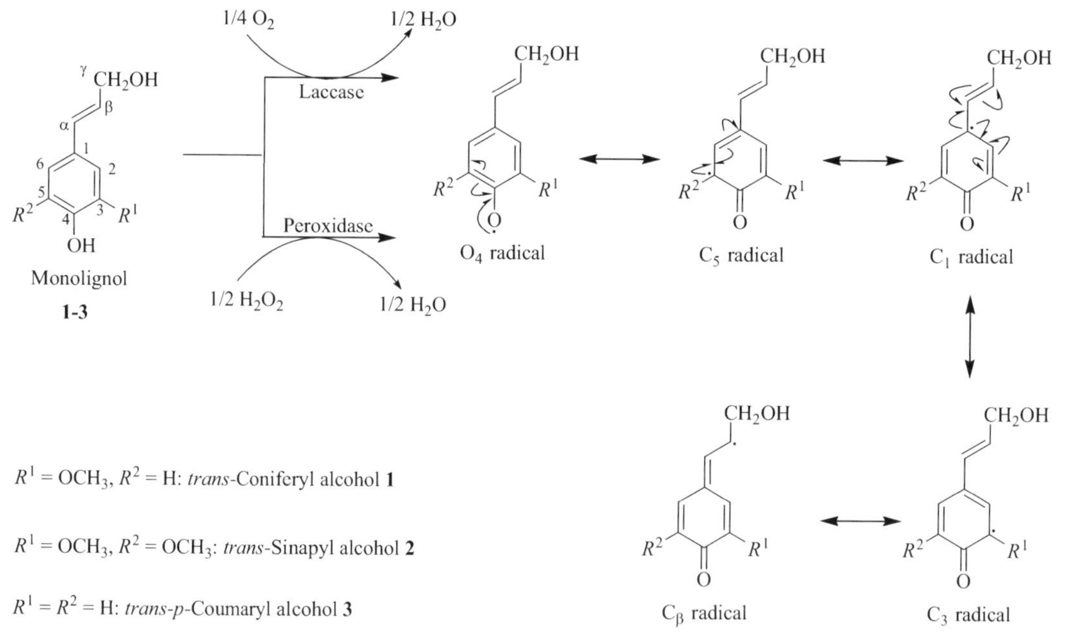
1/4 O_2
1/2 H_2O
Laccase
Peroxidase
1/2 H_2O_2
1/2 H_2O
CH2OH
γ
β
α
OH
Monolignol
1-3
O_4 radical
C_5 radical
C_1 radical
C_β radical
C_3 radical
$R^1 = OCH_3$, $R^2 = H$: *trans*-Coniferyl alcohol **1**
$R^1 = OCH_3$, $R^2 = OCH_3$: *trans*-Sinapyl alcohol **2**
$R^1 = R^2 = H$: *trans-p*-Coumaryl alcohol **3**

Scheme 20.2

Generation of monolignol radicals. Source: *Modified from Heitner, C., Dimmel, D. R., & Schmidt, J. A. (2010).* Lignin and lignans: Advances in chemistry *(pp. 1–7). Boca Raton, FL: CRC Press.*

independently (e.g., laccase) or in combination (e.g., peroxidise in the presence of hydrogen peroxide) cause the oxidation of monolignols (Dean & Erikson, 1994) to produce a phenolic radical at C–4 position. This O_4 radical is resonance stabilized by conjugation to aromatic and allylic double bonds to afford corresponding carbon radicals at C–1, C–3, C–5, and C–β as shown in Scheme 20.2. Among these radicals, C_β radical appears to be the most reactive and abundantly available, as it participates in all major types of interconnections found in lignin.

The principal β–O–4 type of bond (Table 20.1, entry 1) is the most common and abundant linkage found in all lignin types (softwood, hardwood, and grass). It is formed by the free-radical coupling of O_4 and C_β radicals to generate a nonaromatic quinone-methide **16** as given in Scheme 20.3. The quinone-methide **16** is highly reactive and readily accepts the addition of nucleophile at the α-position to attain aromaticity. For example, it undergoes hydroxylation at α-position by nucleophilic attack by water molecule to produce an aromatic dilignol **17**. The dimer **17** participates in oxidative polymerization (Terashima, Nakashima, & Takabe, 1998).

Likewise, α–O–4 linkage (Table 20.1, entry 2) is found in the structure of lignin, although this type of bonding is not present in a large amount. However, if it exists, it is present along with β–O–4 bond. The reaction commences with the initial formation of β–O–4 bond (Heitner, Dimmel, & Schmidt, 2010), where the formed quinone-methide **16** (Scheme 20.2) undergoes coupling with another O_4 radical to generate a trilignol **18** that has an

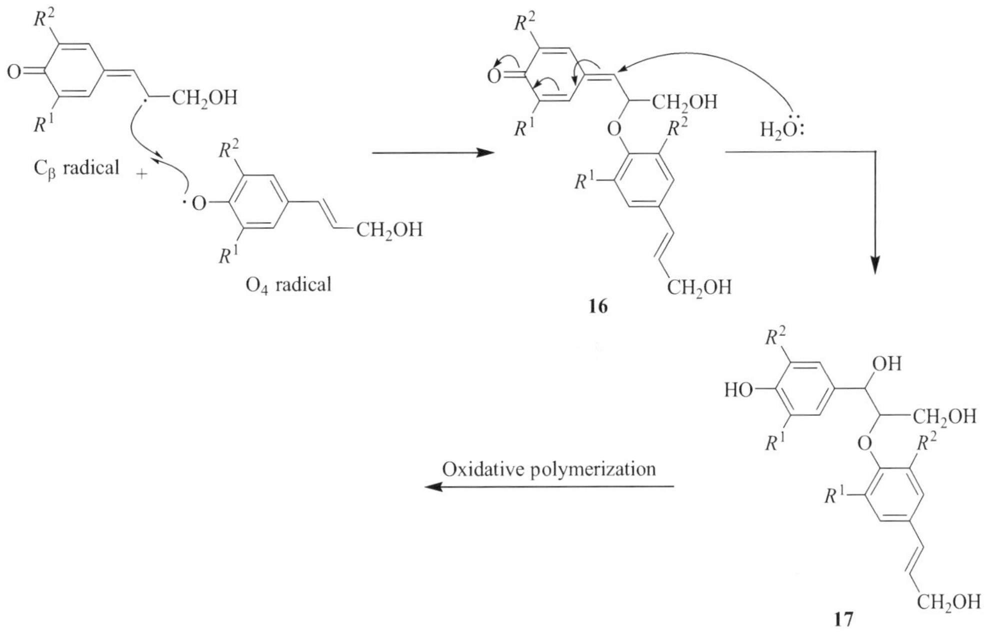

Scheme 20.3
General reaction mechanism for β—O—4 linkage in lignin. Source: *Modified from Heitner, C., Dimmel, D. R., & Schmidt, J. A. (2010). Lignin and lignans: Advances in chemistry (pp. 1—7). Boca Raton, FL: CRC Press.*

α-aryl ether bond as shown in Scheme 20.4. The generated trilignol **18** participates in polymerization till chain-termination to generate macrostructure of lignin.

Besides C—O—C ether linkages (β—O—4 and α—O—4), some C—C linkages of β—5 (Table 20.1, entry 3), β—1 (Table 20.1, entry 4), and β—β (Table 20.1, entry 5) are observed in the structure of lignin. The percentage of any C—C linkages is not very appreciable in any lignin types (Alder, 1977; Heitner et al., 2010). The β—5 type of linkage is more feasible when there is no substituent at C—5 position of C_5-coupling partner or otherwise may need more drastic conditions to release methoxyl substituent R^2 at C—5 position during cyclization. The mechanism of formation of β—5 bond is outlined in Scheme 20.5. The dimer **19** exhibiting β—5 linkage is formed by the free-radical coupling between C_β and C_5 radicals, which rearranges through keto-enol tautomerization to furnish **20**. The subsequent intramolecular Michael addition reaction aromatizes **20** and produces a more stable molecule **21** containing phenylcoumaran structure. Thereafter, **21** participate in the oxidative polymerization.

Another C—C linkage in lignin arises by coupling between C_β and C_1 radicals, and the mechanism is highlighted in Scheme 20.6 (Heitner et al., 2010). The reaction progresses by the initial formation of the diquinone of type **22**, which affords **23** after hydroxylation with

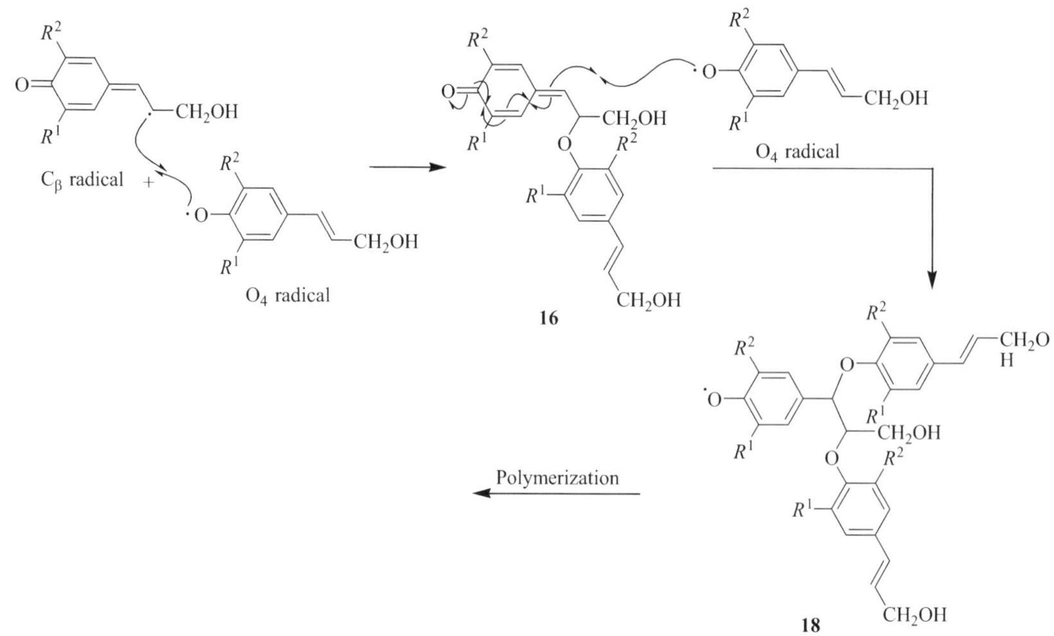

Scheme 20.4
General reaction mechanism for α−O−4 bond in lignin. Source: *Modified from Heitner, C., Dimmel, D. R., Schmidt, J. A. (2010). Lignin and lignans: Advances in chemistry (pp. 1−7). Boca Raton, FL: CRC Press.*

available water molecules. In the subsequent step the appended chain at C−1 position of **23** is cleaved to give an aromatic dilignol of type **24**. This type of cleavage is favorable only with relatively good stability of the aldehyde leaving group.

The general reaction mechanism for analogous nonfeasible C−C bond between C−β and C−3 positions is depicted in Scheme 20.7 (Heitner et al., 2010). This type of linkage is not favorable as there could not be any way for the coupled product **25** to aromatize because the R^1 (=OCH_3) present as C−3 is a poor leaving group and likely reverses back to the precursor radicals C_β and C_3.

The β−β type of bond formation (Table 20.1, entry 5) occurs through the coupling of two C_β radicals and the mechanism is depicted in Scheme 20.8 (Heitner et al., 2010). The structure of formed quinone molecule **26** is such that it can undergo intermolecular Michael addition and rearrangement to yield the resinol type of structure **27**.

The aforementioned discussion concludes that, unlike other biopolymers (Sjostron, 1981), different monolignols in lignin are joined together through different chemical bonds, and there exists a varied proportion of condensed structures, carbon-carbon, and carbon−oxygen bond types. For example, sinapyl alcohol **2** is present mainly in lignins that come from hardwood and these types

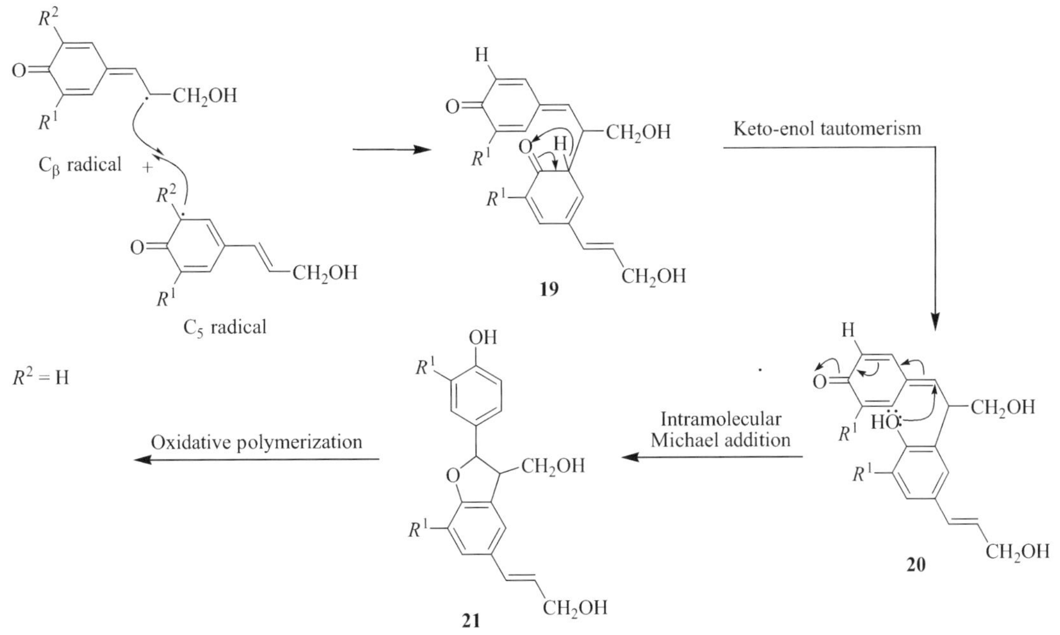

Scheme 20.5

General reaction mechanism for β–5 bond formation in lignin. Source: *Modified from Heitner, C., Dimmel, D. R., & Schmidt, J. A. (2010). Lignin and lignans: Advances in chemistry (pp. 1–7). Boca Raton, FL: CRC Press.*

have relatively less proportion of condensed structures because methoxyl group present at C−3 and C−5 positions will hinder the internal cyclizations. Whereas, softwood lignins have relatively more condensed structures compared to hardwood lignins. The structure of lignins obtained from grasses is more complex because it contains all three types of monolignols **1–3**. Therefore polymerization occurs among all the three types of monomers. Furthermore, the presence of sterically less hindered *p*-coumaryl alcohol **3** in grass lignin favors the reactions of the type shown in Schemes 20.5 and 20.8 that are otherwise not feasible or less favorable in the biosynthesis of softwood or hardwood lignins.

Additionally, the presented discussion explains why the reactivity of one type of lignin quite deviates from the other types and we have seen different types of chemical bonds in different kinds of lignin (Guadix-Montero & Sankar, 2018). In chemical terms the feasibility of a chemical reaction depends on the type of chemical bond under consideration. If the electronegativity difference between the two connecting atoms of a typical bond is large, there is unequal sharing of electron pair forming that bond due to the electronegativity difference, and hence the bond becomes polar and more susceptible to chemical attacks. In general, the more the number of polar heterobonds (such as C−O

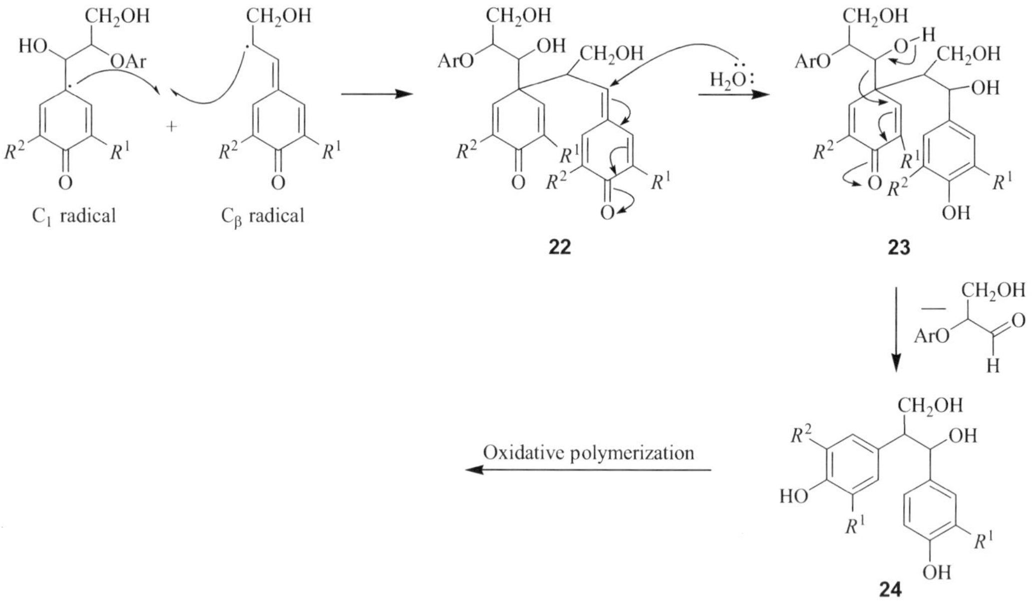

Scheme 20.6
General reaction mechanism for β−1 bond formation in lignin. Source: *Modified from Heitner, C., Dimmel, D. R., & Schmidt, J. A. (2010). Lignin and lignans: Advances in chemistry (pp. 1−7). Boca Raton, FL: CRC Press.*

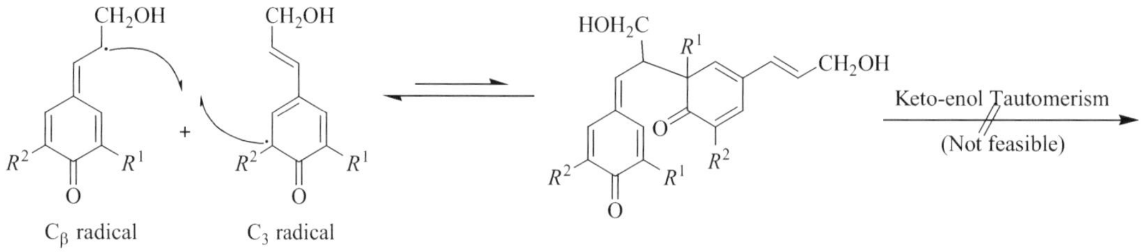

Scheme 20.7
General reaction mechanism for β−3 bond formation in lignin. Source: *Modified from Heitner, C., Dimmel, D. R., & Schmidt, J. A. (2010). Lignin and lignans: Advances in chemistry (pp. 1−7). Boca Raton, FL: CRC Press.*

or O−H) in lignin, the easier would be the chemical modification for that particular lignin.

As discussed, the biosynthesis of lignin has three major steps: firstly, the formation of monolignols, then their radicalization, and lastly the combining of formed radicals. The first two steps are enzyme catalyzed, whereas the recombination of radicals (either the

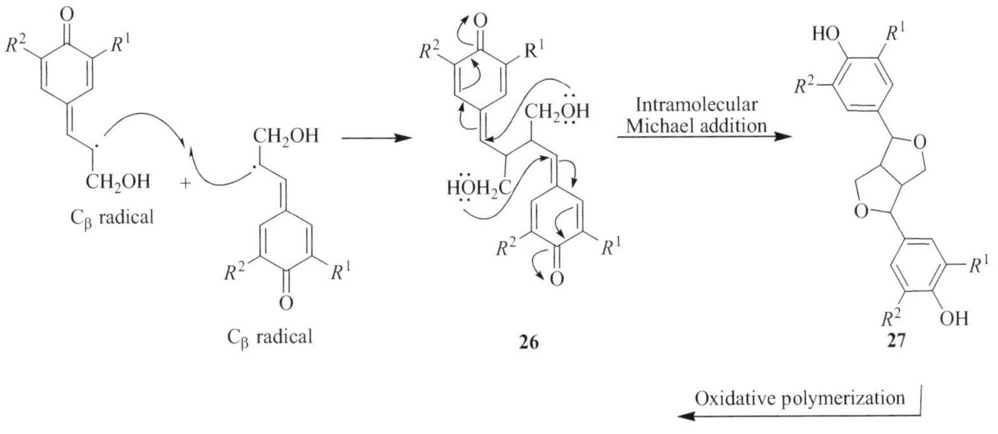

Scheme 20.8
General reaction mechanism β–β bond formation in lignin. Source: *Alder, E. (1977). Lignin chemistry—Past, present and future. Wood Science and Technology, 2, 169–218.*

monolignol radicals or the phenolic end-group radicals of the growing lignin polymer) is nonenzymatic. This means chiral centers generated during the coupling of radicals follow a nonstereoselective pathway to yield a racemic lignin polymer.

20.2 Lignin extraction methods at laboratory and pilot scales

A large number of biomass fractionation methods have been developed (Saleh et al., 1967; Wardrop & Bland, 1959). Lignin could be extracted from biomass through the cleavage of the ester and ether bonds in the lignin–carbohydrate complex (LCC) (Tanaka & Makino, 2009). Some chemical, physicochemical, physical, and biological treatments are known to be effective in separating lignin from biomass (Koukol & Conn, 1962; Neish, 1961; Pearl, 1967); however, all of these treatments have some drawbacks in terms of energy consumption, the complication of multiple-step processes, and contamination. In the following section, some of these methods are discussed.

20.2.1 Lignin extraction using milling methods

Lignin isolated at laboratory scales is mainly used for research purposes that require the well-preserved structure of lignin. Among various types of lignin, milled wood lignin (MWL), cellulolytic enzymatic lignin (CEL), and enzymatic mild acidolysis lignin (EMAL) are known because their structures are close to those of native lignin in biomass due to their mild extracting methods (Ewellyn, Balakshin, Katahira, Chang, & Jameel, 2015). MWL is achieved from biomass mainly by physical milling, while CEL is obtained from a combination of milling and enzymatic treatment, and EMAL is extracted from biomass using milling, acidolysis, and enzymatic treatment (Rui, Wu, Lv, & Guo, 2010). The properties of these lignin types will be discussed further.

Notably, Klason lignin methods (acidolysis using concentrated acid) are widely used for the lignin content determination of biomass (TAPPI T 222 om-02, 2002); but concentrated sulfuric acid renders severe structural damage to lignin, making it improper for generating lignin derivatives. Thus Klason method is basically not considered as a lignin production method. In addition to milling methods, some other methods, such as organosolv, Alcell, soda, and alkaline lignin, could also be carried out at a laboratory scale for lignin-structure analysis. However, they have been successfully applied at the pilot and commercial scales for lignin production. Thus the technical details of these methods and their merits/demerits will be discussed in the next sections.

20.2.1.1 Milled wood lignin

In the MWL method (Fig. 20.1A), intensive milling is applied to extract lignin from biomass at room temperature. Depending on the properties of the milling machines, the milling process lasts for 2–28 days (Anderson, Filpponen, Lucia, & Argyropoulos, 2006). After the milling process, neutral solvent (usually dioxane and water) is used to recover lignin from the milling meal. Acetic acid is used to precipitate lignin in this process and then crude MWL is obtained after drying the precipitated lignin. Crude MWL is further purified via dissolving and precipitation to achieve the final MWL. Rather than oven drying, vacuum drying is used to ensure the low-temperature process to preserve lignin properties (Anderson, Filpponen, Lucia, & Argyropoulos, 2006). Depending on the wood

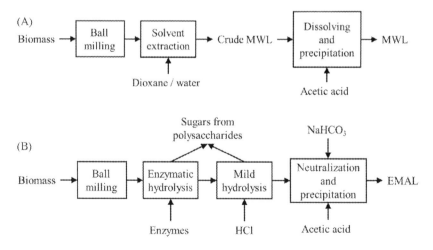

Figure 20.1
Schematic representations of MWL lignin extraction (A) and EMAL lignin extraction (B). *EMAL*, Enzymatic mild acidolysis lignin; *MWL*, milled wood lignin.

Table 20.2: Typical lignin yields, average molecular weights and dispersity indices from various biomass sources isolated by milled wood lignin (MWL), cellulolytic enzymatic lignin (CEL), and enzymatic mild acidolysis lignin (EMAL) (Anderson, Filpponen, Lucia, & Argyropoulos, 2006; Anderson, Filpponen, Lucia, Saquing, et al., 2006).

Biomass	MWL			CEL			EMAL		
	Yield (%)	M_w (g mol^{-1})	Đ	Yield (%)	M_w (g mol^{-1})	Đ	Yield (%)	M_w (g mol^{-1})	Đ
Norway Spruce	11.4	23,500	3.7	23.4	53,850	5.7	44.5	78,400	8.8
Douglas Fir	1.4	7400	3.0	7.1	21,800	4.0	24.8	38,000	5.0
Redwood	15.7	5900	2.5	13.2	23,000	4.2	56.7	30,100	6.4
White Fir	11.3	8300	3.0	11.5	21,700	4.6	42.9	52,000	8.2
E. globulus	34.0	6700	2.6	32.5	17,200	3.1	63.7	32,000	3.7
Southern Pine	11.9	14,900	3.2	12.4	29,600	3.9	56.3	57,600	5.9

Yield determined by Klason method. Đ, Dispersity indices; M_w, average molecular weights.
Source: From Anderson, G., Filpponen, I., Lucia, L. A., & Argyropoulos, D. S. (2006). Comparative evaluation of three lignin isolation protocols for various wood species. Journal of Agricultural and Food Chemistry, 54, 9696–9705; Anderson, G., Filpponen, I., Lucia, L. A., Saquing, C., Baumberger, S., & Argyropoulos, D. S. (2006). Toward a better understanding of the lignin isolation process from wood. Journal of Agricultural and Food Chemistry, 54, 5939–5947.

spices and milling severities, the yield of MWL ranges 5%–15% (Anderson, Filpponen, Lucia, & Argyropoulos, 2006).

MWL is believed to be the best matrix to represent the native lignin in biomass (Tuomela, Vikman, Hatakka, & Itävaara, 2000) and hence is widely used in laboratory studies. Compared with other types of lignin, MWLs have more uniform structures due to the mild extraction protocols. A drawback of MWL is the high content of saccharide contamination because only physical milling cannot effectively cleave the bonds of the LCC (Anderson, Filpponen, Lucia, Saquing, et al., 2006). Meanwhile, lignin is greatly removed alongside polysaccharides, resulting in a low yield of lignin production. As shown in Table 20.2 for most types of biomasses, MWLs have lower purities and yields than CELs and EMALs. Moreover, MWL could only represent the low molecular weight fraction of lignin in biomass. Table 20.2 clearly shows that the molecular weights of MWLs are much lower than CELs and EMALs. The polydispersity indices of MWLs are lower than those of CELs and EMALs, proving that MWLs have more uniform structures.

20.2.1.2 Cellulolytic enzyme lignin

In 1975 CEL was produced using a new method combining the enzymatic degradation of polysaccharides and subsequent dioxane extraction of lignin (Kun, Bauer, & Sun, 2012). The dewaxed wood meal is treated by a ball-milling process, followed by enzyme hydrolysis. The cellulase used in this step could effectively degrade the polysaccharides in biomass and facilitate the lignin extraction process. Then, CEL is obtained after the solvent extraction (ethanol and water) and evaporation (Chang, Cowling, & Wynford, 1975). The yield of CEL ranges 10%−20% for different wood spices (Kun et al., 2012).

As shown in Table 20.2, generally CELs have higher yields, purities, and molecular weights than MWLs because of the application of an enzymatic treatment that effectively removes sugar component from biomass. It is reported that CELs are more representative than MWLs in terms of lignin structure for some wood species, such as birch and maple (Ewellyn et al., 2015). But generally, CEL experiences more structural modification than do MWLs. Furthermore, protein from enzymes becomes another source of lignin contamination (Ibarra et al., 2005).

20.2.1.3 Enzymatic mild acidolysis lignin method

EMAL is separated from biomass using both mild acidolysis and enzymatic treatment (Rui et al., 2010). As shown in Fig. 20.1B, a milling process, like the MWL process, is used in the first step. Then, the milled wood meal is processed by a 48-h enzymatic hydrolysis at 40°C using cellulase and hemicellulose to degrade carbohydrates and expose the fiber structure of biomass. Then, a mild acidolysis using hydrochloric acid is applied to further remove sugars from biomass. EMAL is hence obtained from the mixture after neutralization and precipitation (Wu & Argyropoulos, 2003).

Because the EMAL method is optimized to have the advantages of both acidolysis and CEL methods, a high lignin yield of 70% is achievable without an obvious structural alteration for *Eucalyptus globulus* (southern blue gum) (Wu & Argyropoulos, 2003). For other types of wood, a lignin yield of 20%−50% can still be obtained. Because of the low polysaccharide contamination, the EMAL is widely used for lignin studies, such as the pyrolytic characterization of lignin (Rui et al., 2010). However, EMAL method is one of the most tedious methods for lignin separation at a laboratory scale (Fig. 20.1B). The structure of EMAL is more heterogeneous than those of MWL and CEL according to the molecular weights and polydispersity indices reported in Table 20.2.

20.2.2 Lignin production by novel extraction technologies at pilot scales

Due to an increasing interest in lignin and lignin-derived products, several novel lignin extraction technologies have been developed recently to achieve highly pure lignin. Many of these lignin extraction technologies have been carried out on pilot-scale projects and have opportunities for commercialization. The lignin obtained from these technologies usually has interesting properties that are in favor of downstream applications.

20.2.2.1 Steam-explosion process

DSM-POET (a 50−50 joint venture between Royal Dutch State Mines and POET LLC), Emmetsburg, United States, has reported using steam explosion at a pilot scale for refining

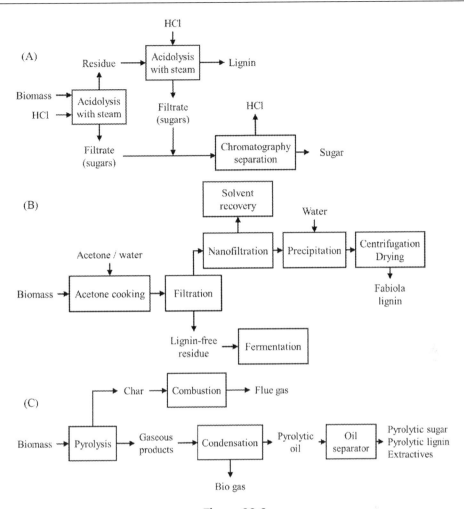

Figure 20.2
Schematic representations of steam-explosion process (A), Fabiola biorefinery process (B), and fast pyrolysis system (C).

biomass producing sugars and lignin (Green Car Congress, 2012). As shown in Fig. 20.2A, after softening the biomass by soaking it in diluted sulfuric acid, the acid-impregnated biomass is treated with high-pressure steam in a steam-explosion reactor. After cooling, the solid from the steam-explosion process is separated by filtration and washed with water. The aqueous hydrolysate solution is rich in hemicellulose-derived sugars and acetic acid. The acid is recycled by reconcentration and the polysaccharides are proposed to be used for hydrogen production or for the fermentation to produce bio-ethanol; meanwhile, lignin is left in the residual solid (POET-DSM Advanced Biofuels, LLC, 2012). The resulting lignin contains little carbohydrates and wood extractive impurities. The yield of lignin is approximately 4%–7% of the dry weight of the sample from birchwood (Green Car Congress, 2012).

The lignin produced in this process resembles the native lignin more than the other produced technical lignins. Pucciariello et al. (2008) succeeded to blend steam-explosion lignin with polycaprolactone (PCL) and claimed that lignin strongly stabilized PCL against UV radiation. However, the molecular weight of the resulting lignin is low because of the treatment at an elevated temperature. It is reported that the average molecular weight of lignin from birchwood using a steam explosion is only 2250–2620 g mol^{-1} with the polydispersity index of 1.5 (Sun & Tomkinson, 2000). Because of the low molecular weight, steam-explosion lignin seems to be a suitable material for phenolic chemical production. As a biorefining process, steam-explosion technology could make full utilization of biomass materials, because the polysaccharides of biomass are used for bioethanol production and the lignin could be applied for the production of aromatics.

20.2.2.2 Sequential liquid-lignin recovery and purification technology

Liquid Lignin Company (South Carolina, United States) designed a pilot-scale lignin recovery process using sequential liquid-lignin recovery and purification (SLRP) (Lake & Blackburn, 2014). This process separates lignin from black liquor of kraft pulping process as a true liquid phase that separates by gravity. This is different from traditional processes that precipitate lignin as small solid particles. Kraft pulping black liquor and carbon dioxide are injected into the carbonation reactor via the injection ports at the top and bottom of the reactor, respectively. This design results in a slurry enriched in lignin instead of lignin particles. Liquid lignin is formed while black liquor reacts with CO_2 at pH 9.0–10.0 (Velez & Thies, 2016). Then, liquid-lignin droplets collide to form a bulk dense slurry, which can be easily separated. To prevent the formation of lignin particles, the pH and temperature in the carbonation reactor are controlled at 2.0–3.0 and 100°C–150°C, respectively (Lake & Blackburn, 2014). The slurry is then transferred into an acidification reactor to react with sulfuric acid at pH 2.0–3.0 for precipitation. Solid lignin is obtained after filtration, washing, and drying. Because the production of SLRP lignin requires a high volume of CO_2, this system has a gas capture system to reduce CO_2 loss and air contamination. The SLRP process could recover 55% lignin in the black liquor (Lake & Blackburn, 2014).

The capital and operating costs are claimed to be much lower than conventional processes, because of small equipment resulting from SLRP's continuous operation. The molecular weight and purity of obtained lignin from this process are the two factors that limit its value in polymeric applications (Lake & Blackburn, 2014; Velez & Thies, 2016). Furthermore, it is reported that the SLRP process results in lignin containing high ash (1%–2%) and sulfur content (2%–3%) (Kihlman, 2015). However, with proper purification to reduce the ash and sulfur contents, SLRP lignin can be still used for bio-fuel applications because of its low production cost in the context of biorefinery.

20.2.2.3 Bio-Flex process

Lixea (previously Chrysalix Technologies, London, United Kingdom), proprietary Bio-Flex process, is a novel pretreatment technology for lignocellulosic biomass offering unprecedented feedstock flexibility. The first pilot Bio-Flex biorefining plant was carried out in Biorenewable Development Centre, York, United Kingdom (https://www.lixea.co/#bioflex-process). Different types of biomass, including agricultural residues, energy crops, as well as forestry residues and even waste construction wood, can be used in this process (https://ubuntoo.com/solutions/bioflex-process). An inexpensive and recyclable ionic liquid separates the different components of biomass, that is, cellulose and lignin, while by-products as well as heavy metals present in waste wood can be recovered. The biomass feedstock is processed with ionic liquid below 200°C and at atmospheric pressure (https://ubuntoo.com/solutions/bioflex-process). Because ionic liquids have low vapor pressure, no harmful emissions are observed in this process (https://www.filtsep.com/chemicals/features/bioflex-turns-wood-into-green-bioproducts-10/). This process has been developed based on the ionosolv biorefining technology, which fractionates any type of lignocellulosic biomass into the biorefining intermediates, cellulose, and lignin via precipitation and filtration. In addition, the compounds containing heavy metal present in biomass can be recovered as a by-product of Bio-Flex process. Lixea plans to design and build an industrial demonstration plant (c. 30,000 t per year) in the future (www.lixea.co/#bioflex-process).

Once isolated individually, these components can be used for a variety of applications, such as biochemicals, precursors for plastics, or as new materials themselves. More specifically, the produced ionosolv lignin can be consumed as a suitable replacement of phenol in resin and adhesive productions (https://www.filtsep.com/chemicals/features/bioflex-turns-wood-into-green-bioproducts-10/). Worth noting, the lignin achieved from Bio-Flex is a sulfur-free material because no sulfur-bearing agent is used in this process (Filtration + Separation, 2019). This makes Bio-Flex lignin more competitive and greener compared with conventional technical lignin. A major drawback of Bio-Flex lignin is the high production cost because synthesis and purification of ionic liquids are challenging and expensive (https://www.filtsep.com/chemicals/features/bioflex-turns-wood-into-green-bioproducts-10/).

20.2.2.4 German lignocellulose feedstock biorefinery project

German lignocellulose feedstock biorefinery (2007–09) was a joint project of 15 partners, aiming to develop a sustainable and economic biorefinery process on a pilot scale to achieve high-quality cellulose, hemicellulose, and lignin from hardwood for biomaterials applications (Michels & Wagemann, 2010). Firstly, biomass extraction is carried out using alcohol to remove the extractives, such as triterpene. Extractive-free biomass is then treated via a pulping process (2–4 h, 170°C–180°C) using ethanol and water (50:50) (Michels & Wagemann, 2010). Lignin and polysaccharides are precipitated from pulping liquor via hot water. The separation of lignin and polysaccharides is achieved by adjusting the pH using NaOH and HCl as following: NaOH solution is used to dissolved lignin, leaving cellulose in solid fraction for filtration; then HCl is added to precipitate lignin from the aqueous phase. It is estimated that approximately 300 kg of wood material can be processed per week, and 80 kg of lignin (27% of feedstock) can be obtained (Michels & Wagemann, 2010).

As the main product of this project, the obtained lignin is a sulfur-free material that is suitable for biofuel and phenolic chemical production. Fraunhofer Institut Chemische Technologie (Pfinztal, Germany) succeeded to produce a polylactic acid product containing 20 wt.% of this lignin (Michels & Wagemann, 2010). Dynea (Lillestrøm, Norway) used this lignin to partially substitute phenol-based resin (up to 30 wt.%) and claimed that obtained resin could be used for wood-construction boards (Michels & Wagemann, 2010). However, this project only focused on the biorefining of hardwood, like beech and poplar. More investigation is required for lignin isolation from softwood using this system. The cost of lignin production is a problem that might undermine its downstream applications. The production cost of lignin in this project was reported to be € 395/t; while the traditional Kraft lignin was reported to be approximately USD 250/t (Abbati de Assis et al., 2018). In addition to lignin, the German lignocellulose feedstock biorefining project could produce fermentable C6/C5 sugars that can be used for bioethanol applications (Michels & Wagemann, 2010).

20.2.2.5 Ammonia fiber explosion lignin

A pilot-scale biomass pretreatment plant in Michigan, United States used an ammonia-based fiber explosion method to produce lignin (Bioeconomy Institute, 2015). In this technology, lignocellulosic biomass is treated with liquid ammonia (1:1 ratio) at low temperature (60°C−90°C) and pressure (3 MPa or above) in a closed vessel for 30−60 min (Teymouri, Laureano-Perez, Alizadeh, & Dale, 2005). The cellulose crystallinity of biomass undergoes a phase change at high temperature and pressure and carbohydrates become active in the presence of ammonia; meanwhile, lignin and a part of hemicellulose are removed. After ammonia treatment, biomass slurry is formed containing both lignin-enriched residues and carbohydrate-enriched hydrolysate, which can be separated via filtration. It is reported that most cellulose and hemicellulose are well preserved without degradation (Moniruzzaman, Dale, Hespell, & Bothast, 1997). On the other hand, lignin can be produced from the unhydrolyzed residues through refluxing (e.g., EtOH:water), washing (hexane and dichloromethane), concentration, and drying. The yield of lignin enriched residue is approximately 38% of the stover feedstock (Bioeconomy Institute, 2015).

A major advantage of the ammonia fiber explosion (AFEX) lignin production is that the cost of AFEX can be reasonable, because at the end of the process, the reaction agents, like ammonia and ethanol, can be recovered and reused, making AFEX industrially attractive. The AFEX lignin is a suitable material for soil fertilizer because ammoxidation could convert nitrogen-free lignin into nitrogenous soil-improving materials (Klinger, Liebner, Fritz, Potthast, & Rosenau,

2013). In addition to lignin, high pure cellulose and hemicellulose can be obtained in the AFEX process. Approximately, 90% of these polysaccharides could be converted into fermentable sugars, making them suitable for bioethanol production (Bioeconomy Institute, 2015). However, alongside this process, it is also possible to form ecotoxic, carbohydrate-derived nitrogenous compounds under ammoxidation conditions (Klinger et al., 2013), which potentially cast a risk in the utilization.

20.2.2.6 FABIOLA lignin

UNRAVEL project, a joint project of 10 European partners, 2018−22, aims to develop advanced pretreatment, separation, and conversion technologies for complex lignocellulosic biomass (UNRAVEL, 2019). As a part of the UNRAVEL project, the scaling-up of the FABIOLA fractionation process on a pilot scale at Fraunhofer Center for chemical−biotechnological processes showed great potential for improving the pretreatment of lignocellulosic biomass. This process is capable of handling various types of biomass, including wheat straw, corn stover, and various hardwoods such as poplar, beech, and birch resulting in cost-effective and high-quality compounds (UNRAVEL, 2019). Fig. 20.2B shows a brief flow diagram of FABIOLA biorefinery (Damen, Smit, Huijgen, & Van, 2017). Hardwood and herbaceous biomass are cooked with acetone at 140°C digester. After filtration, the pulp is separated from the mixture for further fermentation to produce ethanol. The liquid phase is filtrated with a nanofiltration system to separate lignin from the organic solvent. Then, water (4:1 w/w dilution ratio H_2O:liquor) is added into the lignin enriched liquor for lignin precipitation. Lignin is collected by centrifugation and then dried in an oven at 50°C. The acetone solvent is recovered and reused through filtration and distillation. It is reported that the delignification rate of 87% has been achieved in the Fabiola process. With further improvement, it is expected that 95% of lignin can be recovered in this process in the future (UNRAVEL, 2019).

As the Fabiola process uses sulfur-free pulping liquor (Damen et al., 2017), the Fabiola lignin does not contain any sulfur contamination, which is a major advantage for biofuel applications. The structure of Fabiola lignin is very close to native lignin because the cooking temperature of the Fabiola process is relatively low (Fabiola cooking of 140°C, traditional organosolv cooking of 180°C) (Damen et al., 2017). Compared with other types of lignin, Fabiola lignin contains abundant β-O-4 linkages, which is the main interlinkage in native lignin (Damen et al., 2017). As a biorefining process, Fabiola technology also produces xylose (UNRAVEL, 2019), which is a widely used sweetener as a low-calorie sugar in the food industry. A drawback of Fabiola lignin is its high production cost because of the large demand for acetone solvent. The energy consumption of Fabiola lignin production is also a disadvantage as its cooking process lasts at least for 2 h (UNRAVEL, 2019). Although Fabiola lignin has shown desirable properties, the excessive chemical demand and intensive energy consumption of its production may become barriers in its further applications.

20.2.2.7 Proesa lignin

The Proesa technology was designed for the treatment of cellulosic biomass from agricultural residues and energy crops in biorefineries to separate the cellulose and hemicellulose from lignin in a plant located in Crescentino, Italy (https://www.bio.org/sites/default/files/legacy/bioorg/docs/beta%20renewables%20proesa%20technology%20june%202013_bio_michele_rubino.pdf). Biomass is pretreated with steam and water to form biomass slurry via a "Smart Cooking" without adding any corrosive chemicals. Subsequently, enzymes and microorganisms are added into the slurry for hydrolysis and fermentation. The bioethanol is obtained from the produced fermentable sugars and collected via evaporation, leaving lignin in solid residues for value-added applications (Palmisano, 2013). This pilot plant can process 1 t day^{-1} of feedstock, including energy crops, agricultural wastes, woody biomass, and bagasse. Approximately 20% of the product is bioethanol and the rest (80%) is residual lignin and unfermentable sugars (https://www.bio.org/sites/default/files/legacy/bioorg/docs/beta%20renewables%20proesa%20technology%20june%202013_bio_michele_rubino.pdf).

The Proesa lignin is currently simply burnt for energy recovery; however, it is reported that the Proesa lignin is suitable for value-added applications, including bioplastics and phenolic chemical production, such as aromatics, terephthalic acid, and phenols (https://www.bio.org/sites/default/files/legacy/bioorg/docs/beta%20renewables%20proesa%20technology%20june%202013_bio_michele_rubino.pdf). The structure of Prosea lignin is expected to be close to native lignin in biomass because only steam and water are used in the cooking process without any extra use of corrosive chemicals, allowing the structural damage of lignin to be minimum (https://www.bio.org/sites/default/files/legacy/bioorg/docs/beta%20renewables%20proesa%20technology%20june%202013_bio_michele_rubino.pdf). As a biorefining process, the Proesa technology could also produce high-quality bioethanol via the enzyme treatment of the obtained polysaccharides (https://www.bio.org/sites/default/files/legacy/bioorg/docs/beta%20renewables%20proesa%20technology%20june%202013_bio_michele_rubino.pdf). In the future the Proesa claimed to have plans for the production of fatty alcohols, fatty acids, green diesel, and gasoline from cellulosic sugars. On the other hand, the Proesa lignin may have severe sugar and ash contamination because the process liquor containing only water is probably not effective in separating lignin from biomass. Perhaps the energy consumption of Prosea lignin production is also a problem, as long cooking duration is needed due to the unique process. Furthermore, the usage of the enzyme and microorganism increases the production cost of Prosea lignin (https://www.bio.org/sites/default/files/legacy/bioorg/docs/beta%20renewables%20proesa%

20technology%20june%202013_bio_michele_rubino.pdf). These drawbacks may undermine the application of the Proesa lignin.

20.2.2.8 Fast pyrolysis lignin

A novel thermochemical fractionation plant on the pilot scale has been successfully launched by BTG (Biomass Technology Group, the Netherlands), which is an important step toward the commercialization of a fast pyrolysis-based biorefinery (Biomass Technology Group, 2021). It is an innovative two-step conversion process. As shown in Fig. 20.2C, the pyrolysis of biomass is carried out in the reactor chamber at 450°C–600°C to achieve fractionation of the mineral free, liquid product (fast pyrolysis bio-oil) that keeps the key chemical functionalities intact in separate, depolymerized fractions (van de Beld, 2018). Afterward, the pyrolytic gaseous products are transferred into condenser for pyrolytic oil fractionation. The main products of the pyrolytic oil are pyrolytic lignin, pyrolytic sugars, and extractives (Greenovate! Europe, 2019). In addition to pyrolytic oil the biochar and biogas from pyrolysis are combusted to produce flue gas for energy generation. The system is capable of handling all types of biomass resource, meaning that even biomass wastes that are greener than agricultural crops could be utilized in this system for energy, sugar, and lignin production (van de Beld, 2018). It is reported that this plant has the capacity to process 50 tons of biomass feedstock per hour and produce 3.2 t h^{-1} of fast pyrolysis biofuel with the biofuel yield of 6.4% (Biomass Technology Group, 2021).

According to BTG, the fast pyrolysis lignin is more reactive than kraft lignin and native lignin, making it suitable for further modification (Heeres, 2019). The fast pyrolysis lignin is completely free of contamination from external sources because no external chemical is added during pyrolysis (van de Beld, 2018). With further upgrading treatment using commercial catalysts (such as NiMo/ CoMo), fast pyrolysis lignin can partly replace fossil fuel (Biomass Technology Group, 2021). As a biorefining process, fast pyrolysis could also produce pyrolytic sugars suitable for the production of furan-based resins; and extractives quite similar to tall oil liquids to be used as cofeed to produce diesel-like products (van de Beld, 2018). Biochar from pyrolysis is combusted in char combustor to generate flue gas for electricity. In addition, biogas for energy generation is obtained from condenser (van de Beld, 2018). However, a drawback is that pyrolytic lignin may contain sugar contamination from the carbohydrate in biomass, because the produced bio-oil is a mixture of lignin, sugar, and other extractives, and it is not easy to thoroughly separate pyrolytic lignin from the mixed oil product (Biomass Technology Group, 2021).

Notably, the aforementioned methods are still mainly studied at laboratory or pilot scales for research purposes due to their lengthy, energy-consuming protocols, and low lignin yields. However, these lignin separation methods still have potential for commercialization, because they could produce lignin that contains more reactive functional groups and is less contaminated than traditional industrial lignin (Chakar & Ragauskas, 2004). Furthermore, modern technologies have increased the efficiencies of these lignin separation methods. With systematic optimization, these lignin separation methods could be scaled up for commercialization in the future.

20.3 Methods for commercial lignin production

Commercial lignin can be classified as Kraft lignin (from Kraft pulping), lignosulfonate (from sulfite pulping), organosolv lignin (from organosolv pulping), and enzymatic lignin (from the enzymatic ethanol-production process). Industrial lignin is mainly burnt as fuels to recover energy for biomass treatment (Korbag & Mohamed Saleh, 2016). Compared with lignin produced using laboratory methods, industrial pulping processes produce lignin with more severe degradation and higher heterogeneity (Brodin, 2009). Industrial lignin has lower molecular weights and higher dispersity indices than the lignin produced using laboratory methods. For example, the M_w and polydispersity index of MWL from spruce are approximately 23,500 g mol^{-1} and 3.7; on the other hand, those of Kraft lignin from spruce are 4500 g mol^{-1} and 4.5 (Anderson, Filpponen, Lucia, Saquing, et al., 2006; Brodin, 2009). This is because the pulping liquor used in industrial pulping processes could degrade and change the structure of lignin. As a cheap and sustainable material with abundant supplies, industrial lignin with proper modifications is widely used for various value-added applications (Chen & Li, 2000). In the following section, several lignin production methods and examples of value-added downstream commercialization are briefly introduced.

20.3.1 Lignosulfonate production

Lignosulfonates are lignin products generated in the sulfite pulping process. Sulfite pulping was the most commonly used pulping method until the 1940s when Kraft pulping was developed (Cohen, 1987). In a sulfite pulping process, biomass is cooked in fresh liquor containing sulfur dioxide and base solvents to dissolve lignin in biomass. After filtration, the lignin-free residues are screened, washed, and bleached to produce pulp. On the other hand, a basic oxide is added into the spent liquor for the precipitation of lignin. Lignosulfonate is collected after filtration and drying (Aro & Fatehi, 2017). Recycling units are applied in the system to recover SO_2 and the base solvent. In addition to calcium lignosulfonate, sodium/magnesium/ammonium lignosulfonates can be produced in the sulfite process using different base solvent pulping liquors and precipitation chemicals. It is reported that 60%−70% of lignin of corn cob can be recovered following the sulfite pulping (Aro & Fatehi, 2017).

Lignosulfonate has been widely used for binder applications. Borregaard, one of the world major suppliers of lignosulfonate products, has succeeded to convert lignosulfonates into a commercial binder, including Borresperse CA, Borrebond FP, and Norlig A products (Borregaard, 2020). Because of the thermal recalcitrance of lignin, these binders could also work on high-temperature scenario. In this content, Bornstein (1978) used lignosulfonate to produce a lignin-based binder. This binder is compatible with conventional binders for board productions and produces a board with superior stability even under high humidity and temperature. In addition to lignin, ethanol, and sulfite (such as $CaSO_3$ or Na_2SO_3) can be recovered and reused in a sulfite pulping to reduce the production cost. A major drawback of lignosulfonate is its high sulfur content (5–6 wt.%) (Korntner et al., 2018), which may generate an odor during its utilization.

20.3.2 Kraft lignin production

Globally, 85% lignin sold in the market is Kraft lignin, which is the main by-product of Kraft pulping (Chen & Chen, 2015). In the kraft pulping process, biomass is processed by white liquor (NaOH and Na_2S) to cleave the linkage between lignin and fibers at 150°C–170°C (Chakar & Ragauskas, 2004). After filtration, the lignin-free residues are used for pulp production; the spent liquor becomes lignin-enriched black liquor. Sulfuric acid or carbon dioxide is added into black liquor to precipitate Kraft lignin. The production of kraft lignin using the acidification method has been commercialized via LignoForce and LignoBoost technologies.

20.3.2.1 LignoForce technology

LignoForce system, a process developed by FPInnovations (Quebec, Canada) and NORAM (British Columbia, Canada), is designed for lignin recovery from black liquor (Kouisni, Gagné, Maki, Holt-Hindle, & Paleologou, 2014). Today, this technology was used on a commercial scale by West Fraser (Quebec, Canada) (Diels & Browne, 2018). In this system, there are several steps to purify kraft lignin as shown in Fig. 20.3A. Firstly, oxygen is injected into black liquor for oxidation. Then, carbon dioxide is sprayed into the oxidized black liquor for acidification at 70°C–75°C to reach a pH of 9.5–10.0 followed by a coagulation stage. The next step is the precipitation and filtration of the acidified black liquor. Liquor cake enriched in lignin achieved in this step is washed with sulfuric acid and water. LignoForce lignin is achieved after pressing and air drying (Kouisni, Holt-Hindle, Maki, & Paleologou, 2012). LignoForce could recover 60%–62% of lignin in the black liquor with lignin particle size ranging 5–10 μm. A lignin demonstration plant was built by FPInnovations at the Resolute Thunder Bay mill for the production of 12.5 kg h^{-1} of high-quality lignin (Diels & Browne, 2018).

LignoForce lignin has several advantages compared with conventional kraft lignin. LignoForce increases lignin slurry filtration rates without compromising lignin purity or affecting lignin structure. This process reduces the acid demand in precipitation and washing steps (Kouisni et al., 2012). It is also found that the ash content (0.1%–0.7%) is lower than conventional kraft lignin (0.2%–15%) (Kouisni et al., 2012). These facts make

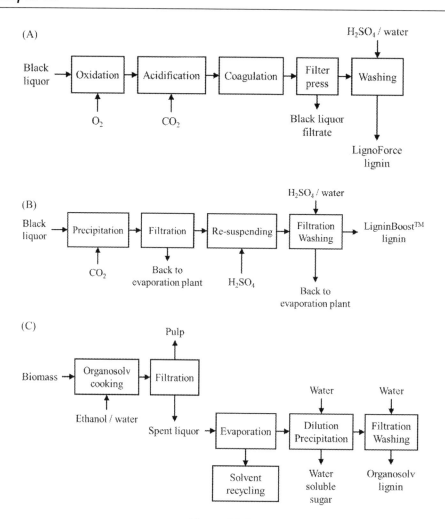

Figure 20.3
Schematic representations of LignoForce Lignin Protocol (A) LignoBoost Lignin Protocol (B), and organosolv pulping (C).

LignoForce lignin proper material for various applications, such as phenolic chemical production and biofuels. A drawback of LignoForce lignin is that the oxidation step may increase the production cost of lignin because of the requirement of oxygen.

20.3.2.2 LignoBoost technology

LignoBoost technology is developed by Chalmers University of Technology (Gothenburg, Sweden) and Innventia (Stockholm, Sweden), aiming to extract lignin from the black liquor of kraft pulping process (Barrett, 2014). This technology was commercialized by Domtar

Inc. in North Carolina, United States (Durruty, 2017; Kong, Wang, Price, Konduri, & Fatehi, 2015). Fig. 20.3B illustrates the protocols of LignoBoost lignin production. The black liquor from kraft pulping is acidified with carbon dioxide to reach a pH of 9.0 to precipitate lignin. After filtration, the obtained lignin is redispersed and acidified again with sulfuric acid. Then, a second filtration step is used to separate lignin from the above mixture. Finally, LignoBoost lignin is produced after acid/water washing and drying (Kong, Wang, et al., 2015). In this case, 75 t of lignin can be produced using LignoBoost lignin each day in a pilot plant (Barrett, 2014).

An advantage of this LignoBoost lignin production is that the pH and temperature of the liquid media of redispersed lignin are nearly equal to those of final washing liquid, thus reducing the concentration gradient during production (Zainab et al., 2018). LignoBoost lignin has been developed for a wide range of applications and markets, including fuels, resins, and biothermoplastics (Barrett, 2014).

20.3.2.3 SunCarbon lignin

To improve the lignin quality from black liquor of kraft pulping process, SunCarbon (Sweden) developed a systematic procedure for the production of lignin oil from biomass (https://suncarbon.se/en/technology/). SunCarbon plans to complete a lignin oil extraction plant at the beginning of 2022 with the capability to produce about 45,000 t of lignin oil per year (Sherrard, 2019). Notably, the feedstock of the system is black liquor obtained from existing pulp mills. Lignin is firstly separated from black liquor via membrane filtration. Afterward, a hydrothermal treatment is applied to process the lignin-rich retentate stream (https://suncarbon.se/en/technology/). During the hydrothermal treatment, the lignin-rich retentate stream is treated in a thermal catalytic cracker at subcritical conditions. Afterward, further purification steps are applied to remove its ash. After additional desalting, the product is a depolymerized lignin-oil, ready for various value-added applications (https://www.hulteberg.com/hulteberg-suncarbon/). In a SunCarbon process, up to 25% of lignin could be recovered from black liquor after a series of purifications.

Compared with conventional pulping lignin, SunCarbon lignin oil is a high-grade lignin product that contains less sugar, ash, and sulfur contamination due to systematic filtration and purification (Sherrard, 2019), and is thus more environmentally friendly for downstream applications, especially for biofuel production. In a SunCarbon process the pulping liquor and catalysts can be recycled through membrane filtration and reused to reduce the production cost. However, the quality of SunCarbon lignin is greatly affected by upstream black liquor production in pulp mills. Furthermore, SunCarbon lignin is expensive in terms of production cost because a series of purification steps are applied for improving the lignin quality (https://suncarbon.se/en/technology/). Another drawback of SunCarbon lignin is its low production yield, which is approximately only 10%−25% of all the lignin produced in the pulp and paper mill (https://suncarbon.se/en/technology/). Because of the above drawbacks, SunCarbon lignin seems not to be economically attractive for value-added applications.

Kraft lignin is a suitable feedstock for various applications and biomaterial production including fertilizers (García, Diez, Vallejo, Garcia, & Cartagena, 1996), plastics (Li & Sarkanen, 2003), binders (Olivares, Guzman, Natho, & Saavedra, 1988), and carbon fibers (Sagues et al., 2019). Ramirez et al. (1997) developed a method to produce sustainable lignin-based multipurpose fertilizers by ammoniation of kraft lignin. In 2007 Eckert & Abdullah (Eckert & Abdullah, 2010) proposed a method for the production of carbon fibers from kraft softwood lignin. Araújo (2008) has investigated the feasibility of vanillin production from kraft lignin and optimized the production system. Considering the increasing global demand for vanillin, kraft lignin, due to its vast availability, has potential for the production of vanillin. However, similar to lignosulfonate, kraft lignin is a sulfur-bearing material. The sulfur content in kraft lignin (1.5−3 wt.%) (Svensson, 2008) may undermine its downstream applications. Compared with other types of industrial lignin, the molecular weight of kraft lignin is relatively low due to its degradation during pulping (Lange, Decina, & Crestini, 2013). This can be a drawback for certain applications, for example, lignin-based flocculants that need large polymers to facilitate the flocculation process.

20.3.3 Organosolv and soda lignin

In organosolv and soda pulping, biomass is cleaved through cooking using sulfur-free pulping liquors. Fig. 20.3C is a brief flow diagram of an organosolv process using ethanol/water pulping liquor. Biomass is cooked using organic solvents, such as acetone and ethanol/water (JhonDutton e-Education, 2021). Then, filtration is applied to obtain the lignin-free residues. The pulp is obtained from the produced solid residues after washing with ethanol and water. On the other hand, the spent liquor after filtration is collected for organosolv lignin production. The organic solvent is recycled through evaporation to reduce the chemical demand (Pan et al., 2006). Then, water is added into the liquor for dilution and precipitation. Organosolv lignin is produced after filtration, washing with water, and drying (da Rosa et al., 2017). In addition, some water-soluble sugar is obtained after precipitation. It is reported that the yield of organosolv lignin is approximately 20% of the dry original biomass feedstock (JhonDutton e-Education, 2021).

The feedstock of soda pulping is usually nonwood biomass, such as straw and bagasse (Doherty & Rainey, 2006). The biomass is cooked using NaOH solvent at 150°C−170°C with anthracenedione and aluminum oxide additives. A filtration process is used to obtain the lignin-free residue for pulp production. The black liquor is blown by flue gas (carbon dioxide) to produce silica gels because feedstock of soda pulping has a high silica content (Doherty & Rainey, 2006). Then, sulfuric acid is added into the black liquor for lignin precipitation at 70°C−80°C (Sameni, Krigstin, dos Santos Rosa, Leao, & Sain, 2014). After filtration, the filtrate is collected for chemical recovery to reduce the production cost. The filtered residues are washed with water and dried to produce soda lignin, which is used for energy recovery and other utilizations. The yield of recovered soda lignin is approximately 4%−10% of the dry original biomass feedstock (Doherty & Rainey, 2006).

As sulfur-free materials, organosolv and soda lignin are environmentally friendly and do not produce odors (Mandlekar, Cayla, & Rault, 2018). Because of less destructive procedures, the structure of sulfur-free lignin is closer to native lignin compared with that of sulfur-bearing lignin. Furthermore, organosolv lignin is the purest lignin among all industrial lignin (Ahmad & Pant, 2018). The low molecular weight of sulfur-free lignin makes it a potential source for platform phenolic compound production. In addition, sulfur-free lignin is a suitable material for resin production. Cook and Hess (1991) developed a method to produce organosolv lignin-phenol aldehyde resin. Due to the unique properties of organosolv lignin, this resin is superior to conventional lignin resin in that it is low in sulfate, ash, and sulfur residues, and thus does not produce harmful sulfur odors during resin preparation or hot-pressing of the adhesive and is unaffected and insolubilized by water. Feng et al. (2016) proposed an approach to produce wood adhesive using soda lignin, which is proven to be environmentally green and a remarkable low cost. In addition to lignin, pulping liquor, such as ethanol, acetone, and methanol, can be recovered, which can be further used as solvents for various processes. A major drawback of organosolv lignin is its relatively high production cost because of the demand for the large excess of organic solvent in the pulping process (Shrotri, Kobayashi, & Fukuoka, 2017). For soda lignin, the high silica and ash contents undermine its downstream application (Mousavioun & Doherty, 2010).

20.3.4 Thermomechanical pulp-bio lignin

FPInnovations (Quebec, Canada) has developed a proprietary TMP-Bio for cellulosic biomass conversion to produce fermentable sugars and hydrolysis lignin (H-lignin) on industrial scales (FPInnovations, 2019). Biomass is transferred to a feed bin and mixed with steam and caustic in mixing screw, and later the mixture is transferred to a pretreatment tower and cooked (Mao et al., 2017). After cooking, the slurry is neutralized and transferred to the screw press, where most of the water is removed from slurry. The biomass is enzymatically hydrolyzed in the reactor. Sugars that are generated during enzymatic hydrolysis are separated from solid biomass (H-lignin) via filtration and later used for fermentable sugar production, like glucose and xylose. On the other hand, hydrolysis lignin (H-lignin) that is separated during filtration is dried using a ring dyer (Mao et al., 2017). It is estimated that in a TMP-Bio process, 1000 kg biomass feedstock could produce 400 kg lignin (40%) and 500 kg sugar products (50%) (FPInnovations, 2019).

Various value-added products can be developed from TMP-Bio lignin, such as activated carbon, resins, and additives (Mao et al., 2017). Worth noting, TMP-Bio lignin is a sulfur-free material that facilitates its utilization in different applications, such as adhesive and biofuel production (Office of Energy Efficiency & Renewable Energy, 2016). The color of traditional pulping lignin is often dark brown or black; however, TMP-Bio lignin has a light yellow color, making it attractive for some applications, such as construction adhesive (Mao et al., 2017). Another advantage is that its properties are close to native lignin because the relatively low-temperature (90°C) cooking is applied in this process. As a biorefining process, the TMP-Bio technology could also produce fermentable sugars and value-added chemicals, such as succinic acid, lactic acid, butanol, ethanol, activated carbon, resins, and additives. The drawback of TMP-Bio is probably its high production cost due to a large demand for various chemicals, including sodium hydroxide, sulfuric acid, enzyme, buffer solution, and water. The energy consumption of TMP-Bio lignin production may be relatively large as there are several heating and cooling cycles in this method (Mao et al., 2017).

20.4 Opportunities and challenges in the commercialization of lignin production

Globally, the production of bio-based chemicals and polymers is estimated to be around 50 million tons per year (https://www.marketwatch.com/press-release/bio-based-chemicals-market-2018-2028-by-regional-analysis-trends-demand-and-leading-companies-2020-01-23). However, most of them are still fossil based. The commercial production of bio-based chemicals could hardly compete with the existing fossil products due to well-grown fossil refineries. However, the fluctuation in the cost of the fossil products and the rising demand for eco-friendly products has boosted the interest in sustainable materials and chemicals derived from renewable resources (Azadi, Inderwildi, Farnood, & King, 2013). In 2007 the US Department of Energy reported the results of an evaluation of opportunities to convert lignin into energy, macromolecules, or chemicals such as methanol, cyclohexane, styrene, and phenol among others. Lignin plays a significant role in increasing the commercial viability of a biorefinery since it constitutes up to 30% of the weight and 40% of the fuel value of biomass (Strassberger, Tanase, & Rothenberg, 2014). However, technologies to convert lignin to macromolecules and aromatic chemicals are under development and represent long-term opportunities.

Currently, the commercially proven technologies for use of lignin do not match the quantity of lignin production (Mordor Intelligence LLP, 2019), implying that the production and extraction of lignin overbalance the applications of value-added lignin-based products. Moreover, the quality of the lignin-based products used in well-established fields of the industry, for example, adhesives, adsorbents, and dispersants, still needs improvement to compete against the existing commercial products (Smolarski, 2012). Therefore, it is essential to develop various applications utilizing technical lignin-based products that can be exploited on an industrial scale. With this aim, various modification reactions have been applied to lignin to improve its overall characteristics to be used in industry, such as sulfomethylation for coal–water slurry and concrete admixture; sulfobutylation for coal–water slurry; carbendazim, carboxymethylation for oil–water emulsions, crude bitumen emulsion, clay, cement, and graphite suspensions; and halogenation for surfactant, animation of cationic surfactants, and cationic asphalt emulsifier productions (Ge, Song, & Li, 2015; Pang, Qiu, Yang, & Lou, 2008). However, the interaction of lignin with organic substances, such as bacteria, antibiotics, proteins, and viruses could be investigated for increasing the number of lignin-based biomedical applications. Using lignin in wound dressing, drug-delivery systems, tissue engineering, and pharmaceutical fields could also help balance the production of lignin with its applications. However, extensive and detailed studies on the toxicity analysis of lignin are required for succeeding with such applications for lignin (Kazzaz, Feizi, & Fatehi, 2019). Opportunities in this field arise sidelong the production of high-quality lignin-based materials that can partially or fully replace fossil-based products. In addition, economic factors relevant to lignin valorization as a part of integrated biorefining processes and its technical problems are barriers in the conversion of lignin into value-added products. The unique structural features of technical lignin as well as the delignification methods, which define its characteristics, are the reasons accounting for the lignin utilization problems (Vishtal & Kraslawski, 2011).

The available lignin sources in the industry vary in structure and characteristics; some low-quality lignin fractions from a biorefinery are suitable for only some specific applications (Narron, Kim, Chang, Jameel, & Park, 2016). Separation of lignin from lignocellulosic biomass can be considered as the main challenge due to the wide distribution of bonds in the various C–O and C–C linkages in the structure of lignin (Jiang et al., 2015). In the case of some lignocellulosic materials and delignification methods, low-molecular-weight lignin fractions have the tendency to undergo condensation reactions that causes lignin to lose its reactive functional groups. Worth noting, such challenges will be substantially diminished as the production in plant is engineered to achieve more reactive lignin with more functional groups for different utilizations (Ragauskas et al., 2014), for example, ammoniated lignin for soil fertilizer (Kouisni et al., 2014) and sulfur-free organosolv lignin for odor-free binder production (Mandlekar et al., 2018).

In addition to energy production, lignin is considered in the production of dispersants, emulsifiers, aromatics, and value-added polymers for its valorization (Dhiman, 2021; Lou & Wu, 2011). With the development of lignin valorization technologies, the quality and value of lignin-based products are continuously improved, making lignin economically more attractive for biorefining. To produce carbohydrate-derived fuels at an industrial scale, the upcoming generation of biorefineries will induce large quantities of lignin as a by-product. Based on the recent economic analysis, lignin valorization has the potential for playing a major role in the overall economic viability of biorefineries (Araujo, Grande, & Rodrigues, 2010; Davis et al., 2013; Vardon et al., 2015). There are also research reports stating that potential methods for the conversion of lignin to value-added chemicals could improve overall economics and sustainability for integrated biorefineries (Davis et al., 2013). For example, the overall economy of a process has been improved and its carbon cost is potentially reduced via the conversion of lignin into 1,4-butanediol and adipic acid according to relevant studies (Araujo et al., 2010; Vardon et al., 2015). Similarly, the coproduction of lignin-based transportation fuel could improve the overall economic viability of a biorefining process for corn ethanol production (Shen, Tao, & Yang, 2019). The production of lignin-based flocculant is also a promising pathway to raise value-added applications for lignin (Couch, Price, & Fatehi, 2016; He, Zhang, & Fatehi, 2016; Kong, Parhiala, Wang, & Fatehi, 2015; Wang, Kong, Fatehi, & Hou, 2018; Wang, Kong, Gao, & Fatehi, 2018). These facts will catch more interest of governments and companies to invest in the commercialization of lignin-based products.

Current macromolecule applications may still be the main commercial uses of lignin, whereas its main potential lies within aromatics. The depolymerization processes of lignin that define the characteristics of aromatics still need development to increase the product yield and selectivity (Pandey & Kim, 2011). The available lignin depolymerization technologies allow only to produce the target chemicals with adequate functional groups. Moreover, a consequent amount of unwanted chemicals with undesired characteristics that must be discarded and a low yield of the products prove that the extent of lignin utilization still needs an improvement (Chen & Wan, 2017). The technology improvement targeting yield, specificity, and costs of the processes appear to be the fastest solution in this regard, though substantial research and efforts are required (Vishtal & Kraslawski, 2011).

20.5 Conclusions and perspectives

This chapter was focused on lignin biosynthesis, lignin production, and its potential applications. The biosynthesis of lignin has three major steps, including the formation of monolignols, their radicalization, and the combining of the formed radicals. Among the various types of interlinkages in lignin, $\beta-O-4$ bond is the most common and abundant linkage. Several methods for lignin production were briefly introduced. Nondestructive milling methods, including MWL, CEL, and EMAL methods, could produce lignin with its structure similar to that of native lignin. Due to an increasing interest in lignin and its utilization, several novel lignin extraction technologies have been developed recently, and many of them have been successfully scaled up to pilot scales, such as a steam explosion, SLRP, Bio-Flex, German lignocellulose feedstock biorefinery, Proesa, and AFEX processes. With further optimization, these processes could be further scaled for industrial production in the future. In addition, several commercial lignin production methods are summarized, including LignoForce, LignoBoost, SunCarbon, and TMP-Bio lignin processes. Currently, the production and value-added application of lignin still face many challenges; however, there are many opportunities for lignin valorization because of the increasing demand for sustainable lignin-based products, such as lignin flocculants, binders, dispersants, and biofuels.

Acknowledgments

The authors would like to thank NSERC, Canada Research Chairs, and the Ontario Centre of Excellence for supporting this research.

References

Abbati de Assis, C., Greca, L. G., Ago, M., Balakshin, M. Y., Jameel, H., Gonzalez, R., & Rojas, O. J. (2018). Techno-economic assessment, scalability, and applications of aerosol lignin micro- and nanoparticles. *ACS Sustainable Chemistry & Engineering, 6*, 11853–11868.

Ahmad, E., & Pant, K. K. (2018). Lignin conversion: A key to the concept of lignocellulosic biomass-based integrated biorefinery. *Waste Biorefinery*, 409–444.

Alder, E. (1977). Lignin chemistry—Past, present and future. *Wood Science and Technology, 1*, 169–218.

Allan, G. G. (1971). Modification reactions. In K. V. Sarkanen, & C. H. Ludwig (Eds.), *Lignins: Occurrence, formation, structure and reactions* (pp. 511–573). New York: Wiley-Interscience.

Anderson, G., Filpponen, I., Lucia, L. A., & Argyropoulos, D. S. (2006). Comparative evaluation of three lignin isolation protocols for various wood species. *Journal of Agricultural and Food Chemistry, 54*, 9696–9705.

Anderson, G., Filpponen, I., Lucia, L. A., Saquing, C., Baumberger, S., & Argyropoulos, D. S. (2006). Toward a better understanding of the lignin isolation process from wood. *Journal of Agricultural and Food Chemistry, 54*, 5939–5947.

Araújo, J. D. P. (2008). *Production of vanillin from lignin present in the Kraft black liquor of the pulp and paper industry* (Doctoral dissertation). Porto: University of Porto.

Araujo, J. D. P., Grande, C. A., & Rodrigues, A. E. (2010). Vanillin production from lignin oxidation in a batch reactor. *Chemical Engineering Research and Design, 88*, 1024–1032.

Aro, T., & Fatehi, P. (2017). Production and application of lignosulfonates and sulfonated lignin. *ChemSusChem, 10*, 1861–1877.

Azadi, P., Inderwildi, O. R., Farnood, R., & King, D. A. (2013). Liquid fuels, hydrogen and chemicals from lignin: A critical review. *Renewable & Sustainable Energy Reviews, 21*, 506–523.

Barrett, A. (2014). *Domtar starts LignoBoost lignin separation based on Metso Technology.* https://bioplasticsnews.com/2014/05/19/domtar-starts-up-commercial-scale-lignoboost-lignin-separation-plant-based-on-metsos-technology/.

Bioeconomy Institute. (2015). *A patent and milestone: Why MBI's AFEX biomass pretreatment is of global importance.* https://bioeconomy.msu.edu/2015/09/25/a-patent-and-milestone-why-mbis-afex-biomass-pretreatment-is-of-global-importance/.

Biomass Technology Group. (2021). *Bio-materials & chemicals.* https://www.btgworld.com/en/rtd/technologies/bio-materials-chemicals.

Boerjan, W., Ralph, J., & Baucher, M. (2003). Lignin biosynthesis. *Annual Review of Plant Biology, 54*, 519–546.

Bornstein, L. F. (Inventor; Georgia Pacific LLC, assignee). (1978). Lignin-based composition board binder comprising a copolymer of a lignosulfonate, melamine and an aldehyde. *U.S. Patent Application No. 4130515*.

Borregaard. (2020). *Sustainable lignin products. Agricultural chemicals.* https://www.lignotech.com/IndustrialApplications/Agriculture/Agricultural-Chemicals.

Branco, R. H. R., Sarafim, L. S., & Xavier, A. M. R. B. (2019). Second generation bioethanol production: On the use of pulp and paper industry wastes as feedstock. *Fermentation, 5*, 1−30.

Brodin, I. (2009). *Chemical properties and thermal behaviour of kraft lignins* (Doctoral dissertation). Stockholm: KTH.

Chakar, F. S., & Ragauskas, A. J. (2004). Review of current and future softwood kraft lignin process chemistry. *Industrial Crops and Products, 20*, 131−141.

Chang, H., Cowling, E. B., & Wynford, B. (1975). Comparative studies on cellulolytic enzyme lignin and milled wood lignin of sweetgum and spruce. *Holzforschung, 29*, 153−159.

Chen, F., & Li, J. (2000). Aqueous gel permeation chromatographic methods for technical lignins. *Journal of Wood Chemistry and Technology, 20*, 265−276.

Chen, H., & Chen, H. (2015). *Lignocellulose biorefinery feedstock engineering. Lignocellulose biorefinery engineering* (1st ed., pp. 37−86). Cambridge: Woodhead Publishing.

Chen, Z., & Wan, C. (2017). Biological valorization strategies for converting lignin into fuels and chemicals. *Renewable and Sustainable Energy Reviews, 73*, 610−621.

Cohen, A. J. (1987). Factor substitution and induced innovation in North American kraft pulping: 1914−1940. *Explorations in Economic History, 24*, 197−217.

Cook, P. M., & Hess, S. L. (Inventors; Eastman Kodak Co, assignee). (1991). Organosolv lignin-modified phenolic resins and method for their preparation. *U.S. Patent Application No. 5010156*.

Couch, R. L., Price, J. T., & Fatehi, P. (2016). Production of flocculant from thermomechanical pulping lignin via nitric acid treatment. *ACS Sustainable Chemistry & Engineering, 4*, 1954−1962.

da Rosa, M. P., Beck, P. H., Müller, D. G., Moreira, J. B., da Silva, J. S., & Durigon, A. M. M. (2017). Extraction of organosolv lignin from rice husk under reflux conditions. *Biological and Chemical Research, 87*−98.

Damen, K. J., Smit, A. T., Huijgen, W. J. J., & Van, H. J. W. (2017). *Fabiola: Fractionation of biomass using low-temperature acetone. 13th International conference on renewable resources & biorefineries*. Petten: ECN.

Davis, R., Tao, L., Tan, E. C., Biddy, M. J., Beckham, G. T., Scarlata, C., et al. (2013). Process design and economics for the conversion of lignocellulosic biomass to hydrocarbons: Dilute-acid and enzymatic deconstruction of biomass to sugars and biological conversion of sugars to hydrocarbons. *Report No. NREL/TP-5100-60223*. Golden, CO: National Renewable Energy Lab.

Dean, J. F. D., & Erikson, K. E. (1994). Laccase and the deposition of lignin in vascular plants. *Holzforschung, 48*, 21−33.

Dhiman, G. (2021). *Lignin biorefinery: An effective biomass conversion to value added product*. http://ttconsultants.com/blog/lignin-biorefinery-an-effective-biomass-conversion-to-value-added-product/.

Diels, L., & Browne, T. (2018 September 18−20). *Lignin international conference*. Edmonton, Alberta. http://www.paptac.ca/communiques/communiquepaptaccentral-lignin2018keynote2.html.

Doherty, W. O. S., & Rainey, T. (2006, May 2−5). Bagasse fractionation by the soda process. In *Proceedings of the Australian Society of Sugar Cane Technologists 2006*. Mackay, Queensland, Australia.

Durruty, J. (2017). *On the local filtration properties of LignoBoost lignin: Studies of the influence of xylan and ionic strength* (Doctoral dissertation). Gothenburg: Chalmers University of Technology.

Eckert, R. C., & Abdullah, Z. (Inventors; Weyerhaeuser NR Co, assignee). (2010). Carbon fibers from kraft softwood lignin. *U.S. Patent Application No. 7678358*.

Ewellyn, C., Balakshin, M., Katahira, R., Chang, H., & Jameel, H. (2015). How well do MWL and CEL preparations represent the whole hardwood lignin? *Journal of Wood Chemistry and Technology, 35*, 17−26.

Feng, M. W., He, G., Zhang, Y., Wang, X., Kouisni, L., & Paleologou, M. (Inventors; FPInnovations, assignee). (2016). High residual content (HRC) kraft/soda lignin as an ingredient in wood adhesives. *U.S. Patent Application No. 15/130,107*.

Filtration + Separation. (2019). *BioFlex uses waste wood instead of crude oil*. https://www.filtsep.com/oil-and-gas/news/bioflex-uses-waste-wood-instead-of-crude-oil/.

FPInnovations. (2019). *FPInnovations and Resolute inaugurate thermomechanical pulp bio-refinery in Thunder Bay*. http://blog.fpinnovations.ca/blog/2019/05/27/fpinnovations-and-resolute-inaugurate-thermomechanical-pulp-bio-refinery-in-thunder-bay/.

García, M. C., Diez, J. A., Vallejo, A., Garcia, L., & Cartagena, M. C. (1996). Use of kraft pine lignin in controlled-release fertilizer formulations. *Industrial & Engineering Chemistry Research, 35*, 245–249.

Ge, Y., Song, Q., & Li, Z. (2015). A Mannich base biosorbent derived from alkaline lignin for lead removal from aqueous solution. *Journal of Industrial and Engineering Chemistry, 23*, 228–234.

Glasser, G. W., & Sarkanen, S. (1989). *Lignin, properties and materials* (pp. 11–28). Washington, DC: American Chemical Society.

Glasser, W. G. (1980). Lignin. In J. P. Casey (Ed.), *Pulp and paper: Chemistry and chemical technology* (pp. 39–111). New York: Wiley-Interscience.

Gougoulias, C., Clark, J. M., & Shaw, L. J. (2014). The role of soil microbes in the global carbon cycle: tracking the below-ground microbial processing of plant-derived carbon for manipulating carbon dynamics in agricultural systems. *Journal of the Science of Food and Agriculture, 94*, 2362–2371.

Green Car Congress. (2012). *POET-DSM contracts with ANDRITZ for biomass pre-treatment process for cellulosic ethanol plant*. https://www.greencarcongress.com/2012/10/andritz-20121025.html.

Greenovate! Europe. (2019). *BTG commissions world-first fast pyrolysis biorefinery pilot plant*. https://cordis.europa.eu/article/id/125435-btg-commissions-worldfirst-fast-pyrolysis-biorefinery-pilot-plant.

Guadix-Montero, S., & Sankar, M. (2018). Review on catalytic cleavage of C–C-inter-unit linkages in lignin model compounds: Towards lignin depolymerisation. *Topics in Catalysis, 61*, 183–198.

He, W., Zhang, Y., & Fatehi, P. (2016). Sulfomethylated kraft lignin as a flocculant for cationic dye. *Colloids and Surfaces A: Physicochemical and Engineering Aspects, 503*, 19–27.

Heeres, H. (2019). *Thermo-chemical fractionation (TCF) of lignocellulosic biomass*. BTG Biomass Technology Group. https://bio4products.eu/wp-content/uploads/2019/11/4.-Hans-Heeres_Thermo-Chemical-Fractionation.pdf.

Heitner, C., Dimmel, D. R., & Schmidt, J. A. (2010). *Lignin and lignans: Advances in chemistry* (pp. 1–7). Boca Raton, FL: CRC Press.

Higuchi, T. (1997). *Biochemistry and molecular biology of wood*. Berlin: Springer-Verlag.

Higuchi, T., Shimada, M., Nakatsubo, F., & Tanahashi, M. (1977). Differences in biosyntheses of guaiacylandsyringyl lignins in woods. *Wood Science and Technology, 11*, 153–167.

Holtzapple, M. T. (2003). Lignin. In B. Caballero, L. C. Trugo, & P. M. Finglas (Eds.), *Encyclopedia of food science and nutrition* (2nd ed., pp. 3535–3542). .

Ibarra, D., José, C., Gutiérrez, A., Rodríguez, I. M., Romero, J., Martínez, M. J., & Martínez, Á. T. (2005). Chemical characterization of residual lignins from eucalypt paper pulps. *Journal of Analytical and Applied Pyrolysis, 74*, 116–122.

Ismail, S. Z., Khandaker, M., Mat, N., & Boyce, A. N. (2015). Effects of hydrogen peroxide on growth, development and quality of fruits: A review. *Journal of Agronomy, 14*, 331–336.

Janusz, G., Pawlik, A., Sulej, J., Swiderska-Burek, U., Jarosz-Wilkolazka, G., & Paszczynski, A. (2017). Lignin degradation: Microorganisms, enzymes involved, genomes analysis and evolution. *FEMS Microbiology Reviews, 41*, 941–962.

Jhon A. Dutton e-Education Institute. (2021). *5.3e: Organic solvation processes*. https://www.e-education.psu.edu/egee439/node/658.

Jiang, Y., Li, Z., Tang, X., Sun, Y., Zeng, X., Liu, S., & Lin, L. (2015). Depolymerization of cellulolytic enzyme lignin for the production of monomeric phenols over Raney Ni and acidic zeolite catalysts. *Energy Fuels, 29*, 1662–1668.

Kazzaz, A. E., Feizi, Z. H., & Fatehi, P. (2019). Grafting strategies for hydroxy groups of lignin for producing materials. *Green Chemistry: An International Journal and Green Chemistry Resource, 21*, 5714–5752.

Kihlman, J. (2015). Acid precipitation lignin removal processes integrated into a kraft mill. In *The 6th Nordic wood biorefinery conference (NWBC) 2015* (pp. 402–410). Helsinki, Finland: Julkaisija-Utgivare.

Kirk, T. K., Higuchi, T., & Chang, H. (1980a). *Lignin biodegradation: Microbiology, chemistry, and potential application* (Vol. 1, pp. 1–231). Boca Raton, FL: CRC Press Inc.

Kirk, T. K., Higuchi, T., & Chang, H. (1980b). *Lignin biodegradation: Microbiology, chemistry, and potential application* (Vol. 2, pp. 1–251). Boca Raton, FL: CRC Press Inc.

Klinger, K. M., Liebner, F., Fritz, I., Potthast, A., & Rosenau, T. (2013). Formation and ecotoxicity of *N*-heterocyclic compounds on ammoxidation of mono-and polysaccharides. *Journal of Agricultural and Food Chemistry, 61*, 9004−9014.

Kong, F., Parhiala, K., Wang, S., & Fatehi, P. (2015). Preparation of cationic softwood kraft lignin and its application in dye removal. *European Polymer Journal, 67*, 335−345.

Kong, F., Wang, S., Price, J. T., Konduri, M. K., & Fatehi, P. (2015). Water soluble kraft lignin-acrylic acid copolymer: Synthesis and characterization. *Green Chemistry: An International Journal and Green Chemistry Resource, 17*, 4355−4366.

Korbag, I., & Mohamed Saleh, S. (2016). Studies on mechanical and biodegradability properties of PVA/lignin blend films. *The International Journal of Environmental Studies, 73*, 18−24.

Korntner, P., Schedl, A., Sumerskii, I., Zweckmair, T., Mahler, A. K., Rosenau, T., & Potthast, A. (2018). Sulfonic acid group determination in lignosulfonates by headspace gas chromatography. *ACS Sustainable Chemistry & Engineering, 6*, 6240−6246.

Kouisni, L., Gagné, A., Maki, K., Holt-Hindle, P., & Paleologou, M. (2014). LignoForce system for the recovery of lignin from black liquor: Feedstock options, odor profile, and product characterization. *ACS Sustainable Chemistry & Engineering, 4*, 5152−5159.

Kouisni, L., Holt-Hindle, P., Maki, K., & Paleologou, M. (2012). The lignoforce system: A new process for the production of high-quality lignin from black liquor. *Journal of Science & Technology for Forest Products and Processes, 2*, 6−10.

Koukol, J., & Conn, E. (1962). Purification and properties of the phenylalanine deaminase of *Hordeum vulgare*. *The Journal of Biological Chemistry, 236*, 2692−2698.

Kun, W., Bauer, S., & Sun, R. (2012). Structural transformation of *Miscanthus × giganteus* lignin fractionated under mild formosolv, basic organosolv, and cellulolytic enzyme conditions. *Journal of Agricultural and Food Chemistry, 60*, 144−152.

Lake, M. A., & Blackburn, J. C. (2014). SLRP—An innovative lignin-recovery technology. *Cellulose Chemistry and Technology, 48*, 799−804.

Lange, H., Decina, S., & Crestini, C. (2013). Oxidative upgrade of lignin—Recent routes reviewed. *European Polymer Journal, 49*, 1151−1173.

Laurichesse, S., & Averous, L. (2014). Chemical modification of lignins: Towards biobased polymers. *Progress in Polymer Science, 39*, 1266−1290.

Li, M., Pu, Y., & Ragauskas, A. J. (2016). Current understanding of the correlation of lignin structure with biomass recalcitrance. *Frontiers in Chemistry, 4*, 1−8.

Li, Y., & Sarkanen, S. (2003). *Biodegradable Kraft lignin-based thermoplastics. Biodegradable polymers and plastics* (pp. 121−139)). Springer.

Lou, R., & Wu, S. (2011). Products properties from fast pyrolysis of enzymatic/mild acidolysis lignin. *Applied Energy, 88*, 316−322.

Lourenco, A., Rencoret, J., Chemetova, C., Gominho, J., Gutierrez, A., del Rio, J. C., & Pereira, H. (2016). Lignin composition and structure differs between xylem, phloem and phellem in *Quercus suber* L. *Frontiers in Plant Science, 7*, 1−14.

Mandlekar, N., Cayla, A., Rault, F., Giraud, S., Salaün, F., Malucelli, G., & Guan, J. P. (2018). *An overview on the use of lignin and its derivatives in fire retardant polymer systems. Lignin—Trends and applications* (pp. 207−231)). InTech Open.

Mao, C., Yuan, Z., Fernando, D. R. L., Wafa, A. D. W., Wong, D., & Browne, T. (2017). TMP-Bio for converting cellulosic biomass to 2nd generation sugar and near-native lignin. In *Proceedings of international forest biorefining conference (IFBC)*. Thunder Bay, Ontario, Canada. https://ifbc2017.files.wordpress.com/2017/02/fpi-ifbc-2017.pdf.

Michels, J., & Wagemann, K. (2010). The German lignocellulose feedstock biorefinery project. *Biofuel, Bioproducts, Biorefining, 4*(3), 263−267. Available from https://www.yumpu.com/en/document/read/41220925/the-german-lignocellulose-feedstock-biorefinery-project.

Moniruzzaman, M., Dale, B. E., Hespell, R. B., & Bothast, R. J. (1997). Enzymatic hydrolysis of high-moisture corn fiber pretreated by AFEX and recovery and recycling of the enzyme complex. *Applied Biochemistry and Biotechnology, 67*, 113–126.

Mordor Intelligence LLP. (2019). Kraft lignin products market: CAGR 7.30% by 2024—New report by mordor intelligence, https://www.prnewswire.com/in/news-releases/kraft-lignin-products-market-cagr-7-30-by-2024-new-report-by-mordor-intelligence-807091727.html.

Mousavioun, P., & Doherty, W. O. S. (2010). Chemical and thermal properties of bagasse soda lignin. *Industrial Crops and Products, 31*(1), 52–58.

Narron, R. H., Kim, H., Chang, H. M., Jameel, H., & Park, S. (2016). Biomass pretreatments capable of enabling lignin valorization in a biorefinery process. *Current Opinion in Biotechnology, 38*, 39–46.

Neish, A. C. (1961). Formation of *m*- and *p*-coumaric acids by enzymatic desamination of corresponding isomers of tyrosine. *Photochemistry, 1*, 1–24.

Office of Energy Efficiency & Renewable Energy. (2016). *Request for information (RFI) DE-FOA-0001615: Cellulosic sugar and lignin production capabilities RFI responses*. https://www.energy.gov/sites/prod/files/2016/10/f33/fpinnovations_point_claire_de_foa_0001615_lignin.pdf.

Olivares, M., Guzman, J. A., Natho, A., & Saavedra, A. (1988). Kraft lignin utilization in adhesives. *Wood Science and Technology, 22*, 157–165.

Palmisano P. (2013). Lignin conversion into bio-based chemicals. In *ECI symposium series, BioEnergy IV: Innovations in biomass conversion for heat, power, fuels and chemicals proceedings*. Basiliani Resort, Otranto, Italy.

Pan, X., Gilkes, N., Kadla, J., Pye, K., Saka, S., Gregg, D., . . . Saddler, J. (2006). Bioconversion of hybrid poplar to ethanol and co-products using an organosolv fractionation process: Optimization of process yields. *Biotechnology and Bioengineering, 94*, 851–861.

Pandey, M. P., & Kim, C. S. (2011). Lignin depolymerization and conversion: A review of thermochemical methods. *Chemical Engineering and Technology, 34*, 29–41.

Pang, Y. X., Qiu, X. Q., Yang, D. J., & Lou, H. M. (2008). Influence of oxidation, hydroxymethylation and sulfomethylation on the physicochemical properties of calcium lignosulfonate. *Colloids and Surfaces A: Physicochemical and Engineering Aspects, 312*, 154–159.

Payane, A. (1838). Memoire sur la composition du tissue propre des plantes et du ligneux. *ComptesRendus, 7*, 1052–1056.

Pearl, I. A. (1967). *The chemistry of lignin* (pp. 136–148). New York: Marcel Dekker.

Phillips, R., Jameel, H., & Chanq, H. M. (2013). Integration of pulp and paper technology with bioethanol production. *Biotechnology for Biofuels, 6*, 1–12.

POET-DSM Advanced Biofuels, LLC. (2012). *POET-DSM makes major technology, process purchase for commercial cellulosic bio-ethanol*. https://poet.com/pr/poet-dsm-makes-major-technology-process-purchase-for-cellulosic-bio-ethanol.

Pucciariello, R., Bonini, C., D'Auria, M., Villani, V., Giammarino, G., & Gorrasi, G. (2008). Polymer blends of steam-explosion lignin and poly(ε-caprolactone) by high-energy ball milling. *Journal of Applied Polymer Science, 109*, 309–313.

Ragauskas, A. J., Beckham, G. T., Biddy, M. J., Chandra, R., Chen, F., Davis, M. F., . . . Wyman, C. E. (2014). Lignin valorization: Improving lignin processing in the biorefinery. *Science, 344*, 1246843.

Ramirez, F., González, V., Crespo, M., Meier, D., Faix, O., & Zúñiga, V. (1997). Ammoxidized kraft lignin as a slow-release fertilizer tested on *Sorghum vulgare*. *Bioresource Technology, 61*, 43–46.

Rui, L., Wu, S., Lv, G., & Guo, D. (2010). Pyrolytic products from rice straw and enzymatic/mild acidolysis lignin (EMAL). *BioResources, 5*, 2184–2194.

Russel, D. W. (1971). The metabolism of aromatic compounds in higher plants. *The Journal of Biological Chemistry, 246*, 3870–3878.

Sagues, W. J., Jain, A., Brown, D., Aggarwal, S., Suarez, A., Kollman, M., . . . Argyropoulos, D. S. (2019). Are lignin-derived carbon fibers graphitic enough? *Green Chemistry: An International Journal and Green Chemistry Resource, 21*, 4253–4265.

Saleh, T. M., Leney, L., & Sarkanen, K. V. (1967). Radioautographic studies of cottonwood, Douglas fir and wheat plants. *Holzforschung, 21*, 116–120.

Sameni, J., Krigstin, S., dos Santos Rosa, D., Leao, A., & Sain, M. (2014). Thermal characteristics of lignin residue from industrial processes. *BioResources, 9*, 725–737.

Sarkanen, K. V., & Hergert, H. I. (1971). Classification and distribution. In K. V. Sarkanen, & C. H. Ludwig (Eds.), *Lignins: Occurrence, formation, structure and reactions* (pp. 43–49). New York: Wiley-Interscience.

Shankar, S., & Rhim, J. W. (2018). *Bionanocomposite films food packaging applications. Reference module in food science* (pp. 1–10). Elsevier.

Shen, R., Tao, L., & Yang, B. (2019). Techno-economic analysis of jet-fuel production from biorefinery waste lignin. *Biofuels, Bioproducts and Biorefining, 13*, 486–501.

Sherrard, A. (2019). *Preem and Sveaskog take stake in SunCarbon—Plans for a lignin oil plant*. https://bioenergyinternational.com/biofuels-oils/preem-and-sveaskog-take-stake-in-suncarbon-plans-for-a-lignin-oil-plant.

Shimada, M., Fushiki, H., & Higuchi, T. (1973). Mechanism of biochemical formation of the methoxyl groups in softwood and hardwood lignin. *Mokuzai Gakkaishi, 19*, 13–21.

Shrotri, A., Kobayashi, H., & Fukuoka, A. (2017). Catalytic conversion of structural carbohydrates and lignin to chemicals. *Advances in Catalysis, 60*, 59–123.

Sjostron, E. (1981). *Fundamentals in carbohydrate chemistry. Wood chemistry fundamentals and applications* (pp. 21–48). New York: Academic Press.

Smolarski, N. (2012). *High-value opportunities for lignin: Unlocking its potential* (p. 15) Paris: Frost & Sullivan.

Strassberger, Z., Tanase, S., & Rothenberg, G. (2014). The pros and cons of lignin valorisation in an integrated biorefinery. *RSC Advances, 4*, 25310–25318.

Sun, R., & Tomkinson, J. (2000). Fractionation and characterization of water-soluble hemicelluloses and lignin from steam-exploded birchwood. *International Journal of Polymeric Materials and Polymeric Biomaterials, 45*, 1–19.

Svensson, S. (2008). *Minimizing the sulphur content in Kraft lignin* (Independent thesis). Västerås and Eskilstuna: Mälardalen University.

Tanaka, A., & Makino, A. (2009). Photosynthetic research in plant science. *Plant & Cell Physiology, 50*, 681–683.

TAPPI T 222 om-02. (2002). *Acid-insoluble lignin in wood and pulp. 2002–2003 TAPPI test methods*. Atlanta, GA: Tappi Press.

Terashima, N. (2000). Formation and ultrastructure of lignified plant cell walls. In K. S. Kim (Ed.), *New horizons in wood anatomy* (pp. 169–180). Gwangju: Chonnam National University Press;.

Terashima, N., Fukushima, K., He, L. F., & Takabe, K. (1993). Comprehensive model of lignified plant cell wall. In H. G. Jung, D. R. Buxton, R. D. Hatfield, & J. Ralph (Eds.), *Forage cell wall structure and digestibility* (pp. 247–270). Madison, WI: The American Society of Agronomy.

Terashima, N., Nakashima, J., & Takabe, K. (1998). *Proposed structure for protolignin in plant cell walls*. ACS symposium series: Vol. 697, Lignin and lignan biosynthesis (pp. 180–193)). Washington, DC: ACS Publications.

Teymouri, F., Laureano-Perez, L., Alizadeh, H., & Dale, B. E. (2005). Optimization of the ammonia fiber explosion (AFEX) treatment parameters for enzymatic hydrolysis of corn stover. *Bioresource Technology, 96*, 2014–2018.

Tuomela, M., Vikman, M., Hatakka, A., & Itävaara, M. (2000). Biodegradation of lignin in a compost environment: A review. *Bioresource Technology, 72*, 169–183.

UNRAVEL. (2019). *Successful scale-up of the FABIOLA™ process in a lignocellulose biorefinery pilot plant*. http://unravel-bbi.eu/successful-scale-up-of-the-fabiola-process-in-a-lignocellulose-biorefinery-pilot-plant/.

van de Beld, B. (2018). Development of fast pyrolysis in the Netherlands: Technology and applications. In *IEA bioenergy triennial summit*. San Francisco, CA: Biomass Technology Group. https://www.ieabioenergy.com/wp-content/uploads/2018/11/902-BTGBiomass-VandeBeld-smaller.pdf.

Vardon, D. K., Franden, M. A., Johnson, C. W., Karp, E. M., Guarnieri, M. T., Linger, J. G., ... Beckham, G. T. (2015). Adipic acid production from lignin. *Energy & Environmental Science*, *8*, 617–628.

Vaughan, P. F. T., & Butt, V. S. (1970). The action of *o*-dihydric phenols in the hydroxylation of *p*-coumaric acid by a phenolase from leaves of spinach beet (*Beta vulgaris* L.). *The Biochemical Journal*, *119*, 89–94.

Velez, J., & Thies, M. C. (2016). Liquid lignin from the SLRPTM process: The effect of processing conditions and black-liquor properties. *Journal of Wood Chemistry and Technology*, *36*, 27–41.

Vishtal, A., & Kraslawski, A. (2011). Challenges in industrial applications of technical lignins. *BioResources*, *6*, 3547–3568.

Wang, S., Kong, F., Fatehi, P., & Hou, Q. (2018). Cationic high molecular weight lignin polymer: A flocculant for the removal of anionic azo-dyes from simulated wastewater. *Molecules (Basel, Switzerland)*, *8*, 2005.

Wang, S., Kong, F., Gao, W., & Fatehi, P. (2018). Novel process for generating cationic lignin based flocculant. *Industrial & Engineering Chemistry Research*, *19*, 6595–6608.

Wardrop, A. B., & Bland, D. E. (1959). The process of lignifications in woody plants. In K. Kratzl, & G. Billik (Eds.), *Biochemistry of wood* (pp. 92–116). New York: Pergamon Press.

Wu, S., & Argyropoulos, D. (2003). An improved method for isolating lignin in high yield and purity. *Journal of Pulp and Paper Science*, *29*, 235–240.

Zainab, A. K., Pradhan, R., Thevathasan, N., Arku, P., Gordon, A., & Dutta, A. (2018). Beneficiation of renewable industrial wastes from paper and pulp processing. *AIMS Energy*, *6*, 880.

Index

Note: Page numbers followed by "*f*" and "*t*" refer to figures and tables respectively.

A

4-Acetamidobenzenesulfonyl chloride, 449
Acetic acid (AA), 328−329
Acetoacetyl-CoA pathway, 522
Acetobacter, 55
   A. aceti, 360−361
   A. hansenii, 55
   A. pasteurianus, 55
   A. xylinum, 55, 360−361, 419−420
Acetylated sugars, 130
Acetylation, 440−442
Achromobacter, 361
Acid pretreatment, 135−136
Acid-adapted preculture approach, 324−325
Acidocalcisomes, 286−287
Acinetobacter sp., 297
   A. calcoaceticus, 288−289
Acrylonitrile−butadiene−styrene, 415−416
Adenosine triphosphate (ATP), 283
Adsorption, 331−332, 380
Aerobacter, 361
Aerobic plant fibrous-bed bioreactor (APFB bioreactor), 264
Aerococcus, 319−320
Aerogels, 89
Aeromonas hydrophila, 233, 288−289, 448
Agar, 36
Agitated/shaking cultivation method, 364−365
Agitation, 538
Agriculture, cellulose for, 64
Agriwaste, 108−110
Agro-food by-products, valorization of, 417−418
Agrobacterium, 55, 361
   A. tumefaciens, 286−287
Agroindustrial by-products fermentation, pullulan production by, 189−192
Air pollution, 393−394
Airlift
   bioreactors, 326−327
   for pullulan production, 195−196
Alcaligenes, 361
   A. latus, 519−520
      ATCC 29713, 526−527
Alcohol precipitation, 108−110
Aliphatic alcohols, 136
Alkaligenes lactus, 498
Alkaline pretreatment, 134−135
α-D-glucose, 76−77
α-PGA, 253
Aluminum oxide, 394−395
Amberlite IRA-420, 331−332
Ammonia fiber explosion (AFEX), 146−147, 568−569
Amorphous polymers, 59
Amylopectin, 32, 32*f*, 34, 76−77, 77*f*
Amylose, 32, 32*f*, 76−77, 77*f*
Anecic earth worms, 23
Anion-exchange membranes (AEMs), 328−329
Antimicrobials
   activity of chitosan, 448−449
   applications of bacterial cellulose, 372−373
   PolyP in, 297
Antisolvents, 529−530
Apiogalacturonan, 103−104
Aquatic environment
   marine, 15−16
   microplastics in, 13−16
   river and lakes, 14−15
Aqueous two-phase system (ATPS), 532
Arabidopsis thaliana, 105, 227
Arabinoglucuronoxylans, 131−132
Arabinose (A), 130
Arabinoxylan, 131
Arabinoxylo oligosaccharides (AXOS), 138
Arenicola marina, 19−20
Argentine Food Code, 118
Aromatic dilignol, 556−558, 557*f*, 560*f*
Artificial cornea and lenses of bacterial cellulose, 375
Artificial ECM, 62
Ascidia sp., 54−55
Aspergillus
   A. aculeatus, 318
   A. nidulans, 483−484
   A. niger, 320
   A. nomius, 483−484
Association of Textile Chemists and Colorists (ATCC), 337
Aureispira marina, 224−226
*Aureobasidium pullulans*CC, 165, 188−189
   cell morphologies of, 189
Autohydrolysis, 137
Azeotropic dehydration, 333
Azotobacter, 361

B

Bacilli, 316
Bacillus sp., 235, 498, 520
  B. amyloliquefaciens, 257
  B. anthracis, 257
  B. atrophaeus, 257
  B. cereus, 257
  B. coagulans, 316
  B. flexus WY2, 524–525
  B. halodurans, 253
  B. licheniformis, 253, 257, 316
  B. megaterium, 223–224, 253, 257, 520, 526–527
  B. mojavensis, 257
  B. mycoides, 525–526
  B. pumilus, 257
  B. stearothermophilus, 316
  B. subtilis, 253, 256–257, 316, 443–444
    B. subtilis 6833, 524–525
  B. thuringiensis, 257
Bacterial cellulose (BC), 7, 61–62, 359–361, 360*f*, 419. *See also* Nanocellulose
  applications, 369–381
    in cosmetics, 378
    electronics, 379
    in food, 376–378
    water purification, 379–380
  biomedical applications, 369–376
    antimicrobial applications, 372–373
    artificial cornea and lenses, 375
    bone tissue engineering, 375–376
    cardiovascular grafts, 374–375
    drug-delivery application, 373–374
    wound-healing applications, 370–372
  commercialization of BC-based products, 381
  production, 361–365
    bacterial strain selection, 361
    cultivation methods, 363–365
    culture medium, 362–363
  structural and functional features of
    biocompatibility, 368–369
    biodegradability, 367–368
    mechanical properties, 365–366
    structure, pore size, and morphology, 366–367
    water holding/release capacity, 366
Bacterial PolyP, 283
Bacterial strains, 316–317
  selection, 361
Bamboo, 401
BASYC, 374
Batch fermentation, 192–193, 235–236, 320–322
β-lactoglobulin (β-lg), 116
β-oxidation, 522
Bio-based materials, 442–444
Bio-based polymers, 66, 75
Bio-Flex process, 567
BIO-Intelligence Service, 108
Bioadimide, 417
Biocompatibility
  of bacterial cellulose, 368–369
  of PHAs, 233–234
Biocomposites, 56, 400, 415–416
  crystallinity of, 429–430
  derived from polyhydroxyalkanoate, 418–420
  derived from polylactic acid, 416–418
  scheme of biocomposite formulation, 415*f*
  synthesis, 414
Biodegradability
  of bacterial cellulose, 367–368
  of PHAs, 234
  of starch granules, 40
  of thermoplastic starch, 39–40
Biodegradable composites, 406
Biodegradable edible films, 118
Biodegradable materials, 75
Biodegradable polymers, 7–8, 54, 54*f*, 342–343, 414
  advantages, 394–395, 409*t*
  commercially available polymer composite in fields, 410*f*
  composites, 393–394, 408
  general commercialization processes, 395–396
  microbial and insect biopolymers, 6–7
  plant-based biopolymers, 4–5
  polymer composites, 394
  potentials and applications, 408–409
  process engineering and commercialization, 8
  types of, 396–408, 398*t*
    carbon nanotube-reinforced composites, 404–406
    double-layer polymer composites, 401–404
    natural fiber composites, 396–401
    petrochemical-based biocomposites, 406–408
Bioethanol, 146–147
Biofuels, 5
Biogenic production of polyphosphate, 287–293
  eukaryotes, 291–293
  prokaryotes, 288–291
Biological adhesives, 272
Biomass, 53, 133, 576–577
  harvesting, 536
  residues, 393–394
  waste, 65
Bioplastic polymers, 63–64
Biopolymers, 3, 5, 39, 53–67, 381, 414, 440–442
  applications of cellulose, 60–66
  biocomposites
    derived from polyhydroxyalkanoate, 418–420
    derived from polylactic acid, 416–418
  biodegradable polymers and polymer composites, 54
  composites, 7–8
  current status and challenges in production of cellulose-based biopolymers, 67
  films, 440–442
  formation, 430–432

Index 591

functional properties of biopolymers and biocomposites
  biopolymer film formation, 430–432
  crystallinity of biocomposites, 429–430
  tensile strength of biopolymer, 427–429
  limitations of biopolymers and overcoming strategies, 66–67
  matrix composites, 54
  packaging, 46
  plant-based cellulose, 55–60
  process engineering and product development, 66
  scheme of biocomposite formulation, 415f
  sources of cellulose, 54–55
  thermal and rheological properties
    polyhydroxyalkanoate and biocomposites, 425–427
    polylactides and biocomposites, 420–425
BioPreferred, 75
Bioproducts obtained from xylan, 140–151
Biopulping treatment, 139–140
Bioreactors, 532–533
  mode of operation for production, 262–264
  for pullulan production, 192–196
Bioremediation, 451
  PGA in, 272–273
Biosorption of heavy metals, 205
Biosynthesis of PGA, 253–254
Biotas, microplastic interaction with, 18–20
Biotechnological applications, PolyP in
  enzyme technology, 297–298
  synthetic biology applications, 298
Biowaste, 8
Bisphenol A (BPA), 21
Blood coagulation, 299

Bone
  formation, 298–299
  tissue engineering, 375–376, 503
Boron nitride (BN), 394–395
Brucella melitensis AUH2, 524–525
Burkholderia
  B. cepacia, 523–524
    B. cepacia AM19 strain, 288–289
  B. sacchari, 523–524

C

C-irradiation depolymerization, 449
C_1 radical, 556f, 557–558
C_3 radical, 556f
C_5 radical, 556f, 557
Caenorhabditis elegans, 19–20
Calcium carbonate ($CaCO_3$), 503
Calcium PolyP fiber, 299
Calcium sulfate ($CaSO_4$), 503
Candida antarctica, 333
Carbon dioxide, 566
Carbon nanotubes (CNTs), 54, 394–395
  carbon nanotube-reinforced composites, 404–406
    polyaniline composites, 404–405
    polypyrrole, 405
    PVA, 405–406
Carbon source, pullulan effect on, 169–171
Carbon-to-nitrogen ratio (C/N ratio), 526–527
Carboxyalkylation, 445–446
Carboxylation, 230
Carboxymethyl pullulan (CMP), 196
Carboxymethylation, 196–198
  sulfation, 198
Carboxymethylcellulose (CMC), 55
Cardiovascular grafts, 374–375
Cardiovascular tissue engineering, 505–506
Carnobacterium, 319–320

Carob, 191
Cartilage tissue engineering, 504
Carya illinoinensis. See Pecan (*Carya illinoinensis*)
Cassava, 79
  starch, 81
Cation-exchange membranes (CEMs), 328–329
Caustic soda (NaOH), 442
Cavitation, 114
Cell immobilization bioreactors, 194–195
Cell morphologies of *Aureobasidium pullulans*, 189
Cellophane, 63
Cellular residues, 82
Cellulolytic enzymatic lignin (CEL), 561–564
Cellulose, 4, 7, 32, 53, 55–56, 129, 359–360, 401–403, 420, 462–463
  applications, 60–66
    for agriculture, 64
    for food industries and packaging applications, 63–64
    for medical applications, 60–61
    for photocatalysis and photovoltaics/electronics, 62–63
    for tissue-engineering applications, 61–62
  biopolymers, 53–67
  cellulose-based biopolymers, 67
  cellulose-based hydrogel, 62
  fibers, 35–36
  mechanical properties of natural fiber, 463t
  nanocomposites, 59
  nanowhiskers/nanocrystals, 420
  plant-based cellulose, 55–60
  polymer, 54
  sources, 54–55
  structure of, 462f
  synthesis, 57
Cellulose nanocrystals (CNCs), 4, 54, 420, 463, 465–467, 482

Cellulose nanofiber (CNF), 4, 85, 463–465
Cellulose nanofibrils (CNFs), 45–46
Cellulose synthase complex (CSC), 57
Cellulose synthase interacting protein 1 (CSI 1), 57
Cellulosic nanofibers, 35–36
Centrifuge, 533
Chemical equilibrium, 537
Chitin, 7, 32, 401, 420, 439–440, 451
 fillers, 85
 nanofibrils, 420
Chitosan, 7, 59, 85, 420, 439–440, 448–449
 applications, 447–453
  bioremediation, 451
  chitosan and composite blends in fields, 447t
  drug delivery, 451–452
  food and packaging, 448–450
  medical, 452–453
  wastewater treatment, 450–451
 biofunctionality, 440–444
  biological process for extraction, 444f
  chemical-processing method, 443f
  extraction and chemical modification of chitosan, 442–444
  structural illustration of chitin and chitosan, 441f
 monomeric units of chitosan and cellulose, 440f
 synthesis of composite blends of, 444–446
  chemical and biological extraction methods, 445t
Chlamydomonas reinhardtii, 286–287
Chlorinated benzenes, 13
Chlorinated diphenyl, 13
Chlorination, 229
Chloroform, 402–403
Cholesterol-bearing pullulans, 199
Citric acid, 84

Citrus pectin, 118–119
Cloisite Na$^+$ (CNa$^+$), 426–427
Cloisite15A (C15A), 426–427
Cluster model, 77
Coconut fiber-PVC/acrylonitrile composites, 399
Coir fiber chemical treatment, 397–398
Collagen, 419–420
 type I, 419–420
Commercial bio-based products, 5
Commercial lignin production, 572–578
 kraft lignin production, 573–576
 lignosulfonate production, 572–573
 organosolv and soda lignin, 576–577
 TMP-Bio lignin, 577–578
Commercial PLA, 428
Commercialization
 of BC-based products, 381
 of biopolymers, 8
Composites, 85
Computational fluid dynamics (CFD), 528
Continuous fermentation, 193–194, 237–239, 323–324
Continuous stirred tank reactor (CSTR), 324–325
Conventional synthetic thermoplastic polymers, 33
Corn, 79
Corneal disease, 375
Corona/plasma treatment, 417
Corynebacterium glutamicum, 235, 270, 290–291, 316
Cosmetics, bacterial cellulose applications in, 378
Cross-linking of pullulan, 198
Crude glycerol, 523
Crude oil, 31
Cryphonectria parasitica, 188
Crystal property of starch, 38–39
Crystalline cellulose, 54–55
Crystalline polymers, 59
Crystalline starch, 37

Crystallinity, 38
 of biocomposites, 429–430
 of PHAs, 232–233
Crystallinity index (X), 478
Crystallization, 429
Crystallization temperature (Tc), 478
Cultivation methods, 363–365
 agitated/shaking cultivation method, 364–365
 static cultivation method for bacterial cellulose production, 363–364
Culture medium of bacterial cellulose, 362–363
Cupriavidus
 C. necator, 427–428, 498, 519–522, 524–527, 539–540
 C. taiwanensis, 525
Cyanobacteria, 318, 520
Cyanobacterial system, 55
Cyclic feast-famine conditions, 520
Cyclophosphates, 286
Cytariahariot i darwinii, 188
C$_\beta$ radical, 555–556, 556f

D

D-galactose (G), 103, 130
D-glucosamine, 442
D-glucose, 77
D-glutamic acid, 254–255
Daphnia magna, 18–19
DdPPK1, 293
DdPPK2, 293
De-structured starch, 32
Deacetylation process, 449
Deep eutectic solvent (DES), 138–139
Degradable biopolymers, 53
Degree of esterification (DE), 103
Dextrose equivalent (DE), 169–170
Diamond, 394–395, 403–404
Dichloromethane, 402–403
Dickeya, 361
Dictyostelium discoideum, 286–287

Index

Dietary antioxidants, 106
Differential scanning calorimetry (DSC), 478
Diffusion dialysis, 328–329
Diffusivity, 328–329
Digestion methods, 530–531
Dimer, 557, 559f
3,5-Dimethoxy-4-hydroxyphenylpropane, 552
Dimethyl carbonate (DMC), 541
1-(2,5-Dimethyl-4-[2,5-dimethyl phenyl] phenyl diazenyl) azonapthalen-2-ol (OilRed O), 501–502
4-Dimethylaminopyridine, 199
Diphosphate. See Pyrophosphates (PPi)
Diquinone, 557–558, 560f
Dissolved oxygen (DO), 525–526
Distillers dried grains with solubles (DDGS), 426
Double-layer polymer composites, 401–404
  graphene composites, 403–404
  polyhydroxyalkanoates composites, 401–403
Downstream processing (DSP), 200–203, 529–531, 536–538. See also Upstream processing (USP)
  agitation, 538
  chemical equilibrium, 537
  digestion methods, 530–531
    PHA-recovery methods, 531t
  equipment size, 538
  fluid dynamics, 537
  nonlinear scale-up, 536–537
  properties of materials, 537
  reaction kinetics, 537
  solvent extraction, 529–530
  steps involved in model development for, 529f
  thermodynamics, 538
Drug delivery, 451–452, 501–502
  application of bacterial cellulose, 373–374
  PHAs in, 241–242
  PolyP application in, 300

Drug-delivery systems (DDSs), 91, 431, 501
  PLA in, 339, 340f
Drug-loading efficiency (DLE), 501

E

Earth worm, 23
Ecoflex, 505
Edible films, 90–91
Electrodialysis, 330–331
Electronics, bacterial cellulose applications in, 379
Electrospinning, 471–472
Enhanced biological phosphate removal method (EBPR method), 295
Enterococcus, 319–320
  E. mundtii, 316
Environmental bioremediation, PolyP in
  enhanced biological phosphate removal, 295
  heavy metal remediation, 295–296
  water softening, 296
Environmental remediation, PLA in, 342–343
Environmentally friendly polymer composites, 408–409
Enzymatic mild acidolysis lignin (EMAL), 561–562, 564
Enzymes, 81
  enzymatic polymerization, 333
  technology, 297–298
Epoxidation, 229
Epoxidized canola oil (eCO), 427
Epoxidized hydroxyl group (eOH), 426–427
Epoxy, 401
Equipment size, 538
Escherichia coli, 118–119, 238–239, 289, 316, 521–522
  LSBJ, 427–428
  strains, 498–499
1-Ethyl-3-(3-dimethylaminopropyl) carbodiimide hydrochloride (EDC), 199

Ethylene/propylene copolymers, 401
Eucalyptus globules, 139
Eukaryotes, 291–293. See also Prokaryotes
  enzymes
    DdPPK1, 293
    DdPPK2, 293
    PPN1 and PPX1, 293
Exopolyphosphatase (PPX), 288
Exopolysaccharide (EPS), 166
Expanded poly(tetrafluoroethylene) (ePTFE), 369
Extracellular cellulose, 360–361
Extracellular matrix (ECM), 61–62
Extrusion-based additive manufacturing technology, 418

F

FABIOLA lignin, 569–570
Fast pyrolysis lignin, 571–572
FDA. See US Food and Drug Administration (FDA)
Fed-batch fermentation, 193–194, 236–237, 322–323
Feed preparation, 535
Feeding strategy, 527
Fermentation, 7
  batch, 320–322
  continuous, 323–324
  fed-batch, 322–323
Fermentative production of biopolymers, 3–4
Fertilizers, 65
Ferulic acid (FEA), 130
Fibers, 420
Fibrous-bed reactors, 327
Filamentous fungi, 315
Film blowing, 42
Film formation, biopolymer, 430–432
Flavors, 115–116
Fluid dynamics, 537
Fluidized-bed reactors (FBRs), 325–326
Foaming, 83
Folates (FOLs), 502

Food
　bacterial cellulose applications in, 376–378
　industry
　　cellulose for food industries and packaging applications, 63–64
　　PGA in, 272
　　pullulan application in, 204–205
　packaging, 31, 448
　pectin in food-processing industries
　　bakery industry, 116–117
　　confectionery, 117–118
　　cosmetics, 119–120
　　dairy industry, 115–116
　　pharmaceutical industry, 118–119
　polyP in food additive, 297
　preservation, 107
　waste, 108–110
　　generation, 108
Food and Agriculture Organization of United Nations (FAO), 108
Fossil fuels, 31
Fossil-based nonbiodegradable materials, 75
Fossil-based polymers, 4
Fourier transform infrared spectroscopy (FTIR), 267
Fullerenes, 403–404
Fungi, 317–318
Fusobacterium nucleatum, 257

G

Galacturonic acid (GalA), 101
γ-polyglutamic acid (γ-PGA), 6, 253–254
　structure, 266
Gelatinization, 33
　process, 78
　temperature of starch, 34, 37
Generally Recognized As Safe (GRAS), 204
Genetic engineering, 521–522
　approaches, 309–310
Genetic modifications, 4, 76
Gentamicin, 241

German lignocellulose feedstock biorefinery project, 567–568
Ghost fishing, 16
Glass transition temperature (T_g), 37, 478
Gluconacetobacter, 359–360
　G. xylinum, 55
　G. xylinus, 359–360
Glucose ($C_6H_{12}O_6$), 553f, 577
Glucosyltransferase, 168
Glucuronic acid (GA), 130
Glucuronic acid A (GlcA), 56
Glucuronoxylan, 131
Glutamate-dependent bacteria, 257–259
Glycerol, 36, 255–256
Glycidyl methacrylate–grafted PHA (PHA-g-GMA) with CSF (PHA-g-GMA/CSF), 426–427
Glycol, 36
Good manufacturing practices (GMPs), 533–535
Good Quality Control Laboratory Practices (GQCLP), 534–535
Graft copolymerization, 446
Graft polysaccharide, 199–200
Grafting, 199–200
Grape pomace, 191–192
Graphene, 394–395, 398t, 401
　composites, 403–404
　graphene-reinforced P3HB4HB, 425–426
Graphene oxide (GO), 380, 403–404
Graphite, 403–404
Greenhouse gases (GHGs), 517
　emission, 8
Guaiacylpropane polymers, 552

H

Haloferax mediterranei, 520, 522, 525, 540
Halolactibacillus halophilus, 316
Halomonas
　H. boliviensis, 525–526
　H. campaniensis, 522
　TD01, 498–499

　TD08, 498–499
Halophile, 520
Hardwood species, 129
Healthcare
　PGA in, 271–272
　pullulan application in, 203–204
Heavy metal remediation, 295–296
Hemicellulases, 81
Hemicellulose, 5, 129–130, 140
Hestrin and Schramm culture medium (HS culture medium), 362
Heterofermentative LAB, 313
Hexachlorocyclohexane, 13
Hexose, 130
HHx. *See* Hydroxydecanoate (HD)
High molecular weight polyphosphate, 286
High-density polyethylene (HDPE), 399
High-impact polystyrene, 415–416
High-intensity ultrasonication, 465
High-pressure homogenization, 463–464
Homofermentative LAB, 313, 316
Homogalacturonan (HGA), 101, 103–104
Homogenization process, 463–464
Homopolymeric PHD, 430
Homoxylan, 131
Human digestive enzymes, 204
Hydra attenuate, 18–19
Hydrodictyon reticulum, 320
Hydrogels, 5, 57–59, 147–148
Hydrogen release compound (HRC), 342–343
Hydrolysis, 39, 467, 523–524
Hydrophilic polymer, 33
Hydrophilic pullulan gels, 205–206
Hydrophobic matrices, 476–477
Hydrophobic modification of pullulan, 199
Hydrothermal pretreatment, 137
4-Hydroxy-3-methoxyphenylpropane, 552

Index

Hydroxyalkanoates (HAs), 223–224
3-Hydroxyalkanoic acids (3HA), 242
3-Hydroxybutyrate (3HB), 496, 499
3-Hydroxybutyric acid, 401–402
10-Hydroxycamptothecin, 199
Hydroxydecanoate (HD), 430, 503
3-Hydroxydecanoate (3HD), 242, 498
3-Hydroxydodecanoate (3HDD), 242, 498
3-Hydroxyhexanoate (3HHx), 242, 496, 498
  nerve channels, 504–505
Hydroxyl groups, 57–59, 462
Hydroxylation, 230–231
3-Hydroxyoctanoate (3HO), 242, 498
Hydroxypropyl methylcellulose (HPMC), 55, 87
Hydroxypropylated starch, 84
Hydroxyvalerate (HV), 402–403, 521
3-Hydroxyvalerate (3HV), 231–232, 498–499, 522
Hylocereus polyrhizus, 113
Hyperbranched polyphosphates (HBPPs), 300

I

Immobilization, 264
Immobilized cell bioreactors, 194–195
In vitro synthesis of biopolymers, 3–4
Inborn cardiovascular defects, 505
Industrial waste, microplastic in, 23–24
Injection blow molding, 42–44
  blowing, 44
  ejection, 44
  injection, 44
Injection stretch blow molding (ISBM), 44–45
  blowing, 45
  conditioning, 44–45
  injection, 44
  stretching, 45

Inoculum preparation, 535
Inorganic PolyPs, 283, 294–295
Insect biopolymers, 6–7
Ionic liquids (ILs), 138–139
IRA-92, 331–332

J

Jatropha seed cake, 192
Jute, 401

K

Kenaf fiber, 399–400
Kiwifruit (KP), 108
Klason lignin methods, 562
Kluyveromyces marxianus, 311
Komagataeibacter, 361
  K. rhaeticus, 361
Kraft lignin production, 573–576
  LignoBoost technology, 574–575
  LignoForce technology, 573–574
  SunCarbon lignin, 575–576
Kraft pulping, 566

L

L-arabinose, 103
L-glutamic acid, 254–255
L-phenylalanine, 553–555, 553f
L-rhamnose, 103
L-tyrosine, 553–555, 553f
Lactic acid (LA), 309–310, 318–319
  microbial production, 311–318
Lactic acid bacteria (LAB), 311
Lactobacillus spp., 80, 319–320, 443–444
  L. acidophilus, 311, 319–320
  L. bifermentans, 319–320
  L. casei, 311
  L. delbrueckii subsp. *bulgaricus*, 311
  L. helveticus, 319–320
  L. lactis, 319–320
  L. maninotivorans, 319–320
  L. paracasei, 319–320
  L. plantarum, 316
  L. rhamnose, 319–320
  L. sanfrancis, 319–320

Lactococcus, 319–320
  L. lactis, 311
Lactose, 170–171
Lakes, microplastics in, 14–15
Layer-by-layer (LBL), 473
Lead Market Initiative, 75
Leuconostoc, 319–320
  L. lactis, 319–320
Life-cycle assessment (LCA), 540–541
Lignin, 5, 129
  extraction methods at laboratory and pilot scales, 561–572
  milling methods, 561–564
  novel extraction technologies at pilot scales, 564–572
  lignin-based flocculant, 579–580
  methods for commercial, 572–578
  occurrence and formation of, 552–561, 554t
  opportunities and challenges in commercialization of, 578–580
  production, 551–552
  valorization, 578–580
LignoBoost technology, 574–575
Lignocellulosic biomass, 7, 129
Lignocellulosic waste fibers, 419
LignoForce technology, 573–574, 574f
Lignosulfonate production, 572–573
Liquid Lignin Company, 566
Liquid–solid ratio (LSR), 134
Littorina, 18–19
Living organisms, 3
Long chain length PHAs (lcl-PHAs), 224–226
Low-density polyethylene (LDPE), 485
Lysinibacillus sphaericus, 257

M

Magnesium-filled PLA biocomposites (Mg-filled PLA biocomposites), 418
Mannans, 5
Manufacturing methods, 399–400

Marine
  ecosystem, 13–16
    classification of polymers found in, 12t
    microplastics in, 15–16
Market demand, 538–539
Mathematical models, 527–528
Matrix-based composites, 397–400
Medical field
  bone formation, 298–299
  drug delivery, 300
  platelets and blood coagulation, 299
  PolyP application in, 298–300
Medical implants, PHAs in, 239–241
Medium-chain length PHAs (mcl-PHAs), 224–228, 429, 496
Melon rinds (MP), 108
Melt mixing, 394
Melt processing, 470–471
Melting temperature (T_m), 37, 478
Membrane filtration, 7, 329–330
Membrane-based processes, 265
Mercerization, 56–57
Mesenchymal stem cells (MSCs), 62
Mesoporous bioactive glass (MBG), 503
Metabolic routes, 524–525
Methanol, 499
2-Methylcitrate synthase, 498–499
Methylobacterium sp, 525–526
  M. extorquens AM1, 499
  M. organophilum, 238–239
Methylotrophic bacteria, 499
Microbacterium aurum TPL18, 524–525
Microbeads, 12–13
Microbial biopolymers, 6–7
Microbial biosynthesis pathway of PGA, 254–255
Microbial consortia, 257–262
  for pullulan production, 188–189
    Aureobasidium pullulans, 188–189
    cell morphologies of *Aureobasidium pullulans*, 189
    strain improvement, 260–262

Microencapsulated phase change materials (MicroPCMs), 426
Microencapsulation of lipophilic materials, 119–120
Microfibrils, 427
Micromonosporaceae, 234
Microorganisms, 3–4, 6, 484
Microparticles, 501–502
MicroPCMs.
  See Microencapsulated phase change materials (MicroPCMs)
Microplastics, 12–13
  in aquatic environment, 13–16
  interaction with biotas, 18–20
  polymers classification in marine ecosystem, 12t
  in soil, 17–18
  in waste, 20–24
Microwave-assisted extraction (MAE), 139
  of pectin, 112–113
Milled wood lignin (MWL), 561–563, 562f
Milling methods, 561–564
  CEL, 563–564
    typical lignin yields, 563t
  EMAL, 564
  MWL, 562–563
minC gene, 499–501
minD gene, 499–501
Miscanthus
  M. sacchariflorus, 137–138
  M. sinensis, 137–138
Mixed microbial consortium (MMC), 519–521
Molasses, 190, 310–311
Monolignols, 552–555, 556f
Monosaccharides, 103, 130
Moving bed biofilm reactor (MBBR), 264
mreB gene, 499–501
Multiwall carbon nanotubes (MWCNTs), 404
Municipal solid wastes (MSWs), 20–21
Mytilus edulis, 19–20

N

N-acetyl-D-glucosamine, 442
Naididae, 15
Nannochlorum sp., 318
Nanocellulose, 4, 375–376, 461, 463–467, 478, 483
  cellulose nanocrystal, 465–467
    variation of sulfuric acid hydrolysis parameters, 467t
  cellulose nanofiber, 463–465, 464f
Nanocellulose-reinforced biocomposites, 461
  barrier property, 480–483
  biodegradation property, 483–485
  processing methods of, 468–473
    electrospinning, 471–472
    fabrication of CNF/thermoplastic starch composites, 471f
    fabrication of nanocellulose-reinforced composites, 469f
    layer-by-layer, 473
    melt processing, 470–471
    schematic diagram of electrospinning technique, 472f
    solvent casting, 468–470
  tensile properties, 473–475
  thermal properties, 475–480
    DSC thermogram, 480f
    TGA curve of CNC/PLA composites compatibilized, 477f
    thermal behavior, 478–480, 479t
    thermal stability, 475–478
Nanoclay, 89, 420
Nanocomposites, 394. *See also* Biocomposites
  materials, 402
  starch-based nanocomposites, 89
Nanofillers, 402–403
Nanomaterials, 420
  reinforcement, 414
Nanoparticles (NPs), 118–119, 501
Nanowhiskers, 420
Natrialba aegyptiaca, 257

Natta-de-coco, 360—361
Natural cellulose, 59
Natural fiber polymer composites (NFPCs), 400
Natural fibers, 394—395, 414
   composites, 395—401
      matrix-based composites, 397—400
      reinforcement-based composites, 400—401
   natural fiber-reinforced composite materials, 399—400
   natural fiber-reinforced plastics, 394—395
   reinforcement, 84—85
Natural polymers, 414
Natural resins, 414
Natural rubber, 65
Naturally occurring polymers, 551—552
Ne-carbobenzyloxy-L-lysine, 453
Nerve tissue engineering, 504—505
Neural stem cell (NSC), 504
Neutral pectin, 107
Nicotinamide adenine dinucleotide phosphate (NADPH), 522
Nile red, 502
Nitrogen source, pullulan effect on, 171—172
Nonbiodegradable petroleum-based polymers, 31
Nonbiodegradable plastics, 223
Nonbiodegradable synthetic polymers, 53
Nondegradable synthetic polymers, 53
NonPHA cell mass (NPCM), 529
Nonrenewable energy use (NREU), 540—541
Nonrenewable fossil fuels, 413
Nostoc muscorum, 520
Novel extraction technologies at pilot scales
   AFEX lignin, 568—569
   bio-flex process, 567
   fabiola lignin, 569—570
   fast pyrolysis lignin, 571—572
   German lignocellulose feedstock biorefinery project, 567—568
   lignin production by, 564—572
   proesa lignin, 570—571
   SLRP, 566
   steam-explosion process, 564—566
Nuclear magnetic resonance (NMR), 230
Nucleoside diphosphates (NDPs), 289
Nucleotide triphosphates (NTPs), 289
Nutrient limitation, 526
Nutrient-feeding methods, 263—264
Nylon nets, 16

O

4-O-methylglucuronic acid (MGA), 130
O_4 radical, 555—556, 556*f*
Oenococcus, 319—320
Organic acids, 135
Organic wastes, 393—394
Organosolv lignin, 576—577
Organosolv pretreatment, 136—137
Oxalic acid (OA), 328—329
Oxygen uptake, 187—188

P

p-hydroxyphenyl pyruvic acid, 553—555, 553*f*
p-hydroxyphenylpropane, 552
Packaging, 148—149
   PLA in packaging and service wares, 339—341, 341*t*
Packed-bed reactors (PBRs), 324—325
Paint particles, 23
Palm fiber-reinforced polystyrene, 399
Papaya pectin, 105—106
Pecan (*Carya illinoinensis*), 417—418
Pectin, 5, 101
   current challenges and future implications, 120—121
   food-processing industries, 115—118
   functional properties of, 105—108
   HGA, 103—104
   polymer, 102*f*
   recent advances in extraction of, 110—115
      microwave-assisted extraction of pectin, 112—113
      ultrasound-assisted extraction of pectin, 113—115
   RG-I, 104
   RG-II, 105
   sources, 108—110
   structure and classification, 102—105
   XGA, 105
Pediococcus, 319—320
Pennisetum purpureum, 418
Pentosanases, 81
Pentose, 130
Peridontal tissue engineering, 505
Personal-care products, 101—102
   PGA in, 272
Petrochemicals, 53
   petrochemical-based biocomposites, 406—408
      polycaprolactone composites, 406—407
      polyester amide composites, 407
      polyurethane composites, 407—408
Petroleum resources, 393—394
Petroleum-based plastics, 517
Petroleum-based polymers, 461
Petroleum-derived plastic material, 413
pH, 524—525
   effect on pullulan, 172—187
*Pha*B1, 522
*Pha*C-PHA, 521—522
phaCAc of *Aeromonas caviae*, 499
phaCMe coding, 499
Pharmaceutical applications, PLA in, 337—338
phbA gene, 521—522
phbB gene, 521—522
phbC gene, 521—522

Phenyl pyruvic acid, 553–555, 553f
Phosphoglucose, 168
Photocatalysis, cellulose for, 62–63
Photodegradation, 39
Photosynthetic cyanobacteria species, 318
Photovoltaics/electronics, cellulose for, 62–63
Phytomonas francai, 286–287
Pineapple leaf fiber (PALF), 399
Plant-based biopolymers, 4–5
Plant-based cellulose, 55–60
Plant fibers, 462
Plastic composite support (PCS), 195
Plasticizers, 33–34, 36, 84
  agents, 4
Plastics, 11, 31, 223. *See also* Microplastics
  mulching, 23
  pellets, 23–24
  pollution, 12
Plasticulture/agriculture, PLA in, 341–342
Platelets, 299
Plug flow reactor, 194
PntAB, 522
Poly-3-hydroxybutyrate-co-3-hydroxy valerate, 342–343
Poly-3-hydroxydecanoate (PHD), 430
Poly-3-hydroxyoctanoate (PHO), 229, 496
Poly-3-hydroxyvalerate (PHBV), 496
Poly-4-hydroxybutyrate (P4HB), 496
Poly-D,L-lactic acid (PDLLA), 310
Poly-D-lactic acid (PDLA), 310
Poly-L-lactic acid (PLLA), 310
Poly-ε-caprolactone (PCL), 501
Poly(3-hydroxybutyrate-co-3-hydroxyhexanoate) (PHBHHx), 505
Poly(3-hydroxybutyrate-co-3-hydroxyvalerate) (PHBV), 470, 521

Poly(3-hydroxybutyrate-co-4-hydroxybutyrate) (P3HB4HB), 425–426
Poly(3-hydroxyoctanoate-co-3-hydroxy-10-undecenoate) (PHOUs), 229–231
Poly(butylene adipate-co-terephthalate) (PBAT), 419
Poly(butylene succinate), 413
Poly(ethylene glycol) (PEG), 242, 501
Poly(ethyleneterephtalate) (PET), 369, 415–416, 473
Poly(hydroxyethyl methacrylate) (PHEMA), 375
Poly(L-lactideco-ε-caprolactone) (PLCL), 504
Poly(methyl methacrylate) (PMMA), 375, 415–416
Poly(N-isopropylacrylamide) (pNIPAM), 199–200
Poly(p-dioxanone), 413
Poly(propylene carbonate) (PPC), 419–420, 505
Poly(ε-caprolactone), 413
Polyaniline (PANI), 398t
  composites, 404–405
Polybutylene succinate, 342–343
Polycaprolactone (PCL), 342–343, 398t, 406
  composites, 406–407
Polycarbonate (PC), 375, 399, 415–416
Polycondensation, 7, 309–310
Polycyclic aromatic hydrocarbons, 13
Polyester
  plastics, 86
  polyester-based DDSs, 501
Polyester amide (PEA), 398t, 406
  composites, 407
Polyglutamic acid (PGA), 6, 253
  applications, 273–274
    bioremediation, 272–273
    food industry, 272
    healthcare, 271–272
    personal-care products, 272
  characterization, 266–267
  commercial production, 267–270
    production cost, 269–270

microbial biosynthesis pathway, 254–255
process parameters for production
  bioreactors mode of operation for production, 262–264
  γ-polyglutamic acid structure, 266
  isolation, analysis, and determination of PGA, 264–266
  microbial consortia, 257–262
  substrate, 255–257
Polyglycolic acid (PGA), 501
Polyhydric alcohols, 136
Polyhydroxyalkanoate composites synthesized with chestnut shell fiber (PHA/CSF), 426–427
Polyhydroxyalkanoates (PHAs), 6, 223–224, 398t, 401, 413, 416f, 495, 500t, 517–518
  applications, 239–243, 518
    in drug delivery, 241–242
    in medical implants and medicines, 239–241
    in tissue engineering, 242–243
  and biocomposites, 418–420, 425–427
  chemical modifications of, 228–231
    carboxylation, 230
    chlorination, 229
    epoxidation, 229
    hydroxylation, 230–231
  chemical structure, 231
  classification, 224–231, 225t
  commercialization, 538–543
    challenges in commercialization of, 543
    commercial production of, 542
    environmental assessment, 540–541
    location of industry, 539
    market demand, 538–539
    social factors affecting, 538–539
    technoeconomic studies, 539–540

Index

composites, 401–403
depolymerases, 498–499
for drug delivery, 501–502
genetic-engineered strains for production of, 496–501
   rational strategies for cost-effective, good-quality large-scale production of, 497–501
   schematic representation for production, 497f
industrial-scale production of, 235–239
   batch fermentation, 235–236
   continuous fermentation, 237–239
   fed-batch fermentation, 236–237
medium chain length, 227–228
PHA-based composites, 8
process development lab scale, 518–532
   downstream processing, 529–531
   upstream processing, 519–528
properties, 231–234
   biocompatibility, 233–234
   biodegradability, 234
   crystallinity, 232–233
   mechanical properties, 233
   thermal properties, 232
scale-up from lab scale to pilot scale, 532–538
short chain length, 227
in tissue engineering, 503–506
   PHAs and biocompatibility studies, 506t
types
   of PHAs for biomedical application, 496
   and properties of, 518
Polyhydroxybutyrate (PHB), 401, 414, 496, 499, 521
   based hydrogels, 502
   based micelles, 502
   based microparticles, 501–502
Polyhydroxyundecenoate (PHU), 230
Polyhydroxyvalerate (PHV), 524

Polyimides, 401
Polylactic acid (PLA), 7, 36, 309–310, 413, 416f, 420–424, 469–470, 501
   commercial strains, 319–320
   commercialization and application, 336–343
      biomedical and pharmaceutical applications, 337–338
      drug-delivery system, 339
      environmental remediation, 342–343
      other applications, 343
      packaging and service wares, 339–341
      plasticulture/agriculture, 341–342
      textiles, 336–337
      tissue engineering, 338–339
   fermentation modes and bioreactors
      batch fermentation, 320–322
      continuous fermentation, 323–324
      fed-batch fermentation, 322–323
   isolation, analysis, and determination technique and process, 327–332
      adsorption, 331–332
      diffusion dialysis, 328–329
      electrodialysis, 330–331
      membrane filtration, 329–330
      reactive extraction, 331
   microbial production, 311–318
      use of bacterial strains, 316–317
      use of cyanobacteria, 318
      use of fungi and yeast, 317–318
   PLA-based biocomposites, 416–418, 420–425, 421t, 428–429
   PLA-based composites, 8
   polymers properties, 335–336
   reactors type used for production, 324–327
   strain improvement, 318–319
   substrate, 310–311

   synthesis and structure of polymers, 332–336
      synthesis, 332–333
      process flow diagram for production, 334–335
      properties of PLA polymers, 335–336
      structure of polymer, 333–334
Polylactides. *See* Polylactic acid (PLA)
Polymer matrix composites (PMCs), 397
Polymerization, 7. *See also* Gelatinization
Polymers, 3–4, 31, 53, 413, 439–440
   composites, 54, 394
   polymer-blend technology, 396
   structure, 333–334
   synthesis of, 332–336
PolyP-dependent nucleoside-diphosphate kinase (PNDK), 290
Polyphosphate (PolyP), 283
   acidocalcisomes, 286–287
   applications, 294–300
      application in medical field, 298–300
      biotechnological applications, 297–298
      in environmental bioremediation, 295–296
      in industry, 296–297
   biogenic production of, 287–293
   challenges associated with polyphosphate production strategies, 300–301
   strategies to improve yield of, 301–302
   structure and types, 284–286
      cyclophosphates, 286
      high molecular weight polyphosphate, 286
      pyrophosphate, 284–286
Polyphosphate kinases (PPKs), 283–284
   PPK1, 289–290
   PPK2, 290–291
   PPK3, 291

Polyphosphates, 6–7
Polypropylene (PP), 195, 393–394, 397–398, 415–416
Polypyrrole (PPy), 398t, 405
Polytunnels, 23
Polyurethane (PU), 398t, 406
  composites, 407–408
  foam, 195
Polyvinyl alcohol (PVA), 398t, 405–406
Polyvinyl chloride (PVC), 399
Pomegranate, 108
Posidonia oceanica, 466
Potassium permanganate ($KMnO_4$), 231
Potatoes, 79
  starch, 82
PPN1, 293
PPX. *See* Exopolyphosphatase (PPX)
PPX1, 293
Prephenic acid, 553–555, 553f
Pretreatment methods for xylan extraction, 133–140
  acid pretreatment, 135–136
  alkaline pretreatment, 134–135
  biopulping treatment, 139–140
  hydrothermal pretreatment, 137
  ILs and DES, 138–139
  MAE, 139
  organosolv pretreatment, 136–137
  steam explosion pretreatment, 137–138
  subcritical or supercritical fluids, 138
Primary microplastics, 12–13, 21
Pristine fiber, 462
Process engineering, 8
  and product development, 66
Process flow diagram for production, 334–335
Processing-induced-type crystallinity, 38
Proesa lignin, 570–571
Prokaryotes, 288–291. *See also* Eukaryotes
  PPK1, 289–290
  PPK2, 290–291
  PPK3, 291

Propionibacterium acidipropionici, 147–148
Protein, 32
Pseudomonas sp., 55, 288–289, 361, 443–444, 498, 520
  P. aeruginosa, 227–228, 498–499
    S164S, 524–525
  P. citronellolis, 427–428
  P. entomophila, 242, 498
  P. fluorescens, 483–484
  P. putida, 227–228, 498
Pseudonocardiaceae, 234
Pullulan, 6, 165
  applications
    food industry, 204–205
    healthcare, 203–204
    miscellaneous applications, 205–206
    waste remediation, 205
  bioreactors and mode of operation for pullulan production, 192–196
    airlift and other fermenters for production, 195–196
    batch fermentation, 192–193
    fed-batch and continuous fermentation, 193–194
    immobilized cell bioreactors, 194–195
  biosynthesis, 166–188, 167f
    factor affecting production of pullulan, 169–188
    physicochemical properties, 168–169
  chemical modification of pullulan and advancement, 196–200, 197t
    carboxymethylation, 196–198
    cross-linking, 198
    grafting, 199–200
    hydrophobic modification, 199
  DSP, 200–203
  microbial consortia for pullulan production, 188–189
  microbial production of, 173t
  production by fermentation of agroindustrial by-products, 189–192
  structure, 166f

Pullularia pullulans. *See Aureobasidium pullulans*
Punica granatum, 114–115
Pure carbon substrates, 522–523
Pyrophosphates (PPi), 283–286

Q

Quality control (QC), 534–535
Quinone molecule, 558, 561f
Quinone-methide, 556, 557f, 558f

R

(R)-3-hydroxyhexanoate, 499
Rabbit aorta smooth muscle cells (RaSMCs), 234
Ralstonia eutropha, 319–320, 498–499, 522, 526–527, 529–530
Random mutagenesis, 319
Rapid prototyping, 418
Raw plastics, 11
Reactive extraction, 7, 331
Reactor
  airlift bioreactors, 326–327
  CSTR, 324–325
  FBRs, 325–326
  fibrous-bed reactors, 327
  PBRs, 325
Reduced graphene oxide (RGO), 403–404
Refractive error (RE), 375
Registration, evaluation, authorization, and restriction of chemical substances Act (REACH Act), 413
Reinforcement-based composites, 400–401
Reinforcement of nanomaterials, 414
Renewable natural polymer, 31
Residual-type crystallinity, 38
Residues of crops, 393–394
Resinol, 558, 561f
Rhamnogalacturonan I (RG-I), 101, 104
Rhamnogalacturonan II (RG-II), 101, 105

Rhizobium, 55, 359–361
Rhizopus
 R. arrhizus, 324–325
 R. oryzae, 315, 317
Rhodamine B, 241
Rhodobacter, 359–361
Rhodospirillum rubrum, 520, 522
Rhodosporidium paludigenum, 188
Rhodotorulabacarum, 188
Ribulose monophosphate pathway (RuMP pathway), 499
Rice, 79
 husk, 425
Rifampicin, 241
Ring-opening polymerization, 7, 309–310
River, microplastics in, 14–15
Rubomycin, 241

S

Saccharomyces cerevisiae, 147, 318
Saccharophagus degradans, 519–520
Saint Lawrence River, 15
Sarcina, 359–360
Scale-up from lab scale to pilot scale, 532–538
 good manufacturing practices, 534–535
 general process validation plan PHA, 534f
 problems and challenges encountered in, 535–538
  biomass harvesting, 536
  downstream processing, 536–538
  upstream processing, 535–536
 process validation, 533–534
 scale-up parameters for equipment, 532–533
  bioreactor, 532–533
  centrifuge, 533
Scanning electron microscope (SEM), 470
Scenedesmus spp., 450
 S. obliquus, 318
Secondary microplastics, 12–13, 21

Sequential batch reactors (SBRs), 521
Sequential liquid-lignin recovery and purification (SLRP), 566
Shewanella oneidensis, 224–226
Shikimic acid, 553–555, 553f
Short-chain-length PHA (scl-PHA), 224–227, 496
Silicate clays, 54
Silkworm silk fiber, 418
Silver (Ag), 118–119
Single-wall carbon nanotubes (SWCNTs), 404
Soda lignin, 576–577
Sodium dodecyl sulfate (SDS), 530
Sodium hydroxide (NaOH), 80–81
Softwoods, 129
Soil
 mesofauna, 23
 microplastics in, 17–18
Solid lignin, 566
Solid waste, microplastic in, 20–23
Solution casting, 394
Solution intercalation, 402–403
Solvent casting, 468–470
Solvent extraction, 529–530
Solvent volatilization method, 432
Solvent-biomass ratio, 114
Sorbitol, 36
Sound waves (SWs), 113–114
Soybean oil, 172
Spartium junceum L., 424–425
Sphingobacterium, 520
Spirillum volutans, 288
Staphylococcus
 S. aureus, 118–119, 297
 S. epidermidis, 257
Starch, 4, 32, 75–76. See also Thermoplastic starch (TPS)
 applications of starch-based materials, 89–93
 granules, 33, 78
 improving mechanical and physicochemical properties, 83–86
 microparticles, 93
 starch-based foaming, 87–88
 starch-based nanocomposites, 89

starch-based polymer blending, 86–87
starch-filled polyethylene, 46
starch-processing techniques, 79–83
starch-restructuring process, 40
structure and properties, 76–79
Starch-based nanoparticles (SNPs), 91
Starchy biomass, 7, 310–311
Static cultivation method for bacterial cellulose production, 363–364
Steam explosion
 pretreatment, 137–138
 process, 564–566, 565f
Stenotrophomonas maltophilia, 297
Stirred tank reactor, 324–325
Strain improvement, 318–319
Streptococcus, 319–320
Streptomycetaceae, 234
Streptosporangiaceae, 234
Styela plicata, 54–55
Subcritical fluids, 138
Substrate, 255–257, 310–311
Sulbactam-cefoperazone, 241
Sulfation, 198
Sulfolobus metallicus, 295–296
Sulfur dioxide (SO_2), 80
Sulperazone, 241
SunCarbon lignin, 575–576
Supercritical fluids (SCF), 138, 532
Supplements, pullulan effect on, 172
Sustainable bio-based renewable alternatives, 413
Synechocystis species, 318
Synthetic biology applications, PolyP in, 298
Synthetic fiber, 24
Synthetic polymers, 53

T

Tartaric acid (TA), 328–329
Technoeconomic studies, 539–540
Teloschistes flavicans, 188
Temperature, 525
 effect on pullulan, 172–187

Tensile strength of biopolymer, 427–429
Terrestrial environment, microplastics in, 13
Tetracycline, 241
Tetragenococcus, 319–320
2,2,6,6-Tetramethylpiperidine-1-oxyl radical (TEMPO), 465
Textiles
 industry materials, 24
 polylactic acid in, 336–337
Thermodynamics, 538
Thermoforming, 45
Thermomechanical pulp-bio lignin, 577–578
Thermomonosporaceae, 234
Thermoplastic matrix composites, 393–394
Thermoplastic polymer, 33
Thermoplastic starch (TPS), 5, 33
 applications of, 45–46
 biodegradability of thermoplastic starch, 39–40
 characterization of, 34–36, 34t
 different source of starch, 33t
 crystal property, 38–39
 mechanical properties, 36
 methods of preparing, 41–45
  film blowing, 42
  injection blow molding, 42–44
  ISBM, 44–45
  thermoforming, 45
 rheological and viscoelastic properties, 37–38
 structure of amylose and amylopectin, 32f
 thermal properties, 37
Thiol-containing chitosan, 446
3D
 cellulose-based scaffolds, 62
 printing, 418
 tissue scaffold, 338
Tissue engineering, 503–506
 bone, 503
 cardiovascular, 505–506
 cartilage, 504
 cellulose for tissue-engineering applications, 61–62

nerve, 504–505
peridontal, 505
PHAs in, 242–243
PLA in, 338–339
Toluene, 402–403
Tommy Atkins, 107
Toxoplasma gondii, 286–287
Trans-5-hydroxyferulic acid, 553–555, 553f
Trans-cafffeic acid, 553–555, 553f
Trans-cinnamic acid, 553–555, 553f
Trans-coniferyl alcohol, 552, 553f
Trans-coumaryl alcohol, 552, 553f
Trans-ferulic acid, 553–555, 553f
Trans-p-hydroxycinnamic acid, 553–555, 553f
Trans-sinapic acid, 553–555, 553f
Trans-sinpyl alcohol, 552, 553f
Transesterification, 333
Transversal flux of polymer, 393–394
Tremella mesenterica, 188
Triacetine (TAC), 430
Tricarboxylic acid cycle (TCA cycle), 253–254
Trilignol, 556–557, 558f
Trypanosoma cruzi, 286–287
Twin-screw extrusion, 394
Tyrosine ammonia lyase, 553–555

U

Ultrasound-assisted extraction of pectin, 113–115
United Nations Environment Program, 13
Unsaturated PHOUs, 230
Upstream processing (USP), 519–528, 535–536. *See also* Downstream processing (DSP)
 application of modeling and simulations in, 527–528
 choice of substrate in fermentation, 522–524
  complex substrates, 523–524
  pure carbon substrates, 522–523

identification of key-process parameters in, 524–527
 C/N ratio, 526–527
 dissolved oxygen, aeration, and agitation, 525–526
 feeding strategy, 527
 nutrient limitation, 526
 pH, 524–525
 temperature, 525
 type of substrate and concentration, 524
selection and type of microbes for production of PHAs, 519–522
 genetically modified microbial strains, 521–522
 MMC, 520–521
 process flow diagram, 519f
 production by pure cultures, 519–520
Uridine diphosphate glucose (UDPG), 166–167
UDPG-pyrophosphorylase, 168
Uridine phosphorylase (UPase), 172
Uronic acids, 130
US Food and Drug Administration (FDA), 204, 233, 310

V

Vagococcus, 319–320
Valorization of agro-food by-products, 417–418
Value-added products, 578
Volatile fatty acids (VFAs), 521

W

Waste
 microplastic in, 20–24
  industrial waste, 23–24
  solid waste, 20–23
 pullulan application in waste remediation, 205
Waste cooking oil (WCO), 523
Wastewater
 systems, 17–18
 treatment, 450–451
Water, 34
 absorption, 36

bacterial cellulose applications in water purification, 379–380
holding/release capacity of bacterial cellulose, 366
softening, 296
water-absorbing hydrogels, 64–65
Weissella, 319–320
Wheat, 79
  starch, 81
  straw, 401
Whey, 190–191, 311
Whiskers, 35–36
Wood plastic composite materials, 399
Worms, 15
Wound-healing applications of bacterial cellulose, 370–372

X

Xylan, 5, 129
  advances and innovation, 151–153
  bioproducts obtained from, 140–151
    bioethanol, 146–147
    enzymes, 150–151
    hydrogels, 147–148
    packaging, 148–149
    xylitol, 142–146
    xylooligosaccharides, 149–150
  environmental aspects, 153
  methods for xylan extraction
  motivation for xylan extraction from different sources, 132–133
  physicochemical characteristics, 132
  pretreatment methods for xylan extraction, 133–140
  sources and types, 130–132
Xylanases, 151
Xylitol, 5, 142–146
Xylogalacturonan (XGA), 103–105
Xyloglucan, 132
Xylooligosaccharides (XOSs), 134, 149–150
Xylose, 170–171

Y

Yeast, 317–318

Z

Z-ring, 499–501
Zymomonas mobilis, 147